Lecture Notes
in Control and Information Sciences 199

Editor: M. Thoma

Guy Cohen and Jean-Pierre Quadrat (Eds.)

11th International Conference on Analysis and Optimization of Systems

Discrete Event Systems

Sophia-Antipolis,
June 15–16–17, 1994

Springer-Verlag London Ltd.

Editors

Guy Cohen, Docteur ès Sciences
Centre Automatique et Systèmes, École des Mines de Paris, 35 rue Saint-Honoré, 77305 Fontainebleau Cedex, France

Jean-Pierre Quadrat, Docteur ès Sciences
INRIA, Domaine de Voluceau, Rocquencourt, BP 105, 78153 Le Chesnay Cedex, France

Honorary Editors

Alain Bensoussan
INRIA - Université Paris IX Dauphine

Jacques-Louis Lions
Collège de France - Académie des Sciences, Paris

ISBN 978-3-540-19896-3 ISBN 978-3-540-39345-0 (eBook)
DOI 10.1007/978-3-540-39345-0

British Library Cataloguing in Publication Data
A catalogue record for this book is available from the British Library

Originally published by Springer-Verlag London Limited in 1994

Typesetting: Camera ready by authors
69/3830-543210 Printed on acid-free paper

11th INTERNATIONAL CONFERENCE ON ANALYSIS AND OPTIMIZATION OF SYSTEMS

DISCRETE EVENT SYSTEMS

SOPHIA-ANTIPOLIS
JUNE 15–16–17, 1994

COORGANIZED BY INRIA AND ÉCOLE DES MINES DE PARIS

HONORARY EDITORS

A. Bensoussan & J.-L. Lions

EDITORS

G. Cohen & J.-P. Quadrat

Foreword

The 11th International Conference on Analysis and Optimization of Systems organized by INRIA is the second on the new format of the series. We recall that this "new format" means that a specific domain is selected for each issue.

This time the field of Discrete Event Systems is covered, following an issue dedicated to Infinite Dimensional Systems. The collection of all conferences will cover the full spectrum of system theory which is much too large to be adressed in a single non specialized meeting.

We would like to express our thanks to the DRET which sponsored this Conference and to École des Mines de Paris which is a coorganizer of this issue and which provided the venue of the Conference in its location of Sophia Antipolis.

We also would like to extend our gratitude to:

- the authors who have kindly accepted all the constraints of the organization,
- the reviewers who have accepted the responsability of selecting papers,
- the chairpersons for having run all the sessions of the Conference with efficiency,
- all the members of the Scientific Committee who took in charge the difficult job of organizing the various sessions,
- our colleagues Guy Cohen who, among many other things, did a wonderful job to improve the quality of the proceedings, and Jean-Pierre Quadrat who was the initiator of the conference,
- the Department of Public Relations of INRIA, and especially Mr. François Tapissier who, by his enthusiasm, challenged the organizers in trying to reach a high standard of efficiency,
- Professor Manfred Thoma and the Publisher Springer Verlag who have, as in the previous series, accepted to publish this series in the Lecture Notes in Control and Information Sciences.

Alain Bensoussan Jacques-Louis Lions

Presentation

The 11th International Conference on Analysis and Optimization of Systems organized by INRIA and École des Mines de Paris can also be considered as the second WODES/ CODES conference (series of WOrkshop or COnferences on Discrete Event Systems).

This Conference has been aimed at engineers and mathematicians working in the field of Automatic Control, Operations Research and Statistics who are interested by the modeling, analysis and optimization of discrete event systems. Many formalisms have been developed to describe such systems. The comparison of the different mathematical approaches and their global confrontation with the applications have been the main goals of the conference.

The Conference has offered eight "key" sessions:

1. The Automata Theoretic Approach under the responsability of W.M. Wonham,
2. The Petri Net Approach under the responsability of B.H. Krogh,
3. The Max-Plus Algebraic Approach under the responsability of G. Cohen, G.J. Olsder and J.-P. Quadrat,
4. Hybrid Systems under the responsability of J.H. van Schuppen,
5. Simulation and Perturbation Analysis under the responsability of P. Brémaud and C. Cassandras,
6. Network Stability under the responsability of F. Baccelli,
7. Large DES under the responsability of G. Fayolle and V. Malyshev,
8. Manufacturing Systems under the responsability of S. Gershwin and R. Suri.

Each session corresponding to a part of these proceedings was composed of three parts:

a. a tutorial lecture giving a survey of the domain,
b. invited lectures developing some aspects introduced in the survey,
c. and finally more specialized contributions.

Each part starts with the survey paper. Invited papers are about ten page long and contributed papers are about seven page long. A few papers are longer since authors of some tutorial and invited lectures decided to provide a joint paper in their area.

Contents

The Petri Net Approach

The Max-Plus Algebraic Approach

Hybrid Systems

Simulation and Perturbation Analysis

Large Discrete Event Systems and Network Stability

Manufacturing Systems

Author Index

The Automata Theoretic Approach

Logical Aspects of Control of Discrete-Event Systems: A Survey of Tools and Techniques

*John G. Thistle**

Département de génie electrique et de génie informatique, École Polytechnique de Montréal, C.P. 6079, succursale Centre-ville, Montréal (Québec), Canada H3C 3A7

1 Introduction

This article surveys recent research on discrete-event control problems that feature qualitative specifications of desired behaviour.

The need for rigourous means of ensuring logical correctness of complex control systems is well recognized; in this regard [BÅ93], for example, offers a view of the potential role of discrete-event control theory in industrial applications. One approach to correctness issues in the control of discrete-event systems (DES) is the *supervisory control theory* initiated by Ramadge and Wonham. Early results in supervisory control studied the solvability of formal control synthesis problems and investigated the management of complexity through modular synthesis and decentralized architecture. (See [RW89] for a survey of the state of supervisory control theory as of January 1989.) Most of this early research was set in a simple, abstract framework of formal languages and finite automata, chosen for conceptual simplicity and generality rather than modelling efficiency or computational tractability. But recent work, in supervisory control and in other settings, has introduced a variety of more structured models, many drawn from computer science and engineering; these offer features such as subsystem aggregation, data structures; real time; formal logic; and control synthesis algorithms of reduced complexity. The purpose of this article is to indicate some of the main directions explored in this recent research.

For the sake of unity, and to highlight issues specific to control, we take as a point of reference the Ramadge-Wonham supervisory control theory, and some of the properties identified within that framework that play key roles in formal control synthesis. We exclude Petri nets from our discussion, as they are the topic of another workshop session. Because of space constraints we also omit a survey of research into the computational complexity of control synthesis.

Section 2 discusses the supervisory control framework. The reader is assumed to be familiar with the original model, and with the properties of *controllability* and *ob-*

* Supported by the Fonds pour la formation de chercheurs et l'aide à la recherche of the Province of Quebec, Canada, Bell-Northern Research Ltd. and the Natural Sciences and Engineering Research Council of Canada, under the Action concertée sur les méthodes mathématiques pour la synthèse de systèmes informatiques.

servability of formal languages and results on modular and decentralized control, as surveyed in [RW89]. Section 3 then discusses hierarchical control, bringing in the property of *hierarchical consistency*, which allows effective implementation of hierarchical control schemes. In §4 we discuss extended system models that include explicit real-time features, and describe the modifications of the fundamental properties of controllability, observability and hierarchical consistency that these entail. This leads to a discussion, in §5, of models based on infinite-string formal languages. The introduction of infinite strings and the corresponding control synthesis algorithms in turn establishes a formal connection with modal logics, often used in the specification of computer systems; in §6 we list some of the applications of formal logic to DES control theory. Section 7 briefly discusses some models that support useful means of aggregation.

2 Supervisory Control

We assume that the reader is familiar with the survey [RW89], which gives a tutorial introduction to the basic supervisory control theory. Here we establish terminology and notation, and mention some recent results based on the paradigms of [RW89].

Discrete-event systems are modelled in [RW89] as *generators* of formal languages. A formal language is a set of strings (usually all finite) of symbols of some *event alphabet* Σ. The set of all finite strings over Σ is denoted Σ^*; this includes the empty string (the string of length zero), denoted by ϵ. A string $s \in \Sigma^*$ is a *prefix* of $t \in \Sigma^*$ if s is an initial substring of t; we write $s \leq t$. The set of prefixes of strings of a formal language $L \subseteq \Sigma^*$ is denoted by $\text{pre}(L)$ or $\overline{L}$. The language L is *(prefix-)closed* if $L = \text{pre}(L) = \overline{L}$. (Note that $L \subseteq \text{pre}(L) = \overline{L}$ always obtains.) The family of prefix-closed subsets of a given language L is denoted by $\mathcal{F}_L$. Power sets are denoted by $\mathcal{P}(\cdot)$.

Associated with a generator G are two languages, a closed language $L(G) \subseteq \Sigma^*$ representing the set of event sequences generated by the DES, and a *marked* language $L_m(G) \subseteq L$ representing completed control tasks. The DES G is *nonblocking* if $\overline{L_M(G)} = L(G)$.

Supervisors are modelled as maps that take the string of past events to the set of currently enabled events, which must include any *uncontrollable* events. The main results surveyed in [RW89] characterize achievable "closed-loop" behaviour in terms of the *controllability* and *observability* of languages, and study the decomposition of control tasks through modular synthesis and decentralized architecture. The characterization of behaviour achievable through decentralized control has recently been extended by Rudie through the definition of *co-observability* [RW92].

3 Hierarchical control

The results on modular and decentralized supervision of [RW89] focus on *horizontal* decomposition of the overall control task, and do not explicitly consider subsupervisors that operate on different time scales or at different levels of logical detail. More recent research has investigated *vertical* decomposition of supervision into tasks performed at different levels of temporal or logical abstraction.

Much of this work is based on the schema of Fig. 1 [ZW90]. In this arrangement, control objectives are formulated with respect to a high-level model, G_{hi}, which represents the plant G_{lo} at an appropriate level of abstraction. Controller C_{hi} is designed to control G_{hi}, using information channel $\mathrm{Inf}_{\mathrm{hi}}$ for observation and control channel $\mathrm{Con}_{\mathrm{hi}}$ for control. However, as G_{hi} is only an abstract plant model, the control channel $\mathrm{Con}_{\mathrm{hi}}$ is in fact a virtual one, and control must actually be implemented by transmitting commands to a low-level controller C_{lo} by way of command channel $\mathrm{Com}_{\mathrm{hilo}}$. Controller C_{lo} controls G_{lo} via $\mathrm{Con}_{\mathrm{lo}}$, on the basis of observation via $\mathrm{Inf}_{\mathrm{lo}}$, taking as its specifications the high-level commands received from C_{hi}. The action of C_{hi} thus only affects G_{hi} when the high-level model is updated to take account of evolution of the low-level plant. Such updating can be considered to be mediated by an information channel $\mathrm{Inf}_{\mathrm{lohi}}$.

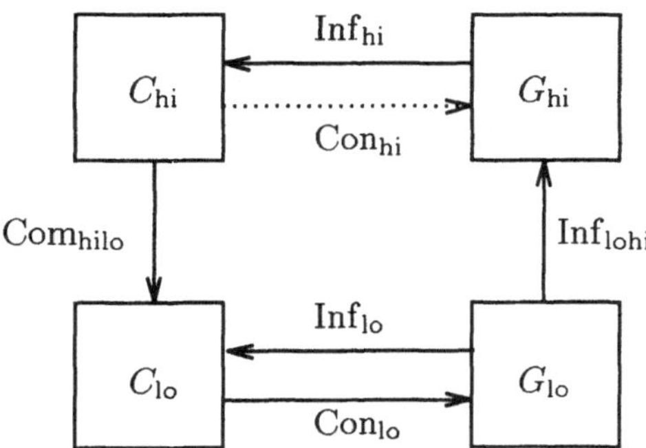

Fig. 1. Hierarchical control scheme

The scheme of Fig. 1 thus divides the overall supervision task into a high-level "managerial" task and a lower-level "operational" one. The effectiveness of such a hierarchical control scheme requires not only that the evolution of the high-level model be consistent with that of the plant, but also that the control technology associated with G_{hi} be consistent with that of G_{lo}, so that the virtual control exercised via $\mathrm{Con}_{\mathrm{hi}}$ can be implemented via $\mathrm{Com}_{\mathrm{hilo}}$. This question of *control consistency* is studied by Zhong and Wonham in [ZW90], and the ideas of [ZW90] are reformulated algebraically and generalized in [WW92].

Suppose that the event alphabets of the low and high levels are Σ and T, respectively. Let the language generated by the low-level plant be L. The information channel $\mathrm{Inf}_{\mathrm{lohi}}$ can be modelled as a map

$$\theta : L \longrightarrow T^*$$

taking strings in L to strings over T. The map θ is interpreted as "summarizing" every sequence of low-level events in the form of a high-level event sequence (which will typically be shorter than the low-level one). We extend θ to a map on sublanguages $\theta : \mathcal{P}(L) \longrightarrow \mathcal{P}(T^*)$ in the natural way. It is assumed that θ is prefix-preserving, meaning that the diagram of Fig. 2 commutes. Thus if $s \in L$ is a prefix of $s' \in L$, then $\theta(s)$ will be a prefix of $\theta(s')$. In general, θ will be many-to-one; thus $t \in T^*$ may represent a high-level "summary" of a subset of low-level strings $\theta^{-1}(t) \subseteq L$.

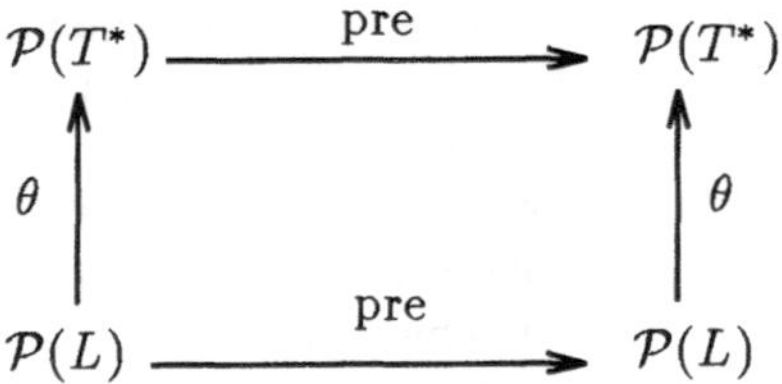

Fig. 2. Prefix-preserving map $\theta : L \longrightarrow T^*$

If the marked language of G_{lo} is $L_m \subseteq L$, then let the languages marked and generated by G_{hi} be $M_m = \theta(L_m)$ and $M = \theta(L)$, respectively. This ensures a certain "dynamical consistency" between the low and high levels. But effective hierarchical control synthesis also requires *control consistency*, meaning that the control mechanism of G_{lo} must admit a consistent high-level counterpart for G_{hi}.

We formalize this notion by bringing in the axiomatic definition of a control mechanism from [WW92]. A map $\mathcal{C} : \mathcal{F}_L \longrightarrow \mathcal{P}^2(L) = \mathcal{P}(\mathcal{P}(L))$ is a *control structure on* L if, for every $H \in \mathcal{F}_L$,

1. $\mathcal{C}(H) \subseteq \mathcal{P}(H)$ is a complete upper semilattice with respect to $\cup$ in $\mathcal{P}(H)$;
2. $\varnothing, H \in \mathcal{C}(H)$;
3. $K \in \mathcal{C}(H) \Longrightarrow \overline{K} \in \mathcal{C}(H)$;
4. for every $F \in \mathcal{F}_L$ with $H \subseteq F, \mathcal{C}(F) \cap \mathcal{P}(H) \subseteq \mathcal{C}(H)$;
5. for every $F \in \mathcal{F}_L$ with $H \in \mathcal{F}_L \cap \mathcal{C}(F), \mathcal{C}(F) \cap \mathcal{P}(H) = \mathcal{C}(H)$.

This definition is satisfied, for example, by the control mechanism postulated by Ramadge and Wonham [RW89], if we let $\mathcal{C}$ map every sublanguage to its set of controllable sublanguages. Still following [WW92], we let

$$\kappa_F : \mathcal{P}(F) \longrightarrow \mathcal{C}(F) : H \longrightarrow \bigcup \{J \subseteq H : J \in \mathcal{C}(F)\}$$

be the *control operator* for $F \in \mathcal{F}_L$ (by Axiom 1). If the control technology is that of [RW89], then $\kappa_F(H) = supC_F(H)$, the union of all controllable sublanguages of H.

Given the hierarchical setup and a control technology $\mathcal{C}_L$ for the low-level model, Wong and Wonham define the triple (L, θ, M) to be *control consistent* if there exists a map $\mathcal{C}_M : \mathcal{F}_M \longrightarrow \mathcal{P}^2(M)$ such that the diagram of Fig. 3 commutes.

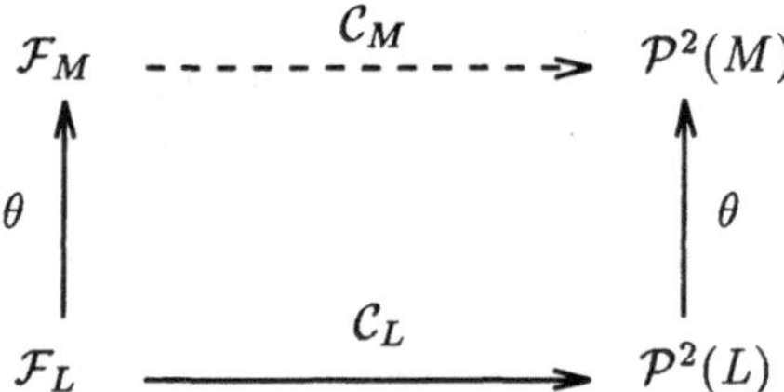

Fig. 3. Control consistency of the triple (L, θ, M)

The commutativity of the diagram intuitively means that if two low-level sublanguages have the same high-level image, then so do their respective "controllable sublanguages." Thus, if the map $\mathcal{C}_M$ constitutes a control structure, it might be considered to represent a suitable high-level counterpart of $\mathcal{C}_L$ in the context of the triple (L, θ, M). Proposition 1 states that $\mathcal{C}_M$ is indeed a control structure, and confirms that it represents an appropriate high-level counterpart to $\mathcal{C}_L$, in the sense that control strategies conceived at the high level can be implemented exactly at the low level.

Proposition 1 (Wong & Wonham [WW92]). *Let $\mathcal{C}_L : \mathcal{F}_L \longrightarrow \mathcal{P}^2(L)$ be a control structure on L and let the triple (L, θ, M) be control consistent. Then the unique map $\mathcal{C}_M$ that makes the diagram of Fig. 3 commute is a control structure on M. Furthermore,*

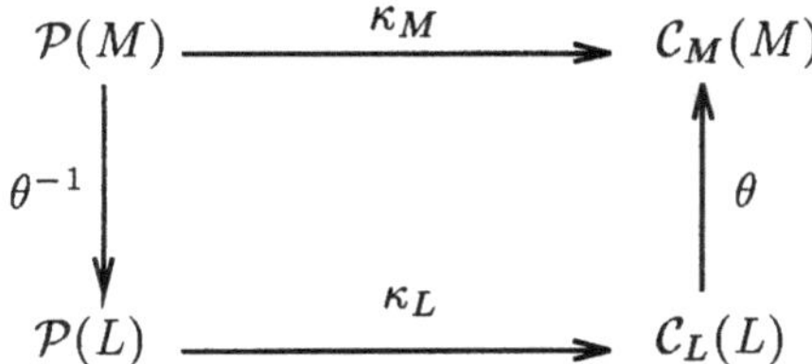

Intuitively, θ^{-1} can be taken to represent the action of command channel $\mathrm{Com}_{\mathrm{hilo}}$, mapping high-level objectives to the low-level. Under this interpretation, the commutativity of the diagram of Proposition 1 means that the effect of transmitting a high-level objective, represented by some sublanguage $E \subseteq M$, to the low level for implementation as the specification $\theta^{-1}(E)$ yields a high-level model behaviour $\theta\kappa_L\theta^{-1}(E)$ that is equivalent to the behaviour $\kappa_M(E)$ that would result from direct control of G_{hi} by $\mathrm{Con}_{\mathrm{hi}}$ through the control structure $\mathcal{C}_M$. In other words, the virtual high-level control is implementable in the low level.

The above notions of control consistency and hierarchical consistency of [WW92] generalize those of [ZW90], which are based on the specific class of control structures of [RW89]. For this standard class of control structures, [ZW90] provides algorithms for testing the appropriate control consistency conditions. In case the conditions fail, procedures are given for enriching the higher-level model and the information transfer between the two levels so as to achieve control consistency. It should be noted that our restriction to two-level hierarchies is not necessary; the above results extend to multiple levels. Nonblocking hierarchical synthesis is studied in [WW92], as is the high-level coordination of modular supervisors. In [WW93], the results on hierarchical consistency are extended to a real-time DES model.

4 Real time

Correctness issues for real-time systems have been the focus of much recent research. The effects of time delays in the implementation of supervisors developed within the original untimed Ramadge-Wonham framework were considered in [LW88]. Extended DES models with explicit real-time features include those introduced in [BH88], where the authors suggest means of separating temporal constraints from issues of event sequencing, and the "timed transition models" of [OW90], in which real-time temporal logic is used for formal verification. In [WTH92], a timed automaton model incorporating real-valued rather than discrete-time clocks is applied to supervisory control. We refer the reader to [Ost92] for a survey of formal models and specifications for real-time systems.

We shall focus here on the model proposed in [BW92] (and inspired by that of [OW90]), which has so far been studied more extensively than others from the standpoint of formal control synthesis. In particular, we shall outline the appropriate extension to this timed model of the properties of controllability, observability and hierarchical consistency, originally defined within the untimed framework.

Let each event σ be equipped with lower and upper time bounds $l_\sigma, u_\sigma \in \mathbb{N}$, respectively, the passage of time being marked by an additional event, tick. Events other than tick are called *activity events*, and the plant generator is called the *activity graph*. Let the set of activity events be denoted by Σ_{act}. An activity event is *enabled* in a given state of the activity graph if there exists a corresponding transition of the activity graph. An enabled activity event $\sigma \in \Sigma_{\text{act}}$ is *eligible* if the number of ticks that have occurred since its enablement lies between l_σ and u_σ. The tick event cannot occur if some activity event σ is *imminent* – that is, if the number of ticks that have elapsed since the enablement of σ is equal to the upper time bound u_σ.

The effect of the above constraints on the occurrence of activity events and the tick event can be captured by a *timed transition graph*. We illustrate the model with a simple example from [BW92]. Figure 4(a) shows a one-state activity graph for a system with two activity events, α and β. If $l_\alpha = u_\alpha = 1$ and $l_\beta = 2$ and $u_\beta = 3$ then the set of event sequences satisfying the model's timing constraints is given by the timed transition graph of Fig. 4(b). The time bounds on the occurrence of enabled α events dictate that the first α must be preceded by a tick, and that subsequently there must be one α between any two consecutive ticks. Event β may occur after the first two ticks, or whenever two ticks have followed the latest β, and must occur at least once for every three ticks. Brandin and Wonham consider only systems whose timed transition graphs are free of *activity loops*, meaning cycles made up only of activity event transitions. Thus activity events may not indefinitely preempt tick. In the finite-state case, any infinite event sequence must therefore contain infinitely many ticks.

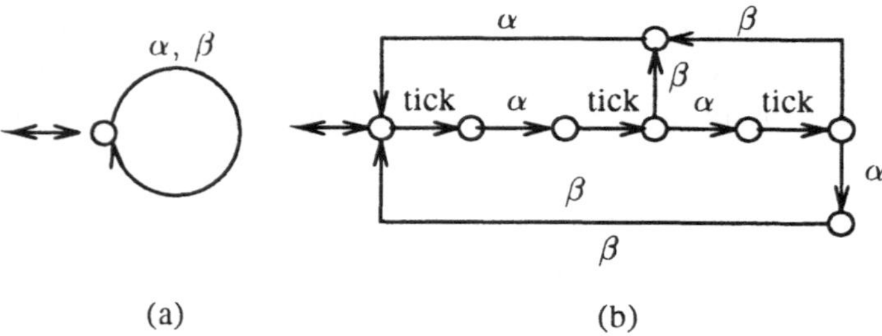

Fig. 4. Activity graph and timed transition graph

To adjoin a control mechanism to the model, define a class of *prohibitible* activity events, $\Sigma_{\text{hib}} \subseteq \Sigma_{\text{act}}$. These will play much the same role as the controllable events of the untimed theory, representing events that can at any given time be prevented from occurring. Activity events that are not prohibitible are called *uncontrollable*; we write $\Sigma_{\text{unc}} := \Sigma_{\text{act}} \setminus \Sigma_{\text{hib}}$. In the timed case, it is natural to consider events that can be *forced* to occur before the advance of the system clock. For this, Brandin and Wonham define a class of *forcible* events that can preempt tick.

The set of events that are eligible after the occurrence of a string s is formally defined by $\text{Elig}_{L(G)}(s) := \{\sigma \in \Sigma : s\sigma \in L(G)\}$, where G is the timed transition graph and $\Sigma = \Sigma_{\text{act}} \cup \{\text{tick}\}$. Note that $\text{Elig}_{L(G)}(s)$ may include tick as well as activity events. A *supervisor* is defined as a map $V : L(G) \longrightarrow \mathcal{P}(\Sigma)$, such that, for all $s \in L(G)$, $V(s) \cap \text{Elig}(s) \neq \varnothing$, and

$$V(s) \supseteq \begin{cases} \Sigma_{\text{unc}} \cup \{\text{tick}\} & \text{if } V(s) \cap \Sigma_{\text{for}} = \varnothing\,, \\ \Sigma_{\text{unc}} & \text{if } V(s) \cap \Sigma_{\text{for}} \neq \varnothing\,. \end{cases}$$

The interpretation is that, as in the untimed case, supervisors never disable uncontrollable events, but if any forcible events are enabled, then tick is disabled – this means that some event (not necessarily forcible) will occur before the next tick of the clock. The definitions of the closed and marked behaviours of the supervised DES V/G, and of nonblocking supervisors, are exactly analogous to those of the untimed theory.

The appropriate extension of controllability to the real-time case is given in [BW92] (see also [WTH92]). First, extend the notion of eligibility by writing, for any $K \subseteq L(G)$

$$\text{Elig}_K(s) := \{\sigma \in \Sigma : s\sigma \in \overline{K}\}, \ \forall s \in \Sigma^*\,.$$

A sublanguage $K \subseteq L(G)$ is then said to be *controllable* with respect to $L(G)$ if, for all $s \in \overline{K}$,

$$\text{Elig}_K(s) \supseteq \begin{cases} \text{Elig}_{L(G)}(s) \cap (\Sigma_{\text{unc}} \cup \text{tick}) & \text{if } \text{Elig}_K(s) \cap \Sigma_{\text{for}} = \varnothing\,, \\ \text{Elig}_{L(G)}(s) \cap \Sigma_{\text{unc}} & \text{if } \text{Elig}_K(s) \cap \Sigma_{\text{for}} \neq \varnothing\,. \end{cases}$$

Thus K is controllable if it allows for the occurrence of any uncontrollable event eligible in $L(G)$, or of tick, if tick is eligible in $L(G)$ and no forcible events are eligible in K. This property plays a role analogous to that of controllability in the original RW theory. In particular, we have

Theorem 2 (Brandin & Wonham [BW92]). *Let $K \subseteq L_m(G)$, $K \neq \varnothing$. There exists a nonblocking supervisor V for G such that $L_m(V/G) = K$ if and only if*

1. *K is controllable with respect to $L(G)$, and*
2. *$K = \overline{K} \cap L_m(G)$.*

In addition, the real-time controllability property is also preserved under arbitrary unions, so a specification of legal timed event sequences admits a maximally permissive supervisor, as in the untimed case.

Lin and Wonham [LW93] have recently extended the results on supervision under partial observations to this real-time model. As in the untimed setting, the event alphabet Σ is partitioned into observable events $\Sigma_o \subseteq \Sigma$ and unobservable events $\Sigma_{uo} = \Sigma \setminus \Sigma_o$.

The effect of partial observation is modelled by a projection $P : \Sigma^* \longrightarrow \Sigma_o^*$ defined by induction on the length of strings:

$$P(\epsilon) = \epsilon \ (\epsilon \text{ is the empty string});$$
$$P(s\sigma) = \begin{cases} P(s) & \text{if } \sigma \in \Sigma_{uo}, \\ P(s)\sigma & \text{if } \sigma \in \Sigma_o. \end{cases}$$

The projection P has the effect of "erasing" all unobservable events. Supervisors are now defined to be maps $V : P(L(G)) \longrightarrow \mathcal{P}(\Sigma)$, such that, for all $s \in L(G)$,

$$V(P(s)) \supseteq \begin{cases} \Sigma_{\text{unc}} \cup \{\text{tick}\} & \text{if } V(P(s)) \cap \Sigma_{\text{for}} = \varnothing, \\ \Sigma_{\text{unc}} & \text{if } V(P(s)) \cap \Sigma_{\text{for}} \neq \varnothing. \end{cases}$$

The closed and marked languages of the supervised DES, denoted by V/G are again defined by direct analogy with the untimed case.

The observability condition necessary for supervision under partial observations is based on a *consistency* relation $\text{consis}_K \subseteq \Sigma^* \times \Sigma^*$, where, for any $K \subseteq L(G)$, and $s, s' \in \Sigma^*$, $\text{consis}_K(s, s')$ if and only if

1. $(\forall \sigma \in \Sigma)\ s\sigma \in \overline{K}\ \&\ s' \in \overline{K}\ \&\ s'\sigma \in L(G) \Longrightarrow s'\sigma \in \overline{K}$, and
2. $(\forall \sigma \in \Sigma)\ s'\sigma \in \overline{K}\ \&\ s \in \overline{K}\ \&\ s\sigma \in L(G) \Longrightarrow s\sigma \in \overline{K}$.

Intuitively, s and s' satisfy $\text{consis}_K(s, s')$ if and only if s and s' are consistent with respect to one-step extension in K. The language $K \subseteq L(G)$ is *observable* with respect to $L(G)$ if and only if, for all $s, s' \in \Sigma^*$,

$$P(s) = P(s') \Longrightarrow \text{consis}_K(s, s').$$

Thus a language is observable if the projection P preserves enough information to allow an appropriate choice of event disablement.

Observability represents the additional property necessary to characterize achievable closed-loop behaviour in the face of partial observations. In particular, we have

Theorem 3 (Lin & Wonham [LW93]). *Let $K \subseteq L_m(G)$, $K \neq \varnothing$. There exists a non-blocking supervisor $V : P(L(G)) \longrightarrow \mathcal{P}(\Sigma)$ such that $L_m(V/G) = K$ if and only if*

1. *K is controllable with respect to G,*
2. *K is observable with respect to $L(G)$, and*
3. *$K = \overline{K} \cap L_m(G)$.*

Besides extending the definition of observability to the real-time model, Lin and Wonham [LW93] discuss its relationship with normality and study the on-line computation of supervisors under partial observations [HL93].

The notions of control consistency and hierarchical consistency discussed in §3 are extended to the real-time case (with complete observations) in [WW93]. This helps to characterize the amount of timing information that must be passed up to the higher level of the scheme of Fig. 1 in order for high-level real-time control schemes to be implementable.

A closed language $L \subseteq \Sigma^*$ is defined to be a *time behaviour* if any strings in L can be extended to a longer string in L, and if all infinite strings of which all finite prefixes belong to L contain infinitely many ticks. Thus a time behaviour is a closed language in which tick occurs infinitely often. This captures a version of the *activity-loop freedom* assumption of [BW92] (§3). To capture the notion that a supervisor never disables all eligible events, a time behaviour $L \subseteq \Sigma^*$ is defined to be *proper* if for all $s \in L$, tick $\notin \mathrm{Elig}_L(s)$ implies $\mathrm{Elig}_L(s) \cap \Sigma_{\mathrm{hib}} \subset \mathrm{Elig}_L(s)$, where the inclusion is strict. In other words, the time behaviour L is proper if whenever tick is ineligible in L, there exists some uncontrollable event that is eligible in L. For $L \subseteq \Sigma^*$, the family of subsets of of L that are time behaviours (respectively, proper time behaviours) is denoted by $\mathcal{TF}_L$ (respectively, $\mathcal{PTF}_L$).

Wong and Wonham define an abstract notion of a control technology inspired by the control mechanism of [BW92]: a map $\mathcal{C} : \mathcal{PTF}_L \longrightarrow \mathcal{P}^2(L)$ is a *time control structure* with respect to $\mathcal{PTF}_L$ if it is a control structure with respect to $\mathcal{PTF}_L$ (that is, if it satisfies the axioms given in §3) and in addition $\mathcal{C}(H) \cap \mathcal{F}_H \setminus \{\varnothing\} \subseteq \mathcal{PTF}_L$, for any $H \in \mathcal{PTF}_L$. Intuitively, a time control structure is a control structure such that control never stops the advance of time and indeed always produces a proper time behaviour.

The hierarchical control scheme of Fig. 1 is modelled formally with a low-level alphabet $\Sigma = \Sigma_{\mathrm{act}} \cup \{\mathrm{tick_{lo}}\}$ and high-level alphabet $T = T_{\mathrm{act}} \cup \{\mathrm{tick_{hi}}\}$, each consisting of activity events and a tick event. Projections p_{lo}, p_{hi} erase all activity events from strings in Σ^*, T^*, respectively. The closed behaviour of the low-level plant is represented by a time behaviour $L \in \mathcal{TF}_{\Sigma^*}$. Information channel $\mathrm{Inf_{lohi}}$ is modelled by a map $\theta : L \cup \mathrm{tick}^*_{\mathrm{lo}} \longrightarrow T^*$ with the following properties:

$$\begin{aligned} \theta \circ \mathrm{pre} &= \mathrm{pre} \circ \theta \,, \\ \theta(\mathrm{tick}^*_{\mathrm{lo}}) &= \mathrm{tick}^*_{\mathrm{hi}} \,, \\ \theta \circ p_{\mathrm{lo}} &= p_{\mathrm{hi}} \circ \theta \,. \end{aligned}$$

In addition, it is assumed that for any infinite string t over T of which all finite prefixes belong to $\theta(L)$, there exists an infinite string s over Σ such that all finite prefixes of s belong to L and the images of the finite prefixes of s are exactly the finite prefixes of t. The first of these properties states simply that θ is prefix-preserving as in §3. The second states that timing information is sent up to the higher level infinitely often, and the third that timing information is summarized independently of other information.

With these preliminary definitions, Wong and Wonham prove a result analogous to Proposition 1 of §3:

Proposition 4 (Wong & Wonham [WW93]). *Let $\mathcal{C}_{\mathrm{lo}} : \mathcal{PTF}_L \longrightarrow \mathcal{P}^2(L)$ be a time control structure with respect to $\mathcal{PTF}_L$ and let $M := \theta(L)$. If there exists a map $\mathcal{C}_{\mathrm{hi}} : \theta\mathcal{PTF}_L \longrightarrow \mathcal{P}^2(M)$ such that $\mathcal{C}_{\mathrm{hi}}\theta(H) = \theta\mathcal{C}_{\mathrm{lo}}(H)$ for $H \in \mathcal{PTF}_L$, then $\mathcal{C}_{\mathrm{hi}}$ is a time control structure with respect to $\theta\mathcal{PTF}_L$, and furthermore,*

$$\kappa_M = \theta \circ \kappa_L \circ \theta^{-1}$$

where κ_M and κ_L are the control operators for $M := \theta(L)$ and L, defined by direct analogy with §3.

Intuitively, the control consistency represented by the existence of the map $\mathcal{C}_{\mathrm{hi}}$ ensures that θ provides sufficient timing information to extend the hierarchical consistency of §3 to the real-time setting.

5 Infinite strings

In the formal definitions of [WW93], the use of infinite strings is in fact only implicit. Nevertheless, the above notions of time behaviours and real-time control consistency illustrate the usefulness of infinite event strings in expressing *liveness* properties, which assert that some event or condition will occur eventually (or, as in the above cases, infinitely often). Infinite-string formal languages (*ω-languages*) and the corresponding automata (*ω-automata*) are standard models for the formal analysis of computer systems. (See for example [MP92] and the article by Kurshan in these proceedings.) Several authors have extended results of supervisory control to the infinite-string setting.

Ramadge [Ram89] has used *Büchi automata* to state assumptions of fairness in the scheduling of concurrent systems. This necessitates an extended notion of controllability. Defining an ω-language to be *topologically closed* if it contains all limit points of its set of finite prefixes (that is, all infinite strings that have infinitely many prefixes in the set), Ramadge defines an ω-language to be controllable if i) its set of finite prefixes is a controllable language (in the usual sense) and ii) it is topologically closed relative to plant behaviour.

This notion of controllable ω-languages captures achievable closed-loop behaviour in Ramadge's setting. Furthermore, Ramadge shows that, if an ω-language S is topologically closed relative to the ω-language generated by the plant generator, then it contains a unique supremal controllable sublanguage. Kumar, Garg and Marcus discuss the computation of this supremal controllable sublanguage, and show that for ω-languages that are closed relative to plant behaviour, the finite-string notions of observability and normality carry over to the infinite-string case in much the same way as controllability [KGM92].

The assumption of topological closure of specification languages in [Ram89, KGM92] means that these languages can only express *safety* properties, which assert that some condition or event must never occur. Indeed, the results of [KGM92] exploit a reduction to the finite-string case. A more general class of specification languages, allowing the expression of some liveness properties, is considered in [YSG92]; closely related to this work are the results of [OW91]. The still more general class of *ω-regular* languages (that is, ω-languages that can be represented by finite automata) is considered in [TW94a].

The extension to more general specification languages reveals some closer parallels with computer science than are evident in the finite-string case. While work in programming theory and computer engineering focusses primarily on verification, a considerable body of research on formal synthesis has been framed in terms of finite ω-automata and a synthesis problem first defined by Church. The infinite-string extension of a basic supervisory control synthesis problem turns out to be equivalent to Church's problem, and the appropriate generalization of controllability – *ω-controllability* – is closely related to notions of *implementability* and *realizability* of computer systems (see [TW94b]).

6 Formal logic

The extension of supervisory control to infinite strings establishes a formal connection with the modal logics widely used for specification and verification of computer systems

[MP92, Ost92]. Several authors have studied the use of one such logical system, that of *temporal logic*, in control of discrete-event systems. We refer the reader to Ostroff's use of real-time temporal logic [OW90] and the articles by Ionescu and Lin and Seow and Devanathan that appear in these proceedings.

In [RW87], logical tools originally developed for the verification of sequential programs are applied to modular synthesis of controllers for control-invariance problems. State subsets are represented by state predicates, and the effects of state transitions are captured by *predicate transformers*. Similar techniques are employed in [KGM93], where control under partial observations is also considered.

The logical predicate used in [KGM93] to represent the current state estimate resembles that used in the COCOLOG theory of Caines and Wang. In COCOLOG, control strategies are expressed as collections of rules in which logical predicates enable control actions. Axioms describe plant dynamics and the evolution of state estimates, and automatic theorem proving is used to implement the control policy. See the article by Caines, Mackling and Wei in these proceedings.

7 Aggregation

The recent work of Brave and Heymann [BH93] seeks to exploit the modelling economy and special structure of a class of *hierarchical state machines* (HSMs). HSMs represent a special case of the statecharts formalism that has been proposed for the efficient modelling of complex computer systems [Har87].

Statecharts provide syntactical support for the aggregation of state-machine models without combinatorial explosion. Among their features are *state-levels*, which represent hierarchical structure or depth, and *orthogonality*, which models concurrency. In [BH93] Brave and Heymann consider systems modelled by a special class of statecharts (asynchronous hierarchical state machines (AHSMs)) in which orthogonal components contain no shared events and are thus completely asynchronous. All interaction among these parallel components occurs either through high-level transitions between superstates of the components or through the action of a controller (see also [Ram89]). This special structure is exploited in solving forbidden state problems.

İnan and Varaiya have proposed a model for discrete-event systems based on an algebra of subsystem composition. *Finitely recursive processes* are defined by recursions employing deterministic choice operators, synchronous and sequential composition, and operators that alter the event alphabet [IV88]. İnan has since applied this framework to supervisory control, representing the interaction between plant and supervisor by the synchronous composition operator: in [İnan92], results on decentralized control are reformulated within the algebraic framework; İnan's contribution to this workshop presents new results on control under partial observations.

8 Conclusion

We have sketched some of the main directions in which theoretical frameworks for control of discrete-event systems have recently been extended and elaborated. While many of the proposed models bear marked similarities to the tools of computer scientists, control theory for DES is distinguished by an emphasis on formal synthesis that is typical of control science. As a consequence, properties such as controllability, observability and hierarchical consistency play a central role; we have discussed the generalization of some of these properties to extended modelling frameworks. In conclusion, we note that recent applications in areas such as semiconductor manufacturing [BHG+93], nuclear reactor safety [Law92] and intelligent vehicle / highway systems [Var93] suggest that tools and techniques for control of DES may be approaching a level of sophistication sufficient for the solution of problems of realistic complexity.

Acknowledgments

It is a pleasure to acknowledge some helpful discussions with Kai Wong.

References

[BÅ93] Albert Benveniste and Karl J. Åström. Meeting the challenge of computer science in the industrial applications of control: An introductory discussion to the special issue. *IEEE Trans. Automatic Control*, 38(7):1004–1010, July 1993.

[BH88] Y. Brave and M. Heymann. Formulation and control of real time discrete event processes. In *Proc. 27th IEEE Conf. on Decision and Control*, pages 1131–1132, December 1988.

[BH93] Y. Brave and M. Heymann. Control of discrete event systems modeled as hierarchical state machines. *IEEE Trans. Automatic Control*, 12:1803–1819, 1993.

[BHG+93] S. Balemi, G. J. Hoffmann, P. Gyugyi, H. Wong-Toi, and G. F. Franklin. Supervisory control of a rapid thermal multiprocessor. *IEEE Trans. on Automatic Control*, 38(7):1040–1059, July 1993.

[BW92] B. A. Brandin and W. M. Wonham. Supervisory control of timed discrete-event systems. Tech. Rpt. 9210, Systems Control Group, Dept. of Electl. and Comp. Engrg., University of Toronto, August 1992.

[Har87] D. Harel. Statecharts: a visual formalism for complex systems. *Science of Computer Programming*, 8:231–274, 1987.

[HL93] M. Heymann and F. Lin. On-line control of partially observed discrete event systems. Tech. Rpt. 9310, Center for Intelligent Systems, Technion – Israel Institute of Technology, Haifa 32000 Israel, March 1993.

[İnan92] K. İnan. An algebraic approach to supervisory control. *Mathematics of Control, Signals and Systems*, 5:151–164, 1992.

[IV88] Kemal İnan and Pravin Varaiya. Finitely recursive processes. In P. Varaiya and A. B. Kurzhanski, editors, *Discrete Event Systems: Models and Applications, IIASA Conference, Sopron, Hungary, Aug. 3 – 7, 1987 (LNCIS, vol. 103)*, pages 1–18, New York, 1988. Springer-Verlag.

[KGM92] Ratnesh Kumar, Vijay Garg, and Steven I. Marcus. On supervisory control of sequential behaviors. *IEEE Trans. Automatic Control*, 37(12):1978–1985, December 1992.

[KGM93] Ratnesh Kumar, Vijay Garg, and Steven I. Marcus. Predicates and predicate transformers for supervisory control of discrete event dynamical systems. *IEEE Trans. on Automatic Control*, 38(2):232–247, February 1993.

[Law92] Mark Stephen Lawford. Transformational equivalence of timed transition models. Tech. Rpt. Systems Control Group, Dept. of Electl. and Comp. Engrg., Univ. of Toronto, January 1992.

[LW88] Y. Li and W. M. Wonham. On supervisory control of real-time discrete-event systems. *Information Sciences*, 46(3):159–183, 1988.

[LW93] F. Lin and W. M. Wonham. Supervisory control of timed discrete event systems under partial observations. Tech. Rpt. 9316, Systems Control Group, Dept. of Electl. and Comp. Engrg., Univ. of Toronto, November 1993.

[MP92] Zohar Manna and Amir Pnueli. *The Temporal Logic of Reactive and Concurrent Systems*, volume 1. Specification. Springer-Verlag, New York, 1992.

[Ost92] J. S. Ostroff. Formal methods for the specification and design of real-time safety-critical systems. *J. Systems and Software*, 18:33–60, April 1992.

[OW90] Jonathan S. Ostroff and W. Murray Wonham. A framework for real-time discrete-event control. *IEEE Trans. Automatic Control*, 35(4):386–397, 1990.

[OW91] Cüynet M. Özveren and Alan S. Willsky. Output stabilizability of discrete-event dynamic systems. *IEEE Trans. Automatic Control*, 36(8):925–935, August 1991.

[Ram89] Peter J. G. Ramadge. Some tractable supervisory control problems for discrete-event systems modeled by Büchi automata. *IEEE Trans. Automatic Control*, 34(1):10–19, January 1989.

[RW87] P. J. Ramadge and W. M. Wonham. Modular feedback logic for discrete event systems. *SIAM J. Control and Optimization*, 25(5):1202–1218, 1987.

[RW89] P. J. Ramadge and W. M. Wonham. The control of discrete event systems. *Proceedings of the IEEE*, 77(1):81–98, January 1989.

[RW92] Karen Rudie and W. Murray Wonham. Think globally, act locally: Decentralized supervisory control. *IEEE Trans. Automatic Control*, 37(11):1692–1708, November 1992.

[TW94a] J. G. Thistle and W. M. Wonham. Control of infinite behaviour of finite automata. To appear in *SIAM J. Control and Optimization* **32** (4), 1994.

[TW94b] J. G. Thistle and W. M. Wonham. Supervision of infinite behaviour of discrete-event systems. To appear in *SIAM J. Control and Optimization* **32** (4), 1994.

[Var93] Pravin Varaiya. Smart cars on smart roads: Problems of control. *IEEE Trans. on Automatic Control*, 38(2):195–207, February 1993.

[WTH92] Howard Wong-Toi and Gérard Hoffmann. The control of dense real-time discrete-event systems. Preprint, 1992.

[WW92] K. C. Wong and W. M. Wonham. Hierarchical and modular control of discrete-event systems. In *Proceedings of 30th Ann. Allerton Conf. on Communication, Control, and Computing*, pages 614–623, September 1992.

[WW93] K.C. Wong and W. M. Wonham. Hierarchical control of timed discrete-event systems. In *Proceedings of Second European Control Conference*, pages 509–512, June-July 1993.

[YSG92] S. Young, D. Spanjol, and V. K. Garg. Control of discrete event systems modeled with deterministic Büchi automata. In *Proceedings of 1992 American Control Conference*, pages 2809–2813, 1992.

[ZW90] Zhong Hao and W. Murray Wonham. On the consistency of hierarchical supervision in discrete-event systems. *IEEE Trans. Automatic Control*, 35(10):1125–1134, October 1990.

Automata-Theoretic Verification of Coordinating Processes

R. P. Kurshan

AT&T Bell Laboratories, Murray Hill, New Jersey 07974

1 Introduction

We address the problem: how to verify mathematically that a system of coordinating components behaves as it should. In this context, the "system" typically is a hardware and/or software implementation of a control algorithm. Examples of systems subject to the type of formal verification addressed here include controllers which implement communication protocols, cache coherency protocols and telephone switches. However, for our purposes, a system may as well be a subcircuit which implements an adder, a state machine implementing a lexical parser, a game such as nim, or a discrete-event economic model. The real-time behavior of systems also may be analyzed in this context, as long as the tested attributes are discrete events.

Given a mathematical model of such a system, one seeks to verify that the model has a given attribute. The attribute could be that all modelled behaviors are consistent with another model. The second model may represent a single aspect of the first model, such as "every message is eventually delivered", "all cache reads are consistent", "add$(a, b) = a + b$", "black wins in 3" and so on.

It is this process which we call formal verification. The expression "formal verification", as it appears in the literature, refers to a variety of (often quite different) methods used to prove that a model of a system has certain specified attributes. What distinguishes *formal* verification from other undertakings also called "verification", is that *formal* verification conveys a promise of mathematical certainty. The certainty is that if a model is formally verified to have a given attribute, then no behavior or execution of the model ever can be found to contradict this. (This is assuming the formal verification itself is mathematically correct!). This is distinct from other types of "verification" which simply test the given model in a (possibly large but not exhaustive) set of cases, wherein the model is said to be "verified" if none of the tested cases contradict the asserted attribute. These types of "verification", more accurately called "testing" (or, in hardware circles, "simulation"), never can serve to prove that the model in all cases has the required attribute. (Typical circuit testing may run billions of tests, requiring more than a year to execute, and yet not cover even close to 1% of the possible circuit states. Moreover, many circuits implement nonterminating models, which cannot be tested exhaustively through circuit execution, in finite time.)

While simulation is the current norm for testing computer hardware (software testing typically is even more primitive), it has become increasingly recognized in industry that this is (even grossly) insufficient for many applications. This realization has given rise to a resent upsurge of interest in formal verification, at least in the hardware community. Although methods for testing hardware are applicable to testing much software as well, *e.g.*, the large control software found in communication protocols and telephone switches, the software community remains about 5 years behind the hardware community in the application of these techniques, early warnings notwithstanding [Yeh83] (*cf.* [Pet91]). Only recently — as the hardware community has begun to adopt formal verification — has the software community begun to adopt hardware simulation methods.

Today, there is a consensus among practitioners of formal verification that from a practical point of view, computer systems are too complex to be verified by hand [Bar89]. As a result, *formal verification* is becoming synonymous with *computer-aided verification* (*cf.* [CAV]). Computer-aided verification means using a computer, for increased speed and reliability, to carry out the steps of the verification. More than simply automating calculations, computer-aided verification has spawned techniques which would be entirely infeasible without a computer. These new techniques are based upon largely new mathematical and logical foundations. They have engendered new algorithms, as well as questions concerning computational complexity, expressive equivalence and heuristics.

Generally speaking, formal verification may be construed as theorem-proving in a given logic. However, in practice, research in formal verification falls within various subcategories. The oldest and most general form of computer-aided formal verification is this: given two formulae f, g in a reasonably expressive logic, prove $f \Rightarrow g$. This most general form has been studied and practiced since before 1960, under the name *automated theorem-proving*. While results in automated theorem-proving are the most general in scope, they are the most limited algorithmically; most often the question "$f \Rightarrow g$?" in such a general context is undecidable. Thus, in this context, computer-aided verification requires extensive user involvement. Dissatisfaction with this largely nonalgorithmic form of verification and its effective inability to address many industrial problems gave rise to highly restricted forms of theorem-proving for which all theorems are decidable. For example, in restricted logics such as linear-time temporal logic (LTL), proving $f \Rightarrow g$ is equivalent to testing the unsatisfiability of the formula $f \wedge \neg g$, which is decidable, albeit in time exponential in the size of the formula. This exponential complexity continued to present a practical barrier, as did expressive limitations of LTL (it cannot adequately express sequentiality, and thus is not useful for defining most system models). Great inroads through these barriers were made in the context of another form of formal verification known as *model-checking*, which for the branching-time logic CTL admits of a linear-time decision procedure. In model-checking as discussed here, the theorems "$f \Rightarrow g$" are restricted to the case where f is a sequential system model and g is a logical formula which is "checked" on f; the relation generally is written $f \models g$ ("f is a model of g"). Between automated theorem-proving and model-checking fall other types of formal verification such as the extensive study of the Turing-expressive Petri nets whose theorems in general are undecidable, but for reasonably broad special cases are decidable and thus machine-checkable.

The sequential system model and linear-time decidability afforded by CTL model-checking represented a major practical advance in formal verification. Recently, its applicability has been broadened dramatically through the utilization of fixed-point procedures implemented with binary decision diagrams [BCM90], [McM93]. This has allowed large sets of states to be analyzed simultaneously, further offsetting the limitation imposed by computational complexity. However, system models typically are expressed in a fashion which is logarithmically succinct as compared with the size of the model state space (a situation often called "state-space explosion"). Thus even these major advances are of no help in the direct model-checking of large classes of problems, the computational complexity of which remain orders of orders of magnitude beyond any conceivable computational resources.

This remaining computational barrier is our reason for focusing on *automata-theoretic verification.* Automata-theoretic verification is a type of model-checking which admits of specific *reduction* algorithms which, although only heuristics, often can break through the computational complexity barrier. Automata also are capable of expressing eventuality (or "fairness") assumptions not expressible in CTL (or even the expressively richer — and computationally more complex — CTL*), whereas CTL can express branching-time properties not expressible by automata. In automata-theoretic verification, "formulae" all are of the form $\mathbf{x} \in \mathcal{L}(P)$ (read "$\mathbf{x}$ is a behavior of model P"). Thus, theorems all are of the form $\mathbf{x} \in \mathcal{L}(P) \Rightarrow \mathbf{x} \in \mathcal{L}(T)$, *i.e.*, $\mathcal{L}(P) \subset \mathcal{L}(T)$. As a form of model-checking, the system is modelled by the automaton P, the "formula" to be checked by the automaton T, and the "check" is whether the language of P is contained in the language of T. Using an automaton P to model the system provides support for eventuality ("fairness") assumptions not possible with pure CTL, while maintaining the expressiveness of model sequentiality. Using the automaton T rather than a CTL formula, to express the property to be checked, sacrifices the expressiveness of branching-time properties and the syntactic flexibility of logic, in exchange for the expressiveness of arbitrary ω-regular sequential properties (with fairness), and most significantly, the opportunity to apply powerful *reduction* algorithms.

The essence of reduction is to replace a computationally expensive test $\mathcal{L}(P) \subset \mathcal{L}(T)$ with a set of collectively cheaper tests $\mathcal{L}(P_i') \subset \mathcal{L}(T_i')$, $i = 1, \ldots, n$, such that $\mathcal{L}(P_i') \subset \mathcal{L}(T_i') \; \forall i \Rightarrow \mathcal{L}(P) \subset \mathcal{L}(T)$. While there is no guarantee that "cheaper" is cheap enough, the reductions developed here have been of considerable practical value in a large variety of problems arising in industrial settings. This has expanded the class of models, and properties of those models, which can be formally verified.

Reduction consists of two steps: *task decomposition* and *task localization.* In the decomposition step, a global system "task" or property T is decomposed into local properties T_i such that

$$\cap \mathcal{L}(T_i) \subset \mathcal{L}(T) , \quad (1)$$

i.e., the local properties taken together imply the global property. Thus, the sought-after conclusion $\mathcal{L}(P) \subset \mathcal{L}(T)$ will follow from

$$\mathcal{L}(P) \subset \mathcal{L}(T_i), \quad \forall i .$$

The localization step replaces each test

$$\mathcal{L}(P) \subset \mathcal{L}(T_i) \quad (2)$$

with a computationally simpler test

$$\mathcal{L}(P_i') \subset \mathcal{L}(T_i') \quad (3)$$

which implies the former. The automaton P_i' is a reduction of P *relative* to T_i. This reduction serves to abstract those portions of P "irrelevant" to the "local" property T_i; the automaton T_i' is an abstraction of T_i consistent with the abstraction of P relative to T_i.

The localization reduction of (2) to (3) and the checking of (1) and (3) all may be done algorithmically. Finding useful heuristics to determine a viable decomposition (1) is one of the foremost open problems in this field. Presently, one uses knowledge of P and T to guess a decomposition (1); a bad guess may fail to satisfy (1), or may not lead to tractable reduction checks (3), but otherwise cannot interfere with the validity of the verification process.

Once (1) is verified, and (3) is verified for all i, (2) follows, and thus so does its consequence $\mathcal{L}(P) \subset \mathcal{L}(T)$.

2 An Example

System models arising in industrial settings often have state spaces of size 10^{100} to 10^{10^6} states. We consider next (for illustrative purposes) a somewhat smaller system model with 8 states.

Consider the example of a system comprising a "crossroad traffic controller", composed of 3 coordinating components: Avenue A, intersecting Boulevard B and traffic Controller C. We model each of these components by a state machine (named A, B and C, respectively), each with inputs and outputs. The outputs of A are 3 tokens:

(A : no_cars)
(A : cars_waiting)
(A : cars_going)

and likewise for B. The outputs of C are also 3 tokens:

(C : go_A)
(C : go_B)
(C : pause) .

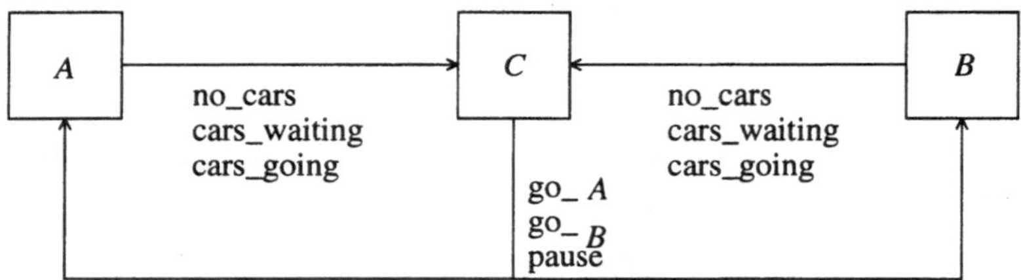

Fig. 1. Data-flow diagram for the Crossroad Traffic Controller, showing the flow of data (control tokens) among its three component processes.

The outputs of the various machines form the inputs to the others. The data-flow (*i.e.*, input/output token-flow) among the 3 component state machines is given by the diagram of Fig. 1.

Each of A, B and C have 2 states, as shown in Fig. 2. At each state (*e.g.*, STOP, GO, go_A, *etc.*) of each state machine there are 2 possible outputs, given inside the associated $\{\cdots\}$; in any execution of the system, the choice of output from each respective state is nondeterministic. For example, in its state STOP, machine A may output either the token (A : no_cars) or the token (A : cars_waiting), the choice of which being nondeterministic. The diagrams below define state transitions in terms of Boolean predicates of the output tokens, described along edges, with $*$ denoting AND, $+$ denoting OR and $\sim$ denoting NOT. The transition structure for B is the same as that for A, but with A and B interchanged throughout. (The thoughtful reader may have trouble interpreting aspects of this example, such as the possibility that (A : cars_waiting) may be followed immediately by (A : no_cars); while a little fanciful thought may help — say the car turned around and left the intersection! — the purpose of the example is to illustrate verification methodologies, not the proper modelling of a crossroad controller!)

The crossroads "system" is modelled by the coordination of the 3 components A, B and C, comprising the system model X. The states of X thus are 3-tuples of states of A, B and C respectively, while the transition conditions for X are given in terms of respective Boolean conjunctions of transition conditions of the components. Thus, for example, letting A(STOP, GO) be the condition for A to move from its state STOP to its state GO, namely,

$$\mathsf{A(STOP, GO)} = \mathsf{A(cars\_waiting)} * \mathsf{(C : go\_A)}$$

and likewise for B, C and X, we get

$$\begin{aligned}&\mathsf{X((STOP, STOP, go\_A),\ (GO, STOP, go\_A))}\\ &\qquad = \mathsf{A(STOP, GO)} * \mathsf{B(STOP, STOP)} * \mathsf{C(go\_A\ ,\ go\_A)}\\ &\qquad = \mathsf{(A : cars\_waiting)} * \mathsf{(B : no\_cars)} * \mathsf{(C : go\_A)}\end{aligned}$$

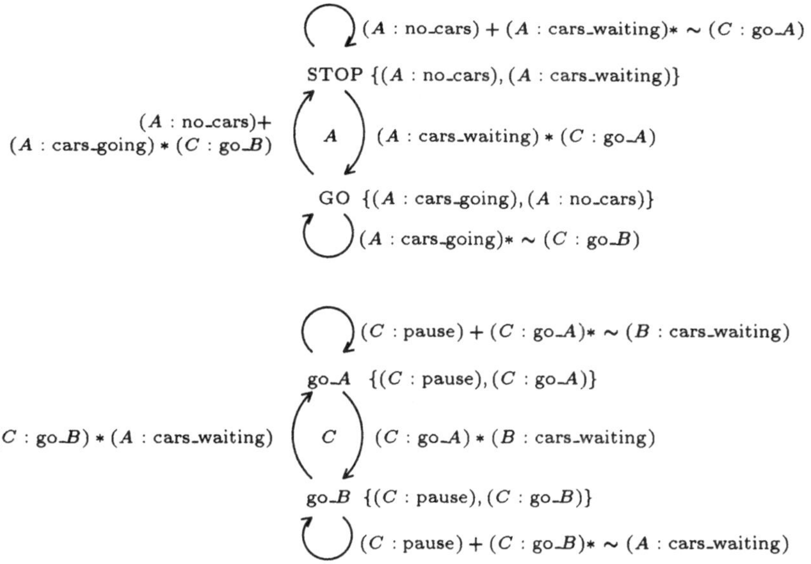

$(A : \text{no\_cars}) + (A : \text{cars\_waiting}) * \sim (C : \text{go\_}A)$

STOP $\{(A : \text{no\_cars}), (A : \text{cars\_waiting})\}$

$(A : \text{no\_cars}) +$ $(A : \text{cars\_going}) * (C : \text{go\_}B)$

A

$(A : \text{cars\_waiting}) * (C : \text{go\_}A)$

GO $\{(A : \text{cars\_going}), (A : \text{no\_cars})\}$

$(A : \text{cars\_going}) * \sim (C : \text{go\_}B)$

$(C : \text{pause}) + (C : \text{go\_}A) * \sim (B : \text{cars\_waiting})$

go_A $\{(C : \text{pause}), (C : \text{go\_}A)\}$

$(C : \text{go\_}B) * (A : \text{cars\_waiting})$

C

$(C : \text{go\_}A) * (B : \text{cars\_waiting})$

go_B $\{(C : \text{pause}), (C : \text{go\_}B)\}$

$(C : \text{pause}) + (C : \text{go\_}B) * \sim (A : \text{cars\_waiting})$

Fig. 2. State transition diagrams for processes A and C.

as is easily computed from the diagrams above.

We add to the definitions of A, B and C a designation of respective *initial* states, say:

$$I(A) = \{\mathsf{STOP}\}\ ,$$
$$I(B) = \{\mathsf{STOP}\}\ ,$$
$$I(C) = \{\mathsf{go\_A}, \mathsf{go\_B}\}\ .$$

This is inherited by X, giving the initial states of X:

$$I(X) = \{(\mathsf{STOP}, \mathsf{STOP}, \mathsf{go\_A}),\ (\mathsf{STOP}, \mathsf{STOP}, \mathsf{go\_B})\}\ .$$

We express the assumption that X is defined by the coordination of A, B and C as above, by writing $X = A \otimes B \otimes C$.

Having modelled the crossroad system by X (defined in terms of its coordinating component models A, B, C as $X = A \otimes B \otimes C$), we are ready to "verify" X. What should we verify? Let us identify two specific properties we wish to verify for X. We may think of these properties as *tasks* which X is intended to perform. Thus, we seek to *verify* that X performs each of these *tasks*. The first task is to ensure that *no cars collide* (an invariant property — it is enough to check that it holds at each state — in some circles called a "safety" property, although "safety" is used somewhat more generally). The second task whose performance we wish to verify is that the system model X ensures that *all the cars on each road eventually get through the intersection.* (This "eventuality" property is intrinsically more complicated than an invariant property, as it cannot directly be seen to be true or false at a given state — sometimes this type of property is called a "liveness" condition.) We may model each of these two tasks by automata.

The invariant task may be modelled by a 2-state automaton T_1 as follows:

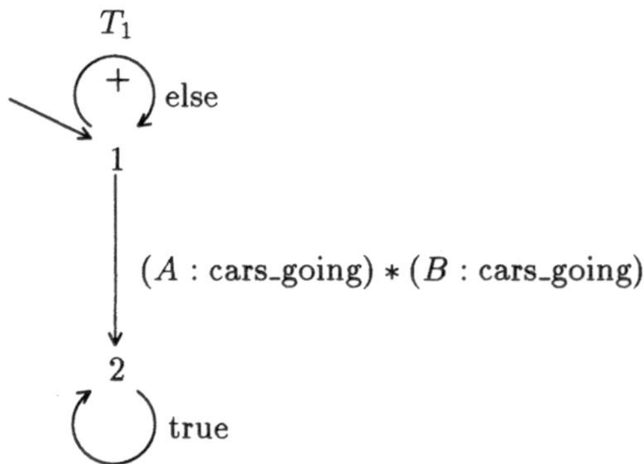

The way to read this is: T_1 has initial state 1, designated by the short arrow entering 1; the "self-loop" transition $1 \to 1$ marked by a '+' is a "recur" edge, meaning that any infinite path through the transition structure of T_1 which crosses the edge $1 \to 1$ infinitely often designates a behavior "accepted" by T_1. It is clear that every behavior (of X) is accepted by T_1 unless it involves **cars_going** simultaneously on both roads (a collision), in which case T_1 moves to its state 2, precluding an infinite number of recur-edge crossings, and thus precluding the acceptance by T_1 of any behavior which includes that collision event. *Every invariant task may be modelled by such a 2-state automaton.* (When implemented in a verification computer program, it is common simply to halt the program — returning an error — when the condition which labels the state transition $1 \to 2$ is enabled.)

The eventuality task, which we already said is more complicated, can be modelled by a 4-state automaton T_2. However, let us instead decompose T_2 into local tasks as in (1). This circumvents computations which involve X and T_2, in favor of less complex computations involving X and simpler subtasks of T_2. It moreover permits the "localization" of X relative to each subtask. Localization replaces X with a system which is reduced relative to the subtask, thereby lessening the computational complexity of verification of each subtask, and hence of T_2 itself. We may decompose this task T_2 into two subtasks: *all cars on A eventually get through* and *all cars on B eventually get through*. This decomposition may be accepted empirically. Alternatively, we may model the subtasks and then prove that all behaviors common to each of the two subtasks are behaviors of T_2, as in (1). Notice that this decomposition need be only approximate, in the sense that (1) involves only language containment, not equality. Notice also that the subtasks are "local", referring to a smaller portion of X than the more "global" property T_2. In practice, it is often the case that global tasks (invariant tasks, as well as eventuality tasks) may be decomposed into several (simpler) local tasks, as in this example. The advantage of localization, demonstrated explicitly below, is that verification may be simplified by collapsing portions of the system model X which are irrelevant to the performance of the subtask.

To model *all cars on A eventually get through*, we may use the 2-state automaton T_A shown in Fig. 3, and likewise for T_B, modelling *all cars on B eventually get through*. (Notice that a behavior is accepted *unless* in A, at some point cars are waiting, and thereafter it is never the case that cars are going or that there are no more cars.)

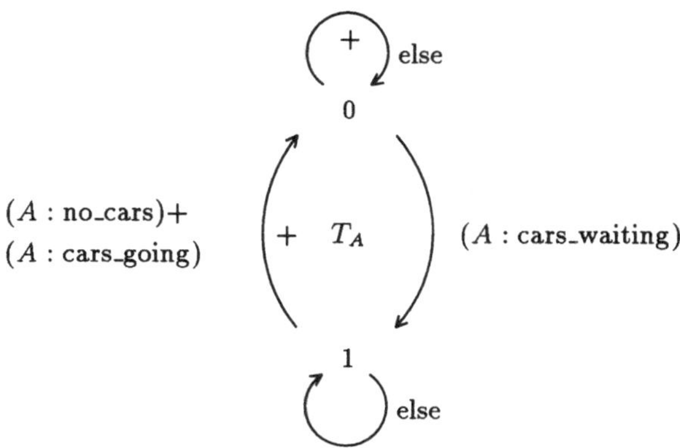

Fig. 3. The automaton T_A, modelling the requirement that all cars on Avenue A eventually get through the intersection.

We now describe the formal context in which we seek to prove that X performs the tasks defined by T_1 and T_2. The "behavior" of X is defined in terms of its (nonterminating) executions. Each behavior of X is by definition an (infinite) sequence $\mathbf{x} = (x_0, x_1, \ldots)$ where each $x_i = x_{Ai} * x_{Bi} * x_{Ci}$, for tokens x_{Ai}, x_{Bi}, x_{Ci} output from A, B and C respectively. Thus, for example, we may have

$$x_0 = (\mathsf{A}:\mathsf{cars\_waiting}) * (\mathsf{B}:\mathsf{cars\_waiting}) * (\mathsf{C}:\mathsf{go\_A}) .$$

In this case, a possible value for x_1 is

$$x_1 = (\mathsf{A}:\mathsf{cars\_going}) * (\mathsf{B}:\mathsf{cars\_waiting}) * (\mathsf{C}:\mathsf{pause})$$

(the reader is invited to find the other seven possible values for x_1).

Let $\mathcal{L}(X)$ (the "language" of X) denote the set of all such behaviors $\mathbf{x}$ of X. Likewise, we define the language $\mathcal{L}(T)$ of an automaton T to be the set of all sequences $\mathbf{x}$ accepted by T, where each x_i is of the form described above. Although each x_{Ai}, x_{Bi}, x_{Ci} is required to be an output token of A, B and C respectively, there is no requirement that $\mathbf{x} \in \mathcal{L}(T)$ satisfy $\mathbf{x} \in \mathcal{L}(X)$.

For example, if for all $i \geq 0$,

$$x_i = (\text{ A : cars\_waiting}) * (\text{B : cars\_going}) * (\text{C : pause}) \ ,$$

then $\mathbf{x} \notin \mathcal{L}(T_A)$ (also $\mathbf{x} \notin \mathcal{L}(X)$ — why?). On the other hand, if for all $i \geq 0$

$$x_{2i} = (\text{A : cars\_waiting}) * (\text{B : cars\_going}) * (\text{C : pause}) \ ,$$
$$x_{2i+1} = (\text{A : cars\_going}) * (\text{B : cars\_waiting}) * (\text{C : pause})$$

then $\mathbf{x} \in \mathcal{L}(T_1) \cap \mathcal{L}(T_A) \cap \mathcal{L}(T_B)$ (but again, $\mathbf{x} \notin \mathcal{L}(X)$).

"Verification" of X (in this context) consists of proving that

$$\mathcal{L}(X) \subset \mathcal{L}(T_1) \cap \mathcal{L}(T_2) \ .$$

We do this by checking the following steps:

$$\mathcal{L}(X) \subset \mathcal{L}(T_1) \ ;$$
$$\mathcal{L}(X) \subset \mathcal{L}(T_A) \ ;$$
$$\mathcal{L}(X) \subset \mathcal{L}(T_B) \ ;$$
$$\mathcal{L}(T_A) \cap \mathcal{L}(T_B) \subset \mathcal{L}(T_2) \ .$$

Consider the first language containment check: although both $\mathcal{L}(X)$ and $\mathcal{L}(T_1)$ are infinite sets, it is possible to check $\mathcal{L}(X) \subset \mathcal{L}(T_1)$ in a finite number of steps. Basically, it works like this: construct the transition system $X \otimes T_1$ (analogous to the earlier definition $X = A \otimes B \otimes C$), and find the cycles in the finite directed graph which $X \otimes T_1$ represents. Although there are an infinite number of cycles, they are all captured by the (finite) set of strongly connected components of this graph. (A strongly connected component of a directed graph is a nonempty set of vertices maximal with respect to the property that any ordered pair of vertices from the set are connected by a directed path from the first to the second.) Since $X \otimes T_1$ is a finite state system, any (infinite) behavior $\mathbf{x} \in \mathcal{L}(X \otimes T_1)$ must describe a trajectory which eventually cycles within some single strongly connected component of the graph. It is enough then to check that each strongly connected component is consistent with the acceptance structure (in this case, the *recur* edges) of T_1. One can write a decision procedure to perform this check.

In order to check $\mathcal{L}(X) \subset \mathcal{L}(T_1)$, we may construct $X \otimes T_1$ and proceed as above. This requires searching a state space roughly of size $|X \otimes T_1| = |A| \cdot |B| \cdot |C| \cdot |T_1| = 2^4$, where $|Z|$ denotes the number of states in Z ("roughly", because some of the states of $X \otimes T_1$ are unreachable from its initial states, and thus do not enter into the decision procedure). As we are interested to be able to perform the language containment check automatically on a computer, we are concerned about the computational complexity of this check: the complexity is (roughly) related to the size $|X \otimes T_1|$ of the state space. We now describe a "reduction" of this language containment test to a search of a smaller state space.

First, we must discuss how we define the language of system components. For the component A of X, we define $\mathcal{L}(A)$ as the set of behaviors $\mathbf{x}$ *over the same output space as* X consistent with the transition structure of A. This takes into account tokens from B, although A "doesn't care" about B. (This output space is an underlying constant, forming the "alphabet" of all automata/state machines.) Thus, $\mathbf{x} \in \mathcal{L}(A)$ has the form $x_i = x_{Ai} * x_{Bi} * x_{Ci}$ described earlier; for $\mathbf{x}$ defined by

$$x_0 = (\mathsf{A : cars\_waiting}) * (\mathsf{B : cars\_waiting}) * (\mathsf{C : go\_A}) ,$$
$$x_i = (\mathsf{A : cars\_going}) * (\mathsf{B : cars\_going}) * (\mathsf{C : go\_A})$$

for all $i > 0$, $\mathbf{x} \in \mathcal{L}(A)$ while $\mathbf{x} \notin \mathcal{L}(B)$, $\mathbf{x} \notin \mathcal{L}(C)$ and $\mathbf{x} \notin \mathcal{L}(X)$. A basic observation concerning $\otimes$ is that $\otimes$ supports the *language intersection property*:

$$\mathcal{L}(A \otimes B \otimes C) = \mathcal{L}(A) \cap \mathcal{L}(B) \cap \mathcal{L}(C) .$$

(The intuition for this is that each component constrains behaviors, and these constraints taken together comprise the constraints of X.) Thus, in order to prove that $\mathcal{L}(X) \subset \mathcal{L}(T_1)$, if we can prove instead that

$$\mathcal{L}(X') \subset \mathcal{L}(T_1)$$

for $X' = A \otimes B \otimes C'$ with $\mathcal{L}(C) \subset \mathcal{L}(C')$, then by the language intersection property, $\mathcal{L}(X) \subset \mathcal{L}(T_1)$ follows. (However, if $\mathcal{L}(X') \subset \mathcal{L}(T_1)$ fails, $\mathcal{L}(X) \subset \mathcal{L}(T_1)$ may or may not hold.) If $|C'| < |C|$, then $|X'| < |X|$ and testing $\mathcal{L}(X') \subset \mathcal{L}(T_1)$ hopefully involves a search of a smaller state space than does testing $\mathcal{L}(X) \subset \mathcal{L}(T_1)$. If in this context $\mathcal{L}(X) \subset \mathcal{L}(X')$, and $|X' \otimes T_1| < |X \otimes T_1|$, we say X' is a *reduction* of X (relative to T_1). Our candidate for X' is obtained by reducing C to a 1-state model C', having (nondeterministic) output from that state of all the 3 tokens of C:

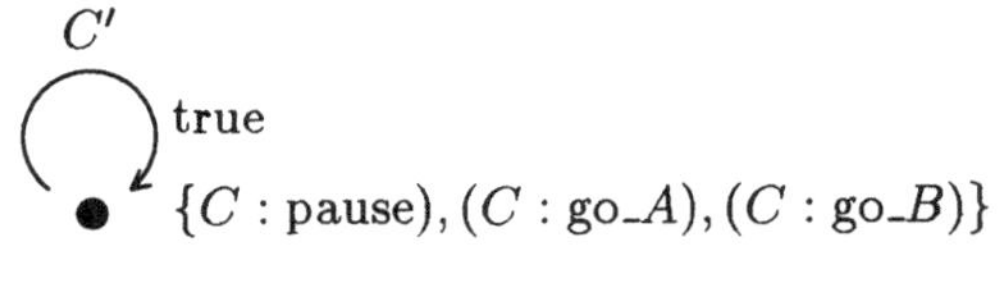

C'

true

$\{C : \text{pause}), (C : \text{go\_}A), (C : \text{go\_}B)\}$

It should be clear upon a moment's reflection that $\mathcal{L}(C) \subset \mathcal{L}(C')$, as C' provides none of the transition constraints that C does. The resulting X' is a particularly trivial form of reduction. In general, C' may have fewer states than C, but more than 1 state; in fact, as we shall see shortly, we may get more powerful reductions by changing the underlying alphabet.

In fact it can be checked that $\mathcal{L}(X') \subset \mathcal{L}(T_1)$ and thus we can conclude indirectly that $\mathcal{L}(X) \subset \mathcal{L}(T_1)$.

Next, let us consider the test

$$\mathcal{L}(X) \subset \mathcal{L}(T_A) .$$

We could apply the same method as before, this time reducing B to a single-state machine B'. However, we can do even better than this. From the perspective of analyzing X for the performance of the task T_A, there is no need to distinguish among the output tokens of B. Thus, let us define an entirely new system $X' =$

$A' \otimes B' \otimes C'$ with associated task T'_A, in which B' has a single output token, say $(\mathsf{B}' : \mathsf{null})$, and A', B', C' and T'_A are derived from A, B, C and T_A respectively by replacing every appearance of $(\mathsf{B} : \mathsf{no\_cars})$, $(\mathsf{B} : \mathsf{cars\_waiting})$ or $(\mathsf{B} : \mathsf{cars\_going})$ with the token $(\mathsf{B} : \mathsf{null})$. We require that T_A be invariant with respect to this transformation: that $T'_A = T_A$, as is obviously true, since T_A does not involve B. After such a simplifying transformation, it is often possible in general to reduce the state spaces of the components, while retaining their respective languages (in the transformed token space). For this example, this transformation gives $A' = A$. However, B' looks like this (after simplification of expressions):

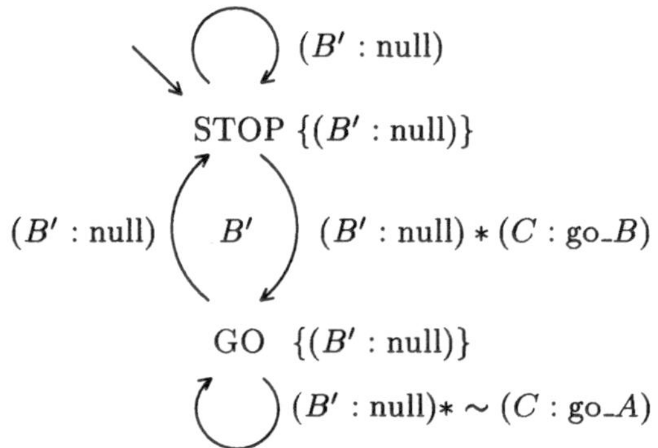

Applying state minimization to B' gives B'' with $\mathcal{L}(B'') = \mathcal{L}(B')$, where B'' looks like this:

B''

true

$\{(B' : \text{null})\}$

(as is readily evident). Although the transition structure of C' is simpler than C, for this transformation we cannot reduce the size of the state space of C' and maintain the same language $\mathcal{L}(C')$. (While we could reduce C to a single state as before, it can be checked that the resulting X' would not perform the task — the reduction, while valid, is too great to prove performance.)

Setting $X'' = A' \otimes B'' \otimes C'$, since $T'_A = T_A$,

$$\mathcal{L}(X'') \subset \mathcal{L}(T'_A) \Rightarrow \mathcal{L}(X) \subset \mathcal{L}(T_A)\ .$$

(In fact, we have what is called a *strongly co-linear* reduction, which implies language containment for the reduction is both *necessary* as well as sufficient for containment of the unreduced problem.) Note that there is no reason to require $\mathcal{L}(B') = \mathcal{L}(B'')$; $\mathcal{L}(B') \subset \mathcal{L}(B'')$ is enough. Examples show that such a more general reduction can be very effective in reducing verification to computationally tractable checks.

Next, we observe an example of a fairly common method for exploiting symmetry in a model. It is common to find system models which contain a fair amount of symmetry such as the symmetry between A and B in X.

Let φ be the map of system output tokens which interchanges the tokens of A and B (so $\varphi(\mathsf{A}:\mathsf{no\_cars}) = (\mathsf{B}:\mathsf{no\_cars})$, *etc*), and $\varphi(\mathsf{C}:\mathsf{go\_A}) = (\mathsf{C}:\mathsf{go\_B})$, $\varphi(\mathsf{C}:\mathsf{go\_B}) = (\mathsf{C}:\mathsf{go\_A})$, $\varphi(\mathsf{C}:\mathsf{pause}) = (\mathsf{C}:\mathsf{pause})$. Then φ induces a map of the transition predicates, which may be extended to a map of the states and hence of the state machines and automata, themselves, giving $\varphi A = B$, $\varphi B = A$, $\varphi C = C$, $\varphi X = X$ and $\varphi T_A = T_B$. Once we have proved that $\mathcal{L}(X) \subset \mathcal{L}(T_A)$, it follows that $\mathcal{L}(X) = \mathcal{L}(\varphi X) \subset \mathcal{L}(\varphi T_A) = \mathcal{L}(T_B)$. Thus, it is unnecessary to perform the specific analysis to prove $\mathcal{L}(X) \subset \mathcal{L}(T_B)$; it is sufficient to prove $\varphi A = B$, $\varphi B = A$ and $\varphi C = C$ (from which it follows that $\varphi X = X$), and to prove that $\varphi T_A = T_B$. Each of these checks is computationally simple, being a check on a component rather than a check on the entire system model X.

Thus, we have reduced the check

$$\mathcal{L}(X) \subset \mathcal{L}(T_2)$$

to the check

$$\mathcal{L}(X') \subset \mathcal{L}(T_A)\ .$$

In fact, this check fails! The trouble can be seen from the following error track which shows a sequence of global states (*i.e.*, states of X) starting from an initial state 1(0) and ending in a "Bad cycle", in this case from 3(1) back to itself. Here $X.A.\$$ is the state of the A-component of X, $X.B.\$$ is the state of the *reduced* B''-component of X and so on; the variables '#' store the values of the respective nondeterministically selected output tokens which determine the transitions from one state to the next.

```
1(0) .X.A.$=STOP .X.B.$=0 .X.C.$=go_B .TA.$=0
     .X.A.#=cars_waiting .X.B.#=null .X.C.#=go_B

Bad cycle at 3(1) back to 3(1):

3(1) .X.A.$=STOP .X.B.$=0 .X.C.$=go_A .TA.$=1
     .X.A.#=cars_waiting .X.B.#=null .X.C.#=pause
```

The problem is seen to begin in global state number 3(1), reachable from the initial state number 1(0) in a single step. At global state 3(1), A has selected **cars_waiting** and C is in its state **go_A**, but C has selected **pause** instead of **go_A**, so A cannot move out of its state STOP. This error track

terminates in an unaccepted cycle (which may repeat forever), in violation of the task T_A. However, we may like to change the model so as to forbid or disregard the possibility that C may "pause" forever. There are a number of ways to do this. For example, we could add it as a restriction to C, or as a relaxation of T_A. Let us relax T_A so as to model the task: **either** *all cars on A eventually get through* **or** *C eventually pauses forever.* To relax T_A (or restrict C) we use a second acceptance structure called a "cycle set": a designated set of states of T_A. If a behavior of X causes T_A eventually to stay inside a cycle set then that behavior also is accepted by T_A.

With this redefinition of T_A (and associated redefinitions of T_B and T_2), it can be checked that $\mathcal{L}(X') \subset \mathcal{L}(T_A)$, from which it follows that $\mathcal{L}(X) \subset \mathcal{L}(T_A)$, $\mathcal{L}(X) \subset \mathcal{L}(T_B)$ and hence $\mathcal{L}(X) \subset \mathcal{L}(T_2)$. We also need to verify that the transformation $X \to X'$ (induced by $B \to B''$) leaves T_A invariant: $T'_A = T_A$, and that $\mathcal{L}(T_A) \cap \mathcal{L}(T_B) \subset \mathcal{L}(T_2)$. The machinery for all of these checks is implemented in COSPAN [HK90].

References

[Bar89] J. Barwise. Mathematical Proofs of Computer System Correctness. *Notices* **36**, pages 884–851, (1989).

[BCM90] J. R. Burch, E. M. Clarke, K. L. McMillan, D. L. Dill, and L. J. Hwang. Symbolic Model Checking: 10^{20} States and Beyond, LICS; Sequential Circuit Verification Using Symbolic Model Checking, IEEE Design Automation Conference (DAC), 1990.

[CAV] Computer-Aided Verification. The proceedings to date have been published as *Lect. Notes in Comput. Sci.* **407** (1989), **531** (1990), **575** (1991), **663** (1992), **697** (1993).

[HK90] Z. Har'El and R. P. Kurshan. Software for Analytical Development of Communications Protocol. *AT&T Tech. J.* **69**, pages 45–59, (1990).

[McM93] K. L. McMillan. *Symbolic Model Checking.* Kluwer, 1993.

[Pet91] I. Peterson. Software Failure: Counting up the Risks. *Science News* **140**, page 388, (1991).

[Yeh83] R. Yeh. Software Engineering. *IEEE Spectrum*, pages 91–94, (1983).

Hierarchical COCOLOG for Finite Machines*

Y.J. Wei and P. E. Caines

Department of Electrical Engineering, McGill University
3480 University Street, Montreal, P.Q, Canada H3A 2A7
peterc@cim.mcgill.edu

1 Introduction

The apparent existence of hierarchical systems in the real world and the perception that hierarchical structures possess certain properties of efficient information processing and control have motivated many attempts to give mathematical formulation to the notion of a hierarchically structured system. In this paper we formulate hierarchical control problems for finite machines in terms of a hierarchical extension of COCOLOG (conditional observer and control logic) control theory (see [CW90], [WeC92]) to the lattice of so-called dynamically consistent partition machines.

Within the automata formulation of control theory for discrete event systems (DES), the research by Zhong and Wonham [ZW90] is devoted to the hierarchical structure of DES; one of the contributions of this work is the creation of aggregation methods which ensure that high level control events are well defined as functions of the aggregated events of a given (low level) automaton. The resulting theory has some points of contact with the present work in that the function of the editing procedure described in [CWe94] is to generate a set of dynamically consistent partition machines. Other work on system decomposition and decentralized control is to be found in Lin and Wonham [LW90] and Rudie and Wonham [RW92]; one principal distinction between that work and the notions presented here is that the state of the overall controlled system in [LW90,RW92] is the product of the states of the various sub-automata, whereas in this paper, the state of the system is a singleton $x \in X$ which falls in the appropriate state of a partition submachine according to the choice of state partition π.

The material in the first part of this paper follows in the line of work of the algebraic theory of automata and input-state-output machines. Notable among many contributions in this domain is the Krohn-Rhodes decomposition theory; they proved that any finite automaton can be decomposed via a covering by a cascade product of permutation-reset automata [G68]. While the Krohn-Rhodes theory is expressed in terms of the covering of an automaton, the hierarchical lattice of a finite machine $\mathcal{M}$, which is presented in [CWe94] and is summarized in §2 of this paper, is a lattice structured decomposition of $\mathcal{M}$. In this context we note that a different notion of a lattice of submachines of a given finite machine is to be found in the book by J.Hartmanis and R. Stearns [HS66].

* Work supported by NSERC grant A1329 and the PRECARN-NCE-IRIS program B5

Sections 3, 4 and 5 of this paper are dedicated to the introduction of the COCOLOG formulation of the notion of the hierarchical structure of a machine and its control theory. A family of multiple sort first-order languages for this purpose is defined in §3. In §4, a family of axiom sets $\left\{\Sigma_k^{\pi}, \Sigma_k^{1}, \ldots, \Sigma_k^{|\pi|}; k \geq 0\right\}$ is introduced to logically express multi-level COCOLOG control theories for $\mathcal{M}$. Section 5 deals with the communication that takes place between the COCOLOG theories at adjacent levels of the hierarchical system; this communication is mediated through two sets of extra-logical rules called, respectively, *the instantiation rule* and *the block membership rule*. The paper concludes with an example of the hierarchical control of an eight state machine.

2 The Hierarchical Lattice of a Finite Machine

2.1 Dynamically Consistent Partitions and Partition Machines

In this paper, a finite system is modeled by a finite machine $\mathcal{M} = < X, U, \Phi >$, where X is a finite set of states, U is a finite set of inputs and $\Phi : X \times U \to X$ is the transition function. Our formulation of the hierarchical structure of a finite machine is based upon the lattice of partitions of the machine's state space; readers are referred to [CWe94] for a detailed presentation of the ideas introduced in this and the subsequent subsections.

A partition π of X is a collection of pairwise disjoint subsets (called blocks) of X such that the union of this collection equals to X. A partition π_1 is said to be *stronger than* a partition π_2,, denoted by $\pi_1 \leq \pi_2$, if every π_1-block is a subset of some π_2-block. The greatest lower bound and the least upper bound of two partitions are defined respectively via *intersection* and *chain union*, for whose standard definitions we refer the readers to [CWe94]. Further, the resulting lattice PAR(X) of all partitions of X is bounded above and below by the trivial partition π_{tr} and the identity partition π_{id} respectively. We shall use the notation $\overline{X}_i$ whenever the set X_i is to be taken as a singleton in a partition set $\pi = \left\{\overline{X}_1, \ldots, \overline{X}_{|\pi|}\right\}$.

To a given state $x \in X$, one may apply a sequence σ of inputs, called a *control sequence*, and so generate a sequence t of states called a *trajectory*. Con$(\mathcal{M})$ and Tra$(\mathcal{M})$ shall denote the set of all legal control sequences (with respect to same specified constraint) and the set of resulting trajectories respectively. Given $\pi = \{X_1, \ldots, X_p\} \in$ PAR(X), A sequence of states $t = x_1 \ldots x_{l-1} x_l \in$ Tra$(\mathcal{M})$ is called an *internal trajectory with respect to* X_i if $t \subset X_i$; it is called a *direct trajectory from* X_i *to* X_j if $x_1 \ldots x_{l-1} \subset X_i$ and $x_l \in X_j$ for some $i \neq j$. Tra$(\pi)_i^i$ and Tra$(\pi)_i^j$ shall denote respectively the set of all internal (respectively direct) trajectories of $\mathcal{M}$ with respect to X_i and X_j. Any trajectory can be decomposed into a sequence of sub-trajectories, say $t_1 t_2 \ldots t_m$ such that each segment t_i is either an internal or a direct trajectory.

For the mapping $\theta : X \times \text{Con}(\mathcal{M}) \to \text{Tr}(\mathcal{M}), \theta((x, \sigma)) = x_0 x_1 x_2 \ldots x_p \in \text{Tra}(\mathcal{M})$, where $\sigma = u_0 u_1 u_2 \ldots u_{p-1} \in \text{Con}(\mathcal{M})$ and $x_0 = x$, $x_i = \Phi(x_{i-1}, u_{i-1})$, $1 \leq i \leq p$, we can define the *block control events* $U_i^i = \theta^{-1}\left(\text{Tra}(\pi)_i^i\right)$ and $U_i^j = \theta^{-1}\left(\text{Tra}(\pi)_i^j\right)$. These sets shall play the role of high level control inputs in the partition machine defined below. Obviously the number of symbols appearing in different elements of U_i^j may be different.

Now we introduce the concept of dynamical consistency. By this we shall mean that a high level transition from a block X_i to X_j is possible only when all states in X_i can be driven to some state in X_j without passing through a third block. A partition such that this consistency obtains is called a dynamically consistent partition.

Definition 1 (Dynamically Consistent Partition Machines — DCPMs). Given $\pi = \{\overline{X}_1, \ldots, \overline{X}_k\} \in \text{PAR}(X)$ and $(\overline{X}_i, \overline{X}_j) \in \pi \times \pi$, a *dynamically consistent condition* (DCC) holds for the set pair (X_i, X_j) if for each element x_i of X_i, there exists at least one direct trajectory from X_i to X_j which has initial state x_i, i.e.

$$\forall x \in X_i, \exists \sigma \in U^*, \forall \sigma' < \sigma \; (\Phi(x, \sigma') \in X_i \wedge \Phi(x, \sigma) \in X_j), \;\; i \neq j, \qquad (1)$$

$$\forall x \in X_i, \exists \sigma \in U^*, \forall \sigma' \leq \sigma \; (\Phi(x, \sigma') \in X_i), \;\; i = j, \qquad (2)$$

where $\sigma' \leq \sigma$ means that σ' is the initial segment of σ. For $\mathcal{U} \triangleq \{\overline{U}_i^j;\; U_i^j = \theta^{-1}\left(\text{Tra}(\pi)_i^j\right), 1 \leq i, j \leq k\}$, the transition function $\Phi^\pi : \pi \times \mathcal{U} \to \pi$ of the *(dynamically consistent) partition machine* (DCPM) $\mathcal{M}^\pi = <\pi, \mathcal{U}, \Phi^\pi>$ is defined by the equality relation $\Phi^\pi(\overline{X}_i, \overline{U}_i^j) = \overline{X}_j$ whenever DCC holds for (X_i, X_j), $1 \leq i, j \leq |\pi|$ and is not defined when DCC fails for (X_i, X_j). PAR$(\mathcal{M})$ shall denote the set of all partition machines of $\mathcal{M}$.

Definition 2 (Partial Ordering of DCPMs of $\mathcal{M}$). Given $\mathcal{M}^{\pi_1} = <\pi_1, \mathcal{U}, \Phi^{\pi_1}>$, and $\mathcal{M}^{\pi_1} = <\pi_1, \mathcal{V}, \Phi^{\pi_2}> \in \text{PAR}(\mathcal{M})$. We say $\mathcal{M}^{\pi_2}$ is *weaker* than $\mathcal{M}^{\pi_1}$, written as $\mathcal{M}^{\pi_1} \preceq \mathcal{M}^{\pi_2}$, if $\pi_1 \leq \pi_2$.

The trajectories in Tra$(\mathcal{M})$ and control sequence in Con$(\mathcal{M})$ can be concatenated in a consistent way. Let $s = x_1 \ldots x_p, t = y_1 \ldots y_q \in \text{Tra}(\mathcal{M})$, then the *dynamically consistent concatenation* $s \circ t$ of the strings s and t is defined to be st if there exists $u \in U$ such that $\Phi(x_p, u) = y_1$, ϵ (empty string) otherwise. For $A, B \subset \text{Tra}(\mathcal{M})$, the dynamically consistent concatenation of A and B is: $A \circ B \triangleq \{s \circ t; s \in A \& t \in B\} \subset AB$. For $w_1 = (x_0, \sigma_1), w_2 = (y_0, \sigma_2) \in X \times \text{Con}(\mathcal{M})$, the *dynamically consistent concatenation of* w_1 *and* w_2, written as $w_1 \circ w_2$ is the element $(x_0, \sigma_1\sigma_2)$ if $\Phi(x_0, \sigma_1) = y_0$, and is (x, ϵ) otherwise. For $C, D \subset X \times \text{Con}(\mathcal{M})$, $C \circ D = \{w \circ v; w \in C, v \in D\}$.

It may be shown that $\theta^{-1}\left(\text{Tr}(\pi)_i^j \circ \text{Tr}(\pi)_j^k\right) = \theta^{-1}\left(\text{Tr}(\pi)_i^j\right) \circ \theta^{-1}\left(\text{Tr}(\pi)_j^k\right)$. For two block control events $\overline{U}_i^j$ and $\overline{U}_j^k$, the symbol $\overline{U}_i^j \circ \overline{U}_j^k$ is interpreted to mean the block control event corresponding to the set $U_i^j \circ U_j^k$. Let $\mathcal{U}^\circ \triangleq \{\overline{U}_{i_1}^{i_2} \circ \overline{U}_{i_2}^{i_3} \cdots \circ \overline{U}_{i_{n-1}}^{i_n};\; \overline{U}_{i_l}^{i_{l+1}} \in \mathcal{U}, 1 \leq l < n, n \geq 1\}$. Then Φ^π will be extended inductively to $\pi \times \mathcal{U}^\circ$ as follows:

$$\Phi^\pi\left(\overline{X}_i, \overline{U}_{i_1}^{i_2} \circ \overline{U}_{i_2}^{i_3} \cdots \circ \overline{U}_{i_{n-1}}^{i_n}\right) \triangleq \Phi^\pi\left(\Phi^\pi\left(\overline{X}_i, \overline{U}_{i_1}^{i_2} \circ \overline{U}_{i_2}^{i_3} \cdots \circ \overline{U}_{i_{n-2}}^{i_{n-1}}\right), \overline{U}_{i_{n-1}}^{i_n}\right).$$

It may be shown [CWe94] that the extension of Φ^π is consistent with the definition of Φ^π and that of the concatenation of elements of $X \times \text{Con}(\mathcal{M})$. In fact, $\overline{U}_{i_1}^{i_2} \circ \overline{U}_{i_2}^{i_3} \cdots \circ \overline{U}_{i_{n-2}}^{i_{n-1}} \circ \overline{U}_{i_{n-1}}^{i_n}$ generates a (necessarily consecutively pairwise dynamically consistent block states) trajectory $\overline{X}_i = \overline{X}_{i_1}, \overline{X}_{i_2}, \ldots, \overline{X}_{i_n} = \overline{X}_j$; this block state trajectory has a set level realization which contains all trajectories starting from all states $x_i \in X_i$ which pass successively through $X_{i_1}, X_{i_2}, \ldots, X_{i_{n-1}}$ and which directly terminate in X_j.

Furthermore it can be shown that if $\pi_1 \preceq \pi_2$, then a π_2-control event is a collection of dynamically consistent concatenations of some π_1- control events.

Definition 3. For $\mathcal{M}^{\pi_1}$ and $\mathcal{M}^{\pi_2} \in$ PAR($\mathcal{M}$), define

$$\mathcal{M}^{\pi_1} \cap \mathcal{M}^{\pi_2} = \mathcal{M}^{\pi_1 \cap \pi_2}, \; \mathcal{M}^{\pi_1} \cup^c \mathcal{M}^{\pi_2} = \mathcal{M}^{\pi_1 \cup^c \pi_2}.$$

It is immediate that HIPAL($\mathcal{M}$) $\underline{\Delta}$ $<$ PAR($\mathcal{M}$), $\cap$, $\cup^c$, $\preceq>$ forms a lattice, which we call the *dynamically consistent hierarchical lattice of* $\mathcal{M}$.

2.2 In-block Controllability Partition Machines and the Associated Lattice

For an analysis of hierarchical system behavior, one needs not only to consider the global dynamics between the state space of submachines treated as singletons but also the local dynamics within each submachine given by the partition element $X_i \in \pi$.

Definition 4 (Controllable Finite Machines). A machine $\mathcal{M}$ is called *controllable* if for any $(x, y) \in X \times X$, there exists a control sequence $\sigma \in U^*$ which gives rise to a trajectory from x to y, i.e.

$$\forall x \in X, \forall y \in X, \exists s \in U^*(\Phi(x, s) = y).$$

For a partition machines we have

Definition 5 (In-block and Between-block Controllability). $\mathcal{M}^\pi$ is called *in-block controllable* if every associated submachine $\mathcal{M}_i = < X_i, U_i, \Phi_{|X_i \times U_i} >$ of the base machine is controllable, i.e.

$$\forall X_i \in \pi, \forall x, y \in X_i, \exists s \in U_i^*, \forall s' \leq s, (\Phi^i(x, s') \in X_i \wedge \Phi^i(x, s) = y).$$

$\mathcal{M}^\pi$ is called *between-block controllable* if

$$\forall \overline{X}_i, \overline{X}_i \in \pi, \exists S \in \mathcal{U}^\circ, (\Phi^\pi(\overline{X}_i, S) = \overline{X}_j).$$

In other words, $\mathcal{M}^\pi$ is controllable as a finite machine. IBCP($\mathcal{M}$) (respectively, BBCP($\mathcal{M}$)) shall denote the set of all in-block controllable (respectively, between-block controllable) partition machines.

Theorem 6. IBCP($\mathcal{M}$) *is closed under chain union.*

From the above theorem, $\mathcal{M}^{\pi_1 \cup^c \pi_2}$ is the least upper bound of $\mathcal{M}^{\pi_1}$ and $\mathcal{M}^{\pi_2}$ in IBCP($\mathcal{M}$). Unfortunately, intersection does not preserve in-block controllability. Hence IBCP($\mathcal{M}$) cannot be a lattice with respect $\cap$ and $\cup^c$.

Definition 7 (Greatest Lower Bound in IBCP($\mathcal{M}$)). Given a finite machine $\mathcal{M}$ and two in-block controllable partitions π_1 and π_2, set

$$\pi_1 \sqcap \pi_2 = \bigcup^c \{\pi'; \; \pi' \preceq \pi_1, \pi' \preceq \pi_2\}, \; \mathcal{M}^{\pi_1} \sqcap \mathcal{M}^{\pi_2} = \mathcal{M}^{\pi_1 \sqcap \pi_2}.$$

Theorem 8. *If* $\mathcal{M}$ *is controllable, then* $<$ IBCP($\mathcal{M}$), $\sqcap$, $\cup^c$, $\preceq>$ *forms a lattice.*

An in-block controllable partition machine $\mathcal{M}^{\pi}$ inherits the controllability of the base machine as stated in:

Theorem 9. *Let $\mathcal{M}^{\pi} \in$ IBCP($\mathcal{M}$), then $\mathcal{M}^{\pi}$ is controllable (i.e. between-block controllable) if and only $\mathcal{M}$ is controllable.*

Definition 10. The tuple $\mathcal{M}_2^{\pi} \triangleq < \mathcal{M}^{\pi}, \mathcal{M}_1, \ldots, \mathcal{M}_p, \psi >$ is called a *two level hierarchical structure of $\mathcal{M}$*, where ψ is the partition function given by $\psi(x) = \overline{X}_i$ if $x \in X_i$, $1 \leq i \leq |\pi|$. $\mathcal{M}^{\pi}$ is called the *high level machine* of $\mathcal{M}_2^{\pi}$ and each $\mathcal{M}_i$ is called a *low level machine* of $\mathcal{M}_2^{\pi}$. ψ is called the *communication function.*

3 Syntax of Hierarchical COCOLOG

In order to present the simplest version of a hierarchical COCOLOG system, we consider throughout the rest of this paper an element $\pi \in$ IBCP($\mathcal{M}$) and the associated two level hierarchical structure $\mathcal{M}_2^{\pi}$. A hierarchical COCOLOG system is a family of first order theories $\{\text{Th}_k^{\pi}, \text{Th}_k^1, \ldots, \text{Th}_k^{|\pi|};\ k \geq 0\}$. Each of these theories has its own syntax, semantics and inference rules where each theory represents the properties of one component in the hierarchical structure. Each one may make its own extra-logical transition to a subsequent theory at the same level through the extra-logical control rules of that level. Each of them has the facility to communicate with theories at an adjacent level through another set of extra-logical rules consisting of instantiation rules and block-membership rules. The full details of standard COCOLOG can be found in [CW90] and [WeC92]. In order to define a two level hierarchical COCOLOG control structure for $\mathcal{M}_2^{\pi}$, we extend the basic COCOLOG languages L_k by the addition of (i) a high level COCOLOG language L_k^{π} and (ii) the attachment of a sorted language for each of components in the structure. These are described briefly below.

The COCOLOG Language of a Hierarchical Structure $k \geq 0$:

$$L_k = L_k^{\pi} \cup L_k^1 \cup L_k^2 \cup \cdots \cup L_k^{|\pi|},$$

where L_k^{π} is a *block-sort COCOLOG language* for $\mathcal{M}^{\pi}$, L_k^i is an *$i-$sort COCOLOG language* for $\mathcal{M}_i$, $1 \leq i \leq |\pi|$. The sets of constant symbols, function symbols and predicate symbols are listed below for each language. For L_k^{π}, $k \geq 0$, $1 \leq i \leq |\pi|$:

$$\text{Const}(L_k^{\pi}) = \{\overline{X}_1, \ldots, \overline{X}_p, \overline{U}_i^j, \ldots, \overline{U}_k^l, \overline{U(0)}, \ldots \overline{U(k-1)}, \overline{0}, \overline{1}, \ldots, \overline{K(p)+1}\};$$
$$\text{Func}(L_k^{\pi}) = \{\Phi^{\pi}, +^{\pi}, -^{\pi}\},$$
$$\text{Pred}(L_k^{\pi}) = \{\text{Rbl}^{\pi}(\cdot,\cdot,\cdot), \text{Eq}^{\pi}(\cdot,\cdot), \text{CSE}_m^{\pi}(\cdot), 1 \leq m \leq k\}.$$

For L_k^i, $1 \leq i \leq p$,

$$\text{Const}(L_k^i) = X_i \cup X_i^e \cup U \cup \{0, 1, \ldots, K(n_i)+1\},$$
$$\text{Func}(L_k^i) = \{\Phi^i, +, -\},$$
$$\text{Pred}(L_k^i) = \{\text{Rbl}^i(\cdot,\cdot,\cdot), \text{Eq}^i(\cdot,\cdot), \text{CSE}_m^i(\cdot), E_j^i(\cdot),\ 1 \leq m \leq k; 1 \leq j \leq |\pi| \& j \neq i\},$$

where the elements of X_i^e are called entry states from block X_i to neighbouring blocks and $X_i \cap X_e^i = \emptyset$ and $\Phi^i : X_i \times U_i \to X_i \cup X_e^i$. $E_j^i(\cdot)$ is the new (with respect to standard COCOLOG language) predicate called the *entry state predicate* from X_i to X_j. WFF(L_k^π), WFF(L_k^i) are the sets of *well formed formulas* generated by the Backus-Naur passing rules with respect to L_k^π and L_k^i respectively.

4 Axiomatization and Proof Theory of a Hierarchical COCOLOG

The axiom set of a hierarchical COCOLOG consists of several parts. Each of them lies in one sorted language only and inference in each component at any given level will be carried out only within the corresponding language and axiom set.

Definition 11. The axiom set Σ_k^π of the high level COCOLOG theory Th_k^π consists of the following for $k \geq 0$:

(1) Block transition axiom set: $\mathrm{AXM}^{\mathrm{dyn}}(L_0^\pi) \underline{\Delta} \{\mathrm{Eq}_\pi(\Phi^\pi(\overline{X}^i, \overline{U}_i^j), \overline{X}^j);\ 1 \leq i, j \leq p\}$, whenever $\Phi^\pi(\overline{X}^i, \overline{U}_i^j)$ is defined.
(2) Block reachability axiom set: $\mathrm{AXM}^{\mathrm{rbl}}(L_0^\pi)$;
(3) Arithmetic function axiom set: $\mathrm{AXM}^{\mathrm{arith}}(L_k^\pi)$;
(4) Equality axiom set: $\mathrm{AXM}^{eq}(L_k^\pi)$;
(5) Logical axiom set: $\mathrm{AXM}^{\log}(L_k^\pi)$;
(6) Size axiom set: $\mathrm{AXM}^{\mathrm{size}}(L_k^\pi)$;
(7) State estimation axiom set: $\mathrm{AXM}^{\mathrm{est}}(L_k^\pi)$ (when $k > 0$).

The reader should refer to [CW90] for complete definitions of the above axiom sets. The inference rules are simply the restricted *modus ponens* and *generalization* rules:

Definition 12 (Set of Inference Rules). $\mathrm{IR}_k^\pi = \{\mathrm{MP}(L^\pi), G(L^\pi)\}$, where

$$\mathrm{MP}(L_k^\pi) : \frac{A, A \to B}{B},\ A, B \in \mathrm{WFF}(L_k^\pi),\ G(L_k^\pi) : \frac{A}{\forall x A},\ A \in \mathrm{WFF}(L_k^\pi),$$

provided $x \in L_k^\pi$ does not occur free in A for rule G.

From the axiom set Σ_k^π one can generate iteratively a new formula A, called a *theorem*, by applying the rules of inference to the axioms and previously generated theorems. In this case, we say A is *provable from* Σ_k^π and this is denoted by $\Sigma_k^\pi \vdash A$. At any instant k, the collection of all such theorems is called a *high level theory*, and this is denoted

$$\mathrm{Th}_k^\pi \underline{\Delta} \left\{A;\ A \in \mathrm{WFF}(L_k^\pi) \& \Sigma_k^\pi \vdash_{\mathrm{IR}_k^\pi} A\right\}.$$

By the completeness of a COCOLOG theory [CW90], Th_k^π contains all the logically true formulas about $\mathcal{M}^\pi$, which are expressible in terms of the first-order language L_k^π.

Definition 13. For $1 \leq i \leq |\pi|, k \geq 0$, the axiom set Σ_k^i of the low level COCOLOG theory Th_k^i consists of the following:

(1) Transition axiom set for $\mathcal{M}_i$: $\mathrm{AXM}^{\mathrm{dyn}}(L_0^i) \underline{\Delta} \{\mathrm{Eq}^i(\Phi^i(x^i, u^q), x^{i_q});\ x^i \in X_i \& u^q \in U \& x^{i_q} \in X_i \cup X_i^e\}$.

(2) Local reachability axiom set: $\mathrm{AXM}^{\mathrm{rbl}^i}(L_k^i)$;
(3) Arithmetic function axiom set: $\mathrm{AXM}^{\mathrm{arith}}(L_k^i)$;
(4) $\mathrm{AXM}^{eq}(L_k^i)$;
(5) $\mathrm{AXM}^{\log}(L_k^i)$;
(6) Size axiom set: $\mathrm{AXM}^{\mathrm{size}}(L_k^i)$;
(7) State estimation axiom set: $\mathrm{AXM}^{\mathrm{est}}(L_k^i)$ when $k > 0$;
(8) Exit state axiom set: $\mathrm{AXM}^{\mathrm{exit}}(L_k^i) \triangleq \{E_j^i(x^l);\ j \neq i \,\&\, x^l \in X_e^i\}$, whenever there exists $x_i \in X_i$, $u \in U$ such that $\Phi(x_i, u) = x^l$.

Definition 14 (Inference Rules $\mathrm{IR}(L_k^i) = \{\mathrm{MP}(L_k^i), G(L_k^i)\}$). These correspond to those in Definition 12.

Similarly, the following set of formulas is called the *low level theory for $\mathcal{M}_i$*:

$$\mathrm{Th}_k^i \triangleq \{A;\ A \in \mathrm{WFF}(L_k^i) \& \Sigma_k^i \vdash_{\mathrm{IR}_k^i} A\}.$$

Theorem 15. *For $k \geq 0, 1 \leq i \leq |\pi|$, Th_k^π and Th_k^i are consistent and decidable.*

The following theorem shows that the set of unrestricted inference rules $\mathrm{IR}_k = \{\mathrm{MP}(L_k), G(L_k)\}$ does not provide more theorems than the set of rules IR_k^i with respect to $\mathrm{WFF}(L_k^i)$.

Theorem 16. *For $F \in \mathrm{WFF}(L_k^i)$, $\Sigma_k^i \vdash_{\mathrm{IR}(L_k^i)} F \Longleftrightarrow \Sigma_k^i \vdash_{\mathrm{IR}(L_k)} F$.*

In terms of semantics, we remark that one defines the model of a hierarchical COCOLOG theory in such way that the algebraic properties of the domain reflect the concepts listed in the previous §1.

5 Extra-Logical Transitions in Hierarchical COCOLOG

For a standard COCOLOG theory, there is defined a set of extra-logical operation called *extra-logical control rules* which make extra-logical transitions from one theory to a subsequent theory. In a hierarchical COCOLOG system, these transitions will take place between theories at a given level. In addition, each theory has at least one more set of rules for passing information from one level to another. An *instantiation rule* passes information from a high level machine to a low level machine by assigning a control objective to the adjacent low level machine. A *block membership rule* passes information from a low level machine to the high level machine by reporting which block the low level machine currently occupies (see Fig.1). The information to be sent down to a low level machine at any instant is task dependent. Hence the construction of the following rules depends upon the overall machine task. In the following, we write out these rules with respect to reachability problems, by which we mean that the overall machine to generate a trajectory from the current state in X to a certain target state in X.

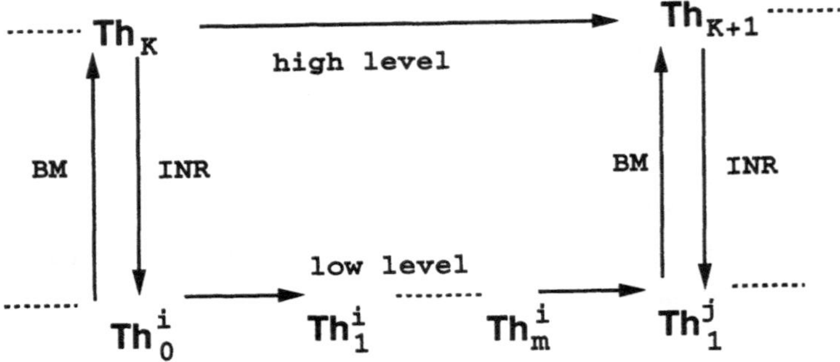

Fig. 1. Extra-logical transitions of a hierarchical COCOLOG system.

High level control rule CCR(L_k^π) For each $\overline{U}_i^j \in \mathcal{U}$:

$$\mathbf{IF} \quad F_i^j(L_k^\pi) \quad \mathbf{THEN} \quad \mathrm{Eq}^\pi(\overline{U(k)}, \overline{U}_i^j), \tag{3}$$

where $F_i^j(L_k^\pi) \in \mathrm{WFF}(L_k^\pi)$ is the control condition formula for the high level control event $\overline{U}_i^j$. The following example allows $\mathcal{M}^\pi$ to simply pick out an arbitrary control that can drive the current block state one step closer to the target block $\overline{X}^T$:

$$F_i^j(L_k^\pi) = \exists l, \mathrm{CSE}_k^\pi(\overline{X}_i) \wedge \mathrm{Rbl}^\pi(\overline{X}_i, \overline{X}^T, \bar{l} +^\pi \bar{1}) \wedge \mathrm{Rbl}^\pi(\Phi^\pi(\overline{X}_i, \overline{U}_i^j), \overline{X}^T, \bar{l}). \tag{4}$$

Low level control rule CCR(L_k^i) **for** $\mathcal{M}_i$ For $u^p \in U_i$

$$\mathbf{if} \quad D_p(L_k^i)(x) \quad \mathbf{then} \quad \mathrm{Eq}^i(U_i(k), u^p), \tag{5}$$

where $D_p(L_k^i)(x) \in \mathrm{WFF}(L_k^i)$ is the control condition for the low level control event u^p and x appears free in $D_p(L_k^i)$. One example of $D_p(L_k^i)(x)$ (see [CW90]) is the following formula asserting that u^p initiates a shortest trajectory from the current state (estimate) y to the target state x:

$$\begin{aligned} D_p(L_k^i)(x) = \forall y\{&\neg\mathrm{CSE}_k^i(y) \vee \mathrm{Eq}^i(y, x) \vee [\exists z \exists l \forall s, \mathrm{Eq}^i(\Phi^i(y, u^p), z) \\ &\wedge \mathrm{Rbl}^i(z, x, l) \wedge \{\neg\mathrm{Eq}^i(s - l + 1, k(N_i) + 1) \vee \neg\mathrm{Rbl}^i(y, x, s)\}]\}, \end{aligned} \tag{6}$$

where $\mathrm{Eq}^i(s - l + 1, k(N_i) + 1)$ is the COCOLOG formula meaning $s < l + 1$. Let $\mathrm{CCF}(L_k^i)(x) \triangleq \{D_p(L_k^i)(x);\ 1 \le p \le |U_i|\}$. Then the following rule substitutes the free variable x by a local target state, which must be one of the exit states x_i^j from X_i to X_j when $\overline{U}_i^j$ is invoked at the high level. In case $\overline{U}_i^i$ is invoked, x shall be substituted by the global target state x^T. We will use the notation $\mathrm{CCF}(L_k^i)(x/x^i)$ to indicate that the substitution is carried for each element in the set.

Instantiation Rule INS$_k$ **for** $\overline{U}_i^j \in \mathcal{U}$

$$\textbf{IF} \quad \text{Eq}^\pi(\overline{U(k)}, \overline{U}_i^j) \quad \textbf{SEND} \quad \text{CCF}(L_k^i)(x/x_i^j), \quad i \neq j. \tag{7}$$

$$\textbf{IF} \quad \text{Eq}^\pi(\overline{U(k)}, \overline{U}_i^i) \quad \textbf{SEND} \quad \text{CCF}(L_k^i)(x/x^T), \quad i = j. \tag{8}$$

Notice the **SEND** parts are sets of control condition formulas for the low level controller, hence the subtask is specified by CCF$(L_k^i)(x/x_i^j)$. The instantiated formulas lie in WFF(L_k^i) and can be proved true or false from Σ_k^i. This gives the construction of the logical controller for $\mathcal{M}_i$ with respect to the task x_i^j. At the time instant k, submachine $\mathcal{M}_i$ starts to operate to achieve this control objective. During the operation of $\mathcal{M}_i$, Th$_k^i$ makes its theory transitions solely as a result of control action chosen to satisfy CCF$(L_k^i)(x/x_i^j)$. In general, it takes several low level control steps to achieve a given control objective. When it is achieved, the following rule reports back to the high level to update the current state of $\mathcal{M}^\pi$:

Block-membership Rule BM$_k^i$ For each $x' \in X_e^i$:

$$\textbf{if} \quad \text{CSE}_k^i(x') \quad \textbf{send} \quad \text{CSE}_{K+1}^\pi(\overline{X}'), \quad \text{where} \quad \psi(x') = \overline{X}'. \tag{9}$$

This rule does not provide any information to the high level machine before the current target state is reached because there is no new information available during that period concerning block position. Once the control objective has been achieved by the low level machine, the block position is changed and the high level COCOLOG theory makes one theory transition step.

Example 1. See Fig. 2. Suppose that the current and target states of $\mathcal{M}_8$ at time $k = 0$ are x^3 and x^4 respectively. Then the current and target states of high level machine $\mathcal{M}^\pi$ are $\overline{X}_1$ and $\overline{X}_2$ respectively. From (3) and (4), either $\overline{U}_1^2$ or $\overline{U}_1^3$ may be invoked; if Eq$^\pi(\overline{U(1)}, \overline{U}_1^2)$ is selected, then from (6) and (7), the task for $\mathcal{M}_1$ is specified by the formulas:

$$D_p(L_1^1)(x/x^5) = \forall y\{\neg\text{CSE}_k^i(y) \vee \text{Eq}^i(y, x^5) \vee [\exists z \exists l \forall s, \text{Eq}^i(\Phi^i(y, u^p), z) \\ \wedge \text{Rbl}^i(z, x^5, l) \wedge \{\neg\text{Eq}^i(s - l + 1, k(N_i) + 1) \vee \neg\text{Rbl}^i(y, x^5, s)\}]\}. \tag{10}$$

Then by (5) and (10), Th$_1^1$ decides upon Eq$^1(U_1(1), b)$ and evolves into Th$_2^1$. After one more transition Th$_2^1 \to$ Th$_3^1$, Th$_3^1$ decides upon Eq$^1(U_1(3), b)$ and the current state becomes x^5. Since $\psi(x^5) = \overline{X}_2$, by (9), Th$_1^\pi$ evolves into Th$_2^\pi$ and the process above repeats within $\mathcal{M}_2$ until x^4 is reached.

References

[CW90] P. E. Caines, S. Wang, COCOLOG: A conditional observer and controller logic for finite machines, *Proceeding of the 29th IEEE CDC*, Hawaii, 1990, pp. 2845–2850, complete version under revision for *SICOPT*.

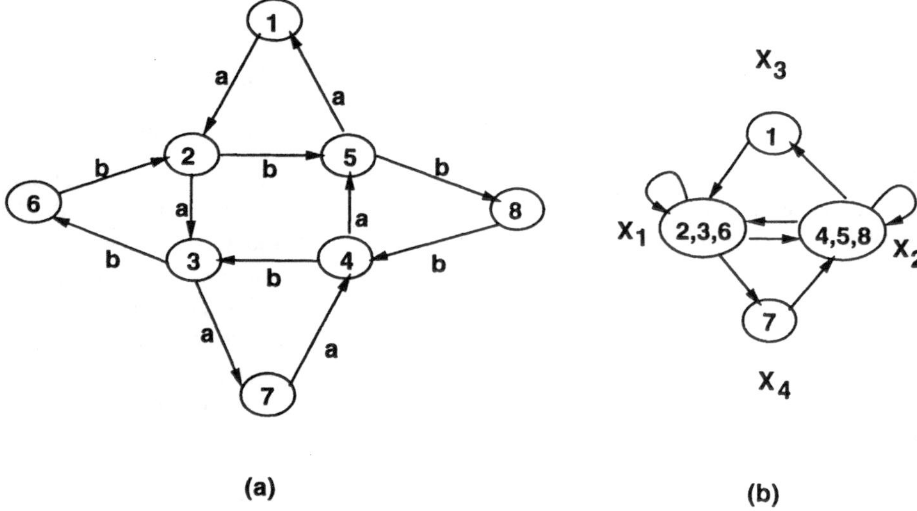

Fig. 2. (a) An 8 state machine $\mathcal{M}_8$. (b) The partition $\pi = \{\overline{2,3,6}, \overline{4,5,8}, \overline{1}, \overline{7}\}$ and its partition machine $\mathcal{M}^{\pi}$.

[CWe94] P.E. Caines, Y.J.Wei The hierarchical lattice of a finite machine, submitted for publication to *System and Control Letters*, Feb, 1994

[G68] A. Ginzburg, *Algebraic Theory of Automata*, Academic Press, New York, 1968

[HS66] J. Hartmanis, R.E. Stearns, *Algebraic Structure Theory of Sequential Machines*, Prentice-Hall, Englewood Cliffs, N.J., 1966.

[LW90] F. Lin, W.M. Wonham, Decentralized control and coordination of discrete-event systems with partial observation, *IEEE Trans. on Automatic Control*, V. 35, Dec. 1990.

[RW92] K. Rudie, W.M. Wonham, Think globally, act locally: decentralized supervisory control, *IEEE Trans. on Automatic Control*, V. 37, Nov. 1992.

[WeC92] Y.J. Wei, P.E. Caines, On Markovian fragements of COCOLOG for logic control systems, *Proceedings of the 31st IEEE CDC*, Tucson, AZ, 1992, pp. 2967–2972.

[ZW90] H. Zhong and W.M. Wonham, On the consistency of hierarchical supervision in discrete-event systems, *IEEE Trans. on Automatic Control*, 35(10) Oct. 1990.

Nondeterministic Supervision under Partial Observations*

Kemal İnan

Electrical & Electronics Eng. Dep., Middle East Technical University, Ankara, Turkey
e-mail : inan@vm.cc.metu.edu.tr

1 Introduction

Supervisory control problem was first formulated by Ramadge and Wonham [RW87b] where control actions for an event driven automaton consisted in blocking a subset of the so called controllable events as a function of the observed string of events. The control objective is given as a legal language defined on a finite event alphabet and control actions are chosen to disallow any deviation from the event sequences of the legal language and allow all the strings to occur within the legal language. The necessary and sufficient conditions for a legal language to be synthesizable under full observation is characterized as the definition of the controllability of a language. When the legal language is not controllable it is demonstrated that the supremum sublanguage of it that is synthesizable (controllable) exists and is effectively computable when both the plant and the specification languages are regular [RW87a]. The practical interpretation of this construct is to view the given legal language as the absolute boundary of safety conditions and try to synthesize an optimal language with the largest degree of freedom that remains within the given legal boundary.

When the observations are partial the necessary and sufficient conditions for synthesizability of a language have been shown to be that it be both controllable and observable [CDFV88, LW88b]. The observability condition for a language boils down to the absence of conflicting situations arising from the partiality of the observations where blocking an event is required for safety reasons but restricts the capability to generate certain legal strings. Unfortunately unlike controllability the property of observability is not closed under unions and therefore the set of controllable and observable sublanguages of a given legal language does not possess a supremum element. One remedy to this problem has been suggested via the concept of normality [LW88a] , a property that is closed under unions. Since normality is a stronger property than observability the corresponding supremal sublanguage constitutes a special suboptimal solution to the problem. Another remedy consists of reformulating the problem by defining legality of a language in terms of an interval of languages under the set inclusion partial order [LW88b]. When legal constraints are relaxed in this way there is no need to compute the optimal (supremal)

* Research partially supported by the Institute for Mathematics and its Applications (IMA) of the University of Minnesota and project EEEAG Yazılım 2 of the Turkish Scientific and Technical Research Council. Some of the results of this paper were reported in the paper by this author titled *Supervisory Control : Theory and Application to the Gateway Synthesis Problem* presented in the Belgium-French-Netherlands summer school on discrete event systems held in Spa, Belgium, 7–12 June, 1993.

solution which indeed does not exist in general under partial observations. On the other hand these partial remedies to the original problem do not constitute a solution to the following projected specification problem which is of primary practical significance : to find the necessary and sufficient conditions for the existence of a synthesizable language M such that $\mathrm{Proj}_{\Sigma_e}(M) = K$, where the Proj_{Σ_e} operator derives the legal language K from M by erasing all the symbols in the strings of M except those in the executive event set Σ_e which are of interest for specification purposes. A primary motivation for this paper has been to bring a solution to the partial observation problem discussed above that also addresses the projected specification problem.

The approach taken in this work involves trying every feasible control strategy under partial observations. Control strategies play a dual and conflicting role : when too restrictive it narrows down the observation error and, in turn, allows less restrictive strategies to be applied next, thanks to sharper observations, and vice-versa. Unless *all* feasible control strategies relative to a given boundary legal language are taken into consideration it is not possible to judge the real-time manoeuvring possibilities for the problem at hand. The way we enhance all the control options is by allowing the control decision at each observation instant to be a nondeterministic one. There are two constraints involved in allowing the control actions to take place within a nondeterministic range of choices : (i) the control action should not generate strings outside the given legal language ; (ii) every string generated should allow completion to a task termination string, that is, supervisory control should be non-blocking. The latter constraint has a special significance since, unlike the deterministic theory, supervision is nondeterministic and deadlock is more natural unless precautions are taken.

In the next section we formulate the automata-theoretic notation and background for the supervisory control problems. In §3 we present the language aspects of nondeterministic control. In particular we define, characterize and illustrate with an example two linguistic concepts : weakly controllable and observable (wco) languages and control and observation closed (co-closed) languages. These properties are both closed under unions and allow for supremal elements to exist. In §4 we characterize wco and co-closed languages constructively as the closed loop languages of nondeterministically controlled systems. In particular we construct the nondeterministic supervisory automaton B that synthesizes the supremal wco sublanguage of a closed language and a nonblocking supervisor N obtained from B that synthesizes the supremal co-closed sublanguage of a given legal language. Finally in §5 we formulate the projected specification problem and state the necessary and sufficient conditions for a solution to exist within the nondeterministic supervision frame.

2 Preliminaries

In this section we develop the notation and the technical preliminaries for the supervisory control problem. We define an *automaton* A as a hextuple $A = (Q_A, \Sigma, \rightarrow_A, \alpha_A, F_A, I_A)$, where Q_A is the set of states, Σ is the universal event alphabet, $\rightarrow_A \subseteq Q_A \times (\Sigma \cup \{\tau\}) \times Q_A$ is the transition relation where τ stands for the special silent event, $\alpha : Q_A \rightarrow 2^{\Sigma}$ is the event control function F_A is the target (final) subset of states and I_A is the nonempty set of initial states. We use the notation $x \xrightarrow{a}_A y$ instead of $(x, a, y) \in \rightarrow_A$ and drop the subscript A of $\rightarrow$ whenever the context is nonambiguous. Note that the first 5 attributes of an automaton define a directed graph where the edges are labeled by events in Σ and the nodes are labeled by points in $2^{\Sigma} \times \{\text{false}, \text{true}\}$ as dictated by α_A and F_A. We shall refer to this underlying graph by the symbol G_A.

Also observe that for a fixed G_A any subset of states can be interpreted as a distinct automaton when I_A is taken as this subset.

An automaton is called *deterministic* if it has no τ transitions, $\rightarrow_A: Q_A \times \Sigma \rightarrow Q_A$ is a partial function and I_A is a singletone set. We shall use the convention where $\rightarrow$ is also used for its transitive closure, that is, $x \overset{s}{\rightarrow} y$ where $s \in (\Sigma \cup \{\tau\})^*$. We use the notation $x \overset{s}{\rightarrow}$ if there exists a $y \in Q_A$ for which $x \overset{s}{\rightarrow} y$. If there is no such y we write $x \overset{s}{\not\rightarrow}$. We derive the relation $\Rightarrow \subseteq Q_A \times \Sigma \times Q_A$ from $\rightarrow$ by letting $x \overset{a}{\Rightarrow} y$ if $x \overset{\tau^* a \tau^*}{\rightarrow} y$. The transitive closure and the other conventions used for $\rightarrow$ extend to $\Rightarrow$. In particular $x \overset{\epsilon}{\Rightarrow} y$ stands for $x \overset{\tau^*}{\Rightarrow} y$ where ϵ stands for the empty string in Σ^*.

It is possible to hide (or internalize) certain event transitions of an automaton A. For a subset Σ_o of Σ we let $A.\Sigma_o$ denote the automaton obtained from A by replacing the graph G_A by $G_A.\Sigma_o$ where the latter is obtained from the former by replacing the labels of the edges of G_A by the symbol τ unless the label is in Σ_o.

Let A be an automaton and let $Q \subseteq Q_A$. We define $L(Q)$, the (prefix) closed language generated by Q, as follows :

$$L(Q) := \{s \in \Sigma^* \mid \exists q \in Q : q \overset{s}{\Rightarrow}\} \ .$$

We define the language $L_m(Q)$, the marked language generated by Q, as

$$L_m(Q) := \{s \in \Sigma^* \mid \exists q \in Q \exists p \in F_A : q \overset{s}{\Rightarrow} p\} \ .$$

We use the notation L_A and L_{mA} to denote $L(I_A)$ and $L_m(I_A)$ respectively. Finally Q/s denotes the set of all reachable states from a subset $Q \subseteq Q_A$ via the string s. Formally

$$Q/s := \{x \in Q_A \mid \exists q \in Q : q \overset{s}{\Rightarrow} x\} \ .$$

Note that Q/s is nonempty iff $s \in L(Q)$.

Next we define the mechanism of synhchronization via the product automaton. Let A and B denote two automata then we define $C = A \times B$ where $Q_C := Q_A \times Q_B$, $(x, y) \in F_C$ iff $x \in F_A$ and $y \in F_B$, $\alpha_C(x, y) := \alpha_A(x) \cup \alpha_B(y)$ and $I_C := I_A \times I_B$. The transition relation $\rightarrow_C$ is defined as the smallest relation that satisfies the following implications :

$$\begin{aligned}
&(x \overset{a}{\rightarrow}_A y \wedge x' \overset{a}{\rightarrow}_B y' \wedge a \neq \tau) \Longrightarrow (x, x') \overset{a}{\rightarrow}_C (y, y') \ , \\
&(x \overset{a}{\rightarrow}_A y) \wedge a \notin \alpha_B(x') \wedge x' \overset{a}{\not\rightarrow}_B \wedge a \neq \tau) \Longrightarrow (x, x') \overset{a}{\rightarrow}_C (y, x') \ , \\
&(x' \overset{a}{\rightarrow}_B y') \wedge a \notin \alpha_A(x) \wedge x \overset{a}{\not\rightarrow}_A \wedge a \neq \tau) \Longrightarrow (x, x') \overset{a}{\rightarrow}_C (x, y') \ , \\
&(x \overset{\tau}{\rightarrow}_A y) \Longrightarrow (x, x') \overset{\tau}{\rightarrow}_C (y, x') \ , \\
&(x' \overset{\tau}{\rightarrow}_B y') \Longrightarrow (x, x') \overset{\tau}{\rightarrow}_C (x, y') \ .
\end{aligned}$$

Observe that if A and B are deterministic automata then so is $A \times B$.

In supervisory control problems we model the plant by an automaton P. In the context of this paper the plant automaton P is deterministic, $F_P = Q_P$ and $\alpha_P(x) = \Sigma$ at all the states x. In order to model supervision we let $\mathcal{S}(\Sigma_c, \Sigma_o)$ denote the set of all deterministic automata S subject to the following control, observation and marking constraints :

(1) $\alpha_S(x) \subseteq \Sigma_c$, for all x in Q_S,
(2) $L_S \subseteq \Sigma_o^*$,

(3) $F_S = Q_S$.

The control problem is to find an $S \in \mathcal{S}$ where we have omitted the arguments of $\mathcal{S}$, such that the closed-loop system $H := P \times S$ satisfies some language constraint in terms of L_H and/or L_{mH}. For example the standard problem is to find the necessary and sufficient conditions for a given sublanguage K of L_P for which there exists $S \in \mathcal{S}$ with the property $L_{(P \times S)} = \bar{K}$ where $\bar{M}$ denotes the closure of a language M. It has been shown that the necessary and sufficient condition is that the language K be controllable and observable [CDFV88].

Our approach to bring a solution to the problem when K is not observable relies on the concept of nondeterministic supervision where all feasible control actions are applied in a nondeterministic manner. The underlying mathematical construct involves enlargening the set $\mathcal{S}$ to $\mathcal{S}_n$ where the latter is defined as the set of all *nondeterministic* automaton with no τ transitions and subject to the same three constraints as in the case of $\mathcal{S}$. In the next section we introduce new linguistic concepts that are relevant to nondeterministic supervision.

3 Language Characterizations

In the following we use the additional notation Σ_{uc} and Σ_{uo} to denote the subsets of uncontrollable and unobservable events respectively. Let L stand for L_P and let $K \subseteq L$. K is defined to be *controllable* if $\bar{K}\Sigma_{uc} \subseteq \bar{K}$ and it is defined to be *observable* if $s, s' \in \bar{K}$, $sa, s'a \in L$ and $s.\Sigma_o = s'.\Sigma_o$ then $sa \in \bar{K}$ iff $s'a \in \bar{K}$ [RW87b, CDFV88] ,where $s.\Sigma'$ denotes the string obtained from s after erasing in it all the symbols that are not in the event subset Σ'. The following definitions are new.

Definition 1. The language K is weakly controllable and observable (wco) iff it is controllable and if for every $s \in \bar{K}$ of the form

$$s = \zeta_0 a_1 \zeta_1 \ldots a_n \zeta_n \ ,$$

where $a_i \in \Sigma_o$ and $\zeta_i \in \Sigma_{uo}^*$, a string $s' \in L$ also holds the property that $s' \in \bar{K}$ where s' is given by

$$s' = \zeta_0' a_1 \zeta_1' \ldots a_n \zeta_n' \ ,$$

with $\zeta_i' \in \Sigma_{uo}^*$ and

$$\mathrm{ev}(\zeta_i') \subseteq \mathrm{ev}(\zeta_i) \cup \Sigma_{uc} \ ,$$

where $\mathrm{ev}(s)$ denotes the set of all events that occur in the string s.

Definition 2. The language K is control and observation closed (co-closed) iff

$$K = \cup\{M \subseteq K \mid M \text{ is co}\} \ .$$

The intuition behind wco languages is simple : if for some $s \in \bar{K}$ there is an $s' \in L$ as given in the definition above, but $s' \notin \bar{K}$, then no controllable and observable sublanguage of K can contain the string s. For any control policy that rules the string $s' \notin \bar{K}$ out it also must rule the string s out since control decisions are to be made after each observation instance and must remain valid until the end of the next observation instance. Therefore all strings of the form s above that violate the definition of wco are to be eliminated in all the co sublanguages of K. Although the converse statement that every string s of a wco language K is also a string of some co sublanguage of K is intuitively appealing it is only correct when K is a closed sublanguage. Consider the following example: $\Sigma = \{a, b, c, A, B, C\}$ where all the events corresponding to upper case letters are observable and the lower case ones are unobservable and $\Sigma_c = \{a, c, B\}$. Let

$$L := \overline{Cc + AbcB^* + AaB} \ ,$$
$$K := Cc + Abc + AaB \ .$$

The language K is controllable but not observable : the strings $s = Abc$, $s' = Aa$ where $sB, s'B \in L$ violates the definition of observability since $s'B \in \bar{K}$ but $sB \notin \bar{K}$. On the other hand K is wco since every string of $\bar{K}$ satisfies the condition of the definition of wco. In fact the only string of $\bar{K}$ that may violate the definition of wco are $s = AaB \in K$ and $s' = AbcB \in L$. But since $\text{ev}(bc) = \{b, c\} \not\subseteq \text{ev}(a) \cup \Sigma_{uc} = \{a, b, A, C\}$, the strings s and s' do not violate the condition of wco. But now note that no sublanguage of K is controllable and observable. Therefore K is not co-closed. On the other hand if we replace K by its closure, that is,

$$K := \overline{Cc + Abc + AaB} \ ,$$

then K is co-closed since

$$K = M \cup M' \ ,$$

where $M := \overline{Cc + Abc}$ and $M' = \overline{C + Ab + AaB}$ which are both co sublanguages of K. Next we present the formal results that substantiate the arguments elaborated by this simple example.

Proposition 3. *Every co-closed sublanguage of L is a wco language.*

The following proposition implies that the converse statement of the previous proposition holds only for closed languages and ,in general, gives equivalent characterizations of closed wco languages.

Proposition 4. *Let K be a closed sublanguage of L, then the following statements are equivalent:*

(i) K is a wco language.
(ii) For each $s \in K$

$$\text{track}(s) := \inf\{W \subseteq L \mid s \in W,\ W \textit{ is closed and co}\} \subseteq K \ .$$

(iii) K is a co-closed language.

It is easily observed that both wco and co-closed sublanguages of any sublanguage K are closed under unions and hence possess supremal elements. What is missing at this stage is that unlike wco languages there does not seem to be a simple and explicit characterization of co-closed languages. In the next section we fill this gap in a constructive way by first constructing a nondeterministic supervisor B that synthesizes the supremal wco sublanguage of the closure of a given language K and then describe an algorithm that computes the nonblocking supervisor N that synthesizes the supremal co-closed sublanguage of K.

4 Nondeterministic Supervision

Our analysis is based on the canonical deterministic automaton C that generates a given controllable sublanguage K of L. Let D be any deterministic automaton for which $L_{mD} = K$ and where $\alpha_D(x) \equiv \Sigma$. We define C as $P \times D$ except for the following modification on the α_C function defined on $Q_C = Q_P \times Q_D$:

$$a \in \alpha_C(y, y') \Longleftrightarrow y \xrightarrow{a} \wedge y' \not\xrightarrow{a} \quad .$$

We thus have $L_C = \bar{K}$ and $L_{mC} = K$ since $K \subseteq L$ and there is full synchronization due to the choice of α_D. Moreover since K is controllable $\alpha_C(x) \subseteq \Sigma_c$ for all $x \in Q_C$ and the final state set F_C of C marks the strings in K. If we modify the automaton C by replacing F_C by Q_C then the resulting automaton, say S, belongs to the set $\mathcal{S}(\Sigma_c, \Sigma)$ and is therefore the supervisor for which $L_{(P\times S)} = \bar{K}$.

Next we construct the nondeterministic supervisor $B \in \mathcal{S}_n(\Sigma_c, \Sigma_o)$ making use of the automaton C. We define Q_B as the following subset of $2^{Q_C} \times 2^{\Sigma_c}$:

$$(X, \Gamma) \in Q_B \Longleftrightarrow \cup_{x\in X}\alpha_C(x) \subseteq \Gamma \quad . \tag{1}$$

The set X is the information available to the supervisor as to the location of the exact current state of C and Γ represent the supervisory control action consisting of blocking all the transitions within the event set Γ. The relation given by (1) is called the *feasability condition* and essentially ensures that all blocking prerequisites to remain within the safety limit $\bar{K}$ are included within the control blocking set Γ.

In order to define the transition relation we need the following notation : let G_A be an underlying graph of an automaton A and $\Sigma' \subseteq \Sigma$. The graph $G_A^{\Sigma'}$ is obtained from G_A after removing all the edges of G_A that have labels in the set Σ'.

The transition relation is given by letting for any $a \in \Sigma_o$, $(X, \Gamma) \xrightarrow{a} (X', \Gamma')$ iff

(i) $a \notin \Gamma$,
(ii) the set $X' \subseteq Q_C$ is given by :

$$X' = \{x' \in Q_C \mid \exists x \in X : x \xrightarrow{a}_C y \wedge y \overset{\epsilon}{\Rightarrow} x' \; in \; G_C^{\Gamma'}.\Sigma_o\} \quad .$$

Note that by the controllability of L_C the pair (X', Σ_c) always satisfies the feasibility condition of Q_B and if for some $x \in X, x \xrightarrow{a}, a \notin \Gamma$ and $(X, \Gamma) \in Q_B$ then $(X, \Gamma) \xrightarrow{a} (X', \Sigma_c)$ where X' is obtained from condition (ii) above with $\Gamma' = \Sigma_c$. That is, by virtue of controllability, the feasibility condition does not give rise to inconsistencies such as a transition being forced into an infeasible pair (X, Γ) - in our formulation a non-state. Also note that the transition relation above defines a nondeterministic automaton with no τ transitions.

In order to complete the construction we take $\alpha_B(X, \Gamma) := \Gamma$,$F_B = Q_B$ and the initial state set I_B is given by the following subset of Q_B :

$$(X, \Gamma) \in I_B \Longleftrightarrow X = \{x \in Q_C \mid I_C \overset{\epsilon}{\Rightarrow} x \in \; G_C^{\Gamma}.\Sigma_o\} \; ,$$

where I_C denotes the singletone initial state set of the automaton C. The associated theoretical results are summarized by the following theorems.

Theorem 5. *The language $L_{P\times B}$ is the supremal wco sublanguage of $\bar{K}$.*

Theorem 6. *A sublanguage M of L is wco if and only if there exists $B' \in \mathcal{S}_n(\Sigma_c, \Sigma_o)$ such that $L_{P\times B'} = \bar{M}$.*

We now proceed to construct the nonblocking supervisor $N \in \mathcal{S}_n$ that synthesizes the supremal co-closed sublanguage of K. First we state the definition of a nonblocking supervisor.

Definition 7. A supervisory automaton $S \in \mathcal{S}_n$ is said to be *nonblocking* relative to a language K if for every reachable state x of the closed loop automaton $P \times S$ via a string $s \in L_{P\times S}$ there exists a string t such that $st \in L_{P\times S} \cap K$ and $x \stackrel{t}{\rightarrow}_{P\times S}$.

Observe from the definition that if a supervisor is nonblocking relative to a language K then it is nonblocking relative to any language K' where $K \subseteq K'$.

Algorithm for computing the nonblocking supervisor N

Step 0 : Set $B_0 = B, i = 0$:
Step 1 : Discard all the states (X, Γ) and the transitions (edges) incident at those states in B_i that do not satisfy the following terminability condition : for all $x \in X$ there exists a string

$$t = \zeta_0 a_1 \zeta_1 \ldots a_n \zeta_n$$

with $a_i \in \Sigma_o$, $\zeta_i \in \Sigma_{uo}^*$ such that $x \stackrel{t}{\rightarrow}_C y$ where $y \in F_C$

$$(X_0, \Gamma_0) := (X, \Gamma) \stackrel{a_1}{\rightarrow} (X_1, \Gamma_1) \stackrel{a_1}{\rightarrow} (X_2, \Gamma_2) \ldots \stackrel{a_n}{\rightarrow} (X_n, \Gamma_n)$$

in B_i and $\mathrm{ev}(\zeta_i) \cap \Gamma_i = \varnothing$ for each i . Call the resulting automaton B_{i+1}. If $B_{i+1} = B_i$ stop and set $N = B_i$; else set $i := i + 1$ and go to Step 2.
Step 2 : Discard all the (inconsistent) states (X, Γ) in B_i and all the transitions incident at those states for which there exists $a \in \Sigma_o$ such that $a \notin \Gamma$, $(X, \Gamma) \stackrel{a}{\nrightarrow}_{B_i}$ and for some $x \in X, x \stackrel{a}{\rightarrow}_C$. Call the resulting automaton B_{i+1}. If $B_{i+1} = B_i$ set $i := i + 1$ and go to Step 1 ; else set $i := i + 1$ and repeat Step 2.

The algorithm above clearly stops after a finite number of steps since states are removed from B at each step. The solution N is the desired (nonempty) automaton provided that it possesses at least one initial state of B. The consequences of the algorithm above are summarized by the following theorems.

Theorem 8. *The language $L_{P\times N} \cap K$ is the supremal co-closed sublanguage of K for which N is a nonblocking supervisor relative to K.*

Theorem 9. *K is a co-closed sublanguage of L if and only if there exists a nonblocking supervisor $N' \in \mathcal{S}_n$ relative to K such that $L_{P\times N'} = \bar{K}$.*

We illustrate these results by an example of nondeterministic supervision. Consider the graphs corresponding to the plant automata P and the corresponding C automaton given by Fig. 1.

The associated languages are given by

$$L = \overline{(ac + b(c + d))^*} \ ,$$
$$K = (acb + bda)((acbd)^* acb + (bdac)^* bda)^* \ .$$

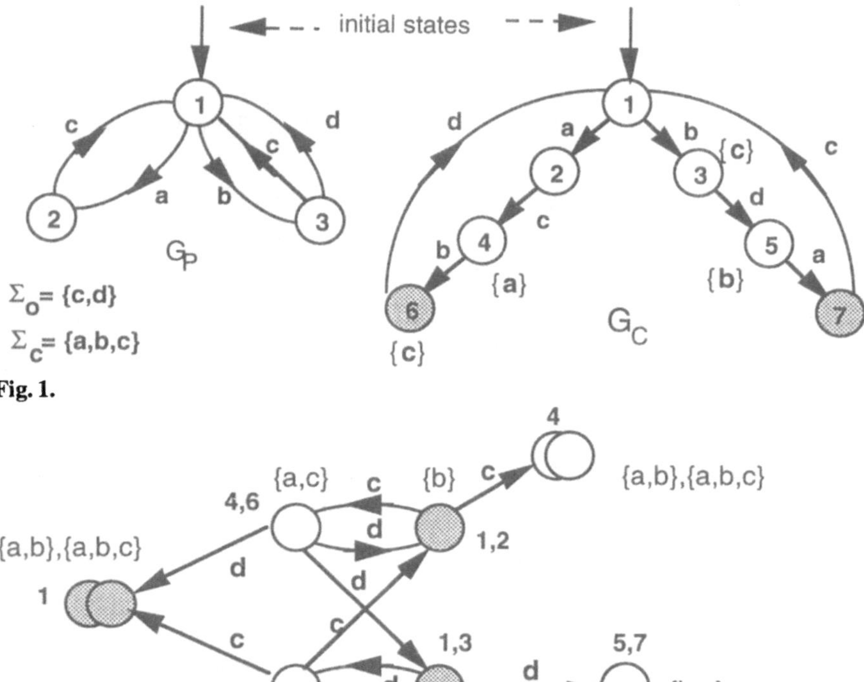

Fig. 1.

Fig. 2. The nondeterministic supervisor B

The nonempty event control sets of C are denoted within curly brackets at the relevant states of C and $F_C = \{6, 7\}$ as seen in the figure. Clearly the language K is controllable but not observable since the strings $s = a, s' = b \in \bar{K}$,$ac \in \bar{K}$ and $bc \in L \setminus K$. It can easily be shown that K is wco. The nondeterministic supervisor automaton B as described above is given for this example in Fig. 2.

The initial states of B are shown with darkened circles and for each state (X, Γ) the elements of Γ are given in curly brackets and the set X is given for a group of states in dark numerals seperated by commas. The closed loop automaton, which itself is necessarily nondeterministic is given by $P \times B$. On the other hand it is not difficult to observe that many of the states of B correspond to dead-end states which will drive the closed loop system to be blocked by the supervisor B. Therefore if some of the states of B corresponding to restrictive control policies are eliminated then these deadlocks could be avoided. This corresponds to applying the algorithm above to B in order to obtain the nonblocking supervisor N which is given by Fig. 3.

For this example the algorithm converges after one application of Step 1, one application of Step 2 which does not remove any states since inconsistency does not occur and the final application of Step 1 which terminates it. The closed loop system given by the automaton $P \times N$ is nondeterministic, generates the language $\bar{K}$, does not suffer from any deadlocks and is given by Fig. 4.

For this example it can be shown that the language K is both wco and co-closed. In

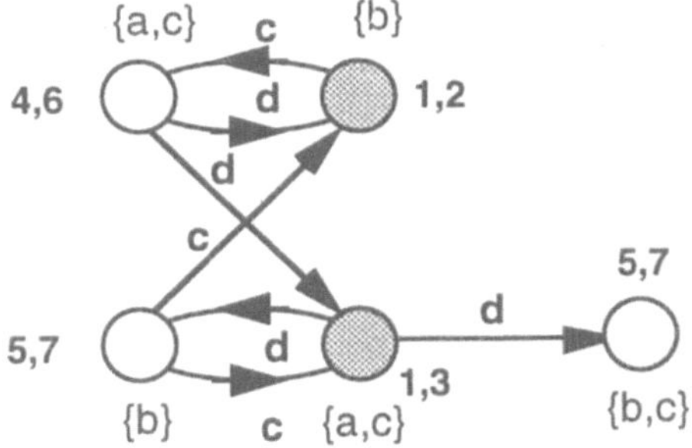

Fig. 3. The nonblocking supervisor N

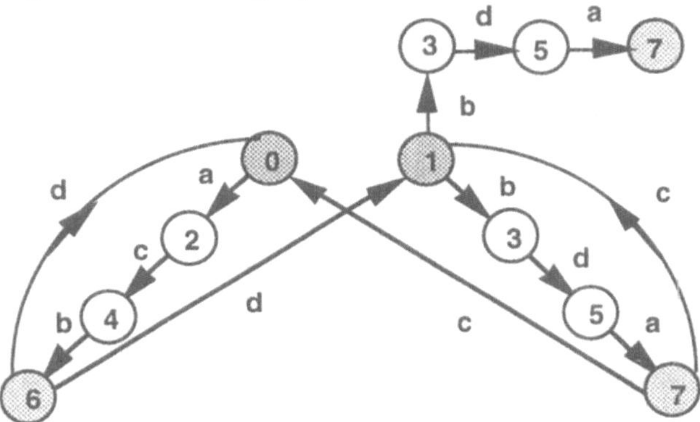

Fig. 4. The closed-loop automaton $P \times N$

fact we may write K as the union of two co languages as below:

$$K = acb((acbd)^*acb)^* + bda((bdac)^*bda)^* \quad .$$

5 The Projected Specification Problem

Let Σ_e be a subset of Σ that signifies the events that are relevant in the specified behaviour of the closed loop system. We search for a nondeterministic supervisor S such that for a given (not necessarily closed) sublanguage K of $\text{Proj}_{\Sigma_e}(L)$ the following specification relation holds :

$$\text{Proj}_{\Sigma_e}(L_{P\times S}) = \bar{K} \tag{2}$$

and the following additional nonblocking condition is satisfied : for every state x of $P \times S$ reached via a string $s \in L_{P\times S}$ there exists a string t such that $x \xrightarrow{t}$ and $st.\Sigma_e \in K$.

Let $M := \text{Proj}_{\Sigma_e}^{-1}(K) \cap L$, that is, M is the largest sublanguage of L such that

$$\text{Proj}_{\Sigma_e}(M) \subseteq K \quad .$$

For any sublanguage R of L let $\text{co}(R)$ denote the supremal co-closed sublanguage of R. The following theorem states the necessary and sufficient conditions for the existence of a nondeterministic supervisor $S \in \mathcal{S}_n$ that solves the projected specification problem given by 2.

Theorem 10. *There exists a nondeterministic supervisor $S \in \mathcal{S}_n$ that solves the projected specification problem if and only if*

$$\mathrm{Proj}_{\Sigma_e}(\mathrm{co}(M)) = K \ .$$

Proof. Let $\mathrm{Proj}_{\Sigma_e}(\mathrm{co}(M)) = K$ then $\mathrm{co}(M)$ is nonempty and co-closed and thus there exists a nonblocking supervisor $S \in \mathcal{S}_n$ relative to $\mathrm{co}(M)$ such that $L_{P\times S} = \overline{\mathrm{co}(M)}$ by Theorem 9. Therefore

$$\mathrm{Proj}_{\Sigma_e}(L_{P\times S}) = \mathrm{Proj}_{\Sigma_e}(\overline{\mathrm{co}(M)}) = \overline{\mathrm{Proj}_{\Sigma_e}(\mathrm{co}(M))} = \bar{K}$$

and S solves equation 2. Since S is nonblocking relative to $\mathrm{co}(M)$, then by the remark following the definition of nonblocking, S is nonblocking relative to M since $\mathrm{co}(M) \subseteq M$. Therefore for each reachable state x of $P \times S$ via a string $s \in L_{P\times S}$ there exists a string t such that $x \xrightarrow{t}_{P\times S}$ and $st \in M$. But this implies by definition of M that $st.\Sigma_e \in K$ and therefore the nonblocking condition of the projected specification problem is also satisfied.

Conversely let S solve the projected specification problem. We claim that S is a nonblocking supervisor relative to $L_{P\times S} \cap M$. To show this let x be a reachable state of $P \times S$ via string s then by hypothesis there exists t such that $x \xrightarrow{t}$ and $st.\Sigma_e \in K$. But the latter condition implies by definition of M that $st \in M$ and hence $st \in L_{P\times S} \cap M$ which establishes the desired result. Next we claim that $\overline{L_{P\times S} \cap M} = L_{P\times S}$. It is enough to show that $L_{P\times S} \subseteq \overline{L_{P\times S} \cap M}$ since the reverse inclusion is obvious. Letting $s \in L_{P\times S}$ and using the nonblocking property of S we stipulate the existence of a string t such that $st \in L_{P\times S} \cap M$ which proves that $s \in \overline{L_{P\times S} \cap M}$. But now by Theorem 9 the language $L_{P\times S} \cap M$ is co-closed and since $L_{P\times S} \cap M \subseteq M$ we conclude that $L_{P\times S} \cap M \subseteq \mathrm{co}(M)$ since $\mathrm{co}(M)$ is the supremal co-closed sublanguage of M. The theorem now follows if we can show that $\mathrm{Proj}_{\Sigma_e}(L_{P\times S} \cap M) = K$. But by the fact that S solves the projected specification problem we have $\mathrm{Proj}_{\Sigma_e}(L_{P\times S}) = \bar{K}$ and for any $s \in K$ it follows that $s \in \bar{K}$ and thus there exists $t \in L_{P\times S}$ such that $t.\Sigma_e = s$. On the other hand by definition of M and the fact that $s \in K$ we must have $t \in M$ and therefore $t \in L_{P\times S} \cap M$ which implies that $s \in \mathrm{Proj}_{\Sigma_e}(L_{P\times S} \cap M)$. Therefore $K \subseteq \mathrm{Proj}_{\Sigma_e}(L_{P\times S} \cap M)$ and the reverse inclusion follows by definition of M since $\mathrm{Proj}_{\Sigma_e}(M) \subseteq K$. □

References

[CDFV88] R. Cieslak, C. Desclaux, A. Fawaz, and P. Varaiya. Supervisory control of discrete event processes with partial observations. *IEEE Trans. Automatic Control*, 33(3):249–260, March 1988.

[LW88a] F. Lin and W.M. Wonham. Decentralized supervisory control of discrete-event systems. *Information Sciences*, 44:199–224, 1988.

[LW88b] F. Lin and W.M. Wonham. On observability of discrete-event systems. *Information Sciences*, 44:173–198, 1988.

[RW87a] P.R. Ramadge and W.M. Wonham. On the supremal controllable sublanguage of a given language. *SIAM Journal on Control and Optimization*, 25:637–659, May 1987.

[RW87b] P.R. Ramadge and W.M. Wonham. Supervisory control of a class of discrete-event processes. *SIAM Journal on Control and Optimization*, 25(1):206–230, January 1987.

Avoiding Blocking in Prioritized Synchronization Based Control of Nondeterministic Systems*

Ratnesh Kumar[1] *and Mark A. Shayman*[2]

[1] Department of Electrical Engineering, University of Kentucky, Lexington, KY 40506-0046
[2] Department of Electrical Engineering and Institute for Systems Research, University of Maryland, College Park, MD 20742

1 Introduction

Majority of the research effort on supervisory control of discrete event systems (DESs) has focused on the control of *deterministic* systems, and relatively little progress has been made towards control of *nondeterministic* systems–systems in which knowledge of the current state and next event is insufficient to uniquely determine the next state. Nondeterminism in the context of supervisory control arises due to unmodeled system dynamics such as suppression of "internal" events, suppression of "ticks of the clock" to obtain the untimed model from a timed model [2], partial observation of the system behavior due to lack of enough sensors, etc.

Most of the prior work on supervisory control such as [11, 7, 1, 3], etc., assume that there are never events which may occur in the supervisor without the participation of the plant. However, this assumption may be unreasonably restrictive for nondeterministic systems. When the plant is nondeterministic, there is generally no way to know a priori whether a command issued by the supervisor can be executed by the plant in its current state.

Heymann introduced an interconnection operator called *prioritized synchronous composition* (PSC) [4], which relaxes the synchronization requirements between the plant and supervisor. Each system in a PSC-interconnection is assigned a *priority set* of events. For an event to be enabled in the interconnected system, it must be enabled in all systems whose priority sets contain that event. Also, when an enabled event occurs, it occurs in each subsystem in which the event is enabled. In the context of supervisory control, the priority set of the plant contains the controllable and uncontrollable events, while the priority set of the supervisor contains the controllable and driven events.

Language models identify systems that have the same set of *traces*. The failures model of Hoare [6] identifies processes that have the same set of so-called *failures*. Failure equivalence refines language equivalence. Heymann showed that failure equivalence is too coarse to support the PSC operator [4]. In other words, there exist two different plants with the same failures model (and hence with the same language model) such that their PSC's with a common supervisor have different language models. Thus, neither the

* This research was supported in part by Center for Robotics and Manufacturing, University of Kentucky, in part by by the National Science Foundation under the Engineering Research Centers Program Grant CDR-8803012, and the Minta Martin Fund for Aeronautical Research, University of Maryland.

language model nor even the failures model retains enough information about a system to do control design using the operation of PSC.

This led Heymann to introduce the *trajectory model*, a refinement of the failures model [4, 5]. The trajectory model is similar to the *failure-trace model* (also called the refusal-testing model) in concurrency theory [10], but differs from this model in its treatment of hidden transitions. The trajectory model treats hidden transitions in a way that is consistent with the failures model. In a previous paper [13], we proved that the trajectory model retains sufficient detail to permit PSC-based controller design.

In [13], we showed that supervisory control of nondeterministic discrete event systems, in the presence of driven events, can be achieved using prioritized synchronous composition as a mechanism of control, and trajectory models as a modeling formalism. The specifications considered in [13] were given by *prefix-closed* languages. In this paper, we extend our previous work to include the notion of markings, by introducing the notion of recognized and generated trajectory sets, so that non-closed specifications and issues such as blocking can be addressed.

The usual notion of non-blocking, referred to as *language model non-blocking* in this paper, requires that each trace belonging to the generated language of a controlled system be extendable to a trace belonging to the recognized language. This property adequately captures the notion of non-blocking in a deterministic setting. However, the execution of a trace belonging to the generated behavior of a controlled nondeterministic system may lead to more than one state. Language model non-blocking only requires that every such trace be extendable to a trace in the recognized behavior from *at least one* such state–as opposed to *all* such states. Thus, a language model non-blocking nondeterministic system can deadlock, as illustrated by the example in the next section. Consequently, there is a need for a stronger type of non-blocking for nondeterministic systems. This leads us to introduce the property of *trajectory model non-blocking*, which requires that each refusal-trace belonging to the generated trajectory set of a controlled nondeterministic system be extendable to a refusal-trace belonging to the recognized trajectory set. We provide conditions for the existence of language model as well as trajectory model non-blocking supervisors.

2 A Motivating Example

In this section, we describe an example that illustrates some of the issues to be addressed in this paper. Figure 1(a) gives a deterministic model for a plant in which parts arrive at a machine from a conveyor and are then processed. The incoming parts are of two types that differ slightly in their widths. The standard width is the wider one. Events a_1 and a_2 denote the arrival at the machine of wide and narrow parts respectively. Events b_1 and b_2 denote the input into the machine of a part with the guides set to wide and narrow respectively. The default setting of the guides is wide, but intervention by a controller can reset them to narrow. A wide part can only be input with the guides set to wide. A narrow part can be input with either guide setting. However, input of a narrow part with the guides set to wide leads to the machine jamming–event d. If a part is input with the correct guide setting, then it can be successfully processed and output–event c. It is assumed that a_1, a_2, c, d are uncontrollable events and that there is no sensor that can distinguish between the two widths of incoming parts–i.e., the observation mask $M(\cdot)$ identifies a_1 and a_2–say $M(a_1) = M(a_2) := a$. A natural control specification is that the supervised plant be non-blocking since this guarantees that continuous operation is possible.

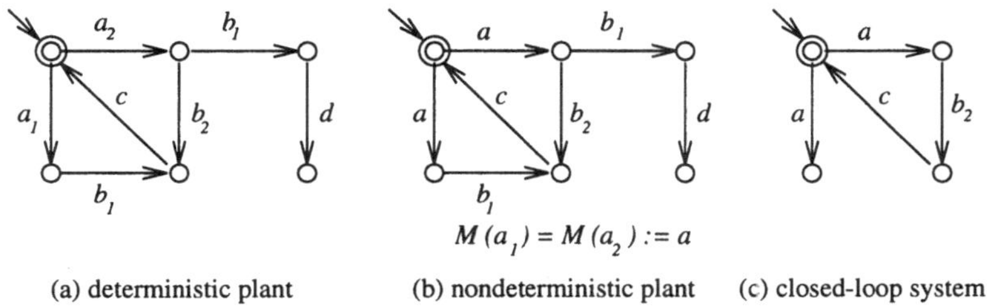

(a) deterministic plant (b) nondeterministic plant (c) closed-loop system

Fig. 1. Diagram illustrating the example of §2

It is clear that the performance specification cannot be met by any supervisor S of the Ramadge-Wonham type that is consistent with the observation mask. To prevent blocking arising from the uncontrollable jamming event d, S would need to disable b_1 following any occurrence of a_2. However, since the mask cannot distinguish between a_1 and a_2, S would also disable b_1 following any occurrence of a_1. But this would give a controlled plant that would deadlock with the arrival of the first wide part.

Suppose we replace the event labels a_1 and a_2 by their common mask value a, thereby obtaining the nondeterministic system shown in Figure 1(b). By so identifying a_1 and a_2, the events are made indistinguishable from the viewpoints of specification, control and observation–whereas in the partially observed deterministic model, they are indistinguishable only from the viewpoint of observation. For this system, however, the nondeterministic model is essentially equivalent to the partially observed one from the viewpoint of control since a_1, a_2 are uncontrollable and hence could not be distinguished in a supervisory control law. However, the non-blocking specification implicitly distinguishes between a_1 and a_2 and consequently forces the definition of a new type of non-blocking appropriate for nondeterministic systems.

Let $\mathcal{P}$ denote the nondeterministic state machine (NSM) depicted in Figure 1(b), and let $L(\mathcal{P})$, $L^m(\mathcal{P})$ denote its generated and recognized languages respectively. Then

$$L^m(\mathcal{P}) = [a(b_1 + b_2)c]^*, \qquad L(\mathcal{P}) = \mathrm{pr}[[a(b_1 + b_2)c]^* ab_1 d],$$

where $\mathrm{pr}(\cdot)$ denotes the prefix-closure operation. Having replaced the original partially observed deterministic model with a completely observed nondeterministic model, let us consider whether the specification can be met by a supervisor of the Ramadge-Wonham type. The closed-loop nondeterministic system $\mathcal{Q}$ obtained by disabling b_1 following any occurrence of a is depicted in Figure 1(c). Since $L(\mathcal{Q}) = \mathrm{pr}((ab_2c)^*) = \mathrm{pr}(L^m(\mathcal{Q}))$, the supervisor is non-blocking from the language model point of view. However, this control design is clearly unsatisfactory since the closed-loop system can deadlock. After all, the nondeterministic plant model is derived from the partially observed deterministic plant model, and there is no non-blocking Ramadge-Wonham type supervisor for that model.

The problem is that the usual language model definition of non-blocking given by $L(\mathcal{P}) = \mathrm{pr}(L^m(\mathcal{P}))$ is not suitable for control specifications in a nondeterministic setting. This motivates us to consider a stronger non-blocking requirement which we refer to as *trajectory model non-blocking* to distinguish it from the usual language model non-blocking condition. Using trajectory models for the plant and supervisor, and PSC as the mode of interconnection, it is possible to design a supervisor so that the closed-loop system meets the stronger non-blocking requirement. The details are given in §5, Example 1.

3 Notation and Preliminaries

Given a finite event set Σ, Σ^* is used to denote the collection of all *traces*, i.e., finite sequences of events, including the zero length sequence, denoted by ϵ. A subset of Σ^* is called a language. Symbols H, K, etc. are used to denote languages. Given a language $K \subseteq \Sigma^*$, the notation $\text{pr}(K) \subseteq \Sigma^*$, called the *prefix-closure* of K, denotes the set of all prefixes of traces from K. K is said to be prefix-closed if $K = \text{pr}(K)$.

The set $2^\Sigma \times (\Sigma \times 2^\Sigma)^*$ is used to denote the collection of all *refusal-traces*, i.e., finite sequences of alternating *refusals* and events [5, 13] of the type:

$$\Sigma_0(\sigma_1, \Sigma_1) \dots (\sigma_n, \Sigma_n),$$

where $n \in \mathbb{N}$. The sequence $\sigma_1 \dots \sigma_n \in \Sigma^*$ is the trace, and for each $i \leq n$, $\Sigma_i \subseteq \Sigma$ is the set of events refused (if offered) at the indicated point. Symbols P, Q, R, S, etc. are used to denote sets of refusal-traces. Refusal-traces are also referred to as *trajectories*.

Given $e \in 2^\Sigma \times (\Sigma \times 2^\Sigma)^*$, we use $|e|$ to denote the length of e, and for each $k \leq |e|$, $\Sigma_k(e) \subseteq \Sigma$ is used to denote the kth refusal in e and $\sigma_k(e) \in \Sigma$ is used to denote the kth event in e, i.e.,

$$e = \Sigma_0(e)(\sigma_1(e), \Sigma_1(e)) \dots (\sigma_k(e), \Sigma_k(e)) \dots (\sigma_{|e|}(e), \Sigma_{|e|}(e)).$$

If $f \in 2^\Sigma \times (\Sigma \times 2^\Sigma)^*$ is another refusal-trace such that $|f| \leq |e|$ and for each $k \leq |f|$, $\Sigma_k(f) = \Sigma_k(e)$ and $\sigma_k(f) = \sigma_k(e)$, then f is said to be a prefix of e, denoted $f \leq e$. For each $k \leq |e|$, e^k is used to denote the prefix of length k of e. If $f \in 2^\Sigma \times (\Sigma \times 2^\Sigma)^*$ is such that $|f| = |e|$ and for each $k \leq |f|$, $\Sigma_k(f) \subseteq \Sigma_k(e)$ and $\sigma_k(f) = \sigma_k(e)$, then f is said to be dominated by e, denoted $f \sqsubseteq e$.

The prefix-closure of $e \in 2^\Sigma \times (\Sigma \times 2^\Sigma)^*$, $\text{pr}(e) \subseteq 2^\Sigma \times (\Sigma \times 2^\Sigma)^*$, is defined as $\text{pr}(e) := \{f \in 2^\Sigma \times (\Sigma \times 2^\Sigma)^* \mid f \leq e\}$, and the dominance-closure of e, $\text{dom}(e) \subseteq 2^\Sigma \times (\Sigma \times 2^\Sigma)^*$, is defined as $\text{dom}(e) := \{f \in 2^\Sigma \times (\Sigma \times 2^\Sigma)^* \mid f \sqsubseteq e\}$. The prefix-closure and dominance-closure maps can be defined for a set of refusal-traces in a natural way. Given a refusal-trace $e \in 2^\Sigma \times (\Sigma \times 2^\Sigma)^*$, the trace of e, denoted $\text{tr}(e) \in \Sigma^*$, is defined as $\text{tr}(e) := \sigma_1(e) \dots \sigma_{|e|}(e)$. The trace map can be extended to a set of refusal-traces in a natural way. Given a set of refusal-traces $P \subseteq 2^\Sigma \times (\Sigma \times 2^\Sigma)^*$, we use $L(P) := \text{tr}(P)$ to denote its set of traces.

Symbols $\mathcal{P}$, $\mathcal{Q}$, $\mathcal{R}$, etc. are used to denote NSM's (with ϵ-moves). Let the 5-tuple

$$\mathcal{P} := (X_\mathcal{P}, \Sigma, \delta_\mathcal{P}, x^0_\mathcal{P}, X^m_\mathcal{P})$$

represent a discrete event system modeled as a NSM, where $X_\mathcal{P}$ is the state set, Σ is the finite event set, $\delta_\mathcal{P} : X_\mathcal{P} \times (\Sigma \cup \{\epsilon\}) \to 2^{X_\mathcal{P}}$ denotes the nondeterministic transition function[3], $x^0_\mathcal{P} \in X_\mathcal{P}$ is the initial state, and $X^m_\mathcal{P} \subseteq X_\mathcal{P}$ is the set of accepting or marked states. A triple $(x_1, \sigma, x_2) \in X_\mathcal{P} \times (\Sigma \cup \{\epsilon\}) \times X_\mathcal{P}$ is said to be a transition if $x_2 \in \delta_\mathcal{P}(x_1, \sigma)$. A transition (x_1, ϵ, x_2) is referred to as a *silent* or *hidden* transition. We assume that the plant cannot undergo an unbounded number of silent transitions.

The *ϵ-closure* of $x \in X_\mathcal{P}$, denoted $\epsilon^*_\mathcal{P}(x) \subseteq X_\mathcal{P}$, is defined recursively as:

$$x \in \epsilon^*_\mathcal{P}(x), \text{ and } x' \in \epsilon^*_\mathcal{P}(x) \Rightarrow \delta_\mathcal{P}(x', \epsilon) \subseteq \epsilon^*(x),$$

[3] ϵ represents both an *internal* or *unobservable* event, and an *internal* or *nondeterministic* choice [6, 9].

and the set of *refusal events* at $x \in X_{\mathcal{P}}$, denoted $\mathfrak{R}_{\mathcal{P}}(x) \subseteq \Sigma$, is defined as:

$$\mathfrak{R}_{\mathcal{P}}(x) := \{\sigma \in \Sigma \mid \delta_{\mathcal{P}}(x', \sigma) = \emptyset, \forall x' \in \epsilon^*_{\mathcal{P}}(x)\}.$$

In other words, given $x \in X_{\mathcal{P}}$, $\epsilon^*_{\mathcal{P}}(x)$ is the set of states that can be reached from x on zero or more ϵ-moves, and $\mathfrak{R}_{\mathcal{P}}(x)$ is the set of events that are undefined at each state in the ϵ-closure of x.

The transition function $\delta_{\mathcal{P}} : X \times (\Sigma \cup \{\epsilon\}) \to 2^{X_{\mathcal{P}}}$ is extended to the set of *traces*, denoted as $\delta^*_{\mathcal{P}} : X \times \Sigma^* \to 2^{X_{\mathcal{P}}}$, and to the set of *refusal-traces*, denoted as $\delta^T_{\mathcal{P}} : X \times (2^{\Sigma} \times (\Sigma \times 2^{\Sigma})^*) \to 2^{X_{\mathcal{P}}}$ [13]. These maps are then used to obtain the language models and the trajectory models of $\mathcal{P}$ as follows:

$$L(\mathcal{P}) := \{s \in \Sigma^* \mid \delta^*_{\mathcal{P}}(x^0_{\mathcal{P}}, s) \neq \emptyset\}, \quad L^m(\mathcal{P}) := \{s \in L(\mathcal{P}) \mid \delta^*_{\mathcal{P}}(x^0_{\mathcal{P}}, s) \cap X^m_{\mathcal{P}} \neq \emptyset\},$$

$$T(\mathcal{P}) := \{e \in 2^{\Sigma} \times (\Sigma \times 2^{\Sigma})^* \mid \delta^T_{\mathcal{P}}(x^0_{\mathcal{P}}, e) \neq \emptyset\},$$
$$T^m(\mathcal{P}) := \{e \in T(\mathcal{P}) \mid \delta^T_{\mathcal{P}}(x^0_{\mathcal{P}}, e) \cap X^m_{\mathcal{P}} \neq \emptyset\}.$$

$L(\mathcal{P}), L^m(\mathcal{P}), T(\mathcal{P}), T^m(\mathcal{P})$ are called the *generated language, recognized language, generated trajectory set, recognized trajectory set*, respectively, of $\mathcal{P}$. It is easy to see that $L(T^m(\mathcal{P})) = L^m(\mathcal{P})$ and $L(T(P)) = L(\mathcal{P})$. The pairs $(L^m(\mathcal{P}), L(\mathcal{P}))$ and $(T^m(\mathcal{P}), T(\mathcal{P}))$ are called the language model and the trajectory model, respectively, of $\mathcal{P}$. Two language models $(K^m_1, K_1), (K^m_2, K_2)$ are said to be equal, written $(K^m_1, K_1) = (K^m_2, K_2)$, if $K^m_1 = K^m_2, K_1 = K_2$; equality of two trajectory models is defined analogously.

4 Trajectory Models and Prioritized Synchronization

We first obtain a necessary and sufficient condition for a given refusal-trace set pair to be a trajectory model. This requires the definition of saturated refusal-traces. Given a refusal-trace set $P \subseteq 2^{\Sigma} \times (\Sigma \times 2^{\Sigma})^*$, we define the *saturation map* on P by $\text{sat}_P : P \to 2^{\Sigma} \times (\Sigma \times 2^{\Sigma})^*$ where

$$\text{sat}_P(e) := \Sigma_0(\sigma_1(e), \Sigma_1) \ldots (\sigma_{|e|}, \Sigma_{|e|}),$$
$$\text{where } \forall k \leq |e| : \Sigma_k := \Sigma_k(e) \cup \{\sigma \in \Sigma \mid e^k(\sigma, \emptyset) \notin \text{dom}(\text{pr}(P))\}.$$

The *saturated refusal-traces* of P, denoted $P_{\text{sat}} \subseteq P$, is defined to be the set of fixed points of $\text{sat}_P(\cdot)$.

Theorem 1. *Given a pair of refusal-trace sets (P^m, P) with $P^m, P \subseteq 2^{\Sigma} \times (\Sigma \times 2^{\Sigma})^*$, it is a trajectory model if and only if (T1) $P \neq \emptyset$; (T2) $P = \text{pr}(P)$; (T3) $P = \text{dom}(P_{\text{sat}})$; (T4) $\forall e \in P, k < |e| : \sigma_{k+1}(e) \notin \Sigma_k(e)$; (T5) $P^m = \text{dom}(P_{\text{sat}} \cap P^m)$.*

Note that given a trajectory model, the trace map can be used to obtain the associated language model. On the other hand, given a language model (K^m, K), it is possible to obtain a *deterministic* trajectory model having (K^m, K) as the language model.

Definition 2. A trajectory model (P^m, P) is said to be *deterministic* if there exists a deterministic state machine $\mathcal{P} := (X_{\mathcal{P}}, \Sigma, \delta_{\mathcal{P}}, x^0_{\mathcal{P}}, X^m_{\mathcal{P}})$ such that $T^m(\mathcal{P}) = P^m$ and $T(\mathcal{P}) = P$.

Given a language model (K^m, K), the trajectory map $\mathrm{trj}_K : K \to 2^\Sigma \times (\Sigma \times 2^\Sigma)^*$ can be used to obtain the associated deterministic trajectory model:

$$\mathrm{trj}_K(s) := \Sigma_0(s)(\sigma_1(s), \Sigma_1(s)) \ldots (\sigma_{|s|}(s), \Sigma_{|s|}(s)) \in 2^\Sigma \times (\Sigma \times 2^\Sigma)^*,$$
$$\text{where } \forall k \leq |s| : \Sigma_k(s) := \{\sigma \in \Sigma \mid s^k\sigma \notin K\}.$$

Proposition 3. *Given a language model* (K^m, K)*, define* $\det(K) := \mathrm{dom}(\mathrm{trj}_K(K))$ *and* $\det^m(K^m, K) := \mathrm{dom}(\mathrm{trj}_K(K^m))$. *Then* $(\det^m(K^m, K), \det(K))$ *is the unique deterministic trajectory model with language model* (K^m, K).

In [4, 5, 13], prioritized synchronous composition (PSC) of systems is used as the mechanism of control. In this setting, each system is assigned a priority set of events. When systems are interconnected via PSC, an event can occur in the composite system only if it can occur in each subsystem which has that event in its priority set. In this way, a subsystem can prevent the occurrence of certain events, thereby implementing a type of supervisory control. Given two NSM's $\mathcal{P}$ and $\mathcal{Q}$ with priority sets $A, B \subseteq \Sigma$, respectively, their PSC is denoted as $\mathcal{P}\ {}_A\|_B\ \mathcal{Q}$. The PSC of the associate trajectory model is denoted by $T^m(\mathcal{P})\ {}_A\|_B\ T^m(\mathcal{Q})$ and $T(\mathcal{P})\ {}_A\|_B\ T(\mathcal{Q})$. For a formal definition of the PSC of two NSM's and associate trajectory models refer to [13].

Theorem 4. $T^m(\mathcal{P})\ {}_A\|_B\ T^m(\mathcal{Q}) = T^m(\mathcal{P}\ {}_A\|_B\ \mathcal{Q})$ *and* $T(\mathcal{P})\ {}_A\|_B\ T(\mathcal{Q}) = T(\mathcal{P}\ {}_A\|_B\ \mathcal{Q})$.

Corollary 5. *Let* $\mathcal{P}_1, \mathcal{P}_2, \mathcal{Q}_1, \mathcal{Q}_2$ *be NSM's with* $T^m(\mathcal{P}_1) = T^m(\mathcal{P}_2)$, $T(\mathcal{P}_1) = T(\mathcal{P}_2)$, $T^m(\mathcal{Q}_1) = T^m(\mathcal{Q}_2)$, $T(\mathcal{Q}_1) = T(\mathcal{Q}_2)$. *Then for any* $A, B \subseteq \Sigma$*:*

1. $T^m(\mathcal{P}_1\ {}_A\|_B\ \mathcal{Q}_1) = T^m(\mathcal{P}_2\ {}_A\|_B\ \mathcal{Q}_2)$ *and* $T(\mathcal{P}_1\ {}_A\|_B\ \mathcal{Q}_1) = T(\mathcal{P}_2\ {}_A\|_B\ \mathcal{Q}_2)$,
2. $L^m(\mathcal{P}_1\ {}_A\|_B\ \mathcal{Q}_1) = L^m(\mathcal{P}_2\ {}_A\|_B\ \mathcal{Q}_2)$ *and* $L(\mathcal{P}_1\ {}_A\|_B\ \mathcal{Q}_1) = L(\mathcal{P}_2\ {}_A\|_B\ \mathcal{Q}_2)$.

The first part of Corollary 5 states that the trajectory model contains enough detail to support the prioritized synchronous composition of NSM's. The second part of Corollary 5 states that the trajectory model serves as a *language congruence* [4] with respect to the operation of prioritized synchronous composition.

Corollary 6. *Let* (P^m, P) *and* (Q^m, Q) *be two trajectory models, and* $A, B \subseteq \Sigma$. *Then* $(P^m\ {}_A\|_B\ Q^m, P\ {}_A\|_B\ Q)$ *is a trajectory model.*

The following corollary states that when the priority sets of two systems are each equal to the entire event set, then the language model of the composed system can be obtained as the intersection of the individual language models.

Corollary 7. *For NSMs* $\mathcal{P} := (X_\mathcal{P}, \Sigma, \delta_\mathcal{P}, x^0_\mathcal{P}, X^m_\mathcal{P})$, $\mathcal{Q} := (X_\mathcal{Q}, \Sigma, \delta_\mathcal{Q}, x^0_\mathcal{Q}, X^m_\mathcal{Q})$*:*

$$L^m(\mathcal{P}\ {}_\Sigma\|_\Sigma\ \mathcal{Q}) = L^m(\mathcal{P}) \cap L^m(\mathcal{Q}); \quad L(\mathcal{P}\ {}_\Sigma\|_\Sigma\ \mathcal{Q}) = L(\mathcal{P}) \cap L(\mathcal{Q}).$$

Corollary 8. *Let* (P^m, P) *and* (Q^m, Q) *be trajectory models. Then*

$$L(P^m\ {}_\Sigma\|_\Sigma\ Q^m) = L(P^m) \cap L(Q^m); \quad L(P\ {}_\Sigma\|_\Sigma\ Q) = L(P) \cap L(Q).$$

Next theorem and corollary assert that PSC is *associative*.

Theorem 9. *Let* $\mathcal{P}, \mathcal{Q}, \mathcal{R}$ *be NSM's and* $A, B, C \subseteq \Sigma$. *Then* $(\mathcal{P}\ {}_A\|_B\ \mathcal{Q})\ {}_{A\cup B}\|_C\ \mathcal{R} = \mathcal{P}\ {}_A\|_{B\cup C}\ (\mathcal{Q}\ {}_B\|_C\ \mathcal{R})$.

Corollary 10. *For trajectory models* (P^m, P), (Q^m, Q) *and* (R^m, R), *and priority sets* $A, B, C\Sigma$:

$$(P^m\ {}_A\|_B\ Q^m)\ {}_{A\cup B}\|_C\ R^m = P^m\ {}_A\|_{B\cup C}\ (Q^m\ {}_B\|_C\ R^m),$$
$$(P\ {}_A\|_B\ Q)\ {}_{A\cup B}\|_C\ R = P\ {}_A\|_{B\cup C}\ (Q\ {}_B\|_C\ R).$$

Next we define *augmentation* by an event set $D \subseteq \Sigma$ to be the PSC of a given NSM with a deterministic state machine $\mathcal{D}$ consisting of a single state having self-loops on events in $D \subseteq \Sigma$. Clearly, $L^m(\mathcal{D}) = L(\mathcal{D}) = D^*$ and $T^m(\mathcal{D}) = T(\mathcal{D}) = \det(D^*)$.

Definition 11. Given an NSM $\mathcal{P} := (X_\mathcal{P}, \Sigma, \delta_\mathcal{P}, x^0_\mathcal{P}, X^m_\mathcal{P})$ and $D \subseteq \Sigma$, the *augmented NSM with respect to D*, denoted $\mathcal{P}^D$, is defined to be $\mathcal{P}^D := \mathcal{P}\ {}_\varnothing\|_\varnothing\ \mathcal{D}$. Given a trajectory model (P^m, P), *the augmented trajectory model with respect to D*, denoted $(P^m, P)^D$, is defined to be $(P^m, P)^D := (P^m, P)\ {}_\varnothing\|_\varnothing\ (\det(D^*), \det(D^*))$.

Using $((P^m)^D, P^D)$ to denote $(P^m, P)^D$, we have $(P^m)^D = P^m\ {}_\varnothing\|_\varnothing \det(D^*)$ and $P^D = P\ {}_\varnothing\|_\varnothing \det(D^*)$. Clearly the trajectory model $(T^m(\mathcal{P}^D), T(\mathcal{P}^D))$ of the augmented NSM $\mathcal{P}^D$ is equal to $((T^m(\mathcal{P}))^D, (T(\mathcal{P}))^D)$.

Next we show that the PSC of two systems is equivalent to the strict synchronous composition (over the union of the two priority sets) of the associated augmented systems. Thus the notion is augmentation is useful in reducing the PSC operation to a strict synchronous composition operation.

Proposition 12. *Let* (P^m, P), (Q^m, Q) *be trajectory models, and* $A, B \subseteq \Sigma$. *Then*

$$P^m\ {}_A\|_B\ Q^m = (P^m)^{B-A}\ {}_{A\cup B}\|_{A\cup B}\ (Q^m)^{A-B}; \quad P\ {}_A\|_B\ Q = P^{B-A}\ {}_{A\cup B}\|_{A\cup B}\ Q^{A-B}.$$

5 Supervisory Control Using PSC

In [13], we studied the supervisory control of nondeterministic systems in the setting of trajectory models and prioritized synchronization under the assumption that all refusal-traces of the plant are marked and the desired behavior is specified by a prefix-closed language; thus the issue of deadlock/blocking was not investigated. In this section, we generalize the results in [13] to include non-closed specifications and arbitrary marking of the plant refusal-traces.

Since trajectory models contain sufficient detail to support the operation of prioritized synchronization, we use trajectory models, rather than NSM's, to represent a discrete event system. Unless otherwise specified, the trajectory model of the plant and that of the supervisor are denoted by (P^m, P) and (S^m, S), and the priority set of the plant and that of the supervisor are denoted by A and B, respectively. In the setting of supervisory control, $A = \Sigma_u \cup \Sigma_c$, where $\Sigma_u, \Sigma_c \subseteq \Sigma$ denote the sets of *uncontrollable* and *controllable* events respectively [11, 12]; and $B = \Sigma_c \cup \Sigma_d$, where $\Sigma_d \subseteq \Sigma$ denotes the set of so-called *driven* [4] or *forcible* [3, 2] or *command* events [1]. The sets Σ_u, Σ_c and Σ_d are pairwise disjoint and exhaust the entire event set. Note that $A \cup B = \Sigma$.

Definition 13. Given a plant (P^m, P) with priority set $A \subseteq \Sigma$, a supervisor, with trajectory model (S^m, S) and priority set $B \subseteq \Sigma$, is said to be *non-marking* if $S^m = S$; it is said to be *language model non-blocking* if $\mathrm{pr}(L(P^m\ {}_A\|_B\ S^m)) = L(P\ {}_A\|_B\ S)$; and it is said to be *trajectory model non-blocking* if $\mathrm{pr}(P^m\ {}_A\|_B\ S^m) = P\ {}_A\|_B\ S$.

Note that if a supervisor is trajectory model non-blocking, then it is also language model non-blocking. On the other hand, if a plant as well as a supervisor are deterministic, and the supervisor is language model non-blocking, then it is also trajectory model non-blocking.

Next we obtain a necessary and sufficient condition for the existence of a non-marking and language model non-blocking supervisor. This result is then used to obtain a necessary and sufficient condition for the existence of a non-marking and trajectory model non-blocking supervisor.

Theorem 14. *Let* (P^m, P) *be the trajectory model of a plant,* $A, B \subseteq \Sigma$ *with* $A \cup B = \Sigma$, *and* $K^m \subseteq L((P^m)^{\Sigma-A})$ *with* $K^m \neq \emptyset$. *Then there exists a non-marking and language model non-blocking supervisor with trajectory model* (S, S) *such that* $L(P^m \ {}_A\|_B \ S) = K^m$ *if and only if*

(Relative-closure) $\mathrm{pr}(K^m) \cap L((P^m)^{\Sigma-A}) = K^m$; *and*
(Controllability) $\mathrm{pr}(K^m)(A - B) \cap L(P^{\Sigma-A}) \subseteq \mathrm{pr}(K^m)$.

In this case, S can be chosen to be $\det(\mathrm{pr}(K^m))$.

Remark. In contrast to the standard conditions [11] for the existence of a non-marking and language model non-blocking supervisor in the absence of driven events, the controllability and relative-closure conditions given in Theorem 14 refer to the language model of the *augmented* plant. It is easily demonstrated [13, Example 2] that this language model depends on the *trajectory model* of the plant and generally cannot be determined if only the language model of the plant is known.

Theorem 15. *Let* (P^m, P) *be the trajectory model of a plant,* $A, B \subseteq \Sigma$ *with* $A \cup B = \Sigma$, *and* $K^m \subseteq L((P^m)^{\Sigma-A})$ *with* $K^m \neq \emptyset$. *Then there exists a non-marking and trajectory model non-blocking deterministic supervisor with trajectory model* (S, S) *such that* $L(P^m \ {}_A\|_B \ S) = K^m$ *if and only if*

(Relative-closure) $\mathrm{pr}(K^m) \cap L((P^m)^{\Sigma-A}) = K^m$, *and*
(Controllability) $\mathrm{pr}(K^m)(A - B) \cap L(P^{\Sigma-A}) \subseteq \mathrm{pr}(K^m)$, *and*
(Trajectory-closure) $P \ {}_A\|_B \ \det(\mathrm{pr}(K^m)) = \mathrm{pr}[P^m \ {}_A\|_B \ \det(\mathrm{pr}(K^m))]$.

In this case, S can be chosen to be $\det(\mathrm{pr}(K^m))$.

With Theorem 15 in hand, we revisit the example of §2.

Example 1. Consider the open-loop NSM $\mathcal{P}$ described in §2 and depicted in Figure 1(b). It is given that a, c, d are uncontrollable. We regard b_1 as a controllable event–i.e., requiring the participation of both the plant and supervisor. However, we regard b_2 as a driven event. This models the possibility that the supervisor may request the input of a part with the guides set to narrow, but that this request may be refused by the plant if the arriving part is wide. Thus, for PSC-based design, the priority sets of the plant and supervisor are $A = \{a, b_1, c, d\}$ and $B = \{b_1, b_2\}$ respectively.

In §2, we indicated that a language model non-blocking Ramadge-Wonham type supervisor could be constructed that gives $K_1^m := (ab_2c)^*$ as the closed-loop recognized language. We also noted that the closed-loop system is unsatisfactory since deadlock can occur. Let us determine whether a PSC-based supervisor can impose the specification K_1^m without permitting deadlock. The augmented plant $\mathcal{P}^{\Sigma-A}$ is shown in Figure 2(a) and has generated language given by

$$L(\mathcal{P}^{\Sigma-A}) = \mathrm{pr}[[b_2^* a(b_2^* b_1 + b_2)b_2^* c]^* b_2^* ab_1 b_2^* db_2^*].$$

It is straightforward to verify that K_1^m satisfies the controllability and relative-closure conditions of Theorem 14. Thus, it follows from Theorem 14 that the non-marking supervisor (S_1, S_1), $S_1 := \det(\mathrm{pr}(K_1^m))$, is language model non-blocking and imposes K_1^m as the closed-loop recognized language.

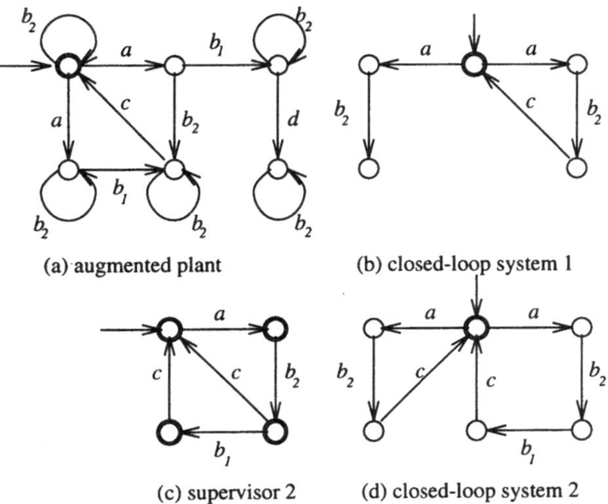

(a) augmented plant (b) closed-loop system 1

(c) supervisor 2 (d) closed-loop system 2

Fig. 2. Diagram illustrating Example 3

To determine whether there is any deterministic non-marking and *trajectory model non-blocking* supervisor that can impose K_1^m, we must check the trajectory-closure condition in Theorem 15. Let $\mathcal{S}_1$ denote the minimal deterministic state machine (with 3 states) with $L^m(\mathcal{S}_1) = L(\mathcal{S}_1) = \mathrm{pr}(K_1^m)$. Let $\mathcal{Q}_1 := \mathcal{P}\ {}_A\|_B\ \mathcal{S}_1$ depicted in Figure 2(b). Letting (P^m, P) denote the trajectory model of $\mathcal{P}$, it follows from Theorem 4 that

$$P\ {}_A\|_B \det(\mathrm{pr}(K_1^m)) = T(\mathcal{Q}_1), \quad P^m\ {}_A\|_B \det(\mathrm{pr}(K_1^m)) = T^m(\mathcal{Q}_1).$$

Since $e := \varnothing(a, \varnothing)(b_2, \{c\}) \in T(\mathcal{Q}_1) - \mathrm{pr}(T^m(\mathcal{Q}_1))$, the trajectory-closure condition fails to hold. Thus, K_1^m cannot be imposed without trajectory model blocking.

Consider the alternative specification $K_2^m := (ab_2(\epsilon + b_1)c)^*$, which also satisfies the controllability and relative-closure conditions of Theorem 14. Let $\mathcal{S}_2$ denote the deterministic state machine with $L^m(\mathcal{S}_2) = L(\mathcal{S}_2) = \mathrm{pr}(K_2^m)$ depicted in Figure 2(c). The resulting closed-loop system $\mathcal{Q}_2 := P\ {}_A\|_B\ \mathcal{S}_2$ is shown in Figure 2(d). Since $T(\mathcal{Q}_2) = \mathrm{pr}(T^m(\mathcal{Q}_2))$, the trajectory-closure condition holds. Thus, the non-marking deterministic supervisor $(\mathcal{S}_2, \mathcal{S}_2)$, $\mathcal{S}_2 := \det(\mathrm{pr}(K_2^m))$, is trajectory model non-blocking and imposes K_2^m as the closed-loop recognized language.

The supervisor implements the following control strategy: When a part arrives, the supervisor requests that it be input with the guides set to narrow. If the part happens to be narrow, the plant accepts this request and executes the event b_2. However, if the part is wide, the plant refuses to execute the event and the transition on b_2 occurs only in the supervisor. The part is then input with the guides set to wide. By following this strategy, a narrow part is never input with the guides set to wide, so jamming does not occur. Allowing for the possibility that a supervisor-initiated event may be refused by the plant is an essential feature of this control design.

6 Conclusion

In this paper, we have extended our earlier work on supervisory control of nondeterministic systems to include markings so that the issue of blocking can be investigated. Since language model non-blocking is inadequate for certain nondeterministic systems, the stronger requirement of trajectory model non-blocking is introduced. Necessary and sufficient conditions are obtained for the existence of language model non-blocking as well as trajectory model non-blocking supervisors that meet given language specifications. In [8] we extend the work presented here by considering design of supervisor under partial observation.

References

1. S. Balemi, G. J. Hoffmann, P. Gyugyi, H. Wong-Toi, and G. F. Franklin. Supervisory control of a rapid thermal multiprocessor. *IEEE Transactions on Automatic Control*, 38(7):1040–1059, July 1993.
2. B. A. Brandin and W. M. Wonham. Supervisory control of timed discrete event systems. Technical Report CSGR 9210, University of Toronto, Toronto, Canada, 1992.
3. C. H. Golaszewski and P. J. Ramadge. Control of discrete event processes with forced events. In *Proceedings of 26th IEEE Conference on Decision and Control*, pages 247–251, Los Angeles, CA, 1987.
4. M. Heymann. Concurrency and discrete event control. *IEEE Control Systems Magazine*, 10(4):103–112, 1990.
5. M. Heymann and G. Meyer. Algebra of discrete event processes. Technical Report NASA 102848, NASA Ames Research Center, Moffett Field, CA, June 1991.
6. C. A. R. Hoare. *Communicating Sequential Processes*. Prentice Hall, Inc., Englewood Cliffs, NJ, 1985.
7. R. Kumar, V. K. Garg, and S. I. Marcus. On controllability and normality of discrete event dynamical systems. *Systems and Control Letters*, 17(3):157–168, 1991.
8. R. Kumar and M. A. Shayman. Supervisory control under partial observation of nondeterministic systems via prioritized synchronization. Technical report, Department of Electrical Enginnering, University of Kentucky, Lexington, KY, 1994.
9. R. Milner. *A Calculus of Communicating Systems*. Springer Verlag, 1980.
10. I. Phillips. Refusal testing. *Theoretical Computer Science*, 50:241–284, 1987.
11. P. J. Ramadge and W. M. Wonham. Supervisory control of a class of discrete event processes. *SIAM Journal of Control and Optimization*, 25(1):206–230, 1987.
12. P. J. Ramadge and W. M. Wonham. The control of discrete event systems. *Proceedings of IEEE: Special Issue on Discrete Event Systems*, 77:81–98, 1989.
13. M. Shayman and R. Kumar. Supervisory control of nondeterministic systems with driven events via prioritized synchronization and trajectory models. *SIAM Journal of Control and Optimization*, May 1994. To Appear.

Supervisory Control for Nondeterministic Systems

Ard Overkamp

CWI, P.O.Box 94079, 1090 GB Amsterdam, Netherlands

1 Introduction

Up till now discrete event systems have been mostly modelled as deterministic systems. The current state and the next event uniquely determine the next state. Deterministic systems are effectively described by the language they can generate. Conditions under which a supervisor exists and the behavior of the controlled system are also stated in a language context. But in case of partial observation (and partial specification) systems are more appropriately modelled as nondeterministic systems.

It is not sufficient to describe nondeterministic systems only by the language they generate. The blocking properties of the systems are also important [5]. Consider the following example. Given two systems, A and B (Fig. 1 a, b). A is nondeterministic and B is its deterministic equivalent. The languages that the systems can generate are the same, but when connected to system C, (Fig. 1 c) $A||C$ can deadlock but $B||C$ can not (Fig. 1 d, e).

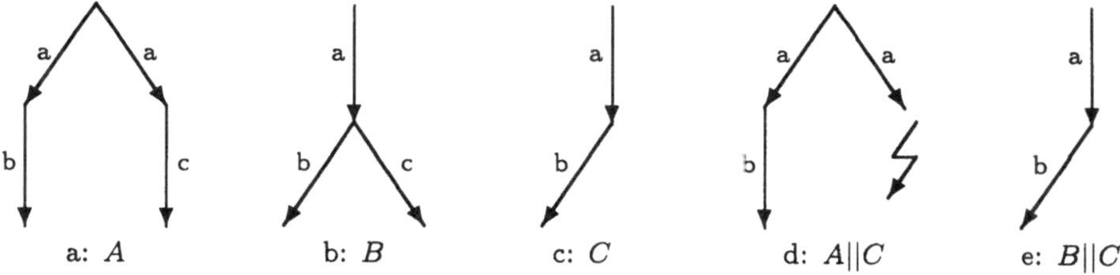

Fig. 1. Illustration of the difference between a nondeterministic and a deterministic system.

The problem that we are trying to solve in this paper is the following. Given a nondeterministic system and a nondeterministic specification, find a supervisor such that the controlled system behaves according to the specification. Here 'behaves like' means that the controlled system may only do what the specification allows, and that it may only block an event if the specification can also block that event. A new aspect in this setup is that the specification may also be nondeterministic. This gives the possibility to specify that some nondeterminism is still allowed in the controlled system.

In §3 an alternative for language semantics will be introduced: failure semantics [1, 4]. It is a commonly used semantics in computer science. We will use it to state necessary and sufficient conditions for the existence of a supervisor such that the controlled system behaves according to the specification. An algorithm to automatically generate a supervisor will be presented. In §5 these results will be extended to deal with uncontrollable events.

2 Notation and Definitions

Definition 1 (Finite State Machine). A (nondeterministic) finite state machine (FSM) is a four tuple $(Q, \Sigma, \delta, \mathcal{Q}_0)$ where,
Q is the state space,
Σ is the set of event labels,
$\delta : Q \times \Sigma \to 2^Q$ is the transition function, and
$\mathcal{Q}_0 \subseteq Q$ is the nonempty set of initial states.

The domain of the transition function can be extended in a natural way to strings and sets of states. Let $A = (Q_a, \Sigma, \delta_a, A_0)$ be a FSM, $\mathcal{A} \subseteq Q_a$, $q_a \in Q_a$, $\sigma \in \Sigma$, and $s \in \Sigma^*$. We define
$\delta_a(q_a, \varepsilon) = q_a$; $\delta_a(q_a, s\sigma) = \delta_a(\delta_a(q_a, s), \sigma)$; $\delta_a(\mathcal{A}, s) = \bigcup_{q \in \mathcal{A}} \delta_a(q, s)$.
For notational convenience we will use the state machine itself as an argument instead of the initial states. $\delta(A, s) = \delta_a(A_0, s)$.

For a deterministic finite state machine (DFSM) $\delta(q, s)$ is always a singleton set. In this case we will write $q' = \delta(q, s)$ instead of $\{q'\} = \delta(q, s)$. Also, we will simply write the initial state instead of a singleton set containing the initial state.

A useful function is the out-function, which describes the outgoing arcs of a state. $\text{out}(q) = \{\sigma \in \Sigma \mid \delta(q, \sigma) \neq \varnothing\}$.

The set of traces, or language, generated by a FSM is defined as $L(A) = \{s \in \Sigma^* \mid \delta(A, s) \neq \varnothing\}$. A language is prefix closed if $sv \in L \Rightarrow s \in L$. All languages generated by a FSM are prefix closed.

Control will not be enforced by a control map as in the Ramadge-Wonham approach, but by synchronization on common events. The controlled system (i.e. the synchronous composition of the plant and the supervisor) can only execute those events that both the supervisor and the plant can execute.

Definition 2 (Synchronous Composition). Let $A = (Q_a, \Sigma, \delta_a, A_0)$ and $B = (Q_b, \Sigma, \delta_b, B_0)$. The synchronous composition of A and B is $A||B = (Q_a \times Q_b, \Sigma, \delta_{ab}, A_0 \times B_0)$ where $\delta_{ab}((q_a, q_b), \sigma) = \delta_a(q_a, \sigma) \times \delta_b(q_b, \sigma)$. Note that for $\mathcal{Q}_a \subseteq Q_a, \mathcal{Q}_b \subseteq Q_b : \varnothing \times \mathcal{Q}_b = \mathcal{Q}_a \times \varnothing = \varnothing$.

The next lemma is a direct result from the definition.

Lemma 3. *For FSMs A, B, and $s \in \Sigma^*$;*

1. $(q_a, q_b) \in \delta(A||B, s) \;\Leftrightarrow\; q_a \in \delta(A, s) \wedge q_b \in \delta(B, s)$.
2. *For* $(q_a, q_b) \in \delta(A||B, s)$ $\text{out}((q_a, q_b)) = \text{out}(q_a) \cap \text{out}(q_b)$.
3. $L(A||B) = L(A) \cap L(B)$.

3 Failure Semantics

In the introduction we already argued that the language alone does not provide enough information for nondeterministic systems. The blocking properties of the systems are also important. In computer science failure semantics is introduced to deal with such problems. In failure semantics a system is not only described by the traces it can generate but also by the set of events that it can block after a trace (fail to execute). We will not elaborate on it here. The reader is referred to [1, 4] for more information. We will only define what we need and show that it fits our interests.

First, let us define what we mean by blocking. In the specification it may be stated that after certain traces the system is allowed to block, but after other traces it may not. So we have to define nonblockingness with respect to an already executed trace.

Definition 4 (Blocking). Let A be a FSM and $s \in L(A)$; A is nonblocking after s if $\forall q_a \in \delta(A, s)\ \text{out}(q_a) \neq \varnothing$. A FSM A is nonblocking if it is nonblocking after all $s \in L(A)$.

As stated in the introduction, the controlled system may only do what the specification allows and it may only block an event if the specification can also block that event. Because the specification is nondeterministic it is possible that sometimes an event is blocked in one branch of a nondeterministic choice and allowed in another branch. So, the controlled system has the choice to either block or execute that event. Because of this freedom the system does not have to be equivalent to the specification. It only has to reduce (implement) it [2, 3].

Definition 5 (Reduction). Let A, B be FSMs; A reduces B ($A \precsim B$) if
i) $L(A) \subseteq L(B)$, and
ii) $\forall s \in L(A),\ \ \forall q_a \in \delta(A, s),\ \ \exists q_b \in \delta(B, s)$ s.t. $\text{out}(q_b) \subseteq \text{out}(q_a)$.

Here, point *i* states that system A may only do what system B allows, and point *ii* states that A may only block what B can also block.

The reduction relation guarantees that the controlled system does not block in an environment if the specification does not block in the same environment. This is formally stated in Theorem 6.

Theorem 6. *Let A, B be FSMs such that $L(A) \subseteq L(B)$. For all FSM C and $s \in \Sigma^*$, $B||C$ nonblocking after s implies $A||C$ nonblocking after s, if and only if $A \precsim B$.*

Proof (if part). if $A \precsim B$ then $\forall q_a \in \delta(A, s)\ \exists q_b \in \delta(B, s)$ s.t. $\text{out}(q_b) \subseteq \text{out}(q_a)$. If $B||C$ is nonblocking after s then, by Lemma 3, $\forall q_c \in \delta(C, s)\ \text{out}(q_b) \cap \text{out}(q_c) \neq \varnothing$. Then also $A||C$ is nonblocking after s because $\text{out}(q_a) \cap \text{out}(q_c) \supseteq \text{out}(q_b) \cap \text{out}(q_c) \neq \varnothing$.

(only if part). We will prove that if A does not reduce B then there exists a FSM C and a string s such that $B||C$ is nonblocking after s, but $A||C$ can block after s. If not $A \precsim B$ then $\exists s \in L(A), \exists q_a \in \delta(A, s)$ s.t. $\forall q_b \in \delta(B, s)\ \text{out}(q_b) \not\subseteq \text{out}(q_a)$. Let C be a DFSM such that $q_c = \delta(C, s)$ with $\text{out}(q_c) = \Sigma - \text{out}(q_a)$. Then $\forall q_b \in \delta(B, s)\ \text{out}(q_b) \cap \text{out}(q_c) = \text{out}(q_b) \cap (\Sigma - \text{out}(q_a)) = \text{out}(q_b) - \text{out}(q_a) \neq \varnothing$. But $\text{out}(q_a) \cap \text{out}(q_c) = \text{out}(q_a) \cap (\Sigma - \text{out}(q_a)) = \varnothing$. □

4 Controller Synthesis

In this section we will first state under what condition there exists a supervisor such that the controlled system reduces the specification. We will call this condition reducibility. Then, an algorithm will be presented that generates such a supervisor. In the following G will denote an uncontrolled system and E a specification.

Definition 7 (Reducibility). Let G, E be FSMs. A language K is reducible (w.r.t. G, E) if

$$\forall s \in K, \forall q_g \in \delta(G, s), \exists q_e \in \delta(E, s) \text{ s.t. } \mathrm{out}(q_e) \subseteq \mathrm{out}(q_g) \cap \{\sigma \in \Sigma \mid s\sigma \in K\}.$$

Lemma 8. *Let G, E be FSMs. Let S be a DFSM such that $L(S)$ is reducible. Then $G||S \precsim E$.*

Proof (Point i of the definition of reduction). If $s \in L(S) \cap L(G)$ then $\delta(G, s)$ is not empty. So, by the definition of reducibility, there exists a $q_e \in \delta(E, s)$. Hence $s \in L(E)$, and $L(G||S) \subseteq L(E)$.

(Point ii of the definition of reduction). For $s \in L(G||S)$ let $(q_g, q_s) \in \delta(G||S, s)$, then, by Lemma 3, $q_g \in \delta(G, s)$ and $q_s = \delta(S, s)$. Because S is deterministic $\{\sigma \in \Sigma \mid s\sigma \in L(S)\} = \mathrm{out}(q_s)$. Then, by reducibility of $L(S)$, we know $\exists q_e \in \delta(E, s)$ such that $\mathrm{out}(q_e) \subseteq \mathrm{out}(q_g) \cap \mathrm{out}(q_s)$. Hence also, by Lemma 3, $\mathrm{out}(q_e) \subseteq \mathrm{out}((q_g, q_s))$. □

Theorem 9. *Let G, E be FSMs. There exists a supervisor S such that $G||S \precsim E$ if and only if there exists a nonempty, prefix closed, and reducible language K.*

Proof (if part). Let S_K be a deterministic state machine generating K. The proof follows directly from Lemma 8.

(only if part). Let $K = L(S||G)$. Then K is nonempty and prefix closed. We will prove that K is reducible. $\forall s \in K$, $s \in L(S||G)$, so $\delta(S||G, s) \neq \varnothing$. Then, by lemma 3, $\delta(G, s) \neq \varnothing$ and $\delta(S, s) \neq \varnothing$. Thus $\forall s \in K$, $\forall q_g \in \delta(G, s)$ $\exists q_s$ s.t. $(q_g, q_s) \in \delta(G||S, s)$. By the definition of reduction $\exists q_e \in \delta(E, s)$ s.t. $\mathrm{out}(q_e) \subseteq \mathrm{out}((q_g, q_s))$. By lemma 3 $\mathrm{out}((q_g, q_s)) \subseteq \mathrm{out}(q_g)$. And by construction of K $\mathrm{out}((q_g, q_s)) \subseteq \{\sigma \in \Sigma \mid s\sigma \in K\}$. So, $\mathrm{out}(q_e) \subseteq \mathrm{out}(q_g) \cap \{\sigma \in \Sigma \mid s\sigma \in K\}$. □

Algorithm 10. The following algorithm will construct a supervisor (if it exists) such that the controlled system reduces the specification.

1. Generate a deterministic state machine $R^0 = (Q_r^0, \Sigma, \delta_r^0, R_0^0)$, where

$$Q_r^0 = 2^{Q_g} \times 2^{Q_e},$$
$$R_0^0 = (G_0, E_0),$$
$$\delta_r^0((\mathcal{G}, \mathcal{E}), \sigma) = \begin{cases} (\delta_g(\mathcal{G}, \sigma), \delta_e(\mathcal{E}, \sigma)), & \text{if } \delta_g(\mathcal{G}, \sigma) \neq \varnothing \wedge \delta_e(\mathcal{E}, \sigma) \neq \varnothing, \\ \text{empty}, & \text{otherwise}. \end{cases}$$

2. Construct R^{i+1} from R^i by removing all reachable states $q_r^i = (\mathcal{G}, \mathcal{E})^i$ (including its in- and out-going transitions) from the state space of R^i that do *not* satisfy the following condition,

$$\forall q_g \in \mathcal{G} \ \exists q_e \in \mathcal{E} \text{ s.t. } \mathrm{out}(q_e) \subseteq \mathrm{out}(q_g) \cap \mathrm{out}(q_r^i) .$$

3. Repeat the previous step until all reachable states satisfy the condition or until there are no more reachable states left. Let R be the last R^i.

Theorem 11. *Let G and E be FSMs. Then Algorithm 10 produces a DFSM R in a finite number of steps. If the state space of R is nonempty then $G||R \precsim E$. If the initial state is removed from R then no supervisor exists such that the controlled system reduces the specification.*

Proof (The algorithm stops in a finite number of steps). If in one step of the algorithm no state is removed, then the algorithm stops because all states satisfy the condition. If in every step at least one state is removed from the state space then the algorithm will eventually halt because the state space is finite.

(The algorithm returns a correct solution). We will proof that when the algorithm finds a solution then $L(R)$ is reducible. And thus, by Lemma 8, $G||R \precsim E$. Let $s \in L(R)$ and $q_r = (\mathcal{G}, \mathcal{E}) = \delta(R, s)$. Then $\mathcal{G} = \delta(G, s)$, $\mathcal{E} = \delta(E, s)$, and because R is deterministic $\text{out}(q_r) = \{\sigma \in \Sigma \mid s\sigma \in L(R)\}$. Then, by step 3 of the algorithm, $\forall q_g \in \delta(G, s)\ \exists q_e \in \delta(E, s)$ s.t. $\text{out}(q_e) \subseteq \text{out}(q_g) \cap \text{out}(q_r) = \text{out}(q_g) \cap \{\sigma \in \Sigma \mid s\sigma \in L(R)\}$.

(The algorithm finds a solution if one exists). Assume there exists a nonempty, prefix closed, and reducible language K. First, we will prove that $K' = K \cap L(G)$ is also nonempty, prefix closed and reducible. Then we will prove that $K' \subseteq L(R)$. Thus the algorithm will return a nonempty solution.

Because $\varepsilon \in K$ and $\varepsilon \in L(G)$, $K \cap L(G) \neq \varnothing$. Because K and $L(G)$ are prefix closed, $sv \in K \cap L(G) \Rightarrow s \in K \wedge s \in L(G) \Rightarrow s \in K \cap L(G)$. So K' is prefix closed. By reducibility of K, $\forall s \in K \cap L(G)$, $\forall q_g \in \delta(G, s)$, $\exists q_e \in \delta(E, s)$ s.t. $\text{out}(q_e) \subseteq \text{out}(q_g) \cap \{\sigma \in \Sigma \mid s\sigma \in K\} = \text{out}(q_g) \cap \{\sigma \in \Sigma \mid s\sigma \in K \cap L(G)\}$. So K' is also reducible.

Now, we will prove by induction on the number of steps of the algorithm that $K' \subseteq L(R^i)$ for all i. So, also $K' \subseteq L(R)$.

Initial step: $\forall s \in K', s \in L(G)$. Then, by reducibility, $\delta(E, s) \neq \varnothing$. So $s \in L(E)$, and $K' \subseteq L(G) \cap L(E) = L(R^0)$.

Inductive hypothesis: $K' \subseteq L(R^i)$.

Let $s \in K'$ and $q_r^i = \delta(R^i, s)$. By the hypothesis, $s\sigma \in K' \Rightarrow s\sigma \in L(R^i) \Rightarrow \delta(R^i, s\sigma) = \delta_{R^i}(q_r^i, \sigma) \neq \varnothing \Rightarrow \sigma \in \text{out}(q_r^i)$. So, $\{\sigma \in \Sigma \mid s\sigma \in K'\} \subseteq \text{out}(q_r^i)$. By reducibility, $\forall q_g \in \mathcal{G}$, $\exists q_e \in \mathcal{E}$ s.t. $\text{out}(q_e) \subseteq \text{out}(q_g) \cap \{\sigma \in \Sigma \mid s\sigma \in K'\} \subseteq \text{out}(q_g) \cap \text{out}(q_r^i)$. So, q_r^i will not be removed from R^i. Hence, $s \in L(R^{i+1})$ and $K' \subseteq L(R^{i+1})$.

□

From the last part of the proof it can be deduced that the algorithm generates the least restrictive supervisor. (The supremal element with respect to language inclusion).

5 Uncontrollable Events

Sometimes a system can generate events that can not be blocked by a supervisor (e.g. machine breakdown). Ramadge and Wonham showed that in the presence of these uncontrollable events we need the condition of controllability to guarantee the existence of a supervisor. We will show that for nondeterministic systems the same condition is needed.

Recall from [6] the definition of controllability. Let G be a FSM, $\Sigma_u \subseteq \Sigma$. A prefix closed language K is controllable (w.r.t. G, Σ_u) if $K\Sigma_u \cap L(G) \subseteq K$. Note that this is equivalent to: $\forall s \in K, \forall q_g \in \delta(G,s)$ $\text{out}(q_g) \cap \Sigma_u \subseteq \{\sigma \in \Sigma \mid s\sigma \in K\}$.

Ramadge and Wonham called a supervisor that always accepts an uncontrollable event complete. We have to adapt the definition of completeness to deal with nondeterministic systems and control by synchronization.

Definition 12 (Completeness). A supervisor S is complete (w.r.t. a FSM G) if $\forall s \in L(S||G)$, $\forall q_s \in \delta(S,s)$, $\forall q_g \in \delta(G,s)$, $\text{out}(q_g) \cap \Sigma_u \subseteq \text{out}(q_s)$.

Theorem 13. *Let G and E be FSMs. There exists a supervisor S such that* $G||S \sqsubseteq_{\sim} E$, *and S complete w.r.t. G if and only if there exists a nonempty, prefix closed, reducible and controllable language K.*

Proof (if part). Let S_K be a DFSM generating K. Then, by lemma 8, $G||S_K \sqsubseteq_{\sim} E$. $\forall s \in L(S_K||G)$, let $q_s = \delta(S_K, s)$. Then, by controllability of $L(S_K)$, $\forall q_g \in \delta(G,s)$ $\text{out}(q_g) \cap \Sigma_u \subseteq \{\sigma \in \Sigma \mid s\sigma \in L(S_K)\} = \text{out}(q_s)$. so S_K is complete.

(only if part). Take $K = L(S||G)$, then by the proof of Theorem 9 (only if part) K is reducible. $\forall s \in L(S||G), \forall q_s \in \delta(S,s), \forall q_g \in \delta(G,s)$ $\text{out}(q_g) \cap \Sigma_u \subseteq \text{out}(q_s) \Rightarrow$ $\text{out}(q_g) \cap \Sigma_u \subseteq \text{out}(q_s) \cap \text{out}(q_g) = \text{out}((q_g, q_s)) \subseteq \{\sigma \in \Sigma \mid s\sigma \in L(S)\}$. So, $L(S||G)$ is controllable. □

Algorithm 14. This algorithm constructs a complete supervisor (if it exists) such that the controlled system reduces the specification.

The algorithm is the same as Algorithm 10, except that the following is added to step 2.

2. ...
   Also, remove those states $q_r^i = (\mathcal{G}, \mathcal{E})^i$ that do *not* satisfy the following condition,

$$\forall q_g \in \mathcal{G} \;\; \text{out}(q_g) \cap \Sigma_u \subseteq \text{out}(q_r^i) \;\; .$$

Theorem 15. *Let G and E be FSMs. Then Algorithm 14 produces a DFSM R in a finite number of steps. If the state space of R is nonempty then* $G||R \sqsubseteq_{\sim} E$ *and R is complete. If the initial state is removed from R then no complete supervisor exists such that the controlled system reduces the specification.*

Proof. The proof goes along the same lines as the proof of Theorem 11. The following has to be added to the different steps of the proof.

(The algorithm returns a correct solution). $\forall s \in L(R||G)$, let $q_r = (\mathcal{G}, \mathcal{E}) = \delta(R,s)$. Then, by step 3 of the algorithm, $\forall q_g \in \mathcal{G} = \delta(G,s)$, $\text{out}(q_g) \cap \Sigma_u \subseteq \text{out}(q_r)$. Hence R is complete.

(The algorithm finds a solution if one exists). After the sentence that starts with 'By reducibility': Also, by controllability, $\forall s \in K', \forall q_g \in \mathcal{G}$, $\text{out}(q_g) \cap \Sigma_u \subseteq \{\sigma \in \Sigma \mid s\sigma \in K'\} \subseteq \text{out}(q_r^i)$. So, q_r^i will not be removed from R^i ... □

6 Conclusions

In case of systems that are partially observed or partially specified one has to realize that the behaviour of a system depends on the nondeterministic properties of that system. This paper has been an attempt to set up a supervisory theory for nondeterministic systems. A condition (reducibility) is found for the existence of a supervisor such that the controlled system behaves like the specification. An algorithm is described which synthesizes a deterministic least restrictive supervisor. These results are extended to deal with uncontrollable events.

What remains to be done is to analyze the consequences of these results for systems with partial specification and partial observation.

References

1. J.C.M Baeten and W.P. Weijland. *Process algebra.* Cambridge University Press, Cambridge, 1990.
2. E. Brinksma, G. Scollo and C. Steenbergen. LOTOS specifications, their implementations, and their tests. *Proc. of IFIP Workshop 'Protocol Specification, Testing and Verification VI'*, pages 349–360, 1987.
3. S.D. Brooks, C.A.R. Hoare, and A.W. Roscoe. A theory for communicating sequential processes. *Journal of the ACM*, 31:560–599, 1984.
4. R.J. van Gladbeek. The linear time - branching time spectrum. *Proc. CONCUR '90*, Amsterdam, Lecture Notes in Computer Science 458, pages 278–297, 1990.
5. M. Heymann. Concurrency and discrete event control. *IEEE Control Systems Magazine*, 10:103–112, 1990.
6. P.J.G. Ramadge and W.M. Wonham. The control of discrete event systems. *Proc. of the IEEE*, 77:81–98, 1989.

Effective Control of Logical Discrete Event Systems in a Trace Theory Setting Using the Reflection Operator

Rein Smedinga

Department of Computing Science, University of Groningen
p.o.box 800, 9700 AV Groningen, the Netherlands
tel. +31 50 633937 , fax +31 50 633800, E-mail: `rein@cs.rug.nl`

Logical discrete event systems can be modelled using trace theory. In this paper we present an effective algorithm to find a controller using an operator (the reflection) that leads to systems that go beyond our scope.
We define a discrete event system (DES) to be a triple, see [Sme93b, Sme93c]:

$$P = \langle \mathbf{a}P, \mathbf{b}P, \mathbf{t}P \rangle$$

with $\mathbf{a}P$ the alphabet (set of events), $\mathbf{b}P \subseteq (\mathbf{a}P)^*$ the behaviour set, and $\mathbf{t}P \subseteq (\mathbf{a}P)^*$ the task set. $\mathbf{a}P$ is a finite set of symbols, $\mathbf{b}P$ and $\mathbf{t}P$ are possibly infinite sets of strings of symbols (traces). For any $x \in \mathbf{t}P$ we assume that, after x, P may stop without performing another event, while after $x \in \mathbf{b}P \backslash \mathbf{t}P$ the system will eventually perform another event or it deadlocks.
We call a DES *realistic*, if the behaviour is prefix-closed: $\mathbf{b}P = \mathbf{pref}(\mathbf{b}P)$ and each completed task is a behaviour: $\mathbf{t}P \subseteq \mathbf{b}P$. An unrealistic DES goes beyond our scope of a discrete event system. Nevertheless, it will play a crucial role in the remainder of this paper.

The restriction of a trace x to some alphabet A, $x\lceil A$, is defined by $\epsilon\lceil A = \epsilon$, and $xa\lceil A = x\lceil A$, if $a \notin A$, and $xa\lceil A = (x\lceil A)a$, otherwise. Here ϵ denotes the empty string. Alphabet restriction can easily be extended to work on trace sets: $T\lceil A = \{x\lceil A \mid x \in T\}$, and on DESs: $P\lceil A = \langle \mathbf{a}P \cap A, \mathbf{b}P\lceil A, \mathbf{t}P\lceil A \rangle$.

For DESs P and R we define the interaction, see [Sme93b, Sme93c], by:

$$\begin{aligned} P \,\|\, R = \langle \mathbf{a}P \cup \mathbf{a}R,\ & \{x \mid x\lceil\mathbf{a}P \in \mathbf{b}P \ \wedge\ x\lceil\mathbf{a}R \in \mathbf{b}R\}, \\ & \{x \mid x\lceil\mathbf{a}P \in \mathbf{t}P \ \wedge\ x\lceil\mathbf{a}R \in \mathbf{t}R\}\rangle \end{aligned}$$

For interactions of systems the common events can be seen as internal events. Such events need no longer be visible outside the interaction. Therefore, we introduce the external interaction operator that deletes the common events:

$$P \,]\![\, R = (P \,\|\, R)\lceil(\mathbf{a}P \div \mathbf{a}R)$$

For systems with equal alphabets we say P is a *subsystem* of R if:

$$P \subseteq R \ \Leftrightarrow\ \mathbf{a}P = \mathbf{a}R \ \wedge\ \mathbf{b}P \subseteq \mathbf{b}R \ \wedge\ \mathbf{t}P \subseteq \mathbf{t}R$$

For DESs P and R with equal alphabets ($\mathbf{a}P = \mathbf{a}R$) we define the difference by

$$P \backslash R = \langle \mathbf{a}P, \mathbf{b}P \backslash \mathbf{b}R, \mathbf{t}P \backslash \mathbf{t}R \rangle$$

From Verhoef [T.V90, T.V91] we have the following definition of the reflection of some DES (in fact the complementary system):

$$\sim P = \langle \mathbf{a}P,\ (\mathbf{a}P)^*\backslash\mathbf{b}P,\ (\mathbf{a}P)^*\backslash\mathbf{t}P\rangle$$

If P and R are realistic, it can be shown that $P \parallel R$, $P\,]\![\,R$, and $P\lceil A$ are also realistic. Notice that, if P is realistic, $\sim P$ need not be. This is why we need unrealistic DESs as well. The *realistic interior* of some DES P is defined by

$$\begin{aligned}\mathbf{real}(P) = \langle \mathbf{a}P,\ &\{x \mid x \in \mathbf{b}P\ \wedge\ (\forall y : y \in \mathbf{pref}(x) : y \in \mathbf{b}P)\},\\ &\{x \mid x \in \mathbf{t}P\ \wedge\ (\forall y : y \in \mathbf{pref}(x) : y \in \mathbf{b}P)\}\rangle\end{aligned}$$

$\mathbf{real}(P)$ is the greatest realistic subsystem of P.

In the sequel we will use $a, b, \ldots$ to denote events, $x, y, \ldots$ to denote strings, and $P, R, \ldots$ to denote DESs. $|A|$ denotes the number of elements in a set A.

1 A control problem

Assume systems P, L_{min}, and L_{max} are given with $L_{min} \subseteq L_{max}$. Our control problem is finding a system R such that $L_{min} \subseteq P\,]\![\,R \subseteq L_{max}$.
In this formulation L_{min} and L_{max} describe minimal and maximal wanted behaviours of the interaction. Mostly, L_{min} describes the minimal acceptable behaviour and L_{max} the legal or admissible behaviour. Notice that $\mathbf{a}L_{min} = \mathbf{a}L_{max}$ and R should be such that $\mathbf{a}R = \mathbf{a}P \div \mathbf{a}L_{min}$. Events from $\mathbf{a}R$ are used to control the order of the remaining events. Earlier versions of this control problem (formulated using trace structures instead of DESs) can be found in [Sme89]. From [Sme92, Sme93c] we know that $F(P, L) = \sim(P\,]\![\sim L)$ may lead to a solution:

Theorem 1. *The control problem has a solution if and only if*

$$L_{min} \subseteq P\,]\![\,F(P, L_{max})$$

and, if it is solvable, the greatest solution (with respect to $\subseteq$*) is* $F(P, L_{max})$*.*
If P*,* L_{min}*, and* L_{max} *are realistic, the greatest possible realistic solution equals* $\mathbf{real}(F(P, L_{max}))$*.*

2 State graphs

In order to have algorithms to compute solutions for our control problem effectively, we introduce so-called state graphs for our DESs and construct a controller, according to theorem 1, in terms of algorithms on these state graphs.

It is well-known that we can associate with a trace structure (language) a (finite) state automaton. Each path in the automaton, starting in the initial state and ending in a final state corresponds to a trace in the trace set. If the trace structure is regular, the number of needed states is finite. A DES is in fact a pair of trace structures, so we could use two automatons to represent one DES. However, in this way we lose the correspondence between behaviour and task. Instead, we use a more general automaton, called a state graph here, containing two kinds of final states:

Definition 2. A state graph is a tuple (A, Q, δ, q, B, T) with A the alphabet, a finite set of labels; Q the state set; $\delta: Q \times A \to Q$ the state transition function; $q \in Q$ the initial state; $B \subseteq Q$ the behaviour state set; and $T \subseteq Q$ the task state set. δ is a total function.

The state set Q need not be a finite set. Because we deal with paths in the graph, we extend δ to $\delta^*: Q \times A^* \to Q$, by: $\delta^*(p, \epsilon) = p$, and $\delta^*(p, xa) = \delta(\delta^*(p, x), a)$.

For a DES P we can construct a state graph using (extended) Nerode equivalence with equivalence classes:

$$[x]_P = \{y \mid (\forall z :: xz \in \mathbf{b}P \Leftrightarrow yz \in \mathbf{b}P \ \wedge \ xz \in \mathbf{t}P \Leftrightarrow yz \in \mathbf{t}P)\}$$

Given some state graph $G = (A, Q, \delta, q, B, T)$ the corresponding DES equals

$$\mathbf{des}(G) = \langle A, \ \{x \mid x \in A^* \ \wedge \ \delta^*(q, x) \in B\}, \ \{x \mid x \in A^* \ \wedge \ \delta^*(q, x) \in T\}\rangle$$

and given some system P a possible state graph is:

$$\mathbf{sg}(P) = (\mathbf{a}P, \{[x]_P \mid x \in (\mathbf{a}P)^*\}, \delta, [\epsilon]_P, \{[x]_P \mid x \in \mathbf{b}P\}, \{[x]_P \mid x \in \mathbf{t}P\})$$

with δ defined by $\delta([x]_P, a) = [xa]_P$.
If the behaviour and the task set of a system are regular sets, the number of equivalence classes is finite and the resulting state graph has only a finite number of states. State graphs can be displayed as is shown in Figure 1.

We have $\mathbf{des}(\mathbf{sg}(P)) = P$, but, in general, $\mathbf{sg}(\mathbf{des}(G)) \neq G$, because more state graphs exist that correspond to the same DES.

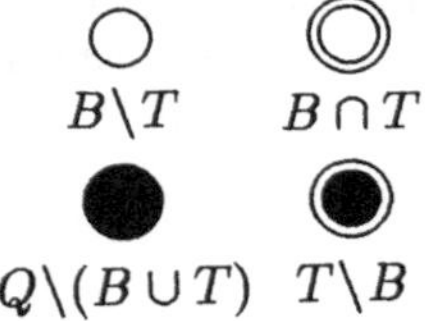

Fig. 1. Displaying of different states

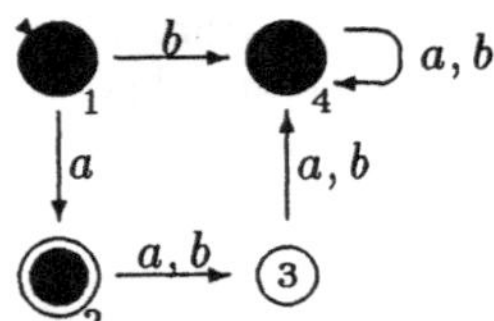

Fig. 2. System from example 1

Example 1. Consider the system $P = \langle\{a, b\}, \{aa, ab\}, \{a\}\rangle$ The following equivalence classes can be found: $p_1 = [\epsilon] = \{\epsilon\}$, $p_2 = [a] = \{a\}$, $p_3 = [aa] = \{aa, ab\}$, and $p_4 = [b] = \{a, b\}^* \setminus (p_1 \cup p_2 \cup p_3)$. In Figure 2 the corresponding graph is shown. $q = [\epsilon]$ is denoted using an extra small arrow.

For the operator $]\![$ we need a nondeterministic graph (nd-graph), because we may have to delete labels. This results in graphs in which δ is no longer a function.

Definition 3. An nd-graph is a tuple $(A, Q, \gamma, q, B, T)_{\text{nd}}$ with A, Q, q, B, and T as in definition 2 and $\gamma: Q \times (A \cup \{\epsilon\}) \to \mathbf{2}^Q$ the state transition map. Again γ is supposed to be total.

The DES $\mathbf{des}(G_{\text{nd}})$ corresponding to an nd-graph G_{nd} is given by

$$\langle A, \ \{x \mid x \in A^* \ \wedge \ \gamma^*(q, x) \cap B \neq \varnothing\}, \ \{x \mid x \in A^* \ \wedge \ \gamma^*(q, x) \cap T \neq \varnothing\}\rangle$$

where $\gamma^*(Q', x)$ represents the set of all states, reachable from a state in $Q' \subseteq Q$ via a path x including zero or more ϵ-transitions. γ^* is the extension of γ.

Example 2. In Figure 5 an nd-graph is given for $P = \langle\{a, b\}, \{\epsilon, a, b, ab\}, \{a, b, ab\}\rangle$

In the remainder of this paper we use state graphs $G_i = (A_i, Q_i, \delta_i, q_i, B_i, T_i)\,(i = , 1, 2)$ and nd-graph $G_{\text{nd}} = (A, Q, \gamma, q, B, T)_{\text{nd}}$. We give algorithms on state graphs for all operators on DESs. We will use the same operator symbol for DESs as well as for state graphs.

3 Algorithms on state graphs

Algorithm 1.

$$G_1 \| G_2 = (A_1 \cup A_2, Q_1 \times Q_2, \delta, (q_1, q_2), B_1 \times B_2, T_1 \times T_2)$$

where
$$\begin{aligned} \delta((p_1, p_2), a) &= (\delta_1(p_1, a), p_2) && \text{if } a \in A_1 \backslash A_2 \\ &= (p_1, \delta_2(p_2, a)) && \text{if } a \in A_2 \backslash A_1 \\ &= (\delta_1(p_1, a), \delta(p_2, a)) && \text{if } a \in A_1 \cap A_2 \end{aligned}$$

Complexity: $\mathcal{O}(|Q_1| \cdot |Q_2| \cdot |\mathbf{a}P \cup \mathbf{a}R|)$
Property: $\mathbf{des}(\mathbf{sg}(P_1) \| \mathbf{sg}(P_2)) = P_1 \| P_2$

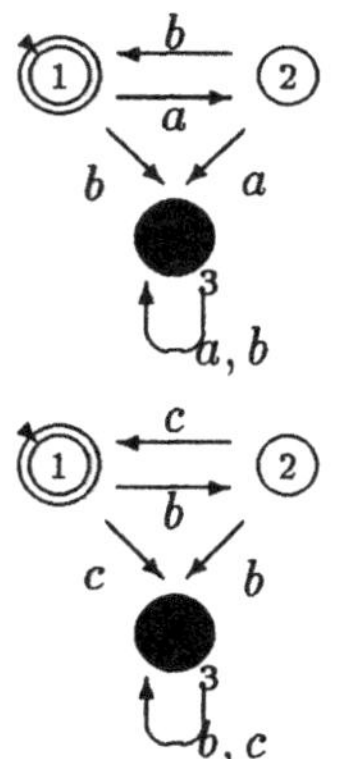

Fig. 3. State graphs G_1 (upper diagram) and G_2 for example 3

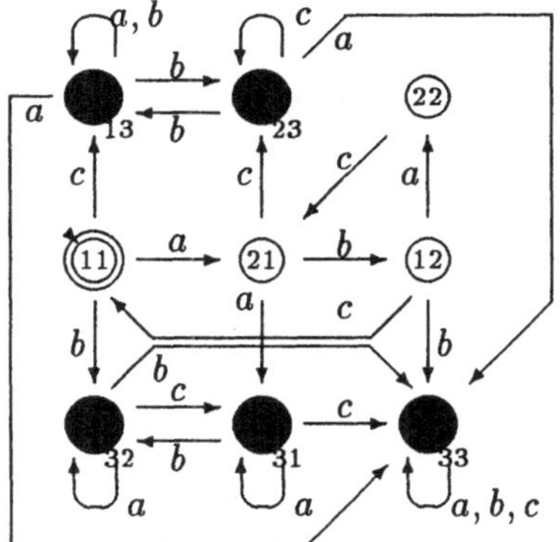

Fig. 4. Interaction of G_1 and G_2

Example 3. Consider the graphs as given in Figure 3. According to the previous property the graph for the interaction is as given in Figure 4. Notice that this graph is not minimal: all non-behaviour states can be replaced by one non-behaviour state.

Algorithm 2.

$$\mathbf{det}(G_{\text{nd}}) = (A, 2^Q, \delta, \overline{q}, \overline{B}, \overline{T})$$

where (for $r \in 2^Q$): $\overline{B} = \{r \mid r \cap B \neq \varnothing\}$ $\overline{q} = \gamma^*(q, \epsilon)$
$\overline{T} = \{r \mid r \cap T \neq \varnothing\}$ $\delta(r, a) = \bigcup_{p \in r} \gamma^*(p, a)$

Property: $\mathbf{des}(\mathbf{det}(G_{\text{nd}})) = \mathbf{des}(G_{\text{nd}})$

The above construction is a generalization of the well-known construction to find the deterministic equivalent of a nondeterministic automaton. Apart from unreachable states, each state in $\mathbf{det}(G_{nd})$ is the set of states that can be reached from another set by doing zero or more ϵ-moves, followed by one normal move, followed by zero or more ϵ-moves.

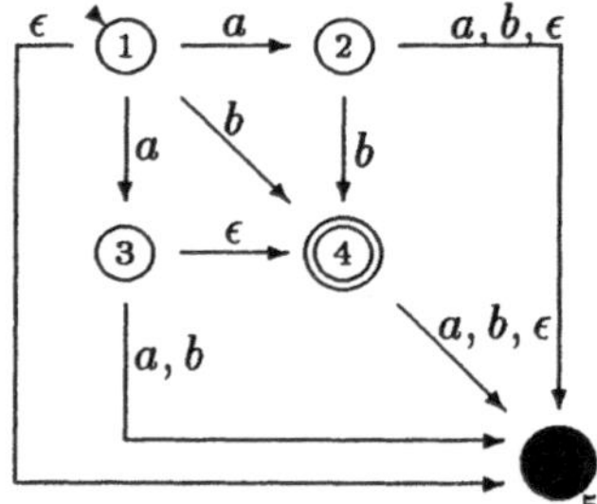

Fig. 5. An nd-graph representing the system from example 2

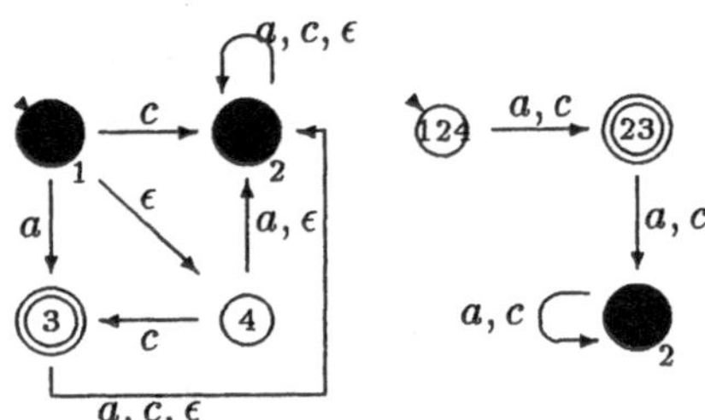

Fig. 6. An nd-graph and its deterministic equivalence for example 4

Example 4. We can use the above construction on the graph of Figure 6 (left) to get a deterministic graph. We find: $\overline{q} = \gamma^*(q, \epsilon) = \{p_1, p_2, p_4\}$, $\gamma^*(\{p_1, p_2, p_4\}, a) = \{p_2, p_3\}$, $\gamma^*(\{p_1, p_2, p_4\}, c) = \{p_2, p_3\}$, $\gamma^*(\{p_2, p_3\}, a) = \{p_2\}$, $\gamma^*(\{p_2, p_3\}, c) = \{p_2\}$, $\gamma^*(\{p_2\}, a) = \{p_2\}$, and $\gamma^*(\{p_2\}, c) = \{p_2\}$, which leads to the graph of Figure 6 (right). Notice that we only have examined the reachable states. The behaviour and task states are $\overline{B} = \{\{p_1, p_2, p_4\}, \{p_2, p_3\}\}$ and $\overline{T} = \{\{p_2, p_3\}\}$. It can easily be checked that this graph also represents the same system.

Algorithm 3.

$$G\lceil A_1 = \mathbf{det}(G_{nd}) \text{ with } G_{nd} = (A \cap A_1, Q, \gamma, q, B, T)_{nd}$$

where: $\gamma(p, \epsilon) = \bigcup_{a \in A \setminus A_1} \delta(p, a)$ $\gamma(p, a) = \{\delta(p, a)\}$ (for $a \in A \cap A_1$)
Complexity: $\mathcal{O}(|A| \cdot |Q|)$ (for computing G_{nd})
Property: $\mathbf{des}(\mathbf{sg}(P)\lceil A) = P\lceil A$

We also have: $\mathbf{des}\big((\mathbf{sg}(P) \,\|\, \mathbf{sg}(R))\lceil(\mathbf{a}P \div \mathbf{a}R)\big) = P\rceil\lceil R$, that shows a way to get a graph for $P\rceil\lceil R$:

Algorithm 4.

$$G_1\rceil\lceil G_2 = (G_1 \,\|\, G_2)\lceil(A_1 \div A_2)$$

Example 5. Consider the system $P = \langle\{a, b, c\}, \{a, b, bc\}, \{a, bc\}\rangle$. Its corresponding graph $\mathbf{sg}(P)$ is given in Figure 7. The graph $\mathbf{sg}(P)\lceil\{a, c\}$ is to be found in Figure 6, constructed using Algorithm 3. Making this graph deterministic leads to a graph representing the system $\langle\{a, c\}, \{\epsilon, a, c\}, \{a, c\}\rangle$ which is equal to $P\lceil\{a, c\}$.

The reflection operator of a graph is simply the graphs complement, i.e., interchange the types of all states:

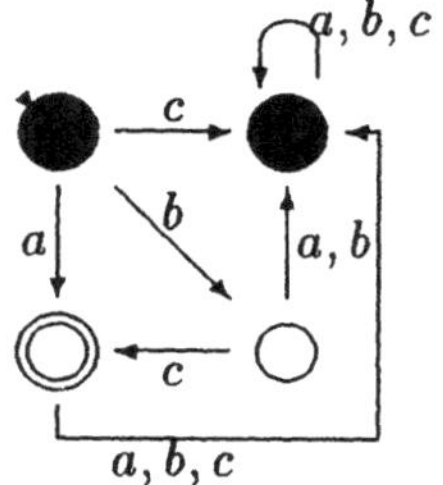

Fig. 7. sg(P) for example 5

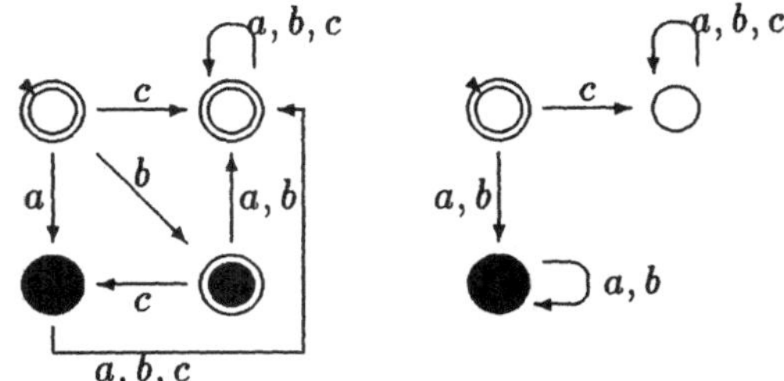

Fig. 8. Graph $\sim$sg(P) (left) and its realistic interior for example 6.

Algorithm 5.

$$\sim G = (A, Q, \delta, q, Q \backslash B, Q \backslash T)$$

Complexity: $\mathcal{O}(|Q|)$
Property: $\mathbf{des}(\sim \mathbf{sg}(P)) = \sim P$

Finding the DES-interior means finding all states reachable from q with paths not going through non-behaviour states:

Algorithm 6.

$$\begin{aligned} \mathbf{real}(G) &= (A, \{q\}, \mathbf{2}, q, \varnothing, \varnothing) && \text{if } q \notin B \\ &= (A, B \cup \{\overline{q}\}, \overline{\delta}, q, B, F \cap B) && \text{otherwise} \end{aligned}$$

where $\overline{q}$ is a fresh state ($\notin B$) and $\overline{\delta}(p, a) = \delta(p, a)$ if $\delta(p, a) \in B$, $\overline{\delta}(p, a) = \overline{q}$ otherwise, and $\overline{\delta}(\overline{q}, a) = \overline{q}$ for all $a \in A$.
Complexity: $\mathcal{O}(|A| \cdot |Q|)$
Property: $\mathbf{des}(\mathbf{real}(\mathbf{sg}(P))) = \mathbf{real}(P)$

The graph $(A, \{q\}, \mathbf{1}, q, \varnothing, \varnothing)$ is a representation for the empty system:

$$\mathbf{des}((A, \{q\}, \mathbf{1}, q, \varnothing, \varnothing)) = \langle A, \varnothing, \varnothing \rangle$$

Algorithm 7.

$$G_1 \backslash G_2 = (A, Q_1 \times Q_2, \delta, (q_1, q_2), B_1 \times Q_2 \backslash B_2, T_1 \times Q_2 \backslash T_2)$$

with $\delta((p_1, p_2), a) = (\delta_1(p_1, a), \delta_2(p_2, a))$.
Complexity: $\mathcal{O}(|A) \cdot |Q_1| \cdot |Q_2|)$
Property: $\mathbf{des}(\mathbf{sg}(P) \backslash \mathbf{sg}(R)) = P \backslash R$

Example 6. Computing $\sim$sg(P) for the system displayed in Figure 7 leads to the system as displayed in Figure 8 (left). In Figure 8 (right) the realistic part of that system can be found.

Example 7. Reconsider the systems as displayed in Figure 3. Computing $G_1 \backslash G_2$ leads to the same graph as in Figure 4 but with other state types: $B = \{p_{13}\}$ and $T = \{p_{12}, p_{13}\}$.

Because we deal with paths in a graph starting in the initial state it has no effect if states ared added that cannot be reached from the initial state. Such states can, if present, also easily be eliminated. Algorithms can be made more efficient if only the states, reachable from the initial state, are really computed. Moreover, we can use an extension of the standard algorithm on automatons (see [HU79]) to minimize state graphs. Also, we can easily extend the operators to work on nd-graphs as well. Therefore we can do without the operator **det** from Algorithm 2.

4 Effectively computable

If P and L are regular, i.e., can be displayed using finite state graphs, we see from the algorithms and properties above, that $\mathbf{real}(F(P, L))$ can be computed in polynomial time. Moreover, to test the condition for having a solution, we can compute the graph equivalence of $L_{\min} \setminus (P]\![\mathbf{real}(F(P, L_{\max})))$. If $L_{\min} \subseteq (P]\![\mathbf{real}(F(P, L_{\max})))$, this results in an empty graph, i.e., $B = \varnothing$ and $T = \varnothing$.

Theorem 4. *Let $G_{\min}$, $G_{\max}$, and G_P be state graphs for $L_{\min}$, $L_{\max}$, and P, respectively. Then is*

$$G_R := \mathbf{real}(\sim(G_P]\![\sim G_{\max}))$$

a state graph for $\mathbf{real}(F(P, L_{\max}))$. It can be computed in polynomial time. If computation of $G_{\min} \setminus (G_P]\![G_R)$ results in an empty graph, the control problem is solvable and G_R represents the largest possible solution. Also this last computation can be done in polynomial time.

5 Conclusions

We have shown that a trace theory based approach leads to an elegant definition of a logical discrete event system and gives a nice algorithm to find a solution for a control problem, where we, temporally, go beyond the scope of a DES. The algorithm can be translated to work on state graphs, leading to an effectively computable solution if the systems itself are regular. Proofs of all properties can be found in Smedinga [Sme92].

References

[BKS93] S. Balemi, P. Kozák, and R. Smedinga, editors. *Discrete Event Systems: Modeling and Control*, volume 13 of *Progress in Systems and Control Theory*. Birkhäuser Verlag, Basel, Switzerland, 1993. (Proceedings of the Joint Workshop on Discrete Event Systems (WODES'92), August 26–28, 1992, Prague, Czechoslovakia).

[HU79] J.E. Hopcroft and J.D. Ullman. *Introduction to automata theory, languages, and computation*. Addison Wesley, 1979.

[Sme89] R. Smedinga. *Control of discrete events*. PhD thesis, University of Groningen, 1989.

[Sme92] R. Smedinga. The reflection operator in discrete event systems. Technical Report CS9201, Department of computing science, University of Groningen, 1992.

[Sme93a] R. Smedinga. Discrete event systems. course-notes, second version, Department of computing science, University of Groningen, 1993.

[Sme93b] R. Smedinga. Locked discrete event systems: how to model and how to unlock. *Journal on Discrete Event Dynamic Systems, theory and applications*, 2(3/4), 1993.

[Sme93c] R. Smedinga. *An Overview of Results in Discrete Event Systems using a Trace Theory Based Setting*, pages 43–56. Volume 13 of Balemi et al. [BKS93], 1993.

[T.V90] T.Verhoeff. Solving a control problem. Internal report, Eindhoven University, 1990.

[T.V91] T.Verhoeff. Factorization in process domains. Internal report, Eindhoven University, 1991.

Diagnosability of Discrete Event Systems

Meera Sampath[1], *Raja Sengupta*[1], *Stéphane Lafortune*[1], *Kasim Sinnamohideen*[2] *and Demosthenis Teneketzis*[1]

[1] University of Michigan, Ann Arbor, MI 48109, USA
[2] Johnson Controls, Inc., Milwaukee, WI 53201 USA

1 Introduction

Failure detection and diagnosis is an important task in the automatic control of large complex systems. With the increased complexity of man-made systems and with increasingly stringent requirements on performance and reliability, one needs sophisticated and systematic means of diagnosing system failures in order to identify their exact location and cause. In this paper, we consider a discrete event systems (DES) approach to the problem of failure diagnosis. This approach to failure diagnosis is applicable to systems that fall naturally in the class of DES; moreover, for the purpose of diagnosis, continuous variable dynamic systems can often be viewed as DES at a higher level of abstraction. The major advantage of this approach is that it does not require detailed in-depth modelling of the system to be diagnosed and hence is ideally suited for diagnosis of large complex systems like HVAC (Heating, Ventilation and Air Conditioning) units, power plants and semiconductor manufacturing lines. We propose a language based definition of diagnosability, discuss the construction of *diagnosers* and state necessary and sufficient conditions for diagnosability.

2 The Notion of Diagnosability

The system to be diagnosed is modelled as a finite state machine (FSM) $G = (X, \Sigma, \delta, x_0)$ where X is the state space, Σ is the set of events or the alphabet, δ is the partial transition function and x_0 is the initial state of the system. The behaviour of the DES is described by the language $L(G)$ generated by G. Henceforth, we shall denote $L(G)$ by L. We assume for simplicity that L is live, i.e., there is a transition defined at each state in X.

The above DES model accounts for the normal *and* failed behaviour of the system. The event set Σ is partitioned as $\Sigma = \Sigma_o \dot{\cup} \Sigma_{uo}$ where Σ_o represents the set of observable events and Σ_{uo} represents the set of unobservable events. The observable events may be one of the following : Commands issued by the controller, sensor readings immediately after the execution of the above commands, or changes of sensor readings. Let $\Sigma_f \subseteq \Sigma$ denote the set of failure events which are to be diagnosed. We assume, without loss of generality, that $\Sigma_f \subseteq \Sigma_{uo}$, since an observable failure event can be trivially diagnosed.

Our objective is to identify the occurrence, if any, of the failure events given that in the traces generated by the system, only the events in Σ_o are observed. In this regard, we partition the set of failure events into disjoint sets corresponding to different failure types: $\Sigma_f = \Sigma_{f1} \cup \Sigma_{f2} \cup \cdots \cup \Sigma_{fm}$. Let Π_f denote this partition. The partition Π_f is motivated by the following considerations: (i). We may not be required to identify uniquely the occurrence of every failure event. We may simply be interested in knowing if one of a set of failure events has happened, as for example, when the effect of the set of failures on the system is the same. (ii). Inadequate instrumentation may render it impossible to diagnose uniquely every possible fault.

We now introduce the notation necessary for the subsequent development. The *projection* operator $P : \Sigma^* \to \Sigma_o^*$ is defined in the normal manner (see, e.g., [4]). P simply 'erases' the unobservable events in a trace. The inverse projection operator P_L^{-1} is then defined as : $P_L^{-1}(y) = \{s \in L : P(s) = y\}$. Let L/s denote the postlanguage of L after s, i.e., $L/s = \{t \in \Sigma^* \mid st \in L\}$ and let $\overline{\omega}$ denote the prefix-closure of $\omega \in \Sigma^*$. Let $\Psi(\Sigma_{fi}) = \{s\sigma_f \in L \mid \sigma_f \in \Sigma_{fi}\}$, i.e., $\Psi(\Sigma_{fi})$ denotes the set of all traces of L that end in a failure event belonging to the class Σ_{fi}.

We now propose the following definition of diagnosability.

Definition 1. A prefix-closed and live language L is said to be **diagnosable** with respect to the projection P and with respect to the partition Π_f on Σ_f if the following holds:

$$(\forall i \in \Pi_f)(\exists n_i \in \mathbb{N})(\forall s \in \psi(\Sigma_{fi}) \wedge \forall t \in L/s)\ \|t\| > n_i \Rightarrow D$$

where the diagnosability condition D is: $D : \omega \in P_L^{-1}[P(st)] \Rightarrow \overline{\omega} \cap \Psi(\Sigma_{fi}) \neq \varnothing$.

The above definition of diagnosability means the following: Let s be any trace generated by the system that ends in a failure event from the set Σ_{fi} and let t be any sufficiently long continuation of s. Condition D then requires that every trace belonging to the language that produces the same record of observable events as the trace st should contain in it a failure event from the set Σ_{fi}. This implies that along every continuation t of s one can detect the occurrence of a failure of the type Σ_{fi} with a finite delay; specifically in at most n_i transitions of the system after s. Alternately speaking, diagnosability requires that every failure event leads to observations distinct enough to enable unique identification of the failure type with a finite delay.

The case of multiple failures from the same set of the partition deserves special attention. When more than one failure of the same type, say, Σ_{fi} occurs along a trace s of L , the above definition of diagnosability does not require that each of these occurrences be detected. It suffices to be able to conclude, within finitely many events, that along s, a failure from the set Σ_{fi} happened. In §4 where we develop necessary and sufficient conditions for diagnosability, we show how this feature distinguishes the case of multiple failures from the case where it is not possible to have more than one failure event of the same type along any system trajectory.

We illustrate by a simple example the above notion of diagnosability. Consider the system represented in Fig. 1. Here, α, β, γ and δ are observable events, σ_{uo} is an unobservable event while σ_{f1}, σ_{f2} and σ_{f3} represent failure events. If one chooses the partition $\Sigma_{f1} = \{\sigma_{f1}, \sigma_{f2}\}$ and $\Sigma_{f2} = \{\sigma_{f3}\}$, i.e., it is not required to distinguish between failures σ_{f1} and σ_{f2}, then the above system is diagnosable with $n_1 = 2$ and $n_2 = 1$.

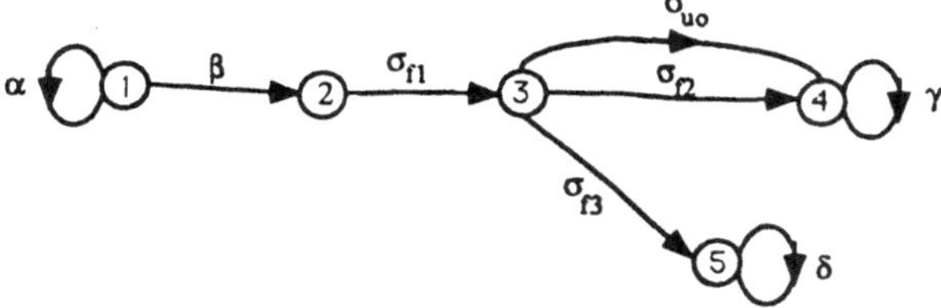

Fig. 1. Example of a system with multiple failures

On the other hand, if the partition is $\Sigma_{f1} = \{\sigma_{f1}\}$, $\Sigma_{f2} = \{\sigma_{f2}\}$, $\Sigma_{f3} = \{\sigma_{f3}\}$, then the system is not diagnosable since it is not possible to deduce the occurrence of failure σ_{f2}.

The problem of diagnosability has received relatively little attention in the literature of DES, until recently. In [3], the author addresses, in a different framework, the problems of off-line and on-line diagnosis. On the other hand, several related notions like observability [1], [2],[4], [5],[8], observability with delay [5] and invertibility [6], [7] have been studied in detail. All of the above, including diagnosability, fall in the general class of problems dealing with partial observations. However, several important differences exist between diagnosability and the problems mentioned above. In [9], we explain in detail the differences between these problems with several illustrative examples.

3 The Diagnoser

Consider the state space X of G and define $X_0 = \{x_0\} \cup \{x \in X : x$ has an observable event into it $\}$. Let $\Delta_f = \{F_1, F_2, \dots, F_m\}$ where $\|\Pi_f\| = m$ and define $\Delta = \{\{N\}, \{A\}\} \cup 2^{\Delta_f}$. Let G_d denote the diagnoser. A state q_d of G_d is of the form $q_d = \{(x_1, \ell_1), \dots, (x_n, \ell_n)\}$ where $x_i \in X_0$ and $\ell_i \in \Delta$ i.e., ℓ_i is of the form $\ell_i = \{N\}$ em "normal" or $\ell_i = \{F_{i_1} F_{i_2}, \dots, F_{i_k}\}$ where $\{i_1, i_2, \dots, i_k\} \in \{1, 2, \dots, m\}$ or $\ell_i = \{A\}$ *"ambiguous"*.

An *observer* for G (see [5]) gives estimates of the current state of the system after the occurrence of every observable event. The diagnoser G_d can be thought of as an *extended observer* in which we append to every state estimate a label of the form mentioned above. The labels attached to the state estimates carry the failure information and failures are diagnosed by checking these labels.

The initial state of the diagnoser is defined to be $(x_0, \{N\})$. Let the current state of the diagnoser (i.e., the set of estimates of the current state of G with their corresponding labels) be q_1 and let the next observed event be σ. The new state of the diagnoser q_2 is computed following a three step process:

Step 1. For every state estimate x in q_1, compute the *reach* due to σ, defined to be $R(x, \sigma) = \{\delta(x, s\sigma)$ where $s \in \Sigma_{uo}^*\}$.

Step 2. Let $x' \in R(x, \sigma)$ with $\delta(x, s\sigma) = x'$. Propogate the label ℓ associated with x to x' according to the following rules: (i). If $\ell = \{N\}$ and s contains no failure events, the label ℓ' associated with x' is also $\{N\}$. (ii). If $\ell = \{A\}$ and s contains no failure events, the label ℓ' is also $\{A\}$. (iii). If $\ell = \{N\}$ or $\{A\}$ and s contains failure events from Σ_{fi}, Σ_{fj}, $\ell' = \{F_i, F_j\}$. (iv). If $\ell = \{F_i, F_j\}$ and s contains failure events from

Σ_{fk}, $\ell' = \{F_i, F_j, F_k\}$.

Step 3. Let $q_2 = \{(x', \ell')$ pairs computed for each (x, ℓ) in $q_1\}$. Replace by (x', A) all $(x', \ell'), (x', \ell'') \in q_2$ where $\ell' \neq \ell''$. That is, if the same state estimate x' appears more than once in q_2 with different labels, we associate with x' the *ambiguous* label A.

Figure 2 illustrates a system G and its diagnoser G_d. Here $\alpha, \beta, \gamma, \delta$ and σ are observable events while σ_{uo}, σ_{f1}, σ_{f2} and $\sigma_{f2'}$ are unobservable. $\Sigma_{f1} = \{\sigma_{f1}\}$ and $\Sigma_{f2} = \{\sigma_{f2}, \sigma_{f2'}\}$.

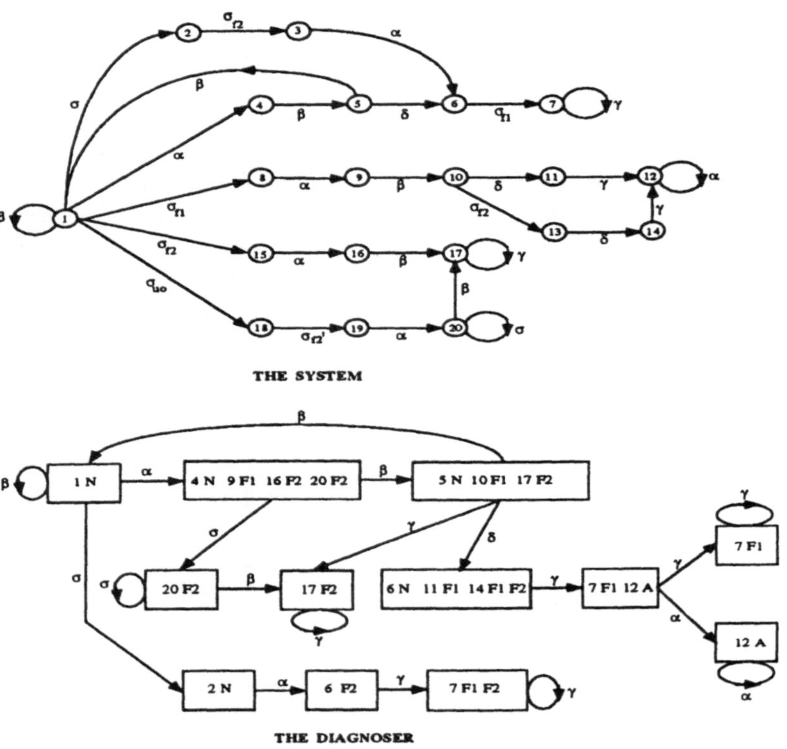

Fig. 2. Example illustrating construction of the diagnoser

For a formal definition of the diagnoser and for its construction procedure, refer to [9].

If L is such that more than one failure from the same set of the partition are possible, we construct a slightly modified diagnoser. We now define $\Delta = \{N\} \cup 2^{\Delta_f}$ i.e., we drop the A label and also omit step 3 of the construction procedure which assigns the A label. Hence, it is now possible to have pairs of the form (x, ℓ), , (x, ℓ') with $\ell \neq \ell'$ in a state of this diagnoser. The reason for this modification will become apparent in the next section.

4 Necessary and Sufficient Conditions

Before stating the necessary and sufficient conditions for diagnosability, we introduce the concepts of *Fi-certain states, Fi-uncertain states, Ambiguous states* and *Fi-indeterminate cycles*.

Definition 2. (i) An Fi*-certain* state of the diagnoser G_d is a state of G_d where all components have the label Fi. (ii) An Fi*-uncertain* state of the diagnoser G_d is a state of G_d where there is at least one component with the label Fi and one component without the label Fi. (iii) An *ambiguous* state of the diagnoser G_d is a state of G_d where there is at least one component with the label A.

(Note that part (iii) above is only relevant in the case of no multiple failures.)

Definition 3. An Fi*-indeterminate cycle* in G_d is a cycle composed exclusively of Fi-uncertain states and for which there exists: 1. a corresponding cycle (of observable events) in G involving only states that carry Fi in their labels in the cycle in G_d and 2. a corresponding cycle (of observable events) in G involving only states that do *not* carry Fi in their labels in the cycle in G_d.

From the above definition of an Fi-certain state, it is obvious that if the current state of the diagnoser is Fi-certain, then we can conclude for sure that a failure of the type Σ_{fi} has occurred, regardless of what the current state of G is. Presence of an ambiguous state in G_d corresponds to the situation where there are two (ambiguous) strings s and s' in L such that the set of all possible continuations of s in L is the same as that of s' (i.e, $\delta(x_0, s) = \delta(x_0, s')$), s contains a failure event of a particular type, say, Σ_{fi} while s' does not and in addition, the strings s and s' produce the same record of observable events. An Fi-indeterminate cycle in G_d indicates the presence in L of two strings s and s' of arbitrarily long length, both of which have the same observable projection, with s containing a failure event from the set Σ_{fi} while s' does not.

We now state the necessary and sufficient conditions for diagnosability. For proofs of these conditions, we refer the reader to [9].

Theorem 1. (i) **Case of No Multiple Failures of the Same Type:** *L is diagnosable iff its diagnoser G_d satisfies the following:* 1. *There are no Fi-indeterminate cycles in G_d, for all failure types i;* 2. *No state of G_d is ambiguous.*
(ii) **Case of Multiple Failures:** *L is diagnosable iff its diagnoser G_d satisfies the following: There are no Fi-indeterminate cycles in G_d, for all failure types i.*

Suppose L contains two traces s and s' which result in an ambiguous state in G_d. If L has to satisfy Definition 1, every $t \in L/s$ should contain in it a failure event from Σ_{fi}. This, however, is possible only in the case of multiple failures. Therefore, in the case of no multiple failures, presence of an A label in G_d immediately implies L is non-diagnosable while on the other hand, in the case of multiple failures, presence of ambiguous strings s and s' does not always imply non-diagnosability.

Note that not all cycles of Fi-uncertain states in the diagnoser qualify as Fi-indeterminate cycles. Figure 3 represents a system with a cycle of Fi-uncertain states in its diagnoser. This, however, is not an Fi-indeterminate cycle and the system is **diagnosable**. Figure 4 on the other hand, illustrates a **non-diagnosable** system with an

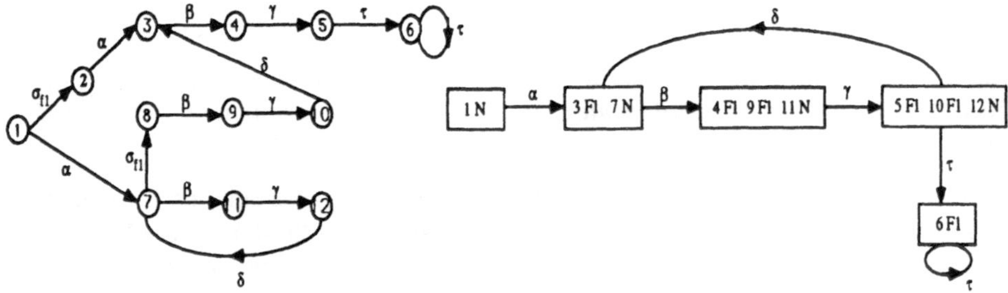

Fig. 3. Example of a diagnosable system with a cycle of Fi-uncertain states in its diagnoser

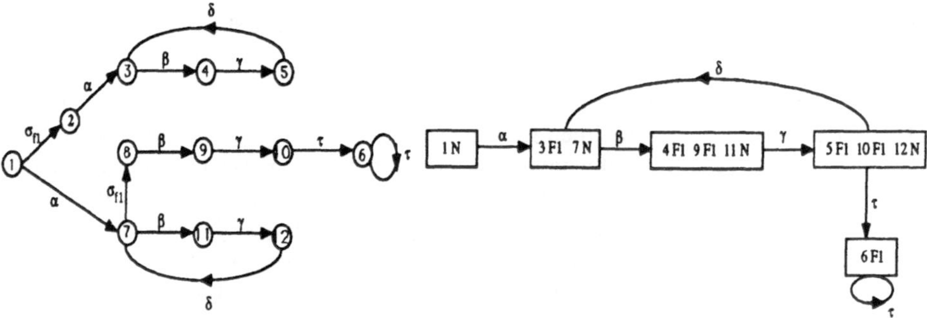

Fig. 4. Example of a non-diagnosable system with an Fi-indeterminate cycle in its diagnoser

Fi-indeterminate cycle in its diagnoser.

Note that the conditions of the theorem imply that given a failure of the type Σ_{fi} has occured, every Fi-uncertain state leads to a Fi-certain state in a bounded number of transitions, which in turn implies that occurrences of failures of the type Σ_{fi} can be diagnosed with a finite delay.

5 Discussion

In this paper, we have presented a language based definition of diagnosability. We have outlined the construction of a diagnoser and provided necessary and sufficient conditions for diagnosability.

In [9], we also propose a relaxed definition of diagnosability which requires the diagnosability condition D to hold not for all traces containing a failure event but only for those in which the failure event is followed by certain "indicator" observable events associated with every failure type. This relaxation is motivated by the following physical consideration. Consider, for example, an HVAC system with a controller unit. Assume that when the controller fails off, it does not sense the presence of any load on the system and hence the system is never switched into operation. Under such conditions, it is obvious that one cannot diagnose any system failure. While such a system would be considered non-diagnosable according to the definition proposed here, the modified definition of diagnosability would require the system to be called into operation before deciding on its diagnosability. This "Start Operation" event would be the indicator event associated with the controller. This notion is discussed in detail in [9]. The theory outlined in this paper has been applied to two different HVAC systems. The diagnosers for these systems and their analysis is presented in [10].

Acknowledgements

This research was supported in part by NSF grants ECS-9057967, ECS-9312134 and NCR-9204419 with additional support from DEC and GE.

References

1. Caines, P., Greiner, R., Wang, S.: Classical and logic based dynamic observers for finite automata. IMA Journal of Math. Control Inform. **8** (1991) 45–80
2. Cieslak, R., Desclaux, D., Fawaz, A., Varaiya, P.: Supervisory control of discrete event processes with partial observations. IEEE Trans. AC **33** (1988) 249–260
3. Lin, F.: Diagnosability of discrete event systems and its applications. Preprint, March, 1993.
4. Lin, F., Wonham, W.M.: On observability of discrete event systems. Inform. Sci. **44** (1988) 173–198
5. Özveren, C.M., Willsky, A.S.: Observability of discrete event dynamic systems. IEEE Trans. AC **35** (1990) 797–806
6. Özveren, C.M., Willsky, A.S.: Invertibility of discrete event dynamic systems. Math. Control Signals Systems **5** (1992) 365–390
7. Park, Y., Chong, E.: On the eventual invertibility of discrete event systems and its applications. Proc. of the 32nd CDC (1993).
8. Ramadge, P.J.: Observability of discrete event systems. Proc. of the 25th CDC (1986) 1108–1112
9. Sampath, M., Sengupta, R., Lafortune, S., Sinnamohideen, K., Teneketzis, D: Diagnosability of discrete event systems. Tech. Rep. CGR 94–2, Dept. of EECS, Univ. of Michigan, MI 48109, USA (1994)
10. Sampath, M., Sengupta, R., Lafortune, S., Sinnamohideen, K., Teneketzis, D: Failure diagnosis using discrete event models. Tech. Rep. CGR 94–3, Dept. of EECS, Univ. of Michigan, MI 48109, USA (1994)

On the Supremal *Lm*-closed and the Supremal *Lm*-closed and *L*-controllable Sublanguages of a Given Language

Roberto M. Ziller and José E. R. Cury

LCMI -- EEL - UFSC - Cx. postal 476 -- 88040-900 - Florianópolis - SC - Brazil

e-mail: ziller@lcmi.ufsc.br, cury@lcmi.ufsc.br

1 Introduction

In the last ten years, the Ramadge-Wonham framework for Discrete Event Systems modeling, analysis and controller (supervisor) synthesis has advanced to an important position among the various models suggested in this research field.

In the basic RW-model, both the system to be controlled and the desired behavior are specified through the use of languages. Solving a synthesis problem amounts to find a controller - called *supervisor* - that restricts the physically possible behavior of the system to be controlled - called *plant* - to the desired one. Solutions are stated in terms of the *supremal Lm-closed and L-controllable sublanguage* of the language representing the target behavior. While several authors have contributed to the computation of the supremal *L*-controllable sublanguage (see especially [WR87] and also [Rudi88], [LVW88], [BGK⁺90] and [KGM91]), no algorithms or formulas for the supremal *Lm*-closed and the supremal *Lm*-closed and *L*-controllable sublanguages of a given language could be found in the surveyed literature.

In this paper we present formulas for these sublanguages. Section 2 recalls some results from RW-theory, section 3 presents the formulas; computational complexity is analyzed in section 4, and an example is given in section 5.

2 Preliminaries

A detailed description of the Ramadge-Wonham framework is given in [RW87] and related articles, the reader being referred to these sources for background knowledge. We only recall some facts needed to present our contribution.

System behavior is represented by a 5-tuple $G = \langle \Sigma, Q, \delta, q_0, Q_m \rangle$ called *generator,* where Σ is a set of event labels, also called the *event alphabet,* Q is a set of states, and $\delta: \Sigma \times Q \to Q$ is a (generally partial) transition function defined at each $q \in Q$ for a subset of the $\sigma \in \Sigma$ so that $q' = \delta(\sigma, q)$ represents the state transition $q \xrightarrow{\sigma} q'$, meaning that the occurrence of event σ takes the system from state q to state q'. $q_0 \in Q$ is the *initial state* and $Q_m \subseteq Q$ is a set of *marker states.*

These are used to mark the termination of certain event sequences, representing the completion of a task by the system.

Each generator G has two associated *languages:* $L(G)$, the language *generated* by G, and $L_m(G)$, the language *marked* by G. These are sets of *words* formed with symbols of Σ. $L(G)$ represents the physically possible behavior of the system, while $L_m(G)$ stands for the tasks it is able to complete.

The alphabet Σ is partitioned into *controllable* and *uncontrollable* events according to $\Sigma = \Sigma_c \cup \Sigma_u$ and $\Sigma_c \cap \Sigma_u = \varnothing$. Control action is performed by an external agent called *supervisor,* which observes the events generated by the plant and applies a *control input* $\gamma \subseteq \Sigma$ to the system in response to them. The events in γ are those specified to be enabled by the supervisor.

This control action restricts the system generated and marked languages. The languages representing the physically possible behavior and the tasks the system may complete under supervision are denoted by $L(S/G)$ and $L_c(S/G)$, respectively.

The prefix-closure $\overline{K}$ of a language K is the set of all prefixes (initial segments) of strings in K. K is said to be *prefix-closed* iff $K = \overline{K}$. Given two arbitrary languages $K, L \subseteq \Sigma^*$ and a partition $\Sigma = \Sigma_c \cup \Sigma_u$, K is said to be *L-closed* iff $K = \overline{K} \cap L$; K said to be *L-controllable* iff $\overline{K}\Sigma_u \cap L \subseteq \overline{K}$.

The main synthesis problem in the RW-model can be stated as follows:

Supervisory Control Problem (SCP): Given a plant G, a target language $E \subseteq L_m(G)$ and a minimal acceptable language $A \subseteq E$, construct a proper supervisor S for G such that

$$A \subseteq L_c(S/G) \subseteq E.$$

The language E is interpreted as the desired behavior under supervision, while A stands for the minimal closed-loop behavior that is still acceptable.

It is shown in [RW87] that the class $C(E)$ of all $L(G)$-controllable sublanguages of E and the class $F(E)$ of all $L_m(G)$-closed sublanguages of E are both non empty and closed under arbitrary unions. Consequently, these classes contain the supremal elements $\sup C(E)$ and $\sup F(E)$, called *supremal $L(G)$-controllable sublanguage of E* and *supremal $L_m(G)$-closed sublanguage of E,* respectively. The same applies to the class $CF(E) = C(E) \cap F(E)$, so there also exists $\sup CF(E)$, the *supremal $L(G)$-controllable and $L_m(G)$-closed sublanguage of E.*

The solution to SCP is then given by the following theorem ([RW87], theorem 7.1, part (ii)):

Theorem 1: SCP is solvable iff $\sup CF(E) \supseteq A$.

3 Computing Solutions

In this section we present formulas for both the supremal $L_m(G)$-closed sublanguage and the supremal $L_m(G)$-closed and $L(G)$-controllable sublanguage of a given language. From this point on, we abbreviate $L(G)$ by L and $L_m(G)$ by L_m whenever no confusion is possible. We need the following lemmas:

Lemma 1: Given a language $K \subseteq L$, if $K = \overline{K}$, then $K \cap L_m$ is L_m-closed.

Proof: ($\subseteq$): $K \cap L_m \subseteq \overline{K \cap L_m}$ and $K \cap L_m \subseteq L_m$, so $K \cap L_m \subseteq \overline{K \cap L_m} \cap L_m$.

($\supseteq$): $K = \overline{K} \Rightarrow K \supseteq \overline{K \cap L_m} \Rightarrow K \cap L_m \supseteq \overline{K \cap L_m} \cap L_m$. ♦

Lemma 2: Given a language $E \subseteq L_m$, if E is L_m-closed, so is $\sup C(E)$.

Proof: Let K be an arbitrary controllable (non necessarily L_m-closed) sublanguage of E, so $K \in C(E)$. Then $K \subseteq E \Rightarrow \overline{K} \subseteq \overline{E} \Rightarrow \overline{K} \cap L_m \subseteq \overline{E} \cap L_m = E$. Since $K \subseteq \overline{K}$ and $K \subseteq E \subseteq L_m$, it is also true that $\overline{K} \cap L_m \supseteq K$. By lemma 1, $\overline{K} \cap L_m$ is L_m-closed. As shown below, this language is also controllable:

$$\left[\overline{\overline{K} \cap L_m}\right]\Sigma_u \cap L(G) \subseteq \left[\overline{K} \cap \overline{L_m}\right]\Sigma_u \cap L(G) \subseteq \overline{K}\Sigma_u \cap L(G) \subseteq \overline{K} \subseteq \overline{\overline{K} \cap L_m}.$$

This means that, for every language $K \in C(E)$ that is not L_m-closed, there is an L_m-closed and controllable language $\overline{K} \cap L_m \supseteq K$ that is also contained in $C(E)$, and so $\sup C(E)$ is L_m-closed. ♦

Lemma 3: Given a language $E \subseteq L_m$, $\sup CF(E) = \sup C[\sup F(E)]$.

Proof: By the definitions of $\sup F(E)$ and $\sup CF(E)$ it follows that

$\sup CF(E) \subseteq \sup F(E)$, so $\sup CF(E) = \sup CF[\sup F(E)]$. By lemma 2,

$\sup C[\sup F(E)] = \sup CF[\sup F(E)]$, and the result is immediate. ♦

We emphasize that this result guarantees that computing the supremal L-controllable sublanguage of the target language is sufficient to solve SCP when the target language is known to be L_m-closed.

In the following development let $\sup P(K)$ denote the *supremal prefix-closed sublanguage of K*, defined as $\sup P(K) = \{s : s \in K \wedge \overline{s} \subseteq K\}$, where $\overline{s}$ stands for $\overline{\{s\}}$, the set of all prefixes of the word s. The class of all prefix-closed sublanguages of K is clearly non empty (since the empty language $\varnothing$ is prefix-closed) and closed under arbitrary unions, so the supremal element defined above is guaranteed to exist. For the arbitrary language K let $\overline{K} - K$ denote the set-theoretic difference $\overline{K} \cap K^c$, where K^c is the complement of K with respect to Σ^*.

The following proposition presents our main result:

Proposition 1: Given a language $E \subseteq L_m$, let $M = \sup P\left[\overline{E} - \left(\overline{E} - E\right) \cap L_m\right]$. Then $\sup F(E) = M \cap L_m$ and $\sup CF(E) = \sup C(M \cap L_m)$.

Proof: First part: $\sup F(E) = M \cap L_m$.

(i) $\sup F(E) \supseteq M \cap L_m$: demonstrating this relation amounts to show that $M \cap L_m$ is L_m-closed and that $M \cap L_m \subseteq E$. $M \cap L_m$ is L-closed by lemma 1 and by the fact that $M = \overline{M}$; $M \cap L_m \subseteq E$ because

$$s \in M \cap L_m \Rightarrow s \in \sup P\left[\overline{E} - \left(\overline{E} - E\right) \cap L_m\right] \cap L_m \Rightarrow s \in \left[\overline{E} - \left(\overline{E} - E\right) \cap L_m\right] \cap L_m \Rightarrow$$

$$\Rightarrow s \in \overline{E} \cap L_m - \left(\overline{E} - E\right) \cap L_m \Rightarrow s \in E \cap L_m \Rightarrow s \in E.$$

(ii) $\sup F(E) \subseteq M \cap L_m$: demonstrating this relation amounts to show that

$\forall s: s \in E \wedge \overline{s} \cap L_m = \overline{s} \cap E \Rightarrow s \in M \cap L_m$

It is immediate that $s \in E \Rightarrow s \in L_m$, since $E \subseteq L_m$. To show $s \in M$ we start from the hypothesis $s \in E \wedge \overline{s} \cap L_m = \overline{s} \cap E$. Then

$$w \in \overline{s} \cap L_m \Rightarrow w \in \overline{s} \cap E \Rightarrow w \notin \overline{E} - \overline{s} \cap E \Rightarrow (\overline{s} \cap L_m) \cap \left(\overline{E} - \overline{s} \cap E\right) = \varnothing \Rightarrow$$

$$\Rightarrow \overline{s} \cap \left(\overline{E} - \overline{s} \cap E\right) \cap L_m = \varnothing.$$

Now $s \in E \Rightarrow \overline{s} \subseteq \overline{E}$, so $\overline{s} \subseteq \overline{E} - \left(\overline{E} - \overline{s} \cap E\right) \cap L_m \subseteq \overline{E} - \left(\overline{E} - E\right) \cap L_m$ and $s \in M$, and thus $\sup F(E) = M \cap L_m$. The second part of the proposition, namely $\sup CF(E) = \sup C(M \cap L_m)$, is immediate by lemma 3, so the proof is complete. ♦

4 Computational Complexity

This section presents an analysis of the computational complexity of the algorithms needed to determine $\sup F(E)$ and $\sup CF(E)$. It is based on the following results from [Rudi88]:

- given two generators G_1 and G_2 of common alphabet Σ, let $|\Sigma| = s$, $|\Sigma_u| = s_u$, let n_i and e_i $(i = 1,2)$ denote the number of states and transitions of generator G_i, respectively. Constructing a generator G such that $L_m(G)$ is the supremal $L(G_1)$-controllable sublanguage of $L_m(G_2)$ is a task of complexity $O(sn_1n_2 + s_ue_1e_2)$;
- constructing a generator G such that $L_m(G) = L_m(G_1) \cap L_m(G_2)$ is a task of complexity $O(n_1n_2 + e_1e_2)$;

• given G with n states, m marker states, e transitions and event set of cardinality s, constructing a generator for $[L_m(G)]^c$ is a task of complexity $O[s(n+e)+nm]$.

The above expressions can be simplified considering (i) that for any deterministic finite automaton, $e \leq sn$ and (ii) that s can be viewed as a parameter, rather than a variable which necessarily increases as does the size of the state set. Also, $m = \alpha n$ for some constant α. This gives the expressions $O(n_1 n_2)$, $O(n_1 n_2)$ and $O(n^2)$ for the complexities of the three algorithms above, respectively.

It is easy to see that, given a generator $D = \langle \Sigma, R, \rho, p_0, R_m \rangle$ such that $L_m(D) = N$, a generator for $\sup P(N)$ is

$D' = \mathrm{Ac}(\langle \Sigma, R_m, \rho | \Sigma \times R_m, r_0, R_m \rangle)$ if $r_0 \in R_m$

and $\Sigma_\varnothing = \langle \Sigma, \varnothing, \varnothing, \varnothing, \varnothing \rangle$ otherwise,

where Ac(.) denotes the accessible component of the operand and $L(\Sigma_\varnothing) = \varnothing$.

An algorithm that produces a generator for $\sup C(K)$ is given in [WR87].

It is now easy to verify that the highest degree of complexity arising in the computation of $\sup F(E)$ and $\sup CF(E)$ is $O[\max(n_H^2, n_G n_H)]$, where n_G and n_H are the cardinalities of the state sets of the generators G and H representing the plant and the target behavior, respectively. The computations are hence quadratic in time.

5 Example

Figure 1 shows an imaginary generator G of alphabet $\{\alpha, \beta, \lambda, \mu\}$, where we assume that $\Sigma_c = \{\alpha, \lambda\}$ and $\Sigma_u = \{\beta, \mu\}$. The target language corresponding to the desired closed-loop behavior is the language E marked by generator H shown in figure 2.

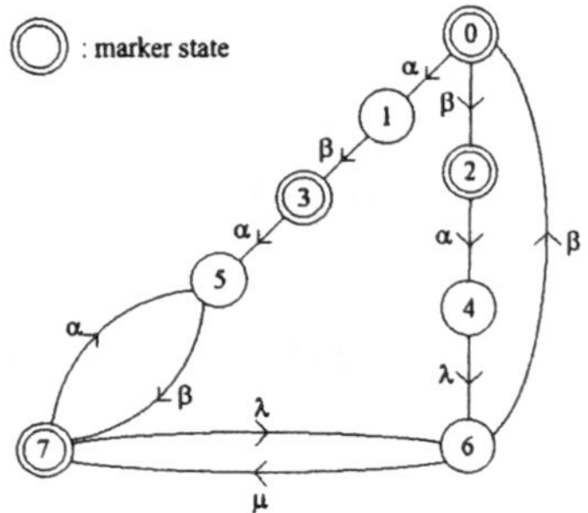

Fig. 1 - Generator *G*: the plant

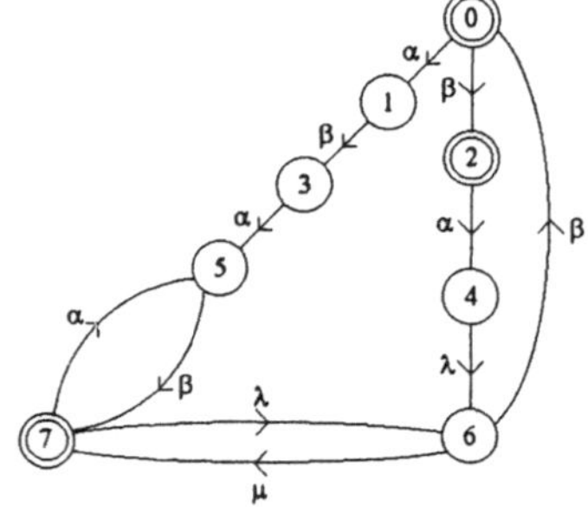

Fig. 2 - Generator *H*: desired behavior

A generator for $\sup F(E)$ is shown in figure 3, where the "bad state" (state 3) that caused E not to be $L_m(G)$-closed has been eliminated. The solution to SCP is given by the supremal $L_m(G)$-closed and $L(G)$-controllable sublanguage of E. A generator for this language is shown in figure 4.

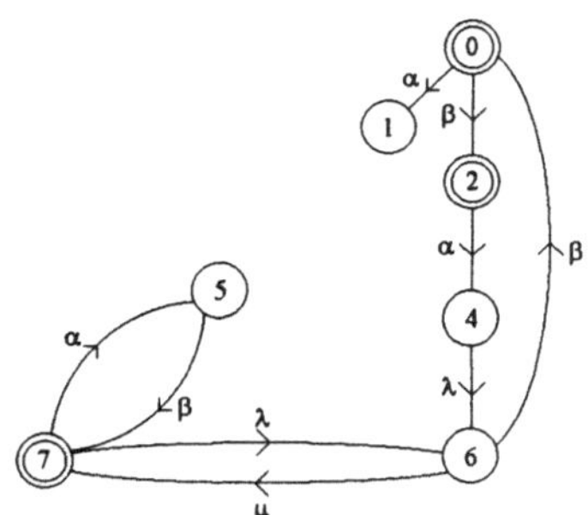

Fig. 3 - Generator for sup$F(E)$

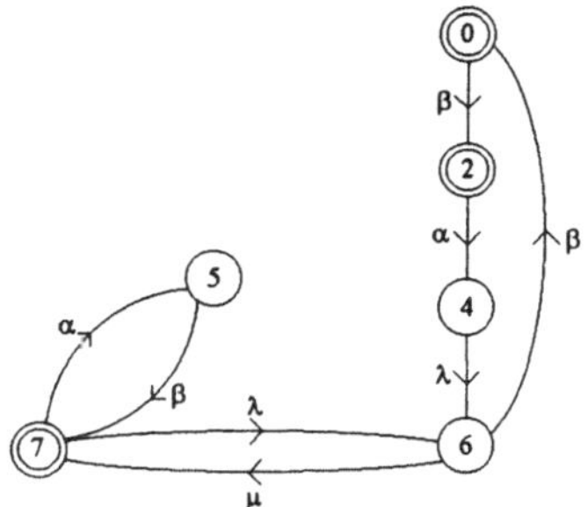

Fig. 4 - Generator for sup$CF(E)$

6 Conclusion

Computing a solution for the *supervisory control problem* (SCP) requires an algorithm for the supremal $L_m(G)$-closed and $L(G)$-controllable sublanguage of the target language E. Only when E is already $L_m(G)$-closed the solution reduces to the computation of $\sup C(E)$. The proposed formula for $\sup CF(E)$ (based on the computation of $\sup F(E)$) allows solving SCP in the general case.

References

[BGK+90] Brandt, R. D.; Garg, V.; Kumar, R.; Lin, F.; Marcus, S. I.; Wonham, W. M.: "Formulas for Calculating Supremal Controllable and Normal Sublanguages". Systems & Control Letters, 15(2):111-117 (1990)

[KGM91] Kumar, R.; Garg, V.; Marcus, S. I.: "On Controllability and Normality of Discrete Event Dynamical Systems". *Systems & Control Letters,* 17:157-168 (1991)

[LVW88] Lin, F.; Vaz, A. F.; Wonham, W. M.: "Supervisor Specification and Synthesis for Discrete Event Systems". *Int. J. Control,* 48(1):321-332 (1988)

[Rudi88] Rudie, Karen G.: "Software for the Control of Discrete Event Systems: A Complexity Study". *M. A. Sc. Thesis,* Department of Electrical Engineering, University of Toronto, Toronto, Canada (1988)

[RW87] Ramadge, P. J.; Wonham, W. M.: "Supervisory Control of a Class of Discrete Event Processes". *SIAM J. Control and Optimization,* 25(1):206-230 (1987)

[WR87] Wonham, W. M.; Ramadge, P. J.: "On the Supremal Controllable Sublanguage of a Given Language". *SIAM J. Control and Optimization,* 25(3):637-659 (1987)

Continuous-Time Supervisory Synthesis for Distributed-Clock Discrete-Event Processes*

S.D. O'Young[1,2]

[1] Department of Electrical and Computer Engineering, University of Toronto, Toronto, Ontario, Canada, M5S 1A4, email: oyoung@control.utoronto.ca

[2] Condata Technologies Ltd, Toronto, Canada

1 Modeling

1.1 Control Mechanism

Figure 1 illustrates a configuration of a real-time supervisor which delays an actuation from the plant controller by disabling its request or expedites an actuation by sending a warning to the low-level plant controller. A unique, nonblocking and least-restrictive supervisor exists and a polynomial time algorithm has been reported in [1, 2] for its computation. The solution is guaranteed to be real-time realizable: all supervisory tasks can be executed within the specified time.

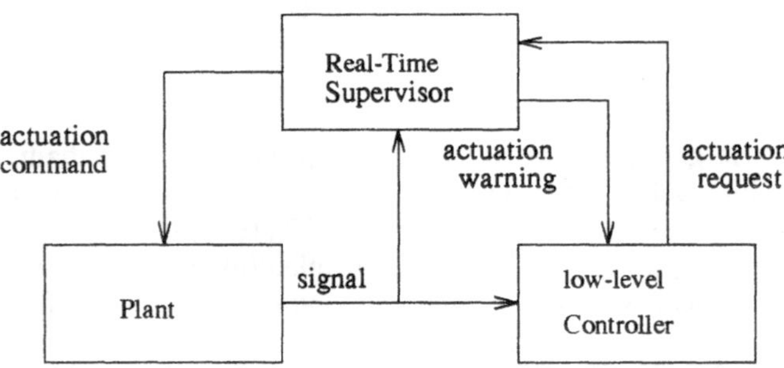

Fig. 1. Real-time Supervisory Control System

Real-time supervisory control using disablement to guarantee safety has been been studied by Golaskewski and Ramadge [3], Ostroff [4, 5] and Ostroff and Wonham [6] using discrete-time models and by Wong-Toi and Hoffmann [7] using an interval continuous (dense) time system. Forcing, as an additional supervisory technology for guaranteeing liveness, has been proposed by Golaskewski and Ramadge [8] in the logical

* The research has been supported by the Information Technology Research Centre of Ontario, Canada

framework and has been extended to the real-time supervisory control by Brave and Heymann [9], O'Young [1] and Brandin et al. [10, 11] for discrete-time models. Our main contribution is in formalizing a continuous-time supervisory control where disablement (delay) and forcing (expediting) are utilized cooperatively into a model with distributed clocks for guaranteed punctuality and satisfaction of hard real-time constraints.

Our continuous-time approach has been motivated by Wong-Toi and Hoffmann [7]. Their development of a synthesis theory is formal-language based whereas ours is state-transition-structure based and we give explicit state and computational data structures [2] for modeling the timed discrete-event processes. For example, an ordered linked list is used for tracking the state transitions of the distributed clocks. We have also introduced an expediting (forcing) mechanism for supervisory control and a timing semantics for expressing hard real-time tasks to continuous-time supervisory synthesis.

1.2 Finite State Representation

A continuous-time discrete-event process (CTDEP) consists of a nontime a Ramadge-Wonham (RW) [12] discrete-event process (Fig. 2), modeling the logical behaviour, and a set of real-valued local clocks tracking the progress of local time. The enablement and resetting of the clocks depend on the state assignment and transition of the logical process, and the eligibility of a logical transition in the logical process, in turn, depends on the clock values and a given set of time bounds. Formally, a CTDEP consists of an uninterpreted logical (nontime) state transition structure modeled by a triplet (Δ, X_m, x_0) with, $\Delta \subseteq \hat{\Delta}$, $X_m \subseteq X$ and $x_0 \in X$ representing the transition set, the marker state set and the initial state of the CTDEP, respectively. A set of time bounds T is added to specify the timing constraints on the transitions in Δ. The reader is referred to [2] for an example interpretation of the effects of T on the nontime structure. It should be noted our synthesis theory is independent of this interpretation.

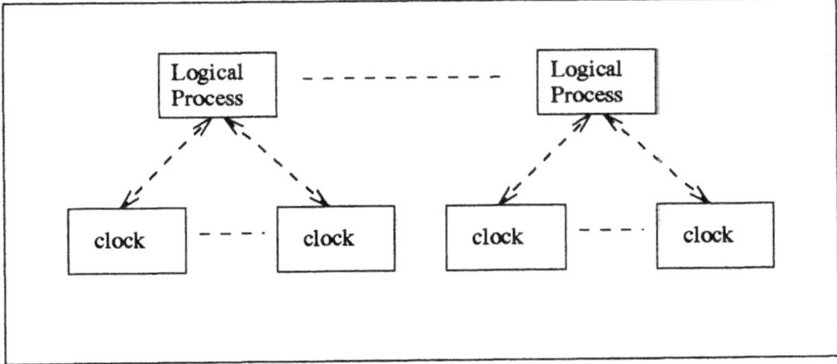

Fig. 2. Distributed-Clock Discrete-Event Processes

1.3 Partition of $\mathbb{R}_+$

Our partition of the state-space of the real-valued clocks within a CTDEP has been motivated by the Alur and Dill's [13] work on timed automaton. Suppose that the time bounds are given in finite precision, and without loss of generality assumed to be in integers. Then, the eligibility of a logical transition in the logical process changes only if some of real-valued clocks cross some integer boundaries. Consider a CTDEP with 2 clocks: c_1 and c_2, with $c_1 = 1.5$ and $c_2 = 3.3$ with time measured in seconds. The integer values of c_1 (= 1) and c_2 (= 3) will not change until 0.5 second has lapsed, and the integer value of c_1 will change to 2 and the integer value of c_2 will remain unchanged at 3. The passage of time, e.g. < 0.5 second, which results in no change of integer values of the clocks is represented by an ϵ transition and passage of time which results in a change of integer values of some or all of the clocks is represented by a τ transition.

The τ transition is a generalization of a *tick* transition in the discrete-time case [4, 5, 6, 1, 10]: *tick* advances all clocks while τ advances a selected subset of clocks. The selection mechanism, e.g. advancing c_1 and not c_2 at the first τ transition in the example, can be coded as a linked list ordering the fractional values of the clocks [2]. This linked list is a finite-state structure for a finite number of local clocks. Due to the occurrence of ϵ transitions which results in no change in the eligibility of the logical transitions, it is possible to partition the $\mathbb{R}^n_+$ space induced from the n real-valued clocks into a finite-state structure for the purpose of finding the states (w.r.t. Nerode equivalence) of a CTDEP. The reader is referred to [2] for a detailed construction of the finite-state structure.

2 Control Synthesis

2.1 A Controlled CTDEP

Let G be a CTDEP. Let the alphabet $\Sigma(G)$ of G be refined into a *delayable* subset $\Sigma_d \subseteq \Sigma(G)$ and an *expeditable* subset $\Sigma_e \subseteq \Sigma(G)$. We make the standing assumption that $\epsilon \notin \Sigma_d \cup \Sigma_e$, i.e. the passage of time cannot be influenced by any control commands.

Definition 1. A transition is *expeditable (delayable)* if it is labeled by an event in expeditable (delayable) set Σ_e (Σ_d) and its occurrence can be expedited (delayed) before (after) the occurrence of the next τ.

Definition 2. A state in G is *singular* if $\delta(\epsilon, x) \neg!$ in G.

The symbol $\delta(\sigma, x)$! in G ($\delta(\sigma, x) \neg!$ in G) denotes a transition labeled by σ exiting from state x is (not) defined in the transition structure of G. A singular state is a zero-time transient state. All singular state assignments have zero support on the time axis and the process spends all measureable (nonzero) time in nonsingular states.

Definition 3. A control pattern $D \in 2^{\Sigma(G)}$ is said to be *executable by* a CTDEP G at state x_G if x_G is nonsingular and $\forall\, \sigma \in D$,

1. ($\sigma \neq \tau$ and $\delta(\sigma, x_G)$! in G) $\Longrightarrow \sigma \in \Sigma_d$, and
2. ($\sigma = \tau$ and $\delta(\tau, x_G)$! in G) $\Longrightarrow$ ($\exists\, \tilde{\sigma} \in \Sigma_e, \tilde{\sigma} \notin D$ and $\delta(\tilde{\sigma}, x_G)$! in G).

Note that an empty control pattern is by definition executable. For a control pattern to be executable, the plant must be in a state where there is time for the execution of the supervisory command.

All transitions labeled by an event σ in D must be prevented from occurring while the supervised CTDEP is in state x_G. If the controller (Fig. 1) requests the actuation of an event σ in the delayable set Σ_d (Condition (1)), the supervisor can defer the actuation of σ until the next τ transition. If $\tau \in D$ (Condition (2)), then the pending τ transition exiting x_G must be preempted by a non-τ transition. This is possible, if there exists a transition $\tilde{\sigma}$ enabled at x_G which can be expedited, i.e. $\tilde{\sigma} \in \Sigma_e$ and $\tilde{\sigma}$ is not prohibited by D.

2.2 Controllability

The role of an optimal real-time supervisor is to exert timely supervision on G in order to retain the largest possible discrete-event subprocess (DESP) $\tilde{G}$ of G for the supervisory technologies which have been built into the plant G. Let $\tilde{G}$ be a reachable DESP of G. For every $x \in X(G)$, the state set of G, construct a control pattern D_x such that

$$D_x := \begin{cases} \{\sigma : \sigma \in \Sigma(G),\ \delta(\sigma, x)\ !\ \text{in } G \text{ and } \delta(\sigma, x) \neg!\ \text{in } \tilde{G}\} & \text{if } x \in X(\tilde{G}); \\ \{\} & \text{otherwise.} \end{cases} \tag{1}$$

Hence, the control pattern D_x contains all the eligible transitions for G which are prohibited by the subprocess $\tilde{G}$. The restricted behaviour $\tilde{G}$ can be realized if the plant accepts and executes a control pattern upon entrance into each new state of $\tilde{G}$.

Definition 4. A DESP $\tilde{G} \subseteq G$ is said to be *controllable* if it is reachable and, for every state $x \in X(\tilde{G})$, the control pattern D_x defined in (1) is executable by G at x.

The symbol $\tilde{G} \subseteq G$ denotes that the transition structure of $\tilde{G}$ is a subset of G. A (unique) largest controllable DESP of P exists if the controllability property is preserved by a union of two controllable DESPs.

Theorem 5. *Suppose G_1 and G_2 are DESPs of G which are controllable. Then $G_1 \cup G_2$ is also controllable.*

Corollary. *There exists a (unique) largest controllable DESP of P with respect to the partial ordering via $\subseteq$.*

2.3 Escapability

A CTDEP may be required to exit from a state before the next τ transition because of some design requirements. We must ensure that the evasive action could be carried out in a finite number of expedited transitions. In particular, we must avoid forcing a CTDEP into a busy loop which can never be broken.

Definition 6. A state x in $X(\tilde{P})$ with $\tilde{P} \subseteq P$ is *exigent* if $\delta(\tau, x)\ !$ in P and $\delta(\tau, x) \neg!$ in $\tilde{P}$.

If the plant is in an exigent state, the pending τ transition must be preempted.

Definition 7. A run $r = x_0, x_1, \cdots, x_n$ in $\tilde{P}$ is *expeditable* if for every j, $1 \leq j \leq n$, there exists $\sigma_j \in \Sigma_e$ and the transition (x_{j-1}, σ_j, x_j) is in $\tilde{P}$.

Definition 8. A state x in $\tilde{P}$ is *escapable in* $\tilde{P}$ if there exist a state $\tilde{x}$ in $X(\tilde{P})$ and an expeditable run r in $\tilde{P}$ such that $\tilde{x}$ is reachable from x via r and $\tilde{x}$ is not exigent.

An exigent state is escapable if the low-level plant controller can be warned to execute a finite number of expeditable transitions (because the state set of a CTDEP is finite) and reach a state where a supervisor-initiated preemption of a pending τ transition is no longer necessary. Note that a non-exigent state is trivially escapable.

In order to guarantee that time be always able to progress, we want to ensure that from every reachable state in $\tilde{P}$, the process can reach a state where a τ transition is defined.

Definition 9. A state x in $\tilde{P}$ is *τ-coreachable* if there exists a state $\tilde{x}$ in $\tilde{P}$ such that $\delta(\tau, \tilde{x})$! in $\tilde{P}$ and $\tilde{x}$ is reachable from x.

A CTDEP $\tilde{P}$ is *τ-coreachable* if every state in $\tilde{P}$ is τ-coreachable.

Proposition 10. *Suppose P_1 and P_2 are τ-coreachable DESPs of P. Then $P_1 \cup P_2$ is also τ-coreachable.*

Proposition 11. *All exigent states in $\tilde{P} \subseteq P$ are escapable if $\tilde{P}$ is τ-coreachable and controllable.*

2.4 State Avoidance

Design constraints can be imposed upon the system P by removing transitions from $\Delta(P)$, the set of transitions of CTDEP P, the transition set of P. Here are some suggestions for editing $\Delta(P)$. If a state $x \in X(P)$ is deemed "bad", remove all transitions entering into x from $\Delta(P)$ such that x will no longer be reachable. If $x \in X(P)$ is deemed "transitory", remove the τ transition, if defined, exiting x such that the system cannot stay in x for any measurable time. We define a DESP $\tilde{P}$ of P to be *state avoiding* if its transition structure does not contain transitions which have been edited from the system P.

Proposition 12. *Suppose P_1 and P_2 are reachable DESPs of P and are state avoiding. Then $P_1 \cup P_2$ is also state avoiding.*

2.5 Continuous-time Optimal Supervision

To ensure that the supervised system is also doing something useful, a nonempty CTDEP must also be nonblocking; i.e. it can always reach a marker state. The nonblocking property can be guaranteed by the trimness of a CTDEP.

Definition 13. A controllable DESP $\tilde{P}$ of P is *State Avoiding Nonblocking and Escapable (SANE)* if it is state avoiding, trim and τ-coreachable.

Recall that the controllability, state-avoiding and τ-coreachability properties are preserved by the union ($\cup$) operation as proven in Theorem 5 and Propositions 12 and 10, respectively. In addition, the trimness property is also preserved by union operation. By Proposition 11, controllability and τ-coreachability imply escapability. We have the main result that the controllable SANE DESPs of the system P with the joint operation being $\cup$ form an upper semi-lattice, and hence a unique supremal element exists.

Theorem 14. *There exists a unique largest DESP of P w.r.t. the ordering via $\subseteq$ which is controllable and SANE.*

3 Bus-Pedestrian Example Revisited

The discrete-time bus-pedestrian example [1] has remodeled as distributed-clock interval-time processes in [2] to demonstrate the continuous-time SANE synthesis method: a pedestrian is crossing a road and his life is threatened by an approaching bus. The objective is to ensure that the pedestrian survives by the timely execution of the supervisory commands for forcing him to jump off the road and/or stopping the bus.

There exists a common set of timing constraints and control settings for both the discrete-time [1, 10] and continuous-time [2] models, where the safety of the pedestrian is only ensured via continuous-time supervision. Hence, a continuous-time system has been demonstrated to be less conservative because it embodies a more precise model of time.

4 Conclusion

The concept of a singular state has made crisp the notion of timed controllability: a system simply cannot be controlled at a singular state because there is no time for the supervisor to exercise control. We have avoided the more complication conditions needed for the discrete-time system such as relating controllability to the present values of the tick counters [1] or imposing a sufficient but not necessary condition on requiring all controllable events to have infinite upper time bounds [10, 11]. Since a continuous-time system admits asychronous and real-valued clocks, and a discrete-time system is restricted to synchronous and integer-valued clocks, a more refined control mechanism can be modeled in continuous time and results in a less restrictive closed-loop solution than the discrete-time counterpart.

References

1. S. D. O'Young. On the synthesis of supervisors for timed discrete event processes. Systems Control Group Report #9107, Dept. Elect. Comp. Engr., Univ. Toronto, 1991. Submitted for Publication.

2. S. D. O'Young. Continuous-time supervisory synthesis for discrete event processes. Systems Control Group Report #9315, Dept. Elect. Comp. Engr., Univ. Toronto, 1993. Submitted for Publication.
3. C. H. Golaszewski and P. J. Ramadge. On the control of real-time discrete event systems. In *Proc. 23rd Conf. on Information Systems and Signals*, pages 98–102, Princeton, NJ, March 1989.
4. J. S. Ostroff. Synthesis of controllers for real-time discrete event systems. In *Proc. 28nd Conf. Decision Contr.*, pages 138–144, Tampa, Fl., December 1989.
5. J. S. Ostroff. *Temporal logic for real-time systems.* John Wiley and Sons, 1989.
6. J. S. Ostroff and W. M. Wonham. A framework for real-time discrete event control. *IEEE Trans. on Automatic Control.*, 35(4), April 1990.
7. H. Wong-Toi and G. Hoffmann. The control of dense real-time discrete event systems. In *Proc. 30th Conf. Decision Contr.*, pages 1527–1528, December 1991.
8. C. H. Golaszewski and P. J. Ramadge. Control of discrete event processes with forced events. In *Proc. 26nd Conf. Decision Contr.*, pages 247–251, December 1987.
9. Y. Brave and M. Heymann. Formulation and control of real time discrete event processes. In *Proc. 27nd Conf. Decision Contr.*, pages 1131–1132, Austin, TX, December 1988.
10. B. A. Brandin and W. M. Wonham. The supervisory control of timed discrete event systems. Systems Control Group Report #9210, Dept. Elect. Comp. Engr., Univ. Toronto, 1992. Submitted for Publication.
11. B. A. Brandin, W. M. Wonham, and B. Benhabib. Manufacturing cell supervisory control – a timed discrete event systems approach. In *Proc. IEEE International Conf. on Robotics and Automaton*, Nice, France, May 1992.
12. P. J. Ramadge and W. M. Wonham. The control of discrete event systems. *Proc. IEEE, Special Issue on Discrete Event Dynamic Systems*, 77(1):81–98, 1989.
13. R. Alur and D. Dill. Automata for modeling real-time systems. In *Lecture Notes in Computer Science*, volume 443. Springer-Verlag, 1990.

Conditions for Optimization of Discrete Event Systems Using Temporal Logic Models

Dan Ionescu

University of Ottawa,Department of Electrical Engineering
Ottawa, Ontario, Canada K1N 6N5

1 Introduction

Many issues in optimal control have been investigated for DESs. Ramadge and Wonham [8] have considered the optimality of the supremal controllable sublanguage of a given language. The graph-theoretic formulation of the optimal control problem for a class of DESs has been given by Sengupta and Lafortune [9], and supervisory optimal control by Lin [3]. Passino and Antsaklis [7] have associated a cost with every state transition of a logical DES and examined optimality with respect to an event cost function. However, among the growing literature that discusses the optimal control for DESs, little work has been done in a temporal logic approach.

In this paper conditions for optimization of DES in a temporal logic framework are shown. For the optimal control problem of DESs, the controller is designed to move the system from the given initial state to the given final state and minimize a given cost function index. As a continuation of the research done in [4, 5], a temporal logic framework is introduced to the optimization of controller design for DESs. Our motivation is to develop an optimal controller synthesis procedure for DESs in a temporal logic framework. We will introduce the temporal logic framework in the next section. In §3, the temporal logic model is given in a measurement space. In §4, the optimal control problem is formulated and the optimal controller synthesis procedure is proposed through heuristic search methods in §5. In §6, an example of applications is shown. Finally, conclusions and topics for further research are suggested.

2 A Temporal Logic Framework

Temporal logic is a logic oriented toward reasoning about time sequences. It has been used for software verification citeClaES:86, ManPn:83a, and has recently been applied to the control problems of discrete event systems [10, 5].

The temporal logic framework we are using here is same as that given by Manna and Pnueli in [6] which has been used in [10] and generalized in [4]. It is a classical logic extended with five temporal operators: $\bigcirc$, $\Diamond$, $\Box$, $\mathcal{U}$, and $\mathcal{P}$; the first three are unary, and the last two binary. Its syntax and semantics are as in Enderton [1] and Lin and Ionescu [4].

A temporal logic model (TLM) is defined by $M = (\mathbf{S}, \mathbf{E}, \mathbf{F}^*, f, s_0, l)$ [4, 5] where $\mathbf{S}$ is the set of states; $\mathbf{E}$ is the set of events, each event being a transition from one state to another; f is a mapping from $\mathbf{E} \times \mathbf{S}$ into $\mathbf{S}$ such that $\forall s \in \mathbf{S}$, $\forall e \in E_s$ which is the set of events that are firing at s, $f(e, s)$ is defined; s_0 is the initial state; and l is a labelling function from $\mathbf{E} \times \mathbf{S}$ to $\mathbf{F}^*$ which is the set of all subsets of the set $\mathbf{F}$ of logical formulas.

With the labeling function l, a TLM is described by a set of formulas. A temporal logic formula $l(e, s)$ has the following form:

$$\Box[s_i(x) \wedge e_{i+1}(x) \Rightarrow \bigcirc s_{i+1}(x)] \ , \tag{1}$$

where x is a local variable, $s(x)$ stands for the predicate representing the state at the present time, $e(x)$ for that representing the event about to occur, and $\bigcirc s'(x)$ for that representing the state at the next time.

An extended state mapping f, defined by $f(e_{k+1}, s_k) = s_{k+1}$ for $k = 0, 1, 2, \ldots, n-1$ is shown to define a semigroup property for DES modeled as TLMs as in the next definition.

The mapping f is extended at state s for a sequence of events $\sigma \in E_s$ and $\forall e \in \mathbf{E}$ such that $\sigma e \in E_s$ by:

$$f(\sigma e, s) \stackrel{\text{def}}{=} f(e, f(\sigma, s)) \ , \tag{2}$$

$$f(\epsilon, s) \stackrel{\text{def}}{=} s \ , \tag{3}$$

where ϵ is a null event.

Equation (3) shows that the extension is consistent.

3 Temporal Logic Models in a Measurement Space

Two sequences result from the execution of a temporal logic model(equation 1): the sequence of states $(s_0, s_1, s_2, \ldots, s_n)$ and the sequence of events $(e_1, e_2, \ldots, e_n)$, which were fired, where n may be infinity. These two sequences are related by the relationship $f(e_{k+1}, s_k) = s_{k+1}$ for $k = 0, 1, 2, \ldots, n$. Both of these sequences provide a record of the execution of the model.

Definition 1. : A finite sequence $\sigma : s_0 \stackrel{e1}{\rightarrow} s_1 \stackrel{e2}{\rightarrow} s_2 \ldots \stackrel{en}{\rightarrow} s_n$ is a finite path or trajectory of a temporal logic model if $e_i \in E_{s_{i-1}}$ and $s_i = f(e_i, s_{i-1})$ for $i = 0, 1, 2, \ldots, n$.

For $\sigma : s_0 s_1 s_2 \ldots s_n$, we define the length of σ by $|\sigma| = n + 1$. For $\sigma_1 : s_0 s_1 \ldots s_k$ and $\sigma_2 : t_1 t_2 \ldots t_n$, we denote the concatenated sequence by $\sigma_1 * \sigma_2$:

$$\sigma_1 * \sigma_2 : s_0 s_1 \ldots s_k t_1 t_2 \ldots t_n \ .$$

Here $|\sigma_1 * \sigma_2| = |\sigma_1| + |\sigma_2|$. If $\sigma = \sigma_1 * \sigma_2$, we write $\sigma_1 < \sigma$ to represent that σ_1 is a prefix of σ; and we write $\sigma_2 = \sigma|^{\sigma_1}$ to represent that σ_2 is obtained by deleting σ_1 from σ.

In order to measure the cost of an event, we need to introduce a measuring function for defining a measurement space associated with the TLM.

Suppose X is an arbitrary non-empty set. Let $\preceq$ be a precedence relation on X and let $m : X \times X \rightarrow \mathbb{R}^+$. For all $x, y, z \in X$, m has the following properties:

(i) $m(x, y) \geq 0$ and $m(x, y) = 0$ iff $x = y$;

(ii) $m(x, z) \leq m(x, y) + m(y, z)$ for $x \preceq y \preceq z$.

The measuring function m is called a measurement on X and the space $\{X; m\}$ is called a measurement space. Obviously, any metric space [2] is a measurement space. So, the definition of measurement space is a generation of that of metric space.

Now we extend the TLM by including the event cost function $\theta : \mathbf{E} \rightarrow \mathbb{R}^+$ with $\theta(e_{ij}) = \theta(s_i, s_j)$ for $e_{ij} \in E_{s_i}$ such that $s_j = f(e_{ij}, s_i)$. The event cost function is defined iff the event $e_{ij} \in E_{s_i}$. Obviously, $\theta(\epsilon) = 0$ where ϵ is a null event. If $e_{1k} = e_1 e_2 \ldots e_k$, then $\theta(e_{1k}) = \sum_{i=1}^{k} \theta(e_i)$. Thus we have obtained an enhanced temporal logic model $M = (\mathbf{S}, \mathbf{E}, \mathbf{F}^*, f, s_0, l, \theta)$.

The execution of an enhanced TLM results in three sequences: a sequence of states, a sequence of events, and a sequence of the costs of events. It can be represented explicitly by a Θ-graph:

- $\mathbf{S} = \{s_0 s_1 s_3 \ldots\}$ is the non-empty set of nodes;
- $\mathbf{E} = \{e_{ij}\} = \{(s_i, s_j);\ s_i, s_j \in \mathbf{S}\}$ is the non-empty set of directed arcs with e_{ij} pointing from s_i to s_j;
- $\Theta = \{\theta_{ij}\} = \{\theta_{ij} = \theta(e_{ij});\ e_{ij} \in \mathbf{E}\}$ is the non-empty set of costs associated with each arc and $\theta_{ij} \geq 0$ for all $\theta_{ij} \in \Theta$.

This Θ-graph is an extension of the graphical representation of a TLM given in [5].

4 Optimal Control Problem for DESs

This section will establish the first step towards developing the foundations for an optimal control theory for DESs in a temporal logic approach.

In the previous section, temporal logic model has been enhanced by defining a cost function index in terms of the costs for events to occur. In addition, for a given final state s_g, a DES plant is modelled by

$$M = (\mathbf{S}, \mathbf{E}, \mathbf{F}^*, f, s_0, l, \theta, s_g) ,$$

where $\mathbf{S}$ is the set of states; $\mathbf{E}$ is the set of events; $f : \mathbf{E} \times \mathbf{S} \rightarrow \mathbf{S}$ is the next-state function; s_0 is the initial state; $l : \mathbf{E} \times \mathbf{S} \rightarrow \mathbf{F}^*$ is a labelling function; $\mathbf{F}^*$ is the set of all subsets of the set of formulas $\mathbf{F}$; $\theta : \mathbf{E} \rightarrow \mathbb{R}^+$ is the event cost function; s_g is the final state which is reachable [13] from s_0; and $\mathbb{R}^+$ is the set of positive real numbers.

The controller is modelled by

$$M_c = (\mathbf{Q}, \mathbf{E}_c, \mathbf{F}_c^*, f_c, q_0, l_c) ,$$

where $\mathbf{Q}$ is the set of controller states; $\mathbf{E}_c \subset \mathbf{E}$; $f_c : \mathbf{E}_c \times \mathbf{Q} \rightarrow \mathbf{Q}$ is the controller next-state function; q_0 is the initial controller state; $l_c : \mathbf{E}_c \times Q \rightarrow \mathbf{F}_c^*$ is a labelling function; and $\mathbf{F}_c^*$ is the set of all subsets of the set of formulas $\mathbf{F}_c$.

The closed-loop DES is formed by synchronizing the events in the plant M and events in the controller M_c.

Thus, the output of the plant is the input of the controller; and the output of the controller is the input of the plant. The optimization forces the controller to generate a sequence of events to move the plant from s_0 to s_g along the optimal trajectory and minimize the cost function index.

Let E_σ denote the set of all events needed to the trajectory $\sigma \in S^*$ that can be generated by M, i.e.

$$E_\sigma = \{e \mid e \in E_s, s \text{ is a state in } \sigma\} \ .$$

Then the cost function index is defined by

$$J(\sigma) = \sum_{e \in E_\sigma} \theta(e) \ ,$$

for all $\sigma \in S^*$. Obviously, $J(\sigma) = 0$ if $\sigma = s_0$.

For finite state DESs, there are only a finite number of trajectories of finite length. Hence

$$J(\sigma^*) = \min_{\sigma \in M(s_0, s_g)} \sum_{e \in E_\sigma} \theta(e) \ . \tag{4}$$

The optimal controller synthesis problem is to find a controller that will generate a sequence of events that drives the system M along the trajectory $\sigma^* \in M(s_0, s_g)$ such that (5) holds.

5 Optimization of Controller Synthesis

To solve the above controller problem the A^* algorithm will be used.

The Θ-graph of the enhanced TLM defined above can be used to the heuristic graph-search procedure. An implicit representation of the Θ-graph G is given by a set of initial nodes and a successor operator $Z : \mathbf{S} \to 2^{\mathbf{S} \times \Theta}$ defined by $Z(s_i) \stackrel{\text{def}}{=} \{(s_{i+1}, \theta_{i,i+1}) \mid s_{i+1} = f(e_{i,i+1}, s_i),\ \theta_{i,i+1} = \theta(e_{i,i+1}),\ e_{i,i+1} \in E_{s_i}\}$. When Z is applied to a node s, the graph is expanded. A subgraph G_s from any $s \in \mathbf{S}$ can be obtained by applying Z to the initial node s. Each node in G_s is reachable [5] from s. There exists a path from s_i to s_j iff s_j is reachable from s_i. Each path has a cost which is obtained by adding the costs of each arc $\theta_{i,i+1} \in \Theta$. An optimal path from s_i to s_j is a path having the smallest cost over the set of all paths from s_i to s_j. Denote this cost by $J(s_i, s_j)$ and an estimate of it by $\hat{J}(s_i, s_j)$.

Let $J(s)$ be the cost of an optimal path constrained to go through s from s_0 to s_g, and let $J(s) = J(s_0, s) + J(s, s_g)$ where $J(s_0, s)$ is the cost of an optimal path from s_0 to s and $J(s, s_g)$ is the cost of an optimal path from s to s_g. Since $J(s)$, $J(s_0, s)$ and $J(s, s_g)$ are not known, the estimates $\hat{J}(s)$, $\hat{J}(s_0, s)$, and $\hat{J}(s, s_g)$ are used. Hence,

$$v(s) = \hat{J}(s) = \hat{J}(s_0, s) + \hat{J}(s, s_g) \tag{5}$$

is chosen. The function $\hat{J}(s, s_g)$ is called the heuristic component of the evaluation function, the heuristic function. If it satisfies the following conditions: i)$0 \leq \hat{J}(s, s_g) \leq$

$J(s, s_g)$ for all $s \in \mathbf{S}$, and ii) $\hat{J}(s_i, s_g) - \hat{J}(s_j, s_g) \leq J(s_i, s_j)$ for all $s_i, s_j \in \mathbf{S}$, and if s_g is reachable from s_0 then the A^* algorithm is said to be admissible.

The following theorem says that if the heuristic function $\hat{J}$ is a measuring function and is a low bound on the cost function θ for the given optimal control problem of DESs, then we can find an admissible and consistent A^* algorithm that will generate an optimal control to move the system from the initial state to the final state with the minimum of cost.

Theorem 2. *Given the optimal control problem described by*

$$\begin{cases} M = (\mathbf{S}, \mathbf{E}, \mathbf{F}^*, f, s_0, l, \theta, s_g) \ , \\ J(\sigma) = \min_{\sigma \in M(s_0, s_g)} \sum_{e \in E_\sigma} \theta(e) \ . \end{cases}$$

Suppose that the final state s_g is reachable from s_0. If the heuristic function $\hat{J}$ is a measuring function and satisfies $\hat{J}(s_i, s_j) \leq \theta(e_{ij})$ for all $s_i, s_j \in \mathbf{S}$ such that $s_j = f(e_{ij}, s_i)$, then there exists an admissible and consistent A algorithm that can select a sequence of events to move the system from the initial state s_0 to the final state s_g with the minimal cost.*

The above theorem givessufficient conditions to the optimal control problem of DESs by using A^* algorithm.

6 An Example

Consider an automated machineshop in a factory [10] which can process parts of two different classes: custom and stock. The example is specified in the temporal logic framework in [9]. Using the same symbols as in the above the specifications of the plant are as follows:

$$\Box\left[\vee_{i=1}^{N}(\delta = \alpha_i \vee \delta = \beta_i \vee \delta = \gamma_i) \vee \delta = \mu \vee \delta = \nu \vee \delta = \lambda \vee \delta = \epsilon\right] \quad (P1)$$

$$\Box\left\{\wedge_{i=1}^{N}\left[\delta = \alpha_i \Rightarrow x_i = O \wedge (\bigcirc x_i) = W \wedge \left(\wedge_{i \neq j=1}^{N}(\bigcirc x_j) = x_j\right) \wedge (\bigcirc n) = n\right]\right\} \quad (P2)$$

$$\Box\left\{\wedge_{i=1}^{N}\left[\delta = \beta_i \Rightarrow x_i = W \wedge (\bigcirc x_i) = P \wedge \left(\wedge_{i \neq j=1}^{N}(\bigcirc x_j) = x_j\right) \wedge (\bigcirc n) = n\right]\right\} \quad (P3)$$

$$\Box\left\{\wedge_{i=1}^{N}\left[\delta = \gamma_i \Rightarrow x_i = P \wedge (\bigcirc x_i) = O \wedge \left(\wedge_{i \neq j=1}^{N}(\bigcirc x_j) = x_j\right) \wedge (\bigcirc n) = n\right]\right\} \quad (P4)$$

$$\Box\left[\delta = \mu \Rightarrow (\bigcirc n) = n + 1 \wedge \left(\wedge_{j=1}^{N}(\bigcirc x_j) = x_j\right)\right] \quad (P5)$$

$$\Box\left[\delta = \nu \Rightarrow n > 0 \wedge (\bigcirc n) = n - 1 \wedge \left(\wedge_{j=1}^{N}(\bigcirc x_j) = x_j\right)\right] \quad (P6)$$

$$\Box\left[\delta = \lambda \Rightarrow (\bigcirc n) = n \wedge \left(\wedge_{j=1}^{N}(\bigcirc x_j) = x_j\right)\right] \quad (P7)$$

$$\Box[O \neq W \wedge O \neq P \wedge P \neq W] \quad (P8)$$

$$\wedge_{i=1}^{N} x_i = O \wedge n = 0 \quad (P9)$$

A software package, which will be the subject of a further publication will be used to obtain the reachability graph, the restrictions on it and the optimization of the controller. The reachability graph is given in Fig. 1.

According to rules $P1$-$P4$, expressions of the desired properties of the closed-loop system are formally given below.

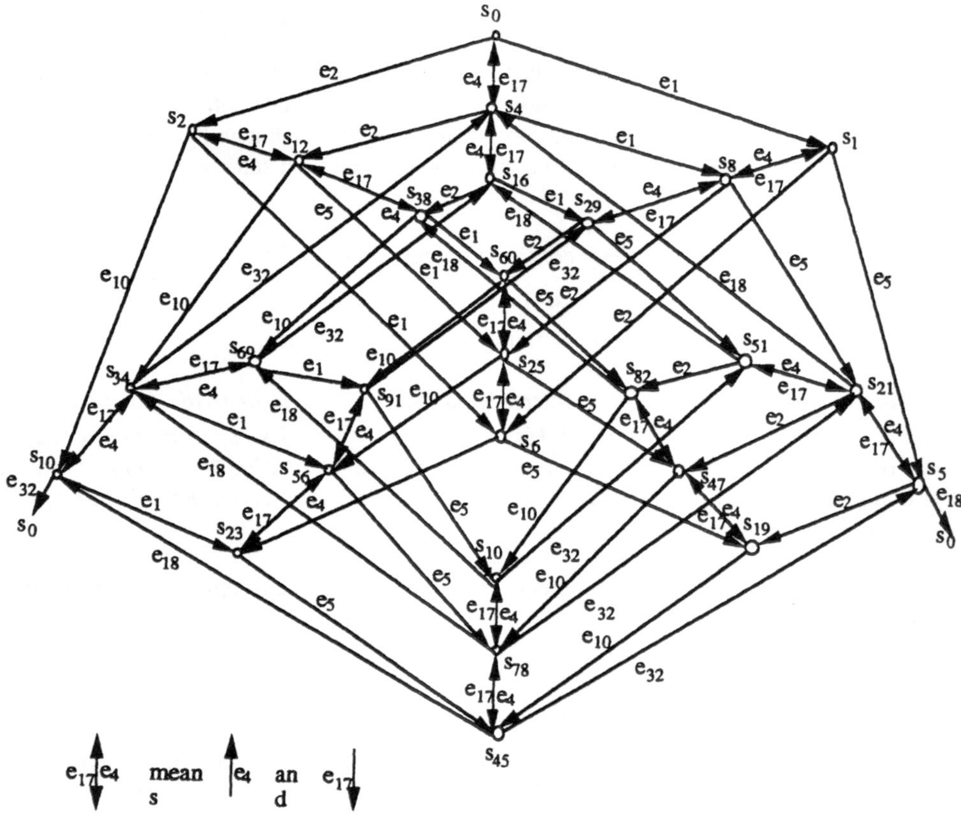

Fig. 1. Reachability graph of the plant.

$$\Box\left[\wedge_{i=1}^{N}\left(x_i = P \Rightarrow \wedge_{i\neq j=1}^{N} x_j \neq P\right)\right] \tag{CL1}$$

$$\Box\left[\vee_{i=1}^{N}(x_i = P) \Rightarrow n = 0\right] \tag{CL2}$$

$$\Box\left[\vee_{i=1}^{N}(x_i = W) \Rightarrow \delta \neq \mu)\right] \tag{CL3}$$

$$\Box\left[\wedge_{i,j=1;i\neq j}^{N}(x_i \neq W \Rightarrow (\delta = \alpha_i \mathcal{P} \delta = \alpha_j \Rightarrow \delta = \beta_i \mathcal{P} \delta = \beta_j))\right] \tag{CL4}$$

$$\Box\left[(\wedge_{j=1}^{N}(x_i = O) \Rightarrow \delta \neq \lambda\right] \tag{CL5}$$

According to the specifications of the required behavior, there are many undesirable states in the uncontrolled system. For instance, at state $s21 = (P, O, 1)$, custom part 1 is being processed while there is one stock part being processed. This violates the specification ($CL2$).

Figure 1 shows that there are more than one path from one state to another state, for instance, from $S0$ to $S19$ and to $S54$, from $S10$ to $S6$, from $S16$ to $S19$, from $S23$ to $S88$. For this manufacturing system, it is obvious that the costs might be different through different paths from the same initial state to the same final state. Therefore, it is of great significance to find out the path which has the minimum of the costand then to generate the corresponding sequence of events that drives the system along that path.

Table 1. Costs of events.

events	e_1	e_2	e_3	e_4	e_5	e_{10}	e_{17}	e_{18}	e_{32}
costs	1	2	1	2	1	1	2	5	5

The costs of events for this example are listed in Table 1. To use the A^* algorithm to solve the optimization problem, the heuristic index will be needed. The state preference indecies of the manufacturing system are given in Table 2.

Table 2. The index of states.

events	S_0	S_1	S_2	S_4	S_5	S_6	S_{10}	S_{16}	S_{19}	S_{23}	S_{44}	S_{54}	S_{75}	S_{88}
costs	0	1	2	2	1	2	3	3	3	3	4	4	5	5

Let $h(s_i)$ be the preference index of the state s_i. Then the heuristic function

$$\hat{h}(s_i, s_j) \stackrel{\text{def}}{=} |h(s_i) - h(s_j)|$$

is a semi-metric function and there holds

$$\hat{h}(s_i, s_j) = \left|\sum_{k=i}^{j-1}[h(s_k) - h(s_{k+1})]\right|$$
$$\leq \sum_{k=i}^{j-1} |h(s_k) - h(s_{k+1})| = \sum_{k=i}^{j-1} \theta(e_{k,k+1}, s_k) = \theta(e_{ij}, s_i) \ .$$

We choose some states from the reachability set of the closed-loop system as the initial states and the final states.

Figure 2 gives the optimal trajectories and corresponding optimal sequences of events for the initial state S_0 and the final state S_{19}, for the initial state S_0 and the final state S_{54}, and for the initial state S_{16} and the final state S_{19}, respectively.

7 Conclusions

The optimization problem of DESs supervisors has been shown to have a solution if sufficient conditions are met. A measurement space and function have been define in order to properly formulate this problem. The A^* algorithm generates the control such that the system move from the initial state to the final state with the least cost if the heuristic function is a measuring function and is a low bound of the event cost function.

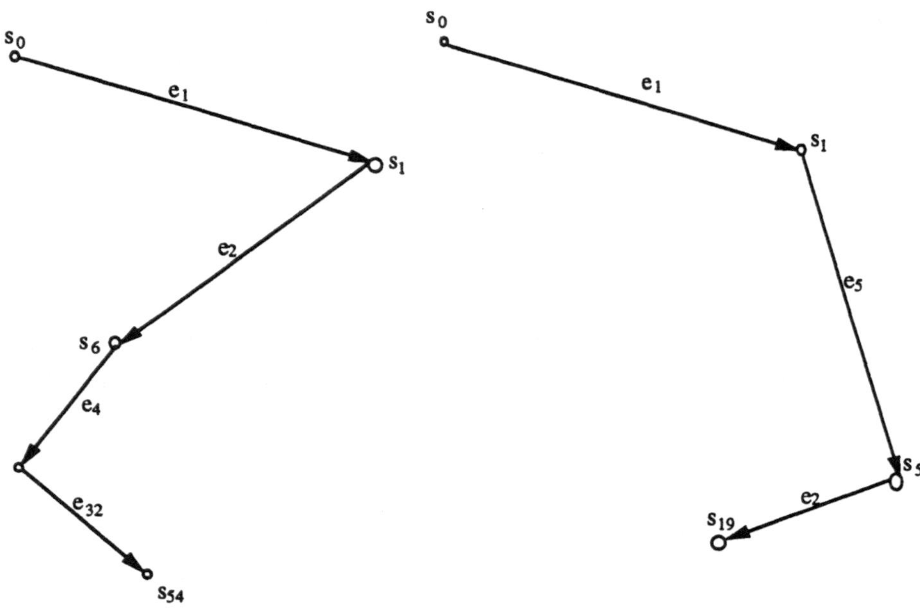

Fig. 2. Optimal paths.

References

1. H.A. Enderton, *A Mathematical Introduction to Logic*, Academic Press, New York, 1972.
2. E. Kreyzig, *Introductory Functional Analysis with Applications*, John Wiley, New York, 1978.
3. F. Lin, "A note on optimal supervisory control," *Proc. 1991 IEEE Int. Symposium on Intelligence Contr.*, Arlington, Virginia, Aug.13–15, 1991, pp.227–232.
4. J.-Y. Lin and D. Ionescu, "Verifying a class of nondeterministic discrete event systems in a generalized temporal logic framework," *IEEE Trans. Systems, Man and Cybernetics*, vol.22, no.6, pp.1461–1469, 1992.
5. J.-Y. Lin and D. Ionescu, "A Reachability Synthesis Procedure for Discrete Event Systems in A Temporal Logic Framework," to appear in vol.23 of *IEEE Trans. Systems, Man and Cybernetics*.
6. Z. Manna and A. Pnueli, "Verification of concurrent programs: A temporal proof system," *Foundations of Computer Science IV*, Mathematical Centre Tracts 159, Mathematish Centrum, Amsterdam, pp.163–255, 1983.
7. K.M. Passino and P.J. Antsaklis, "On the optimal control of discrete event systems," *Proc. 28th IEEE Conf. Decision and Control*, pp.2713–2718, 1989.
8. P.J. Ramadge and W.M. Wonham, "Supervisory control of a class of discrete event process," *SIAM J. Contr. Optimiz.*, **25**, pp.206–230, 1987.
9. R. Sengupta and S. Lafortune, "Optimal control of a class of discrete event systems", *Preprints of IFAC Int. Symposium on Distributed Intelligence Systems*, Arlington, Virginia, August 13–15, 1991, pp.25–30.
10. J.G. Thistle and W.M. Wonham, "Control problems in a temporal logic framework," *Int. J. Control*, **44**, pp.943–976, 1986.

Partial Difference Equation Extensions to Automata Regulator Theory

Qingsheng Yuan and Albert D. Baker

University of Cincinnati, Cincinnati, Ohio 45221, USA

1 Introduction

Gatto and Guardabusi [GG1] established the regulator theory for finite automata using primarily graph–theoretic arguments. We provide an algebraic extension to Gatto and Guardabusi's results using a partial difference equation (PDE) model [BA1]. In the approach followed here, algebraic procedures are established to determine the controllability as well as to find an optimal control sequence for finite automata. A feedback control law is obtained under the assumption of full state observation. The conditions for the existence of a linear control law are developed. These ideas have been applied to three case studies which are briefly discussed.

2 The Regulator Problem in Partial Difference Equation Form

Gatto and Guardabusi [GG1] studied the regulator problem for finite automata and provided conceptual definitions. They defined controllability, the conditions for which there exits a feedback control law, how to find the feedback control law, etc. This paper follows in the same manner as Gatto and Guardabusi's, but it uses the partial difference equation model developed by Baker [BA1].

Without loss of generality, let the system under control be a finite automaton $\mathbf{S}$ represented by m concurrent finite state machines. Each machine has $n_1, n_2, \ldots, n_m$ states respectively. The whole system state is represented as a $(n_1+n_2+n_3+\cdots+n_m)\times 1$ state vector. We refer to the ordered set of active states in each state machine as the *list of active states* of $\mathbf{x}_i$ written $(i_1, i_2, \ldots, i_m)$. Thus for example, if a discrete–event dynamic system (DEDS)'s list of active states is (2,5,8) and it has altogether 9 states, then $\mathbf{x}_i = [0\ 1\ 0\ 0\ 1\ 0\ 0\ 1\ 0]^T$, where $i_1 = 2, i_2 = 5$ and $i_3 = 8$. Furthermore, $\mathbf{c}$ is defined to be a sentence product and ϕ to be the state transition matrix over $\mathbf{c}$.

Definition 1. A state $\mathbf{x}_j$ is *controllable* to state $\mathbf{x}_i$ if there exists a state transition matrix $\phi(\mathbf{c})$ such that $\mathbf{x}_i = \phi(\mathbf{c})\mathbf{x}_j$.

Definition 2. An automaton $\mathbf{S}$ is controllable to state $\mathbf{x}_i$ if there exists a state transition matrix $\phi(\mathbf{c}_j)$ such that $\mathbf{x}_i = \phi(\mathbf{c}_j)\mathbf{x}_j$ for every state $\mathbf{x}_j$ of the automaton $\mathbf{S}$. The sequence of events $\mathbf{c}_j$ is called a *regulating sentence* to state $\mathbf{x}_i$.

The regulator problem is to find a regulating sentence $\mathbf{c}_j$ to a specific state $\mathbf{x}_i$ from any initial state $\mathbf{x}_j$. This is basically to control $\mathbf{S}$ such that the state of $\mathbf{S}$ is kept at a suitable preassigned value and led back to this value when the system is affected by any perturbation. Under the assumption of full state observation, the problem becomes a state feedback control problem for a DEDS. In other words, we want a feedback control law $\mathbf{K}(\cdot) : \mathbf{X} \rightarrow \mathbf{E}$ such that all states $\mathbf{x}_j$ of $\mathbf{S}$ can be homed to a specific state $\mathbf{x}_i$ by feeding back $\mathbf{e} = \mathbf{K}(\mathbf{x}_j)$.

Using the partial difference equation model, a closed–form solution to determine whether the automaton $\mathbf{S}$ is controllable to a specific state $\mathbf{x}_i$ can be obtained. First, $\mathbf{A}'_k$ matrices are formed by replacing all 1's in the state matrices $\mathbf{A}_k$ by the corresponding event for all k from 1 to n, where n is the total number of different events. The matrix $\mathbf{A}'$ is then defined as:

$$\mathbf{A}' = \mathbf{A}'_1 + \mathbf{A}'_2 + \mathbf{A}'_3 + \cdots + \mathbf{A}'_n \ . \tag{1}$$

Then, $\mathbf{A}'$ is modified by adding all the possible self loops to the diagonal elements, this forms the matrix $\mathbf{A}$.

For example, consider a DEDS modeled by the finite state machines shown in Fig. 1. In this example, the $\mathbf{A}'$ matrix is of the form:

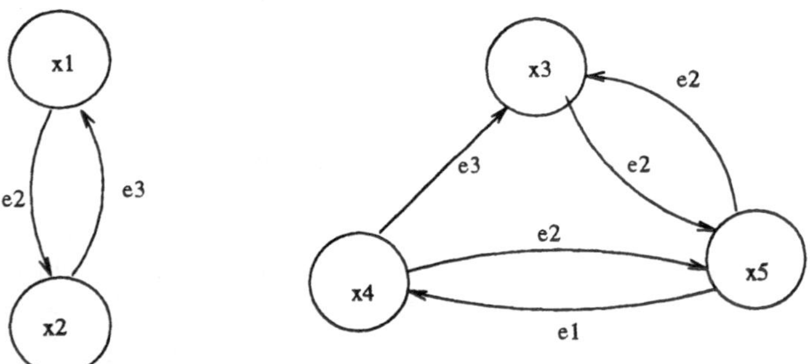

Fig. 1. Concurrent finite state machine model of an example discrete–event system

$$\mathbf{A}' = \begin{bmatrix} \mathbf{A}'_1 & \mathbf{0} \\ \mathbf{0} & \mathbf{A}'_2 \end{bmatrix} \tag{2}$$

where $\mathbf{A}'_1$ and $\mathbf{A}'_2$ are for each of the finite state machines. After replacing all state–matrix 1's by the corresponding events and adding all self–loops to the matrix, the $\mathbf{A}$ matrix is formed :

$$\mathbf{A} = \begin{bmatrix} \mathbf{A}_1 & \mathbf{0} \\ \mathbf{0} & \mathbf{A}_2 \end{bmatrix} \tag{3}$$

where

$$\mathbf{A}_1 = \begin{bmatrix} \mathbf{e}_1 + \mathbf{e}_3 & \mathbf{e}_3 \\ \mathbf{e}_2 & \mathbf{e}_1 + \mathbf{e}_2 \end{bmatrix}$$

and

$$\mathbf{A}_2 = \begin{bmatrix} \mathbf{e}_1 + \mathbf{e}_3 & \mathbf{e}_3 & \mathbf{e}_2 \\ \mathbf{0} & \mathbf{e}_1 & \mathbf{e}_1 \\ \mathbf{e}_2 & \mathbf{e}_1 & \mathbf{e}_3 \end{bmatrix}$$

and by '+' we mean, for example, by $\mathbf{e}_1 + \mathbf{e}_3$ that either $\mathbf{e}_1$ or $\mathbf{e}_3$ will cause the state transition.

The (i, j)-th element in the $\mathbf{A}$ matrix specifies that state j in the list of active states will be replaced by state i when the event specified by the (i, j)-th element occurs. However, this is only a 1-step event. If we calculate $\mathbf{A}^2, \mathbf{A}^3, \ldots$, we will get the 2-step, 3-step, ... event sequences from one state to another. The reachability matrix $\mathbf{R}$ is defined as [HNC1]:

$$\mathbf{R} = \mathbf{A} + \mathbf{A}^2 + \mathbf{A}^3 + \cdots + \mathbf{A}^l \tag{4}$$

where $l = \max(n_1, n_2, \ldots, n_m) - 1$.

State $\mathbf{x}_j$, whose list of active states is $(j_1, j_2, \ldots, j_n)$, is controllable to state $\mathbf{x}_i$, whose list of active states is $(i_1, i_2, \ldots, i_m)$, if the $(i_1, j_1), (i_2, j_2), \ldots, (i_m, j_m)$ elements of $\mathbf{R}$ have common event sequences. Note, each element of $\mathbf{R}$ will contain zero or more event sequences separated by the 'or' operator '+'. Furthermore, the reverse ordering of that common event sequence is the regulating sentence from $\mathbf{x}_j$ to $\mathbf{x}_i$. If all states of an automaton are controllable to $\mathbf{x}_i$, then the automaton is controllable to $\mathbf{x}_i$.

Consider the previous example. In this case, $l = 2$, and

$$\mathbf{A}^2 = \begin{bmatrix} \mathbf{A}_1^2 & \mathbf{0} \\ \mathbf{0} & \mathbf{A}_2^2 \end{bmatrix} \tag{5}$$

where

$$\mathbf{A}_1^2 = \begin{bmatrix} (\mathbf{e}_1 + \mathbf{e}_3)^2 + \mathbf{e}_3\mathbf{e}_2 & (\mathbf{e}_1 + \mathbf{e}_3)\mathbf{e}_3 + \mathbf{e}_3(\mathbf{e}_1 + \mathbf{e}_2) \\ \mathbf{e}_2(\mathbf{e}_1 + \mathbf{e}_3) + (\mathbf{e}_1 + \mathbf{e}_2)\mathbf{e}_2 & \mathbf{e}_2\mathbf{e}_3 + (\mathbf{e}_1 + \mathbf{e}_2)^2 \end{bmatrix},$$

and

$$\mathbf{A}_2^2 = \begin{bmatrix} (\mathbf{e}_1 + \mathbf{e}_3)^2 + \mathbf{e}_2\mathbf{e}_2 & (\mathbf{e}_1 + \mathbf{e}_3)\mathbf{e}_3 + \mathbf{e}_3\mathbf{e}_1 + \mathbf{e}_2\mathbf{e}_2 & (\mathbf{e}_1 + \mathbf{e}_3)\mathbf{e}_2 + \mathbf{e}_3\mathbf{e}_1 + \mathbf{e}_2\mathbf{e}_3 \\ \mathbf{e}_1\mathbf{e}_2 & \mathbf{e}_1\mathbf{e}_1 + \mathbf{e}_1\mathbf{e}_2 & \mathbf{e}_1\mathbf{e}_1 + \mathbf{e}_1\mathbf{e}_3 \\ \mathbf{e}_2(\mathbf{e}_1 + \mathbf{e}_3) + \mathbf{e}_3\mathbf{e}_2 & \mathbf{e}_2\mathbf{e}_3 + \mathbf{e}_2\mathbf{e}_1 + \mathbf{e}_3\mathbf{e}_2 & \mathbf{e}_2\mathbf{e}_2 + \mathbf{e}_2\mathbf{e}_1 + \mathbf{e}_3\mathbf{e}_3 \end{bmatrix}.$$

Note that two events being multiplied, for example, $\mathbf{e}_1\mathbf{e}_2$ means that $\mathbf{e}_2$ happens first followed by e_1. Thus, multiplication is not assumed to be commutative here, it is important to preserve the order.

So, if we want to know whether state $\mathbf{x}_j = [1\ 0\ 1\ 0\ 0]^T$ whose list of active states is (1,3) is controllable to state $\mathbf{x}_i = [0\ 1\ 0\ 1\ 0]^T$ whose list of active states is (2,4), the $(2, 1)$-th and $(4, 3)$-th elements in $\mathbf{R}$ are examined. In fact, this is equivalent to examining the $(2, 1)$-th and $(4, 3)$-th elements in $\mathbf{A}, \mathbf{A}^2, \ldots, \mathbf{A}^l$, and if there are any common elements the search can be stopped. If there is no common sentence until the maximum number l, then $\mathbf{x}_j$ is not controllable to $\mathbf{x}_i$.

In this example, the $(2, 1)$-th, $(4, 3)$-th elements in $\mathbf{A}^2$ are $\mathbf{e}_2(\mathbf{e}_1 + \mathbf{e}_3) + (\mathbf{e}_1 + \mathbf{e}_2)\mathbf{e}_2$ and $\mathbf{e}_1\mathbf{e}_2$ respectively. The common element is $\mathbf{e}_1\mathbf{e}_2$, whose reverse ordering $\mathbf{e}_2\mathbf{e}_1$ is the regulating sentence from state $\mathbf{x}_j$ to $\mathbf{x}_i$. This is easily verified from Fig. 1.

Based on the above discussion, we have the following theorem for the controllability of an automaton modeled by partial difference equations which is similar to Theorem 1 in Gatto and Guardabusi's :

Theorem 3. *The automaton* $\mathbf{S}$ *is controllable to state* $\mathbf{x}_i$ *if every state* $\mathbf{x}_j$ *in* $\mathbf{S}$ *is controllable to* $\mathbf{x}_i$*. A state* $\mathbf{x}_j$ *in* $\mathbf{S}$ *is controllable to* $\mathbf{x}_i$ *if the* (i_1, j_1)*-th,* (i_2, j_2)*-th* $\ldots$ *,* (i_m, j_m)*-th elements of* $\mathbf{R}$ *have common event sequences. The reverse ordering of those events is the regulating sentence.*

The above ideas have been extended to solve the *weighted event* problem, and the *weighted state* problem [YQ1]. The *weighted event* problem is the case where the occurrence of each event may have a different weight (e.g., different cost) and a regulating sentence is sought so that the automaton goes to the desired state at a minimum weight (cost). In this case, the cost w_i is appended to each e_i. After finding the sequences using the above method, the cost is also known for each sequence. The sequence with the smallest cost is chosen as the regulating sentence.

The *weighted states* problem is where there is a cost associated with visiting each state. A minimum–cost regulating sentence is sought. If the cost is only for a single state element, the same method can be used to find the minimum cost sequence as for the weighted event case. However, if the cost is for the whole state of the system (certain combinations of state elements), then we can not use the concurrent finite state machine model. There is no means for one state element to know the activity of another state element though these elements together determine the cost. In this case, we need to unify the concurrent state machine. An unified state machine is one whose states are the Cartesian product of all possible states in the concurrent state machines. Then the same methods can be applied to get the minimum cost sequence.

3 Feedback Control Law Synthesis

Once it has been determined that the system is controllable to a state $\overline{x}$, a closed–form solution for the state feedback control law can be developed.

Once the regulating sentence is known, the leftmost event is picked as the *feedback event* for the state from which the system is going.

We want to synthesize a feedback control law in the following form:

$$\mathbf{e} = \mathbf{K}(\mathbf{x}) \tag{6}$$

where $\mathbf{x}$ is the system state and $\mathbf{e}[i] = 1$, all other elements are zero, if the i-th event is the feedback event.

For a controllable single finite state machine, there will always exits a linear feedback control law in the form of $\mathbf{e} = \mathbf{K}\mathbf{x}$ where $\mathbf{K}$ is a matrix. Suppose the finite state machine has m states and the total number of events is n. Then the $\mathbf{e}$ vector will be $n \times 1$, $\mathbf{x}$ will be $m \times 1$ and $\mathbf{K}$ will be $n \times 1$. Because we have a single finite state machine, all but one element in the state vector $\mathbf{x}$ are zero, the remaining element is one.

The $\mathbf{K}$ matrix is given as:

$$\mathbf{K} = [k_1 \ k_2 \ \ldots \ k_n]^T \tag{7}$$

where

$$\mathbf{k}_i = \sum_{\mathbf{x} \in \mathbf{x}_i} \mathbf{x}$$

and $\mathbf{x}_i$ is set of all state vectors whose feedback event is e_i.

If the system is modeled by m concurrent finite state machines and each machine has $n_i, i = 1, \ldots, m$ states, it can be unified to a single finite state machine and a linear–form feedback control law can be obtained. In this way, the state vector will be $n_1 \times n_2 \times \cdots \times n_m$ long. Equation (7) can also be performed directly on a concurrent finite state machine model using a few modifications. A state vector which is $n_1 + n_2 + \cdots + n_m$ long is constructed. At all times, there will be m 1's in this vector, one for each machine. For this representation, a linear controller exits *iff all possible pairs of states with common 1's in their state vectors have the same feedback event.* This is because we can add all the vectors with the i-th feedback event and the i-th row of the $\mathbf{K}$ matrix will be the transpose of this vector divided by m. If this is not the case, i.e., we have a different feedback event for each vector, and we can not guarantee that we get a one in an event element with all others being zero. In this case, we need nonlinear feedback.

The nonlinear feedback control law is formulated as follows: Let $\mathbf{e} = [e_1 \;\; e_2 \ldots \;\; e_n]^T$ be the event vector. For a specific event e_j which is the feedback event for p states, we have :

$$e_j = \mathbf{k}_{j1}\mathbf{x} + \mathbf{k}_{j2}\mathbf{x} + \cdots + \mathbf{k}_{jp}\mathbf{x} \tag{8}$$

where $\mathbf{k}_{ji}$ is the transpose of the i-th state vector in p state vectors which is then divided by m, the number of state machines.

Therefore, for the whole system, Equation (6) becomes:

$$\mathbf{e} = \lfloor(\mathbf{K}_1\mathbf{x})\rfloor + \lfloor(\mathbf{K}_2\mathbf{x})\rfloor + \cdots + \lfloor(\mathbf{K}_p\mathbf{x})\rfloor \tag{9}$$

where the j-th row of $\mathbf{K}_i$ is $\mathbf{K}_{ji}$ for $j = 1, \ldots, n$ and $i = 1, \ldots, p$.

The operator $\lfloor\cdot\rfloor$ is the nonlinear floor operator defined as:

$$\lfloor x \rfloor := \max\{i \in \mathbb{N} \mid i \leq x\}$$

where $\mathbb{N}$ is the set of integers. In our special case, this operator reduces to a Boolean function:

$$\lfloor x \rfloor = \begin{cases} 1 \text{ if x=1} \\ 0 \text{ if x} \leq 1 \end{cases}$$

when $\lfloor\cdot\rfloor$ operates on a vector $\mathbf{x} = [x_1, x_2, \ldots x_n]^T$, the following result is obtained:

$$\lfloor(\mathbf{x})\rfloor := [\lfloor(x_1)\rfloor, \lfloor(x_2)\rfloor, \ldots, \lfloor(x_n)\rfloor]^T \tag{10}$$

For the earlier example, if we want the system to go from state $\mathbf{x}_j$ whose list of active states is (1,3) to state $\mathbf{x}_i$ whose list of active states is (2,4), we know the regulating sentence is e_2e_1, thus the state feedback is e_2. So, in Equation (7) for event e_2, there will be a $\mathbf{k}_2$ vector of $\left[\frac{1}{2} \;\; 0 \;\; \frac{1}{2} \;\; 0 \;\; 0\right]$.

The advantage of the concurrent finite state machine representation is that we use a $n_1 + n_2 + \cdots + n_m$ vector instead of a $n_1 \times n_2 \times \cdots \times n_m$ vector to represent the system state. But we need more matrices than just one as is used with an unified machine. So the conclusion is: for a state machine with a small number of states, we may use the unified machine representation and a linear controller can be obtained. In case of a large number of states, the concurrent finite state machine representation may be employed and only in very restricted situations will a linear controller exist. In most applications, a nonlinear controller is found and the number of matrices needed will be larger than that used in the unified state machine case though each of the matrices will be smaller.

4 Case Studies

The above ideas have been successfully applied to two case studies and unsuccesfully applied to another [YQ1]. One succesful application is to a manufacturing system. In this case, the controller, the plant as well as the control console are modeled by 6, 9 and 8 finite state machines respectively. For the plant, there are altogether 36 possible states 27 of which are forbidden states. There are 13 possible events. Based on this model, an optimal regulating sentence under an emergency stop is found. Another successful application is network protocol analysis [SD1]. In this case, we prove what the regulating sequence would be for each state the system might enter thus conclude that this protocol is deadlock free. We attempted to apply the above ideas to a VLSI tesing problem with limited success because of the presence of an internal feedback loop. We are able to formulate a reachability matrix but the derived regulating sequences contain both external and internal–feedback events. We can not control the internal-feedback events without knowing the evolving state of the system. Thus, we have not been able to guarantee that any derived regulating sequence can be executed.

5 Summary

An algebraic solution to the regulator problem for the finite automata based on the partial difference equation model is proposed. It gives closed form solutions by matrix manipulations. Case study applications for this model have been developed.

References

[GG1] M. Gatto and G. Guardabassi: The Regulator Theory for Finite Automata. Information and Control. **31** (1976) 1–16

[BA1] A.D. Baker: Modeling Discrete–Event Systems Using Partial Difference Equations. UC ECE TR 138/8/1992. (Paper not accepted for publication in these proceedings. Available by anonymous ftp to ftp.ece.uc.edu as pub/biblio/papers/TR138_8_92.ps).

[SD1] Deepinder P. Sidhu: Authentication Protocols for Computer Networks:I. Computer Network and ISDN Systems. **11** (1986), 297–310.

[YQ1] Qingsheng Yuan: A Partial Difference Equation Extension to the Automata Regulator Problem. MS thesis, University of Cincinnati, Electrical and Computer Engineering Department, in preparation.

[HNC1] Frank Harary, Robert Z. Norman and Dorwin Cartwrighto: Structural Models: An Introduction to the Theory of Directed Graphs. New York, Wiley (1965).

An Algebraic Temporal Logic Approach to the Forbidden State Problem in Discrete Event Control

KiamTian Seow and R. Devanathan

School of Electrical and Electronic Engineering
Nanyang Technological University
Nanyang Avenue S(2263) Republic of Singapore

1 Introduction

In this paper, propositional linear time temporal logic of Manna and Pneuli [1] is adapted for analysis of DES and its control. While research efforts on the use of temporal logic have been initiated (e.g. [2, 3]), the primary focus has been on the verification issues of control systems for a given controller design. Existing works on logical invariants (e.g. [4, 5]) basically employ the predicate and predicate transform theory [6] in the setting of the R-W framework [7]. The fundamental contribution of our approach over existing works is the mechanical synthesis of a higher level feedback control logic (meta-control logic) for a controllable forbidden state specification in the setting of strict concurrency [8].

2 Discrete Event Control & Temporal Logic

2.1 Model Definition

Let **G** be any discrete event process (DEP) given by a 5-tuple model $G = (\mathbf{P}, \Phi, \Theta, \Sigma, \delta)$ where $\mathbf{P}$ is a set of states of the process, Φ is a set of activity variables representing the different states of the process such that $\phi \in \Phi$ is a mapping $\phi : \mathbf{P} \mapsto \{0, 1\}$, $\Theta \in \Phi$ is the initial condition, Σ is an event set, and δ is the state transition function. $\Sigma = \Sigma_c \cup \Sigma_u$ and $\Sigma_c \cap \Sigma_u = \varnothing$, where Σ_c refers to the set of controllable events and Σ_u refers to the set of uncontrollable events. δ is a mapping $\delta : \mathbf{P} \times \Sigma \mapsto \mathbf{P}$. Note that there is a one-to-one and onto correspondence between the elements $p \in \mathbf{P}$ and $\phi \in \Phi$.

A set of DEPs $\mathbf{G_i}$ each modelled by : $G_i = (\mathbf{P}_i, \Phi_i, \Theta_i, \Sigma_i, \delta_i)$ constitute a (N concurrent DEP) discrete event system DES **G** modelled by $G = (Q, \Pi, \Sigma, \Theta, \delta) = G_1 \parallel \cdots \parallel G_i \parallel \cdots \parallel G_N$ such that $Q = \mathbf{P}_1 \times \cdots \times \mathbf{P}_i \times \cdots \times \mathbf{P}_N$, every system state variable $\pi \in \Pi$ is a mapping $\pi : Q \mapsto \{0, 1\}$ according to $\pi(p_1, p_2, \ldots, p_i, \ldots, p_N) = \prod_{i=1}^{N} \phi_i$, where $p_i \in \mathbf{P}_i$ correspond to $\phi_i \in \Phi_i$, $\Sigma = \bigcup_{i=1}^{N} \Sigma_i$, $\Theta = \prod_{i=1}^{N} \Theta_i \in \Pi$ and $\delta : Q \times 2^{\Sigma} \mapsto Q$.

In order to control a DES **G**, we adjoin a means of control such that the controlled discrete event system (CDES) $\mathbf{G_c}$ is given by a 6-tuple model $G_c = (\Pi, Q, \Gamma, \Sigma, \Theta, \delta_c)$, where every $\gamma \in \Gamma$ is a control mapping $\gamma : 2^{\Sigma} \mapsto \{0, 1\}$ according to $\gamma(\Sigma_s) = \prod_{k=1}^{\|\Sigma_s\|} \gamma(\sigma_k)$, where $\sigma_k \in \Sigma_s \subseteq \Sigma$ and $\|\Sigma_s\|$ is the number of distinct nonempty elements in Σ_s. δ_c is a mapping $\delta_c : \Gamma \times 2^{\Sigma} \times Q \mapsto Q$ according to

$$\delta_c(\gamma, \Sigma_s, q) = \begin{cases} \delta(\Sigma_s, q) & \text{if } \delta(\Sigma_s, q) \text{ is defined and } \gamma(\Sigma_s) = 1, \\ \text{undefined} & \text{otherwise.} \end{cases}$$

Γ in the CDES model G_c implicitly assumes the existence of an external agent or controller, also known as a supervisor. The controller can enable or disable the appropriate events via the respective event-controls $\gamma(\sigma)$ (also written as f_σ) in Γ at each discrete state q of G.

2.2 Model Semantics

Interpretations in Temporal Logic An interpretation, in the context of DES **G**, refers to a possible execution sequence or initialized legal trajectory (*legal* in the sense that it corresponds to the actual behaviour of DES **G** and *initialized* in the sense that it begins from the initial state) of its model G that 'interprets' (that is, makes true or satisfies) the temporal formula describing the desired behaviour.

Definition 1 (Interpretation I). An interpretation I is a 3-tuple $(G,\mathcal{A},s)$, where s is a temporal sequence of states in model or structure G, starting from its initial state q_0, and $\mathcal{A}$ assigns a logic value for every global variable in G.

The global variables (expressions) do not change as the DES evolves through a sequence of states. Quantification will thus be allowed only on global variables. In our abstraction, since the structure G and $\mathcal{A}$ are fixed for a given DES **G**, for convenience, an interpretation may also be written as (G, s) or simply s.

Definition 2 (Suffix). For k $\geq$ 0, the k-truncated suffix of a temporal sequence $s = q_0 \to q_1 \to q_2 \to \dots$ (written : $q^{(k)}$) is the sequence $q_k \to q_{k+1} \to q_{k+2} \to \dots$.

$I^{(k)}$ is the interpretation obtained from I, by replacing q with $q^{(k)}$.

Now, for all interpretations I, we say that I satisfies w (written : $\models^I w$) where the satisfaction relations of some operators are defined inductively on formulae as follows :

$\models^I \Box w$ iff for all k $\geq$ 0, $\models^{I^{(k)}} w$,
$\models^I \Diamond w$ iff there exists a k $\geq$ 0 such that $\models^{I^{(k)}} w$.

Any formula that is not qualified by a temporal operator is assumed to be true only in the initial state q_0.

Definition 3 (G-validity). A formula w is said to be G-valid if and only if every legal trajectory in G satisfies w, that is, notationally,

$$G \models w \text{ iff } (\forall I \in G) : \models^I w.$$

For convenience, we use the general theorem notation $\vdash w$ to denote $G \models w$ unless otherwise stated.

2.3 Basic Control Formulation

In the rest of this paper, in any proof, Theorems quoted as T[n] are taken from [1]. Any propositional rule or theorem quoted as PR is taken from [9].

DESC Global Axioms and Theorem

$$\text{DESC-GA1}: \vdash \neg(\sigma \in \Sigma_c) \leftrightarrow (\sigma \in \Sigma_u).$$

DESC-GA1 is an alternative way of asserting that $\Sigma = \Sigma_c \cup \Sigma_u$ such that $\Sigma_c \cap \Sigma_u = \varnothing$.

$$\text{DESC-GA2}: \vdash \neg f_\sigma \rightarrow (\sigma \in \Sigma_c).$$

Intuitively, DESC-GA2 may be paraphrased as "If an event can be disabled, it must be controllable".

$$\text{DESC-GT1}: \vdash (\sigma \in \Sigma_u) \rightarrow f_\sigma.$$

Proof. It follows from DESC-GA, DESC-GA1 and PR. □

Intuitively, DESC-GT1 may be paraphrased as "If an event is uncontrollable, it will always remain enabled".

DESC Axiom and Theorem In the context of a given DEP **G**, consider a state $p \in \mathbf{P}$. Let $F \in \Phi$ be a state variable corresponding to $p \in \mathbf{P}$.

Define a meta-control function $R : \mathbf{P} \mapsto \{0, 1\}$ for $p \in \mathbf{P}$ such that

$$R_p(\cdot) = \left[F \vee \sum_{j=1}^{\|\Sigma_p\|} F_j \wedge f_{\sigma_j} \right], \tag{1}$$

where $\sigma_j \in \Sigma_p \subseteq \Sigma$, Σ_p corresponds to the set of diligent (i.e. derived from events in Σ) transitions incident on $p \in \mathbf{P}$ and $F_j \in \Phi$ is the state variable corresponding to $p_j \in \mathbf{P}$ at which $\sigma_j \in \Sigma_p$ is defined such that $\{p\} \cap \{p_j\} = \varnothing$.

$$\text{Then DESC-A1}: \vdash \Box[\neg R_p(\cdot) \rightarrow \bigcirc\neg F].$$

Intuitively, control axiom DESC-A1 may be paraphrased as "At every system state q, if an event $\sigma \in \Sigma_p$ defined at state q is disabled, the system will not enter a next system state where the process state variable F holds". DESC-A1 characterizes the fundamental property of f_σ, namely, it is *permissive* in the sense that while a disabled σ is definitely prevented from happening, an enabled σ is not necessarily forced to occur.

For a set of concurrent DEPs $\mathbf{G}_i$, $i = 1, 2, \ldots, N$, $N \geq 1$, let $F_{p_c} = \prod_{i=1}^{M} F_{k_i}$, $1 \leq M \leq N$, $k_i \in \{1, 2, \ldots, N\}$, $i = 1, 2, \ldots, M$, $k_i \neq k_j$ for $i \neq j$, $\Sigma_{p_c} = \bigcup_{i=1}^{M} \Sigma_{p_{k_i}}$, $p_{k_i} \in \mathbf{P}_{k_i}$. Generalizing the control function, let $R_{p_c} : Q \rightarrow \{0, 1\}$ such that

$$R_{p_c}(\cdot) = \prod_{i=1}^{M} R_{p_{k_i}}(\cdot) = \prod_{i=1}^{M} \left[F_{k_i} \vee \sum_{j=1}^{\|\Sigma_{p_{k_i}}\|} F_{k_i j} \wedge f_{\sigma_{k_i j}} \right], \tag{2}$$

where $\sigma_{k_i j} \in \Sigma_{p_{k_i}} \subseteq \Sigma_{k_i}$, $\Sigma_{p_{k_i}}$ corresponds to the set of diligent transitions incident on $p_{k_i} \in \mathbf{P}_{k_i}$ and $F_{k_i j} \in \Phi_{k_i}$ is the state variable corresponding to $p_{k_i j} \in \mathbf{P}_{k_i}$ at which $\sigma_{k_i j} \in \Sigma_{p_{k_i}}$ is defined such that $\{p_{k_i}\} \cap \{p_{k_i j}\} = \varnothing$.

Finally, let $F_{p_d} = \sum_{l=1}^{L} \prod_{i=1}^{M_l} F_{r_i^l k_i^l}$, $1 \le M_l \le N$, $l = 1, 2, \ldots, L$, $L \ge 1$, $r_i^l \in \mathbf{N}$, $r_i^l \ne r_j^l$ for $i \ne j$, $k_i^l \in \{1, 2, \ldots, N\}$, $i = 1, 2, \ldots, M_l$, $k_i^l \ne k_j^l$ for $i \ne j$, $\Sigma_{p_d} = \bigcup_{l=1}^{L} \bigcup_{i=1}^{M_l} \Sigma_{p_{r_i^l k_i^l}}$, $p_{r_i^l k_i^l} \in \mathbf{P}_{k_i^l}$. Let $R_{p_d} : Q \to \{0, 1\}$ such that

$$R_{p_d}(\cdot) = \sum_{l=1}^{L} \prod_{i=1}^{M_l} R_{p_{r_i^l k_i^l}}(\cdot) = \sum_{l=1}^{L} \prod_{i=1}^{M_l} \left[F_{r_i^l k_i^l} \vee \sum_{j=1}^{\left\|\Sigma_{p_{r_i^l k_i^l}}\right\|} F_{r_i^l k_i^l j} \wedge f_{\sigma_{r_i^l k_i^l j}} \right], \tag{3}$$

where $\sigma_{r_i^l k_i^l j} \in \Sigma_{p_{r_i^l k_i^l}} \subseteq \Sigma_{r_i^l k_i^l}$, $\Sigma_{p_{r_i^l k_i^l}}$ corresponds to the set of diligent transitions incident on $p_{r_i^l k_i^l} \in \mathbf{P}_{k_i^l}$ and $F_{r_i^l k_i^l j} \in \Phi_{k_i^l}$ is the state variable corresponding to $p_{r_i^l k_i^l j} \in \mathbf{P}_{k_i^l}$ at which $\sigma_{r_i^l k_i^l j} \in \Sigma_{p_{r_i^l k_i^l}}$ is defined such that $\{p_{r_i^l k_i^l}\} \cap \{p_{r_i^l k_i^l j}\} = \varnothing$.

It should be obvious by the definitions of F_{p_d} and $R_{p_d}(\cdot)$ that :

$$\vdash \Box(F_{p_d} \to R_{p_d}(\cdot)) \tag{4}$$

which equivalently means :

$$\vdash \Box(\neg R_{p_d}(\cdot) \to \neg F_{p_d}). \tag{5}$$

By temporal reasoning based on axiom DESC-A1, it follows easily that :

$$\text{DESC-T1} : \vdash \Box[\neg R_{p_d}(\cdot) \to \bigcirc \neg F_{p_d}].$$

3 Forbidden State Problem

Controllability Axioms Under the conditions that

1. for every $p_{r_i^l k_i^l j} \in \mathbf{P}_{k_i^l}$, there is (at least) one initialized trajectory passing through $p_{r_i^l k_i^l j} \in \mathbf{P}_{k_i^l}$ first before reaching its state satisfying F_{p_d} and
2. it is possible to enter the forbidden state F_{p_d} if all the events defined at $p_{r_i^l k_i^l j} \in \mathbf{P}_{k_i^l}$ are enabled,

the axioms that formally characterize the (invariance) controllability of $\Box\neg F_{p_d}$ with respect to DES $\mathbf{G}$ are as follows :

$$\text{DESC-ICA1} : \vdash \neg F_{p_d} \quad \text{DESC-ICA2} : \vdash \Box(\neg F_{p_d} \to \neg R_{\Sigma_{p_d}}(\cdot)).$$

Intuitively, DESC-ICA2 may be paraphrased as "Whenever the current state q of the system does not satisfy F_{p_d}, control can be imposed at state q by disabling at least an event in every $\Sigma_{p_c} \subseteq \Sigma_{p_d}$ which is defined at q to prevent the system from entering any state satisfying F_{p_d} in the next instance". Together with DESC-ICA1, the axioms mean that the system can be directed to follow any trajectory that does not pass through any state satisfying F_{p_d}; hence the existence of optimal or maximally permissive control solution is implied.

Event Identification In the context of invariance control of $\neg F_{p_d}$ whenever it is satisfied, we can deduce that (at least) an uncontrollable event $\sigma \in \Sigma_{p_d}$ exists provided there is a state in a sequence $s \in G$ where $R_{\Sigma_{p_d}}(\cdot)$ holds when it should never be, resulting in a possibility of entering a state satisfying F_{p_d} which we want to avoid. This is precisely captured by the following event identification axiom :

$$\text{DESC-ICEIA}: \vdash \exists(s \in G) : \Diamond(\neg F_{p_d} \wedge R_{\Sigma_{p_d}}(\cdot)) \leftrightarrow \exists(\sigma \in \Sigma_{p_d}) : (\sigma \in \Sigma_u).$$

Control Logic Equation Rewrite DESC-ICA2 as an Invariance P-Control Equation (IPC-Eqn) :

$$C(\Pi, \Lambda) = 1 \text{ (rel to } P) \tag{6}$$

such that :

$$C(\Pi, \Lambda) = \neg R_{\Sigma_{p_d}}(\cdot), \text{ where } \Lambda \subseteq \Gamma \text{ and } P = \neg F_{p_d}. \tag{7}$$

Hence, the optimal control solution in Λ of (IPC-Eqn) exists that ensures the invariance of P if $\Box P$ is (invariance) controllable.

4 Main Results

On Controllability

Lemma 4. $\vdash \exists(s \in G) : \neg\Box\,(R_{\Sigma_{p_d}}(\cdot) \to F_{p_d}) \equiv \{\exists(\sigma \in \Sigma_{p_d}) : (\sigma \in \Sigma_u)\}$.

Proof. It follows from T5, DESC-ICEIA and PR. □

Theorem 5 (Invariance State Control Theorem ISCT).
$\vdash \Box(\neg F_{p_d} \to C) \equiv \vdash \{\forall(\sigma \in \Sigma_{p_d}) : (\sigma \in \Sigma_c)\}$.

Proof. It follows from Lemma 4, T40, DESC-GA1, definition of G-validity and PR. □

Corollary 6. $\Box\neg F_{p_d}$ *is controllable if and only if the following conditions hold :*

$\vdash \neg F_{p_d}$,
$\vdash \{\forall(\sigma \in \Sigma_{p_d}) : (\sigma \in \Sigma_c)\}$.

Proof. It follows from axiomatic definition of controllability, Theorem 5 and PR. □

On Event-Control The supremal σ-friend of P [4] can be characterized using the concept of Boolean difference which is defined as follows :

Definition 7. Given a Boolean function $F = f(x_1, \ldots, x_i, \ldots, x_n)$, the Boolean difference with respect to its logic variable x_i, as given in [10], is defined as follows :

$$\frac{dF}{dx_i} = f(x_1, \ldots, x_{i-1}, \text{false}, x_{i+1}, \ldots, x_n) \oplus f(x_1, \ldots, x_{i-1}, \text{true}, x_{i+1}, \ldots, x_n).$$

Theorem 8 (The Forbidden State Event-Control Theorem FSECT). *Let P be a forbidden state formula. Then if □P is controllable, the supremal σ-friend of P (denoted by* sup $F_\sigma(P)$*) for* $\sigma \in \Sigma_c$ *is given by:*

$$\sup F_\sigma(P) = \neg \frac{dC}{df_\sigma} \ (\mathit{rel\ to\ P})$$

where C is the invariance P-State control logic.

Proof. By expression (3),

$$R_{p_d}(\cdot) = \sum_{l=1}^{L} \prod_{i=1}^{M_l} R_{p_{r_i^l k_i^l}}(\cdot) = \sum_{l=1}^{L} \prod_{i=1}^{M_l} \left[F_{r_i^l k_i^l} \vee \sum_{j=1}^{\left\| \Sigma_{p_{r_i^l k_i^l}} \right\|} F_{r_i^l k_i^l j} \wedge f_{\sigma_{r_i^l k_i^l j}} \right].$$

Rewriting $R_{p_d}(\cdot)$ in terms of a distinct $\sigma \in \Sigma_{p_d}$, we have :

$$R_{p_d}(\cdot) = (A \vee B \wedge f_\sigma) \tag{8}$$

where A and B are logic combinations of process state variables and other event-control logics (excluding f_σ) in **positive** (or normal, i.e. without any 'negation' connective) form.

Therefore, if $\Box P$ is controllable, by expressions (7) and (8), we have : $C = \neg(A \vee B \wedge f_\sigma)$. Taking the Boolean difference of C with respect to event-control logic f_σ for a distinct $\sigma \in \Sigma_{p_d}$, where $\sigma \in \Sigma_c$, we have : $\frac{dC}{df_\sigma} = A \oplus [A \vee B] = (\neg A) \wedge B$. Now, if $\frac{dC}{df_\sigma}$ = true (or Logic 1), i.e. $A = 0$, $B = 1$, then $C = \neg(A \vee B \wedge f_\sigma) = \neg f_\sigma$. Therefore, $\frac{dC}{df_\sigma} \rightarrow (C \equiv \neg f_\sigma)$. Since $C = 1$ (rel to P), we have : $\frac{dC}{df_\sigma} \rightarrow \neg f_\sigma$ (rel to P) . By PR, $f_\sigma \rightarrow \neg \frac{dC}{df_\sigma}$ (rel to P) . Hence, the supremal expression of f_σ (denoted by sup $F_\sigma(P)$) for $\sigma \in \Sigma_c$ is given by sup $F_\sigma(P) = \neg \frac{dC}{df_\sigma}$ (rel to P) . □

5 Example

A manufacturing system example [4] as shown in Fig. 1 is used to illustrate the ideas developed.

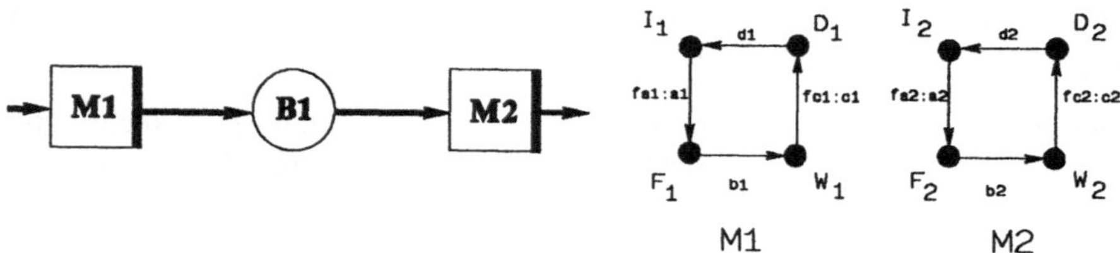

Fig. 1. Example : system flow configuration and state-graphs of the individual CDEPs

For each process, each process state variable[1] represents an activity (process state), while diligent transitions between states, that is, events, mark the completion of one activity and the start of another. Only the event-controls f_σ against each controllable event σ are shown. In detail, in state I_i, $\mathbf{M_i}$ is "idle"; in state F_i, $\mathbf{M_i}$ is "fetching" a part from its input buffer ($\mathbf{B_1}$ in the case of $\mathbf{M_2}$, and a buffer (not shown in Fig. 1) assumed not empty at all times in the case of $\mathbf{M_1}$); in state W_i, $\mathbf{M_i}$ is "working"; and in state D_i, $\mathbf{M_i}$ is "depositing" a part in its output buffer ($\mathbf{B_1}$ in the case of $\mathbf{M_1}$, and a buffer (not shown in Fig. 1) assumed not full at all times in the case of $\mathbf{M_2}$).

Consider a requirement due to mutually exclusive use of $\mathbf{B_1}$ given by $\Box P_1$, where $P_1 = \neg(D_1 \wedge F_2)$. By Theorem 6, $\Box P_1$ is controllable. Hence, by expression (6), we have :

$$C_1 = \neg([D_1 \vee W_1 \wedge f_{c_1}] \wedge [F_2 \vee I_2 \wedge f_{a_2}])$$

and $C_1 = 1$ (rel to P_1) is the control logic equation.
Now, consider the supremal control of event a_2 of P_1. By Theorem 8, we have:

$$\sup F_{a_2}(P_1) = \neg\frac{dC_1}{df_{a_2}} \text{ (rel to } P_1) = \neg[I_2 \wedge (D_1 \vee W_1 \wedge f_{c_1})] \text{ (rel to } P_1).$$

This is the supremal a_2-control (relative to P_1) for the concurrent manufacturing system where the simultaneity of event occurrences is possible. However, if we assume that the system operation is serial (one event per system transition), then it does not depend on c_1-control f_{c_1} and therefore specializes (by assigning $f_{c_1} = 0$) to :

$$\sup F_{a_2}(P_1) = \neg(I_2 \wedge D_1) \text{ (rel to } P_1).$$

This is equivalent to the supremal a_2-control (relative to P_1) for the same example of Ramadge and Wonham [4, pp. 1215], wherein it is given as :

$$f_{1a_2} = [(q_1, q_2) \neq (D_1, I_2)] \text{ (rel to } P_1).$$

6 Conclusion

In this paper, we present an axiomatic approach to the forbidden state problem in discrete event control for a class of concurrent DES using temporal logic. A simple example illustrates the main ideas developed.

References

1. J. S. Ostroff, "Appendix A : Formal overview of temporal logic and Appendix B : Temporal logic theorems and rules," *Temporal logic for Real Time Systems*, pp. 155–171 and 172–183, New York: Research Studies Press Ltd. John Wiley & Sons Inc, 1989. Advanced Software Development Series.
2. J. G. Thistle and W. M. Wonham, "Control problems in a temporal logic framework," *International Journal of Control*, vol. 44, no. 4, pp. 943–976, September 1986.

[1] For convenience, the terms *process state* and its *state variable* are used interchangeably.

3. Jing-Yue Lin and Dan Ionescu, "Verifying a class of nondeterministic discrete event systems in a generalized temporal logic," *IEEE Transactions on Systems, Man and Cybernetics*, vol. 22, no. 6, pp. 1461–1469, November/December 1992.
4. P. J. Ramadge and W. M. Wonham, "Modular feedback logic for discrete event systems," *SIAM Journal of Control and Optimization*, vol. 25, no. 5, pp. 1202–1218, September 1987.
5. Ratnesh Kumar, Vijay K. Garg, and Steven I. Marcus, "Using predicate transformers for supervisory control," *Proceedings of the 30th IEEE International Conference on Decision and Control*, (Brighton, England), pp. 98–103, December 1991.
6. E. W. Dijkstra, *A Discipline of Programming*. Prentice Hall, Inc. Englewood Cliffs, New Jersey, 1985.
7. P. J. Ramadge and W. M. Wonham, "Supervisory control of a class of discrete event processes," *SIAM Journal of Control and Optimization*, vol. 25, no. 1, pp. 206–230, January 1987.
8. Yong Li and W. M. Wonham, "Strict concurrency and nondeterministic control of discrete event systems," *Proceedings of the 28th IEEE International Conference on Decision and Control*, (Tampa, Florida, U.S.A), pp. 2731–2736, December 1989.
9. K. A. Ross and C. R. B. Wright, *Discrete Mathematics*, ch. 2.2 : Propositional Calculus. Prentice Hall, Inc. Englewood Cliffs, New Jersey, 2nd ed., 1988. Tables on Logical Equivalences, Logical implications and Rules of inference.
10. R. P. Trueblood and A. Sengupta, "Dynamic analysis of the effects access rule modifications have upon security," *IEEE Transactions on Software Engineering*, vol. SE-12, no. 8, pp. 866–870, August 1986.

Automata Timing Specification

D. Delfieu[1] *and A.E.K. Sahraoui*[1,2]

[1] Laboratoire d'Automatique et d'Analyse des Systèmes du C.N.R.S.
7, avenue du colonel Roche, 31077 Toulouse Cedex, France
e-mail : delfieu@laas.fr, kader@laas.fr

[2] École Nationale d'Ingénieurs de Tarbes, chemin d'Azereix, 65016 Tarbes, France

1 Introduction

Reactive systems are discrete systems which need particular attention during their design because they manipulate a large amount of events and they are subject to hard temporal constraints. Many works deal with reactive systems, Synchronous language for design phase, languages with real-time features for implementation phase, timing tools for integration phase, ... we focus our work on the specification phase of the software development cycle. We propose an approach that consists in building a temporal acceptor which will allow to check in a separate way the temporal aspects of a specification that takes into account the temporal and the functional aspects.

Some methods of specification based on "Process Algebras" [5] mix temporal and functional specification. Time is introduced by mean of time stamps associated to events [4]. We retain this idea, introducing a special event which occurrence marks the elapsing of time. We object two arguments for not using these algebras. Firstly, these formalisms (CCS, Lotos, ...) are based on Finite Automata which are expressively equivalent to regular expressions which is a limited formalism that cannot express some sequence properties [2]. Secondly, a language equivalence is sufficient for the comparison of two temporal behaviors. These two last points have lead us to go back to grammar based approach [3] that we will describe in next section, then we will present an axiomatic method to compose statically Temporal Constraints (TC).

The general context of our approach is first to extract TC from requirement, by the mean of a grammar formalism. A representation is then obtained, by an incremental process of composition, for which we will present the first step of the algorithm. The final representation will serve to test temporal properties. This paper is structured in two parts : the expression and the composition of temporal constraints.

2 The Temporal Expression

2.1 Basic Temporal Constraints

In the process of formalization of TC, we need to define a basic form of TC upon which all calculation will be made. Basic TC are constraints on which every TC can be expressed in terms of conjunction of this type. In [1] the author has characterized TC in three categories, minimum, maximum and durational. Examining these constraints we extract the entities which define it :

- O: an event representing the origin of time upon which the date is expressed,
- E: the expected event,
- D: an integer representing the duration expressed in some unit of time,
- Type = {minimum, maximum}. Minimum means "no sooner than" and maximum means "no later than".

A basic TC (BTC) is represented by the 4-tuple (O, E, D, T). The processus of formalization of TC, is described by the following **algorithm** :

- The **extraction** phase where all the temporal constraints are extracted from the requirements. At this step every TC is converted in basic TC. This phase will not be addressed in this paper.
- The **composition** phase allow to produce the conjunction of the temporal constraints.
  - First step (BTC_i and BTC_j are Basic TC)
    IF there exits some dependences between BTC_i and BTC_j
    THEN $C_1 \leftarrow \text{BTC}_i * \text{BTC}_j$
    ELSE $C_1 \leftarrow \text{BTC}_i \parallel \text{BTC}_j$
  - Current step (composition of C_n with BTC)
    IF there exits some dependences between C_n and BTC
    THEN $C_{n+1} \leftarrow C_n * \text{BTC}$
    ELSE $C_{n+1} \leftarrow C_n \parallel \text{BTC}$

Definition 1. We say that there is a dependence between two basic temporal constraints BTC_1 and BTC_2, if the expected or origin event of BTC_1 is equal to the origin and the expected event of BTC_2.

Definition 2. We say that there is a dependence between a basic temporal constraints BTC and a temporal constraint TC, if it exists one BTC_i composing TC which is in dependence with BTC.

2.2 Definition of a Grammar

A grammar is formally defined by the 4-tuple $G = (T, NT, R, S)$ where T is the set of the terminal symbols which represent the events implied in the temporal constraint ; NT represent the set of non terminal event, they are symbols allowing recursive definitions ; R is the set of rules, a rule is composed of two members, the left member is the rewriting of the right member ; S is the axiom, it indicates the first rule to apply. (see [3] for more detailed explanation).

2.3 Grammar of Basic TC

minimum : (o, e, d, min)
$G_{\text{min}} = (T_1 = \{o, e, h\}, NT_1 = \{S, T\}, R_1, S_1)$,

where R_1 is the set of the following equations :

$$S \to o.h^d.T \ ,$$
$$T \to h.T + e \ .$$

The first rule expresses that after event o, the system waits for d units of time (the system has to receive d occurrences of h noted h^d) then there is another phase (Recursive rule T) where e can occur at every instant. ‘

maximum : $(o, e, d, \min)$
$G_{\max} = (\{o, e, h\}, \{S\}, R, S)$ where R is :

$$S \to \sum_{i=0}^{d} o.h^i.e \ .$$

This grammar give all the acceptable traces according to the unit of time defined by the event h : after o, e can occur synchronously ($i = 0$) or after one unit of time ($i = 1$) and so on . . . .

3 Composition

Composing two TC allows to find the set of traces that are compatible with the two grammars. This set of traces can be also characterized by a grammar. Our goal is to then to define a method in order to obtain the resulting grammar. We present in the following for all the different cases of dependences the composition of two basic TC. First, we illustrate, a composition by an example.

3.1 Example

"Today, (i) John must phone no later than 3'o clock and (ii) he cannot leave his work, no sooner than 5'o clock."

The extraction phase reveals two TC (i) and (ii) which can be expressed by :

$C_1 = (bd, p, 3, \max) \ , \quad C_2 = (bd, l, 5, \min) \ ,$

where bd is the event marking the beginning of the day, p for phoning and l for leaving. The grammars accepting the behavior are :

- for C_1 (constraint on the phoning event):

$$S \to bd.\left(\sum_{i=0}^{3} h^i.p\right) \ .$$

- for C_2 (constraint on the leaving event):

$$S \to bd.h^5.T \ ,$$
$$T \to h.T + l \ .$$

The resulting expression which takes into account the two constraints is :

$$S \rightarrow bd.\left(\sum_{i=0}^{3} h^i.p.h^{5-i}\right).T \ ,$$
$$T \rightarrow h.T + l \ .$$

The next subsection presents the general forms of basic compositions for every type of basic TC.

3.2 Same Origin

The conjunction of these constraints represents a behavior where two events are bounded to the same origin.

$C_1 = (o_1, a_1, d_1, T_1)$, $C_2 = (o_1, a_2, d_2, T_2)$:

i) $T_1 = \min$ and $T_2 = \max$ and $d_1 < d_2$:

$$S \rightarrow o_1\left[\left(\sum_{i=0}^{d_2} h^i.a_2.h^{\max(d_1-i,0)}\right)T + h^{d_1}.\left(\sum_{i=0}^{d_2-d_1} h^i.a_1\right).\left(\sum_{j=i}^{d_2} h^j.a_2\right)\right] \ ,$$
$$T \rightarrow h.T + a_1 \ .$$

ii) $T_1 = \min$ and $T_2 = \max$ and $d_1 \geq d_2$:

$$S \rightarrow o_1\left[\left(\sum_{i=0}^{d_2} h^i.a_2.h^{d_1-i}\right).T\right] \ ,$$
$$T \rightarrow h.T + a_1 \ .$$

iii) $T_1 = T_2 = \max$ and $d_1 \geq d_2 \geq 1$:

$$S \rightarrow o_1\left[\sum_{i=0}^{d_2} h^i.\left(a_1.\left(\sum_{j=0}^{d_2-i} h^j.a_2\right) + a_2.\left(\sum_{j=0}^{d_1-i} h^j.a_1\right)\right)\right] \ .$$

We got the same formula for $d_2 \geq d_1 \geq 1$ replacing respectively a_1 by a_2 and d_1 by d_2.

iv) $T_1 = T_2 = \min$ and $d_1 > d_2$:

$$S \rightarrow o_1.h^{d_2}\left[h^{d_1-d_2}.T + \left(\sum_{i=0}^{d_1-d_2} h^i.a_2.h^{d_1-d_2-i}\right)V\right] \ ,$$
$$T \rightarrow a_1.U + a_2.V + h.T \ ,$$
$$U \rightarrow h.U + a_2 \ ,$$
$$V \rightarrow h.V + a_1 \ .$$

We got the same formula for $d_2 \geq d_1 \geq 1$ replacing respectively a_1 by a_2 and d_1 by d_2.

3.3 The Expected Event Same as the Origin

The composition gives the expression of the concatenation of the constraints, one being prior to the other.

$C_1 = (o_1, a_1, d_1, T_1)$, $C_2 = (a_1, a_2, d_2, T_2)$:

i) $T_1 = T_2 = \min$:

$$\begin{aligned} S &\to o_1.h^{d_1}T \ , \\ T &\to a_1.h^{d_2}.U + h.T \ , \\ U &\to h.U + a_2 \ . \end{aligned}$$

ii) $T_1 = T_2 = \max$:

$$S \to o_1.\left(\sum_{i=0}^{d_1} h^i.a_1\right)\left(\sum_{i=0}^{d_2} h^i.a_2\right) \ .$$

iii) $T_1 = \min$ and $T_2 = \max$:

$$\begin{aligned} S &\to o_1.h^{d_1}T \ , \\ T &\to a_1.\left(\sum_{i=0}^{d_2} h^i.a_2\right) + h.T \ . \end{aligned}$$

iv) $T_1 = \max$ and $T_1 = \min$:

$$\begin{aligned} S &\to o_1.\left(\sum_{i=0}^{d_1} h^i.a_1\right).h^{d_2}.T \ , \\ T &\to a_2 + h.T \ . \end{aligned}$$

3.4 Same Expected Event

An event is bounded to two references.

$C_1 = (o_1, a_1, d_1, T_1)$, $C_2 = (o_2, a_1, d_2, T_2)$:

i) $T_1 = T_2 = \min$ and $d_2 > d_1$:

$$\begin{aligned} S &\to o_1.T + o_2.U \ , \\ T &\to h.T + o_2.h^{d_2}.V \ , \\ U &\to h^{d_2-d_1}.W + \left(\sum_{i=0}^{d_2-d_1-1} h^i.o_1.h^{d_2-i}\right).V \ , \\ W &\to h.W + o_1.h^{d_1}.V \ , \\ V &\to h.V + a_1 \ . \end{aligned}$$

We got the dual formula for $d_2 \geq d_1 \geq 1$.

ii) $T_1 = T_2 = \max$:

$$S \to o_1.\left[h^{d_1}.o_2.a_1 + \sum_{i=0}^{d_1-1} h^i.o_2.\left(\sum_{j=0}^{\min(d_1-i,d_2)} h^j.a_1\right)\right]$$

$$+ o_2.\left[h^{d_2}.o_1.a_1 + \sum_{i=0}^{d_2-1} h^i.o_2.\left(\sum_{j=0}^{\min(d2-i,d_1)} h^j.a_1\right)\right] .$$

iii) $T_1 = \min$, $T_2 = \max$ and $d_1 > d_2$:

$$S \to o_1.h^{d_1-d_2}.\left[h^{d_2}.U + \sum_{i=1}^{d_2} h^{d_2-i}.o_2.\left(\sum_{j=i}^{d_2} h^j.a_1\right)\right] ,$$

$$U \to h.U + o_2.\left(\sum_{i=0}^{d_2} h^i.a_1\right) .$$

iv) $T_1 = \min$, $T_2 = \max$ and $0 < d_1 \le d_2$:

$$S \to o_1.\left[h^{d_1}.U + \sum_{i=0}^{d_1-1} h^i.o_2.\left(\sum_{j=d_1-i}^{d_2} h^j.a_1\right)\right]$$

$$+ o_2.\left[\sum_{i=0}^{d_2-d_1} h^i.o_1.h^{d_1}.\left(\sum_{j=0}^{d_2-d_1-i} h^j.a_1\right)\right] ,$$

$$U \to h.U + o_2.\left(\sum_{i=0}^{d_2} h^i.a_1\right) .$$

v) $T_1 = \min$, $T_2 = \max$ and $d_1 = 0$:

$$S \to o_1.U + o_2.\left[\sum_{i=0}^{d_2} h^i.o_1.\left(\sum_{j=0}^{d_2-i} h^j.a_1\right)\right] ,$$

$$U \to h.U + o_2.\left(\sum_{i=0}^{d_2} h^i.a_1\right) .$$

3.5 Same Origin and Same Expected Event

The composition express the notion of tightening of constraint.

$C_1 = (o_1, a_1, d_1, T_1)$, $C_2 = (o_2, a_1, d_2, T_2)$:

i) $T_1 = \min$, $T_2 = \min$ and $d_1 > d_2$: (C_1 and C_2) is equivalent to C_1,
ii) $T_1 = \max$, $T_2 = \max$ and $d_1 > d_2$ (C_1 and C_2) is equivalent to C_1,
iii) $T_1 = \max$, $T_2 = \min$ and $d_2 < d_1$:

$$S \to o_1.h^{d_2}.\left(\sum_{i=0}^{d_1-d_2} h^i\right).a_1 .$$

In the last case ($T_1 = \max$, $T_2 = \min$ and $d_2 > d_1$), the constraints are not compatible.

4 Validation

A first validation is made step by step. Composing TC, if the intermediate result give an empty set of traces, the TC is rejected and the requirement step is doubted over.

A second aspect of the validation is that the final temporal behavior obtained, can be used to check the temporal aspect of an executable specification representing the requirement. This checking will be made by putting these two elements in a "multi-way synchronization" [6], which signifies that the synchronization is made on a predefined set of events.

Conclusion

We have presented in this paper some procedures of an algorithm which provide from the requirements a syntactic expression representing the temporal behavior of a specification. This expression can be used as a syntactic analyzer, to recognize valid traces. The composition process allows also to reject incompatible traces.

A software tool supporting this work is being implemented. The perspective of the present work is to develop a formal basis for every type of constraints that will allow to build automatically a language acceptor for valid temporal constraints. All this part is used to extend actual software engineering method as SA-RT and Statecharts for the specification of reactive systems.

References

1. B. Dasarathy. Timing constraints of real_time systems: construct for expressing them, methods of validating them. *IEEE transaction Software Engineering*, 11:80–86, 1985.
2. D. Delfieu and A.E.K Sahraoui. Expression and verification of temporal constraints for real-time systems. In *7th Annual European Computer Conference*. IEEE, 1993.
3. J. E. Hopcroft and J. D. Ullman. *Introduction to automata theory, languages and computation*. Addison Wesley, 1979.
4. C. Miguel, A.Fernandez, and L. Vidaller. Extended lotos towards performance evaluation. In *FORTE'92*, 1992.
5. R. Milner. *A Calculus of Communicating Systems*. Lecture Notes in Computer Science, 1980.
6. J. Quemada et al. Introduction to lotos. Comett Alice, may 1991.

Synthesis of Static Controllers for Forbidden States Problems in Boolean C/E Systems Using the Boolean Differential Calculus

Stefan Kowalewski

Process Control Group (AST), Chemical Engineering Department,
University of Dortmund, D-44221 Dortmund, Germany.
email: stefan@astaire.chemietechnik.uni-dortmund.de

1 Introduction

The successful application of model based discrete controller synthesis methods depends very much on the easy development of clear and verifiable models for complex systems. We believe that the Condition/Event system framework developed by Sreenivas and Krogh [SK91] with its ability for block diagram and hierarchial modeling, and especially Boolean C/E (BC/E) systems [KK93] as an effective computational representation can serve as a platform for this approach. In this paper we therefore present first ideas about controller synthesis based on BC/E models for a special class of problems, namely the static controller synthesis for forbidden states problems. We will use two main tools: the mathematical theory of the so-called Boolean Differential Calculus (BDC) and the Boolean analysis software XBOOLE. Both tools together offer the possibility of concise formulation and effective realization of the synthesis algorithms.

The paper is organized as follows. The next section gives a review of BC/E systems and XBOOLE. Section 3 provides a short introduction to the BDC. In §4 we define the control problem we are looking at and split it into two subproblems: the synthesis of an observer and of the actual controller. The solution to the first subproblem is addressed in §5, the second one in §6. Section 7 presents an example.

2 Boolean Condition/Event Systems and XBOOLE

In [KK93] Boolean C/E (BC/E) systems were introduced as a multi-input/multi-output extension of the original C/E model [SK91]. The binary valued in-/output signal components u_i, v_i, y_i, z_i and the state variables x_i of a BC/E system are collected in five vectors u, v, x, y, z of the dimensions n_u, n_v, n_x, n_y, n_z, resp., where u represents the condition input, v the event input, y the condition output, z the event output, and x the state of the system. The behavior can then be described by three Boolean functions: The state transition function F (1), and two output functions G and H for each kind of signal (2), (3). Together with the set X_0 of possible initial states $x(0)$, u, v, x, y, z and F, G,

H form a complete model of a BC/E system.

$$F\big(x(t^-), u(t^-), v(t), x(t)\big) = 1 \ , \tag{1}$$

$$G\big(x(t), u(t), y(t)\big) = 1 \ , \tag{2}$$

$$H\big(x(t^-), x(t), v(t), z(t)\big) = 1 \ , \quad \text{with } t^- = \lim_{\Delta t \to 0} (t - \Delta t). \tag{3}$$

In [KK93] BC/E systems are represented and analysed using the software library XBOOLE [BS91]. XBOOLE uses so-called *ternary vector lists* (TVLs) as data structures for solution sets of Boolean functions. A TVL is a table of entries from the set {0, 1, −}, where "−" is the "don't care" symbol indicating that the appropriate value can be 0 and 1. A row of a TVL is called *ternary vector* and represents one or more binary vectors, depending on the number of "-"-components. A TVL is called *orthogonal*, if no binary vector is represented by more than one ternary vector. Orthogonality will be an useful notion for observability analysis of BC/E systems. XBOOLE offers various operations on TVLs, e.g. set algebraic operations or orthogonalization. A complete XBOOLE representation of a BC/E system consists of five SVs for the vectors u, v, x, y, z and four TVLs for the functions F, G, H and the set X_0.

3 Boolean Differential Calculus

The BDC can be regarded as the approach to establish a differential calculus for Boolean functions by keeping an analogy to the "classical" differential calculus for real-valued vector spaces. The initial work goes back to the 1950s and was performed by Akers [Ak59] and Talanzev [Ta59]. Since then the BDC has been developed into a well founded mathematical theory which is comprehensively represented in [Th81] and [BP81]. It has found its applications mainly in the field of switching circuits, but also in control [SW91]. The BDC includes a lot of mathematical notions, reaching for example from binary differences to vector derivatives. We will concentrate on the so-called *partial differential operations*. They are defined as follows:

Definition 1 [BP81]. Let $f(x)$ with $x = (x_1, \ldots, x_i, \ldots, x_n)$ be a Boolean function of n variables. Then

$$\frac{\partial f(x)}{\partial x_i} = f(x_1, \ldots, x_i, \ldots, x_n) \oplus f(x_1, \ldots, \overline{x_i}, \ldots, x_n), \tag{4}$$

is called the *partial derivative of f over x_i*, where $\oplus$ symbolizes the XOR or antivalence operation,

$$\min_{x_i} f(x) = f(x_1, \ldots, x_i, \ldots, x_n) \wedge f(x_1, \ldots, \overline{x_i}, \ldots, x_n), \tag{5}$$

is called the *minimum of f over x_i*, and

$$\max_{x_i} f(x) = f(x_1, \ldots, x_i, \ldots, x_n) \vee f(x_1, \ldots, \overline{x_i}, \ldots, x_n), \tag{6}$$

is called the *maximum of f over x_i*.

All three operators can be applied successively for different variables of f. This leads to the *k-times partial derivative, minimum and maximum over $x_1, \ldots, x_k$*. These operations are also available as XBOOLE functions.

4 Problem Formulation

We look at a special class of synthesis problems, called the *static controller synthesis problem for forbidden states specifications*. The major restriction is that the controller $\mathcal{C}$ has to be a static system. In the BC/E system framework a static system is exposed by the fact that it has only one state $x^c(t) \equiv x^c_{\text{only}}$ [1]. We also assume that $\mathcal{C}$ has no event input signals and therefore cannot generate or feed through events: $z^c \equiv 0$. Such systems are completely defined by their G-function (7). $\mathcal{C}$ is a permissive controller: it will not force events.

$$G^c(x^c_{\text{only}}, u^c(t), y^c(t)) = 1 \ . \tag{7}$$

We suppose, that the information about the current state of the plant $\mathcal{P}$ is carried by the condition input signals u^c. If these signals are not provided by the plant, they have to be produced by an additional BC/E system $\mathcal{O}$, the observer. We assume that $\mathcal{O}$ is a static system, too. For the plant $\mathcal{P}$ we distinguish between control input signals and disturbance input signals: $u^p(t) = (u^p_c(t), u^p_d(t)), \ v^p(t) = (v^p_c(t), v^p_d(t))$.

To analyse the controlled system, we have to connect $\mathcal{P}$, $\mathcal{C}$ and $\mathcal{O}$. Building new systems by connecting mutiple BC/E systems is described in [KK93]. A connection is specified by two mappings YUMAP and ZVMAP describing which output signal is connected to which input signal. Here, we choose the following mappings. Fig. 1 illustrates the resulting structure.

$$\text{YUMAP}: \ \left((u^p_d, u^p_c, y^p) = u^o, \ y^o = u^c, \ y^c = u^p_c\right), \ \ \text{ZVMAP}: \ \left(z^c = v^p_c\right). \tag{8}$$

We can now formulate our controller synthesis problem:

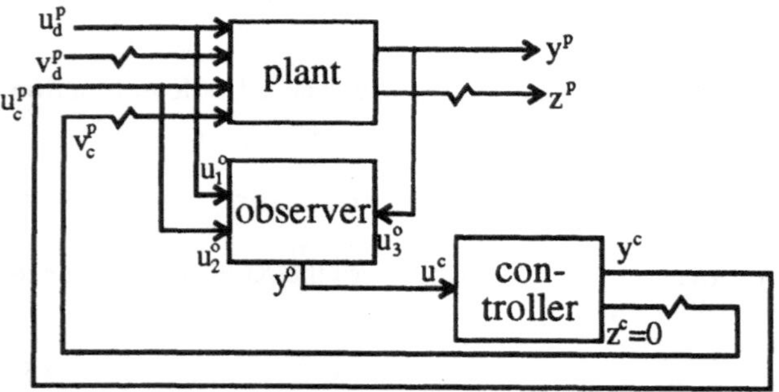

Fig. 1. Block diagram of control structure.

Given a BC/E system $\mathcal{P}$ and a set of forbidden states $X^p_f \subset X^p$:

1. *Find a static BC/E system $\mathcal{O}$, the observer, which will identify each possible state of the $\mathcal{P}$ by a distinct condition signal value.*

[1] We use superscripts to identify the systems: c for controller, p for plant, and o for observer.

2. *Find a static BC/E system C, the controller, according to (7), so that $\mathcal{P}$ as a subsystem of the system resulting from the connection according to (8), will never reach a forbidden state.*

In contrast to the BDC-based approach in [SW93], here, the concept of signals is used explicitly, and thus, the observable states and preventable transitions have to be determined and are not known a priori.

5 Observer Synthesis

Given the condition in- and output signal values, the state x^p is uniquely determinable, if $G^p(u^p, x^p, y^p) = 1$ can be transformed into $x^p = \gamma(u^p, y^p)$, with $\gamma : \{0, 1\}^{n_{up}} \times \{0, 1\}^{n_{yp}} \longrightarrow \{0, 1\}^{n_{xp}}$ being a unique mapping. The BDC offers a criterion for the solvability of this problem [BP81], but it requires $\min^k_{u^p, y^p} \overline{G}(u^p, x^p, y^p) = 0$, which means that at least one value of x^p corresponds to each possible value of $(u^p, y^p) \in \{0, 1\}^{n_{up}} \times \{0, 1\}^{n_{yp}}$. This is generally not the case for a BC/E system, because there may be impossible values of (u^p, y^p). If, for example, a valve is used to control a tank level, the signals u^p_{c1} ="valve open" and u^p_{c2} ="valve closed" can never be 1 both at the same time. For this reason we will use another criterion based on TVL representation: When G^p is represented by an orthogonal TVL, there will be exactly one state for each value of (u^p, y^p), if no value of (u^p, y^p) appears in more than one row, because otherwise the corresponding states in each of these rows would be different. This can be checked by the algorithm in Table 1, using again the orthogonality property.

Table 1. XBOOLE algorithm to check observability.

`G:=ORTH(G)`	First `G` is orthogonalized.
`Guy:=DCO(G,x)`	Delete columns corresponding to x.
`ans:=ORTHTEST(Guy)`	Test whether `Guy` is orthogonal. (The function `ORTHTEST` is not an XBOOLE function, but can be realized easily with the help of other XBOOLE functions.) If `ans`=yes, the state is determinable, and `G` represents the desired observer.

6 Controller Synthesis

For the controller synthesis we need to know which pure transitions (no self-loops) can happen independently from u^p_d and v^p_d and for $v^p_c \equiv 0$. They are represented by the function ϕ: (We abbreviate: $x^- = x(t^-)$ and $u^- = u(t^-)$.)

$$\phi(x^{p-}, u^{p-}_c, x^p) = \max^k_{v^p_c, v^p_d, u^{p-}_d} \left(F^p(x^{p-}, u^{p-}_c, u^{p-}_d, 0, v^p_d, x^p) \wedge \overline{F^p(x^p, u^{p-}_c, u^{p-}_d, v^p_c, v^p_d, x^p)} \right) . \quad (9)$$

We have to determine all transitions which cannot be disabled by appropriate values of u_c^p, i.e. those transitions which are a solution of ϕ for every possible value of u_c^p. If there are no restrictions on the possible values of u_c^p, the solution can be obtained easily by applying the minimum operator (prt stands for *preventable transitions*):

$$p_{\text{prt}}(x^{p-}, x^p) = \min_{u_c^{p-}}{}^k \phi(x^{p-}, u_c^{p-}, x^p) \ . \tag{10}$$

But in general we have to consider a restricted value set for u_c^p, as illustrated in §5 by the valve example. Thus, these values have to be added as solutions of ϕ before the minimum is performed:

$$p_{\text{prt}}(x^{p-}, x^p) = \min_{u_c^{p-}}{}^k \Big(\phi(x^{p-}, u_c^{p-}, x^p) \vee \Big(\max_{u_c^{p-}}{}^k \phi(x^{p-}, u_c^{p-}, x^p) \wedge \overline{\max_{x^{p-}, x^p}{}^k \phi(x^{p-}, u_c^{p-}, x^p)} \Big) \Big) . \tag{11}$$

We can now determine the extended forbidden states set X_{fe} (represented by $p_{fe}(x) = 1$) containing all states which are either forbidden or from which forbidden states can be reached via unpreventable transitions. This is done by reachability analysis (cf. [KK93]) backwards along the unpreventable transitions. At this point the following has to be valid: $x_0^p \notin X_{fe}$. Otherwise there exists no controller matching the specification.

The admissable plant behaviour under control is given by all transitions beginning and ending in states $x^p \notin X_{fe}$ and being a solution of ϕ. Building the maximum over x^p and renaming variables yields a controller output function G^c:

$$\begin{aligned} G^c(x^c_{\text{only}}, u^c, y^c) &= G^c(x^c_{\text{only}}, y^o, u_c^{p-}) = G^c(x^c_{\text{only}}, x^{p-}, u_c^{p-}) \\ &= \max_x{}^k \left(\phi(x^{p-}, u_c^{p-}, x^p) \wedge \overline{p_{fe}(x^{p-})} \wedge \overline{p_{fe}(x^p)} \right) \ . \end{aligned} \tag{12}$$

G^c determines all admissible control inputs u_c^{p-}. In general, there is not a unique u_c^{p-} for each x^{p-}, and arbitrary values have to be chosen. It also has to be checked, whether G^c presents at least one control input for every reachable state of the controlled plant. If this is the case, G^c is the desired static controller.

7 Example

We consider a batch reactor as an example. It is equipped with a cooling device which can be switched on and off. As a safety feature, a stopper liquid can be dropped in, which will bring the reaction immediately to an end. The reaction is exothermal and can explode, if the cooling is not sufficient. It will die down irreversably, if the temperature gets too low. Table 2 shows the signals and states (incl. the forbidden ones). The TVLs for the BC/E model and the algorithm results are presented in Table 3. The algorithm of Table 1 generates a TVL G′, which is orthogonal. Thus, the desired observer exists and is represented by the TVL G^o. The controller synthesis algorithm first determines X_{fe} = {explosion, dead, temperature critical high}, and then generates the controller G^c, in which the "-" has to be replaced by an arbitrary value.

Table 2. Signals, states and variable names for the reactor example.

Signal vector	Components	XBOOLE variable name
u:	cooling on/off	`cool`
v:	stopper liquid drops in	`stop`
x:	exploding (forbidden)	`expl`
	temperature critically high	`crit`
	temperature high	`high`
	temperature nominal	`nom`
	temperature low	`low`
	reaction dead (forbidden)	`dead`
y:	measured temperature nominal or above	`anom`
	measured temperature nominal or below	`bnom`
	measured temperature critically high	`dang`
	measured temperature low	`tlow`
z	reaction dies	`dies`

Table 3. TVLs F, G and H of the BC/E model for the reactor (self-loops in F are omitted) and the resulting TVLs for observer and controller.

```
F: ech  d c                  G:                       H: erh  d
   xrinle o s eih  d            ech  d c abdt            xrinle ech  d s d
   pigooa o t xrinle            xrinle o nnal            pigooa xrinle t i
   lthmwd l o pigooa            pigooa o oono            lthmwd pigooa o e
   ------ - p lthmwd            lthmwd l mmgw            ------ lthmwd p s
   ====== = = ======            ====== = ====            ====== ====== = =
   010000 - 1 000001            100000 - 0000            010000 000001 - 1
   001000 - 1 000001            010000 - 1010            001000 000001 - 1
   000100 - 1 000001            001000 - 1000            000100 000001 - 1
   000010 - 1 000001            000100 - 1100            000010 000001 - 1
   000001 - 1 000001            000010 - 0101            000010 000100 - 0
   000010 0 0 000100            000001 - 0100            000100 001000 - 0
   000100 0 0 001000                                     001000 010000 - 0
   001000 0 0 010000                                     010000 100000 - 0
   010000 - 0 100000
   010000 1 0 001000
   001000 1 0 000100
   000100 1 0 000010
   000010 1 0 000001
```

```
G':               G^o: x yyy  y u uuuu     G^c: x ucu  u y
                       o oooyyo o oooo          c cccuuc c
   c abdt              o echood c abdt          o echccd c
   o nnal              n xrinle o nnal          n xrinle o
   o oono              l pigooa o oono          I pigooa o
   l mmgw              y lthmwd l mmgw          y lthmwd l
   = ====              = ====== = ====          = ====== =
   - 0000              1 100000 - 0000          1 001000 1
   - 1010              1 010000 - 1010          1 000100 -
   - 1000              1 001000 - 1000          1 000010 0
   - 1100              1 000100 - 1100
   - 0101              1 000010 - 0101
   - 0100              1 000001 - 0100
```

Acknowledgements

The author gratefully acknowledges helpful discussions with Prof. Bruce H. Krogh and Dr. Hans-Michael Hanisch. He also likes to thank Reiner Scheuring who provided an interpreter for the XBOOLE software.

References

[Ak59] Akers, S. B.: On a theory of Boolean functions. Journal of Soc. Indust. Appl. Math. **7**(4), (1959) 487–498.

[BP81] Bochmann, D., Posthoff, C.: Binäre dynamische Systeme. Oldenbourg-Verlag, Munich (1981), (German).

[BS91] Bochmann, D., Steinbach, B.: Logikentwurf mit XBOOLE. Verlag Technik, Berlin (1991), (German).

[KK93] Krogh, B. H., Kowalewski, S.: Boolean condition/event systems: computational representation and algorithms. Preprints IFAC 12th World Congress, Sidney, Australia (1993) 327–330.

[SW91] Scheuring, R. and H. Wehlan: On the design of discrete event systems by means of the Boolean Differential Calculus. Preprints 1st IFAC Symp. on Design Methods of Cont. Systems, Zurich, (1991).

[SW93] Scheuring, R. and H. Wehlan: Control of discrete event systems by means of the Boolean Differential Calculus. In *Balemi et al.: Discrete Event Systems: Modeling ad Control.* Birkhäuser Verlag, Basel (1993).

[SK91] Sreenivas, R. and B. H. Krogh: On condition/event systems with discrete state realizations. Discrete Event Dynamic Systems: Theory and Applications, **1**(2) (1993) 209–236.

[Ta59] Talanzev, A.: On the analysis and synthesis of certain electrical circuits by means of special logical operators. Automat. i. telemeh., vol.20 (1959). (Russian, referenced by [BP81])

[Th81] Thayse, A.: Boolean Calculus of Differences. Lecture Notes in Computer Science, vol. 101, Springer Verlag (1981).

The Petri Net Approach

Analysis of Autonomous Petri Nets with Bulk Services and Arrivals*

Manuel Silva and Enrique Teruel

Dpt. Ing. Eléctrica e Informática. CPS, Univ. Zaragoza
María de Luna 3, E-50015 Zaragoza, Spain
Fax: + 34 76 512932 e-mail: silva@cc.unizar.es

1 Introduction

The mathematical description of reality is one of the basis of the impressive developments of technology. In the case of concurrent systems, a formalism able to adequately mathematize the behaviour of a collection of agents possibly interacting, cooperating and competing in performing their own tasks is crucial. *Petri nets* are a mathematical formalism to deal with concurrent systems, in the same way that differential equations deal with continuous dynamical systems. More precisely, Petri nets are a graph based formalism specially well suited to model and analyze *discrete event dynamic systems (DEDS)* which exhibit parallel evolutions and whose behaviours present synchronization and resource sharing phenomena. They have been used in a wide range of fields, including communication networks [21], computer systems [2, 55] and discrete part manufacturing systems [22, 56].

Although they can represent very complex behaviours, they consist of a few simple objects, relations and rules. A net model consists of two parts: (1) A *net structure*, an inscribed bipartite directed graph, that represents the static part of the system; and (2) a *marking*, representing a distributed state on the structure. The places correspond to the state variables of the system and the transitions to their transformers, or actions. The inscriptions may be very different, leading to different families of nets. If the inscriptions are simply natural numbers, named weights, *Place/Transition (P/T) nets* are obtained. In this case, the weights permit the modelling of bulk services and arrivals, while the marking represents the values of the state variables. A historically and conceptually interesting subclass of P/T nets is obtained when all weights are one. The nets are said to be *ordinary*, and they represent a straightforward but important generalization of automata models. Although ordinary and weighted nets have the same theoretical expressive power, weighted ones produce more compact and natural models of many systems. More elaborate inscriptions on the net structure and marking lead to the so called *High Level Petri Net (HLPN)* formalisms [31]. For finite state space (i.e. bounded) net systems, high level descriptions allow more compact models, but again the theoretical

* This work is supported by the projects CICYT ROB-91-0949, and Esprit BRA Project 7269 (QMIPS) and W.G. 6067 (CALIBAN).

modelling power is the same of ordinary P/T nets. For the purpose of this work, which focuses on basic concepts, the P/T level is considered as the reference level.

The intricate behaviours that can be described by P/T net systems make crucial the consideration of well founded *analysis* and *synthesis* techniques. The efficient treatment of P/T models may be very difficult, but for a few subclasses it is reasonably well understood and efficiently solved. Subclasses can be defined either restricting the behaviour or the structure. A possible way of obtaining *structural subclasses* is restricting the inscriptions (e.g. ordinary nets are a subclass of P/T nets) or the topology, usually aiming at limiting the interplay of conflicts and synchronizations. Historically, subclasses of ordinary nets have received special attention. Powerful results were early obtained for them [17, 28] originating an active branch of net theory [5, 6, 20, 25, 35, 63].

Graphs allow different *interpretations*. Among many others, the following are classic graph interpretations: state diagrams and algorithmic state machines representing sequential switching systems, state transition rate diagrams representing continuous time homogeneous Markov Chains, resource allocation graphs, PERT (Program Evaluation and Review Technique) graphs, etc. An interpretation precises the semantics of objects and their behaviour, eventually making explicit the connection of the model to the external world. As these classic graph models, Petri nets can be interpreted, leading to models describing different perspectives of a given system along its life-cycle. Nevertheless, for the purpose of this work, we mainly consider *autonomous* (i.e. uninterpreted) net models.

This survey is organized as follows: The formal definitions together with the description of some modelling features are contained in §2. Basic analysis methods are overviewed in §3. Among them, several structural techniques are further detailed in §4.

2 Modelling of Systems Using Nets

Formally, a P/T net is an inscribed bipartite directed graph: $\mathcal{N} = (P, T, F, W)$, where P and T are the disjoint sets of places and transitions, $F \subseteq (P \times T) \cup (T \times P)$ defines the support of the flow relation, and $W : F \mapsto \mathbb{N}$ are the inscriptions. The functions Pre, Post : $P \times T \mapsto \mathbb{N}$ can be defined as $\text{Pre}(p, t) = \textbf{if } (p, t) \in F \textbf{ then } W(p, t) \textbf{ else } 0$, and $\text{Post}(p, t) = \textbf{if } (t, p) \in F \textbf{ then } W(t, p) \textbf{ else } 0$. Thus, a P/T net can be alternatively defined as $\mathcal{N} = (P, T, \text{Pre}, \text{Post})$. The incidence functions are usually represented by matrices. If, without loss of generality, $\forall p \in P, t \in T : \text{Pre}[p, t] \cdot \text{Post}[p, t] = 0$, then the incidence matrix, $C = \text{Post} - \text{Pre}$, contains all the flow information. A P/T system is a pair $(\mathcal{N}, M_0)$ where $\mathcal{N}$ is a P/T net and $M_0 : P \mapsto \mathbb{N}$ is the initial marking, that can be represented in vector form. Transition t is enabled at marking M iff $M \geq \text{Pre}[\cdot, t]$. Being enabled it may fire, yielding a reached marking $M' = M + \text{Post}[\cdot, t] - \text{Pre}[\cdot, t]$. Firing a sequence σ of transitions from M yields $M' = M + C \cdot \vec{\sigma}$ (this is known as the *state equation* of the system), where $\vec{\sigma}$ is the firing count vector of sequence σ, i.e. $\forall t \in T : \vec{\sigma}[t] = \#(t, \sigma)$. The set of all reachable markings (states) from the initial one is denoted by $R(\mathcal{N}, M_0)$. (For more detailed presentations see [7, 40, 42, 49, 53].)

Net systems, like automata, are able to model sequential systems: transitions c and d (see the leftmost net in Fig. 1) fire in *sequence*, while a *conflicting* situation appears between transitions c and e. Combining sequences and decisions *iterations* are obtained:

fire e and then f until firing c. But net systems also permit modelling *true concurrency*: transitions b and c may fire in any order or even "simultaneously" (stronger concurrency notion that interleaving). Also the notion of self-concurrency can be represented: after firing a, the two tokens in place 2 permit two concurrent firings of transition b. The basic *synchronization* construct is a transition having more than one input place, such as d: the transition cannot fire until enough tokens have arrived to all the input places. But also arcs from a place to a transition with a weight other than one are a sort of synchronization, because several tokens must assemble in order to enable the transition. These net basic primitives allow to conveniently model typical synchronization schemas, like rendez-vous, semaphores, symmetric and asymmetric rendez-vous implemented by semaphores, fork-join, subprogram flow control, shared resources, guards, etc. [42, 53].

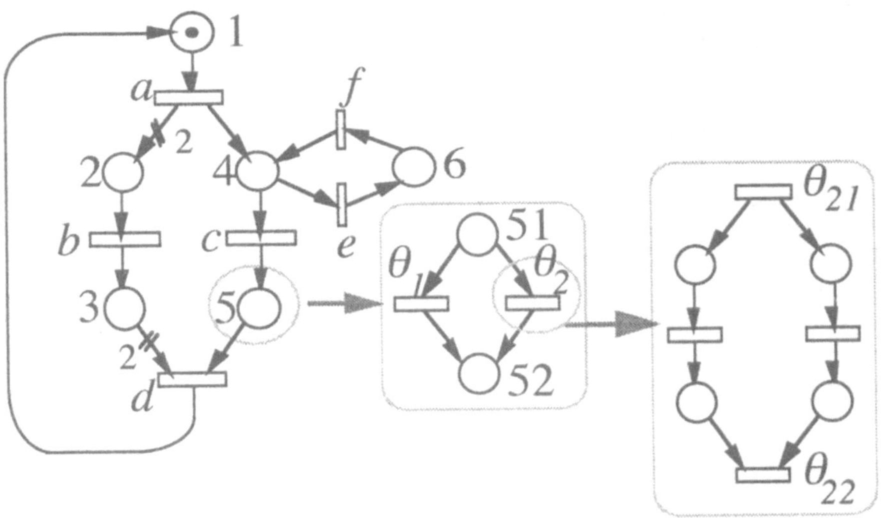

Fig. 1. A P/T net system. Illustration of state and action refinements.

Thanks to the bipartite nature of nets, state variables (places) and their local transformers (transitions) are considered on equal footing. This facilitates the consideration of diverse top-down and bottom-up design methodologies. Figure 1 shows how both state variables, like 5, and actions, like θ_2, can be *refined*. On the other hand, the leftmost net in Fig. 1 could have been built in a *compositional* way. For instance, it can be seen as the synchronization of two components: one defined by places $\{1, 2, 3\}$ together with their adjacent transitions and the other by $\{1, 4, 5, 6\}$ with theirs, merging the common transitions a and d. Or it can be seen as the fusion of two nets defined by their transitions: $\{a, b, c, d\}$ and $\{e, f\}$, merging the common place 4. The design of systems by refinements and compositions is surveyed in [3, 9], and is frequently used in practice [22, 68].

Despite their simplicity, net systems may model extremely complex and intricate behaviours. A classical approach to the solution of complex problems is to study restricted versions of them. This is the aim of defining subclasses. The typical restrictions which are imposed aim at limiting the interplay between synchronizations and conflicts. On one hand, these restrictions facilitate the analysis (we shall discuss some results for a few subclasses in §4). On the other, some modelling capabilities are lost. The designer must find a compromise between modelling power and availability of powerful analysis tools, while one of the theoretician's goals is obtaining better results for increasingly larger subclasses.

The relationship of the system modelled by a net to the external world is usually explicited by the interpretation, that gives a meaning to the abstract objects which form the net. A *state diagram* like interpretation leads to *marking diagrams*, which are well suited to model concurrent switching automatisms. For instance, a *Moore* marking diagram can be constructed associating events to transitions (the transition fires if it is enabled and the corresponding event is true) and unconditional actions to places (the action occurs whenever the associated place is marked). Another interesting family of interpretations is obtained when time and probabilities are associated to the model. For instance, some transitions may have a duration, and a probabilistic conflict resolution conflict may be specified. Timed and stochastic models are frequently used to evaluate performance issues. They can be seen as simple and systematic extensions of *Queueing Networks* (e.g. Queueing Networks with Fork/Join, with passive resources, etc.). This variety of interpretations yields a major advantage of modelling with nets: Net systems provided with appropriate interpretations can be used along the design and operation of systems, using a *single family of formalisms*, and basic concepts and results are shared (can be reused) by the different interpretations, leading to some economy in the analysis and synthesis of models.

In the sequel, only autonomous net systems will be considered.

3 Correctness Verification. An Overview

A major goal of the mathematization of systems is the *verification of their (functional) correctness*. Different concepts of correctness exist. Basically, *a system is said to be correct when two models, namely the* specification *and the* implementation, *are* equivalent *[9, 43, 66], or when the system exhibits a set of* desirable properties, *either expressed as formulas of a* logic *[13, 37], or selected from a given* kit *(e.g. boundedness, liveness, etc. See below)*. Henceforth, a possible criterion to classify analysis methods is the underlying correctness concept.

Another important criterion to classify analysis methods, specially in the case of concurrent systems, is whether the reasoning is performed directly on the description of the behaviour or not. *Behavioural* techniques consider the reachable states, firable sequences, etc. For bounded systems, this typically allows to decide correctness, but leads to extremely inefficient algorithms due to the *state space explosion problem* [29, 33, 42]. On the other hand *structural* techniques try to obtain information about the behaviour by reasoning on the underlying structure and the initial configuration of resources. This approach typically produces more efficient algorithms but does not always permit deciding correctness, but for some particular subclasses and properties.

Among the properties that are usually part of the specification of (reactive) systems, we want to highlight a few, putting them in terms of P/T net systems. A P/T system is *bounded* when every reachable marking is finite, and it is *live* when every transition can ultimately occur from every reachable marking. Boundedness is necessary whenever the system is to be implemented, while liveness is often required to preclude the existence of actions that can become permanently disabled, or dead, even if there is still some activity in the system. They are so important that the name *well-behaved* has been coined for live and bounded systems. A net $\mathcal{N}$ is *structurally (str.) bounded* when $(\mathcal{N}, M_0)$ is bounded for *every* M_0 and it is *str. live* when *there exists* an M_0 such that $(\mathcal{N}, M_0)$ is live. Consequently, if a net $\mathcal{N}$ is str. bounded and str. live there exists some marking M_0 such that $(\mathcal{N}, M_0)$ is well-behaved. In such case, non well-behavedness is exclusively imputable to the marking, and we say that the net is *well-formed.*

Other interesting properties follow. A weaker property related to liveness is *deadlock-freeness*, i.e. existence of at least one enabled transition at every reachable marking. A reachable marking M is a *home state* in $(\mathcal{N}, M_0)$ iff it is reachable from every reachable marking, and $(\mathcal{N}, M_0)$ is *reversible* iff M_0 is a home state. Finally, transition firing dependencies can be expressed and even measured by a series of *synchronic concepts* [50, 54].

Conventionally, analysis methods of autonomous Petri net models are classified as follows [7, 40, 49]:

Enumeration Techniques. The first step is to compute the *reachability graph*: nodes represent the markings while there is an arc labelled t from M_1 to M_2 whenever firing a transition labelled t which is enabled at marking M_1 yields the marking M_2. If the system is bounded, the construction of the reachability graph can be regarded as an exhaustive simulation of the system, which can be used as the *computational model* for a *proof system*, which verifies formulas from a given (temporal) logic, or for *decision procedures* and *tools for automatic verification* [13, 37]. A major problem of this approach is the size of the state space of a concurrent system. To palliate the state space explosion problem, techniques to disregard markings which are not interesting for a given analysis [64, 65] or to take advantage from symmetries [48] have been proposed, obtaining significant practical improvements on the performance of these methods. It is also possible to deal with unbounded systems by using the *coverability graph* [27, 34], but this construction does not allow to decide important properties such as liveness and reachability [42].

Structural Techniques. The basic idea is to obtain useful information about the behaviour reasoning on the structure of the net and the initial marking. Two crucial advantages of this approach are the deep understanding of the system behaviour that is gained, and the eventual efficiency of the algorithms. Two intimately related families of techniques have extensively been used:

Graph Theory. Objects like paths, circuits, handles, bridges, siphons, traps, etc. and their relationships are investigated. Typically only ordinary nets are considered, and the major results are obtained for particular properties, mainly boundedness, liveness and reversibility, and particular subclasses [5, 17, 24, 28, 63].

Linear algebra and convex geometry. Let $(\mathcal{N}, M_0)$ be a P/T system with incidence matrix C. The state equation gives a *necessary* condition for a marking to be reachable: a vector $M \in \mathbb{N}^{|P|}$ such that $\exists \vec{\sigma} \in \mathbb{N}^{|T|} : M = M_0 + C \cdot \vec{\sigma}$ is said to be *potentially reachable*, denoted by $M \in PR(\mathcal{N}, M_0) \supseteq R(\mathcal{N}, M_0)$. The *potential reachability graph* is defined like the reachability graph, but the set of nodes is now $PR(\mathcal{N}, M_0)$ instead of $R(\mathcal{N}, M_0)$. From the state equation it is easy to see that annullers of the incidence matrix lead to linear marking invariants and potential cyclic behaviours (firing invariants), which are useful for reasoning on the behaviour of the system. As an example, a well-known polynomial necessary condition for well-formedness, based on purely structural properties, is *conservativeness* and *consistency*, i.e. existence of positive left and right annullers of the incidence matrix. (For convenience sake, a conservative and consistent net will be called *well-structured*.) These invariants can be used to prove properties like boundedness, mutual exclusion, liveness, etc. More generally, the state equation can be used as a basic description of the system in order to prove or disprove the existence of markings or firing sequences fulfilling some given conditions, eventually expressed as logic formulas [14, 23]. Typically results for general P/T net systems are obtained [15, 36, 39, 59], some of which become specially powerful when restricted to particular subclasses combining graph theory based arguments [11, 25, 58, 60, 61, 62].

Transformation techniques. To facilitate the analysis of a large and complex system it can be transformed (typically reduced) preserving the properties to be analyzed [4, 16, 49]. Transformation rules somehow preserve the behaviour while they are often supported by structural arguments as simple, and efficient, sufficient conditions.

We have outlined the *analysis approach* to the design of correct systems, consisting on verifying correctness by different techniques. It encompasses a "trial and error" methodology. We want to mention to conclude this section that an interesting alternative is the *synthesis approach*, generating correct models by construction (either implementing a given specification or satisfying some desirable properties) by the application of *refinement* and *composition* transformations [38]. Behaviour-based equivalence notions are important to apply these principles [9, 43, 66], but also structural techniques can be used. For instance, a top-down methodology is easily derived reversing the application of reduction rules, and bottom-up methodologies can be obtained by careful study of structural properties. Again restricted subclasses of systems permit the best results [25, 26, 68].

4 Structural Analysis of P/T Net Systems

We concentrate on the structural analysis of selected properties, but we also consider semidecision of properties expressed as formulas of a first order logic.

In order to be able to deal with non ordinary P/T net systems, the use of linear algebra based techniques appears to be more convenient than the use of conventional graph theoretic tools because the matrix representation is exactly the same for ordinary and non ordinary nets, while important graph objects, like siphons and traps, are meaningful only in the ordinary case. Nevertheless the intimate relationship of both families of techniques is often used through the technical development of results, and some objects that were originally conceived from the graph theoretic perspective for ordinary nets can be generalized and are useful for weighted nets [62]. Conversely, results originally conceived from a linear algebraic perspective can be re-interpreted in graph theoretic terms, perhaps leading to more efficient algorithms for particular classes of ordinary nets.

The aim of this section is to overview some of the results that have been recently obtained for non ordinary P/T net systems, either by the use of linear algebra and convex geometry based arguments alone, or by the use of linear algebra combined with graph arguments. The material in this section is organized in two subsections: the first one deals with the analysis of general P/T net systems, and the second is concerned with the analysis of some subclasses.

4.1 Structural Analysis of General P/T Net Systems

The key idea leading to a structurally based verification of properties of a general P/T system is simple, and it has been already commented in the previous section: Let $(\mathcal{N}, M_0)$ be a P/T system with incidence matrix C. If M is reachable from M_0 by firing sequence σ, then $M = M_0 + C \cdot \vec{\sigma}$. Therefore the set of natural solutions, $(M, \vec{\sigma})$, of the state equation $M = M_0 + C \cdot \vec{\sigma}$ defines $PR(\mathcal{N}, M_0)$. This set can be used to analyze properties like marking and submarking reachability and coverability, firing concurrency, conflict situations, deadlock-freeness, mutual exclusion, k-boundedness, existence of frozen tokens, synchronic relations, etc. To do so, the properties are expressed as formulas of a first order logic having linear inequalities as atoms, where the reachability or firability conditions are relaxed by satisfiability of the state equation. This formulas are verified checking existence of solutions to systems of linear equations that are automatically obtained from them [14]. For instance, if $\forall M \in PR(\mathcal{N}, M_0) : M[p] = 0 \vee M[p'] = 0$ then places p and p' are in mutual exclusion. This is verified checking absence of (natural) solutions to $\{M = M_0 + C \cdot \vec{\sigma} \wedge M[p] > 0 \wedge M[p'] > 0\}$. *Integer Linear Programming Problems* [41] where the state equation is included in the set of constraints can be posed to express optimization problems, like the computation of marking bounds, synchronic measures, etc. [14, 54]. For instance, an upper bound for the marking of place p is $M[p] \leq \max\{M[p] \in PR(\mathcal{N}, M_0)\}$. This approach is a generalization of the classical reasoning using linear invariants [36, 39], and it deeply bridges the domains of net theory and convex geometry resulting in a unified framework to understand and enhance structural techniques [14].

Unfortunately, it usually leads to only semidecision algorithms because, in general, $R(\mathcal{N}, M_0) \subset PR(\mathcal{N}, M_0)$. The undesirable solutions are named *spurious*, and they may be removed by the application of certain structural techniques, consequently improving the quality of the linear description of the system [16]. Anyway the algorithms do decide in many situations, and they are relatively efficient, specially if the integrality of variables is disregarded. (This further relaxation may spoil the quality, although in many cases it does not [20, 54].)

In some cases the formulas are complex, like in deadlock-freeness analysis. The formula to express that a marking is a deadlock consists of a condition for every transition expressing that it is disabled at such marking. This condition consists of several inequalities, one per input place of the transition (expressing that the marking of such place is less than the corresponding weight) linked by the "$\vee$" connective (because lack of tokens in a single input place disables the transition). This can be verified checking absence of solutions to a certain (often large) set of linear systems. Nevertheless it is possible to perform simple net transformations preserving the set of solutions to the state equation and reducing the number of systems to solve, basically by obtaining an equivalent wrt. deadlock-freeness net system where the enabling conditions of transitions can be expressed linearly [59]. As a result, deadlock-freeness of a wide variety of net systems can be proven by verifying absence of solutions to a single system of linear inequalities. Even more, in some subclasses it is known that there are no spurious solutions being deadlocks, so the method decides on deadlock-freeness [61].

In temporal logic terms, the above outlined approach is well suited for *safety* properties ("some bad thing never happens"), but not so much for *liveness* properties ("some good thing will eventually happen"). For instance, the formula expressing reversibility would be $\forall M \in PR(\mathcal{N}, M_0) : \exists \vec{\sigma}' \gneq 0 : M_0 = M + C \cdot \vec{\sigma}'$, but this is neither necessary nor sufficient for reversibility. The general approach to linearly verify these liveness properties is based on the verification of safety properties that are necessary for them to hold, together with some inductive reasoning [32]. For instance, deadlock-freeness is necessary for transition liveness, and the existence of some *decreasing potential function* proves reversibility [51].

Another important contribution of linear techniques to liveness analysis has been the derivation of *ad hoc* simple and efficient semidecision conditions. We have already mentioned that conservativeness and consistency are necessary for well-formedness. This condition has been significantly improved by adding a rank upper bound, which was originally conceived when computing the *visit ratios* in certain subclasses of net models [10]. To state this important result, some conflict relations between transitions must be defined first. Two transitions, t and t', are in *Choice relation* iff $t = t'$ or they share some input place. This relation is not transitive. Its transitive closure is named *Coupled Conflict relation*, and the equivalence classes in T are *Coupled Conflict sets*. Another stronger equivalence relation than this is the *Equal Conflict relation*: $t \sim t'$ iff $t = t'$ or $\text{Pre}[P, t] = \text{Pre}[P, t'] \neq 0$. Note that whenever a transition t is enabled, all the transitions in the equivalence class (or *Equal Conflict set*) of t are also enabled. A slightly modified statement of the result in [15] is: *if a* P/T *net $\mathcal{N}$ is well-formed, then $\mathcal{N}$ is well-structured and* $\text{rank}(C) < |\mathcal{E}|$, *where $\mathcal{E}$ is the set of all Equal Conflict sets of $\mathcal{N}$.* This condition has been proven to be also sufficient for some subclasses of nets [18, 62]. The proof in [18], for (ordinary) Free Choice nets, uses elements of the classical theory (Commoner's theorem and the decomposition theorem), so it was of little help to work out non ordinary subclasses, but remarkably it lead to a general *sufficient* condition for ordinary P/T nets [19]: *if an ordinary* P/T *net $\mathcal{N}$ is well-structured and* $\text{rank}(C) < |\mathcal{C}|$, *where $\mathcal{C}$ is the set of all Coupled Conflict sets of $\mathcal{N}$, then $\mathcal{N}$ is well-formed.* Observe that, even for ordinary nets, we do not have a complete characterization of well-formedness. Since $|\mathcal{C}| \leq |\mathcal{E}|$, there is still a range of nets which satisfy neither the necessary nor the sufficient condition for well-formedness!

4.2 Structural Analysis of Subclasses of P/T Net Systems

The development of the structure theory for subclasses of (weighted) P/T net systems is somehow parallel to that for the ordinary case. The syntactical restrictions to facilitate the analysis aim at limiting the interplay between synchronizations and conflicts. The first results were obtained for subclasses where conflicts are structurally forbidden, namely *Weighted T-systems (WTS)* [58] and, more generally, *Choice-free (CF) systems* [60], which are nets where every place has at most one output transition. (The *reverse-dual* subclass, interchanging places and transitions and reversing arcs, forbids synchronizations, and its purely structural results can be obtained automatically transforming those for CF nets.) The next step was allowing both synchronizations and conflicts in a restricted fashion, such that choices are autonomous or free, thus generalizing (Extended) Free Choice net systems. The formal definition of *Equal Conflict (EC) systems* [61] requires that all Coupled Conflict sets are Equal Conflict sets, i.e. $\mathcal{E} = \mathcal{C}$. Basically we are interested in simple and efficient structural characterizations of well-formedness and well-behavedness, that is, correctness of the net and the system respectively. We also consider home states and other properties.

Choice-free Systems [60]. These nets strictly generalize the ordinary *Marked Graphs (MG)* or *T-systems* [5, 17], where every place has one input and one output transition, in two ways: weights and attributions are allowed (hence both restrictions on the inscriptions and the topology are relaxed). They still enjoy a crucial property indicating the absence of undeterminism: they are *persistent* [35] for any initial marking. For str. bounded nets, determinism is reflected at the structural level by *the existence of a unique minimal T-semiflow in strongly connected CF nets*. Well-formedness is characterized easily: *a CF net is well-formed iff it is strongly connected and consistent*, while MG are well-formed iff they are strongly connected, because they are always consistent.

The characterization of liveness of a MG in terms of the circuits which compose the net has its analogue in the case of CF systems, where more complex subnets than circuits have to be considered, but it is not so efficient. Nevertheless *it suffices for well-behavedness (and to find a home state) to fire a sequence greater than or equal to the minimal T-semiflow*.

There are many other results for this subclass. Some of them are technical developments to be used in larger subclasses that can be seen as composed by CF subnets. Some others are interesting on their own. For instance, it is proven that all spurious solutions of a well-behaved and reversible CF system can be removed by adding certain linear constraints to the state equation, so *the reachability set can be described by a system of linear inequalities over the naturals*. This result provides a unified framework for verifying properties of reversible and well-behaved CF systems using the approach outlined in §4.1, which in this case *decides* because no spurious solutions are left. A slight adaptation of this result leads to a method to compute optimal initial markings making the system well-behaved and reversible, plus other custom (linear) constraints. This result, together with liveness monotonicity on the marking, provides a systematic way to construct well-behaved systems from well-formed nets.

Historically, a more restricted subclass than CF systems was considered first, namely WTS [58], where only the inscriptions are generalized wrt. MG. In that case some further results hold. In particular, every well-behaved WTS is reversible, what extremely simplifies the linear characterization of the state space because it leads to absence of spurious solutions to the state equation of a well-behaved WTS.

Equal Conflict Systems [61, 62]. These nets strictly generalize the ordinary *Extended Free Choice (EFC) systems* [5, 28], where every two transitions sharing some input place have exactly the same pre-incidence function, by allowing arbitrary weights except for the output arcs from choice places, whose weights must be equal to keep choice freeness.

Well-formedness is characterized in polynomial time by well-structuredness and the rank condition $\text{rank}(C) = |\mathcal{E}| - 1$. Assuming well-formedness, well-behavedness is characterized in terms of certain subnets which are the analogue of P-components [62]. Unfortunately this condition cannot be checked as efficiently as in the EFC case, where it reduces to prove that there is no minimal P-semiflow whose support is unmarked by the initial marking. Anyway, liveness of a strongly connected and bounded EC system is equivalent to deadlock-freeness [61], so it can be decided using the approach outlined in §4.1 [59].

Other important results for this subclass are: Existence of home states for well-behaved EC systems [61], which generalizes the result in [6] using state equation based arguments. Decomposition and duality theorems for well-behaved systems [62], which generalize the results in [28, 63] after the generalization of the concepts of *components* and *allocations*.

Extensions: Modular Subclasses. Once some basic subclasses are reasonably well understood, their properties can be used for the analysis and synthesis of other subclasses which can be considered as *essentially* similar to them. Among such subclasses, specially challenging and appealing candidates are those which are defined in a *modular* way, i.e. as the composition of smaller models, because this is a most desirable feature from a system design perspective [3]. With respect to the expressiveness versus manageability compromise that was mentioned in §2, the first has often been the motivation for the definition of important modular subclasses (see [3, 68]). At present, some classes of models that have been considered, and for which there exists a rather rich structure theory, are: EC systems, after their decomposition results [62]. We hope that a synthesis theory analogue to that for EFC [25] will be developed soon. (Increasingly larger versions of) *Deterministic Systems of Sequential Processes* [44, 57, 11], where sequential processes cooperate by restricted message passing. *Systems of Simple Sequential Processes with Resources* [26], where restricted sequential processes compete for shared resources represented by monitor places.

5 Conclusion

Our intention was to briefly survey a rich area of net theory, namely the analysis of autonomous models, where the variety of approaches is perhaps the most salient feature. We concentrated on structural methods, which are a peculiar approach originated within the Petri nets field.

Generally speaking, the more restricted the systems the stronger the results. This principle motivates the study of subclasses, where expressiveness is sacrificed to obtain powerful tools. The study of subclasses had been restricted mainly to ordinary systems, where the representation of bulk services and arrivals is cumbersome. We described several extensions to weighted systems, which have larger expressive power than their ordinary counterparts.

The analysis of autonomous models is interesting on its own, but also it is important for the study of interpreted models. In fact, this nice feature of net models that they can be used with different interpretations has led to mutual beneficial influence between functional and performance analysis [10, 11, 12, 52], and it promises to keep on doing so.

References

1. M. Ajmone Marsan, editor. *Application and Theory of Petri Nets 1993*, volume 691 of *Lecture Notes in Computer Science*. Springer Verlag, 1993.
2. M. Ajmone-Marsan, G. Balbo, and G. Conte. *Performance Models of Multiprocessor Systems*. MIT Press, 1986.
3. L. Bernardinello and F. DeCindio. A survey of basic net models and modular net classes. In [46], pages 304–351.
4. G. Berthelot. Checking properties of nets using transformations. In G. Rozenberg, editor, *Advances in Petri Nets 1985*, volume 222 of *Lecture Notes in Computer Science*, pages 19–40. Springer Verlag, 1986.
5. E. Best and P. S. Thiagarajan. Some classes of live and safe Petri nets. In [67], pages 71–94.
6. E. Best and K. Voss. Free Choice systems have home states. *Acta Informatica*, 21:89–100, 1984.
7. G. W. BRAMS. *Réseaux de Petri: Théorie et Pratique*. Masson, 1983.
8. W. Brauer, editor. *Net Theory and Applications*, volume 84 of *Lecture Notes in Computer Science*. Springer Verlag, 1979.
9. W. Brauer, R. Gold, and W. Vogler. A survey of behaviour and equivalence preserving refinements of Petri nets. In [45], pages 1–46.
10. J. Campos, G. Chiola, and M. Silva. Properties and performance bounds for closed Free Choice synchronized monoclass queueing networks. *IEEE Trans. on Automatic Control*, 36(12):1368–1382, 1991.
11. J. Campos, J. M. Colom, M. Silva, and E. Teruel. Functional and performance analysis of cooperating sequential processes. Research Report GISI-RR-93-20, DIEI. Univ. Zaragoza, Sept. 1993.
12. J. Campos and M. Silva. Steady-state performance evaluation of totally open systems of Markovian sequential processes. In M. Cosnard and C. Girault, editors, *Decentralized Systems*, pages 427–438. North-Holland, 1990.
13. E. M. Clarke, E. A. Emerson, and A. P. Sistla. Automatic verification of finite state concurrent systems using temporal logic specifications. *ACM Trans. Prog. Lang. Syst.*, 8:244–263, 1986.
14. J. M. Colom. *Análisis Estructural de Redes de Petri. Programación Lineal y Geometría Convexa*. PhD thesis, DIEI. Univ. Zaragoza, June 1989.
15. J. M. Colom, J. Campos, and M. Silva. On liveness analysis through linear algebraic techniques. In *Procs. of the AGM of Esprit BRA 3148 (DEMON)*, 1990.
16. J. M. Colom and M. Silva. Improving the linearly based characterization of P/T nets. In [45], pages 113–145.

17. F. Commoner, A. W. Holt, S. Even, and A. Pnueli. Marked directed graphs. *Jnl. Comput. System Sci.*, 5:72–79, 1971.
18. J. Desel. A proof of the rank theorem for Extended Free Choice nets. In [30], pages 134–153.
19. J. Desel. Regular marked Petri nets. Presented at "Graph Theoretical Concepts in Computer Science", Utrecht, 1993.
20. J. Desel and J. Esparza. Reachability in cyclic Extended Free Choice systems. *Theoretical Computer Science*, 114:93–118, 1993.
21. M. Diaz. Petri net based models in the specification and verification of protocols. In W. Brauer et al., editors, *Petri Nets: Applications and Relationships to Other Models of Concurrency. Advances in Petri Nets 1986, Part II*, volume 255 of *Lecture Notes in Computer Science*, pages 135–170. Springer Verlag, 1987.
22. F. DiCesare, G. Harhalakis, J. M. Proth, M. Silva, and F. B. Vernadat. *Practice of Petri Nets in Manufacturing*. Chapman & Hall, 1993.
23. J. Esparza. Model checking based on branching processes. In volume 668 of *Lecture Notes in Computer Science*, pages 613–628. Springer Verlag, 1993.
24. J. Esparza and M. Silva. Circuits, handles, bridges and nets. In [45], pages 210–242.
25. J. Esparza and M. Silva. On the analysis and synthesis of Free Choice systems. In [45], pages 243–286.
26. J. Ezpeleta and J. Martinez. Synthesis of live models for a class of FMS. In *Procs. of the IEEE Int. Conf. on Robotics and Automation*, pages 557–563, 1993.
27. A. Finkel. The minimal coverability graph for Petri nets. In [47], pages 210–243.
28. M. H. T. Hack. Analysis of production schemata by Petri nets. Master's thesis, MIT, 1972. (Corrections in *Computation Structures Note* 17, 1974).
29. M. Jantzen and R. Valk. Formal properties of Place/Transition nets. In [8], pages 165–212.
30. K. Jensen, editor. *Application and Theory of Petri Nets 1992*, volume 616 of *Lecture Notes in Computer Science*. Springer Verlag, 1992.
31. K. Jensen and G. Rozenberg, editors. *High Level Petri Nets*. Springer Verlag, 1991.
32. C. Johnen. Algorithmic verification of home spaces in P/T systems. In *Procs. IMACS 1988, 12th World Congress on Scientific Computation*, pages 491–493, 1988.
33. N. D. Jones, L. H. Landweber, and Y. E. Lien. Complexity of some problems in Petri nets. *Theoretical Computer Science*, 4:277–299, 1977.
34. R. M. Karp and R. E. Miller. Parallel program schemata. *Jnl. Comput. System Sci.*, 3:147–195, 1969.
35. L. H. Landweber and E. L. Robertson. Properties of conflict-free and persistent Petri nets. *Jnl. of the ACM*, 25(3):352–364, 1978.
36. K. Lautenbach. Linear algebraic techniques for Place/Transition nets. In W. Brauer et al., editors, *Petri Nets: Central Models and their Properties. Advances in Petri Nets 1986, Part I*, volume 254 of *Lecture Notes in Computer Science*, pages 142–167. Springer Verlag, 1987.
37. Z. Manna and A. Pnueli. The anchored version of the temporal framework. In J. W. de Bakker et al., editors, *Linear Time, Branching Time and Partial Order in Logics and Models of Concurrency*, volume 354 of *Lecture Notes in Computer Science*, pages 201–284. Springer Verlag, 1989.
38. K. M.Chandy and J. Misra. *Parallel Program Design*. Addison-Wesley, 1988.
39. G. Memmi and G. Roucairol. Linear algebra in net theory. In [8], pages 213–223.
40. T. Murata. Petri nets: Properties, analysis and applications. *Proceedings of the IEEE*, 77(4):541–580, 1989.
41. G. L. Nemhauser and L. A. Wolsey. *Integer and Combinatorial Optimization*. Wiley, 1988.
42. J. L. Peterson. *Petri Net Theory and the Modeling of Systems*. Prentice-Hall, 1981.
43. L. Pomello, G. Rozenberg, and C. Simone. A survey of equivalence notions for net based systems. In [46], pages 410–472.

44. W. Reisig. Deterministic buffer synchronization of sequential processes. *Acta Informatica*, 18:117–134, 1982.
45. G. Rozenberg, editor. *Advances in Petri Nets 1990*, volume 483 of *Lecture Notes in Computer Science*. Springer Verlag, 1991.
46. G. Rozenberg, editor. *Advances in Petri Nets 1992*, volume 609 of *Lecture Notes in Computer Science*. Springer Verlag, 1992.
47. G. Rozenberg, editor. *Advances in Petri Nets 1993*, volume 674 of *Lecture Notes in Computer Science*. Springer Verlag, 1993.
48. K. Schmidt. Symmetries of Petri nets. *Petri Net Newsletter*, 43:9–25, 1993.
49. M. Silva. *Las Redes de Petri: en la Automática y la Informática*. AC, 1985.
50. M. Silva. Towards a synchrony theory for P/T nets. In [67], pages 435–460.
51. M. Silva. Logical controllers. In *IFAC Symposium on Low Cost Automation*, pages 157–166. 1989.
52. M. Silva. Interleaving functional and performance structural analysis of net models. In [1], pages 17–23.
53. M. Silva. Introducing Petri nets. In [22], pages 1–62.
54. M. Silva and J. M. Colom. On the computation of structural synchronic invariants in P/T nets. In G. Rozenberg, editor, *Advances in Petri Nets 1988*, volume 340 of *Lecture Notes in Computer Science*, pages 387–417. Springer Verlag, 1988.
55. M. Silva and J. M. Colom. Petri nets applied to the modelling and analysis of computer architecture problems. *Microprocessing and Microprogramming. The EUROMICRO Jnl.*, 38(1-5):1–11, 1993.
56. M. Silva and R. Valette. Petri nets and flexible manufacturing. In G. Rozenberg, editor, *Advances in Petri Nets 1989*, volume 424 of *Lecture Notes in Computer Science*, pages 374–417. Springer Verlag, 1988.
57. Y. Souissi and N. Beldiceanu. Deterministic systems of sequential processes: Theory and tools. In *Concurrency 88*, volume 335 of *Lecture Notes in Computer Science*, pages 380–400. Springer Verlag, 1988.
58. E. Teruel, P. Chrzastowski, J. M. Colom, and M. Silva. On Weighted T-systems. In [30], pages 348–367.
59. E. Teruel, J. M. Colom, and M. Silva. Linear analysis of deadlock-freeness of Petri net models. In *Procs. of the 2nd European Control Conference*, volume 2, pages 513–518. North-Holland, 1993.
60. E. Teruel, J. M. Colom, and M. Silva. Modelling and analysis of deteministic concurrent systems with bulk services and arrivals. In *Procs. of the ICDDS '93*, pages 93–104. North-Holland, 1993.
61. E. Teruel and M. Silva. Liveness and home states in Equal Conflict systems. In [1], pages 415–432.
62. E. Teruel and M. Silva. Well-formedness and decomposability of Equal Conflict systems. Research Report GISI-RR-93-22, DIEI. Univ. Zaragoza, Nov. 1993.
63. P. S. Thiagarajan and K. Voss. A fresh look at Free Choice nets. *Information and Control*, 61(2):85–113, 1984.
64. A. Valmari. Stubborn sets for reduced state space generation. In [45], pages 491–515.
65. A. Valmari. Compositional state space generation. In [47], pages 427–457.
66. K. Voss. Interface as a basic concept for systems specification and verification. In [67], pages 585–604.
67. K. Voss et al., editors. *Concurrency and Nets*. Springer Verlag, 1987.
68. M. C. Zhou and F. DiCesare. *Petri Net Synthesis for Discrete Event Control of Manufacturing Systems*. Kluwer Academic Publishers, 1993.

Dependability and Performability Analysis Using Stochastic Petri Nets

Kishor S. Trivedi[1], *Gianfranco Ciardo*[2], *Manish Malhotra*[3], *Sachin Garg*[1]

[1] Department of Electrical Engineering, Duke University
[2] Department of Computer Science. College of William and Mary
[3] AT&T Bell Laboratories, Holmdel, NJ 07733
[4] 308 W. Delaware, Urbana, IL, 61801

1 Introduction

Performance, reliability, availability, and safety etc. are the key factors that need to be modeled to understand the dynamic behavior of a system. Reliability, availability, safety and related measures are collectively known as *dependability*. In classical modeling, performance of a system has been modeled independent of the faulty (or fault-tolerant) behavior of the system. In many computer and communication systems, however, even if some components fail, the system continues to function correctly, although at a degraded performance level. Analysis of Fault-tolerant systems from pure performance viewpoint tends to be optimistic since it ignores failure-repair, and gracefully degrading behavior of the system. On the other hand, pure dependability analysis tends to be too conservative since performance considerations are not accounted for [40]. To capture such behavior, a composite measure called *performability* has been defined by Beaudry [2] and Meyer [27, 28]. The approach is very useful in modeling graceful degradation of a system. Pure performance and reliability analyses are then special cases of performability analysis.

1.1 Modeling Methods

Factors which govern the choice of a suitable modeling method are: properties of the system being modeled, measures to be obtained, and the tradeoff between accuracy and ease of modeling. *Reliability block diagrams* and *fault trees* offer concise description and very efficient solution techniques and are primarily used for dependability evaluation of systems [43]. However, their use is limited because of their inadequacy in modeling the dependencies between system components [43]. *Task graphs* are widely used for performance analysis of concurrent programs [37]. The implicit assumption of unlimited resources, however, does not hold in many cases. *Product form Queueing Networks* can represent contention for resources. However, the product form is violated when concurrency within a job, synchronization, and server failures are considered.

Markov Models are free from such limitations. They are equally effective in modeling dependencies within a system, contention for resources, synchronization, concurrency, fault tolerance, and reconfiguration, etc. Furthermore, by considering an extension to Markov reward models, degradable performance can also be modeled. One major drawback is the largeness of their state space even for small systems, which results in tedious specification and inefficient solution techniques. For a comprehensive discussion on the hierarchy of dependability model types, the reader is referred to [19, 23].

1.2 Modeling with SPN

Stochastic Petri nets offer a concise, high-level representation for modeling a system's behavior. The reachability graph of the stochastic Petri net, which describes the dynamic behavior of the system, has been shown to be isomorphic to a continuous time Markov chain [29]. Thus their evaluation involves generating and solving the underlying CTMC. Methods for solving a CTMC are well known and software tools exist for automated generation and solution of the underlying CTMC, given the specifications of the net [15, 6]. It should be noted that although SPN's offer a concise specification framework, their modeling power is the same as that of Markov models.

The paper is organized as follows: In §2, we formally define a stochastic Petri net and mention some structural extensions that facilitate compact specification. In §3, we identify and illustrate typical behavioral properties of a system that can be modeled. In §4, we outline and define the types of analyses. The importance of reward specification to be integrated with the SPN is motivated. Typical dependability and performability measures are defined along with the respective reward assignments. A simple example illustrating the above procedure is given. In §5, we discuss the tractability of analytic model solution. Problems are identified and a brief summary of current research in avoiding and tolerating such problems is presented.

2 SPN-formalism

The basic SPN [15] is defined as a 7-tuple $\mathbf{A} = \{P, T, D^-, D^+, \mu_0, \lambda, w\}$ where:

- $P = \{p_1, ..., p_{|P|}\}$ is a finite set of places, each containing a non-negative number of tokens.
- $T = \{t_1, ..., t_{|T|}\}$ is a finite set of transitions ($P \cap T = \varnothing$).
- D^- is the set of input arcs $D^- \in \{0, 1\}^{|P \times T|}$. There is an input arc from place p to transition t iff $D^-_{p,t} = 1$.
- D^+ is the set of output arcs $D^+ \in \{0, 1\}^{|P \times T|}$.There is an output arc from transition t to place p iff $D^+_{p,t} = 1$.
- μ_0 is the initial marking. It defines the initial number of tokens at each place.
- $\forall t \in T, \lambda_t : \mathbb{N}^{|P|} \rightarrow \mathbb{R}^+ \cup \{\infty\}$ is the rate of the exponential distribution for the firing time of the transition t. A marking μ is said to be *vanishing* if there is a marking-enabled transition t in μ such that $\lambda_t(\mu) = \infty$; μ is said to be *tangible* otherwise. Define $\mathcal{T}$ and $\mathcal{V}$ as the sets of tangible and vanishing markings, respectively. Then, $\mathcal{T} \cup \mathcal{V} = \mathcal{S}$. where $\mathcal{S}$ is the reachability set of the SPN.
- $\forall t \in T, w_t : \mathbb{N}^{|P|} \rightarrow \mathbb{R}^+$ describes a weight assigned to a transition t, whenever its rate $\lambda_t = \infty$. The probability of firing transition t enabled in a vanishing marking μ is:

$$\frac{w_t(\mu)}{\sum\limits_{t_i : \mu \xrightarrow{t_i}} w_{t_i}(\mu)} .$$

This definition of SPN corresponds exactly to that os the GSPN by Marson et. al. [25]. If $\forall t \in T, \lambda_t^{-1} > 0$, then the reachability graph contains only tangible markings, i.e. $\mathcal{V} = \emptyset$, and $\mathcal{T} = \mathcal{S}$. The underlying stochastic process is then a CTMC whose state space is isomorphic to the reachability graph. If $\mathcal{V} \neq \emptyset$, the reachability graph needs to be transformed into a CTMC by elimination of the vanishing markings. The CTMC obtained is then solved for necessary analysis.

Structural extensions to the SPN/GSPN have been proposed in order to further facilitate conciseness of model specification [15]. These extensions do not enhance the modeling power of the SPN but they often result in more compact nets. These include guards, marking dependent arc multiplicities, a more general approach to the specification of priorities, and the ability to decide in a marking dependent fashion whether the firing time of a transition is exponentially distributed or zero. For a formal definition of the extended SPN, the reader is referred to [15].

3 Behavioral Modeling with Petri Nets

For correct functional model of a system, the SPN must correctly specify all relevant events, the *pre-conditions* that hold when an event occurs as well as the *post-conditions* that prevail after the event has occurred [33]. Some fundamental properties of events which have temporal or structural relations and that occur widely in many systems are identified and their SPN models are given below.

- *Dependency:* Firing of a transition t_p depends on firing of transitions t_i and t_j (Figure 1a). This property is widely made use of in software modeling [12, 39, 1]. The data flow, task graph models of computation are typical examples.

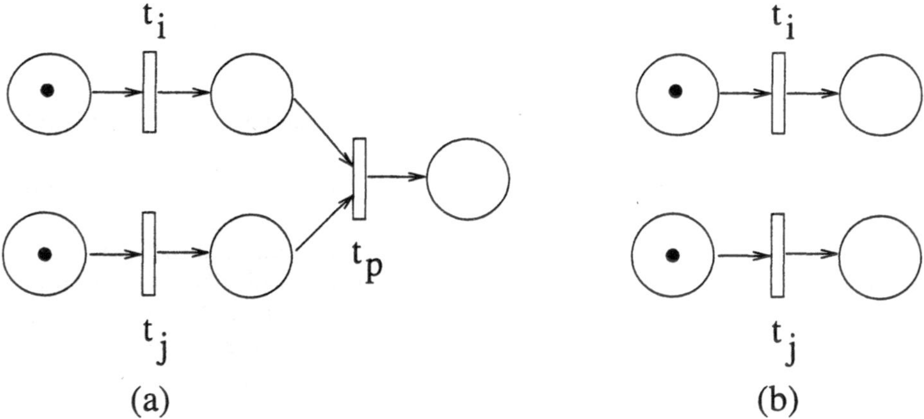

Fig. 1. Dependency and Concurrency

- *Concurrency:* Two or more events (transitions t_i and t_j in figure 1b) are simultaneously enabled and do not interfere or interact with each other. This property applies to modeling of multiprocessor systems [31], concurrent software [12, 1], and so forth.

- *Synchronization:* In a parallel program, concurrently executing tasks may need to synchronize after they finish. This behavior can be modeled by using an immediate transition t_0 that synchronizes the transitions $t_1, \ldots, t_n$. Figure 2(a) shows the synchronization model.

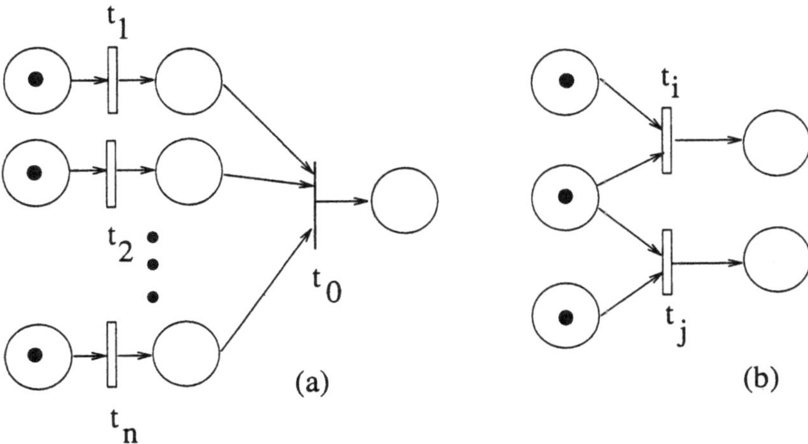

Fig. 2. Synchronization

- *Conflict:* In a conflict situation, two or more transitions are enabled at one time and firing of one disables the others. This property can be used in modeling contention for resources such as in client-server systems [7]. Figure 2(b) shows the model of a conflict situation.

4 SPN Analysis

4.1 Stochastic Reward Nets

In a Markov Reward Model(MRM), a real variable called the *reward rate* is associated with each state of the underlying state space. This technique facilitates the computation of performability measures of a system such as total work done by the system in a finite time interval [3]. This corresponds to computing the accumulated reward in that interval. Although CTMCs have equivalent modeling power, such analysis is cumbersome. Moreover, pure reliability analysis such as computing the time to system failure is a special case of evaluating the MRM with reward rate of 1 and 0 assigned to the system *"UP"* and *"DOWN"* states respectively [41]. A Stochastic reward net provides a high level specification for an MRM. In an SRN, reward rates may be attached to markings and impulse rewards may be associated with firing of transitions. We restrict ourselves to the former case in this paper.

4.2 Analysis of SRN

Modeling and analysis by an SRN is a 3 stage process.

- Specifying the system with a compact SRN (potentially by the use of marking dependent arc cardinalities, guards and other structural extensions) with assignments of transition rates for timed transitions, transition probabilities for immediate transitions, priorities and weights [26] to resolve conflicts among simultaneously enabled transitions,
- Generating the state space of the underlying MRM. This requires constructing the reachability graph and then transforming it to the corresponding MRM by eliminating the vanishing markings [25, 15]. An alternative method is to preserve the vanishing markings and convert the stochastic process into a DTMC, which is then solved. This approach will not be elaborated upon in this paper.
- Solving the generated MRM for the measures of interest. We elaborate on the analysis types and typical measures obtainable in the next 3 sub-sections.

Automated tools and techniques are available for stage 2 [15, 6, 25] and stage 3 [15].

4.3 Analysis Types

Analysis of a stochastic process can be classified in four categories.

1. Steady State Analysis:
2. Transient Analysis:
3. Cumulative Transient Analysis:
4. Sensitivity Analysis:

Let $\{\Theta(t), t \geq 0\}$ be a continuous-time finite-state homogeneous Markov reward model with the state space Ψ and reward r_i assigned to state i. Let Q be the infinitismal generator matrix and $P(t)$ be the state probability vector of the model, where $P_i(t)$ is the instantaneous probability that the MRM is in state i at time t. The transient behavior of the model, given the initial probability vector $P(0)$, is described by the Kolmogorov differential equation:

$$\frac{dP(t)}{dt} = P(t)Q \ . \tag{1}$$

The steady state analysis consists of evaluating $\pi = \lim_{t\to\infty} P(t)$ assuming that the limit exists. It is independent of the initial probability vector $P(0)$ and is given by:

$$\pi Q = 0, \quad \sum_{i\in\Psi} \pi_i = 1 \ . \tag{2}$$

Cumulative transient analysis consists of evaluating the total expected time spent in state i during the interval $[0, t)$. Let $L_i(t)$ denote this quantity. In vector terms, $L(t) = \int_0^t P(x)dx$. This quantity can be evaluated by solving:

$$\frac{dL(t)}{dt} = L(t)Q + P(0) \ . \tag{3}$$

Assume that the entries of Q are functions of some parameter vector θ. Sensitivity analysis consists of evaluating variation in the state probability vector with respect to the model parameters i.e. $\frac{dP(t)}{d\theta_i}$.

We are now in a position to identify typical dependability, and performability measures obtainable from the underlying MRM of the SRN.

4.4 Dependability Measures

In a dependability model, a reward rate of 1 is assigned to all the up (operational) states and a rate of 0 is assigned to all the down (failure) states. In the SRN specification, corresponding reward rates are assigned to markings that represent the up and down states. Let $\Upsilon(t) = r_{\Theta(t)}$ be the instantaneous reward rate of the underlying MRM. The accumulated reward in the time interval $[0, t)$ is given by:

$$\Phi(t) = \int_0^t \Upsilon(x)dx = \int_0^t r_{\Theta(x)}\, dx \ . \tag{4}$$

The instantaneous availability of the system is obtained as expected instantaneous reward rate at time t and is given by:

$$E[\Upsilon(t)] = \sum_{i\in\Psi} r_i P_i(t) \ . \tag{5}$$

and the expected availability when the system has reached steady state is given by $\lim_{t\to\infty} E[\Upsilon(t)]$:

$$E[\Upsilon_{ss}] = \sum_{i\in\Psi} r_i \pi_i \ . \tag{6}$$

The total time the system is available in the interval $[0, t)$ is computed as the expected accumulated reward in that interval and is given by:

$$E[\Phi(t)] = \sum_{i\in\Psi} r_i L_i(t) \ . \tag{7}$$

The interval availability (averaged over time) is given by $\frac{1}{t}E[\Phi(t)]$. If the underlying MRM has absorbing states, mean time to failure (MTTF) needs to be computed. With the binary reward assignment, MTTF is computed as the expected accumulated reward until absorption and is given by:

$$E[\Phi(\infty)] = \sum_{i\in\Psi_T} r_i \tau_i \ , \tag{8}$$

where $\tau_i = \lim_{t\to\infty} L_i(t)$. Often times, it is desirable to compute the distribution of accumulated reward. For instance, the distribution of time to complete a job that requires r units of time on a system can be computed by:

$$P[\Gamma(r) \leq t] = 1 - P[\Phi(t) < r] \ , \tag{9}$$

where $\Gamma(r)$ is the random variable that denotes the time to accumulate reward r.

Note that for computing reliability measures, such as *MTTF*, all the system down states must be absorbing states. Conversely, steady state measures can be computed only when none of the system states are absorbing states [40].

4.5 Performability Measures

Performance in the presence of failures can be evaluated by assigning appropriate reward rates to different states of the underlying MRM and using above equations. For example, consider a multiprocessing system, which has N processing units. Each of the unit is subject to failures. The failures occur independently with a rate λ. There is a single facility that does the repair with a rate μ. Figure 3(a) shows the SPN which models the system. The state of the underlying CTMC is defined as the number of processing units currently functioning. Figure 3(b) shows the underlying continuous time Markov chain, which is a finite birth-death process.

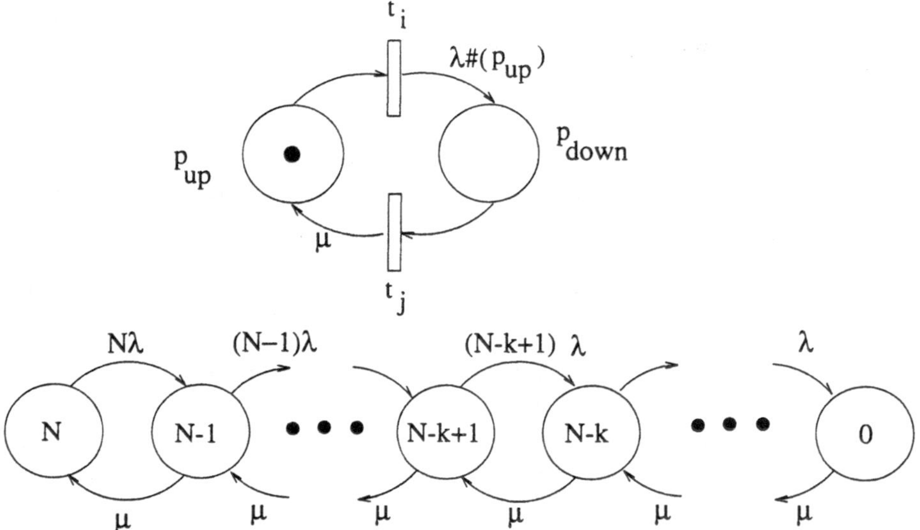

Fig. 3. SPN and underlying CTMC for the multiprocessor

The processing (service) rate of each unit is σ. Then the reward rate assigned to state k, $0 \leq k \leq N$ is $k\sigma$. In the SRN specification, reward associated with a marking equals $\sigma\#(p_{up})$, where $\#p_{up}$ is the number of tokens in the place p_{up} in that marking. To compute the total processing capacity available in time (0,t], we use Eq. 7. Instantaneous capacity is computed by Eq. 5 and steady state capacity by equation 6. Furthermore, if there is no repair facility, state 0 is the absorbing state. We can then calculate the total capacity delevered until absorption by Eq. 8.

5 Problems and Remedies in Model Solution

In this section, we describe the three main problems that arise in analytic modeling. They are *"largeness"* and *"stiffness"* of the model and the need to model *"non-exponential"* distributions. We also describe the methods to overcome these problems and give some simple illustrations.

5.1 Largeness

Solution of the underlying stochastic process is often faced with many problems. The simpler case of steady state analysis of a CTMC involves solving the system of linear equations $\pi = \pi Q$, under the constraint $\sum_i \pi_i = 1$. For modeling even a small "real-world" system, the state space is potentially of the order of hundred thousand, if not more. This poses difficulties in both specification and analysis. The specification problem is handled by the use of SPN framework, whereby the user specifies only the SPN and the underlying CTMC/MRM is automatically generated and stored for analysis. Other frameworks for this purpose also exist and are outlined in [17].

Largeness Avoidance

Largeness needs to be avoided when the model's state space is too large to be stored and efficiently solved. This is done through using approximation techniques such as truncation, lumping, decomposition and the use of fluid models. These techniques are discussed and illustrated below.

Truncation: Continuing with our example of the multiprocessor system, if the number of processing units is large and the failure rate is very small, i.e. $N >> \lambda$, then the probability that the system would reach the state k, $k << N$ is negligible. This fact can be used to eliminate all the states i, where $i < k$ from the MRM. The truncated MRM is shown in the figure 4.

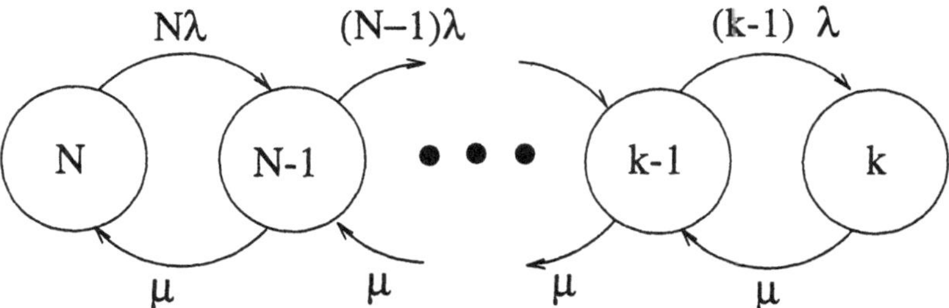

Fig. 4. strict truncation

Formally, given a reachability graph $(\mathcal{S}, \mathcal{A})$, the state truncation results in a truncated reachability graph $(\mathcal{S}', \mathcal{A}')$. If $(\mathcal{S}', \mathcal{A}') \subset (\mathcal{S}, \mathcal{A})$, the method is called "strict truncation".

Alternatively, $(\mathcal{S}', \mathcal{A}')$ might be a subgraph of $(\mathcal{S}, \mathcal{A})$ augmented with one or more states and arcs. In our example, in the truncated model, we might add a new state u and an arc from the state with k failed component to u. This new state corresponds to all further failures.

Note that u is an absorbing state. Figure 5 shows the new model. Since such an approach involves coalescing many states, it is called the "aggregation-truncation" method.

It is interesting to note that the above two approximations allow us to compute the lower and upper bounds on performability measures. For example, we can evaluate the expected instantaneous computational capacity for both models as follows:

Let the transient probability vectors for strict and aggregation truncated models be $P^s(t)$ and $P^a(t)$ respectively. The reward assignment for the strict truncated model is given by ρ^s, where $\rho^s_p = p\mu$, $N - k \leq p \leq N$. The reward assignment for aggregated truncated

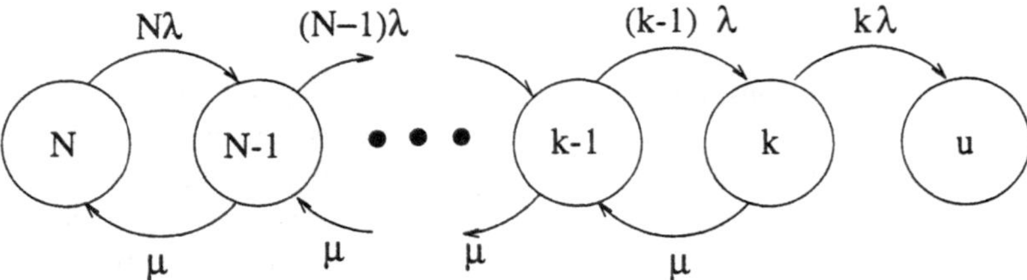

Fig. 5. Aggregation truncation

model is given by ρ^a, where $\rho^a_p = p\mu$, $N - k \leq p \leq N$ and $\rho^a_u = 0$. The expected instantaneous computational power(EICP) for the strict truncated model is given by

$$C^s(t) = \sum_{i \in \{N,\ldots,N-k\}} \rho^s_i P^s_i \ ,$$

and the EICP for the aggregated truncated model is given by

$$C^a(t) = \sum_{i \in \{N,\ldots,N-k,u\}} \rho^a_i P^a_i \ .$$

It can then be shown that

$$C^s(t) \geq C(t) \geq C^a(t) \ .$$

Note that discrete-event simulation, in a probabilistic sense, is a form of state truncation, since the most likely states are visited frequently while the unlikely states may not be visited at all.

Lumping: In our example, if we modify the failure/repair behavior to be different for each unit, then the state space of the model increases from N to 2^N. In this technique, we model each of the units independently. Each processor can be in one of the two states, *up* and *down*. These models are then composed together to obtain the lumped model. We describe the general algorithm to obtain the lumped state space for a system composed of N independent subsystems [15].

1. Generate the reachability graph for a single subsystem.
2. Transform this graph into a SPN as follows: for each marking i, add a place p_i, initially empty; for each arc from state i labeled by transition t, add a transition t_i with marking dependent rate equal to the number of tokens in place p_i times the rate of t_i in marking i for a single subsystem, an input arc from p_i to t_i, and an output arc from t_i to p_j.
3. Set the initial marking of the SPN as follows: For each sub-system, if its initial state is i, add a token to p_i.
4. Generate the CTMC/MRM underlying this SPN.

In general, if there are N subsystems with K states each, the size of the state space with and without lumping is

$$K^N vs. \binom{N+K-1}{K}.$$

Thus lumping always results in the state space reduction. It is substantial especially if N is large and K is small. For our multiprocessor example, the state space reduces from 2^N to $\frac{1}{2}N(N+1)$.

If the subsystems have interactions, then application of lumping requires care. If the interaction is limited to "rate dependence" [14] where the transition rates in a subsystem depend on the number of subsystems in certain states, but not on their identity, this algorithm can still be applied with a different specification of the firing rates for the resulting SPN.

Decomposition: When there is a large difference in rates of performance related events such as task/job arrival rate compared with the failure rates which are very rare, it is safe to make the assumption that the system attains a (quasi) steady state with respect to the performance related events between successive occurrences of failure events. Thus we can compute the performance measures for the system in each of these (quasi) steady states. The overall system can then be characterized by weighting these performance measures by the structure state probabilities. This technique is called *hierarchical* or compositional modeling. A detailed illustration is given in [30]. In general, instead of a monolithic model, a set of submodels are solved independently and the results are combined to account for the interactions among the submodels.

Depending on the type of interaction among the submodels hierarchical composition techniques differ not only in the way the model is constructed but also in the way it is solved . If a cyclic dependence exists among the submodels, a fixed point iteration scheme is used to obtain the models' solutions [38, 9, 14]. If a strict hierarchy exists, the solution techniques are non-iterative [36, 22]. For a unified view of different approaches to hierarchical composition, refer to [23].

Fluid Models: As the number of tokens in a place becomes large, the size of the underlying CTMC grows. It may be possible to approximate the number of tokens in the place as a non-negative real number. It is then possible to write the differential equations for the dynamic behavior of the model and in some cases, provide solutions. Trivedi and Kulkarni have proposed fluid stochastic Petri nets for this purpose [42].

5.2 Stiffness

Model stiffness is caused by extreme disparity between the transition rates. This is especially true of monolithic performability models where the transition rates for performance related events are far greater than failure occurrence rates. In a typical computer system, the rate of arrival of jobs may be as high as 10^9 times the failure rates.

A linear system of differential equations such as Eq. (1) is said to be stiff if the solution has components whose rates of change (decay or gain) differ greatly. The rate of change of each solution component is governed by the magnitude of an eigenvalue of the generator matrix Q. Formally, this system is considered stiff if for $i = 2, ..., m$, $\text{Re}(\lambda_i) < 0$ and

$$\max_i |\text{Re}(\lambda_i)| >> \min_i |\text{Re}(\lambda_i)| \ ,$$

where λ_i are the eigenvalues of Q. The rate of change of a solution component is defined relative to the solution interval. With a CTMC, the solution attains a numerical steady-state after some finite time t_{ss}. Stiffness in the CTMC context may be defined as: "A

CTMC characterized by a system of equations as in 1 is stiff in the interval $[0, t)$ if there exists a solution component of the system which has variation in the interval that is large compared to $1/\min\{t, t_{ss}\}$. The large difference in transition rates of the CTMC approximately translates into large differences in the magnitude of the eigen values of the generator matrix.

Stiffness causes computational problems which adversely affects the stability, accuracy and efficiency of a numerical solution method unless it is specifically designed to handle it. There are two basic methods to overcome stiffness: stiffness avoidance and stiffness tolerance.

Stiffness Avoidance: Stiffness avoidance is linked with largeness avoidance in the sense that the same approximation schemes can be applied. Hierarchical modeling is a potential stiffness avoidance technique. Bobbio and Trivedi [4] have designed an approximation technique based on aggregation.

Stiffness Tolerance: Special numerical methods for the solution of the system of differential equations is used in this approach. Two most common methods are *uniformization* and numerical *ODE* solution methods. Uniformization is advantageous because it involves only additions and multiplications. It is not subject to severe round-off errors like the Taylor series expansion of the matrix exponential. One of the main problems with uniformization is its $O(\eta q t)$ complexity that renders it inefficient [24, 34] for stiff Markov chains as a large value of qt characterizes stiffness. Here, $Q = [q_{ij}]$ is the generator matrix, $q \geq \max_i |-q_{ii}|$, and η is the number of non-zero entries in Q. A modified method has been proposed by Muppala and Trivedi [32] that uses detection of the steady state of the underlying DTMC. The nicety of this approach is that the computation time is controlled by the sub-dominant eigenvalue of the DTMC matrix

$$Q^* = \frac{Q}{q} + I \ ,$$

where $q > \max_i |q_{ii}|$, rather than by qt.

Uniformization methods, however, remain inefficient in comparison to the L-stable ODE methods [21]. Among these, second-order TR-BDF2 [35] is efficient for low accuracy requirements and third order implicit Runge-Kutta method [24] is efficient for high accuracy requirements.

5.3 Modeling Non-exponential Distributions

In modeling with SPN's, the assumption that the transition firing times have exponential distributions is a major constraint. Non-exponentially distributed processing durations often occur in many practical problems. For instance, deterministic times in network flow control, time outs in real time systems and lognormal distribution for repair times.

Phase Approximations: The basic idea in this approach is to replace a non-exponential distribution in a model by a set of states and transitions between those states such that the holding time in each state is exponentially distributed. Then the overall CTMC is generated and solved. The main advantage is that the approach is conceptually simple and the entire range of CTMC solution methods is applicable. Its main drawbacks are the increase in state space, fitting a non-exponential distribution to a set of exponential phases and the inaccuracy of fit, especially for deterministic durations. More discussion on this approach can be found in [5, 20].

Stochastic extensions: The DSPN (Deterministic and stochastic Petri net) allows for deterministic transition times in addition to exponentially distributed times. However, for the model to be analytically tractable, at most one deterministic transition may be enabled at any time. [10, 13] provide the solution equations for steady state and transient behavior of a DSPN model. [11] provides the method for parametric sensitivity analysis of such models. A class of Petri nets called Extended stochastic Petri nets (ESPN) have been proposed by Dugan et al. that allow generally distributed transition times [16]. Under some restrictions, the underlying stochastic process of an ESPN is a SMP and thus analytically solvable. In general, discrete event simulation is used to solve such models.

Choi et. al. recently proposed a new class of stochastic Petri nets called the Markov regenerative stochastic petri nets (MRSPN) [8]. The underlying stochastic process has been shown to be an MRGP. MRSPN class is a superset of SPN, GSPN, DSPN and ESPN classes. It allows for generally distributed transition times including zero firing delay. The analysis is based on the property that not all state changes need be regeneration points. Specifically, [8] shows that an SPN is an MRSPN if

- at most one *GEN*(generally distributed, timed) transition is enabled in a marking
- the firing of the *GEN* transition is sampled at the time the transition is enabled; the firing time distribution may depend upon the marking at the time the transition is enabled. Firing time cannot change until the transition either fires or is disabled.

The solution method for steady-state and transient analysis of MRSPN is given in [8] and the method for parametric sensitivity analysis is given in [18].

References

1. H.H. Ammar, S.M.R. Islam, and S. Deng. Performability analysis of parallel and distributed algorithms. In *Proceedings of Petri Nets and Performance Models*, pp. 240–248, kyoto, Japan, Dec. 1989.
2. M. Beaudry. Performance related reliability for computer systems. *IEEE Transactions on Computers*, C-27:540–547, June 1978.
3. Andrea Bobbio. Stochastic Reward Models in Performance/Reliability Analysis. In R. Puigjaner and D. Poitie (eds.) *Modeling Techniques and Tools for Computer Performance Evaluation*, pp. 353–365, Plenum Press, 1989
4. A. Bobbio and K.S. Trivedi. An aggregation technique for the transient analysis of stiff Markov chains. *IEEE Transactions on Computers*, C-35(9):803–814, Sept. 1986.
5. S.C. Bruel, P.Z. Chen, and G. Balboa. Alternative methods for incorporating non-exponential distributions into stochastic timed Petri nets. In *Proc. of the 3rd Int'l. Workshop on Petri Nets and Performance Models (PNPM89)*, Kyoto, Japan, 1989.
6. J.A. Carrasco. Automated construction of compound Markov chains from generalized stochastic high-level Petri nets. In *Proc. of the 3rd Int'l. Workshop on Petri Nets and Performance Models (PNPM89)*, Kyoto, Japan, 1989.
7. O.C. Ibe, H. Choi, and K.S. Trivedi. Stochastic Petri net models of client-server systems, *IEEE Transactions on Parallel and Distributed Systems*. to appear

8. H. Choi, V.G. Kulkarni and K.S. Trivedi. Markov Regenerative Stochastic Petri nets. In *16th IFIP W.G. 7.3 Int'l Sym. on Computer Performance Modeling, Measurement and Evaluation (Performance'93)*, Rome, Italy, Sep. 1993.
9. H. Choi and K.S. Trivedi. Approximate Performance Models of Polling Systems using Stochastic Petri Nets. In *Proceedings of the IEEE INFOCOM 92*, Florence, Italy, May 1992.
10. H. Choi, V.G. Kulkarni, and K.S. Trivedi. Transient analysis of deterministic and stochastic Petri nets, *Lecture Notes in Computer Science*, M.A. Marson (ed.), Springer Verlag, 1993.
11. H. Choi, V. Mainkar, and K.S. Trivedi. Sensitivity analysis of deterministic and stochastic petri nets. In *Proc. of MASCOTS'93, the Int'i Workshop on Modeling, Analysis and Simulation of Computer and Telecommunication Systems*, pages 271–276, San Diego, USA, Jan. 1993.
12. G. Ciardo, J.K. Muppala, and K.S. Trivedi. Analyzing concurrent and fault-tolerant software using stochastic Petri nets. *Journal of Parallel and Distributed Computing*. Vol. 15, pp. 255–269, 1992.
13. G. Ciardo and C. Lindemann. Analysis of deterministic and stochastic Petri nets. In *Proceedings of Petri nets and Performance Models 1993*, pp. 160–169, Toulouse, France, Oct. 19–22, 1993.
14. G. Ciardo and K.S. Trivedi. A decomposition approach for stochastic reward net models. In *Proceedings of the Fourth Intl. Workshop on Petri Nets and Performance Models (PNPM91)*, Melbourne, Australia, Dec. 1991.
15. G. Ciardo, A. Blakemore, P.F. Chimento, J.K. Muppala and K.S. Trivedi. Automated generation and analysis of Markov reward models using stochastic reward nets. In *Linear Algebra, Markov Chains, and Queueing Models*, Carl Meyer and R.J. Plemmons (eds.), IMA Volumes in Mathematics & its Applications, Vol. 48, pp 145–191, Springer Verlag, Heidelberg, 1993.
16. J.B. Dugan, K.S. Trivedi, R.M. Geist, and V.F. Nicola. Extended stochastic Petri nets: Applications and analysis, *Performance '84*, E. Gelenbe (ed.), North-Holland.
17. B.R. Haverkort and K.S. Trivedi. Specification and generation of Markov reward models. *Discrete Event Dynamic Systems: Theory and Applications 3*, pp.219–247, 1993.
18. V. Mainkar, H. Choi, and K.S. Trivedi. Sensitivity analysis of Markov regenerative stochastic Petri nets. In *Proceedings of Petri nets and Performance Models 1993*, pp. 180–191, Toulouse, France, Oct. 19–22, 1993.
19. M. Malhotra and K.S. Trivedi. Power-hierarchy of dependability model types. To appear in *IEEE Transactions on Reliability*.
20. M. Malhotra and A. Reibman. Selecting and implementing phase approximations for semi-Markov models. To appear in *Stochastic Models*, 1993.
21. M. Malhotra, J.K. Muppala, and K.S. Trivedi. Stiffness-tolerant methods for transient analysis of stiff Markov chains. *Technical Report DUKE-CCSR-92-003*, Center for Computer Systems Research, Duke University, 1992.
22. M. Malhotra and K.S. Trivedi. Reliability analysis of redundant arrays of inexpensive disks. *Journal of Parallel and Distributed Computing*, 17:146–151, Jan. 1993.
23. M. Malhotra and K.S. Trivedi. A methodology for formal expression of hierarchy in model solution. In *Proceedings of Petri nets and Performance Models 1993*, pp. 258–267, Toulouse, France, Oct. 19–22, 1993.
24. M. Malhotra. Specification and solution of dependability models of fault-tolerant systems. *Ph.D. Thesis, Department of Computer Science, Duke University*, April, 1993.
25. M.A. Marson and G. Conte. A class of generalized stochastic Petri nets for the performance evaluation of multiprocessor systems. *ACM Trans. Comp. Systems*, Vol 2. No. 2, May 1984, pp. 93–122
26. M.A. Marson, G. Balbo, G. Chiola, and G. Conte. Generalized stochastic Petri nets revisited; Random switches and priorities. In *Proceedings of Petri Nets and Performance Models*, pp

44–53, Wisconsin, Madison, August, 1987.

27. J.F. Meyer. On evaluating the performability of degradable computer systems. *IEEE Trans. on Computers*. C-29(8):720–731, 1980.
28. J.F. Meyer. Closed-form solutions of performability. *IEEE Trans. on Computers*. C-31(7):648–657, 1982.
29. M. Molloy. Performance analysis using stochastic Petri nets. *IEEE Transactions on Computers*, C-31(9):913–917, Sept. 1982.
30. J.K. Muppala, and K.S. Trivedi, Composite Performance and Availability Analysis using a Hierarchy of Stochastic Reward Nets, *Proc. fifth Intl. Conf. Modeling Techniques and Tools for Computer Performance Evaluation*, Torino, Italy, Feb. 1991.
31. J.K. Muppala, A.S. Sathaye, R.C. Howe, and K.S. Trivedi. Dependability modeling of a heterogeneous multiprocessor system using stochastic reward nets. In D. Avresky (ed.), *Hardware & Software Fault Tolerance in Parallel Computing Systems*, pp. 33–59, Ellis Horwood Ltd. 1992.
32. J.K. Muppala and K.S. Trivedi. Numerical transient analysis of finite Markovian queueing systems. In U. Bhat and I. Basawa, editors, *Queueing and Related Models*, pp. 262–284. Oxford University Press, 1992.
33. J.L. Peterson. Petri net theory and the modeling of systems, *Prentice Hall*, 1981.
34. A. Reibman and K.S. Trivedi. Numerical Transient analysis of Markov Models. *Computers and Operations Research*, 15(1):19–36, 1988.
35. A. Reibman and K.S. Trivedi. Transient analysis of cumulative measures of Markov model behavior. *Stochastic Models*, 5(4):683–710, 1989.
36. R.Sahner and K.S. Trivedi. A software tool for learning about stochastic models. *IEEE Transactions on Education*, 36(1):56–61, Feb. 1993.
37. R. Sahner, and K.S. Trivedi. Performance analysis using directed acyclic graphs. *IEEE Transactions on Software Engg.*, 14(10):1105–1114, Oct. 1987.
38. L. Tomek and K.S. Trivedi. Fixed-Point iteration in availability modeling. In M.Dal Cin (ed.), *Informatik-Fachberichte*, Vol. 91: *Fehlertolerierende Rechensystems*. Springer Verlag, Berlin, 1991.
39. L.A. Tomek, J.K. Muppala, and K.S. Trivedi. Modeling Correlation in Software Recovery Blocks. *IEEE Transactions on Software Engg.* (sp. issue on software reliability), 19(11):1071–1086, November 1993.
40. K.S. Trivedi, J.K. Muppala, S.P. Woolet, and B.R. Haverkort. Composite performance and dependability analysis, *Performance Evaluation* 14(1992) 197–215, North Holland.
41. K.S. Trivedi and M. Malhotra. Reliability and performability techniques and tools: A survey. In *Proc. 7th ITG/GI Conference on Measurement, Modeling, and Evaluation of Computer & Communication Systems*. Aachen, pp. 159–172, Sept. 1993.
42. K.S. Trivedi, and V.G. Kulkarni. FSPNs: Fluid Stochastic Petri Nets, *Lecture Notes in Computer Science* 691, Springer Verlag, 1993.
43. K.S. Trivedi. Probability and Statistics with Reliability , Queueing, and Computer Science Applications. *Prentice Hall*, Engelwood Cliffs, NJ, USA, 1982.

Controlled Petri Nets: A Tutorial Survey

L. E. Holloway[1] *and B. H. Krogh*[2]

[1] Center for Manufacturing Systems and Dept. of Electrical Engineering, University of Kentucky, Lexington, KY 40506-0108 USA, email: holloway@engr.uky.edu

[2] Department of Electrical and Computer Engineering, Carnegie Mellon University, Pittsburgh, PA 15213-3890 USA, email: krogh@ece.cmu.edu

1 Introduction

This paper surveys recent research on the synthesis of controllers for discrete event systems (DESs) modeled by *controlled Petri nets* (CtlPNs). Petri nets have been used extensively in applications such as automated manufacturing [6], and there exists a large body of tools for qualitative and quantitative analysis of Petri nets [28]. For control, Petri nets offer a structured model of DES dynamics that can be exploited in developing efficient algorithms for controller synthesis.

The following section presents the basic untimed CtlPN model, discusses special classes of CtlPNs, and gives an overview of the various control problems that have been formulated and solved in the literature. Section 3 describes state feedback controllers for CtlPNs and the general conditions that must be satisfied for a state feedback control policy to exist to prevent the CtlPN from reaching a given set of forbidden markings. In §4, control of the sequential behavior of a CtlPN is considered. We summarize methods for converting event-based sequential control problems into state feedback problems. We then present two general approaches to computing state feedback policies, namely, the linear integer programming approach of Li and Wonham (§5) and the path-based approach introduced by Holloway and Krogh (§6). The paper concludes with a discussion of several directions for further research in §7.

2 Controlled Petri Nets

Controlled Petri nets (CtlPNs) are a class of Petri nets with external enabling conditions called *control places* which allow an external controller to influence the progression of tokens in the net. CtlPNs were first introduced by Krogh [20] and Ichikawa and Hiraishi [19]. Formally, a CtlPN $\mathcal{G} = (\mathcal{P}, \mathcal{T}, \mathcal{C}, \mathcal{E}, \mathcal{B})$ consists of a finite set $\mathcal{P}$ of *state places*, a finite set $\mathcal{T}$ of *transitions*, a finite set $\mathcal{C}$ of *control places*, a set $\mathcal{E} \subseteq (\mathcal{P} \times \mathcal{T}) \cup (\mathcal{T} \times \mathcal{P})$ of directed arcs connecting state places to transitions and vice versa, and a set $\mathcal{B} \subseteq (\mathcal{C} \times \mathcal{T})$ of directed arcs connecting control places to transitions. The sets $\mathcal{P}, \mathcal{T}, \mathcal{C}$ are mutually disjoint. For a transition $t \in \mathcal{T}$, we denote the set of input control places as ${}^{(c)}t := \{c \mid (c,t) \in \mathcal{B}\}$, and for a control place $c \in \mathcal{C}$ we denote the set of output transitions as $c^{(t)} := \{t \mid (c,t) \in \mathcal{B}\}$. Similarly, we denote the set of input (output) state places for transition t as ${}^{(p)}t := \{p \mid (p,t) \in \mathcal{E}\}$ ($t^{(p)} := \{p \mid (t,p) \in \mathcal{E}\}$). The set of input (output) transitions for a state place p is defined as ${}^{(t)}p$ ($p^{(t)}$). A transition t is said to be a *controlled transition* if its set of control inputs ${}^{(c)}t$ is nonempty.

The state of a CtlPN is given by its *marking*, which is the distribution of *tokens* in the state places. Formally, a marking is a function $m : \mathcal{P} \longrightarrow \mathbb{N}$, where $m(p)$ is the number of tokens in place p. $\mathcal{M}$ denotes the set of all markings. A set of transitions $T \subseteq \mathcal{T}$ is *state enabled* if for all $p \in \mathcal{P}$, $m(p) \geq |p^{(t)} \cap T|$.

A *control* for a CtlPN is a function $u : \mathcal{C} \longrightarrow \{0, 1\}$ associating a binary value to each control place. The set of all such controls is denoted by $\mathcal{U}$. A set of transitions $T \subseteq \mathcal{T}$ is said to be *control enabled* if for all $t \in T$, $u(c) = 1$ for all $c \in {}^{(c)}t$. A control $u \in \mathcal{U}$ is said to be *as permissive as* control $u' \in \mathcal{U}$ if $u(c) \geq u'(c)$ for all $c \in \mathcal{C}$. Control u is said to be *more permissive than* control u' if u is as permissive as u' and $u(c) > u'(c)$ for some $c \in \mathcal{C}$. The most permissive control is $u_{\text{one}} \equiv 1$, and the least permissive control is $u_{\text{zero}} \equiv 0$.

Letting $\mathcal{T}_e(m, u) \subseteq 2^{\mathcal{T}}$ denote the collection of sets of transitions that are both state enabled by $m \in \mathcal{M}$ and control enabled by $u \in \mathcal{U}$, any set of transitions $T \in \mathcal{T}_e(m, u)$ can *fire*, thereby changing the marking of the net to the marking m' defined by

$$m'(p) = m(p) + |{}^{(t)}p \cap T| - |p^{(t)} \cap T| \ . \tag{1}$$

In words, firing a set of transitions $T \in \mathcal{T}_e(m, u)$ causes one token to be removed from each $p \in {}^{(p)}t$, and one token to be added to each $p \in t^{(p)}$, for each $t \in T$.

Given a marking $m \in \mathcal{M}$ and control $u \in \mathcal{U}$, the set of *immediately reachable markings*, $R_1(m, u)$, is given by

$$R_1(m, u) = \{m\} \cup \{m' \in \mathcal{M} \mid m' \text{ is given by (1) for some } T \in \mathcal{T}_e(m, u)\} \ . \tag{2}$$

The set of *reachable markings* under an arbitrary number of transition firings from a given marking $m \in \mathcal{M}$ with a constant control $u \in \mathcal{U}$ is denoted $R_\infty(m, u)$.

An uncontrolled *subnet* (P, T, E) of $\mathcal{G}$ where $P \subseteq \mathcal{P}$, $T \subseteq \mathcal{T}$, and $E = ((P \times T) \cup (T \times P)) \cap \mathcal{E}$ is called a *marked graph structure* if all state places $p \in \mathcal{P}$ have at most one input transition and one output transition in the subnet. Marked graph structures can model synchronization of concurrent processes: tokens in places which share an output transition must progress synchronously. A subnet (P, T, E) is a *state graph structure* if all transitions $t \in \mathcal{T}$ have at most one input state place and one output state place in the subnet. State graph structures with a single token are analogous to finite-state automata. Section 6 examines a family of control synthesis methods which exploit state graph and marked graph structures in CtlPNs to compute state feedback policies.

It is sometimes useful to write the state transition equation (1) as a linear matrix-vector equation by introducing indices for the places and transitions, $\mathcal{P} := \{p_1, \ldots, p_n\}$ and $\mathcal{T} := \{t_1, \ldots, t_m\}$, and, with a slight abuse of notation, redefining the marking as a column vector, $m \in \mathbb{N}^n$, where $m_i := m(p_i)$. Given a sequence of sets of transitions $\{T_0, \ldots, T_k\} \subseteq (2^{\mathcal{T}})^k$, the *firing vector* $v(\{T_0, \ldots, T_k\}) \in \mathbb{N}^m$ has components $v_j(\{T_0, \ldots, T_k\})$ defined as the number of times transition $t_j \in \mathcal{T}$ occurs in the sequence of sets $T_0, \ldots, T_k$. Given a marking $m \in \mathcal{M}$ and control $u \in \mathcal{U}$, if $T \in \mathcal{T}_e(m, u)$, the state transition equation (1) can then be written as

$$m' = m + Ev(T) \ , \tag{3}$$

where $E \in \{-1, 0, 1\}^{n \times m}$ is the *incidence matrix* for the CtlPN graph [28].

Thus, Petri nets dynamics can be viewed from a linear algebraic perspective, as described in §5. This approach has been pursued in a control context by Giua, DiCesare, and Silva [11], and by Li and Wonham (which they call *vector discrete event systems*) [26, 27].

In much of the Petri net literature it is assumed that only a single transition can fire at any instant. We will refer to this case as the *no concurrency (NC) assumption*, under which the state transition equation (1) applies only to singleton sets. Given a CtlPN $\mathcal{G}$ and an initial marking $m_0 \in \mathcal{M}$, under the NC assumption $L(\mathcal{G}, m_0) \subseteq \mathcal{T}^*$ denotes the *language of valid sequences of transition firings* for $\mathcal{G}$ starting from m_0 and the control u_{one}, where $\mathcal{T}^*$ is the set of all finite-length strings of elements of $\mathcal{T}$. The firing of a transition corresponds to an *event* in the usual DES terminology. In general, Petri net languages are defined in terms of a separate set of event labels which can be assigned to some or all of the transitions. By identifying the transitions with the event labels, each transition is a distinct event, which corresponds to a so-called *free-labeled* Petri net in the theory of Petri net languages [30]. The notion of the language of a CtlPN is useful for specifying and analyzing problems of sequential control which are considered in §4.

3 State Feedback and State Specifications

A *state feedback policy* for a CtlPN is a function $U : \mathcal{M} \longrightarrow 2^{\mathcal{U}}$. The state feedback policy is *deterministic* if $U(m)$ is a singleton for all markings $m \in \mathcal{M}$. We extend the notation for immediately reachable markings from a marking $m \in \mathcal{M}$ for a feedback policy U by defining $R_1(m, U) = \bigcup_{u \in U(m)} R_1(m, u)$. Similarly, the set of markings reachable under an arbitrary number of firings from a marking m for a state feedback policy U is denoted $R_\infty(m, U)$. Extending the concept of relative permissiveness of controls to state feedback policies, we say state feedback policy U_1 is as permissive as state feedback policy U_2, denoted by $U_1 \geq U_2$, if for each $m \in M$, $U_1(m) \supseteq U_2(m)$. It follows that $U_1 \geq U_2$ implies $R_\infty(m, U_1) \supseteq R_\infty(m, U_2)$ for any marking $m \in M$.

State feedback policies for CtlPNs have been investigated by a number of researchers [15, 21, 2, 26, 18, 1, 16, 35]. In most cases the fundamental problem is to design a state feedback policy that guarantees the system remains in a specified set of allowed states, or, equivalently, that the marking of the CtlPN is never in a specified set of *forbidden markings*. Given a CtlPN $\mathcal{G}$ with initial marking m_0, let $\mathcal{M}_F$ denote the set of forbidden markings. The objective is to find a state feedback policy U_F for which: (1) $R_\infty(m_0, U_F) \cap \mathcal{M}_F = \varnothing$; and (2) U_F is more permissive than any other state feedback policy satisfying (1). We call a state feedback policy satisfying these two conditions a *maximally permissive state feedback policy* for the given forbidden state specification $\mathcal{M}_F$.

From the general theory of controlled DES developed by Ramadge and Wonham, a necessary and sufficient condition for the existence of a maximally permissive state feedback policy is determined by an analysis of the CtlPN behavior under the control u_{zero}. Specifically, define the set of *admissible markings* for a CtlPN $\mathcal{G}$ with respect to a set of forbidden markings $\mathcal{M}_F$ as

$$\mathcal{A}(\mathcal{M}_F) = \{m \in \mathcal{M} \mid R_\infty(m, u_{\text{zero}}) \cap \mathcal{M}_F = \varnothing\} \ . \tag{4}$$

Necessary and sufficient conditions for the existence of a state feedback control policy that keeps a CtlPN out of a given set of forbidden markings is then given by the following theorem.

Theorem 1 ([21]). *Given a CtlPN $\mathcal{G}$ with initial marking m_0 and a forbidden marking specification $\mathcal{M}_F$, a unique maximally permissive state feedback policy exists if and only if $m_0 \in \mathcal{A}(\mathcal{M}_F)$.*

The *unique* maximally permissive state feedback policy in Theorem 1 is *nondeterministic* because the firing rule (1) allows multiple transitions in the CtlPN to fire simultaneously [20, 25]. In general, there will not be a unique *deterministic* maximally permissive policy because the set of controls $U_F(m)$ does not have necessarily a unique maximal element. On the other hand, under the NC assumption where the firing rule (1) is restricted to singleton sets, a unique *deterministic* maximally permissive state feedback policy exists [26].

When the initial marking for a CtlPN satisfies the condition $m_0 \in \mathcal{A}(\mathcal{M}_F)$ in Theorem 1, the maximally permissive state feedback policy can be described simply as the policy which does not allow any state transitions to markings outside $\mathcal{A}(\mathcal{M}_F)$.

Theorem 2 ([21]). *Given a CtlPN $\mathcal{G}$ with a forbidden marking specification $\mathcal{M}_F$ and an initial condition $m_0 \in \mathcal{A}(\mathcal{M}_F)$, if $m \in R_\infty(m_0, U_F)$, then*

$$U_F(m) = \{u \in \mathcal{U} \mid R_1(m, u) - \mathcal{A}(\mathcal{M}_F) = \varnothing\} \ .$$

One approach to compute the set of admissible controls for a given marking is to simply create the equivalent controlled automaton for the CtlPN, which is a matter of generating the *reachability graph* for the Petri net structure with the associated control information. Since the reachability graph can grow exponentially with respect to the size of the CtlPN model [22], alternative methods are desirable for computing feedback policies. Sections 5 and 6 present the two approaches that have been developed to use the structure of the CtlPN model directly, thereby avoiding the generation of the equivalent controlled automaton.

We conclude this section with a brief summary of various extensions and generalizations of state feedback policies considered in the literature. Baosheng and Haoxun consider partial observability in the context of distributed state feedback control systems [13]. Control capability is distributed among several predefined controllers, and each controller has only limited observations of the net marking. Li and Wonham consider partial state observability for centralized controllers [26], and prove that a maximally permissive state feedback policy exists if and only if the given predicate (set of admissible states) is both controllable and observable [26]. They also consider *modular state feedback* policies where the overall specification for the admissible states for the closed-loop system are given as the conjunction (intersection) of a collection of *subspecifications*, similar to the forbidden marking specifications considered in §6.

4 Event Feedback and Sequential Specifications

Event feedback control for a CtlPN is based on observations of the transition firings rather than the marking. Under the NC assumption (which is assumed throughout this section), an *event feedback policy* is a function $\mathcal{V} : \mathcal{T}^* \longrightarrow \mathcal{U}$ mapping each sequence of transitions onto a control. $L(\mathcal{V} \mid \mathcal{G}, m_0) \subseteq \mathcal{T}^*$ denotes the language generated by the net $\mathcal{G}$ from initial marking m_0 under an event feedback policy $\mathcal{V}$. The goal of event feedback control is to restrict the operation of the Petri net such that it produces a desired (prefix-closed) language $K \subseteq L(\mathcal{G}, m_0)$. From the general Ramadge and Wonham theory, there exists an event feedback policy $\mathcal{V}$ such that $L(\mathcal{V} \mid \mathcal{G}, m_0) = K$ if and only if the language K is *controllable*, where K is defined to be controllable with respect to $L(\mathcal{G}, m_0)$ if $K \subseteq L(\mathcal{G}, m_0)$ and $KT_u \cap L(\mathcal{G}, m_0) \subseteq K$ [32]. [3]

[3] KT_u denotes the set of all strings of the form $\omega\sigma$ such that $\omega \in K$ and $\sigma \in T_u$.

As with state feedback control policies, one approach to synthesizing event feedback controllers for a CtlPN $\mathcal{G}$ is to create the equivalent controlled automata model for $\mathcal{G}$ and apply the supervisor synthesis algorithms from the standard Ramadge and Wonham theory. As an alternative to automata-based supervisors, some researchers have considered using Petri nets as supervisors. This increases the class of controllable languages that can be realized by event feedback policies, but computational complexity becomes a problem for more general Petri net languages. Indeed, Sreenivas has shown that controllability is undecidable for the most general Petri net languages, but it is decidable if the specification language is accepted by a free-label Petri net [34]. Giua and DiCesare obtained necessary and sufficient conditions for the existence of a Petri net supervisor when the controllable language specification can be accepted by a conservative Petri net [8, 9].

An alternative to constructing dynamic event feedback controller directly in the form of an automaton or Petri net is to augment the dynamics of the plant with an additional automaton or Petri net that "encodes" the desired sequential behavior of the system in the state of the augmented system. One can then apply state feedback to the augmented system to achieve the desired sequential behavior. This approach has been developed by Kumar and Holloway [23] using a Petri net to augment the plant dynamics, and by Li and Wonham [26] using an automaton called a *memory* to augment the plant dynamics. In both cases it is shown that the maximally permissive state feedback control for avoiding an appropriately defined predicate or set of forbidden states on the augmented state space results in the *supremal controllable sublanguage* $K\uparrow$ for a given sequential specification K. Since event feedback control problems for sequential specifications can be converted into state feedback control problems, we turn our attention in the following two sections to methods for computing state feedback policies for CtlPNs.

5 Linear Integer Programming Approach

Linear algebraic methods and linear integer programming have been standard tools for solving many problems in the Petri net literature based on the matrix-vector representation of the CtlPN state transition equation (3) [28]. In the context of control, Giua, DiCesare, and Silva use linear integer programming to evaluate properties of DESs controlled by a class of Petri net controllers [10, 11]. In this section we describe the linear integer programming method proposed by Li and Wonham [24, 27] for computing state feedback policies.

Under the NC assumption and the assumption each control place is connected to a single controlled transition, Li and Wonham consider the synthesis of maximally permissive feedback policies when the allowable states are specified by a *linear predicate* P of the form

$$P = \{m \in \mathcal{M} \mid a^T m \leq b\} \ , \tag{5}$$

where a is an n-vector and b is a scalar. The control objective is to guarantee the linear constraints on the marking are satisfied for all markings reachable under the control.

The set of admissible markings corresponding to a predicate P of the form (5) is denoted as $[P]$ by Li and Wonham, and the unique maximally permissive control input for a given $m \in [P]$ is denoted by $u_{[P]}(m)$. From the general theory in §3, computing $u_{[P]}(m)$ is a matter of determining if markings of the form $\hat{m} = m + Ev(c^{(t)})$ are in $[P]$. Li and Wonham show that this problem can be reduced to solving a linear integer program provided the CtlPN satisfies a particular structural condition, namely, the uncontrolled portion of the CtlPN must be *loop free*. To describe this result, we use the language notation for strings of admissible transitions to express $[P]$ as:

$$[P] = \{m \in \mathcal{M} \mid a^T m + a^T E v^*(m)) \leq b\} \ , \tag{6}$$

where $v^*(\omega)$ is the solution to the following optimization problem

$$\max_{\omega \in L(u_{\text{zero}} | \mathcal{G}, m)} \quad a^T E v(\omega) \ . \tag{7}$$

The reduction of the optimization problem (7) to a *linear integer program* is based on the following general result for Petri nets (which we state for CtlPNs).

Theorem 3 ([19]). *Given a* loop-free *CtlPN $\mathcal{G}$ and a marking $m \in \mathcal{M}$, $\omega \in L(\mathcal{G}, m)$ if and only if $\omega \in \mathcal{T}^*$ and*

$$m + E v(\omega) \geq 0 \ . \tag{8}$$

The significance of this theorem is that any firing vector satisfying (8) corresponds to a valid firing sequence. Giua and DiCesare have called Petri nets with this property *easy Petri nets* [7]. To apply this result to the optimization problem (7), observe that the language $L(u_{\text{zero}} \mid \mathcal{G}, m)$ is the same as the language for the Petri net obtained when all controlled transitions are removed from $\mathcal{G}$, which Li and Wonham denote by $\mathcal{G}_u$. Thus, if $\mathcal{G}_u$ is loop free, (7) becomes a *linear integer program*.

The maximally permissive feedback control can be computed using this linear integer program by observing that $u_{[P]}(m)(c) = 1$ when $c^{(t)}$ is state enabled if and only if

$$a^T \hat{m} + a^T E v^*(\hat{m}) \leq b \ ,$$

where $\hat{m} = m + E v(c^{(t)})$.

Li and Wonham develop this basic approach in several ways, including the generalization to multiple linear predicates (modular synthesis) and developing a closed-form expression for the maximally permissive control under further structural assumptions. They also consider sequential specifications in the form of linear predicates on the firing vector which they call *linear dynamic specifications*. For sequential specifications, they convert the problem to a state feedback problem using a *memory* as described in the previous section.

6 Path-Based Algorithms

Path control algorithms decompose the forbidden state control problem into the regulation of tokens in individual paths in the CtlPN. For some classes of CtlPNs it has been shown that this exploitation of the net structure can lead to significant gains in computational efficiency for on-line control synthesis [22]. Path control has been used to address a variety of control specifications, including state control [15, 21], sequence control [18], and distributed control [13].

Path control assumes the set of forbidden states is expressed in the form of *forbidden conditions*, which are specifications of sets of forbidden markings based on linear inequalities on the marking vectors [21, 2]. A forbidden condition is represented by the triple (F, v, k), where $F \subseteq \mathcal{P}$ is a subset of places, $v : F \longrightarrow \mathbb{N}$ is a *weighting function* over the places in F, and k is the *threshold* of the forbidden condition. The forbidden condition specifies a set of forbidden markings as

$$M_{(F,v,k)} := \left\{ m \in \mathcal{M} \mid \sum_{p \in F} m(p)v(p) > k \right\} . \tag{9}$$

Given a collection $\mathcal{F}$ of set conditions, the set $M_{\mathcal{F}}$ of forbidden markings for $\mathcal{F}$ is defined as the union of $M_{(F,v,k)}$ for all $(F, v, k) \in \mathcal{F}$.

In general, arbitrary forbidden marking sets cannot be represented by forbidden conditions; however, for the special case of cyclic controlled marked graphs, any subset of live and safe markings can be defined by a set $\mathcal{F}$ of forbidden set conditions [14]. Previous work has considered specific cases of the generalized forbidden condition specification. Holloway and Krogh consider live and safe cyclic controlled marked graphs where for each (F, v, k), $v(p) = 1$ for all $p \in F$, and $k = |F| - 1$ in [15] and $1 \leq k < |F|$ in [21]. Boel et al. consider a class of controlled state graphs without restrictions on v or k [2]. Holloway and Guan consider a general class of Petri nets with forbidden conditions where $v(p) = 1$ for all $p \in F$ and $k = |F| - 1$ [17].

Path control methods characterize the reachability of forbidden conditions in terms of the markings of directed paths in the CtlPN. We consider only paths which begin with a transition and end with a state place. The following notation is used in this section. For a given path $\pi = (t_0 p_0 \ldots t_n p_n)$, the starting transition t_0 is denoted t_π. We define set operations on paths to be over the set of places and transitions in the path. Thus, $\pi \cap \mathcal{P}$ is the set of places in the path, and $\pi \cap \mathcal{T}$ is the set of transitions in the path. We extend the marking notation such that $m(\pi)$ is the sum of $m(p)$ over all $p \in \pi$. $\Pi_c(p)$ denotes the set of all *precedence paths* for a place p, where $\pi = (t_0 p_0 \ldots t_n p_n) \in \Pi_c(p)$ implies that: (1) path π ends at place p; (2) t_π is a controlled transition or has ${}^{(p)}t_\pi \cap \pi \neq \varnothing$; (3) all other transitions $t \in (\pi \cap \mathcal{T}) - \{t_\pi\}$ are uncontrolled transitions; and (4) for any $0 \leq i, j \leq n$ then $p_i \neq p_j$ and if $t_i = t_j$ then either $i = 0$ or $j = 0$. The last condition (4) restricts a path to not include a cycle.

The key question in path control is: For a given marking m, does there exist an uncontrollably reachable marking, $m' \in R_\infty(m, u_{\text{zero}})$, for which $m'(p) \geq k$? This problem is referred to as the *uncontrollable k-coverability problem.* The role of precedence paths in determining the uncontrollable k-coverability problem is most easily illustrated by considering the case where $k = 1$. For a marking m and a precedence path π, define the predicate

$$\Lambda_m(\pi) := \begin{cases} 1 & \text{if } m(p) \geq 1 \text{ for some } p \in \pi \ , \\ 0 & \text{else.} \end{cases} \tag{10}$$

Theorem 4 ([15]). *Given a CtlPN $\mathcal{G}$ and place p such that $\Pi_c(p) = \{\pi_1, \pi_2, \ldots \pi_n\}$ is a marked graph structure, for a marking m there exists an uncontrollably reachable marking $m' \in R_\infty(m, u_{\text{zero}})$ with $m'(p) \geq 1$ if and only if*

$$\Lambda_m(\pi_1) \wedge \Lambda_m(\pi_2) \wedge \cdots \Lambda_m(\pi_n) = 1 \ . \tag{11}$$

Theorem 5 (follows from both [2] and [17]). *Given a CtlPN $\mathcal{G}$ and place p such that $\Pi_c(p) = \{\pi_1, \pi_2, \ldots \pi_n\}$ is a state graph structure, for a marking m there exists an uncontrollably reachable marking $m' \in R_\infty(m, u_{\text{zero}})$ with $m'(p) \geq 1$ if and only if*

$$\Lambda_m(\pi_1) \vee \Lambda_m(\pi_2) \vee \cdots \Lambda_m(\pi_n) = 1 \ . \tag{12}$$

The above theorems show that the marked graph structure and state graph structure lead to the complimentary characterizations of the uncontrolled 1-coverability of a place. These conditions are generalized by Holloway and Guan to cases where the uncontrolled region of the net leading to a given place may have a mixture of both marked graph structures and state graph structures [17]. We note that conditions for uncontrolled k-coverability do not have the same convenient Boolean characterization as the uncontrolled 1-coverability problem has. However, for state graph structures, Boel et al. show that uncontrolled k-coverability can be characterized through the sum of tokens among precedence paths leading to a place [2].

Given the characterization of the uncontrollably reachable markings in the above theorems, we now describe a method for avoiding forbidden states by controlling the markings of individual paths. Our discussion is primarily based on the method presented in [15] and [21] for controlled marked graphs. The reader is referred to [2] and [17] for control laws for other net structures. For the remainder of this section, we assume controlled marked graphs with binary markings, i.e. *safe* markings [28], where all cycles within the net contain at least one marked place under any initial marking. To be able to achieve the decomposition of control upon which path control methods depend, we also require the following assumption on the interaction of paths for places in a forbidden condition: Given a set $\mathcal{F}$ of forbidden conditions, for any $(F, v, k) \in \mathcal{F}$ and any $p, p' \in F$ with $p \neq p'$, if $\pi \in \Pi_c(p)$ then $p' \notin \pi$ and $p' \notin {}^{(p)}t_\pi$. This is referred to as the Specification Assumption (SA).

We consider a set $\mathcal{F}$ of forbidden conditions of the form (F, v, k), where $v(p) = 1$ for all $p \in F$ and $1 \leq k < |F|$. For a forbidden condition $(F, v, k) \in \mathcal{F}$ and marking m, define $\Lambda_m(p)$ from (11) for each $p \in F$ as the conjunction of $\Lambda_m(\pi)$ for all $\pi \in \Pi_c(p)$. Furthermore, define

$$L_F(m) := \{p \in F \mid \Lambda_m(p) = 1\} \ . \tag{13}$$

From Theorem 4 and the SA, it can be shown that $L_F(m)$ is the set of places in F which can become marked uncontrollably from the marking m [15]. From the definition of the admissible marking set $\mathcal{A}(\mathcal{M}_\mathcal{F})$ in §3, it then can be shown that for any given marking m, $m \in \mathcal{A}(\mathcal{M}_\mathcal{F})$ if and only if $L_F(m) \leq k$ [21].

To prevent a forbidden marking from being reachable, a control must be enforced to ensure that $|L_F(m')| \leq k$ for all $(F, v, k) \in \mathcal{F}$ for all immediately reachable markings under the control. To determine a control using path control techniques, we first examine the markings of individual paths to determine which places in a condition F could potentially join the set $L_F(m')$ after the next transition set firing.

Define the predicate $\Delta_m(\pi) = 1$ iff t_π is state enabled under m, and define

$$\Delta_m(p) := \begin{cases} 1 & \text{if } \Lambda_m(p) = 0 \text{ and } \Lambda_m(\pi) \vee \Delta_m(\pi) = 1 \text{ for all } \pi \in \Pi_c(p) \ , \\ 0 & \text{else,} \end{cases} \tag{14}$$

$$B_F(m) := \{p \in F \mid \Delta_m(p) = 1\} \ . \tag{15}$$

It can be shown that $B_F(m)$ is the set of places in F which could become uncontrollably marked following the next transition set firing unless some control is enforced. From the definition of $\Delta_m(p)$ above and from Theorem 4, we note that keeping at least one path $\pi \in \Pi_c(p)$ unmarked is sufficient to ensure that p is an element of $B_F(m')$ (and thus not an element of $L_F(m')$) for any immediately reachable marking m'. This can be accomplished by disabling the controlled transition t_π for some path $\pi \in \Pi_c(p)$

which is unmarked but for which t_π is state enabled. From this observation we have that

$$D_F(m,u) := \{p \in B_F(m) \mid u(c) = 0 \text{ for some } c \in {}^{(c)}t_\pi \text{ for some } \pi \in \Pi_c(p) \text{ with } \Delta_m(\pi) = 1 \text{ and } \Lambda_m(\pi) = 0\}$$

is the set of places in B_F which will be prevented from joining $L_F(m')$ for a next reachable marking m' under the control u. The number $|B_F(m)| - |D_F(m,u)|$ thus represents the maximum number of places in F that could join $L_F(m')$ after the next transition set firing. In order to ensure that $|L_F(m')| \leq k$ for all immediately reachable markings $m' \in R(m,u)$, the net control u then must satisfy the following equation.

$$|L_F(m)| + |B_F(m)| - |D_F(m,u)| \leq k \ . \tag{16}$$

From Theorem 2, the following result can be obtained:

Theorem 6 ([21]). *Given a CtlPN $\mathcal{G}$ with a forbidden marking specification $\mathcal{F}$ satisfying SA, let $U_{\mathcal{F}}$ be the policy such that for each $m \in \mathcal{A}(\mathcal{M}_{\mathcal{F}})$, $U_{\mathcal{F}}(m)$ is the set of controls such that (16) is satisfied for each $(F, v, k) \in \mathcal{F}$. The control policy $U_{\mathcal{F}}$ is the maximally permissive state feedback policy.*

7 Directions for Future Research

This paper surveys research on feedback control policies for discrete event systems using Petri net models. The primary objective in this research is to develop modeling, analysis and synthesis procedures that take advantage of the structural properties of Petri nets to reduce computational complexity. Toward this end there are several open directions for further research.

It would be of interest to determine how more classes of Petri net structures such as free-choice nets can be exploited for feedback control. The computational complexity of Petri net methods versus unstructured automata-based approaches also needs futher investigation. One comparison is presented in [22] where it is shown that the complexity of computing maximally permissive state feedback policies for controlled marked graphs (CtlMGs) is polynomial in the number of transitions and the number of set conditions in the forbidden state specifications, whereas the complexity grows exponentially for the equivalent automata-based models. Although there are efficient methods for solving similar problems independent concurrent automata [31], it has been shown that forbidden state problems for synchronized concurrent systems (of which CtlMGs are a subset) are in general computationally intractable [12]. Thus the boundary between tractable and intractable problems needs to be explored more deeply.

Recently a number of researchers have been interested in developing methods for extending methods for synthesizing feedback control policies for untimed (logical) models of DESs to models and specifications which include explicit representations of real time [4, 29, 3]. Petri nets offer an attractive framework for developing these extensions for controlled DESs because there are established methods for introducing and analyzing timing in uncontrolled Petri net models [28]. Two extensions of the methods discussed in this survey to *controlled time Petri nets* (CtlTPNs) are the work of Sathaye on synthesis of dynamic supervisors for sequential specifications for CtlPNs [33], and the work of Brave and Krogh on extensions of the path-based approach for forbidden marking specifications for the special case of *controlled time marked graphs*(CtlTMGs) [5]. Given the complexity of timed DESs, we believe research in this direction should be guided by an understanding of problems arising in specific applications.

Acknowledgments: L. E. Holloway has been supported in part by NSF grant ECS-9308737, NASA grant NGT-40049, Rockwell International, and the Center for Robotics and Manufacturing Systems at the University of Kentucky. B. H. Krogh has been supported by Rockwell International and the Alexander von Humboldt Foundation.

References

1. Z. A. Banaszak and B. H. Krogh. Deadlock avoidance in flexible manufacturing systems with concurrently competing process flows. *IEEE Transactions on Robotics and Automation*, 6(6), December 1990.
2. R.K. Boel, L. Ben-Naom, and V. Van Breusegem. On forbidden state problems for a class of controlled Petri nets. *IEEE Transactions on Automatic Control*, 1994. to appear.
3. B.A. Brandin and W.M. Wonham. Supervisory control of timed discrete-event systems. In *Proc. 27th IEEE Conf. on Decision and Control*, pages 3357–3362, Austin, TX, Dec 1988.
4. Y. Brave and M. Heymann. Formulation and control of real-time discrete event processes. In *Proc. 27th IEEE Conf. on Decision and Control*, pages 1131–1132, Austin, TX, Dec 1988.
5. Y. Brave and B.H. Krogh. Maximally permissive policies for controlled time marked graphs. In *Proceedings of 12th IFAC World Congress*, Sydney, July 1993.
6. F. Dicesare, G. Harhalakis, J. M. Proth, M. Silva, and F. B. Vernadat. *Practice of Petri Nets in Manufacturing*. Chapman and Hall, London, 1993.
7. A. Giua and F. DiCesare. Easy synchronized Petri nets as discrete event models. In *Proc. 29th IEEE Conf. on Decision and Control*, pages 2839–2844, Honolulu, Dec 1990.
8. A. Giua and F. DiCesare. Supervisory design using Petri nets. In *Proceedings of the 30th IEEE Conference on Decision and Control*, Brighton, UK, December 1991.
9. A. Giua and F. DiCesare. Blocking and controllability of Petri nets in supervisory control. *Transactions on Automatic Control*, 39(2), Feb. 1994. to appear.
10. A. Giua and F. DiCesare. Petri net structural analysis for supervisory control. *IEEE Transactions on Robotics and Automation*, April 1994. to appear.
11. A. Giua, F. DiCesare, and M. Silva. Generalized mutual exclusion constraints on nets with uncontrollable transitions. In *Proceedings of 1992 IEEE International Conference on Systems, Man, and Cybernetics*, pages 974–979, Chicago, October 1992.
12. C.H. Golaszewski and P.J. Ramadge. Mutual exclusion problems for discrete event systems with shared events. In *Proceedings 27th IEEE Conf. on Decision and Control*, Austin, Texas, Dec 1988.
13. C. Haoxun and H. Baosheng. Distributed control of discrete event systems described by a class of controlled Petri nets. In *Preprints of IFAC International Symposium on Distributed Intelligence Systems*, Arlington, Virginia, August 1991.
14. L. E. Holloway. Feedback control synthesis for a class of discrete event systems using distributed state models. Technical Report LASIP-88-17, Laboratory for Automated Systems and Information Processing, Dept of Electrical and Computer Engineering, Carnegie Mellon University, Pittsburgh, PA, Sept. 1988.
15. L. E. Holloway and B. H. Krogh. Synthesis of feedback control logic for a class of controlled Petri nets. *IEEE Transactions on Automatic Control*, 35(5):514–523, May 1990. Also appears in *Discrete Event Dynamic Systems: Analyzing Complexity and Performance in the Modern World*, edited by Y.C. Ho, IEEE Press, New York, 1992.
16. L. E. Holloway and B. H. Krogh. On closed-loop liveness of discrete event systems under maximally permissive control. *IEEE Transactions on Automatic Control*, 37(5), May 1992.
17. L.E. Holloway and X. Guan. A generalization of state avoidance policies for controlled Petri nets. In *Proceedings of 32nd IEEE Conference on Decision and Control*, San Antonio, Texas, December 1993.

18. L.E. Holloway and F. Hossain. Feedback control for sequencing specifications in Controlled Petri Nets. In *Third International Conference on Computer Integrated Manufacturing*, pages 242–250, Troy, New York, May 1992.
19. A. Ichikawa and K. Hiraishi. Analysis and control of discrete event systems represented by Petri nets. *Discrete Event Systems: Models and Applications*, IIASA Conference, Sopron, Hungary, August 3-7, 1987, 1988. Springer-Verlag, New York.
20. B. H. Krogh. Controlled Petri nets and maximally permissive feedback logic. *Proceedings of 25th Annual Allerton Conference*, Sept. 1987. University of Illinois, Urbana.
21. B. H. Krogh and L. E. Holloway. Synthesis of feedback control logic for discrete manufacturing systems. *Automatica*, July 1991.
22. B. H. Krogh, J. Magott, and L. E. Holloway. On the complexity of forbidden state problems for controlled marked graphs. In *Proceedings of the Conference on Decision and Control*, Brighton, UK, December 1991.
23. R. Kumar and L. E. Holloway. Supervisory control of Petri net languages. In *Proceedings of the 31st IEEE Conference on Decision and Control*, pages 1190–1195, Tucson, Arizona, December 1992.
24. Y. Li. *Control of Vector Discrete-Event Systems.* PhD thesis, Systems and Control Group, Department of Electrical Engineering, University of Toronto, July 1991.
25. Y. Li and W. M. Wonham. Strict concurrency and nondeterministic control of discrete-event systems. In *Proceedings of IEEE Conference on Decision and Control*, pages 2731–2736, Tampa, Florida, December 1989.
26. Y. Li and W.M. Wonham. Control of vector discrete-event systems I – the base model. *IEEE Transactions on Automatic Control*, 38(8):1214–1227, August 1993.
27. Y. Li and W.M. Wonham. Control of vector discrete-event systems II – controller synthesis. *IEEE Transactions on Automatic Control*, 39(3), March 1994. to appear.
28. T. Murata. Petri nets: Properties, analysis and applications. *Proceedings of the IEEE*, 77(4):541–580, April 1989.
29. J. S. Ostroff and W. M. Wonham. A framework for real-time discrete event control. *IEEE Transactions on Automatic control*, 35(4), April 1990.
30. J. L. Peterson. *Petri Net Theory and the Modeling of Systems.* Prentice-Hall, Englewood Cliffs, NJ, 1981.
31. P. J. Ramadge. Some tractable supervisory control problems for discrete event systems. *Symposium on the Mathematical Theory of Networks and Systems*, June 1987.
32. P. J. Ramadge and W. M. Wonham. Supervisory control of a class of discrete-event processes. *SIAM J. on Control an Optimization*, 25, Jan 1987.
33. A.S. Sathaye. *Logical Analysis and Control of Real-Time Discrete Event Systems.* PhD thesis, Carnegie Mellon University, 1993.
34. R.S. Sreenivas. A note on deciding the controllability of a language K with respect to a language L. *IEEE Transactions on Automatic Control*, pages 658–662, April 1993.
35. T. Ushio. Maximally permissive feedback and modular control synthesis in Petri nets with external input places. *IEEE Transactions on Automatic Control*, 35(7), July 1990.

Functional and Performance Analysis of Cooperating Sequential Processes*

E. Teruel, M. Silva, J.M. Colom and J. Campos

Dpt. Ing. Eléctrica e Informática. CPS, Univ. Zaragoza
María de Luna 3, E-50015 Zaragoza, Spain
Fax: + 34 76 512932 e-mail: eteruel@cc.unizar.es

1 Introduction

The design of concurrent and distributed systems is a complex task, compelling the use of formal methods. A major trend in the modelling of such systems is the use of a single formalism during the entire design process [9], which should provide: (1) Basic modelling features; (2) A well founded logical theory providing the definition and validation of functional properties; (3) A natural representation of time and the possibility of quantitative analysis of performance properties. A modelling paradigm satisfying the above requirements are Petri nets [3, 10, 13], extended with the corresponding time representations.

In this paper we concentrate on systems obtained by the application of a simple modular design principle: several *sequential processes* execute concurrently and *cooperate using asynchronous communication by message passing through a set of buffers*. As soon as a process S, the *sender*, needs to communicate with another one R, the *receiver*, S deposits *messages* in a buffer, which are taken by R when it is ready. The possible information contained in messages is disregarded, paying attention to the control flow only: messages are considered as *authorizations*. The restrictions imposed to the connectivity of buffers prevent competition and the possibility that buffers condition the resolution of (private) conflicts of a sequential process. Application domains where this class of systems appear include computer networks, information systems, operating systems, real-time systems, nonsequential programming languages, and discrete part manufacturing systems.

Several works exist concerning the analysis of these Petri net models, named *Deterministic Systems of Sequential Processes*: Some aspects of modelling and functional analysis can be found in [11, 12]; Efficient (with polynomial time complexity on the net size) performance analysis for some subclass is presented in [6]. In this paper we try to bridge qualitative and quantitative aspects, with the goal of obtaining benefits in both the validation of functional properties and the evaluation of performance indices of such net systems [14]. The paper is organized as follows: in §2 the classes under consideration are defined. Section 3 deals with some aspects concerning the functional analysis, and

* This work is supported by the projects CICYT ROB-91-0949, and Esprit BRA Project 7269 (QMIPS) and W.G. 6067 (CALIBAN).

§4 considers performance properties. The proofs and further details had to be omitted, but the interested reader can find them in [4].

2 Deterministic Systems of Sequential Processes

In this section we define the class of Deterministic Systems of Sequential Processes as a subclass of Petri net systems, after recalling some definitions and notations about Petri nets [3, 10, 13].

A *P/T net* is a 4-tuple $\mathcal{N} = (P, T, \text{Pre}, \text{Post})$, where P and T are disjoint sets of *places* and *transitions* ($|P| = n$, $|T| = m$), and Pre (Post) is the *pre- (post-) incidence function* representing the input (output) arcs. *Ordinary* nets are those whose pre- and post-incidence functions take values in $\{0, 1\}$. The incidence function of an arc in a non-ordinary net is called *weight* or *multiplicity*. A P/T net can be seen as a bipartite directed graph in which places and transitions are the two kinds of nodes (e.g. the dot notation is used to refer to the pre- or post-set of a node). The *pre- and post-incidence matrices* of the net are defined as $\text{PRE} = [\text{Pre}(p_i, t_j)]$ and $\text{POST} = [\text{Post}(p_i, t_j)]$. The *incidence matrix* of the net is $C = \text{POST} - \text{PRE}$. Right and left natural annullers of C are called *T-* and *P-semiflows* respectively. Some important properties are defined in terms of semiflows and similar vectors (*str.* stands for *structurally*): if $\exists X \geq 1$ such that $C \cdot X = (\geq)0$, then $\mathcal{N}$ is consistent (str. repetitive); If $\exists Y \geq 1$ such that $Y \cdot C = (\leq)0$, then $\mathcal{N}$ is conservative (str. bounded).

A function $M\colon P \to \mathbb{N}$ (usually represented in vector form) is called *marking*. A *P/T system*, $(\mathcal{N}, M_0)$, is a P/T net $\mathcal{N}$ with an *initial marking* M_0. A transition $t \in T$ is *enabled* at marking M iff $\forall p \in P\colon M(p) \geq \text{Pre}(p, t)$. A transition t enabled at M can *fire* yielding a new marking M' defined by $M'(p) = M(p) - \text{Pre}(p, t) + \text{Post}(p, t)$. It is denoted by $M \xrightarrow{t} M'$. A sequence of transitions $\sigma = t_1 t_2 \ldots t_n$ is a *firing sequence* in $(\mathcal{N}, M_0)$ iff there exists a sequence of markings such that $M_0 \xrightarrow{t_1} M_1 \xrightarrow{t_2} M_2 \ldots \xrightarrow{t_n} M_n$. Marking M_n is said to be *reachable* from M_0 by firing σ, and this is denoted by $M_0 \xrightarrow{\sigma} M_n$. The *reachability set* $R(\mathcal{N}, M_0)$ is the set of all markings reachable from the initial marking. A place $p \in P$ is said to be *k-bounded* iff $\forall M \in R(\mathcal{N}, M_0)$, $M(p) \leq k$. A P/T system is said to be (marking) k-bounded iff every place is k-bounded, and bounded iff there exists some k for which it is k-bounded. A P/T system is *live* when every transition can ultimately occur from every reachable marking, and it is *deadlock-free* when at least one transition is enabled at every reachable marking. M is a *home state* in $(\mathcal{N}, M_0)$ iff it is reachable from every reachable marking, and $(\mathcal{N}, M_0)$ is *reversible* iff M_0 is a home state. The name *well-behaved* has been coined for live and bounded systems. A net $\mathcal{N}$ is *str. bounded* when $(\mathcal{N}, M_0)$ is bounded for *every* M_0, and it is *str. live* when *there exists* an M_0 such that $(\mathcal{N}, M_0)$ is live. Consequently, if a net $\mathcal{N}$ is str. bounded and str. live there exists some marking M_0 such that $(\mathcal{N}, M_0)$ is well-behaved. In such case, non well-behavedness is exclusively imputable to the marking, and we say that the net is *well-formed*. For convenience sake, a str. bounded and str. repetitive net will be called *well-structured*. Some well-known relations between these concepts are summarized as follows [3, 5, 10, 13]: Let $(\mathcal{N}, M_0)$ be a connected P/T system. If $\mathcal{N}$ is well-formed, then it is well-structured, which is equivalent to consistent and conservative. If $\mathcal{N}$ is well-structured, then it is strongly connected. If $(\mathcal{N}, M_0)$ is well-behaved, then $\mathcal{N}$ is strongly connected and consistent.

State Machines (SM) are ordinary nets such that $\forall t \in T$: $|{}^\bullet t| = |t^\bullet| = 1$ [3, 10, 13]. Strongly connected SM are, essentially, the Petri net counterpart of classical closed monoclass queueing networks.

Definition 1. A P/T system $((P_1 \cup \ldots \cup P_q \cup B, T_1 \cup \ldots \cup T_q, \text{Pre}, \text{Post}), M_0)$ is a *Deterministic System of Sequential Processes (DSSP)* iff:

1. $\forall i, j \in \{1, \ldots, q\}, i \neq j$: $P_i \cap P_j = \emptyset$, $T_i \cap T_j = \emptyset$, $P_i \cap B = \emptyset$,
2. $\forall i \in \{1, \ldots, q\}$: $(SM_i, M_{0i}) = (P_i, T_i, \text{Pre}_i, \text{Post}_i, M_{0i})$ is a strongly connected and 1-bounded SM (where Pre_i,Post_i, and M_{0i} are the restrictions of Pre, Post, and M_0 to P_i and T_i),
3. the set B of buffers is such that $\forall b \in B$:
   (a) $|{}^\bullet b| \geq 1$ and $|b^\bullet| \geq 1$,
   (b) $\exists i \in \{1, \ldots, q\}$, such that $b^\bullet \subset T_i$,
   (c) $\forall p \in P_1 \cup \ldots \cup P_q$: $t, t' \in p^\bullet \Rightarrow \text{Pre}(b, t) = \text{Pre}(b, t')$.

Items 1 and 2 of the above definition state that a DSSP is composed by a set of SM $(SM_i, i = 1, \ldots, q)$ and a set of buffers (B). By item 3(a), buffers are neither source nor sink places. The output-private condition is expressed by condition 3(b). Requirement 3(c) justifies the word "deterministic" in the name of the class: the marking of buffers does not disturb the decisions taken by a SM, i.e. *choices in the SM are free*. In [12] buffers are required to have not only a single output SM but also a single input SM (input-private). From a queueing network perspective, DSSP are a mild generalization of *Fork-Join Queuing Networks with Blocking* [2], allowing bulk services and arrivals, and having complex servers (1-bounded SM with a rich connectivity to buffers).

Another interesting subclass of P/T nets are *Equal Conflict nets* [16]. Let $\mathcal{N}$ be a P/T net. Two transitions, $t, t' \in T$, are in *Equal Conflict relation* iff $t = t'$ or ${}^\bullet t \cap {}^\bullet t' \neq \emptyset \Rightarrow \forall p \in P$: $\text{Pre}(p, t) = \text{Pre}(p, t')$. This is an equivalence relation on the set of transitions of a net, and every equivalence class is called an *Equal Conflict (set)*. The set of all Equal Conflict sets is denoted by $\mathcal{E}$. $\mathcal{N}$ is an *Equal Conflict (EC) net* iff $\forall t, t' \in T$: ${}^\bullet t \cap {}^\bullet t' \neq \emptyset \Rightarrow \forall p \in P$: $\text{Pre}(p, t) = \text{Pre}(p, t')$. EC nets are such that every choice is free, so they generalize the ordinary subclass of Free Choice nets. Many nice results from the Free Choice theory have been recently extended to EC net systems [16, 17].

Observe that, in a DSSP, any conflict of a SM is Equal by the "deterministic" assumption. All the remaining transitions are elementary Equal Conflict sets. On the sequel, for easy presentation, Equal Conflict sets will be supposed to have at most two transitions (otherwise serialize choices into binary ones). Neither DSSP are a subclass of EC nets nor the converse. Nevertheless, if, for instance, it happens that all buffers of a DSSP, $(\mathcal{N}, M_0)$, have exactly one output transition ($\forall b \in B$: $|b^\bullet| = 1$), that is, if they are *strictly* output-private, then obviously $\mathcal{N}$ is EC, and it is said to be an *Equal Conflict DSSP (EC-DSSP)*. Naturally, EC-DSSP inherit all the nice properties of EC nets and systems. In fact, by means of several *transformations* preserving, among other properties, boundedness, liveness and the existence of home states [4], the results that are valid for the EC-DSSP subclass can be extended to many non EC nets. Observe, though, that *not* all DSSP can be transformed into EC.

Regarding time, we consider net systems with time independent random variables associated to the firing of transitions, which define their *service time*. For the modelling of conflicts *immediate transitions* t_i with the addition of marking and time independent *routing rates* $r_i \in \mathbb{Q}^+$ are used [1]. Consequently, routing is decoupled from duration of activities. The only restriction that this decoupling imposes is that *preemption* cannot be modelled with two timed transitions (in conflict) competing for the tokens. (This is equivalent to the use of a *preselection policy* for the resolution of conflicts among timed transitions.) Assuming the above described time interpretation, the timed model has almost surely the *fair progress* property, that is, no transition can be permanently enabled without firing. Additionally, it has the *local fairness* property, that is, all output transitions of a shared place simultaneously enabled at infinitely many markings will fire infinitely often.

3 Functional Analysis of DSSP

For DSSP, there is an special necessary condition for well-formedness, that we conjecture to be also sufficient (actually, it is proven to be for EC-DSSP [17]), and a simple algebraic sufficient condition for liveness, based on the equivalence of liveness and deadlock-freeness (which is also proven to be necessary for EC-DSSP [16]).

Theorem 2 ([4]). *Let* $(\mathcal{N}, M_0)$ *be a DSSP.*

1. *If* $\mathcal{N}$ *is well-formed, then* $\mathcal{N}$ *is well-structured and* $\text{rank}(C) = |\mathcal{E}| - 1$.
2. *If* $(\mathcal{N}, M_0)$ *is strongly connected and bounded, it is live iff it is deadlock-free.*

If $\mathcal{N}$ is not well-structured, then it cannot be well-formed. If $\mathcal{N}$ is well-structured but $\text{rank}(C) \geq |\mathcal{E}|$, then it is not str. live (Theorem 2.1) and for every initial marking the system would deadlock (Theorem 2.2). In performance terms, in both cases, assuming boundedness, the system would present null throughput for any initial configuration of resources/customers due to a problem that is rooted on the net structure. The problem can be detected in polynomial time (both well-structuredness and the rank are analyzed in polynomial time), so this test should be applied prior to any other.

In case $\mathcal{N}$ is well-formed, the problem is determining whether the initial marking makes the system live or not. To achieve this, Theorem 2.2 can be used, so only deadlock-freeness needs to be proven, instead of liveness. In [15] a general sufficient condition for deadlock-freeness in terms of the absence of integer solutions to a set of systems of linear inequalities is presented. In the particular case of DSSP, among other subclasses, such algebraic condition can be expressed as a *single system of linear inequalities*. This general sufficient condition for deadlock-freeness is also necessary in the case of EC systems [16], and we conjecture that it is so also for DSSP (1-boundedness of the SM is necessary here.)

The existence of home states is a strong property, sometimes appearing in the specification of systems that require states to return to from any point in order to be able to resume their activity. It also facilitates the analysis since the reachability graph is known to be rather particular: it has a unique terminal strongly connected component. It is proven in [16] that well-behaved EC systems have home states, so:

Theorem 3. *Let $(\mathcal{N}, M_0)$ be a well-behaved DSSP-EC. There exists a home state $M \in R(\mathcal{N}, M_0)$.*

Although the result is stated for EC-DSSP, again the result can be extended to many non EC nets by the corresponding net transformations preserving the existence of home states. We conjecture that general DSSP have also home states.

4 Performance Analysis of DSSP

It is well-known that for any net under (possibly marking dependent) exponentially distributed random variables associated to the firing of transitions and Bernouilli trials for the successive resolutions of each conflict, the underlying *Continuous Time Markov Chain (CTMC)* is isomorphous to the reachability graph of the untimed net model [1]. Thus, the existence of home states leads to ergodicity of the marking process for bounded net systems with exponential firing times. After Theorem 3, using the classical expansion of Coxian variables into their exponential stages it follows that:

Theorem 4 ([4]). *Let $(\mathcal{N}, M_0)$ be a well-behaved EC-DSSP with Coxian random variables associated to the firing of transitions and Bernouilli trials for the successive resolutions of each conflict. The underlying CTMC is ergodic.*

The *visit ratio* of transition t_i with respect to t_j, $v_i^{(j)}$, is the average number of times t_i is visited (fired) for each visit to (firing of) the reference transition t_j. To prove that vector $\vec{v}^{(j)}$ depends neither on the marking, provided it allows infinite behaviours, nor on the service times, observe that: (1) if $\mathcal{N}$ is well-structured, the visit ratio vector normalized for t_j, $\vec{v}^{(j)}$, should be a T-semiflow of $\mathcal{N}$; (2) the conflicts in the SM of a DSSP are free, because the buffers do not condition the conflict resolution. Structurally speaking, these conflicts correspond to Equal Conflicts at the net level. Let t_a and t_b be in Equal Conflict relation. Then $r_b\ v_a^{(j)} - r_a\ v_b^{(j)} = 0$. An equation like this holds for every (binary) Equal Conflict. Rewritten in vector form: $(r_{ab}) \cdot \vec{v}^{(j)} = 0$, which for the set of all Equal Conflicts leads to $R \cdot \vec{v}^{(j)} = 0$ where R is a matrix with $m - |\mathcal{E}|$ (number of binary Equal Conflicts) rows and m columns. (In other words, $m - |\mathcal{E}|$ is the number of independent linear relations fixed by the routing rates at binary Equal Conflicts, so rank$(R) = m - |\mathcal{E}|$.)

Theorem 5 ([4]). *Let $\mathcal{N}$ be a well-formed DSSP net. The system of equations:*

$$\begin{pmatrix} C \\ R \end{pmatrix} \cdot \vec{v}^{(j)} = 0, \quad v_j^{(j)} = 1,$$

has only one solution (i.e. the vector of visit ratios depends neither on the marking, provided it allows infinite behaviours, nor on the service times).

Now we present some *insensitive* (i.e. holding for any probability distribution function for the firing times) performance bounds, for the case of marking independent service times. Basically, throughput upper bounds (i.e. mean interfiring time lower bounds) are computed by finding the slowest isolated subnet among those generated by P-semiflows of the net.

Theorem 6 ([7]). *Let* $(\mathcal{N}, M_0)$ *be a DSSP and* $\vec{D}^{(j)}$ *the vector of marking independent average service demands from transitions, with components* $D_i^{(j)} = s_i\, v_i^{(j)}$. *A lower bound for the mean interfiring time* $\Gamma^{(j)}$ *of transition* t_j *is:*

$$\Gamma^{(j)} \geq \max \left\{ Y \cdot \text{PRE} \cdot \vec{D}^{(j)} \mid Y \cdot C = 0 \wedge Y \cdot M_0 = 1 \wedge Y \geq 0 \right\}. \tag{1}$$

We remark that the computation of the above bound has polynomial time complexity on the net size (the computation of $\vec{D}^{(j)}$ is polynomial by Theorem 5 and also linear programming problems can be solved in polynomial time). If the solution of (1) is unbounded then non-liveness (i.e. infinite interfiring time) can be assured. If the visit ratios of all transitions are non-null, the unboundedness of (1) implies that a total deadlock is reached by the system. This result has the following interpretation: if (1) is unbounded then there exists an unmarked P-semiflow, and the system is non-live.

Concerning throughput lower bounds, provided the net system is live, they can be derived by adding the service time of all transitions, weighted by the visit ratios. This computation implies a complete sequentialization of all the activities represented in the model.

Theorem 7 ([7]). *Let* $(\mathcal{N}, M_0)$ *be a well-behaved DSSP. An upper bound for the mean interfiring time* $\Gamma^{(j)}$ *of transition* t_j *is:* $\Gamma^{(j)} \leq \sum_{i=1}^{m} s_i\, v_i^{(j)} = \sum_{i=1}^{m} D_i^{(j)}$.

We remark that the above bound can also be computed in polynomial time, since the vector of visit ratios can be computed in polynomial time (Theorem 5).

Bounds for other performance indices can be computed using classical formulas in queuing theory such as Little's formula (see [7]).

5 Conclusions and Future Work

We have introduced a structured subclass of Petri nets which generalizes the Deterministic Systems of Sequential Processes (DSSP) presented in [12]. It permits the modelling of some cooperating sequential processes: the processes are modelled by 1-bounded State Machines while their cooperation is represented by places called buffers. The output-private and the deterministic assumptions, together with 1-boundedness of the State Machines, preclude competition.

We have presented some results concerning the functional and performance analysis of these models, and we have outlined some extensions and conjectures, which are known to hold for the subclass of Equal Conflict net systems, and which we are working out at present.

References

1. M. Ajmone-Marsan, G. Balbo, and G. Conte. *Performance Models of Multiprocessor Systems.* MIT Press, 1986.
2. H. Ammar and B. Gershwin. Equivalence relations of queueing models of Fork/Join networks with blocking. *Performance Evaluation*, 10 (1989) 233–345.
3. G.W. BRAMS. *Réseaux de Petri: Théorie et Pratique.* Masson, 1983.

4. J. Campos, J.M. Colom, M. Silva, E. Teruel. Functional and performance analysis of cooperating sequential processes. *GISI RR-93-20*. Univ. Zaragoza. 1993.
5. J.M. Colom, J. Campos, and M. Silva. On liveness analysis through linear algebraic techniques. In *Procs. of the AGM of ESPRIT BRA 3148 (DEMON)*, 1990.
6. J. Campos and M. Silva. Steady-state performance evaluation of totally open systems of Markovian sequential processes. In *Decentralized Systems*, 427–438. North-Holland, 1990.
7. J. Campos and M. Silva. Structural Techniques and Performance Bounds of Stochastic Petri Net Models. In *LNCS* 609: 352–391. Springer-Verlag, 1992.
8. E. Gelenbe and G. Pujolle. *Introduction to Queuing Networks*. Wiley, 1987.
9. U. Herzog. Performance evaluation and formal description. In *Procs. of the IEEE Conference CompEuro 91*, 750–756. Bologna, 1991.
10. T. Murata. Petri nets: Properties, analysis, and applications. *Proceedings of the IEEE*, 77(4):541–580, 1989.
11. W. Reisig. Deterministic buffer synchronization of sequential processes. *Acta Informatica* 18, 117–134, 1982.
12. Y. Souissi and N. Beldiceanu. Deterministic Systems of Sequential Processes: Theory and tools. In *LNCS* 335: 380–400. Springer-Verlag, 1988.
13. M. Silva. *Las Redes de Petri: en la Automática y la Informática*. AC, 1985.
14. M. Silva. Interleaving functional and performance structural analysis of net models. In *LNCS* 691: 17–23. Springer-Verlag, 1993.
15. E. Teruel, J.M. Colom and M. Silva. Linear analysis of deadlock-freeness of Petri net models. In *Procs. of ECC '93*, 2:513–518. North-Holland, 1993.
16. E. Teruel and M. Silva. Liveness and home states in Equal Conflict systems. In *LNCS* 691: 415–432. Springer-Verlag, 1993.
17. E. Teruel and M. Silva. Well-formedness and decomposability of Equal Conflict systems. *GISI RR-93-22*. Univ. Zaragoza. 1993.

Hierarchically Combined Queueing Petri Nets

Falko Bause, Peter Buchholz and Peter Kemper

University of Dortmund, Informatik IV, D-44221 Dortmund, Germany
e-mail: {bause,buchholz,kemper}@ls4.informatik.uni-dortmund.de

1 Introduction

Combining Petri Nets with Queueing Networks yields a modelling formalism which allows a convenient description of queues and scheduling strategies as well as conflicts and synchronisation aspects. In this paper we combine Petri Nets and Queueing Networks to a two-level hierarchy, hence called hierarchically combined Queueing Petri Nets. This hierarchical modelling formalism supports an efficient technique for quantitative analysis based on Markov chains. The analysis technique allows analysis of models with a state space about 10 times larger than conventional algorithms do. If the model description reveals symmetries this factor grows towards several orders of magnitude. The benefits of the approach are illustrated by a non-trivial example.

2 The QPN modelling formalism

Queueing Petri Nets (QPNs) [BB1, Ba1] combine Coloured Generalized Stochastic Petri Nets (CGSPNs) [AB1] with Queueing Networks (QNs) by hiding stations in special places of the CGSPN which are called timed queueing places. Here we regard a natural generalisation of QPNs that allows to integrate a QN into a place of the CGSPN. This specification can be regarded as a two-level description of a system where the QNs of all timed queueing places form the low level part of the model and the Petri Net structure is seen as the top level.

Figure 1 i) depicts such a timed queueing place and its pictorial representation. Here a QN with two stations is integrated into the place of a CGSPN. Such a timed queueing place consists of two components, the QN and a depository for tokens having completed their service at this QN. Tokens, when fired onto a timed queueing place by any of its input transitions, are inserted into one of the queues of the embedded QN. Tokens within a QN are not available for the QPN transitions. After completion of its service at a QN station, a token moves to a different one or leaves the QN part. Tokens that leave the QN part move to the depository of the timed queueing place. Here tokens are available for all output transitions of the timed queueing place. An enabled timed transition will fire after a certain exponentially distributed delay according to a race policy like in GSPNs. Enabled immediate transitions will fire according to relative firing frequencies. The

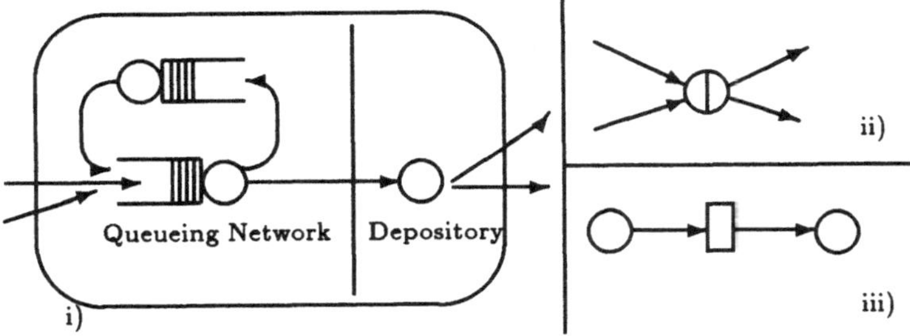

Fig. 1. i) Example timed queueing place of a QPN, ii) its pictorial representation, and iii) its simplified interpretation from a Petri Net point of view

firings of immediate transitions have priority over those of timed transitions. Every QPN describes a stochastic process which can be analysed by Markovian analysis techniques [Ba1].

3 Generation of the Markov Chain for a QPN

Apart from a more convenient specification the hierarchical structure of a QPN can be exploited for solution purposes. The solution is based on the analysis of the Markov chain (MC) underlying a QPN. The MC has a structure which is strongly related to the model structure and therefore allows a very efficient computation of state space and generator matrix. Furthermore the structure can be exploited for various analysis techniques as shown below. However, due to space limitations we can only give a very brief description of the approach, for a more comprehensive introduction the reader is referred to [Bu1, Bu2].

On the PN level of the QPN each timed queueing place behaves comparable to two places connected by a timed transition. The first place includes tokens actually in the QN underlying the timed place, the second place includes the tokens on the depository, which are ready to leave, see Fig. 1 iii. Let P_{QN} be the set of all places including QNs and T_{QN} the set of newly introduced transitions specifying the output behaviour of the embedded QNs. We impose the following restrictions onto the embedded QNs to simplify the description of the analysis technique: Tokens do not change their colour inside a QN and whenever there is at least one colour c token inside a QN this token can be the next to leave.

These restrictions ensure that transitions from T_{QN} fire at a given time one token from the first to the second place without changing the token colour. A transition $t \in T_{QN}$ does not describe an exponentially distributed delay, however, it behaves qualitatively like an ordinary (exponential) timed transition due to the infinite support of exponential distributions. Note that this point of view just considers qualitative aspects concerning the state space of the PN part. The state space of the PN can be computed using standard methods for GSPNs (see [AB1]) without taking care of the embedded QNs. Vanishing markings can be eliminated locally in the PN part of the model using known techniques from GSPN analysis [AB1] without effecting any embedded QN. Let M_0 be the marking space of tangible markings resulting from that net. M_0 is, of course, not isomorphic to the state space of the underlying MC, but describes an aggregated view on this state space by hiding the detailed states of all included QNs. Let Q_0 be the transition matrix of the PN part. Each non-zero in Q_0 is originated by a timed transition in the PN or an entity leaving a QN. If it is originated by a timed transition, then $Q_0(n, m) = \lambda_{n,m}$, where $\lambda_{n,m}$ is given by the firing rate of the timed transition possibly multiplied with the probabilities of subsequent firing immediate transition to yield marking m. If it is originated by a transition from T_{QN}, then $Q_0(n, m) = (p_{n,m}, i)$, where $p_{n,m}$ is given by the probabilities of firing immediate transitions to yield marking m and i is the number of the QN where the customer (token) leaves. Furthermore define $Q_0(n, n) = \lambda_n = -\sum_{m \neq n} \lambda_{n,m}$.

The markings from M_0 determine the possible population vectors in the embedded QNs. Let QN_j be the QN embedded in timed queueing place p_j and N_j be the set of all possible population vectors in QN_j. By interpreting the tokens of the PN as customers in the embedded QNs and colours of tokens as customer chains we have $n \in N_j$ if $n(k) = m_j(k)$ for some $m \in M_0$.

The set N_j defines the possible interaction of QN_j with its environment realised by the rest of the model. Thus it is possible to specify QN_j in combination with a finite capacity source and a sink to compute transition matrices and state space. Let the population in QN_j be n, the finite capacity source possibly generates a chain k entity, if $n + e_k \in N_j$. If customers of chain k can be generated, then the interarrival time is exponentially distributed with a fictive rate $\lambda > 0$. Inside QN_j customers of chain k can have an additional class identity from a set of classes $\mathcal{K}_k$ ($\mathcal{K}_k \cap \mathcal{K}_l = \emptyset$ for $k \neq l$). A customer of chain k leaving QN_j looses his class identity and is absorbed by the sink in the isolated specification.

In combination with the source/sink pair QN_j can be interpreted as an isolated model which is mapped on the underlying MC. Let M_j be the state space and Q_j be the corresponding generator matrix. M_j is decomposed into disjoint subsets $M_j(n)$ ($n \in N_j$) according to the actual population in QN_j. In the same way Q_j is decomposed into submatrices $Q_j^{n,m}$ including all transitions between states from $M_j(n)$ and $M_j(m)$. The following interpretation can be used for the different submatrices:

- If $n = m$, then $Q_j^{n,n} = Q_j^n + \sum_{k \in K} \delta(n(k))\lambda I$ and Q_j^n includes all rates of transitions locally in QN_j and in the main diagonal the negative sum of transition rates of possible transitions originated in QN_j.
- If $n - e_k = m$, then $Q_j^{n,m} = S_j^{n,m}$ includes the departure rates of chain k customers from QN_j.
- If $n + e_k = m$, then $Q_j^{n,m} = \lambda U_j^{n,k}$ and $U_j^{n,k}$ includes conditional arrival probabilities of chain k customers into QN_j.
- In all other cases $Q_j^{n,m} = 0$.

Let M be the state space of the complete model, M can be decomposed into disjoint subset $M(n)$ according to the state $n \in M_0$. Thus each state is described by a state of the PN and a state for each embedded QN. M is composed from the isolated state spaces as follows.

$$M = \bigcup_{m \in M_0} M(m) = \bigcup_{m \in M_0} \times_{j \in P_{QN}} M_j(m_j) \tag{1}$$

The size of the overall state space grows with the product over the sizes of the subspaces of the included QNs which yields the well known state explosion. Nevertheless

M has a very regular structure and can be generated efficiently from the isolated parts. The next step is to compute the generator matrix Q of the MC underlying the QPN. This matrix can be computed from the isolated matrices following the ideas of state space generation. Q is decomposed into submatrices $Q^{n,m}$ according to the state of the PN part. Before we describe the computation of submatrices some additional functions have to be defined. The following two functions are used to map the state of a QN into the state of the complete model.

$$l_j(n) = \prod_{i<j, p_i \in P_{QN}} \|M_i(n_i)\| \ , \qquad u_j(n) = \prod_{i>j, p_i \in P_{QN}} \|M_i(n_i)\| \tag{2}$$

where $\| \ldots \|$ denotes the cardinality of a set.

To describe the behaviour of QNs embedded in the PN we have to specify the behaviour after the arrival of a batch of customers which is possible from the PN but has not been considered in the isolated specification. Define for $n, m \in N_j$ with $n(k) \leq m(k)$ an ordered set $\Delta_j(n, m) = \{(n^1, k^1), \ldots, (n^X, k^X)\}$ such that $n^x = n^{x-1} + e_k$ for some $k \in K, \ 0 \leq x \leq X$, where $n^0 = n$ and $n^X = m$. The difference $m - n$ defines the arriving batch of customers, $\Delta_j(n, m)$ is used to define an ordering on this set, thus the batch is handled like a sequence of customers arriving at the same time.

$$U_j^{n,m} = \prod_{(n',k) \in \Delta_j(n,m)} U_j^{n',k} \tag{3}$$

Furthermore let $A(n, m)$ for $Q_0(n, m) \neq 0$ $(n, m \in M_0)$ be the set of QNs which receive new customers when the marking of the PN changes from n to m. Elements from $A(n, m)$ are ordered according to the numbering of the involved QNs. Now we can compute the submatrices $Q^{n,m}$ as follows (see [Bu1, Bu2] for further details).

$$Q^{n,m} = \begin{cases} \lambda_n I_{\|M(n)\|} + \sum_{j \in P_{QN}} I_{l_j(n)} \otimes Q_j^{n_j} \otimes I_{u_j(n)} & \text{if } n = m \\ \lambda_{n,m} \prod_{j \in A(n,m)} I_{l_j(m)} \otimes U_j^{n_j,m_j} \otimes I_{u_j(n)} & \text{if } Q_0(n,m) = \lambda_{n,m} \\ p_{n,m} I_{l_i(n)} \otimes S_i^{n_j,m_j} \otimes I_{u_i(n)} \prod\limits_{j \in A(n,m)} I_{l_j(m)} \otimes U_j^{n_j,m_j} \otimes I_{u_j(n')} & \\ & \text{if } Q_0(n,m) = p_{n,m}, i \\ 0 & \text{if } Q_0(n,m) = 0 \end{cases} \tag{4}$$

where I_x is the identity matrix of order x and $\otimes$ is the tensor product; $n'_j = n_j$ if $j \neq i$ and $n'_i = m_i$.

With the above formula the generator matrix can be computed efficiently, but the size of the matrix for more complex models is often too large to be stored completely using todays computers. In the next section we consider very briefly analysis methods exploiting the structure of the MC to deal with large models.

4 Analysis of QPNs

Analysis of complex MCs faces two problems: time and space requirements. The state space of MCs underlying realistic models often contains more than 10^6 states. Although generator matrices are very sparse, they can exceed storage capacities and even the generation of the state space is not possible. Thus memory requirements are often the most important problem. The use of virtual memory to increase space is often of no help since access time to virtual memory is slow increasing the overall solution time dramatically. Our main goal here is to manage the space problem which often implicitly attacks time complexity.

State space generation as defined above is very efficient since only small subspaces are generated using the conventional algorithms which need to compare a newly computed state with the states already generated. The complete state space and generator is computed using well defined operations which are very efficient. However, this does not help for the solution. The solution requires the determination of the steady state distribution π given as the solution $\pi Q = 0$ normalised to 1, or the transient distribution at time t $\pi(t)$ given as the solution of $\pi(t) = \pi(0)e^{Qt}$. For both solutions iterative solvers are used, e.g. the iteration

$$p^{k+1} = p^k + 1/q_{\max} p^k Q \text{ where } q_{\max} = 1.001 \max(|Q(i,i)|) \tag{5}$$

converges for irreducible MC and $k \to \infty$ against π. The vector $\pi(t)$ is given by $\sum_{k=0}^{\infty} p^k e^{q_{\max} t} (q_{\max} t)^k / k!$, where $p^0 = \pi(0)$ and the vectors p^k are computed as shown above. Other iterative approaches, in particular for stationary analysis, exist, but they all follow similar ideas.

To use the structure of the MC in the solution we decompose all vectors into subvectors according to the state of the PN, i.e. π_n is a $\|Z(n)\|$-dimensional vector including the steady state probabilities of the states from $Z(n)$. With this decomposition the above iteration is equivalent to:

$$p_n^{k+1} = p_n^k + 1/q_{\max} \sum p_m^k Q^{m,n} \tag{6}$$

Each product $p_m^k Q^{m,n}$ is the sum or product of matrices of the form $I \otimes A \otimes I$. Such a matrix has a very regular structure and the multiplication of a vector with a matrix of this type can be performed using only A and the dimension of the identity matrix. A procedure performing the multiplication is given in [Bu2]. Thus it is possible to iterate using only Q_0 and the QN matrices, the huge matrix Q never needs to be computed and stored as a whole. Compared with conventional techniques, computing first Q, MCs can be solved which are approximately larger by the order of a magnitude. With the new approach the iteration and solution vector becomes the limiting factor of an analysis and not the generator matrix, thus we reach some natural limit for analysis approaches which compute the state vector as a whole. However, the use of tensor products only reduces the space complexity, the number of operations is similar to the conventional approach. Nevertheless, the additional effort for some address transformations is rather small for non-trivial QNs due to the large amount of data locality given by the QN matrices.

To decrease also the time complexity of a solution, a special structure of the model is required or an approximation error has to be introduced. A special structure is given, if the PN is symmetric and the embedded QNs in symmetric parts are identical, then an aggregated MC resulting from lumpability can be generated and used for the analysis (see [Bu3]).

5 Example

This example illustrates the modelling and analysis of distributed systems using the QPN formalism. The system architecture is given by $n + 1$ processors (proc$_j$, manager) with local memory (LM). One processor (manager) is dedicated to a special task, collecting results and informing each processor of the new status. This processor is a multiprocessor system consisting of two processors with separate queues and arriving jobs are inserted into the shortest queue. All processors are connected via a bidirectional bus which must be used exclusively. This architecture is employed for the following iteration problem:

$$f(x^k) = x^{k+1} \text{ where } f(x) = (f_1(x), \ldots, f_t(x)); x, x^k, x^{k+1} \in \mathbb{R}^m \tag{7}$$

f is a vectorial function and each processor proc$_j$ calculates the part $f_i(x^k)$ of f for a given x^k where $i = j + np$ for some $p \in \mathbb{N}_0$. We assume that calculation of the fixed point takes a huge number of iterations, and that the complexity of computing $f_i(x^k)$ requires several local memory accesses at each local processor. Furthermore we assume that $t > n$, i.e. the size of the problem is greater than the number of available processors. The whole system behaves as follows:

After computation of $f_i(x^k)$ processor j transmits the result (x_i^{k+1}) to the manager who updates its local copy of x^k. If the manager has received the results of all processors, i.e. the value $f_i(x^k)$ for all i, it distributes the updated vector x^{k+1} to all processors via broadcasting. Figure 2 shows the corresponding QPN model using the expression representation for Coloured Petri Nets [Je1]. Each processor is modelled by a timed queueing place whose QN part is as in Fig. 1 i). The components x_i^k and $f_i(x^k)$ are represented by a token of colour f_i. After computation proc$_j$ transmits $f_i(x^n)$ to the manager provided the bus is available. The manager updates its local copy. The synchronisation aspect of the iteration is represented by t_2 which consumes t tokens from manager if the bus is available. Broadcasting is modelled by t_1 firing tokens of colour f_i.

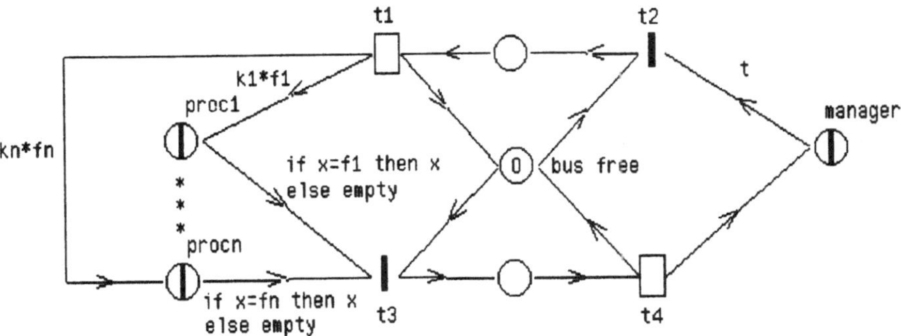

Fig. 2. QPN with $k_i = (t \text{ div } n) + g(i, n, t)$ and $g(i, n, t) := 1$ if $i \leq t \bmod n$ and 0 else.

The example model has been analysed for $n = 5$ processors and an increasing number of tasks. Table 1 summarises the performance of the analysis approaches. All experiments have been performed on a workstation with 24MB main memory and 42MB

Table 1. Sizes and solver performance for different configurations

		conventional		hierarchical		reduced		
t	$\|M\|$	non-zeros	time	non-zeros	time	$\|M_{red}\|$	non-zeros	time
9	6.42e+4	5.41e+5	3.80e+3	4.64e+3	7.45e+2	6.00e+3	6.33e+2	1.89e+2
13	1.18e+6	–	–	2.37e+4	2.14e+4	7.76e+4	2.02e+3	2.32e+3
20	–	–	–	–	–	1.71e+6	7.62e+3	2.05e+5

virtual memory. The first column of the table includes the value of t, the second column the number of states of the MC, column 3 and 4 show the number of non-zero elements and the elapsed time in seconds which is needed to compute the solution vector with an estimated accuracy of 0.01% including the times to generate state spaces and matrices from the model specification for the conventional approach. Columns 5 and 6 include the corresponding quantities for the hierarchical approach. For the conventional analysis immediate transitions are eliminated on net level. The last three columns of the table show the number of states, non-zero elements in the submatrices and analysis time for a reduced MC resulting from an exploitation of the identities of processors. The technique used for reduction has been introduced in [Bu3]. The largest MC which can be generated and analysed with the conventional approach results from the model with $t = 9$, for $t = 10$ we were not even able to generate the state space in an acceptable amount of time (i.e. less than 24 hours). With the new approach we can analyse the model for $t = 13$ which results in a MC with a state space twenty times larger than the one to be analysable with conventional techniques. Additionally the new approach is faster for large models since paging is not necessary. Exploitation of symmetries yields a significant increase in the size of models which are solvable. Combining this reduction with the new analysis technique allows us to analyse models with t up to 20 which results in a MC about 3 orders of a magnitude larger than the MC solvable with the conventional approach.

References

[AB1] Ajmone-Marsan, M., Balbo, G., Conte, A.: Performance models of multiprocessor systems. MIT Press, London, 1986.

[BB1] Bause, F., Beilner, H.: Eine Modellwelt zur Integration von Warteschlangen- und Petri-Netz-Modellen. Springer IFB 218, 1989, pp. 190-204.

[Ba1] Bause, F.: Queueing Petri nets: a Formalism for the Combined Qualitative and Quantitative Analysis of Systems. 5th International Workshop on Petri Nets and Performance Models, Toulouse (France), 1993.

[Bu1] Buchholz, P.: Hierarchies in Colored GSPNs. Springer LNCS 691, 1993.

[Bu2] Buchholz, P.: A Class of Hierarchical Queueing Networks and Their Analysis. to appear in QUESTA.

[Bu3] Buchholz, P.: Hierarchical Markovian Models -Symmetries and Reduction-. Computer Performance Evaluation '92, Edinburgh University Press, 1993.

[Je1] Jensen, K.: Coloured Petri Nets. Springer EATCS, 1992.

Optimizing structural analysis of extended Petri nets models

Luca Ferrarini [(1)] *and Massimo Trioni* [(2)]

[1] Dipartimento di Elettronica e Informazione, Politecnico di Milano, P.za L. da Vinci n° 32, 20133 Milano (ITALY), Fax. (+39) 2 2399 3587, ferrarin@ipmel2.elet.polimi.it

[2] former student at Politecnico di Milano, now graduated

1 Introduction

The paper deals with the problem of the modeling, analysis and synthesis of the controllers for discrete event systems [2,8,11]. The adopted approach is quite general, and may be exploited in a number of fields. In particular, it has been successfully applied in the design of programmable logic controllers [3,8,11] and of manufacturing systems [5,6].

The proposed approach is based on Petri nets [2,9,10] modules combined in a bottom-up fashion, as opposed to other, still effective, approaches in the manufacturing field [12]. The designer defines *Elementary Control Tasks* (*ECTs*), modeled as strongly connected state machines with just one token, and then to connect these tasks with one another with three kinds of connections, namely *self-loops*, *inhibitor arcs* and *synchronizations*. These connections lead to bounds for the behavior of an ECT, called *connection bundles*. No particular restrictive rule is given for the connections. Thus, it is often the case that the resulting net is not live or that the initial marking is not reversible.

In [5] it was shown that the net liveness and the reversibility of initial marking may be studied in terms of structural properties, that is, in term of the underlying graph of a Petri net. In [6] the case of ECTs with a particular structure recalling that of the if-the-else one used in programming languages was considered, with a special emphasis on the incremental approach. The reversibility and reachability problems for such nets were considered in [4].

One of the original aspects in the proposed approach is the way in which the analysis can be carried out. The two basic notions are those of *module* and *loop*. The module constitutes the first step of the decomposition of the net and is composed of an ECT and one input and one output connection bundle. The loop is a determined by a set of modules such that the output connection bundle of a module is the input of another one.

The present work describes how the analysis for liveness and reachability/reversibility can be efficiently performed. In particular, this work shows that there are a number of modules, whose characteristics can be quickly decided, that can be neglected in the analysis. This allows one to avoid first the construction of all the possible loops containing such modules and then the application of checking algorithms to those loops. In sections 2, 3 and 4 the model definition and the main results relating to the liveness and reversibility analysis will be briefly recalled, to make the comprehension of the remainder of the paper possible.

2 Definition of the Petri net model [4,5]

The model used in this paper to describe a discrete event system is conceived as decomposed into tasks, called Elementary Control Tasks (ECTs). If the model represents a logic controller, such structures may model the cyclic sequential tasks devoted to the control of a specific device as well as a work cell.

Definition 1. A strongly connected state machine (P,T,F,M_o) whose initial marking is such that only one place is marked with just a token while all the other places are unmarked (*unitary marking*) is said to be an Elementary Control Task (ECT).

Strongly connected state machines with an initial unitary marking are typically used to model cyclic sequential tasks. They will always have a unitary marking, are live and reversible, are covered by p-invariants and t-invariants (their properties are easy to obtain, see [9,10,12]).

The connections between different ECTs are now introduced.

Definition 2. Let p* and p** be places of an ECT E_1, and t* a transition of an other ECT E_2. The following connections between E_1 and E_2 are defined:

1. self-loop (SL), starting from place p* and with arrival transition t* (See Fig. 1); transition t* may be enabled only if the place p is *marked*, while holding the usual enabling conditions for the places connected to t* with normal arcs.
2. inhibitor arc (IA), starting from place p** and with arrival transition t* (see Fig. 1); transition t* may be enabled only if the place p is *unmarked*, while holding the usual enabling conditions for the places connected to t* with normal arcs.

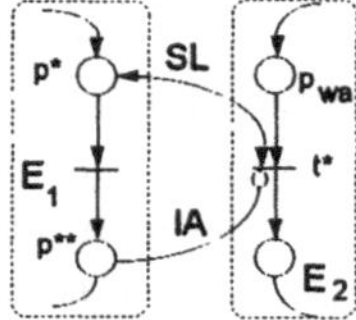

Fig. 1 - SL and IA

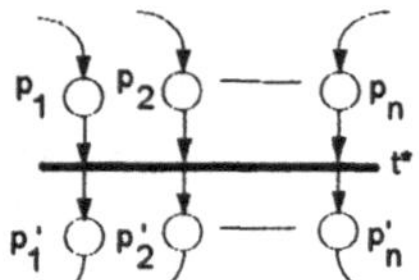

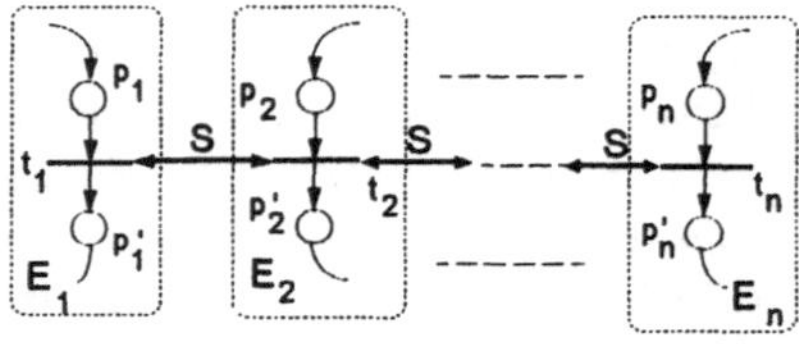

Fig. 2 - Synchronization (a) and its graphic symbol (b)

3. (simple) synchronization (S) is the merging of two or more transitions, each of them belonging to a different ECT, into one called t*, as shown in Fig. 2a; the graphic symbol for a synchronization is depicted in Fig. 2b and consists in a double arrow line between any pair of graphically consecutive transitions involved by this connection.

The usage of a self-loop (an inhibitor arc) from E_1 to E_2 allows to influence the behavior of E_2 with the *presence* (*absence*) of a token in a place of E_1 *without modifying* the behavior of E_1; a synchronization represents a *rendez-vous* between two ECTs.

Remark 1: Henceforth, when referring to a net N without any specification, it is meant that N is made by ECTs connected by the connections given in Def. 2. Thus, the only acceptable markings are those in which just one place in each ECT is marked with just one token and the other places are unmarked.

Finally, a transition at which at least one connecting arc (self-loop, inhibitor arc, synchronization) arrives is called conditioned transition. It is apparent from what has been exposed till now that eventual liveness problems are necessarily associated with conditioned transitions.

3 Basic Structures For The Analysis [5]

Definition 3. Let E_1 and E_2 be two ECTs of a net; the set of connections from E_1 to a transition t* of E_2 is said to be the Connection Bundle (CB) from E_1 to t*.

Connection bundles are worth defining, because the influence of an ECT E_1 on another one E_2, due to the connections starting from E_1 and arriving at a conditioned transition t* of E_2, may be dealt with as a whole in terms of *markings enabling the transition t**. Thus, any connection of a net belongs to at least a CB and a connection bundle can be made of one or more connections. Actually, a *unilateral* connection such as SL or IA gives rise to a single CB, while a *bilateral* connection such as a synchronization S between E_1 and E_2, produces two different CBs: one from E_1 to E_2 and the other one from E_2 to E_1. We stress that two or more connections make a single connection bundle when they start from just one ECT and arrive at just one transition t*.

Remark 2. A synchronization S among two or more ECTs, as shown in Fig. 2b, leads to a couple of CBs *between any pair of ECTs involved in S* even though, for the sake of brevity, only few arcs really introduced by an S connection are graphically represented.

A CB C_1 from an ECT E_1 to an ECT E_2 may be simply characterized by two entities. The first one is the transition it conditions, say t*, in E_2, or, equivalently, the place in the preset of t*, called wait place (WA(t*)). The second one is the set of enabling places, i.e., the set of places in E_1 that necessarily must be marked in all the markings enabling t*. A marking in which all the marked places are wait places relative to some transition is said to be a wait marking. As an example, consider the net composed of the ECT E_1 and E_2 and connection S shown in Fig. 2b. In such a net there are *two* CBs: C_1 consisting of the synchronization S in which E_2 influences E_1 (the wait place relative to t_1 is p_1) and C_2 also consisting of S in which E_1 influences E_2 (the wait place relative to t_2 is p_2). The only wait marking of N is the one where p_1 and p_2 are marked. Finally, the set of the enabling places for t_1 (this relates to C_1) is made just by p_2 and, similarly, the set of the enabling places for t_2 is $\{p_1\}$ (this relates to C_1).

Remark 3: When the set of the enabling places of E relative to t* is empty, the conditioned transition of that bundle can never fire; since these cases are trivial for analysis and represent just a few cases, listed in [6], they will not be considered any longer.

Now, to analyze a net, three structures of ECTs and the CBs will be defined in the following.

Definition 4. In a net N, a set of ECTs and CBs among them is said to be a net of bundles (B-net) of N.

The decomposition of a net into B-nets allows the development of an analysis theoretical body for the identification of the minimum number of ECTs and CBs which are responsible for the liveness loss of the net. In addition, particular B-nets, called *loop*, can successfully be exploited to find not reversible markings. As an example, consider the net composed of the ECT E_1 and E_2 and connection S shown in Fig. 2b, where there are two CBs, C_1 and C_2, introduced by the S connection, it is possible to identify three different B-nets that are: $B_1=\{E_1, E_2, C_1\}$; $B_2=\{E_1, E_2, C_2\}$; $B_3=\{E_1, E_2, C_1, C_2\}$.

Definition 5. Let N be a net with at least two ECTs. A module of N consists of: *i)* an ECT E of N; *ii)* a connection bundle CB_{in} from *any* other ECT of N to a transition t~ of E; *iii)* a connection bundle CB_{out} from E to a transition t of *any* other ECT of N.

Observe that, given an ECT E_1, there may be more than one input connection bundle to one or more transitions of E_1 from other ECTs and more than one output connection bundle from E_1 to some other ECT. Thus, one ECT may originate many modules, since a module is composed of only one input connection bundle and only one output connection bundle. In each module MOD it is possible to identify: one conditioned transition t~(MOD), which is the only conditioned transition of MOD, one wait place of the module, which is WA(t~), and the set of enabling places of CB_{out}.

To analyze a net, the properties of a single module have been investigated. In particular, given a module MOD, it has been studied how the presence of CB_{in}(MOD) conditions the firing of the arrival transition of CB_{out}(MOD). Starting from the wait marking of the module M_{wa}(MOD) the CB_{out} is said to be locked if no marking M* where an enabling place of MOD is marked (called enabling marking of MOD) is reachable from M_{wa}(MOD) without the firing of the transition t~ conditioned by CB_{in}; CB_{out} is said to be free if M_{wa} is an enabling marking ; CB_{out} is finally, said to be quasi-locked if it is not free but exists an enabling marking M* reachable from M_{wa}(MOD) without the firing of t~.

Definition 6. Consider, in a net N, n (n>1) modules MOD_i of N, i=1,...n, such that: 1) each module contains a different ECT; 2) $CB_{out}(MOD_i)=CB_{in}(MOD_{i+1})$ for i=1,...n-1; 3) $CB_{out}(MOD_n)= CB_{in}(MOD_1)$.

Then, B-net containing the ECTs and CBs of MOD_i, i=1,...n, ; is said to be a loop of n-th order of N.

Since for each ECT of a loop there exists just one module with that ECT, the locution "modules of a loop", frequently used in the sequel, can not determine any misunderstanding. In particular, observe that for every ECT of a loop there is just one wait place and just one conditioned transition. Loops and modules may be given a simple graphical representation, called net graph. The net graph is a graph with directed arcs, which is associated to a B-net B_1 according to this rule: a node in the net graph is inserted for each ECT of B_1, and an arc is inserted between two nodes iff there exists a CB between the corresponding ECTs. An example of net graph is reported in the following (Fig. 5b).

Finally, according to the types of CB_{out} of modules of a loop, we may distinguish three cases: a) a loop whose modules all have the CB_{out} locked is said to be locked; b) a loop that is not locked and whose modules all have the CB_{out} locked or quasi-locked is said to be quasi-locked; c) a loop that has at least one module with the CB_{out} free is said to be free.

Properly exploiting the structures just introduced, a theoretical body has been proposed and developed for both liveness and reversibility/reachability properties. The analysis results and method are briefly and informally recalled in the next paragraph to give prominence to the problems due to their application and to understand the advantages of the optimization analysis criteria here stated and proved.

4 Analysis of a Net

Liveness Analysis. In [5] it has been proved that two are the fundamental causes for which a net not containing a dead connection bundle is not live: the presence of particular B-nets, called essentially locked B-nets (ELB-net) and the presence of locked transitions.

An ELB-net represents a subset of the ECTs and connections of a net, in which each conditioned transition of each module can not be enabled if a wait marking is chosen as the initial marking. Thus, an ELB-net defines a mutual wait (*locking condition*) among many ECTs, variably connected, thus extending the classical concept of circular wait. In addition, an ELB-net ELB_1 is a B-net whose CBs all are *essential* for the *locking* of ELB_1, in the sense that every B-nets obtained from ELB_1 neglecting one of the CB of ELB_1 is no more locked (*essentiality condition*).

Basically, it can be proved that the wait marking M_{WA} of an ELB-net ELB_1 and any marking $M \in [M_{WA}>$ is not live markings for a net N containing the same ECTs of ELB_1 and all the connections that contribute to the CBs of ELB_1. In addition, call Σ the set of the markings M above discussed, and N* a generic net containing ELB_1. Consider now a marking M* of N* from which a marking M' is reachable whose restriction to the ECTs of ELB_1 is a marking belonging to Σ. If M* is the initial marking of N*, then N* is not live and ELB_1 is said to be *active* with M*.

The second cause of liveness loss is due to the presence of a locked transition. This situation is due to the impossibility to mark at the same time an enabling place relative to each CB arriving at such a transition, despite that both its wait place and its enabling places can be marked in some marking M reachable from the initial marking.

One of the most important results in the liveness analysis is the proof that the two mentioned causes of loss of liveness are independent of one another ; another crucial property is that if a net is not live, then at least an *active* ELB-net or a locked transition occurs in the net.

A method to easily identify ELB-net has been proposed. Actually, the definition of ELB-net, even if conceptually simple, is not effective. Investigating the topology of ELB-nets, it has been possible to identify some structural properties of such B-nets. Such properties are extremely useful to identify all the ELB-nets of a given net. The most relevant results are the following: 1) an ELB-net ELB_1 is completely covered by loops (that is any ECT and any CB belonging to ELB_1 belongs to at least a loop included in ELB_1); 2) an ELB-net either coincides with a locked loop or is covered by two or more quasi-locked loops and does not contain any free or locked loop.

Using the topological properties 1 and 2 an algorithm to find the ELB-nets of in a given net N has been developed in [5]. Finally, observe that an ELB-net *does not* represent a problem if it is not active.

Remark 4: Consider a connection bundle CB* to a transition t_1, which is involved in a synchronization with another transition t_2. By definition of synchronization, t_1 and t_2 *are* the same transition. Thus, to find all the loops of a net, CB* should be considered as a connection to t_2, as well. This fact clearly shows that, when there is a connection bundle CB* to a transition involved in a synchronization, the number of eventual loops may significantly increase, thus making liveness analysis more time-consuming.

Reversibility and reachability analysis [4]

Definition 7. Let MOD be a module of a net, p* one of its places and M_{p*} the marking of MOD such that M*(p)=1 for p=p* and M*(p)=0 otherwise. In addition, let {EN(MOD)} represents the set of enabling places of the module MOD. Then:

a) If CB_{out}(MOD)≠S, p* is called <u>external</u> iff there exists an enabling place $p_E \in$ {EN(MOD)} such that M_{p*} is reachable from the marking M_E with $M_E(p_E)$=1 without the firing of the transition t~(MOD).

b) If CB_{out}(MOD)=S on a transition t*, with t*≠t~, p* is called <u>external</u> iff, defined M' as that marking in which only the place of MOD in the postset of t* is marked, $M^* \in M'[\sigma >$, where σ is a sequence of transitions in which t~(MOD) does not appear.

c) If CB_{out}(MOD)=S on the transition t~, then no place p* is called external.

d) p* is called <u>internal</u> iff it is not external.

Even if the definition of external places is split in three parts, the idea underlying Def. 7 is unique. External places represent those places of a module that can be reached from an enabling place, without the firing of t~(MOD), when the arrival transition of the output connection bundle of the module is set free to fire.

According to the previous definition, the markings of a loop can be partitioned into two complementary sets: the <u>unrepeatable</u> and <u>repeatable</u> markings. An unrepeatable marking with respect to a loop is a marking where the marked place in any ECT of the loop is internal. The fundamental results on reachability are now stated.

i) In a net coinciding with a loop of n-th order, whose ECTs *are* connected by a single synchronization, then all the markings of the loop are reachable from an initial one and all the markings reachable from an initial one are reversible.

ii) Let N be a net containing a generic loop LP of n-th order, *not* constituted by a unique synchronization among n modules. Let M_{unrep} and M_{rep} be two markings of N, respectively unrepeatable and repeatable with respect to LP. Then $M_{unrep} \notin [M_{rep}>$. In addition, if the net is live, every unrepeatable marking is not reversible. These results can be directly exploited to derive criteria to choose the initial marking of the net.

5 Optimization of Structural Analysis

Optimization of analysis algorithms can be basically achieved through the reduction of the number of modules to be considered. As a matter of fact, this reduction immediately results in a reduction of the loops to build and analyze. Modules and loops may be neglected in the analysis according to two basic criteria. One criterion is based on the consideration that modules and loops may be neglected because they convey no information at all for the analysis. The other criterion allows to neglect modules and loops because the information they convey can be obtained analyzing other modules and loops.

According to the first criterion, the modules such that all their places are external and with a free CB_{out} can be neglected in both liveness and reachability analysis. Such modules are called *superfluous*. In effect, any loop containing a superfluous module is free (thus, it does not generate any non-live markings, as mentioned in section 4.1). In addition, if all the places of a module are external, then there exists no unrepeatable marking with respect to any loop containing that module (by definition. of unrepeatable marking). An example of superfluous module is given in Fig. 3, where the wait place is also enabling.

Another significant reduction can be obtained according to the second reduction criterion above mentioned. This leads to the definition of *trivial modules*.

Definition 9. A module MOD is said to be a trivial module iff CB_{out}(MOD) is determined by a synchronization and is free.

To realize the importance of trivial modules, consider now those loops containing trivial modules. They are free, so they give no information for liveness analysis. As for reachability analysis, it is useful to distinguish the following cases: *i*) the loop is formed only by trivial modules; *ii*) the loop is not formed only by trivial modules. In case *i*) all the places of every module is internal and, as proved in [4], such a loop is composed of ECTs connected by a unique synchronization, and thus it is not associated with non reversible markings. In case *ii*) in the loop some unrepeatable marking, which is not reversible, can be found. Thus, such a loop can not *a priori* be neglected. What we will prove is that, given a loop not constituted only by trivial modules, another loop constituted just by non trivial modules can be found which has the same information as the original one relating to the non reversible markings. Thus we will conclude that those loops containing at least one trivial module are not necessary for the analysis.

To that aim, we will use the following notations: two modules will be said to be equivalent iff they have the same ECT, the same conditioned transition and the CB_{out} is determined by the same connection(s). It is easy to prove that equivalent modules have the same internal (external) places and have the same type (locked, quasi-locked and free) of CB_{out}.

Prop. 1. *Let N be a net containing a loop L with m≥2 non trivial modules and at least a trivial module. Then in N there exists just one loop L' of m-th order constituted by non trivial modules which are equivalent to the non trivial modules of L (have the same internal and external places and the same type of CB_{out} of the non trivial modules of L).*

Proof. The loop L contains some trivial modules by hypothesis. Within the loop L, such modules may be adjacent or not. Let *r* be the number of modules belonging to a group of adjacent trivial modules of L and MOD_i the first of these trivial modules. Without loosing generality, those adjacent modules will be labeled with consecutive numbers (MOD_i, MOD_{i+1}, ... , MOD_{i+r-1}). Since the CB_{out} of a trivial module is determined by a synchronization from the conditioned transition t~ of the module, and since $CB_{out}(MOD_k)=CB_{in}(MOD_{k+1})$, all the conditioned transitions of the *r* trivial modules and the transition t~(MOD_{i+r}) are connected with the same synchronization, say S_1. On the contrary, $CB_{in}(MOD_i)$, say CB*, can not be the synchronization S_1, being MOD_i the first of the *r* considered trivial modules. Thus, MOD_{i-1} (and similarly MOD_{i+r}) is certainly not trivial. This situation is sketched in Fig. 4.

Now, recalling Remark 4, since CB* is a connection bundle to a transition involved in a synchronization (S_1), CB* should be duplicated to any transition synchronized with S_1, that is to t~(MOD_k), $i \leq k \leq i+r$. Let CB' be that connection bundle, originated by the duplication of CB*, conditioning t~(MOD_{i+r}). Then, with CB' two new modules can be defined: MOD^*_{i-1} (with the same ECT and CB_{in} of MOD_{i-1}, but with CB_{out}=CB*) and MOD^*_{i+r} (with the same ECT and CB_{out} of MOD_{i+r}, but with CB_{in}=CB*). Observe that MOD^*_{i-1} (MOD^*_{i+r}) is equivalent to MOD_{i-1} (to

MOD_{i+r}), and thus they have the same internal and external places and, also the same type (locked, quasi-locked and free) of CB_{out}, by construction.

The net containing the loop L contains a loop L* made of the same modules of L except for the trivial modules of L MOD_k, $i \le k \le i+r-1$ and of the modules MOD^*_{i-1} and MOD^*_{i+r} instead of MOD_{i-1} and MOD_{i+r}. Now, if also L* contains trivial modules, the reasoning can be iterated. At the end of the process, a loop L** is obtained which is made of just non trivial modules that are equivalent to the non trivial modules of L.

L** exists since at least MOD_{i-1} and MOD_{i+r} are not trivial and must be different, otherwise the connection bundle CB* (and CB' as well) would not make sense. L** is also unique. Let, by absurd, L' be another loop whose modules are not trivial and equivalent to the non trivial modules of L. Then, the order of L' and L** must be the same. In addition, since the modules of L** are equivalent to those of L, they are also equivalent to those of L'. This means that the 3-tuples of equivalent modules of L, L** and L' have the same conditioned transitions and their CB_{out} is determined by the same connection. Since this hold for all the modules of L', it means that also the CB_{in} of the above 3-tuple is determined by the same connection. Thus, L' and L** coincides. (EOP)

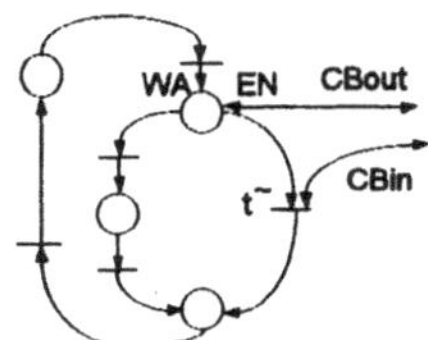

Fig. 3 - Superfluous modules

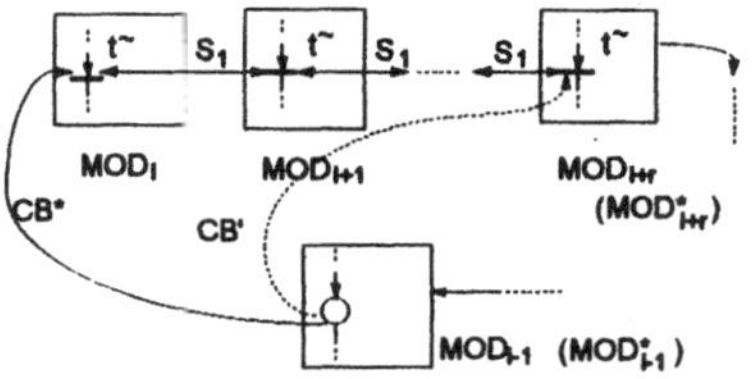

Fig. 4 - Loop with trivial modules

Prop. 2. *Let N be a net containing a loop L with m≥2 non trivial modules and at least a trivial module and let L' be the loop of m-th order whose modules are not trivial and equivalent to the non trivial ones of L. Then, a marking M that is unrepeatable with respect to L is also unrepeatable with respect to L'.*

Proof. Given the loop L, the loop L' exists and is unique. Let M be a marking unrepeatable with respect to L. Then, all the places of L marked under M are internal. Consider now the marking M/L' (the marking M restricted to the places of the ECT of the modules of the loop L'). That marking is definitely internal with respect to L', since the modules of L' are equivalent to those of L. (EOP)

Theorem *1. A loop L containing at least a trivial module can be discarded in the liveness and reachability analysis.*

Proof. Each loop L containing a trivial module is free, by definition; thus, it can be neglected in the analysis of the net liveness. If the loop L is constituted just by trivial modules, then every CB of L is due to a unique synchronization. In this case, no unrepeatable marking can be defined with respect to L.

If the loop L is not constituted just by trivial modules, then not every CB_{out} of the modules of L is the same synchronization. However, for Prop. 1, given L, there exists one and only one loop L' constituted of modules that are equivalent to the non trivial modules of L . In addition, Prop. 2 states that a marking unrepeatable with respect to L is also unrepeatable with respect to L'. Thus, the unrepeatable markings that can be obtained from L, can be obtained from L'. Of course, L' may have "its own" unrepeatable markings (EOP)

6. Example and Concluding Remarks

Let us consider the net N shown in fig. 5a. N is composed of three ECTs, E_1, E_2 and E_3 connected with two synchronizations S_1 and S_2. In fig. 5b the largest B-net belonging to N is described using the synthetic net-graph formalism. In this way *all* the CBs introduced by S_1 and S_2 are depicted, and the identification of all the modules and all the loops belonging to N can be easily carried out simply looking at the picture.

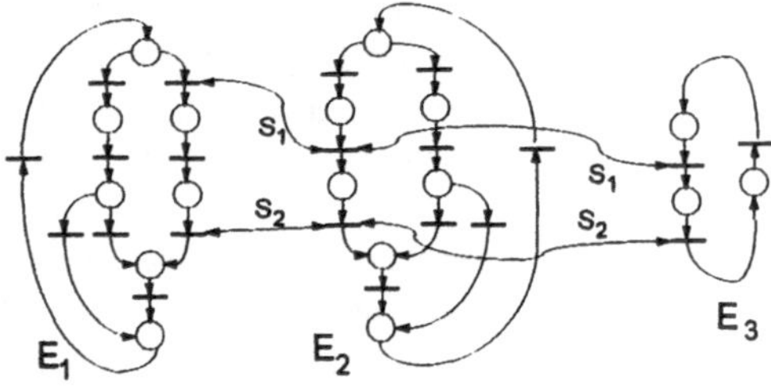

Fig. 5a - Example with 3 ECT

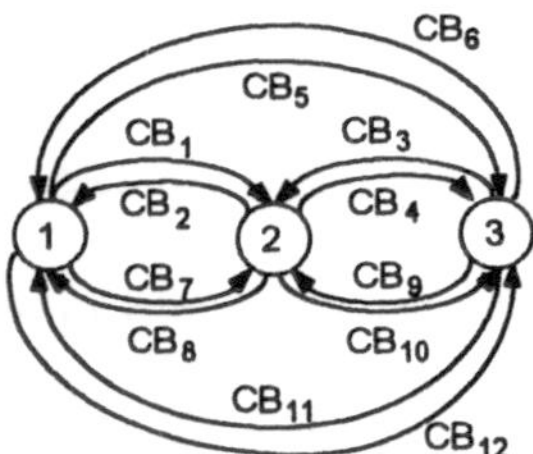

Fig. 5b - Net graph of Fig. 5a

The CBs CB_1 to CB_6 are due to S_1, while the others are due to S_2. In the net N there are 24 modules and 28 loops (12 loops of the 2nd order and 16 loops of the 3rd order). Without using the optimization criteria it is necessary to build and analyze all the 28 loops, while neglecting the trivial modules only *six* loops will be built with consequent reduction of time consumption. The six loops that must be analyzed are: $\{E_1,E_2,CB_1,CB_8\}$; $\{E_1,E_2,CB_2,CB_7\}$; $\{E_1,E_3,CB_6,CB_{12}\}$; $\{E_1,E_3,CB_5,CB_{11}\}$; $\{E_2,E_3,CB_3,CB_{10}\}$; $\{E_2,E_3,CB_4,CB_9\}$. Notice that the gains in applying optimization criteria increases with the number of synchronizations and with the number of ECTs involved in them.

As it is clear from the above example, the criteria illustrated in the present paper allow a sensible reduction of the complexity analysis of nets, especially in the case when synchronizations are used. As illustrated in [6], synchronizations are one of the most used way of describing the interaction among parallel running tasks, as in the case of manufacturing systems. Properly exploiting the notions of trivial and equivalent module, it is possible to develop a checking algorithm that does not a priori compute a great number of loops, since they do not carry information for the analysis. Other reduction techniques will be investigated more directly related to *the order* in which the various loops are constructed.

Acknowledgment. The work has been supported by MURST project "Ingegneria del Controllo".

References

[1] K. P. Brand and J. Kopainski, "Principles and Engineering Process Control with Petri Nets," *IEEE Trans. Aut. Contr.*, vol. 33, no. 2, pp. 138-149, Feb. 1988.

[2] R. David, "Modeling of Dynamic Systems by Petri Nets," *Proc. European Control Conf.* (Grenoble, Fr), July 1991.

[3] L. Ferrarini, "An incremental approach to logic controller design with Petri nets," *IEEE Trans. Syst Man Cybern.*, vol. SMC-22, no. 3, pp.461-473, 1992.

[4] L. Ferrarini, "On the reachability and reversibility problem in a class of Petri nets," accepted for pubblication in *IEEE Trans. Syst Man Cybern.*

[5] L. Ferrarini, A. Lecci and M. Trioni, "Incremental design of logic controllers," Int. Rep. n°93049 Dipartimento di Elettronica e Informazione, Politecnico di Milano, Milano (Italy).

[6] L. Ferrarini, M. Narduzzi and M. Tassan-Solet, "A new approach to modular liveness analysis conceived for large logic controllers' design," to be published on *IEEE Trans. on Rob. & Automation*, (special issue on Computer integrated manufacturing, Part II)

[7] M. D. Jeng and F. DiCesare, "A synthesis method for Petri net modeling of automated manufacturing systems with shared resources" *Proc. of 31st Conf. on Dec. and Control* (Tucson, AZ) 1992 pp.1184-1189.

[8] C. Maffezzoni and L. Ferrarini, "Designing Logic Controllers with Petri Nets," *IFAC Symposium on Computer Aided Design in Control Systems*, (Swansea, GB), July 1991.

[9] T. Murata, "Petri nets: properties, analysis and applications," *Proc. of the IEEE*, vol. 77, n°4, pp 541-580, Apr. 1989.

[10] W. Reisig. *Petri Nets: an Introduction*. Springer Verlag, 1982.

[11] M. Silva, "Logic controllers, " *Proc. Symp. Low Cost Automat.*, (Milan, I), vol. 2, pp. F157-F166bis, 1989.

[12] M. C. Zhou, F. DiCesare and A. Desrochers, "Hybrid methodology for synthesis of Petri nets models for Manufacturing Systems, " *IEEE Trans. Robotics. Autom.*, vol. 8, no.3, pp. 350-361, 1992.

Reduced State Space Generation of Concurrent Systems Using Weak Persistency

Kunihiko Hiraishi

School of Information Science
Japan Advanced Institute of Science and Technology, Hokuriku
15 Asahi-dai, Tatsunokuchi, Nomi-gun, Ishikawa 923-12 Japan (email: hira@jaist.ac.jp)

1 Introduction

State space explosion is a serious problem in analyzing discrete event systems that allow concurrent occurring of events. The purpose of this paper is to propose a new method to reduce the number of states to examine.

Several mathematical models have been proposed for describing discrete event systems, such as automata model, Petri nets, process algebra, etc. Given such a model, its state space is generated in the form of a transition system. To avoid the state space explosion, several methods have been proposed [Wol]. Among these methods, reduced state space generation will be studied here. A reduced state space is constructed by restricting some state transitions of the original state space, and preserves the properties we intend to verify.

The method shown in this paper is an improvement of Valmari's stubborn sets method [Val90, Val92]. Stubborn sets are based on persistency, which is a behavioral property of systems, and is defined as a relationship between a pair of transitions. We are using weak persistency, which is defined as a relationship among a set of transitions. Using Petri nets as a model of discrete event systems, we will show a method for generating reduced state spaces that preserve liveness, livelocks and terminal states.

2 Preliminaries

Let $\mathbb{N}$ denote the set of nonnegative integers. For a finite set A, we identify a function $f : A \to \mathbb{N}$ with a $|A|$ dimensional vector of nonnegative integers and write $f \in \mathbb{N}^A$.

Definition 1. *A state space* is a transition system $TS = (S, E, \tau)$, where S is a nonempty set of states, E is a finite set of *labels*, and $\tau \subset S \times E \times S$ is *a transition relation*.

Let E^* denote the set of all sequences (including the empty sequence λ) over E. Each element of E^* is called *an occurrence sequence*. Let $\psi : E^* \to \mathbb{N}^T$ be a function such that $\psi(\sigma)(e)$ denote the number of occurrences of label e in σ, e.g., $\psi(abaabbc)(a) = 3$. For an occurrence sequence $\sigma \in E^*$, let $\mathrm{supp}(\sigma) = \{e \in E \mid \psi(\sigma)(e) > 0\}$.

For a transition relation τ, let $\tau^* \subset S \times E^* \times S$ denote the smallest set satisfying (i) $\forall s \in S : (s, \lambda, s) \in \tau^*$ and (ii) $(s, e, s') \in \tau \wedge (s', \sigma, s'') \in \tau^* \Rightarrow (s, e\sigma, s'') \in \tau^*$. We write $(s, \sigma >\in \tau^*$ to indicate $\exists s' \in S : (s, \sigma, s') \in \tau^*$. We say that state s' is *reachable* from state s iff $\exists \sigma \in E^* : (s, \sigma, s') \in \tau^*$, and that an occurrence sequence σ is *enabled* at state s iff $(s, \sigma >\in \tau^*$. For a state $s \in S$, let $TS(s)$ denote the subspace consists of all states reachable from s.

A state s is called *terminal* iff there is no label e such that $(s, e >\in \tau$. A subset C of S is called *a strong connected component* iff there is a state $s \in C$ such that $C = \{s' \in S \mid \exists \sigma \in E^* : (s, \sigma, s') \in \tau^* \wedge \exists \sigma' \in E^* : (s', \sigma', s) \in \tau^*\}$. A strong connected component C is called *nonterminal* iff there exist $s \in C$, $s' \in S - C$ and a label e such that $(s, e, s') \in \tau$. Otherwise C is called *terminal*. Terminal strong connected components correspond to livelock situation. A label e is called *live* iff for any $s \in S$, there is an occurrence sequence $\sigma \in E^*$ such that $(s, \sigma e >\in \tau^*$. A state space is called *live* iff every label is live.

3 Weakly Persistent Sets

The main idea of reduced state space generation is selective search, i.e., only some of enabled transitions are executed at each state. For a given state space $TS = (S, E, \tau)$, a reduced state space will be constructed as a transition system $\underline{TS} = (S, E, \underline{\tau})$ such that

(i) $\tau \subset \underline{\tau}$, and
(ii) $\underline{TS}$ preserves liveness, terminal states and terminal strong connected components.

Definition 2. Let $TS = (S, E, \tau)$ be a state space. A subset E_s of E is called *a weakly persistent set* (*W. P. set*, for short) at state s iff

$$\forall e \in E_s : [(s, \sigma e, s') \in \tau^* \wedge (s, e >\in \tau \wedge \mathrm{supp}(\sigma) \cap E_s = \varnothing \\ \Rightarrow \exists \sigma' \in E^* : (s, e\sigma', s') \in \tau^* \wedge \psi(\sigma') = \psi(\sigma)].$$

We note that *persistency* is defined as a property that $(s, e_1, s_1) \in \tau \wedge (s, e_2, s_2) \in \tau \Rightarrow (s_1, e_2, s') \in \tau \wedge (s_2, e_1, s') \in \tau$ [Lan].

Let $TS = (S, E, \tau)$ be a state space. A subset E_s of E is called *a prior set* at state s iff:

C1. E_s is a W. P. set at s.
C2. $\forall e \in E_s : [(s, \sigma e >\in \tau^* \wedge \mathrm{supp}(\sigma) \cap E_s = \varnothing \Rightarrow (s, e >\in \tau]$.
C3. $\exists e \in E_s : [(s, \sigma >\in \tau^* \wedge \mathrm{supp}(\sigma) \cap E_s = \varnothing \Rightarrow (s, \sigma e >\in \tau^*]$ (Such label e is called *a key* of E_s).

Using C1 and C2, we can easily show the following.

Lemma 3. *Let $TS = (S, E, \tau)$ be a state space and let E_s be a prior set at state s. Suppose that $(s, \sigma e, s') \in \tau^*$,* $\mathrm{supp}(\sigma) \cap E_s = \varnothing$, *and $e \in E_s$. Then there exists an occurrence sequence σ' such that $(s, e\sigma', s') \in \tau^* \wedge \psi(\sigma') = \psi(\sigma)$.*

Given a fixed prior set E_s for each state s, the reduced state space is constructed as a transition system $\underline{TS} = (S, E, \underline{\tau})$, where $\underline{\tau} \subset \tau$ is defined by $(s, e, s') \in \underline{\tau}$ iff $(s, e, s') \in \tau \wedge e \in E_s$.

Lemma 4. *Let $TS = (S, E, \tau)$ be a state space. Suppose that $(s, \sigma, s') \in \tau^*$ and s' is a terminal state of TS. Then there exists a label $e \in \mathrm{supp}(\sigma) \cap E_s$ such that $(s, e > \in \tau$.*

Proof. Since a prior set contains a key, $\mathrm{supp}(\sigma) \cap E_s \neq \emptyset$ holds. Let e be the label in E_s that occurs first in σ. By C2, it follows that $(s, e > \in \tau$. □

By Lemma 3 and Lemma 4, we can show that the reduced state space preserves terminal states.

Corollary 5. *Let $TS = (S, E, \tau)$ be a state space and let $\underline{TS} = (S, E, \underline{\tau})$ be the reduced state space.*

(i) Suppose that $(s, \sigma, s') \in \tau$ and s' is a terminal state of TS. Then there exists an occurrence sequence σ' such that $(s, \sigma', s') \in \underline{\tau}^*$ and $\psi(\sigma) = \psi(\sigma')$. In addition, s' is a terminal state of $\underline{TS}$.
(ii) Suppose that $(s, \sigma, s') \in \underline{\tau}^*$ and s' is a terminal state of $\underline{TS}$. Then s' is a terminal state of TS.

The reduced state space defined by C1 $\sim$ C3 preserves sufficient information for systems that are intend to terminate. In addition, this reduced state space preserves existence of infinite occurrence sequences.

Lemma 6. *Let $TS = (S, E, \tau)$ be a state space. Suppose that $(s, \sigma > \in \tau^*$. Then there exist $e \in E_s$ and an occurrence sequence σ' such that $(s, e\sigma' > \in \tau^* \wedge \psi(e\sigma') \geq \psi(\sigma)$.*

Proof. Case 1: $\mathrm{supp}(\sigma) \cap E_s \neq \emptyset$. Let e be the label in E_s that occurs first in σ. Suppose that $\sigma = \sigma_1 e \sigma_2$ and $(s, \sigma_1 e, s_1) \in \tau^*$. By C1 and C2, there exists an occurrence sequence σ_3 such that $(s, e\sigma_3, s_1) \in \tau^* \wedge \psi(\sigma_3) = \psi(\sigma_1)$. Hence the lemma holds for $\sigma' = \sigma_3\sigma_2$.
Case 2: $\mathrm{supp}(\sigma) \cap E_s = \emptyset$. Let e be a key of E_s. By C3, $(s, \sigma e > \in \tau$ holds. By C1 and C2, there exists an occurrence sequence σ' such that $(s, e\sigma' > \in \tau^* \wedge \psi(\sigma') = \psi(\sigma)$. □

Corollary 7. *Let s_0 be any initial state. There is an infinite occurrence sequence in the reduced state space $\underline{TS}(s_0)$ iff there is an infinite occurrence sequence in the ordinary state space $TS(s_0)$.*

For the *'ignoring problem'* [Val90, Wol], we can obtain the same result as shown in the stubborn set method. Thus we obtain the following.

Theorem 8. *Let $TS = (S, E, \tau)$ be a state space and let $\underline{TS} = (S, E, \underline{\tau})$ be the reduced state space. Assume that ignoring does not occur. Then for any initial state $s_0 \in S$, the following holds:*

(1) (Occurrences of labels) A label occurs in $TS(s_0)$ iff it occurs in $\underline{TS}(s_0)$.
(2) (Preserving liveness) A label is live in $TS(s_0)$ iff it is live in $\underline{TS}(s_0)$.
(3) (Preserving livelocks) Assume that S is finite.
   a. Let $C \subset S$ be a terminal strong connected component of $TS(s_0)$. There is a terminal strong connected component $\underline{C}$ in $\underline{TS}(s_0)$ such that $\underline{C} \subset C$ and for every label e, $\exists s, s' \in C : (s, e, s') \in \tau$ iff $\exists \underline{s}, \underline{s'} \in \underline{C} : (\underline{s}, e, \underline{s'}) \in \underline{\tau}$.
   b. Let $\underline{C} \subset \underline{S}$ be a terminal strong connected component of $\underline{TS}(s_0)$. There is a terminal strong connected component C in $TS(s_0)$ such that $\underline{C} \subset C$ and for every label e, $\exists \underline{s}, \underline{s'} \in \underline{C} : (\underline{s}, e, \underline{s'}) \in \underline{\tau}$ iff $\exists s, s' \in C : (s, e, s') \in \tau$.

4 Weakly Persistent Sets on Petri Nets

In this section we will consider Petri nets as a model of discrete event systems, and will study conditions on the model corresponding to C1 $\sim$ C3.

Persistency and weak persistency are defined as behavioral properties on state spaces. Therefore, if we intend to know whether a state space satisfies C1 $\sim$ C3, then we have to examine all states and all occurrence sequences in the state space. In Petri nets, conditions for persistency and weak persistency have already been obtained [Lan, Hir], i.e., if a net satisfies the condition, then the generated state space is persistent (weakly persistent). These conditions are defined on net structure, and are easy to verify. Using these conditions, we can efficiently compute prior sets.

Definition 9. *A net* is a triple $N = (P, T, A)$, where P and T are disjoint finite sets, and $A : (P \times T) \cup (T \times P) \rightarrow \{0, 1\}$ is a function. Each element in P is called *a place*, and each element in T is called *a transition.*

We note that the considering nets are ordinary (single-arc). Let $N = (P, T, A)$ be a net. As usual for each $x \in P \cup T$, let ${}^\bullet x = \{y \mid (y, x) \in A\}$ and $x^\bullet = \{y \mid (x, y) \in A\}$. For $X \subset P \cup T$, let ${}^\bullet X = \cup_{x \in X} {}^\bullet x$ and $X^\bullet = \cup_{x \in X} x^\bullet$. Each function $s : P \rightarrow \mathbb{N}$ is called *a marking*, which is a state of the net. Let U be a subset of $P \cup T$. Then $N|U$ denote the subnet consists of U and all the nodes connected to U.

Let $\tau_N \subset \mathbb{N}^P \times T \times \mathbb{N}^P$ denote the transition relation defined by

$$(s, t, s') \in \tau_N \text{ iff } \forall p \in P : [s(p) \geq A(p, t) \wedge s'(p) = s(p) + A(t, p) - A(p, t)].$$

The state space of the net N is defined as a transition system $TS_N = (\mathbb{N}^P, T, \tau_N)$. TS_N has the following property:

$$(s, \sigma, s') \in \tau_N^* \wedge (s, \sigma', s'') \in \tau_N^* \wedge \psi(\sigma) = \psi(\sigma') \Rightarrow s' = s''.$$

A sequence $c = x_1 x_2 \ldots x_n$ over $P \cup T$ is said to be a *circuit* if $A(x_i, x_{i+1}) > 0$ $(i = 1, \ldots, n-1)$ and $x_1 = x_n$. Let P_c denote the set of places on circuit c. A subset U of places is called *a trap* iff $U^\bullet \subset {}^\bullet U$, and is called *a deadlock* iff ${}^\bullet U \subset U^\bullet$.

Now we show conditions corresponding to C1 $\sim$ C3. As a structural condition corresponding to weak persistency, we here use TC-nets(trap-circuit net) [Ich, Mur]. A net is said to be *a TC-net* iff the set of places on each circuit is a trap. To check whether a given net is a TC-net or not, we can use the following fact: A net N is a TC-net iff for any transition t, $N|P - t^\bullet$ has no strongly connected components containing place(s) of ${}^\bullet t$. There is a linear time algorithm that finds strongly connected components of a given graph [Aho].

Using the notion of TC-nets, we define a net structure corresponding to W. P. set. Let $N = (P, T, A)$ be a net. A subset T_s of T has *TC-property* at marking s iff for each $t \in T_s$: if $(s, t >\in \tau_N$, then $N|t \cup (T - T_s)$ is a TC-net. If T_s has TC-property, then it is a W. P. set. The proof is shown in Appendix. Hence we can use the following condition N1 instead of C1.

N1. T_s has TC-property at s.

Since every subnet of a TC-net is also a TC-net, $N|(T - T_s)$ has to be a TC-net. Let *a TC-violation* be defined as a set V of transitions such that $N|(T - V)$ is a TC-net. Since TC-violations are independent of the state s, we can compute all minimal TC-violations before finding prior sets. Thus the set T_s satisfying N1 can be obtained as follows. Initially, give a set T_s containing some TC-violation. If T_s contains an enabled transition t such that $N|t \cup (T - T_s)$ is not a TC-net, then add transitions of $T - T_s$ to T_s until $N|t \cup (T - T_s)$ becomes to be a TC-net.

For each transition t and each place p, we define the following set:

$$\mathrm{up}_t(p) = \{t' \in T \mid A(t', p) > A(p, t') < A(p, t)\}.$$

Transitions in $\mathrm{up}_t(p)$ can add a token to place $p \in {}^\bullet t$. Then the condition corresponding to C2 is obtained as follows:

N2. $\forall t \in T_s[(s, t> \notin \tau_N \Rightarrow \exists p \in P : s(p) < A(p, t) \wedge \mathrm{up}_t(p) \subset T_s]$.

For each transition t and each place p, we define the following set:

$$\mathrm{down}_t(p) = \{t' \in T \mid A(p, t') > A(t', p) < A(p, t)\}.$$

Transitions in $\mathrm{down}_t(p)$ can remove a token from place $p \in {}^\bullet t$. Then the condition corresponding to C3 is obtained as follows:

N3. $\exists t \in T_s : [(s, t> \in \tau_N \wedge \forall p \in P : \mathrm{down}_t(p) \subset T_s]$.

Example 1. For the net in Fig. 1, minimal TC-violations are obtained as $\{a, c\}$ and $\{b, c\}$. Given an initial marking $s_0 = [1, 0, 0, 1, 0, 0]$, the state space is computed as shown in Fig.
/2. The reduced state space is indicated by bold arrows. Prior sets are obtained as follows:

$$T_{s_0} = \{a, d, c\}, T_{s_1} = \{b, c\}, T_{s_2} = \{a, c, g\}, T_{s_3} = \{a, b, c, f\}, T_{s_4} = \{a, b, c, g\},$$
$$T_{s_5} = \{a, b, c, e\}, T_{s_6} = \{a, b, c, g\}, T_{s_7} = \{a, b, c, d\}, T_{s_8} = \{a, b, c, g, e\}.$$

For example, at marking $s_1 = [0, 1, 0, 0, 1, 0]$ first c is chosen as a key. Then $\mathrm{down}_c(p_2) = \{b\}$ is added by the requirement of N3. Since $\{b, c\}$ is a TC-violation, N1 is satisfied. Both b and c are enabled, and therefore N2 holds. Hence $T_{s_1} = \{b, c\}$ is a prior set.

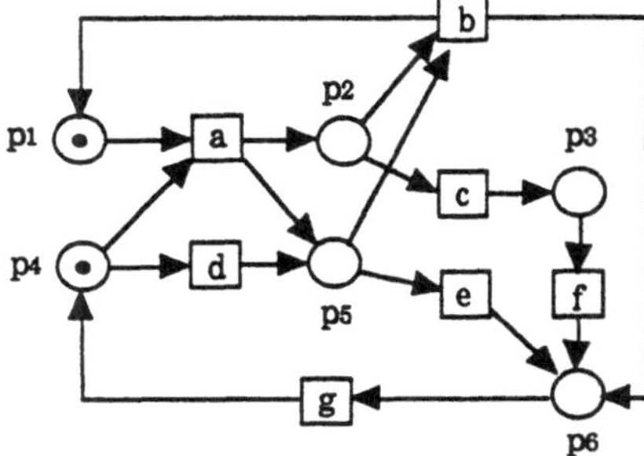

Fig. 1. An example net

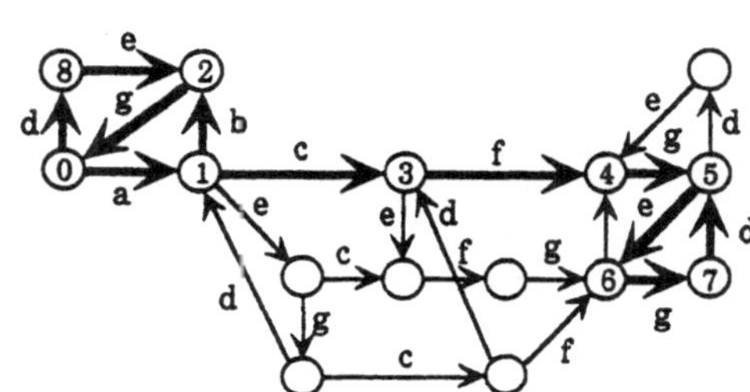

Fig. 2. The state space

5 Comparison with Stubborn Set Method

Now we compare the proposed method with stubborn set method. Stubborn sets are defined so that the following property holds [Val90].

Theorem 10. *Let $T_s \subset T$ be semistubborn at state s, and there is an occurrence sequence $\sigma = t_1 t_2 \dots t_{n-1} t_n$ such that $(s, \sigma, s') \in \tau^*$, $t_1, \dots, t_{n-1} \notin T_s$ and $t_n \in T_s$. Then the occurrence sequence $\sigma' = t_n t_1 \dots t_{n-1}$ satisfies that $(s, \sigma', s') \in \tau^*$.*

It is easy to observe that our condition (Lemma 3) is weaker than that of the stubborn set method. For Petri nets, stubborn sets are defined as follows. Subset of transitions T_s is stubborn (in the weak sense) at state s iff:

S1. $\forall t \in T_s : [(s, t> \in \tau_N \Rightarrow \forall p \in P : A \vee B]$, where
A : $\forall t' \in T - T_s : \min(A(t, p), A(t', p)) \geq \min(A(p, t), A(p, t'))$,
B : $\forall t' \in T - T_s : \min(A(t, p), A(p, t')) \geq \min(A(p, t), A(t', p))]$.
S2. $\forall t \in T_s[(s, t> \notin \tau_N \Rightarrow \exists p \in P : s(p) < A(p, t) \wedge \mathrm{up}_t(p) \subset T_s]$.
S3. $\exists t \in T_s : [(s, t> \in \tau_N \wedge \forall p \in P : \mathrm{down}_t(p) \subset T_s]$.

S2 and S3 are the same as N2 and N3, respectively. It is not difficult to prove that T_s satisfying S1 is a W. P. set. Hence,

Proposition 11. *If T_s is a stubborn set defined by S1 $\sim$ S3, then it is a prior set.*

In Example 1, $T_{s_1} = \{b, c\}$ is a prior set, but not a stubborn set. Considering transition b and place p_5, S1-A does not hold for transition e because $A(b, p_5) = A(e, p_5) = 0$ and $A(p_5, b) = A(p_5, e) = 1$; Similarly S1-B does not hold for transition d. $\{b, c, e\}$ is a stubborn set.

At marking s_1 sequence $egdb$ is enabled. After the occurring of b sequence gde is enabled. Sequence gde is a rearrangement of egd, but egd itself is not enabled. This fact shows the difference between conditions of Lemma 3 and Theorem 10.

References

[Aho] A. V. Aho, J. E. Hopcroft, J. D. Ullman: The Design and Analysis of Computer Algorithms. Addison-Wesley (1974)

[Hir] K. Hiraishi, A. Ichikawa: On Structural Conditions for Weak Persistency and Semilinearity of Petri Nets. Theoretical Computer Science, Vol. 93, 185-199 (1992)

[Ich] A. Ichikawa, K. Hiraishi: Analysis and Control of Discrete Event Systems Represented by Petri Nets. Lecture Notes in Control and Information Sciences, No. 103, 115-134, Springer-Verlag (1987)

[Lan] L. Landweber, E. Robertson: Properties of Conflict-free and Persistent Petri Nets. J. of ACM, 25-3, 352-364 (1978)

[Mur] T. Murata: Petri Nets: Properties, Analysis and Applications. Proc. of the IEEE, Vol. 77, No. 4 (1989)

[Rei] W. Reisig: Petri Nets. EATCS Monographs on Theoretical Computer Science, Vol. 4, Springer-Verlag (1985)

[Val90] A. Valmari: Stubborn Sets for Reduced State Space Generation. Lecture Notes in Computer Sience, No. 483, Springer-Verlag, 491-515 (1990)

[Val92] A. Valmari: Alleviating State Explosion During Verification of Behavioral Equivalence. Dept. of Comuter Science, Univ. of Helsinki, Report A-1992-4 (1992)

[Wol] P. Wolper, P. Godefroid, D. Pirottin: A Tutorial on Partial-Order Methods for the Verification of Concurrent Systems. CAV93 Tutorial (1993).

Appendix.

Lemma A1. [Rei] *Let N be a net. If N has no enabled transitions at a marking s, then the set $\{p \mid s(p) = 0\}$ is a token free deadlock.*

Theorem A2. *Let $N = (P, T, A)$ be a net. Suppose that a subset T_s of T has TC-property at a marking s, then T_s is a W. P. set at s.*

Proof. Suppose that $(s, \sigma t, s_1) \in \tau_N^* \wedge (s, t, s') \in \tau_N$. We use the induction on the length of σ. It is clear when the length $|\sigma| = 0$. We consider the case that $|\sigma| > 0$. Assume that no transition in supp(σ) is enabled at s'. Then by Lemma A1, $N|$supp(σ) contains a token free deadlock at s'. Since $(s, \sigma >\in \tau_N^*$, $N|$supp(σ) contains no token free source places at s', where a place is called a source place if it has no input transitions. Therefore $N|$supp(σ) contains a circuit c such that P_c is a token free deadlock at s'. There are two cases:

(i) P_c is a deadlock of $N|$supp(σt): By the TC-property of T_s, $t \notin P_c^\bullet - {}^\bullet P_c$ holds. Therefore P_c is token free at s. This implies that transitions in $P_c^\bullet$ cannot be enabled at every marking reachable from s. This contradicts $(s, \sigma >\in \tau_N^*$.

(ii) P_c is not a deadlock of $N|$supp(σt): This implies that $t \in {}^\bullet P_c - P_c^\bullet$, and therefore P_c must have token(s) at s'. Contradiction.

Hence Nsupp(σ) has at least one enabled transition. Let $t' \in$ supp(σ) be an enabled transition and let $\sigma = \sigma_1 t' \sigma_2$. By the induction hypothesis, there exists σ_1' such that $(s', t'\sigma_1' >\in \tau_N^*$ and $\psi(\sigma_1') = \psi(\sigma_1)$. Hence we have $(s', t'\sigma_1'\sigma_2 >\in \tau_N^*$. □

A New Approach to Discrete Time Stochastic Petri Nets

Robert Zijal * *and Reinhard German* †

Technische Universität Berlin, Institut für Technische Informatik, Franklinstr. 28/29, 10587 Berlin, F.R.Germany, e-mail: bob@cs.tu-berlin.de, rge@cs.tu-berlin.de

1 Introduction

Several classes of *stochastic Petri nets* (SPNs) have been proposed in order to define a unified framework for the modeling and analysis of discrete event dynamic systems. Most commonly, in a SPN the ordinary Petri net is augmented by continuous time random variables which specify the transition firing times. As a consequence, a stochastic process is underlying the SPN and can be studied in order to obtain quantitative measures of the model. Our main objective is to model technical sytems where the duration of most activities are constant (e.g. transfer time in a communication system, repair time in a fault-tolerant system) and some are better modeled by a probability distribution (e.g. failures). The class of *deterministic and stochastic Petri nets* (DSPNs) [1] contains transitions which fire after an exponentially distributed or deterministic delay. Under the structural restriction that at most one transition with a deterministic delay is enabled in each marking the steady-state solution can be obtained by considering an embedded Markov chain. In order to tackle the structural restriction it was shown in [4], that state equations describing the dynamic behaviour of DSPNs without that restriction can be derived by the method of supplementary variables. The state equations contain partial differential equations and are difficult to solve.

Apart from continuous time also discrete time has been considered. Molloy developed the class of *discrete time stochastic Petri nets* (DTPNs) in [7], where the transition firing times are specified by a geometric distribution with the same time step. Holliday and Vernon introduced *generalized timed Petri nets* (GTPNs) in [6]. In both models the underlying stochastic process is a *discrete time Markov chain* (DTMC). Since a deterministic firing time is a special case of a geometrically distributed firing time, a deterministic transition can be represented in a SPN with discrete time. In order to represent distributions with different time steps a discretization of the time step can be used. In a DTPN the user has to model the discretization with net constructs (a series of places and transitions). As a consequence, conflicts have to be resolved manually by the user. In a GTPN geometric distributions have to be represented by net constructs consisting of deterministic transitions and a looping arc, which becomes more complicated in case of conflicts.

* Robert Zijal was supported by a doctoral fellowship from the German National Research Council under grant Ho 1257/1-2.

† Reinhard German was supported by the Siemens Corporate Research and Development.

In this paper we define the class of *discrete time deterministic and stochastic Petri nets* (dtDSPNs). In a dtDSPN transitions fire either without time (immediate transitions) or after a geometrically distributed time. Transitions with a deterministic firing delay are a special case. Arbitrary time steps are allowed. All types of transitions can be mixed *without* structural restrictions. Moreover, since a discrete time scale is used, simultaneous firings of transitions in the same instant of time are possible. An algorithm is presented which resolves all conflicts automatically and computes the one-step transition probability matrix of the underlying DTMC. The transient and stationary solution can then be obtained by standard techniques (see e.g. [5, 2]). Although the state space may become large in case the greatest common divisor of the time steps of all firing time distributions is small, the approach is well suited to model technical systems where the most activities have a constant duration and failures can be modeled by random variables.

The rest of the paper is organized as follows. In §2 the considered class of dtDSPNs is introduced. In §3 a general description of the solution technique is presented and in §4 a numerical example is given. Concluding remarks are given in §5.

2 The Considered Class of Stochastic Petri Nets

A dtDSPN is formally given by a nine-tuple

$$\text{dtDSPN} = (P, T, I, O, H, \mathcal{P}, \mathbf{m}_0, F, W).$$

P is the set of places. Places may contain tokens. The vector $\mathbf{m} \in \mathbb{N}_0^{|P|}$ of number of tokens in each place is refered to as *marking*. T is the set of transitions and can be partitioned into the set T_{timed} of timed transitions firing after a certain delay and into the set T_{imm} of immediate transitions firing without consuming time. I, O, and H represent the input, output, and inhibitor arcs of the transitions, respectively. The arc multiplicities may be marking-dependent: $I : P \times T \times \mathbb{N}_0^{|P|} \to \mathbb{N}_0$, $O : T \times P \times \mathbb{N}_0^{|P|} \to \mathbb{N}_0$, $H : P \times T \times \mathbb{N}_0^{|P|} \to \mathbb{N}_0$. $\mathcal{P} : T \to \mathbb{N}_0$ assigns a priority to each transition. Timed transitions have lowest priority equal to zero and immediate transitions have priorities greater than zero. $\mathbf{m}_0$ is the initial marking of the net.

The usual firing rules are assumed. The places connected by input (output) arcs to a transition t are refered to as input (output) places of t. t is *enabled* if all of its input places contain at least the number of tokens equal to the multiplicity of the input arc and all places connected by inhibitor arcs contain a number of tokens less than the multiplicity of the inhibitor arcs. Additionally, no transition with a higher priority may fulfil this condition. An enabled transition may *fire* by removing tokens from the input places and by adding tokens to the output places according to the multiplicities of the arcs. All markings reachable from the initial marking constitute the *reachability set* of the net. The *reduced reachability set* is given by all *tangible markings* enabling only timed transitions.

$F : T_{\text{timed}} \to \mathcal{F}$ assigns a geometric firing time distribution to each timed transition. $\mathcal{F}$ is the set of all geometric distributions, denoted by $\text{Geom}(p, n\omega)$, given by the probability p and the time step $n \cdot \omega$ ($n \in \mathbb{N}$, $\omega \in (0, \infty)$). n may be arbitrarily chosen for each transition, wheras ω has to be kept fixed. ω is the greatest common divisor of the time steps of all firing time distributions and therefore the underlying time step of the model. Transitions may fire only at the time instances which are multiples of ω. The probability mass function of the firing time X is given by

$$\Pr\{X = i \cdot n \cdot \omega\} = p \cdot (1 - p)^i,$$

and the mean is given by $\frac{n\omega}{p}$. A deterministic firing time, denoted by Const($n\omega$), is a special case of a geometric distribution: setting $p = 1$ leads to Const($n\omega$) = Geom($1, n\omega$). In a dtDSPN, transitions with deterministic and geometrically distributed firing time can be mixed *without* structural restrictions. The use of a discrete time scale allows the transitions to fire in the same instant of time. Since this may lead to conflicts, weights are needed. $W : T \rightarrow (0, \infty)$ assigns a weight to each transition. The default weight is one. Weights are used to resolve conflicts between timed and between immediate transitions.

3 Description of the Solution Technique

3.1 Definition of the DTMC

The stochastic process underlying a dtDSPN is a DTMC. The time step of the DTMC is given by ω. Since the time steps of the firing time distributions may be multiples of ω, the state space of the DTMC is represented by the tangible markings of the net and by the vector of *remaining firing times* (RFTs) of the transitions. A RFT $r > 0$ of a transition represents that the time step of the transition is reached after the time $r \cdot \omega$ has elapsed. For a disabled transition the RFT is set to zero. Let the set of all tangible markings of the net be denoted as $\mathcal{M} \subset \mathbb{N}_0^{|P|}$, and let the set of all possible vectors of RFTs be denoted as $\mathcal{R} \subset \mathbb{N}_0^{|T_{\text{timed}}|}$. The set of states of the DTMC is then denoted as: $\mathcal{S} \subseteq \mathcal{M} \times \mathcal{R}$. We consider only the case of a finite state space.

3.2 Creating the DTMC

The states and the one-step transition probabilities of the DTMC can be determined by starting with the initial state and by visiting all reachable states. In the following we assume that the initial marking is tangible [1]. The initial state is given by $s_0 = (\mathbf{m}_0, \mathbf{r}_0)$. The entries of $\mathbf{r}_0$ corresponding to enabled transitions are set to their predefined RFT and to zero otherwise.

For each state $s = (\mathbf{m}, \mathbf{r})$ the set $\mathcal{S}'$ of states reachable at the next time step can be determined. The states of $\mathcal{S}'$ and the corresponding one-step transition probabilities can be computed by the following four steps:

Step 1. Determine the set of the *firing enabled* transitions T_{fe}. These transitions may fire at the next time step. The set T_{fe} in state $s = (\mathbf{m}, \mathbf{r})$ is given by

$$T_{\text{fe}} = \{t \in T_{\text{timed}} \mid t \text{ enabled in } \mathbf{m}, \text{ RFT of } t \text{ equal to } 1\}.$$

Step 2. The transitions of T_{fe} are partitioned into *independent groups*. For each group *firing events* are identified. The probabilities of the firing events can be computed for each group in isolation.

First, a binary relation *direct conflict* $\mathcal{DC} \subset T_{\text{fe}} \times T_{\text{fe}}$ between transitions is defined. $\mathcal{DC}$ represents the competition of transitions for the tokens in the input places. Two transitions $t_i, t_j \in T_{\text{fe}}$ are in direct conflict in marking $\mathbf{m}$, if they share a common input place:

$$t_i \ \mathcal{DC} \ t_j \quad \Leftrightarrow \quad \exists p \in P : I(p, t_i, \mathbf{m}) > 0 \wedge I(p, t_j, \mathbf{m}) > 0.$$

[1] Otherwise the set of tangible states reachable from the initial marking has to be considered.

Note that this definition differs from the ordinary definition of conflicts in Petri nets. Since only enabled transitions and a discrete time scale is used, the condition is sufficient. $\mathcal{DC}$ can be determined very easily. Performing the reflexive, symmetric, and transitive closure of $\mathcal{DC}$ leads to the equivalence relation *indirect conflict* $\mathcal{IC} = \mathcal{DC}^*$. The equivalence classes of $\mathcal{IC}$ are refered to as *independent groups* and represent sets of transitions for which the firing of transitions is not independent. In the following we distinguish between three group types: $1 : n$ -, $m : 1$ -, and $m : n$ - groups. For each group type the set of possible exclusive *firing events* is identified, and it is shown how the probabilities of the events can be computed.

$1 : n$ - group. The group consists of one single transition t connected with n input places. Two events are possible: Either t fires or not. Let p denote the probability that t will fire in isolation. t fires with probability p and t does not fire with probability $1 - p$.

$m : 1$ - group. The group consists of m transitions with at least one common input place. $m + 1$ events are possible. Either one of the m transition fires or no transition fires. The probabilities of the events can be computed as follows. Let the transitions of the group be enumerated by t_i, $i = 1, ..., m$. Furthermore, let p_i be the probability that transition t_i will fire in isolation and let w_i denote the weight of transition t_i. Define the index sets $I = \{1, ..., m\}$, $I_i = I \setminus \{i\}$. The probability P_{t_i} of the event that t_i fires, is given by:

$$P_{t_i} = p_i \cdot \sum_{I_i' \subseteq I_i} \Big(\prod_{j \in I_i'} p_j \cdot \prod_{j \in I_i \setminus I_i'} (1 - p_j) \cdot \frac{w_i}{w_i + \sum_{j \in I_i'} w_j} \Big),$$

and the probability P_N of the event, that no transition fires is given by:

$$P_N = \prod_{j \in I} (1 - p_j)$$

$m : n$ - group. The group consists of m transitions connected with n input places, additionally there is no common input place for all transitions. In a $m : n$ - group at least $m + 1$ events are possible. As in a $m : 1$ - group each single transition may fire or no transition may fire. Furthermore, the simultaneous firing of more than one transition may be possible. The computation of the event probabilities is demonstrated for an example. Figure 1 shows a $3 : 2$ - group. Let p_i denote the probability of firing of transition t_i in isolation and let w_i denote the weight of transition t_i, $i = 1, 2, 3$. Possible events are the single firing of t_1, t_2 or t_3, the simultaneous firing of t_1 and t_3, and that no transition fires. The event probabilities are denoted as: P_{t_1}, P_{t_2}, P_{t_3}, $P_{t_1 t_3}$ and P_N, respectively. They are given by:

$$P_{t_1} = p_1(1-p_2)(1-p_3) + p_1 p_2(1-p_3)\frac{w_1}{w_1+w_2}, \quad P_{t_1 t_3} = p_1 p_3(1-p_2) + p_1 p_3 p_2 \frac{w_1+w_3}{w_1+w_2+w_3}$$

$$P_{t_2} = p_2(1-p_1)(1-p_3) + p_2 p_1(1-p_3)\frac{w_2}{w_1+w_2} + p_2(1-p_1)p_3\frac{w_2}{w_2+w_3} + p_1 p_2 p_3 \frac{w_2}{w_1+w_2+w_3}$$

$$P_{t_3} = p_3(1-p_1)(1-p_2) + p_3(1-p_1)p_2\frac{w_3}{w_2+w_3}, \qquad P_N = \prod_{i=1}^{3}(1 - p_i)$$

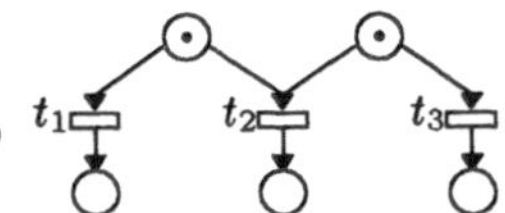

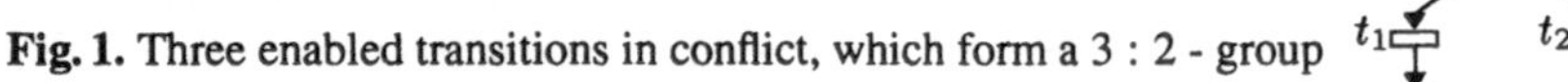

Fig. 1. Three enabled transitions in conflict, which form a 3 : 2 - group

Step 3. The events of the independent groups take place simultaneously. We refer to the simultaneous events as *product events*. Let the groups be enumerated by 1, ..., k and denote the set of events of group i as E_i. A product event is then an element of the product of the event sets: $E_1 \times \cdots \times E_k$. The probability of a product event is given by the product of the single event probabilities. All product event probabilities sum to one.

Step 4. After a product event has taken place in state $s = (\mathbf{m}, \mathbf{r})$, immediate transitions may be enabled. After the possible firing of immediate transitions a new state $s' = (\mathbf{m}', \mathbf{r}')$ is reached. $\mathbf{m}'$ is determined by the firing of the transitions. $\mathbf{r}'$ is determined as follows. Each entry of $\mathbf{r}$ greater than zero is decremented. After that each entry corresponding to a newly enabled transition in marking $\mathbf{m}'$ is set to the predefined RFT of the transition. And finally, the entries of disabled transitions in $\mathbf{m}'$ are set to zero. The set $\mathcal{S}'$ of all reachable states from s is therefore given by the states reachable by product events combined with the possible firing of immediate transitions. The transition probabilities of the DTMC are thus given by the product of the probabilities of the product events and of the probabilities of the immediate transition firing sequences. The probabilities of the immediate transition firing sequences can be determined by the algorithm presented in [2].

3.3 The Geometric Distribution and Deterministic Firing Times

A deterministic firing time is a geometric distribution with probability $p = 1$. As a consequence, the probability of firing events involving transitions with a deterministic firing time may reduce to zero. The RFT is needed in order to discretize the firing time and to represent different deterministic firing times in the underlying DTMC. In case of a geometric distribution an arbitrary average firing time can be achieved for an arbitrary time step. In order to reduce states, a time step equal to ω should be used for a geometric distribution, whenever possible. The mean firing time is then given by $\frac{\omega}{p}$.

3.4 Extracting the Results from the DTMC

In §3.2 it was shown how the one-step transition probability matrix $\mathbf{P}$ of the DTMC can be constructed. The transient and stationary solutions of the DTMC can be obtained with standard techniques (e.g. [5, 2]). Since $\mathbf{P}$ is usually a sparse matrix, sparse numerical techniques should be used. Having obtained the probability distribution over states of the DTMC, $\pi_s,\ s \in \mathcal{S}$, the probability distribution over markings can be obtained by summing up all corresponding probabilities of the DTMC:

$$\pi_{\mathbf{m}} = \sum_{r \in \mathcal{R}} \pi_{(\mathbf{m},\mathbf{r})}, \quad \mathbf{m} \in \mathcal{M}.$$

The probability distribution of tokens in the places of the Petri net and other measures can be derived from the probability distribution over markings.

4 Example

In this section an example is given in order to illustrate the modeling power of the class of dtDSPNs. Consider a flexible manufacturing system with a facility for loading and processing of raw parts. The system has a buffer for K parts. Both the loading and the processing facilities work with a deterministic speed but are subject to failures and repairs. Figure 2 shows a dtDSPN model of the system. The model represents a D/D/1/K-queueing system with failures and repairs. Transition t_1 and t_2 correspond to the loading and processing of parts. Transitions t_3 (t_5) and t_4 (t_6) represent failure and repair of the loading (processing) facilities. Tokens in p_1 model free buffer places, tokens in p_2 model loaded parts. A token in p_3 (p_5) represents a failed loading (processing) facility and tokens in p_4 (p_6) represent a working loading (processing) facility, respectively. Arcs with a small circle at their destination are inhibitor arcs. The loading time is Const(τ_l), the processing time Const(τ_p). Both repair times are Const(τ_r). Failure times for the loading and processing facilities are Geom($p_{f_{\text{load}}}$, 1) and Geom($p_{f_{\text{proc}}}$, 1), respectively. We assume the following parameters: $K = 5$, $\tau_l = 1$, $\tau_p = 2$, $\tau_r = 3$, $p_{f_{\text{load}}} = 10^{-3}$. The underlying time step is therefore given by $\omega = 1$. Note that in most states two transitions with a deterministic firing time are enabled. In the following experiments the failure probability $p_{f_{\text{proc}}}$ is varied from 10^{-4} to 0.5. The model has 24 markings and 104 states. Figure 3 shows the throughput of the processing facility.

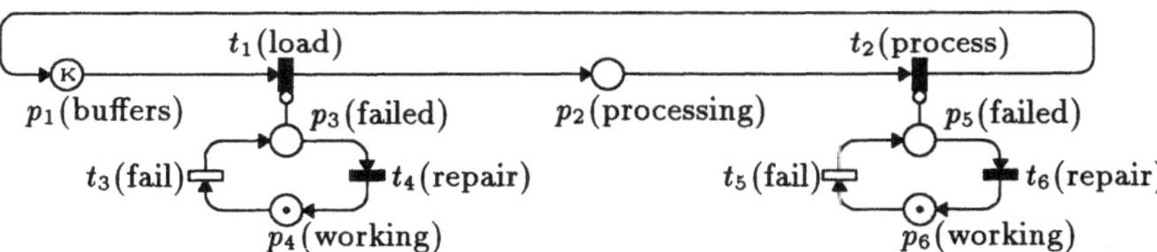

Fig. 2. D/D/1/K-queueing system with failures and repairs

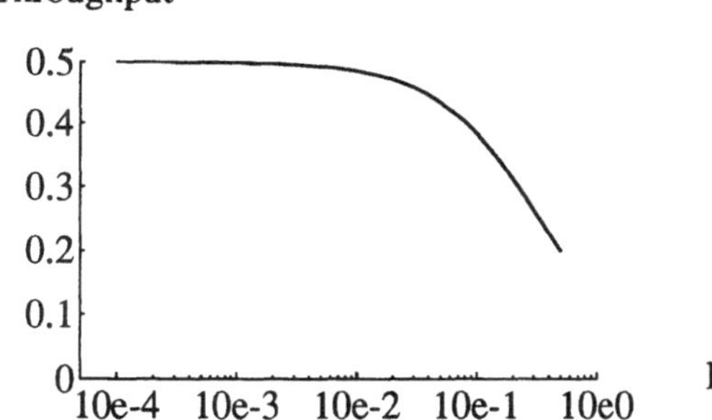

Fig. 3. Throughput of the processing facility

5 Conclusions

In this paper we defined the class of discrete time deterministic and stochastic Petri nets (dtDSPNs) for the modeling of systems comprising concurrent activities with either deterministic or geometrically distributed duration. No restrictions are imposed on the structure of the net. An algorithm was presented which computes the state space of the model and the one-step probabilities of the underlying DTMC. All conflicts can be resolved automatically by the algorithm. Transient and stationary measures can be computed by analyzing the DTMC. Although the state space may become large in case the greatest common divisor of the time steps of all firing time distributions is small, the approach is well suited to model technical systems where the most activities have a constant duration and failures can be modeled by random variables. Numerical results were presented for a D/D/1/K queueing system with failures and repairs.

In future research we intend to investigate whether the costs can be reduced by considering an embedded Markov chain and by normalizing the results [8, 3].

References

1. M. Ajmone Marsan, G. Chiola. "On Petri Nets with Deterministic and Exponentially Distributed Firing Times". In *G. Rozenberg (Ed.) Advances in Petri Nets 1987, Lecture Notes in Computer Science 266*, pp. 132-145, Springer 1987.
2. G. Ciardo, A. Blakemore, P. F. J. Chimento, J. K. Muppala, and K. S. Trivedi. "Automated generation and analysis of Markov reward models using Stochastic Reward Nets". In C. Meyer and R. J. Plemmons, editors, *Linear Algebra, Markov Chains, and Queueing Models*, volume 48 of *IMA Volumes in Mathematics and its Applications*, pages 145-191. Springer-Verlag, 1993.
3. G. Ciardo, R. German, C. Lindemann. "A characterization of the stochastic process underlying a stochastic Petri net". In *Proc. 5th Int. Conf. on Petri Nets and Performance Models*, Toulouse, France, October 1993.
4. R. German, C. Lindemann. "Analysis of Stochastic Petri Nets by the Method of Supplementary Variables". In *Proc. of the PERFORMANCE '93*, Rome, Italy, Sept.-Oct. 1993.
5. D. Gross, C. M. Harris. *Fundamentals of Queueing Theory*. 2nd Edition, John Wiley & Sons, 1985.
6. M. A. Holliday, M. K. Vernon. "A Generalized Timed Petri Net Model for Performance Analysis". In *IEEE Transactions on Software Engineering*, Vol.SE-13, No.12, pp. 1297-1310, Dezember 1987.
7. M. K. Molloy. "Discrete Time Stochastic Petri Nets". In *IEEE Transactions on Software Engineering*, Vol. 11, No. 4, pp. 417-423, April 1985.
8. W. L. Zuberek. "Timed Petri Nets, Definitions, Properties, and Applications". In *Microelectronics and Reliability*, Vol. 31, pp. 627-644, 1991.

Analysis of Timed Place/Transition Nets Using Minimal State Graphs

Hans-Michael Hanisch

Otto-von-Guericke-Universität Magdeburg
Department of Electrical Engineering
PF 4120, D-39016 Magdeburg, Germany

1 Introduction

The paper presents a method to compute a state graph for Place/Transition nets with deterministic timing of arcs directing from places to transitions. Nets with timed places or timed transitions can be easily transformed into subclasses of nets with timed arcs. The basic idea of the state graph is that a new state is generated only by firing of transitions and not by any tick of a clock. Thus, the resulting state graph has a minimal number of states. The method can also be applied to nets in which untimed (causal) firings of transitions are mixed with delayed firings.

Qualitative properties (deadlocks, liveness, conflicts etc.) as well as the quantitative behaviour of the modeled system. As an example, an optimal scheduling problem for a manufacturing system is discussed.

2 Place/Transition Nets with Timed Arcs

A Place/Transition net with timed arcs is a tuple ATPN = (P,T,F,V,m_0,ZB,D_0) where: P is the set of places p, T is the set of transitions t, $F \subseteq P \times T \bigcup T \times P$ is the set of directed arcs, $V : F \rightarrow \mathbb{N}^+$ denotes the arc weights and $m_0 : P \rightarrow \mathbb{N}$ is the initial marking. $\mathbb{N}$ denotes the set of integers. The difference $\Delta t = t^+ - t^-$ describes the change of the marking when t fires. Each arc from a place to a transition is labeled with a time interval $ZB(p,t)$:

$$ZB : (P \times T) \cap F \rightarrow \mathbb{N} \times (\mathbb{N} \cup \{\infty\}) \tag{1}$$

that describes the permeability of this arc. The *permeability* of an arc $ZB = [ZB_r, ZB_l]$ is a tuple of two integer values, where the first value $ZB_r(p,t)$ denotes the begin of the permeability and $ZB_l(p,t)$ denotes the end of the permeability of the arc relative to the local clock of p.

The state $z = (m, D)$ of a timed net consists of two components, the marking of the net and local clocks D denoting the age of the marking of the places:

$$D : P \rightarrow \mathbb{N} \ . \tag{2}$$

A transition t is called *marking-enabled* if all places $p \in Ft$ carry enough tokens according to the token weight of the arc:

$$\underline{0} \leq m + \Delta t, \quad \underline{0} \ldots \text{null vector} \ , \tag{3}$$

and called *time-enabled* if all arcs directing to the transition are permeable:

$$\max\{\operatorname{sgn} ZB_r(p,t) - D(p)) \mid p \in Ft\} \leq \min\{ZB_l(p,t) - D(p) \mid p \in Ft\} \ . \tag{4}$$

A transition is called *state-enabled* if it is both marking-enabled and time-enabled. Let set K be the set of all state-enabled transitions in a given state z. The earliest firing time $\text{eft}(t)$ for a marking-enabled transition is the time the transition becomes time-enabled and is given as follows:

$$\text{eft}(t) = \max\{\operatorname{sgn}(ZB_r(p,t) - D(p)) \mid p \in Ft\} \ . \tag{5}$$

We use the earliest firing rule with the consequence that maximal steps must be fired but no multiple firing of a transition is allowed.

The step delay $\Delta\tau$ denotes the earliest time a state-enabled transition can fire:

$$\Delta\tau = \min\{\text{eft}(t) \mid t \in K\} \ . \tag{6}$$

The set E is a subset of the enabled transitions and contains the transitions which fire exactly after $\Delta\tau$:

$$E = \{t \mid t \in K \wedge f(t) = \Delta\tau\} \ . \tag{7}$$

Since the earliest and maximal firing rule is used, the set E must be partitioned into maximal subsets of transitions which fire simultaneously after $\Delta\tau$ time units. Such subsets are called *maximal steps* u^*. For each maximal step u^* the vector Δu^* of dimension $|P|$ denotes the change of marking by firing u^*:

$$\Delta u^* = \sum_{t \in u^*} \Delta t \ . \tag{8}$$

The firing of a maximal step u^* after $\Delta\tau$ time units creates a new state (abbreviated: $z\left[(u^*, \Delta\tau) > z'\right)$ by changing the marking and the local clocks of the timed net:

$$m' := m + \Delta u^* \ , \tag{9}$$

$$D'(p) := \begin{cases} D(p) + \Delta\tau, & \text{if } m(p) > 0, \wedge\, p \notin Fu^* \cup u^*F \ , \\ 0, & \text{otherwise} \ . \end{cases} \tag{10}$$

The concept includes the special cases:

1. no delay of permeability ($ZB_r(p,t) = 0$),
2. no limitation of permeability ($ZB_l(p,t) = \omega$).

If all arcs have a time interval $ZB = [0 \ldots \omega]$, then the net behaves like a causal Petri net under the maximal firing rule.

3 The State Graph for Place/Transition Nets with Timed Arcs

Similar to the definition of the reachability graph of classical, causal Petri nets [1], we can define the state graph for arc-timed Place/Transition nets as follows. Let G be the set of all state transitions:

$$G \subseteq \mathbf{P}(T) \times \mathbb{N} \ , \qquad \mathbf{P}(T) \ldots \text{power set of } T \ . \tag{11}$$

The abbreviation $z\left[g > z'\right.$ denotes that state z' can be reached from state z by state transition g, where g is a tuple of the maximal step u^* and the step delay $\Delta\tau$. The set of all possible firing sequences of state transitions g is the set of words $W_{\mathrm{ATPN}}(G)$. We call a state z' *reachable from state* z (abbreviated: $z\left[^* > z'\right.$) if there is a sequence $q = g_1, g_2, \ldots, g_n$ with

- q is a word
- q transforms state z into state z':

$$z[* > z' :\iff \exists q\ (q \in W_{\mathrm{ATPN}}(G)\ \wedge\ z[q > z') \ . \tag{12}$$

Let $R_{\mathrm{ATPN}}(z)$ be the set of all states in the arc-timed Petri net which are reachable from state z. We can then define the state graph (dynamic graph) of an arc-timed Petri net as $DG_{\mathrm{ATPN}} = (R_{\mathrm{ATPN}}(z_0), A_{\mathrm{ATPN}})$ with

$$A_{\mathrm{ATPN}} = \left\{(z, g, z') \mid z, z' \in R_{\mathrm{ATPN}}(z_0)\ \wedge\ g \in G\ \wedge\ z[g > z'\right\} \ . \tag{13}$$

Each element of A_{ATPN} describes an arc directing from node z to node z' and labeled with a maximal step $u^*(g)$ and the corresponding step delay $\Delta\tau(g)$. The concept is described in detail in [2] and is implemented in the tool ATNA [3]. Based on the computation of the dynamic graph we are able to check the formal properties of the net as well as to optimize the performance of the modeled system.

4 Performance Evaluation and Optimal Control

Modeling of Control Decisions

The dynamic graph of a persistent Petri net model contains at most one path or cycle. Such a system cannot be influenced by an external control since all conflicts are resolved internally. Since the dynamic graph analysis is applied to compute control strategies, the models which describe the behaviour of the uncontrolled process and which are analysed must contain conflicts. Such conflicts represent *decisions of a control* (see §6). Let $T_s \subseteq T$ be the set of controlled transitions. To be controllable, each of this transitions must be in a dynamic conflict with another controlled transition in at least one state:

$$\forall t \in T_s \quad (\exists t' \in T_s,\ \exists z \in Z(t' \in \mathrm{cfl}(t, z))) \ , \tag{14}$$

where $\mathrm{cfl}(t, z)$ denotes the set of transitions which are in conflict with t in state z. A *control strategy* q_s is a word which only contains controlled transitions and therefore describes a sequence of conflict resolutions (control decisions). If $T_s \subseteq T$ is the set of controllable transitions, the control strategy $q_s(q, T_s)$ for any word $q = (g_1, \ldots, g_i, \ldots, g_k)$ can be determined:

$$u^*(g_{is}) := u^*(g_i) \cap T_s \ , \qquad \Delta\tau(g_{is}) = \Delta\tau(g_i) \ . \tag{15}$$

Non-Cyclic Behaviour

Obviously, any path in the dynamic graph represents a discrete process. A process π is an ordered set of tuples:

$$\pi = ((z_1, g_1), \ldots, (z_i, g_i), \ldots, (z_k, g_k) \mid z_i[g_i > z_{i+1} \\ \wedge \forall i, j \in \{1 \ldots k\} (i \neq j \rightarrow z_i \neq z_j)) \ . \quad (16)$$

The process time is the sum of all step delays in the process:

$$t_{\text{Proc}}(\pi) = \sum_{g \in q(\pi)} \Delta\tau(g) \ . \quad (17)$$

Cyclic Behaviour

As a basis for the analysis and evaluation of the cyclic stationary behaviour, all cycles c of a dynamic graph must be determined. The analysis, evaluation and optimization of the cyclic behaviour is described in detail in [2, 4]. However, the cyclic behaviour is not of interest for the example in §sec5 and shall therefore not be described here.

Time Optimal Control

A control strategy for startup or shutdown procedures must ensure that the system reaches a desired state from an initial one in minimal time. Let Π be the set of all paths (processes) of a dynamic graph and $\Pi(z, z')$ the set of all paths (processes) from state z to state z'. Then the set $\Pi_{\text{opt}}(z, z')$ of optimal processes from z to z' can be found by determing all paths:

$$\Pi_{\text{opt}}(z, z') = \left\{ \pi \mid \pi \in \Pi(z, z') \wedge t_{\text{Proc}}(\pi) = \min_{x \in \Pi(z, z')} \{t_{\text{Proc}}(\pi)\} \right\} \ . \quad (18)$$

5 Transformation of Nets with Timed Places or Timed Transitions to Nets with Timed Arcs

Nets with timed arcs were originally used to model batch production systems in the process industry [4]. Such production systems are very similar to FMS but are more dangerously to operate because any disturbance may cause hazardous situations. So a concept with larger modeling power was needed which can represent disruptions of the operations before their normal duration is reached. This is beyond the modeling capabilities of timed places or timed transitions. Although we do not need the whole modeling power of arc-timed nets in §6, we can apply the analysis and optimization techniques and the tool to scheduling problems in manufacturing systems.

Nets with timed places are defined in [5, 6]. A method to compute the state graph for nets with timed places is described in [6]. Timed places can be easily transformed into timed arcs. If a place p of a net with timed places is timed with a duration $d(p)$, all arcs from p to transitions $t \in pF$ of the corresponding arc-timed net have the same time interval $[d(p) \ldots \omega]$. This ensures that a token must stay at least $d(p)$ time units at place p before it can be removed by firing a transition $t \in pF$.

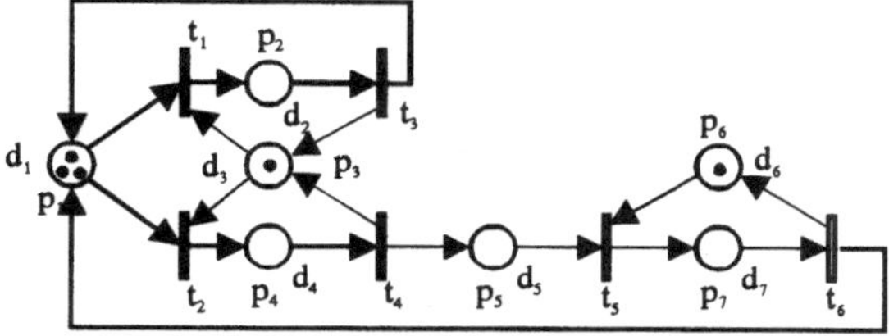

Fig. 1. Petri net model with timed places of a manufacturing systems [5]

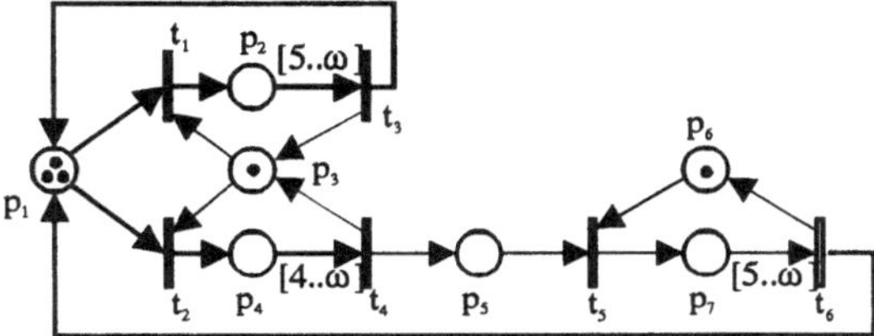

Fig. 2. Arc-timed Petri net model of the manufacturing system

Example 1. Figure 1 shows a Petri net with timed places which models a manufacturing system [5]. Figure 2 shows the corresponding arc-timed Petri net.

Nets with timed transitions were defined very early [7]. A method to compute a state graph is given in [1]. The method requires that any transition must have a time duration $d(t) > 0$ which is often too restrictive for the modeling of production processes. The method to compute the state graph generates a new state by any tick of a clock and therefore more or at least as much states are generated than by the computation of the dynamic graph of the corresponding arc-timed net. Nets with timed transitions can be transformed into arc-timed nets by splitting each transition into a "begin"-transition and an "end"-transition and a place between which is marked during the firing of the transition and has the duration $d(t)$. This timed place can be transformed into a timed arc like described above.

Example 2. Figure 3 shows the Petri net with timed transitions which corresponds to the timed Petri nets in Figure 1 and Figure 2.

P	Interpretation	d(p)	T	Interpretation
p_1	Pallet buffer	0	t_1	Loading P1 on M1
p_2	M1 processes P1	5	t_2	Loading P2 on M1
p_3	M1 available	0	t_3	Unloading P1 from M1
p_4	M1 processes P2	4	t_4	Unloading P2 from M1
p_5	P2 in M1 buffer	0	t_5	Loading P2 on M1
p_6	M2 available	0	t_6	Unloading P2 from M1
p_7	M2 processes P2	5		

Table 1. Interpretation and timing of the net in Figure 1

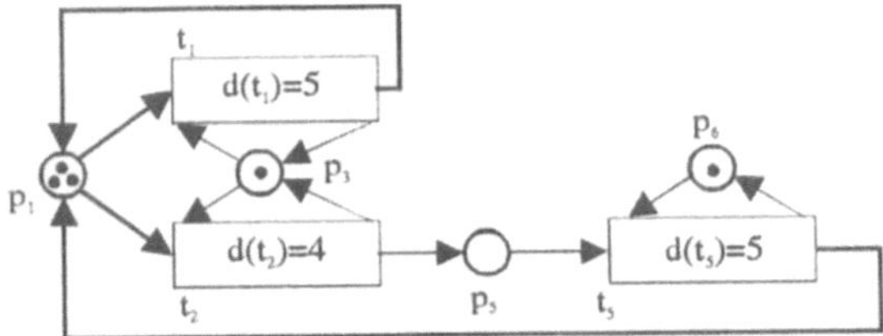

Fig. 3. Petri net model with timed transitions of the manufacturing system

6 Scheduling of a Manufacturing System

We study the manufacturing system which is described in [5] and modeled in Figures 1 to 4.

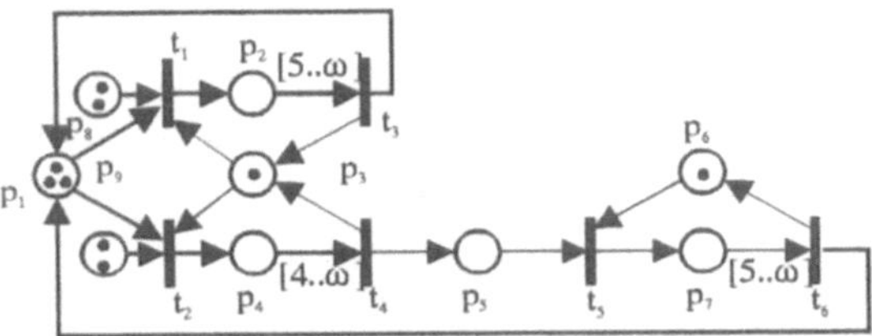

Fig. 4. Non-cyclic model

The interpretation of the model is given in Table 1. We assume in correspondence with [5] that two pieces of product P1 and product P2 are to be manufactured in a minimal time. Since we want to analyse the Petri net model, this must be expressed in the Petri net by the places p_8 and p_9. Figure 4 shows the appropriate model. The model has a non-cyclic behaviour because all paths end at a deadlock marking $m_d(p_i) = (3, 0, 1, 0, 0, 1, 0, 0, 0)$. This marking denotes the end of the manufacturing process.

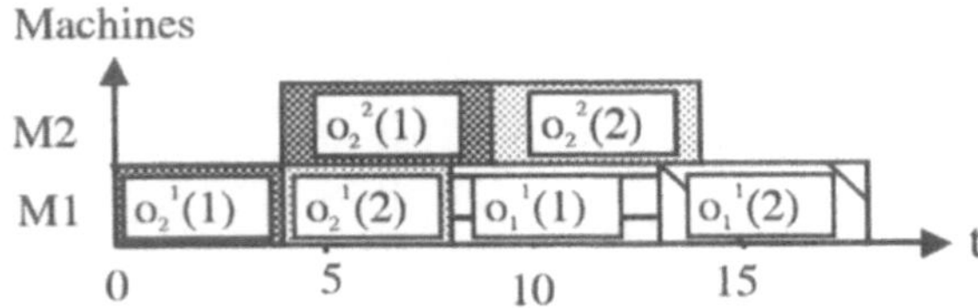

Fig. 5. Optimal schedule 1

The dynamic graph which was computed and analysed by means of ATNA has 39 states and two paths from the initial marking to the deadlock with the minimal process time of 18 time units. One path describes the same control strategy as given in [5]. The corresponding schedule is shown in Figure 5. $O_j^i(k)$ denotes the i-th operation

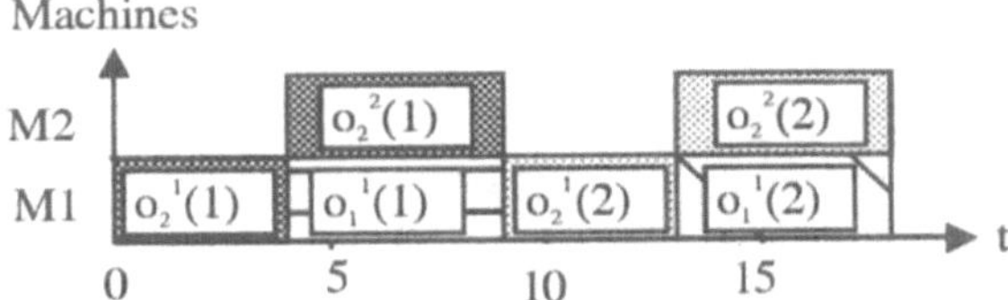

Fig. 6. Optimal schedule 2

for product j with the k-th piece. The firing sequence of maximal steps is q_1=(t_2(0), t_4(4), t_2, t_5(0), t_4(4), t_1(0), t_6(1), t_5(0), t_3(4), t_1(0), t_6(1), t_3(4)) and the control strategy is q_{1s}=(t_2(0), (4), t_2(0), (4), t_1(0), (5), t_1(0), (5)). The empty steps represent the uncontrolled state transitions in the system. The second path represents a schedule which is shown in Figure 6. The firing sequence is q_2=(t_2(0), t_4(4), t_1, t_5(0), t_3, t_6(5), t_2(0), t_4(4), t_1, t_5(0), t_3, t_6(5)). It is also time optimal. The control strategy is q_{2s}=(t_2(0), (4), t_1(0), (5), t_2(0), (4), t_1(0), (5)). If we finally assume that the number of pallets used by the system shall be minimal, then an analysis of the model shows that strategy 2 is better than strategy 1. Strategy 1 requires that three pallets are available from time unit 8 to 9. Strategy 2 needs at most two pallets. Hence, strategy 2 is optimal both in time and in the number of needed resources.

7 Industrial Application

The method was applied to an existing plant of typical industrial size (six polymerization reactors in three production lines coupled by limited resources such as metering tanks and mains for water, vacuum etc.). The problem which was studied was to find the bottlenecks of the plant and to provide a control strategy (i.e. start times for the single reactors) for the cyclic behaviour to maximize the throughput of the system. An optimal control strategy was found by analysis of the dynamic graph of the timed Petri net model. Details about this application will be presented in [8].

Acknowledgement

This work was supported by the Deutsche Forschungsgemeinschaft under grant Ha 1886/1-1. The author wishes to thank Prof. P. Starke for the cooperation in the development of ATNA.

References

1. P. Starke: Analyse von Petri-Netz-Modellen. B. G. Teubner, Stuttgart, 1990.
2. H.-M. Hanisch: Analysis of Place/Transition Nets with Timed Arcs and its Application to Batch Process Control. LNCS, Vol. 691, pp. 282–299.
3. P. Starke: ATNA- Arc Timed Net Analyser. Petri Net Newsletter 37, Dec. 1990, pp. 27–33.
4. H.-M. Hanisch: Petri-Netze in der Verfahrenstechnik. Oldenbourg-Verlag, MÅnchen-Wien, 1992.

5. J. Long and B. Descotes-Genon: Control Synthesis of Flexible Manufacturing Systems Modeled by a Class of Controlled Timed Petri Nets. Preprints of the 12th IFAC World Congress, Sydney, July 1993, Vol. 1, pp. 245–248.
6. H.-M. Hanisch: Dynamik von Koordinierungssteuerungen in diskontinuierlichen verfahrenstechnischen System. at-Automatisierungstechnik 38 (1990), 11, pp. 399–405.
7. C. Ramchandani: Analysis of Asynchronous Concurrent Systems by Timed Petri Nets. MIT, Project MAC, Technical Report 120, Feb. 1974.
8. U. Christmann and H.-M. Hanisch: Modeling, Analysis and Simulation of a Polymer Production Plant by Means of Arc-Timed Petri Nets. CIMPRO '94, New Brunswick, April 1994.

An Algebraic Description of Processes of Timed Petri Nets*

Józef Winkowski

Instytut Podstaw Informatyki PAN, 01-237 Warszawa, ul. Ordona 21, Poland

1 Motivation and Introduction

Petri nets are a widely accepted model of concurrent systems. Originally they were invented for modelling those aspects of system behaviours which can be expressed in terms of causality and choice. Recently a growing interest can be observed in modelling real-time systems, which implies a need of a representation of the lapse of time. To meet this need various solutions has been proposed known as timed Petri nets.

For the usual Petri nets there exist precise characterisations of behaviours (cf. [DMM 89], for example). In the case of timed Petri nets the situation is less advanced since the existing semantics either do not reflect properly concurrency (cf. [GMMP 89], for a review) or they oversimplify the representation of the lapse of time (as in [BG 92]).

In this paper we try to fill up this gap by representing the behaviour of a timed net by an algebra of structures called concatenable weighted pomsets. These structures correspond to concatenable processes of [DMM 89] with some extra information about the lapse of time. If the lapse of time is represented only in terms of delays between situations then we call such structures free time-consuming processes. If also the time instants at which situations arise are given then we call them timed time-consuming processes.

There is a homomorphisms from the algebra of timed time-consuming processes of a net to the algebra of its free time-consuming processes, and a homomorphism from the algebra of free time-consuming processes to an algebra whose elements reflect how much time the respective processes take. More precisely, to each free time-consuming process there corresponds a table of delays between its data and results (a delay table) such that the tables corresponding to the results of operations on processes (a sequential and a parallel composition) can be obtained by composing properly the tables corresponding to components.

An important property of free time-consuming processes and their delay tables is that they do not depend on when the respective data appear. Due to this property one can compute how a process of this type proceeds in time for any given combination of appearance times of its data. The combination which is given plays here the role of a

* This work has been partially supported by the Polish grant no. 2 2047 92 03

marking. This marking is timed in the sense that not only the presences of tokens in places, but also the respective appearance times (which need not be the same) are given.

The possibility of computing how a free time-consuming process applies to a given timed marking allows us to find the corresponding timed process. The possibility of finding timed processes of a net allows us to see which of them can be chosen and to define the possible firing sequences of the net.

Proofs of results described in this paper will be published elsewhere.

2 Concatenable Weighted Pomsets

Processes of timed nets and their delay tables will be represented by partially ordered multisets (pomsets) with some extra arrangements of minimal and maximal elements (similar to those in concatenable processes of [DMM 89]), and with some extra features (weights) and properties.

A *concatenable weighted pomset* (or a *cw-pomset*) over a set V is an isomorphism class α of structures $\mathcal{A} = (X, \leq, d, e, s, t)$, where $(X, \leq)$ is a finite partially ordered set with a subset $X_{\min}$ of minimal elements and a subset $X_{\max}$ of maximal elements such that each maximal chain has an element in each maximal antichain, $d : X \times X \to \{-\infty\} \cup [0, +\infty)$ is a *weight function* such that $d(x, y) = -\infty$ iff $x \leq y$ does not hold, $d(x, x) = 0$, and $d(x, y)$ is the maximum of sums $d(x, x_1) + \cdots + d(x_n, y)$ over all maximal chains $x \leq x_1 \leq \cdots \leq x_n \leq y$ from x to y whenever $x \leq y$, $e : X \to V$ is a *labelling function*, $s = (s(v) : v \in V)$ is a family of enumerations of the sets $e^{-1}(v) \cap X_{\min}$ (an *arrangement of minimal elements*), and $t = (t(v) : v \in V)$ is a family of enumerations of the sets $e^{-1}(v) \cap X_{\max}$ (an *arrangement of maximal elements*). Each such a structure is called an *instance* of α, we write α as $[\mathcal{A}]$, and we use subscripts, $X_{\mathcal{A}}$, $\leq_{\mathcal{A}}$, $d_{\mathcal{A}}$, $e_{\mathcal{A}}$, $s_{\mathcal{A}}$,$t_{\mathcal{A}}$, when necessary.

In this definition by an enumeration of a set we mean a sequence of elements of this set in which each element of this set occurs exactly once, and by an isomorphism from one structure to another we mean a bijection which preserves the order, the weight function, the labelling function, and the arrangements of minimal and maximal elements.

The restriction of $\mathcal{A}$ to $X_{\min}$ with t replaced by s and that to $X_{\max}$ with s replaced by t are instances of cw-pomsets written respectively as $\partial_0(\alpha)$ and $\partial_1(\alpha)$ and called the *source* and the *target* of α. If $X = X_{\min} = X_{\max}$ then $\leq$ reduces to the identity and we call α a *symmetry*. If also $t = s$ then $\alpha = \partial_0(\alpha) = \partial_1(\alpha)$ and α becomes a *trivial symmetry* or, equivalently, a *multiset* of elements of V with the multiplicity of each $v \in V$ given by cardinality$(e^{-1}(v) \cap X)$.

Graphical representations of cw-pomsets can be found in Fig. 1 and 2. Elements are represented by occurrences of their labels. The order and the respective weights are represented by annotated arrows, omitting the arrows which follow from transitivity. The arrangements of minimal and maximal elements are represented by endowing the labels of minimal elements with subscripts and tha labels of maximal elements with superscripts denoting the positions of the corresponding elements in the respective sequences.

For cw-pomsets α and β such that $\partial_0(\beta) = \partial_1(\alpha)$ there exists a unique cw-pomset $\alpha; \beta$ whose each instance $\mathcal{C}$ can be decomposed by a maximal antichain Y into an instance $\mathcal{A}$ of α and an instance $\mathcal{B}$ of β, where $\mathcal{B}$ follows $\mathcal{A}$ in the sense that $X_{\mathcal{A}} = \{x \in$

$X_\mathcal{C} : x \leq_\mathcal{C} y$ for some $y \in Y\}$ and $X_\mathcal{B} = \{x \in X_\mathcal{C} : y \leq_\mathcal{C} x$ for some $y \in Y\}$. The correspondence $(\alpha, \beta) \mapsto \alpha; \beta$ is an associative partial operation such that $\partial_0(\alpha; \beta) = \partial_0(\alpha)$, $\partial_1(\alpha; \beta) = \partial_1(\beta)$, and $\partial_0(\alpha); \alpha = \alpha; \partial_1(\alpha) = \alpha$. We call it a *sequential composition* (see Fig. 1).

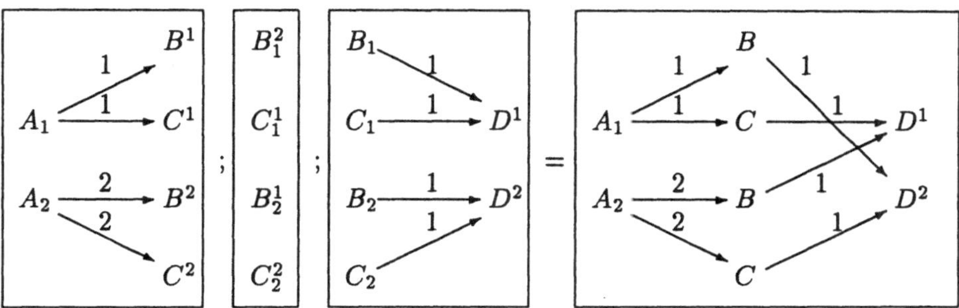

Fig. 1.

For arbitrary cw-pomsets α and β there exists a unique cw-pomset $\alpha \otimes \beta$ whose each instance $\mathcal{C}$ can be splitted into an instance $\mathcal{A}$ of α and an instance $\mathcal{B}$ of β in the sense that $X_\mathcal{C} = X_\mathcal{A} \cup X_\mathcal{B}$ with $X_\mathcal{A} \cap X_\mathcal{B} = \emptyset$ and $x \in X_\mathcal{A}$ incomparable with $y \in X_\mathcal{B}$, each $s_\mathcal{C}(v)$ is $s_\mathcal{A}(v)s_\mathcal{B}(v)$ (the concatenation of $s_\mathcal{A}(v)$ and $s_\mathcal{B}(v)$), and each $t_\mathcal{C}(v)$ is $t_\mathcal{A}(v)t_\mathcal{B}(v)$. The correspondence $(\alpha, \beta) \mapsto \alpha \otimes \beta$ is an associative (but not commutative) operation such that $\partial_0(\alpha \otimes \beta) = \partial_0(\alpha) \otimes \partial_0(\beta)$ $(= \partial_0(\beta) \otimes \partial_0(\alpha))$, $\partial_1(\alpha \otimes \beta) = \partial_1(\alpha) \otimes \partial_1(\beta)$ $(= \partial_1(\beta) \otimes \partial_1(\alpha))$, and $(\alpha; \beta) \otimes (\gamma; \delta) = (\alpha \otimes \gamma); (\beta \otimes \delta)$ whenever $\alpha; \beta$ and $\gamma; \delta$ are defined. We call it a *parallel composition* (see Fig. 2).

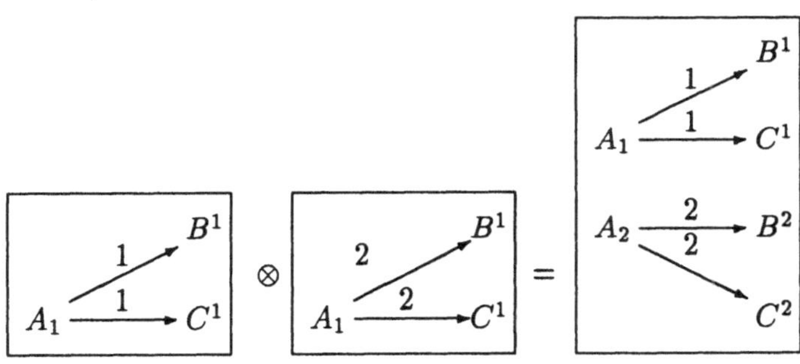

Fig. 2.

Finally, for arbitrary trivial symmetries (multisets) a and b there exists a unique symmetry $I_*(a, b)$ whose each instance $\mathcal{C}$ can be partitioned into an instance $\mathcal{A}$ of a and an instance $\mathcal{B}$ of b such that $X_\mathcal{C} = X_\mathcal{A} \cup X_\mathcal{B}$ with $X_\mathcal{A} \cap X_\mathcal{B} = \emptyset$, each $s_\mathcal{C}(v)$ is $s_\mathcal{A}(v)s_\mathcal{B}(v)$, and each $t_\mathcal{C}(v)$ is $t_\mathcal{B}(v)t_\mathcal{A}(v)$ $(= s_\mathcal{B}(v)s_\mathcal{A}(v))$. The correspondence $(a, b) \mapsto I_*(a, b)$ is an operation such that

$$I_*(a,b);\ I_*(b,a) = a \otimes b,$$
$$(I_*(a,b) \otimes c);\ (b \otimes I_*(a,c)) = I_*(a, b \otimes c),$$
$$I_*(\partial_0(\alpha), \partial_0(\beta));\ \alpha \otimes \beta = \beta \otimes \alpha;\ I_*(\partial_1(\alpha), \partial_1(\beta)).$$

We call it an *interchange* (see Fig. 1, where the second component of the left hand side can be represented as $I_*(B, B) \otimes C \otimes C$).

The set of cw-pomsets over V and the operations thus defined constitute an algebra CWPOMSETS(V) which is a symmetric strict monoidal category in the sense of category theory.

Each cw-pomset α can be obtained with the aid of compositions and interchanges from *atomic* cw-pomsets of two types: *one-element* cw-pomsets, one for each label $v \in V$ with one-element instances whose elements are labelled by v, and *prime* cw-pomsets with instances $\mathcal{P}$ such that $X_\mathcal{P} = (X_\mathcal{P})_{\min} \cup (X_\mathcal{P})_{\max}$, $(X_\mathcal{P})_{\min}$ and $(X_\mathcal{P})_{\max}$ are nonempty and disjoint, and $x \leq_\mathcal{P} y$ for all $x \in (X_\mathcal{P})_{\min}$ and $y \in (X_\mathcal{P})_{\max}$. Moreover, in order to obtain α one needs always the same number, $|\alpha|(\pi)$, of copies of each prime cw-pomset π, that is the same multiset, $|\alpha|$, of prime cw-pomsets.

Each atomic cw-pomset α is *symmetrical* in the sense that $\sigma; \alpha = \alpha; \sigma' = \alpha$ for all symmetries σ, σ' such that the respective compositions are defined.

By restricting an instance $\mathcal{A}$ of a cw-pomset α to the subset of its minimal and maximal elements we obtain an instance $\mathcal{B}$ of a cw-pomset β. As the order of $\mathcal{B}$ reduces to a relation between elements of $(X_\mathcal{A})_{\min}$ and those of $(X_\mathcal{A})_{\max}$, we may regard $\mathcal{B}$ as the matrix

$$(d_\mathcal{A}(x, y) : x \in (X_\mathcal{A})_{\min}, y \in (X_\mathcal{A})_{\max})$$

in which rows and columns are labelled as specified by $e_\mathcal{A}$ and arranged according to $s_\mathcal{A}$ and $t_\mathcal{A}$, respectively. When considered up to a bijective renaming of rows and columns, such a matrix can be identified with the cw-pomset β. We write it as tab(α) and call a *table* over the respective set of labels, V.

The set of tables over V and the operations defined in this set by $\partial_0'(\alpha) = \partial_0(\alpha)$, $\partial_1'(\alpha) = \partial_1(\alpha), \alpha;'\beta = \text{tab}(\alpha;\beta), \alpha \otimes' \beta = \alpha \otimes \beta$, and $I_*'(a,b) = I_*(a,b)$, constitute an algebra TABLES(V) which is a symmetric strict monoidal category. The correspondence $\alpha \mapsto \text{tab}(\alpha)$ is a homomorphism from the algebra CWPOMSETS(V) to TABLES(V). When expressed in terms of instances, the operation $(\alpha, \beta) \mapsto \alpha;'\beta$ is similar to matrix multiplication: for $\mathcal{A} \in \alpha$, $\mathcal{B} \in \beta$, and $\mathcal{C} \in \alpha;'\beta$ such that $X_\mathcal{A} \cap X_\mathcal{B} = Y$ for a maximal antichain Y we have $d_\mathcal{C}(x, y) = \max(d_\mathcal{A}(x, z) + d_\mathcal{A}(z, y) : z \in Y)$. Similarly, for the operation $(\alpha, \beta) \mapsto \alpha \otimes' \beta$: for $\mathcal{A} \in \alpha$, $\mathcal{B} \in \beta$, and $\mathcal{C} \in \alpha \otimes' \beta$ such that $X_\mathcal{C} = X_\mathcal{A} \cup X_\mathcal{B}$ with disjoint $X_\mathcal{A}$ and $X_\mathcal{B}$ we have either $d_\mathcal{C}(x, y) = d_\mathcal{A}(x, y)$, or $d_\mathcal{C}(x, y) = d_\mathcal{B}(x, y)$, or $d_\mathcal{C}(x, y) = -\infty$.

3 Processes of Timed Nets

Let $N = (\text{Pl}, \text{Tr}, \text{pre}, \text{post}, D)$ be a timed place/transition Petri net with a set Pl of places of infinite capacities, a set Tr of transitions, input and output functions pre, post : $\text{Tr} \to \text{Pl}^+$, where Pl^+ denotes the set of multisets of places, and with a duration function

$D : \mathrm{Tr} \to [0, +\infty)$. The multiset $\mathrm{pre}(\tau)$ represents a collection of tokens, $\mathrm{pre}(\tau, p)$ tokens in each place p, which must be consumed in order to execute a transition τ. The multiset $\mathrm{post}(\tau)$ represents a collection of tokens, $\mathrm{post}(\tau, p)$ tokens in each place p, which is produced by executing τ. The non-negative real number $D(\tau)$ represents the duration of each execution of τ. We assume that $\mathrm{pre}(\tau) \neq 0$, $\mathrm{post}(\tau) \neq 0$, $D(\tau) \neq 0$ for all transitions τ, and that $\mathrm{pre}(\tau)$, $\mathrm{post}(\tau)$, $D(\tau)$ determine τ uniquely.

A distribution of tokens in places is represented by a marking $\mu \in \mathrm{Pl}^+$, where $\mu(p)$, the multiplicity of p in μ, represents the number of tokens in p. If many executions of transitions are possible for the current marking but there is too few tokens to start all these executions then a conflict which thus arises is resolved in an indeterministic manner. We assume that it takes no time to resolve conflicts: when an execution of a transition can start, it starts immediately, or it is disabled immediately. Finally, we admit many concurrent nonconflicting executions of the same transition.

The behaviour of N can be described by characterizing the possible processes of N, where a process is either an execution of a transition, or a presence of a token in a place, or a combination of such processes. The processes are considered as *time-consuming*, or *tc-processes*, that is together with the lapse of time. Each such a process may be viewed as a cw-pomset $\alpha = [\mathcal{A}]$, where $\mathcal{A} = (X, \leq, d, e, s, t)$, elements of X represent the tokens which take part in the process, the partial order $\leq$ specifies the causal succession of tokens, the weight function d specifies the delays with which tokens appear after their causal predecessors, the labelling function e specifies the meanings of tokens, and s and t are respectively an arrangement of the tokens which the process receives from its environment and an arrangement of the tokens which the process delivers to its environment. It may be given either without specifying when its tokens appear and then called a *free tc-process*, or together with the respective appearance times and then called a *timed tc-process*. In the first case the labelling function e specifies only the proper meaning of each token x (played by the place where x appears). In the second case e specifies an extended meaning $e(x)$ of each token x, where $e(x)$ consists of the proper meaning $e_{\mathrm{proper}}(x)$ and the respective appearance time $e_{\mathrm{time}}(x)$ such that $e_{\mathrm{time}}(x) = \max(e_{\mathrm{time}}(y) + d(x, y) : y \leq x, y \neq x)$.

For each place $p \in \mathrm{Pl}$ we have the free tc-process of presence of a token in p. This process, $ftcp(p)$, may be viewed as the one-element cw-pomset with the label p. For each transition $\tau \in \mathrm{Tr}$ we have the free tc-process of executing τ. This proces, $ftcp(\tau)$, may be viewed as the prime cw-pomset $[\mathcal{A}]$ with $\mathrm{cardinality}(e_{\mathcal{A}}^{-1}(p) \cap (X_{\mathcal{A}})_{\min}) = \mathrm{pre}(\tau, p)$ and $\mathrm{cardinality}(e_{\mathcal{A}}^{-1}(p) \cap (X_{\mathcal{A}})_{\max}) = \mathrm{post}(\tau, p)$ for all $p \in \mathrm{Pl}$, and with $d_{\mathcal{A}}(x, y) = D(\tau)$ for all $x \in (X_{\mathcal{A}})_{\min}$ and $y \in (X_{\mathcal{A}})_{\max}$. The possible combinations of free tc-processes of the above two types may be viewed as cw-pomsets which can be obtained from the respective atomic cw-pomsets of the forms $ftct(p)$ and $ftcp(\tau)$ with the aid of compositions and interchanges. Thus we obtain a set $\mathrm{fbeh}(N)$ of all possible free tc-processes of N. This set defines a subalgebra $\mathrm{FBEH}(N)$ of the algebra $\mathrm{CWPOMSETS(Pl)}$, called the *algebra of free tc-processes* of N.

To each free tc-process $\alpha \in \mathrm{fbeh}(N)$ there corresponds the table $\mathrm{tab}(\alpha)$. For each instance $\mathcal{A}$ of α and the respective instance

$$(d_{\mathcal{A}}(x, y) : x \in (X_{\mathcal{A}})_{\min}, y \in (X_{\mathcal{A}})_{\max})$$

of this table, each item may be interpreted as the delay with which a token represented by y appears after one represented by x. Consequently, we call tab(α) the *delay table* of α. The correspondence $\alpha \mapsto$ tab(α) is a homomorphism from the algebra of free tc-processes of N to the algebra of delay tables over Pl.

Timed tc-processes of N can be characterized as follows.

For each place $p \in$ Pl we have a family $ttcp(p)$ of timed tc-processes, each of them viewed as a one-element cw-pomset with the label consisting of p and of the respective appearance time. For each transition $\tau \in$ Tr we have a family $ttcp(\tau)$ of timed tc-processes, each of them viewed as a prime cw-pomset whose instance $\mathcal{B}$ can be obtained from $\mathcal{A} \in ftcp(\tau)$ by replacing the labelling $e_\mathcal{A}$ by $e_\mathcal{B}$, where $(e_\mathcal{B})_{\text{proper}} = e_\mathcal{A}$ and $(e_\mathcal{B})_{\text{time}}$ is such that, for all $y \in (X_\mathcal{A})_{\max}$, we have

$$(e_\mathcal{B})_{\text{time}}(y) = \max((e_\mathcal{B})_{\text{time}}(x) + D(\tau) : x \in (X_\mathcal{A})_{\min}).$$

The possible combinations of timed tc-processes of the above two types may be viewed as cw-pomsets which can be obtained from the respective members of sets of the forms $ttcp(p)$ and $ttcp(\tau)$ with the aid of compositions and interchanges. Thus we obtain a set tbeh(N) of all possible timed tc-processes of N. This set defines a subalgebra TBEH(N) of the algebra CWPOMSETS(Pl $\times (-\infty, +\infty)$), called the *algebra of timed tc-processes* of N.

To each timed tc-process α there corresponds a free tc-process free(α) which can be obtained from α by reducing the labelling function of an instance of α to its proper part. The correspondence $\alpha \mapsto$ free(α) is a homomorphism from TBEH(N) to FBEH(N).

For each free tc-process α and each family $M = (M(p) : p \in \text{Pl})$ of finite sequences $M(p)$ of time instants such that the length of $M(p)$ coincides with the multiplicity of p in $\partial_0(\alpha)$ we have a timed tc-process $timed(\alpha, M)$ whose instance $\mathcal{B}$ can be obtained from an instance $\mathcal{A}$ of α by replacing the labelling function $e_\mathcal{A}$ by $e_\mathcal{B}$, where $(e_\mathcal{B})_{\text{proper}}(x) = e_\mathcal{A}(x)$, $(e_\mathcal{B})_{\text{time}}(x) = (M(p))(i)$ for $x = (s_\mathcal{A}(p))(i)$, and

$$(e_\mathcal{B})_{\text{time}}(x) = \max((e_\mathcal{B})_{\text{time}}(y) + d_\mathcal{A}(x, y) : y \leq_\mathcal{A} x, y \neq x)$$

for $x \in X_\mathcal{A} - (X_\mathcal{A})_{\min}$. The source of this process may be viewed as the multiset of pairs of the form (p, t), where p is a place and t is a time instant, and the multiplicity of (p, t) coincides with the number of occurrences of t in $M(p)$. Such a multiset, called a *timed marking*, represents a set of tokens, each token represented by the pair (p, t) consisting of its place p and appearance time t.

Each timed tc-process is of the form timed(α, M) for some α and M. Thus the relatively small algebra of free tc-processes of N determines uniquely the much larger algebra of timed tc-processes.

The algebra of timed tc-processes of N allows to reflect how processes exclude each other due to an earlier enabling of a transition and to reconstruct clasical firing sequences.

In order to describe details let us consider any timed process α, any instance $\mathcal{A}$ of α, and any instant u of time, and define $X(u)$ as the set of $x \in X_\mathcal{A}$ such that $(e_\mathcal{A})_{\text{time}}(y) \leq u$ whenever $y = x$ or y is a direct predecessor of x. Then the maximal elements of $X(u)$ constitute a maximal antichain $Y(u)$ and by restricting $\mathcal{A}$ to $X(u)$ we obtain a set $\alpha|u$ of timed tc-processes, a timed marking $\mu_{\alpha,u}$, where $\mu_{\alpha,u}(p, t)$ is the number of elements of $Y(u)$ with the label (p, t), and a multiset $\Theta_{\alpha,u}$ of prime free tc-processes, where $\Theta_{\alpha,u}(\pi)$ is the number of copies of π in free(α') for any $\alpha' \in \alpha|u$. The set $\alpha|u$ and the multisets

$\mu_{\alpha,u}$, $\Theta_{\alpha,u}$ do not depend on the choice of instance of α, and we may regard $\Theta_{\alpha,u}$ as a multiset of transitions rather than of prime free tc-processes corresponding to transitions. Consequently, we have a chain $-\infty = u_0 < u_1 < \cdots < u_n < u_{n+1} = +\infty$ of time instants such that $\alpha|u$, $\mu_{\alpha,u}$, $\Theta_{\alpha,u}$ are constant and respectively equal to some α_i,μ_i,Θ_i on each interval $[u_i, u_{i+1})$. In this manner to α a sequence $fs(\alpha) = \mu_0[\Theta_1\rangle\mu_1 \ldots [\Theta_n\rangle\mu_n$ there corresponds which may be regarded as a candidate for a possible firing sequence of N.

Whether indeed $fs(\alpha)$ is a possible firing sequence depends on whether the process α cannot or can be excluded by another process due to an earlier enabling of a transition, and it can be reflected with the aid of concepts of dominance and admissibility.

Given two timed tc-processes α and β, we say that β *dominates* α if there exists a time instant u_0 such that $\Theta_{\alpha,u} = \Theta_{\beta,u}$ and $\Theta_{\alpha,u} = \Theta_{\alpha,u_0}$ and $\Theta_{\beta,u} < \Theta_{\beta,u_0}$ for $u < u_0$. Given any set P of timed tc-processes, a member α of P is said to be *admissible* in this set if there is no $\beta \in P$ which dominates α.

Thus we obtain a set of timed tc-processes of N which are admissible in tbeh(N) and firing sequences of N can be defined as $fs(\alpha)$ for α in this set.

References

[BG 92] Brown, C., Gurr, D., *Timing Petri Nets Categorically* , Springer LNCS 623, Proc. of ICALP'92, 1992, pp.571-582

[DMM 89] Degano, P., Meseguer, J., Montanari, U., *Axiomatizing Net Computations and Processes*, in the Proceedings of 4th LICS Symposium, IEEE, 1989, pp.175-185

[GMMP 89] Ghezzi, C., Mandrioli, D., Morasca, S., Pezze, M., *A General Way to Put Time in Petri Nets*, Proc. of the 5th Int. Workshop on Software Specifications and Design, Pittsburgh, May 1989, IEEE-CS Press

The Max-Plus Algebraic Approach

Dioids and Discrete Event Systems*

Guy Cohen[1,2]

[1] Centre Automatique et Systèmes, École des Mines de Paris, 35 rue Saint-Honoré, 77305 Fontainebleau Cedex, France, e-mail: cohen@cas.ensmp.fr
[2] INRIA, B.P. 105, 78153 Le Chesnay Cedex, France

1 Introduction

System theory builds upon a limited number of concepts, of models and of basic problems. With each category of models, a certain category of mathematical tools is associated. For example, linear systems appeal to linear vector spaces and to rational functions of a complex variable. Smooth nonlinear systems have been studied with the help of power series in noncommutative variables, Lie algebra of differential operators, and differential (or difference) algebra. These mathematical tools are also used in other branches of applied mathematics.

On the side of applications, theory should have some relevance even if models are kept simple compared with the complexity of reality. For example, there are almost *no* truly linear systems in the real world (e.g. because state constraints are always involved), yet linear systems are helpful in understanding and in solving certain problems.

In the last ten or twelve years, a new branch of system theory has emerged under the name of Discrete Event (Dynamic) Systems (DES). There have been already a wide variety of models proposed and of issues addressed by these models. Here, the focus is on a class which, in the framework of Petri nets, is referred to as 'Timed Event (or Marked) Graphs' (TEGs). The orientation is towards performance evaluation and optimization. It has been realized that such models are amenable to linear systems provided the so-called 'algebra of dioids' prevails over the more conventional algebra or arithmetics everybody is familiar with since the elementary school.

The main purposes of this paper is to introduce the mathematical tools upon which the theory dwells, and to review the various system-theoretic questions addressed so far. A number of concepts and problems can be studied in a way which parallels similar notions of conventional linear system theory, revealing a very striking analogy. Another side topic is to mention other areas in which dioids are also useful. This encompasses continuous system theory, dynamic optimization, etc. Indeed, this will serve as an introduction to some other invited papers at this conference.

Most of the material of this paper is based on a twelve-year work of a working group mainly active at INRIA under the name of 'Max Plus'; the names of the researchers who contributed to this work, in the past or recently, appear in the references quoted here, and more completely in the bibliography of [2]. Our exposition is restricted to deterministic systems. Stochastic TEGs have been studied by F. Baccelli et al. (see [2] and [8]).

* This work was partially supported by a grant of the French Ministry of Education, Direction of Research and Doctoral Studies.

2 Where the Max-Plus or the Min-Plus Algebra May Pop Up

2.1 Timed Event Graphs (TEGs)

This topic is now almost classic [2]. Consider the TEG of Fig. 1. The black dots are the

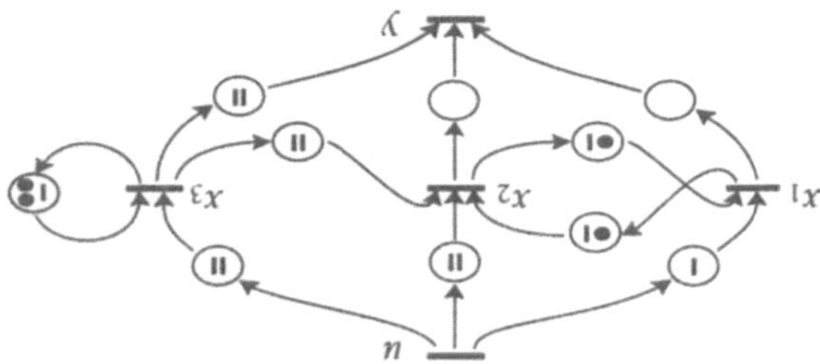

Fig. 1. A timed event graph

tokens of the initial marking as usual, whereas the bars indicates the holding times of places (in time units) and transition firings have no duration (without loss of generality). For a transition with generic name x, $x(t)$ is the number of the first transition firing which occurs at or next to time t (firings are numbered consecutively starting with number 0 for the first firing at or after time 0). Then the following equations are obtained for the example of Fig. 1:

$$x_1(t) = \min\left(x_2(t-1)+1, u(t-1)\right) , \qquad x_2(t) = \min\left(x_1(t-1)+1, x_3(t-2), u(t-2)\right) ,$$
$$x_3(t) = \min\left(x_3(t-1)+2, u(t-2)\right) , \qquad y(t) = \min\left(x_1(t), x_2(t), x_3(t-2)\right) .$$

2.2 A continuous system: linear or nonlinear?

Fig. 2 (right-hand side) represents a continuous system and it should be self-explanatory.

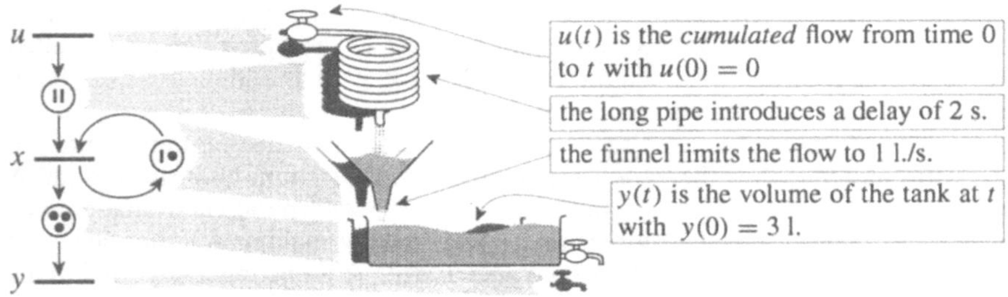
u
x
y
u(t) is the cumulated flow from time 0 to t with u(0) = 0
the long pipe introduces a delay of 2 s.
the funnel limits the flow to 1 l./s.
y(t) is the volume of the tank at t with y(0) = 3 l.

Fig. 2. A continuous system and its equivalent DES

This system obeys the inequality $y(t) \leq \min(u(t-2)+3, y(t-1)+1)$. Note that this inequality is also satisfied by the TEG shown on the left-hand side of the same figure. More generally, it can be proved that the largest solution of the above inequality in continuous time is given by

$$y(t) = \inf_{\tau \in \mathbb{R}} \left(h(t-\tau) + u(\tau)\right) \quad \text{with} \quad h(t) = \begin{cases} 3 & \text{for } t \leq 2 \ ; \\ 1+t & \text{otherwise.} \end{cases} \tag{1}$$

In the form (1), this system should be considered as a *nonlinear*, and even *nonsmooth* system. However, using the min-plus algebraic notation ($\oplus$ for min over a finite set, $\oint_{t\in A}$ for $\inf_{t\in A}$ when A is an infinite, possibly continuous, set, and $\otimes$ for $+$; note that $\otimes$ is generally omitted), (1) is converted into

$$y(t) = \oint_{\tau\in\mathbb{R}} h(t-\tau)\otimes u(\tau) \ , \tag{2}$$

which looks like the input-output relation of a 'linear' system. Note that $h(\cdot)$ may be called the 'impulse response' in that it is the trajectory followed by $y(\cdot)$ if one puts an infinite amount of fluid at the inlet of the pipe at time 0.

This system satisfies the axioms of min-plus linearity, namely, if $u(\cdot)\mapsto y(\cdot)$ and $v(\cdot)\mapsto z(\cdot)$, then $\min(u(\cdot), v(\cdot))\mapsto \min(y(\cdot), z(\cdot))$ (min is pointwise) and $a+u(\cdot)\mapsto a+y(\cdot)$ for any constant a. A systematic development of the theory of min-plus linear systems was undertaken in [2, Chapter 6]. Such systems can be combined serially ($\rightsquigarrow$ inf-convolution), in parallel (mixing of outputs in prescribed proportions $\rightsquigarrow$ pointwise min), in feedback, etc.

2.3 Dynamic Programming as a Linear Process

Consider the very simple optimal control problem

$$\min\sum_{t=0}^{T-1} c(u(t)) + \phi(x(T)) \quad \text{s.t.} \quad x(t+1) = x(t) - u(t) \ , \quad x(0)=\xi \ .$$

The corresponding dynamic programming (DP) equation reads

$$V(x,t) = \inf_y\big(c(x-y) + V(y,t+1)\big) = \oint_y c(x-y)V(y,t+1), \quad V(x,T)=\phi(x) \ .$$

Hence, the Bellman function V appears as the result of the iterated inf-convolution of the 'initial' value ϕ by the 'kernel' c. As such, V is a min-plus linear function of ϕ.

The Cramer transform (see [1]) converts conventional convolutions into inf-convolutions. Iterated (conventional) convolutions occur when iteratively summing up independent random variables (in order to compute the probability distribution of the random sum). Classical asymptotic theorems (law of large numbers, central limit theorem) are available in this situation. J.P. Quadrat [10] obtained similar results for dynamic programming, based on this homomorphism. In [1], the parallelism between probability theory and stochastic processes on the one hand (of conventional algebra), and optimization theory and decision processes on the other hand (of min-plus algebra) is further developed. Note that through the Cramer transform, a conventional linear system is mapped to a min-plus linear system.

2.4 Asymptotic Exponentials

Let $\overline{\mathbb{R}} \stackrel{\text{def}}{=} \mathbb{R}\cup\{-\infty\}\cup\{+\infty\}$ and $\overline{\mathbb{R}}^+$ denote the subset of nonnegative numbers (including $+\infty$). Consider functions $f:\mathbb{R}\to\overline{\mathbb{R}}^+$ and the mapping $c: f\to \limsup_{x\to+\infty}\log(f(x))/x$. Consider now the subset of functions f which are finite sums of exponentials: a generic element of this subset is written $\sum_{i\in I}\alpha_i\exp(a_i x)$ where I is a finite index set, $\alpha_i \in \mathbb{N}$ or $\mathbb{R}^+$, $a_i \in \overline{\mathbb{R}}$ (if $a_i = -\infty$, $\exp(a_i x)$ is identified to the zero constant function, and if $a_i = +\infty$, it is identified to the function everywhere equal to $+\infty$). This subset of functions is closed for ordinary pointwise sum and product, and, for f and g in this subset, one has that

$$c(f) = \max_{i \in I} a_i \ , \quad c(f+g) = \max(c(f), c(g)) \ , \quad c(f \times g) = c(f) + c(g) \ .$$

Therefore, in some sense, the mapping c transforms the conventional algebra into the max-plus algebra.

It turns out that this mapping c is of interest in several branches of applied mathematics. For example, when drawing Bode plots in linear system theory, one indeed generally considers *asymptotic* Bode plots. Also, in the theory of large deviations in probability theory, such considerations are basic. The message conveyed here is that max-plus algebra is also essential in the study of asymptotic phenomena. Maslov and co-workers [9] were very instrumental in this direction.

2.5 Max-Plus Algebra and Convexity Theory

Consider a polynomial in one variable x with coefficients in the max-plus algebra, namely, $p(x) = \bigoplus_i a_i x^i$. If we rewrite this expression in conventional algebra, since x^i is interpreted as $x \otimes \cdots \otimes x$, i times, which is $x + \cdots + x = i \times x$, it is realized that $p(x) = \max_i (i \times x + a_i)$. Hence, as a numerical function of x, $p(x)$ is a (piecewise linear and nondecreasing) convex function. We can further extend this class to encompass nonmonotonic piecewise linear functions by considering 'rational' functions in the max-plus algebra (i.e. pointwise conventional differences of the previous convex functions). We will not elaborate any further on this topic in this survey but it should be realized that an algebraic point of view may be adopted to reconsider the manipulations of certain interesting classes of functions (see [3]).

3 Dioid Theory

Min-plus or max-plus algebra has proved useful to model TEGs and other kinds of systems too. Concentrating from now on on TEGs, it will be seen that several representations are possible and that these representations appeal sometimes to more sophisticated structures than the above mentioned algebras. However all these algebraic structures obey a common set of axioms which defines what has been called 'dioids' already for some time (see [7] and references therein). Other names are also used to designate similar, albeit slightly different, structures (e.g. the word 'semiring' is often used in conjunction with various adjectives, but the following axiom of idempotency of addition precludes the possibility of embedding dioids into rings — see below). Overall, dioids are structures with a double personality:

- on the one hand, dioids resemble conventional linear algebra with which they share several combinatorial axioms (associativity, distributivity, etc.); this explains common results which are true in both frameworks (e.g. the Cayley-Hamilton theorem); and it also allows one to mimic some reasonings or algorithms (e.g. Gaussian elimination to solve some linear systems of equations);
- on the other hand, dioids are lattice-ordered structures and other aspects of properties or proofs strongly rely upon the existence of a natural order associated with the idempotent addition.

3.1 Axiomatics

Definition 1. A dioid is a set $\mathfrak{D}$ endowed with two inner operations $\oplus$ and $\otimes$ such that:

1. the addition $\oplus$ is associative, commutative, has a zero element ε;
2. multiplication[3] $\otimes$ is associative, has an identity element e;
3. $\otimes$ is distributive with respect to $\oplus$ (to the left and to the right);
4. ε is absorbing for $\otimes$: $\forall a \in \mathfrak{D}, a \otimes \varepsilon = \varepsilon \otimes a = \varepsilon$;
5. $\oplus$ is idempotent: $\forall a \in \mathfrak{D}, a \oplus a = a$.

Moreover,

a. $\mathfrak{D}$ is said *commutative* if $\otimes$ is commutative;
b. $\mathfrak{D}$ is said *complete* if it is closed for all infinite sums and if $\otimes$ is distributive with respect to infinite sums.

As already mentioned, the fact that $\oplus$ is idempotent precludes any possibility of embedding a (nontrivial) dioid into a ring. Indeed, suppose that a has an opposite element b such that $a \oplus b = \varepsilon$; then $a \oplus a \oplus b = a$ but the left-hand side is equal to $a \oplus b$ and $a \oplus b$ is just ε, hence $a = \varepsilon$ which is thus the only element admitting a symmetric element. Despite this impossibility, the reader is referred to [2, §3.4–5] for an attempt of symmetrization of some kind in dioids.

3.2 Order Structure

Due to the fact that $\oplus$ is idempotent, an order relation can be derived from this addition.

Theorem 2. *In a dioid $\mathfrak{D}$, $\{a = a \oplus b\}$ is equivalent to $\{\exists c : a = b \oplus c\}$. These equivalent statements define a (partial) order relation denoted $\succeq$ as follows: $a \succeq b \Leftrightarrow a = a \oplus b$. This order relation is compatible with addition and multiplication, that is, $a \succeq b$ implies that, for all c, $a \oplus c \succeq b \oplus c$ and $ac \succeq bc$ (and $ca \succeq cb$). For any two elements a and b in $\mathfrak{D}$, $a \oplus b$ is their least upper bound.*

Therefore, a dioid is a sup-semilattice having a 'bottom' element, namely ε. If the dioid is complete, which implies that any (finite or infinite) subset also admits a least upper bound, then this sup-semilattice can be completed to become a lattice by the following classical construction of the *greatest lower bound* $a \wedge b$ of any two elements a and b:

$$a \wedge b = \bigoplus_{\substack{x \preceq a \\ x \preceq b}} x \ .$$

In a complete dioid $\mathfrak{D}$, $\top = \bigoplus_{x \in \mathfrak{D}} x$ is well defined and it is the greatest element of $\mathfrak{D}$ (called 'top'). It can be proved that $\top$ is absorbing for multiplication (i.e. $\top \otimes x = \top, \forall x \neq \varepsilon$) if the dioid is *Archimedian*, that is, $\forall x, y, \exists z : x \otimes z \succeq y$ (and the same is required for left multiplication by some z' if multiplication is not commutative).

As is the case for a lattice, a complete dioid is said *distributive* if the least upper bound $\oplus$ is distributive with respect to the greatest lower bound $\wedge$ and conversely (distributivity is required for infinite subsets of elements). However, even in this case, $\oplus$ and $\wedge$ do not play symmetric roles in general as long as distributivity of $\wedge$ with respect to $\otimes$ is not always granted (see Example 3 below).

[3] the sign $\otimes$ is often omitted

Example 1. The simplest nontrivial dioid is the Boolean dioid, denoted $\mathbb{B}$, which consists of $\{\varepsilon, e\}$ solely.

Example 2. Let $\overline{\mathbb{R}}_{\max} = (\overline{\mathbb{R}}, \max, +)$, that is, $\oplus = \max$, $\otimes = +$ and the set is $\overline{\mathbb{R}}$. This is a totally ordered and complete dioid in which $\preceq$ coincides with the natural order, $\varepsilon = -\infty$, $e = 0$, $\top = +\infty$ and $\wedge$ is min. Similar examples with $\overline{\mathbb{Z}}$ replacing $\overline{\mathbb{R}}$ and/or min replacing max are also of interest for the remainder of this paper. Notice that, in $\overline{\mathbb{Z}}_{\min}$ and $\overline{\mathbb{R}}_{\min}$, $\varepsilon = +\infty$ and $\top = -\infty$, and moreover, the order defined by Theorem 2 is just reversed with respect to the natural order. Also $(-\infty) + (+\infty) = \varepsilon \otimes \top = \varepsilon = -\infty$ in $\overline{\mathbb{R}}_{\max}$ and $\overline{\mathbb{Z}}_{\max}$, whereas $(-\infty) + (+\infty) = +\infty$ in $\overline{\mathbb{Z}}_{\min}$ and $\overline{\mathbb{R}}_{\min}$ (guess why!).

Example 3. Let $2^{\mathbb{R}^n}$ denotes the collection of subsets of $\mathbb{R}^n$ (including $\varnothing$ and $\mathbb{R}^n$ itself). Then $\left(2^{\mathbb{R}^n}, \cup, +\right)$ is a complete dioid (here $+$ denotes the vector sum of subsets). It is only partially ordered and $\preceq$ coincides with $\subset$, $\wedge$ with $\cap$, $\varepsilon = \varnothing$ and $\top = \mathbb{R}^n$. This dioid is also Archimedian and distributive, but $\wedge$ does not distribute with respect to $\otimes$ (see [2, Example 4.31] for a counterexample).

Example 4. $(\overline{\mathbb{R}}, \max, \min)$, $\left(2^{\mathbb{R}^n}, \cup, \cap\right)$, and more generally any set of real-valued functions over $\mathbb{R}^n$ with pointwise max and min as addition and multiplication respectively, are examples of dioids in which $\wedge$ coincides with $\otimes$ and which are not Archimedian. This type of dioids may be useful for example in fuzzy set theory.

Example 5. Subsets of real-valued functions over $\mathbb{R}$ with pointwise min (or inf) for addition and inf-convolution as multiplication are also of interest for linear system theory in the min-plus algebra.

3.3 Matrix, Polynomial and Related Dioids

Matrix Dioids. Given a dioid $\mathcal{D}$, a matrix dioid can be derived by considering *square* matrices of a fixed but arbitrary dimension, and by endowing this set with usual matrix sum and product defined after the 'scalar' operations $\oplus$ and $\otimes$. The zero and identity elements in the matrix dioids are also denoted ε and e (the latter is the diagonal matrix with e everywhere on the diagonal and ε elsewhere).

Remark. Some authors (e.g. [12]) consider the analogue of vector spaces over dioids that they sometimes call moduloids. For our purpose, we can always embed vectors into square matrices obtained by adding as many zero columns as necessary: there is in general no mixing between the significant part and the arbitrary part of those square matrices in most operations we are interested in. Hence, it is enough for us to speak of matrix dioids (to which all results of noncommutative abstract dioids apply) and there is no need to introduce other algebraic structures.

Matrices and Graphs. Square $n \times n$ matrices $A = \left(A_{ij}\right)$ can be interpreted in a graph-theoretic point of view: the corresponding graph $\mathcal{G}(A)$, called 'precedence graph' in [2], has n nodes and there is a directed arc with weight A_{ij} from node j to node i if $A_{ij} \neq \varepsilon$; otherwise, there is no arc from j to i. With nonsquare matrices $n \times m$, a bi-partite graph $\mathcal{T}(A)$, called 'transition graph', can also be associated: $\mathcal{T}(A)$ has m source nodes and n sink nodes and arcs from the j-th source node to the i-th sink node has weight A_{ij} (there is no arc if $A_{ij} = \varepsilon$). Therefore, the precedence graph of a square matrix can be seen as the transition graph of the same matrix once this has been folded so has to make the i-th source and the i-th sink nodes coincide.

Algebraic operations on matrices can now be interpreted in terms of corresponding operations on their associated weighted digraphs: addition corresponds to 'parallel composition' and multiplication amounts to 'serial composition' (see [2]). More precisely, $\mathfrak{G}(A \oplus B)$ keeps the least upper bound $\oplus$ of weights of parallel arcs of $\mathfrak{G}(A)$ and $\mathfrak{G}(B)$ (i.e. those having the same end nodes in the two graphs — for a pair of nodes, it suffices that one such arc exists in either $\mathfrak{G}(A)$ or $\mathfrak{G}(B)$ to get one in $\mathfrak{G}(A \oplus B)$; $\mathfrak{G}$ can everywhere be replaced by $\mathcal{G}$ for square matrices). The bi-partite graph $\mathfrak{G}(A \otimes B)$ connects any source node of $\mathfrak{G}(B)$ to any sink node of $\mathfrak{G}(A)$ provided that there is a path through any pair of nodes of the form $\{k\text{-th sink node of } \mathfrak{G}(B), k\text{-th source node of } \mathfrak{G}(A)\}$ — given the necessary compatibility of dimensions, note that this is the only possible pairing of nodes a priori. Then the weight for an arc $j \to i$ is the least upper bound of weights of such parallel paths (going through all possible nodes k) knowing that the weight of a path is the product $\otimes$ of the weights of the composing arcs (given the absorbing property of ε, a 'missing' arc in a 'path' results in a weight ε for the whole path). For square matrices, using the precedence graph, A^p encodes the 'maximal' (indeed, least upper bounds of) weights of paths with (exact) length p in $\mathcal{G}(A)$ (a length of a path is the number of arcs which composes the path). Finally, $A^{(p)} \stackrel{\text{def}}{=} A \oplus A^2 \oplus \cdots \oplus A^p$ provides the same information for paths of length $\ell \leq p$. The following definitions are also useful (and are always defined for elements or matrices A in any complete dioid):

$$A^+ = \bigoplus_{p=1}^{+\infty} A^p \ , \qquad A^* = e \oplus A^+ \ .$$

Polynomials and Power Series. In system theory over dioids, we are also led to manipulate power series in one or several variables[4] $z_1, \ldots, z_m$ with coefficients in a (complete) dioid $\mathfrak{D}$ and with exponents in $\mathbb{N}$ or in $\mathbb{Z}$. Endowed with the usual sum and product of power series, these algebraic sets are also dioids. The notation $\mathfrak{D}[\![z_1, \ldots, z_m]\!]$ is used (most of the time exponents are assumed to lie in $\mathbb{Z}$); the notation $\mathfrak{D}[z_1, \ldots, z_m]$ is reserved for the subdioid of polynomials (with a finite number of nonzero coefficients).

The main unusual fact with polynomials (or power series) over dioids is that *formal* polynomials are *not* isomorphic to their associated *numerical functions*. Let us illustrate this point with help of an example.

Example 6. Consider the two formal polynomials $\mathfrak{p}_1 = 1 \oplus z \oplus z^2$ and $\mathfrak{p}_2 = 1 \oplus z^2$ in $\overline{\mathbb{R}}_{\max}[z]$. Formally, they are not equal. But considering the numerical functions $z \mapsto 1 \oplus z \oplus z^2$ and $z \mapsto 1 \oplus z^2$ from $\overline{\mathbb{R}}_{\max}$ to itself, it is easy to prove that

$$\forall z \in \overline{\mathbb{R}} \ , \quad \max(1, z, 2 \times z) = \max(1, 2 \times z)$$

(hint: draw the graphs of these functions). Therefore, the mapping $\mathcal{E}$ which associates numerical functions with formal polynomials is an homomorphism (for pointwise $\oplus$ and $\otimes$ operations in the set of numerical functions) but is not an isomorphism since distinct formal polynomials may be mapped to the same numerical function (here, $[\mathcal{E}(\mathfrak{p}_1)](z) = [\mathcal{E}(\mathfrak{p}_2)](z)$, $\forall z$, whereas $\mathfrak{p}_1 \neq \mathfrak{p}_2$).

Another important remark from the system-theoretic point of view is the following.

[4] mostly in the situation of *commutative* variables — see [5] for various usages

Remark. Consider an infinite sequence or 'trajectory' $\{a(t)\}_{t\in\mathbb{Z}}$ in $\overline{\mathbb{R}}_{\min}$ and the associated formal power series obtained by the classical z-transform, namely $A \stackrel{\text{def}}{=} \bigoplus_{t\in\mathbb{Z}} a(t)z^{-t}$. Consider now the associated numerical function

$$z \mapsto \left[\mathcal{E}(A)\right](z) = \bigoplus_{t\in\mathbb{Z}} a(t)z^{-t} = \inf_{t\in\mathbb{Z}}(a(t) - t \times z) = -\sup_{t\in\mathbb{Z}}(t \times z - a(t)) \ . \quad (3)$$

The last expression is nothing but the discrete Fenchel transform of the mapping $k \mapsto a(k)$, up to a change of sign. Note also that, for two such trajectories $a(\cdot)$ and $b(\cdot)$, the product $A \otimes B$ of their z-transforms is the z-transform of the inf-convolution of $a(\cdot)$ and $b(\cdot)$. This is consistent with the fact that the Fenchel transform converts inf-convolutions into pointwise conventional sums [11] which are pointwise products here. This fact is analogous to the fact that the Laplace transform converts convolutions into products.

4 Descriptions of TEGs

4.1 Counter and Dater Descriptions

For the TEG of Fig. 1, variables $u(t)$, $x_i(t)$, $y(t)$ attached to transitions were introduced with the meaning of 'number given to the first firing occurred at the corresponding transition at or after time t'. The mapping $t \mapsto v(t)$ is called a *'counter'*; counters obey min-plus linear equations. There exists a dual description in terms of *'daters'*: a dater is a mapping $k \mapsto \overline{v}(k)$ in which $\overline{v}(k)$ is the date at which the transition named v incurred its firing numbered k.

Remark. Very soon, we will drop the overbar on dater symbols and we will use the same letter v for both the counter and the dater associated with a given transition, and v will be the name of the transition too. What will then distinguish counters from daters will rather be the name of their argument (or domain): it will be either t (counter over the time domain) or k (dater over the event domain).

The mappings $t \mapsto v(t)$, from the *time domain* to the *event domain*, and $k \mapsto \overline{v}(k)$, from the event domain to the time domain, are in some sense reciprocal of each other. There is indeed a rigorous theory to account of this fact: this is the theory of *residuation*. It is beyond the scope of this paper to develop this theory here in some details and the reader is referred to [2] and the references therein. Let us just say that because $k \mapsto \overline{v}(k)$ is nondecreasing, it can be 'inverted' to yield $t \mapsto v(t)$: the problems arise with horizontal parts of the graph of $\overline{v}(\cdot)$ (this occurs when several events, i.e. transition firings, occur simultaneously at the transition — which is a priori not impossible) and with vertical jumps (this occurs when several time instants separate two successive events). There is indeed two extreme ways to define reciprocal functions $t \mapsto v(t)$ of $k \mapsto \overline{v}(k)$ in this case (such that none is 'multivalued'), and this corresponds to residuated and dually residuated mappings.

Generally speaking, given an isotone — i.e. nondecreasing — mapping $x \mapsto f(x)$ between two lattice-ordered sets, residuation deals with the problem of finding the least upper bound of the subset $\{x \mid f(x) \preceq y\}$ for a given y, and/or the greatest lower bound of the subset $\{x \mid f(x) \succeq y\}$. Under some conditions, and especially a 'semi-continuity' property of f (semi-continuity with respect to the order relations), these bounds will themselves belong to the corresponding subsets (they may then be called 'greatest subsolution' and 'least supersolution' of the equation $f(x) = y$ respectively).

In our particular case, there are thus two different definitions for a counter $t \mapsto v(t)$: $v(t)$ is the number of the *last* firing occurred *before* or *at* time t, or the *first* to occur *at* or *after* t. It turns out that the latter is more convenient mathematically speaking (see [2]) though it is perhaps less attractive intuitively. This latter definition corresponds to dual residuation of the dater, which is thus the residual of this counter. Whatever definition is adopted, the equations satisfied by counters are linear in $\overline{\mathbb{Z}}_{\min}$. While the transformation from counters to daters is rather 'nonlinear'[5], it turns out that daters also satisfy linear equations, but in $\overline{\mathbb{Z}}_{\max}$. For the example of Fig. 1, these equations are

$$x_1(k) = \max(x_2(k-1)+1, u(k)+1) \quad , \quad x_2(k) = \max(x_1(k-1)+1, x_3(k)+2, u(t)+2) \quad ,$$
$$x_3(k) = \max(x_3(k-2)+1, u(k)+2) \quad , \quad y(k) = \max(x_1(k), x_2(k), x_3(t)+2) \quad .$$

We invite the reader to formulate the rules to pass from the counter equations of §2.1 to the dater equations above, to write matrix equations in the corresponding dioids, to compare maximum and minimum delays found for x_i and u in the right-hand sides of both descriptions, etc.

4.2 γ- and δ-Transforms

In a way analogous to the z-transform of conventional system theory, we can introduce the γ-transform of daters $v(k)$ and the δ-transform of counters $v(t)$ in the following way

$$V(\gamma) = \bigoplus_{k \in \mathbb{Z}} v(k)\gamma^k \; ; \qquad V(\delta) = \bigoplus_{t \in \mathbb{Z}} v(t)\delta^t \; .$$

Note that γ and δ can be interpreted as *backward* (rather than *forward* for z in (3)) shift operators in the corresponding domains (roughly speaking, e.g. $\gamma v(k) = v(k-1)$). With these power series in $\overline{\mathbb{Z}}_{\max}[\![\gamma]\!]$, respectively $\overline{\mathbb{Z}}_{\min}[\![\delta]\!]$, one obtains the following equations for the TEG of Fig. 1

$$X(\gamma) = \begin{pmatrix} \varepsilon & 1\gamma & \varepsilon \\ 1\gamma & \varepsilon & 2 \\ \varepsilon & \varepsilon & 1\gamma^2 \end{pmatrix} X(\gamma) \oplus \begin{pmatrix} 1 \\ 2 \\ 2 \end{pmatrix} U(\gamma) \; , \quad X(\delta) = \begin{pmatrix} \varepsilon & 1\delta & \varepsilon \\ 1\delta & \varepsilon & \delta^2 \\ \varepsilon & \varepsilon & 2\delta \end{pmatrix} X(\delta) \oplus \begin{pmatrix} \delta \\ \delta^2 \\ \delta^2 \end{pmatrix} U(\delta) \; ,$$
$$Y(\gamma) = \begin{pmatrix} e & e & 2 \end{pmatrix} X(\gamma) \; , \qquad Y(\delta) = \begin{pmatrix} e & e & \delta^2 \end{pmatrix} X(\delta) \; .$$

4.3 2D-Domain Description

By definition of $\overline{\mathbb{Z}}_{\max}[\![\gamma]\!]$, an expression such as $(m \oplus p)\gamma^n$ means $\max(m,p)\gamma^n$. In $\overline{\mathbb{Z}}_{\min}[\![\delta]\!]$, the same expression turns out to be $n(\delta^m \oplus \delta^p)$ which does not obviously simplify to $n\delta^{\max(m,p)}$. Conversely, $(m \oplus p)\delta^n = \min(m,p)\delta^n$ whereas $n(\gamma^m \oplus \gamma^p)$ is not readily found to be $n\gamma^{\min(m,p)}$. However, a little thinking should show that, if $X(\delta)$ (resp. $X(\gamma)$) is the δ- (resp. γ-) transform of a counter (resp. a dater) $x(\cdot)$, which is a *nondecreasing* function of its argument, then, indeed,

$$n(\delta^m \oplus \delta^p)X(\delta) = n\delta^{\max(m,p)}X(\delta) \quad (\text{resp. } n(\gamma^m \oplus \gamma^p)X(\gamma) = n\gamma^{\min(m,p)}X(\gamma)).$$

[5] in any intuitive sense, inversion is nonlinear

A natural idea is thus to decide that expressions such as $m\gamma^p \oplus n\gamma^q$ and $p\delta^m \oplus q\delta^n$ will now be written simply $\gamma^p\delta^m \oplus \gamma^q\delta^n$ and that the following simplification rules will apply

$$\gamma^m \oplus \gamma^p = \gamma^{\min(m,p)} \ ; \quad \delta^m \oplus \delta^p = \delta^{\max(m,p)} \ . \tag{4}$$

A formal construction of the appropriate dioid consists in starting first from $\mathbb{B}[\![\gamma,\delta]\!]$ (power series with Boolean coefficients and two variables) and then to define an equivalence (indeed a dioid congruence) relation between two elements $X(\gamma,\delta)$ and $Y(\gamma,\delta)$ in the following way

$$X(\gamma,\delta) \equiv Y(\gamma,\delta) \iff \gamma^*\left(\delta^{-1}\right)^* X(\gamma,\delta) = \gamma^*\left(\delta^{-1}\right)^* Y(\gamma,\delta) \text{ in } \mathbb{B}[\![\gamma,\delta]\!] \ .$$

This equivalence is denoted as an equality in the quotient dioid which is called $\mathcal{M}_{in}^{ax}[\![\gamma,\delta]\!]$.

The role of the above congruence is to 'filter out' nondecreasing trajectories: in the plane $\mathbb{Z}^2$ (the x-axis is the event domain, the y-axis is the time domain), an element X of $\mathbb{B}[\![\gamma,\delta]\!]$ can be represented as a collection of points (a point with coordinates (k,t) exists if the monomial $\gamma^k\delta^t$ has coefficient e in X); if the upper hull of this collection of points is not a monotonic trajectory, then this will be the case for $\gamma^*\left(\delta^{-1}\right)^* X$ which is, in some sense, the 'best monotonic approximation of X from above'. Two elements of $\mathbb{B}[\![\gamma,\delta]\!]$ are 'equal' in $\mathcal{M}_{in}^{ax}[\![\gamma,\delta]\!]$ if they have the same such approximation. For example, e, γ^*, $\left(\delta^{-1}\right)^*$ and $\gamma^*\left(\delta^{-1}\right)^*$ are all equal in $\mathcal{M}_{in}^{ax}[\![\gamma,\delta]\!]$, although they are not the same formal power series. We refer the reader to [2, Chapter 5] for a rigorous development of this construction, and for a nice 'informational' interpretation of the simplification rules (4).

Finally, returning once again to the TEG of Fig. 1, the following equations stand in $\mathcal{M}_{in}^{ax}[\![\gamma,\delta]\!]$

$$X(\gamma,\delta) = \begin{pmatrix} \varepsilon & \gamma\delta & \varepsilon \\ \gamma\delta & \varepsilon & \delta^2 \\ \varepsilon & \varepsilon & \gamma^2\delta \end{pmatrix} X(\gamma,\delta) \oplus \begin{pmatrix} \delta \\ \delta^2 \\ \delta^2 \end{pmatrix} U(\gamma,\delta) \ , \quad Y(\gamma,\delta) = \begin{pmatrix} e & e & \delta^2 \end{pmatrix} X(\gamma,\delta) \ .$$

4.4 Transfer Matrices

All representations obtained in previous subsections can be casted into the same formal equations, namely

$$X = AX \oplus BU \ ; \quad Y = CX \oplus DU \tag{5}$$

(for our example, D was zero). In §4.1, X (for example) can be considered as an infinite column vector obtained by concatenation of the $x(k)$ or $x(t)$ vectors for k or t ranging from $-\infty$ to $+\infty$. Consequently, A is an infinite matrix too[6]. In §4.2, X is a vector with dimension equal to the number of internal transitions[7] and with components in $\overline{\mathbb{Z}}_{\max}[\![\gamma]\!]$ or $\overline{\mathbb{Z}}_{\min}[\![\delta]\!]$. Finally, in §4.3, the components belong to $\mathcal{M}_{in}^{ax}[\![\gamma,\delta]\!]$.

Whatever interpretation is given to Eqs. (5), according to a basic theorem in (complete) dioids, the *least* solution (according to the dioid order) of the first implicit equation is given by $X = A^*BU$, and thus $Y = (CA^*B \oplus D)U$. The expression $H = CA^*B \oplus D$ plays the role of the *transfer matrix* of the TEG. Note that the least solution corresponds,

[6] Note that this representation encompasses 'nonstationary' systems too: for example, in the dater representations, holding times of places may vary with the token numbers.

[7] those which have both predecessors and successors

in all interpretations, as the one which will occur if transitions fire as soon as possible and if the most favorable 'initial conditions' take place (which means that all tokens of the initial marking are available since $-\infty$ — a discussion of other initial conditions can be found in [2, §5.4.4.2]). For our example, the calculations performed in $\mathcal{M}_{in}^{ax}[\![\gamma, \delta]\!]$ ([8]) yield $H(\gamma, \delta) = \delta^4(\gamma\delta)^*$. It is then realized that such a transfer matrix (actually transfer function in this SISO case) is also that of the TEG depicted in Fig. 3a. This TEG is remarkably simpler than that of Fig. 1 and this demonstrates all the power of algebra.

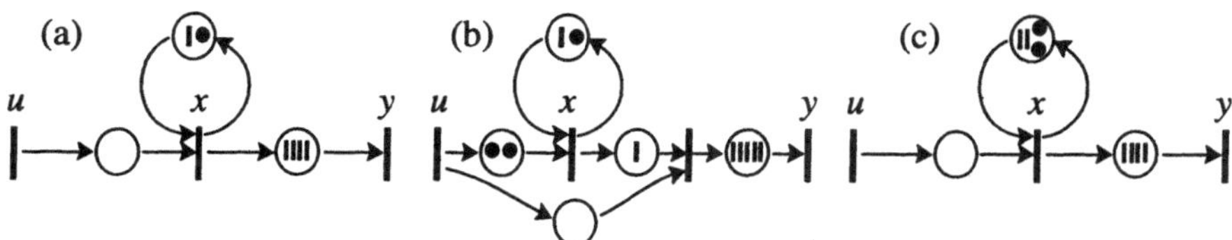

Fig. 3. Reduced event graphs

These simplifications are not at all intuitive and could not be obtained by reasoning directly on the graph. For example, suppose that the holding time of the output place between x_3 and y is changed from 2 into 3. Then the same calculations yield $H(\gamma, \delta) = \delta^5 \left(e \oplus \gamma^2\delta(\gamma\delta)^*\right)$ which results in the more complex TEG shown in Fig. 3b ([9]). Instead of the previous change , if a token is added in the output place between x_1 and y, then $H(\gamma, \delta) = \delta^4 \left(\gamma^2\delta^2\right)^*$ (TEG of Fig. 3c).

5 A Quick Review of Some System-Theoretic Results for TEGs

5.1 Autonomous Systems, Asymptotic Behavior and Eigenvalues

Conventional linear systems have 'modes' which are reached asymptotically (for stable systems) and which are related to their eigenstructures. Similar notions exist for *autonomous* TEGs (i.e. TEGs without input transitions, or with input transitions supplied with an infinity of tokens). Possibly at the price of augmenting the state vector dimensionality, such systems obey (dater) dynamic equations in the standard form $x(k+1) = Ax(k)$ in $\overline{\mathbb{Z}}^n_{\max}$. For simplicity, assume that the precedence graph $\mathcal{G}(A)$ is strongly connected (which will be seen as a sufficient condition of 'stability' later on). Then, such systems reach a *'periodic'* regime within a *finite* number of stages, namely,

$$\exists\lambda \in \mathbb{Q} \ , \quad \exists c \in \mathbb{N} \ , \quad \exists K \in \mathbb{Z} : \quad \forall k > K \ , \quad x(k+c) = \lambda^c x(k) \ ,$$

which means that, for all i, $x_i(k+c) = x_i(k) + c \times \lambda$ in conventional notation. When $\mathcal{G}(A)$ is strongly connected, λ is unique and, it is given by the maximum 'cycle mean' of $\mathcal{G}(A)$, that is, the maximum ratio, with respect to all circuits of $\mathcal{G}(A)$, of the total of weights (holding times) along the circuit over the length (number of arcs) of the same circuit. Algebraically, this can be summarized by the formula

$$\lambda = \bigoplus_{j=1}^{n} (\mathrm{trace}(A^j))^{1/j}$$

[8] with help of the MAX software due to Stéphane Gaubert (INRIA)

[9] However, it can be checked that the two transfer functions differ only by the first term of their developments.

(note that $a^{1/j}$ is a/j in conventional notation). If the equations are kept in their initial and general form $x(k) = \bigoplus_{l=0}^{m} A_l x(k-l)$ (for a TEG with places containing from 0 to m tokens in the initial marking), λ is the maximum ratio of the total of holding times over the total of tokens. For the TEGs depicted in Fig. 1 or 3, λ equals 1.

Indeed, λ is the unique eigenvalue of A (in the standard form), that is, there exists an eigenvector x such that $Ax = \lambda x$. Explicit formulæ are available for eigenvectors and the structure of 'eigenspaces' is fully understood (see [2]). As for the *'cyclicity'* c, it is given by the length (number of arcs or, more generally, of tokens) of the critical circuit[10], *if this circuit is unique*. Otherwise, more complex formulæ are available [2]. In the case when the TEG is made up of several strongly connected components, uniqueness is not preserved in general, but the number of eigenvalues depends on the ordering of eigenvalue magnitudes of components from upstream to downstream.

5.2 Stability and Stabilizability by Dynamic Output Feedback

A completely observable and controllable (conventional) system can be stabilized by dynamic output feedback. For TEGs, stability essentially means that tokens do not accumulate indefinitely inside the graph. A sufficient condition is that the whole system is synchronized, i.e. it consists of a single strongly connected component. A TEG is *structurally controllable* (resp. *observable*) if every internal transition can be reached by a directed path from at least one input transition (resp. is the origin of at least one directed path to some output transition). Then, it is easy to figure out that a structurally controllable and observable TEGs can be stabilized by *output feedback* in that the graph can be made strongly connected by adding appropriate arcs from output to input transitions. The next step is to get the equivalent of a notion of *pole placement*. For a TEG, this means that the new circuits created by feedback, whose primary purpose is to synchronize, hence stabilize, the system, will not damage its performance (in terms of its throughput, which is given by the inverse of the eigenvalue of the closed-loop TEG in the dater setting). This latter condition can always be ensured by *dynamic* feedback: it suffices to place enough tokens in the initial marking of places located on feedback arcs (which results into delays between outputs and inputs in the dater setting). Indeed, so doing, the new circuits resulting from feedback arcs necessarily cease being critical (compared to circuits of the open-loop TEG) when the marking of feedback places exceed some threshold. The purpose of *resource optimization* [6] is to optimize those thresholds given a desired performance level and the unit costs of resources (tokens) to be added in some places.

5.3 Transfer Functions, Rationality and Periodicity

In conventional system theory, a necessary and sufficient condition for an input-output system to admit a rational transfer function (matrix) is that it be realizable by a finite dimensional time-invariant linear system. For $\mathcal{M}_{in}^{ax}[\![\gamma, \delta]\!]$ transfer matrices, an even stronger result holds true since the following three properties are equivalent:

1. the transfer matrix can be realized by a TEG with constant (nonnegative) holding times;
2. the transfer matrix is rational (and causal);
3. the transfer matrix is periodic (and causal).

[10] the one which realizes the maximum cycle mean; for example, $c = 1$ for Fig. 3a whereas $c = 2$ for Fig. 3c

The second property means that each entry of the matrix belongs to the closure of $\{\varepsilon, e, \gamma, \delta\}$ by finitely many $\oplus$, $\otimes$ and * operations; the third property means that each entry can be written as an expression of the form $p \oplus q r^*$ in which p and q are polynomials in (γ, δ) which represent the transient behavior and the repeated pattern respectively, whereas r is a monomial $\gamma^k \delta^t$ which reproduces the pattern q along the 'slope' t/k. Needless to say, for a TEG with strongly connected internal transitions, this slope is nothing but the unique eigenvalue (in the dater representation). For the TEG of Fig. 3b, one has that $p = \delta^5$, $q = \gamma^2 \delta^6$ and $r = \gamma\delta$.

5.4 Frequency Responses

In conventional linear system theory, sine functions of any frequency are eigenfunctions of rational transfer functions $H(s)$, i.e., up to a transient behavior, the output 'mimics' the input with amplification and phase shift. The amplification gain and the phase shift at the frequency ω are computed by replacing the formal operator s by the numerical value $j\omega$ in the expression of $H(s)$. For TEGs, the analogues of sine functions are a certain class of periodic inputs of any rational 'slope' in the plane $\mathbb{Z}^2$ where the x-axis is the event domain and the y-axis is the time domain. The outputs caused by such inputs ('frequency responses') are identical to the inputs, up to the fact that they are shifted along the two axes. Shifts can be evaluated using, in some sense, the slope as a numerical argument of the transfer function (see [2, §5.8] for more detailed explanations).

For example, for the slope 2 (1 event every 2 time units), the eigenfunction $\ell_2(\gamma, \delta)$ is $(\gamma^{-1}\delta^{-2})^*(\gamma\delta^2)^*$ ([11]). Pick $g = -2$ and $d = 1$ (such that the slope is equal to $-g/d$) and evaluate $H(g, d)$ as if H was a *numerical* function over $\mathbb{Z}^2_{\max}$. For example, with $H(\gamma, \delta) = \delta^4(\gamma\delta)^*$, $H(-2, 1) = (1)^4 \otimes ((-2) \otimes 1)^* = (4 \times 1) + \sup(0, -2 + 1, -4 + 2, -6 + 3, \ldots) = 4 = 0 \times g + 4 \times d$, meaning that the shift is equal to 0 in the event domain and to 4 in the time domain. Indeed, it can directly be verified that $H(\gamma, \delta)\ell_2(\gamma\delta) = \delta^4 \ell_2(\gamma, \delta)$. These shifts become infinite whenever the slope is smaller than the eigenvalue λ (in the dater representation): this corresponds to a situation of instability because the input is faster than what the TEG can process, hence tokens accumulate inside the system and delays get infinite. In other words, TEGs exhibit a 'low-pass' feature.

5.5 Co-State Equations and Second-Order Theory

In conventional optimal control, Pontryagin's minimum principle introduces a backward equation for a vector ξ called 'co-state' or 'adjoint state'. In the linear theory of TEG, a similar notion arises about the following problem: given an output (dater) trajectory $\{y(k)\}$, find the *latest (greatest)* input trajectory $\{u(k)\}$ which yields an output trajectory less (earlier) than the given one. This is again a typical problem in the theory of residuation already alluded to at §4.1: indeed, if $H(\gamma)$ is the transfer function, then the problem is to find the greatest $U(\gamma)$ such that $H(\gamma)U(\gamma) \leq Y(\gamma)$. The solution of this problem exists, and it is denoted $U(\gamma) = H(\gamma) \backslash Y(\gamma)$, where, generally speaking, in a complete dioid, $y \mapsto a \backslash y$ ('left division') denotes the residuated mapping of $x \mapsto ax$ ('left multiplication'). It can be proved ([2, §5.6]) that, for the system with dater equations

$$x(k+1) = Ax(k) \oplus Bu(k) \;; \quad y(k) = Cx(k) \;,$$

[11] The purpose of the 'negative tail' is to remove the transient behavior by starting at $-\infty$.

the solution is explicitly given by the backward recursive equations

$$\xi(k) = (A \backslash \xi(k+1)) \wedge (C \backslash y(k)) \quad ; \quad u(k) = B \backslash \xi(k) \ ,$$

and, moreover, $(A \backslash b)_i = \min_k (b_k - A_{ki})$, with the conventions $+\infty - (+\infty) = +\infty$ and $(-\infty) - (-\infty) = +\infty$.

The (nonnegative) differences $\xi_i(k) - x_i(k)$ represent the 'spare time' or the 'margin' which is available at transition x_i for the firing number k; in other words, an exogenous event may delay this event by this spare time without preventing the future deadlines to be met. By returning to dioid notation, differences such as $\xi(k) - x(k)$ emerge from the calculation of the matrix $P(k) = \xi(k) / x(k)$ (here we use the 'right division', i.e. the residuated mapping of right multiplication; indeed, $p(k) = x(k) \backslash \xi(k)$, as a scalar, carries less information). In conventional system theory, $\xi = Px$ where P is a Riccati matrix. Very formally, then, $P = \xi / x$. Unfortunately, for the time being, no Riccati equation has been found for this matrix in the context of TEG (see [4] for some results).

Similarly, differences between input and output daters (resp. counters) of a place or a group of places have obviously a practical interest in terms of evaluation of the sojourn time of tokens (resp. accumulated stock of tokens) in that part of the TEG. We consider such quantities as the analogue of 'correlations' manipulated in the *second-order theory* of conventional linear systems. We refer the reader to [2, §6.6] and [6] for further developments.

References

1. Akian, M., Quadrat, J.-P., Viot, M.: Bellman Processes. This conference (1994).
2. Baccelli, F., Cohen, G., Olsder, G.J., Quadrat, J.P.: Synchronization and Linearity: an Algebra for Discrete Event Systems. John Wiley and Sons, New York (1992).
3. Cuninghame-Green, R.A., Meijer, P.F.J.: An algebra for piecewise-linear minimax problems. Discrete Applied Mathematics, **2** (1980) 267–294.
4. Cohen, G., Gaubert, S., Quadrat, J.P.: From first to second-order theory of linear discrete event systems. 12th IFAC World Congress (1993), Sydney, Australia.
5. Gaubert, S.: Rational series over dioids and discrete event systems. This conference (1994).
6. Gaubert, S.: Théorie Linéaire des Systèmes dans les Dioïdes. Thèse, École des Mines de Paris, Paris, (1992).
7. Gondran, M., Minoux, M.: Graphs and Algorithms. John Wiley and Sons, New York (1986).
8. Mairesse, J.: Stochastic linear systems in the (max,+) algebra. This conference (1994).
9. Dobrokhotov, S.Y., Kolokoltsov, V.N., Maslov, V.P.: Quantization of the Bellman equation, exponential asymptotics and tunneling. In: Maslov, V.P., Samborski, S.: Idempotent Analysis. AMS, Adv. in Sov. Math. **13** (1992).
10. Quadrat, J.P.: Théorèmes asymptotiques en programmation dynamique. Note CRAS Paris **311** (1990) 745–748.
11. Rockafellar, R.T.: Convex Analysis. Princeton University Press, Princeton, N.J. (1970).
12. Wagneur, E.: Subdirect sum decomposition of finite dimensional pseudomodules. This conference (1994).

On Structural Properties of Min-Max Systems

Geert Jan Olsder

Department of Mathematics
Delft University of Technology
P.O. Box 5031, 2600AG Delft, the Netherlands

1 Introduction: systems to be considered

A variety of problems in operations research, performance analysis, manufacturing, communication networks, etc. can be modelled as discrete event systems with minimum and maximum timing constraints. The algebra underlying such systems is based on the three operations maximization, minimization and addition. In this paper some structural properties of such systems will be studied. Systems which are based on the two operations maximization and addition (or based on the two operations minimization and addition), called max-plus systems, are well understood now. The corresponding theory gives conditions under which timed event graphs (which are related to Petri nets) reach a periodic behavior and it characterizes their periodicities. It can also be viewed as a spectral theory of max-plus matrices.

One can distinguish structural and nonstructural properties. The existence of an eigenvalue of a max-plus matrix for instance is a structural property; it does not depend on numerical values (as long as they remain finite) characterizing the matrix. In this paper the existence of (structural) fixed points for min-max systems will be studied specifically. The class of bipartite systems to be introduced (and their decomposition into elementary bipartite systems) turns out to be a very useful aid. Relationships with stability issues, eigenvalues will be briefly indicated and some conjectures will be given.

Though this paper is not a survey, certain known results will also be dealt with in order to complement and relate various results and to briefly overview the concept of stability for min-max systems. Complementary and related results can be found in [9], [5], [4], [2], [1], [8], [10], [6]. In the following, the symbol $\vee$ stands for maximization and $\wedge$ for minimization.

Definition 1. A min-max expression, f, is a term in the grammar

$$f := z_1, z_2, \ldots, | f + a | f \wedge f | f \vee f,$$

where $z_1, z_2, \ldots,$ are variables and where $a \in \mathbb{R}$ is a parameter.

If f is a min-max expression, then it can be placed in the so-called conjunctive normal form (see [7]):

$$f = f_1 \wedge f_2 \wedge \cdots \wedge f_m,$$

where each f_i is a max-plus expression, i.e.

$$f_i = a_1 + z_1 \vee a_2 + z_2 \vee \cdots \vee a_n + z_n.$$

It will be assumed that in max-plus expressions the parameters, here a_i, belong to $\mathbb{R} \cup \{\varepsilon\}$, where $\varepsilon = -\infty$. The quantity ε is the neutral element with respect to maximization; if it occurs in a max-plus expression, the term concerned can be left out without changing the numerical value of this expression. If f is a min-max expression, then it can equally well be placed in the so-called disjunctive normal form:

$$f = f_1 \vee f_2 \vee \cdots \vee f_p,$$

where each f_i is a min-plus expression, i.e.

$$f_i = b_1 + z_1 \wedge b_2 + z_2 \wedge \cdots \wedge b_n + z_q.$$

Here it is assumed that $b_i \in \mathbb{R} \cup \{\top\}$, where $\top = +\infty$. The quantity $\top$ has been added because it is the neutral element with respect to minimization.

Definition 2. A min-max function of dimension p is any function, $\mathcal{M} : \mathbb{R}^p \to \mathbb{R}^p$, each of whose components, $\mathcal{M}_i$, is a min-max expression of p variables $z_1, \ldots, z_p$.

Some properties of $\mathcal{M}$ are:

- $\mathcal{M}$ is continuous in its arguments;
- $\mathcal{M}$ is monotone, i.e. $\bar{z} \leq z \Rightarrow \mathcal{M}(\bar{z}) \leq \mathcal{M}(z)$ (these inequalities must be interpreted as the vectorial notation for inequalities of the individual components);
- $\mathcal{M}$ is homogeneous, in the sense that for any scalar λ, $\mathcal{M}(z + \lambda\mathbf{1}) = \mathcal{M}(z) + \lambda\mathbf{1}$ (the notation $\mathbf{1}$ refers to a vector of appropriate size of which all components are 1).

Definition 3. A min-max system is a system with state $z(k) = (z_1(k), \ldots, z_q(k))'$, where $'$ denotes transposition, which evolves according to $z(k+1) = \mathcal{M}(z(k))$, $k = 0, 1. 2, \ldots$, where $\mathcal{M}$ is a min-max function.

Definition 4. A max-plus (dually, min-plus) system is a min-max system where all $\mathcal{M}_i$ components are max-plus only (respectively, min-plus only) expressions.

Rather than writing a max-plus system as $z(k+1) = \mathcal{M}(z(k))$, one often writes

$$z(k+1) = A \otimes z(k), \tag{1}$$

which is short-hand notation for

$$z_i(k) = \max(a_{i1} + z_1(k), \ldots, a_{ip} + z_p(k)), \quad i = 1, 2, \ldots, p.$$

Definition 5. A separated (min-max) system is a min-max system with mixed constraints: $\mathcal{M}_i$ is a max-plus only expression for $i = 1, \ldots, m$ and $\mathcal{M}_i$ is a min-plus only expression for $i = m+1, \ldots, p$. A separated system is called nondegenerated if $1 \leq m < p$.

Definition 6. A separated min-max system is called bipartite if the max-plus only expressions (say $\mathcal{M}_i$ for $i = 1, \ldots, m$) map $z_{m+1}, \ldots, z_p$ into z_i, $i = 1, \ldots, m$, and if the min-plus only expressions (the remaining $\mathcal{M}_i$; i.e. $\mathcal{M}_i$ for $i = m + 1, \ldots, p$) map $z_1, \ldots, z_m$ into z_i, $i = m + 1, \ldots, p$.

Remark 7. Bipartite systems have some flavor of systems described by means of Petri nets (see [1]); indeed, Petri nets are defined by means of bipartite graphs. It is, however, not difficult to show that there is no (timed) Petri net representation, consisting of a finite number of places and transitions, which corresponds to a bipartite system. Said differently, bipartite systems and Petri nets describe different classes of systems; there is no equivalence of these two classes.

If one writes $z(k + 2) = \mathcal{M}(z(k + 1)) = \mathcal{M}(\mathcal{M}(z(k)))$, one sees that $z(k + 2)$ can be expressed as a min-max expression of the variables $z_i(k)$. These expressions can be written in their disjunctive normal form;

$$z_i(k+2) = \bigvee_j \left(\bigwedge_l (z_l(k) + a_{ijl}) \right), \tag{2}$$

for some parameters a_{ijl}. The expression $\mathcal{M}(\mathcal{M}(z))$ will be written more compactly as $\mathcal{M}^2(z)$. This notation extends to $\mathcal{M}^k(z)$, with k being a natural number, in a natural way. If one now defines

$$y_{ij}(k+1) = \bigwedge_l z_l(k) + a_{ijl}, \quad x_i(k+2) = \bigvee_j y_{ij}(k+1),$$

the latter equation representing (2), one sees that $z(k+2) = \mathcal{M}^2(z(k))$ can be written as a bipartite system. Actually, Eq. (2) describes two such systems: one system for k even, say $k = 2\bar{k}, \bar{k} = 1, 2, \ldots$, and with the state composed of $x(2\bar{k})$ and $y(2\bar{k} - 1)$; the other system for k odd, say $k = 2\bar{k}+1, \bar{k} = 0, 1, \ldots$, and with the state composed of $x(2\bar{k}+1)$ and $y(2\bar{k})$. Please note that these two systems behave completely independently one from the other. Please also note that while on the one hand bipartite systems form a subclass of min-max systems, on the other hand each min-max system can be written as a bipartite system (more precisely: as two independent bipartite systems; in the one the states of the original min-max system appear with odd numbered arguments, in the other the states of the same original min-max system appear, but now with even numbered arguments).

2 Stability, cyclicity and eigenvalues

Definition 8. The mapping $\mathcal{M}$ has a fixed point, $z \in \mathbb{R}^p$, if $\mathcal{M}(z) = z + \lambda\mathbf{1}$, for some $\lambda \in \mathbb{R}$. One also says that z is an eigenvector of $\mathcal{M}$ and that λ is the eigenvalue corresponding to z. One talks about a structural fixed point (and structural eigenvalue) if the existence of a fixed point does not depend on the numerical values, as long as they remain finite, of the parameters through which $\mathcal{M}$ is defined.

It should be emphasized that only finite changes in finitely-valued parameters are allowed in the above definition on a structural fixed point.

Theorem 9. *A max-plus system $z(k+1) = A \otimes z(k)$ has a structural fixed point if and only if the precedence graph of A is strongly connected.*

This follows directly from [2] or [1]. The essence is that the precedence graph is invariant with respect to finite changes in the parameter values of A.

Theorem 10. *A nondegenerated separated system with both the max-part and the min-part strongly connected (see [9]) does not have a structural fixed point.*

This follows directly from [9].

Definition 11. The vector z is a periodic point of $\mathcal{M}$ if for some $\mu \in \mathbb{R}$, $\mathcal{M}^d(z) = z+\mu\mathbf{1}$. The least d with this property is called the period of z. If $d = 1$, z is a fixed point of $\mathcal{M}$. The sequence $z(k)$, $k = 0, 1, \ldots$, with $z(0)$ being a periodic point and $z(k+1) = \mathcal{M}(z(k))$, $k \geq 0$, is called a periodic behavior of $\mathcal{M}$.

Definition 12. The min-max system defined by the mapping $\mathcal{M}$ is called stable if, starting from an arbitrary initial condition $z(0)$ with only finite components, a number L exists such that $|z_i(k) - z_j(k)| < L$ for all $k = 0, 1, \ldots$, and for all $i, j = 1, \ldots, p$. If the existence of stability does not depend on the numerical values of the finitely-valued parameters which determine $\mathcal{M}$, then one talks about structural stability.

Theorem 13. *The max-plus system (1) is structurally stable if and only if the precedence graph of A is strongly connected. Each solution converges in a finite number of steps to a periodic behaviour.*

It is conjectured that a min-max system is (structurally) stable if and only if it has a (structural) fixed point. Moreover, it is conjectured that the solution of a stable min-max system converges in a finite number of steps to a periodic behaviour. Some results in this direction have already been shown to be true, see for instance [4].

3 Bipartite systems

Bipartite systems are of the form

$$x(k+1) = A \otimes y(k) \ , \tag{3}$$

$$y(k+1) = B \odot x(k) \ . \tag{4}$$

The symbol $\otimes$ refers to the matrix vector product in the max-plus algebra; the symbol $\odot$ refers to the matrix vector product in the min-plus algebra. The juxtaposition of the vectors $x = (x_1, \ldots, x_m)'$ and $y = (y_1 \ldots, y_n)'$ is the state of the system. The quantities A and B are matrices of order $m \times n$ and $n \times m$ respectively. One writes $A = \{a_{ij}\}$ and $B = \{b_{ij}\}$. We assume on the one hand that the coordinates of x and the entries of A are elements of $\mathbb{R} \cup \{\varepsilon\}$ and on the other hand that the coordinates of y and the entries of B are elements of $\mathbb{R} \cup \{\top\}$. One has the convention that $\top \otimes \varepsilon = \varepsilon$ and $\top \odot \varepsilon = \top$. The conventional notation of (3) and (4) is:

$$x_i(k+1) = \max(a_{i1} + y_1(k), \ldots, a_{in} + y_n(k)), \quad i = 1, \ldots, m,$$
$$y_i(k+1) = \min(b_{i1} + x_1(k), \ldots, b_{im} + x_m(k)), \quad i = 1, \ldots, n.$$

Remark 14 (The nondegeneracy assumption). It will be assumed for the remainder of the paper that for each i there exists at least one j (which may depend on i) such that $a_{ij} > -\infty$. This condition prevents x_i from being 'degenerated'. Similarly it will be assumed that for each i there exists at least one j such that $b_{ij} < \infty$. Without loss of generality we also assume that for each j there exist i's such that $a_{ij} > -\infty$ and $b_{ij} < \infty$. In terms of the precedence graph of the system (3), (4), this condition states that each node has at least one incoming and one outgoing arc. Nodes which do not have an outgoing arc are 'endstations' of the graph; the behaviour of the system in these nodes is completely determined by the other nodes. All these assumptions together can succinctly be stated as: each row and each column of A and B contains at least one finite element. Such systems are called nondegenerate.

If an initial condition $(x(0), y(0))$ is given, one can calculate the evolution of the equations (3) and (4) uniquely, i.e. one can calculate $(x(k), y(k))$ for $k = 1, 2, \ldots$. The combination of (3) and (4) will, not surprisingly, compactly be written as

$$z(k+1) = \mathcal{M}(z(k)), \tag{5}$$

where $z' \stackrel{\text{def}}{=} (x', y')$.

Theorem 15. *A nondegenerate bipartite system, characterized by matrices A and B as in (3) and (4), has a structural fixed point if and only if for each $i \in \{1, 2, \ldots, m\}$ and each $j \in \{1, 2, \ldots, n\}$ at least one of the two entries a_{ij}, b_{ji} is finite (i.e. $a_{ij} \neq \varepsilon$ and/or $b_{ji} \neq \top$).*

Proof.
Sufficiency. An initial condition $(\bar{z})$ for (5) will be constructed such that $\mathcal{M}(\bar{z}) = \bar{z} + \lambda \mathbf{1}$ for some finite real λ. The vector $\bar{z}$ is then a fixed point of $\mathcal{M}$. Towards this end consider the differential equations

$$\dot{p}(t) = u(t), \tag{6}$$

for the control function $u(t) \in \mathbb{R}^{m+n}$ and the state vector $p(t) \in \mathbb{R}^{m+n}$; the dot refers to derivative with respect to time t. The initial condition for (6) is $p(0) = z(0) \stackrel{\text{def}}{=} (x'(0), y'(0))'$, being also the initial conditions of (3) and (4). The only assumption here on $x(0)$ and $y(0)$ is that all the coefficients are finite. It will be shown that the control function u can be chosen in such a way that for a finite time T we have $p(T) = \bar{z}$.

For $t \geq 0$, define

$$\begin{aligned} c(t) &= \mathcal{M}(p(t)) - p(t), \\ \bar{c}(t) &= \max_i c_i(t), \quad \underline{c}(t) = \min_i c_i(t), \\ \mathcal{B}(t) &= \{i \mid c_i(t) = \underline{c}(t)\}, \\ \mathcal{B}^{c}(t) &= \{i \mid c_i(t) \neq \underline{c}(t)\}. \end{aligned}$$

If $p(t)$ is finite, then because of the nondegeneracy assumption one also has that $c(t)$ is finite and hence $\bar{c}(t)$ and $\underline{c}(t)$ exist. Since $p(0) = z(0)$ is finite by assumption, it will follow from the construction of the control function $u(t)$, $t \geq 0$, to be given shortly, that $p(t)$ remains finite for $t \geq 0$. For $i \in \mathcal{B}^{c}(t)$ choose $u_i(t) = 0$. The other components of $u(t)$, i.e. those $u_i(t)$ for which $i \in \mathcal{B}(t)$, will be chosen in such a way that $\mathcal{B}(t)$ is a nondecreasing function of t. The latter means that once a component $c_j(t)$ is at the 'bottom of the basket of all $c(t)$-components', i.e. $c_j = \underline{c}(t)$, it will remain at the bottom for all larger values of t. As t increases this bottom can only catch more c-components. Towards this end consider

$$\dot{c}_i = \sum_{j \in \mathcal{B}(t)} \frac{\partial \mathcal{M}_i}{\partial p_j} u_j - u_i, \quad i \in \mathcal{B}(t). \tag{7}$$

At this point we should note that function $\mathcal{M}$ is not differentiable everywhere. Since, however, $\mathcal{M}$ is monotonous and continuous, it is differentiable almost everywhere. At points where $\mathcal{M}$ is not differentiable, $\partial \mathcal{M}_i / \partial p_j$ is not uniquely defined: it will be a point to set mapping. In the analysis to follow, simply take an arbitrary element of this set (if necessary). We will not pay any attention to this (non)differentiability anymore. Remark 16 after the proof will view this item from an algorithmic point of view.

The functions $\partial \mathcal{M}_i / \partial p_j$, $i, j = 1, 2, \ldots, m+n$, are written as elements g_{ij} of an $(m+n) \times (m+n)$ matrix G. Now we write (7) concisely as

$$\dot{\tilde{c}} = (\tilde{G} - I)\tilde{u},$$

where I is the identity matrix, $\tilde{c}$ ($\tilde{u}$) is the vector of c_i- (u_i-) components for which $i \in \mathcal{B}(t)$, and where $\tilde{G}$ is an appropriate submatrix of G. Due to the properties of $\mathcal{M}$ mentioned in Section 1,

$$g_{ij} \geq 0, \quad \sum_{j \in \mathcal{B}(t)} g_{ij} \leq 1,$$

and therefore the moduli of the eigenvalues of $\tilde{G}$ are ≤ 1. Now we distinguish two cases, viz.,

- The case that all these eigenvalues lie strictly within the unit circle. Choose $\tilde{u}$ as

  $$\tilde{u} = -\frac{(I - \tilde{G})^{-1}\mathbf{1}}{\|(I - \tilde{G})^{-1}\mathbf{1}\|_\infty} = -\frac{(I + \tilde{G} + \tilde{G}^2 + \ldots)\mathbf{1}}{\|(I + \tilde{G} + \tilde{G}^2 + \ldots)\mathbf{1}\|_\infty}.$$

  Clearly $\|\tilde{u}\|_\infty = \|u\|_\infty = 1$ and $u_i \leq 0$, $i = 1, 2, \ldots, m+n$.
- At least one of the eigenvalues lies on the unit circle. Due to the theorem of Perron-Frobenius [3], there is the real eigenvalue 1. Consider therefore

  $$\tilde{G}\tilde{u} = \tilde{u} \Rightarrow (\tilde{G} - I)\tilde{u} = 0.$$

  According to [3] this eigenvector $\tilde{u}$ can be chosen such that $\tilde{u} \leq 0$, $\tilde{u} \neq 0$. This choice is then normalized such that $\|\tilde{u}\|_\infty = 1$.

As a result, all components of $\dot{\tilde{c}}$ are nonnegative and identical. This shows that indeed 'once a component $c_j(t)$ is at the bottom of the basket of all $c(t)$-components, it will remain at the bottom for all larger values of t'.

It has been shown that, whatever the eigenvalues of $\tilde{G}$ are, $\underline{c}(t)$ is a nondecreasing function. All $c_i(t)$ with $i \in \mathcal{B}^c(t)$ are nonincreasing and hence $\bar{c}(t)$ is nonincreasing also. Moreover, the number of elements in $\mathcal{B}(t)$ is nondecreasing. It remains to be shown that a finite T exists for which $\underline{c}(T) = \bar{c}(T)$. If so, then $p(T) = \bar{z}$ is the vector we were looking for. Moreover, $\bar{c}(T)$ can be interpreted as an eigenvalue of the mapping $\mathcal{M}$. Suppose that not such finite T exists and we will obtain a contradiction.

Since $\|u(t)\|_\infty = 1$ for $t \geq 0$ and $u_i(t) \leq 0$, also for $t \geq 0$, at least one of the p_i-components, say p_{i_1}, approaches $-\infty$ as $t \to \infty$. We also have $\mathcal{B}^c(t) \neq \emptyset$ for all $t \geq 0$ and hence there is at least one p_i-component, say p_{i_2}, which remained a member of $\mathcal{B}^c(t)$ for all $t \geq 0$ and which thus has never changed its value and thus remains finite. Now we are going to use the assumption that for all i, j the values a_{ij} and/or b_{ji} are finite. It then follows that, if p_{i_1} is a component of the vector x then all y components must be equal to $-\infty$; if on the other hand p_{i_1} is a component of the vector y then all x components must be equal to $-\infty$. Repeating this argument one obtains that all components of both x and y must be equal to $-\infty$ which contradicts the fact that at least one p component remained finite.

Necessity. Here we give a simple counterexample. Suppose $m = n = 2$,

$$A = \begin{pmatrix} 5 & 3 \\ \varepsilon & 1 \end{pmatrix}, \quad B = \begin{pmatrix} 5 & \top \\ 3 & 1 \end{pmatrix}. \tag{8}$$

We have that neither a_{21} nor b_{12} is finite. If we assume the initial conditions $x_i(0) = y_i(0) = 0, i = 1, 2$, then $x_i(k) = 5k$ and $y_i(k) = k$ for $k = 1, 2, \ldots$, and this contradicts the assumed existence of an eigenvalue.

Remark 16. The proof above is in principle a constructive one. If in such an algorithm one works with finite differences rather than with differentials, one does not encounter the problem with (non)differentiability as briefly mentioned in the proof. An adjustment that one has to make, if one works with finite differences, is that one must redefine $\mathcal{B}(t)$ as $\mathcal{B}(t) = \{i \mid c_i(t) \leq \underline{c}(t) + \epsilon\}$, where ϵ is a small positive parameter related to the stepsize of the difference equations; see [9] for details.

Remark 17. It is believed that Theorem 15 can also be formulated in a Boolean framework.

We will now study the structure and properties of the system described by (3) and (4), which satisfies the nondegeneracy assumption (see Remark 14), if at least for one pair (i, j) neither a_{ij} nor b_{ji} is finite. The set $\{1, 2, \ldots, m\}$ is split up into two nonoverlapping subsets I_1 and I_2 such that $I_1 \cup I_2 = \{1, 2, \ldots, m\}$ and the set $\{1, 2, \ldots, n\}$ is split up into two nonoverlapping subsets J_1 and J_2 such that $J_1 \cup J_2 = \{1, 2, \ldots, n\}$. Moreover, the sets $\{I_1, I_2\}$ and $\{J_1, J_2\}$ are chosen in such a way that $a_{i_1 j_2} = \varepsilon$ and $b_{j_2 i_1} = \top$ for all $i_1 \in I_1$ and for all $j_2 \in J_2$. At least one such a pair (i_1, j_2) exists by the assumption of the existence of at least one pair (i, j) such that neither a_{ij} nor b_{ji} is finite. Thus one has split up the original system characterized by the x_i variables with $i = 1, 2, \ldots, m$ and by the y_j variables with $j = 1, 2, \ldots, n$ into two subsystems; one is characterized by the x_i variables with $i \in I_1$ and by the y_j variables with $j \in J_1$ and the other one is characterized by the x_i variables with $i \in I_2$ and by the y_j variables with $j \in J_2$. If there are any connections between these two subsystems, then such connections are realized by finite a_{ij} and/ or b_{ji} variables with $j \in J_1$ and $i \in I_2$. This procedure of splitting up systems can be repeated but now with respect to each of the subsystems. An assumption made at this point is that each of the two subsystems must be nondegenerate (see Remark 14) again. If this assumption is not fulfilled, one deletes those x_i and/or y_j variables which do not have an incoming arc (within the same subsystem) or an outgoing arc (also within the same subsystem) such that the subsystems become nondegenerated again. This procedure can be continued till one has only subsystems in each of which each x variable is connected to each y variable by means of a finite a and/or b element. Such subsystems are the basic building blocks for the total system; they are called elementary systems. These elementary systems can be thought of to be connected by a tree structured graph; the root is the total system, and each node corresponds to a subsystem and has two branches at the end of which one has the two 'subsubsystems' (which are 'twins'), except for the end-nodes (the 'leaves'), which correspond to the elementary systems.

Remark 18. The decomposition of a bipartite system into elementary systems is not unique, as is easily shown by examples.

An elementary system has an eigenvalue, as shown by the theorem above. The question now is whether the combination of two 'twinned' elementary systems also has an eigenvalue. The answer to this question is not always positive as shown by the example characterized by the A and B matrix of (8). If these matrices are changed somewhat, however, say to

$$A = \begin{pmatrix} 5 & \varepsilon \\ 3 & 1 \end{pmatrix}, \quad B = \begin{pmatrix} 5 & 3 \\ \top & 1 \end{pmatrix}, \tag{9}$$

then an eigenvalue does exist (and it equals 3 as is shown easily).

Conjecture 19. *We are given two (nondegenerated) subsystems, S_1 and S_2, the states of which are respectively x_{i_1}, $i_1 \in I_1$, y_{j_1}, $j_1 \in J_1$ for S_1, and x_{i_2}, $i_2 \in I_2$, y_{j_2}, $j_2 \in J_2$ for S_2, where I_1 and I_2 are nonoverlapping subsets of $\{1, 2, \ldots, m\}$ and where J_1 and J_2 are nonoverlapping subsets of $\{1, 2, \ldots, n\}$. There are no finite a and/or b values which connect the sets J_2 and I_1. Suppose both subsystems have an eigenvalue; λ_1 and λ_2 respectively. The following statements can be made:*

- *If there are no finite a and/or b values which connect the sets J_2 and I_1, then S_1 and S_2 are disconnected and both systems will behave independently.*
- *If there are no finite a values which connect the sets J_1 and I_2, but finite b values which connect the sets J_1 and I_2 do exist, then the following holds.*
  - *If $\lambda_1 < \lambda_2$, then an eigenvalue of the combination $\{S_1, S_2\}$, with the b-valued connections, does not exist.*
  - *If $\lambda_1 \geq \lambda_2$, then an eigenvalue λ of the combination $\{S_1, S_2\}$, with the b-valued connections, might exist and, if so, it satisfies $\lambda = \lambda_1$. This eigenvalue does exist if (sufficiency condition) each element of J_1 is directly connected to an element of I_2 by means of an arc corresponding to a finite b value.*
- *If there are no finite b values which connect the sets J_1 and I_2, but finite a values which connect the sets J_2 and I_1 do exist, then the following holds.*
  - *If $\lambda_1 < \lambda_2$, then an eigenvalue of the combination $\{S_1, S_2\}$, with the a-valued connections, does not exist.*

  - *If $\lambda_1 \geq \lambda_2$, then an eigenvalue λ of the combination $\{S_1, S_2\}$, with the a-valued connections, might exist and, if so, it satisfies $\lambda = \lambda_2$. This eigenvalue does exist if (sufficiency condition) each element of I_2 is directly connected to an element of J_1 by means of an arc corresponding to a finite a value.*
- *If the sets J_1 and I_2 are connected by both finite a and finite b values, then the following holds.*
  - *If $\lambda_1 < \lambda_2$, then an eigenvalue of the combination $\{S_1, S_2\}$, with the a- and b-valued connections, does not exist.*
  - *If $\lambda_1 \geq \lambda_2$, then an eigenvalue λ of the combination $\{S_1, S_2\}$, with the a- and b-valued connections, might exist and, if so, it satisfies $\lambda_1 \leq \lambda \leq \lambda_2$. This eigenvalue does exist if (sufficiency condition) each element of I_2 is directly connected to an element of J_1 by means of an arc corresponding to a finite a value and if each element of J_1 is directly connected to an element of I_2 by means of an arc corresponding to a finite b value.*

It is believed that a proof of this conjecture can be given along the lines of Lemmas 4.1–4.6 of [9]. This conjecture provides a tool for calculating the eigenvalue of a system (if it exists), starting from the decomposition in elementary subsystems and working up the tree towards larger subsystems. If during the decomposition procedure some x_i and/or y_j variables were deleted because of degeneracy, these must be added again for calculating the eigenvalue of the combination of two subsystems (the formulation of the conjecture above must be adjusted somewhat to include these earlier deleted variables; they must be added again for the calculation of the eigenvalue). Even if the composition of two subsystems does not have a (unique) eigenvalue, one can still talk about a lower bound and upper bound of the speeds of the two subsystems. It is possible that 'higher up the tree', where combinations with other subsystems are constructed, a unique eigenvalue will result again (because for instance the eigenvalue of the lastly added subsystem 'overrules' the lower and upper bound obtained earlier).

4 Conclusion

This paper is a stepping stone in showing that many properties which are known to be true for max-plus systems, also hold true for the more general class of min-max systems. The rewriting of a min-max system as a bipartite system turned out useful in proving results. Though in this paper only deterministic aspects have been dealt with, it is also believed that many properties of stochastic max-plus systems carry over to stochastic min-max systems (see [6]).

References

1. F. Baccelli, G. Cohen, G.J. Olsder, and J.P. Quadrat. *Synchronization and Linearity*. Wiley, 1992.
2. G. Cohen, D. Dubois, J.P. Quadrat, and M. Viot. A linear system-theoretic view of discrete event processes and its use for performance evaluation in manufacturing. *IEEE Transactions on Automatic Control*, AC-30:210–220, 1985.

3. F.R. Gantmacher. *The Theory of Matrices*. Chelsea Publishing Company, New york, 1959.
4. Jeremy Gunawardena. Cycle times and fixed points of min-max functions. Technical report, Department of Computer Science, Stanford University, Stanford, CA 94305, USA, 1993.
5. Jeremy Gunawardena. Periodic behaviour in timed systems with (and,or) causality. part i: systems of dimension 1 and 2. Technical report, Department of Computer Science, Stanford University, Stanford, CA 94305, USA, 1993.
6. A. Jean-Marie and G.J. Olsder. Analysis of stochastic min-max systems: results and conjectures. Technical Report 93-94, of the Faculty of Technical Mathematics and Informatics, Delft University of Technology, 1993.
7. P.T. Johnstone. *Stone Spaces*, volume 3 of *Studies in Advanced Mathematics*. Cambridge University Press, 1982.
8. G.J. Olsder. Descriptor systems in min-max algebra. In *Proceedings of the 1-st European Control Conference*, pages 1825–1830. Hermès, Paris, 1991. Grenoble, France.
9. G.J. Olsder. Eigenvalues of dynamic min-max systems. *Journal of Discrete Event Dynamic Systems*, 1:177–207, 1991.
10. G.J. Olsder. Analyse de systemes min-max. Technical Report 1904, INRIA, Sophia-Antipolis, France, 1993.

Rational Series over Dioids and Discrete Event Systems

Stéphane Gaubert

INRIA, Domaine de Voluceau, BP 105, 78153 Le Chesnay Cédex, France.
e-mail: Stephane.Gaubert@inria.fr

1 Introduction

We survey the different kinds of rational series which arise in the study of Discrete Event Systems (DES) and of certain related Markov Decision Processes. The use of rational series over fields is classical, e.g. as transfer functions, generating series of finite Markov chains, skew Ore series in Difference and Differential Algebra, commutative multivariable series for linear PDE with constant coefficients, Fliess' noncommutative generating series for bilinear systems. It turns out that all these more or less familiar classes of series admit useful counterparts for DES, when the scalars belong to some *dioids* [5] such as the $(\max, +)$ semiring. The main interest of this series theoretical point of view consists in introducing some efficient algebraic techniques in the study of these dynamical systems. Since this paper is obviously too short for such a program, we have chosen to propose an introductive guided tour. A more detailed exposition will be found in our references and in a more complete paper to appear elsewhere.

2 Rational Series in a Single Indeterminate

2.1 Rational Series as Transfers of $(\max, +)$-Linear Systems

We consider systems of recurrent linear equations in the $(\max, +)$ algebra of the form

$$x(n) = Ax(n-1) \oplus Bu(n), \qquad y(n) = Cx(n) \ , \tag{1}$$

where[1] $x(n) \in \mathbb{R}_{\max}^p$, $u(n) \in \mathbb{R}_{\max}$, $y(n) \in \mathbb{R}_{\max}$. As it is well known [2, 5], this class of systems encompasses in particular the dater equations of SISO Timed Event Graphs (TEG). A straightforward argument shows that the least solution[2] of (1) is given by

$$y(n) = \bigoplus_{k \in \mathbb{N}} CA^k Bu(n-k) \ , \tag{2}$$

[1] $\mathbb{R}_{\max}$ denotes the "(max,+) semiring" $(\mathbb{R} \cup \{-\infty\}, \max, +)$. The reader is referred to [5] for the general notation about dioids. In particular, $\oplus$ and $\otimes$ denote the sum and the product, ε denotes the zero and e the unit (in $\mathbb{R}_{\max}$, $a \oplus b = \max(a, b)$, $a \otimes b = a + b$, $\varepsilon = -\infty$, $e = 0$).

[2] which corresponds for a TEG to the earliest behavior [5].

hence, the input-output behavior of the system is completely determined by the following *transfer series*[3]

$$H = \bigoplus_{k \in \mathbb{N}} C A^k B X^k = C(AX)^* B \in \mathbb{R}_{\max}[[X]] \ , \qquad (3)$$

where[4] $M^* \stackrel{\text{def}}{=} \bigoplus_{k\in\mathbb{N}} M^k$. A series is *realizable* if it admits a representation of the form (3) for some finite dimensional triple (A, B, C). The celebrated Kleene-Schützenberger Theorem [3] states that *realizable* series coincide with *rational* series defined as follows.

Definition 1. Let $\mathcal{S}$ denote a dioid. The dioid of *rational series* over $\mathcal{S}$ in the indeterminate X is the least subset of $\mathcal{S}[[X]]$ containing the polynomials and stable by the operations $\oplus$, $\otimes$ and $*$.

Thus, *the transfer series* $H = C(AX)^*B$ *of a stationary finite dimensional* (max, +) *linear system is rational.*

2.2 Generating Series of Bellman Chains

Let us consider a Markovian maximization problem with finite state $S = \{1, \ldots, n\}$, finite horizon k, final reward b and transition reward $i \stackrel{A_{ij}}{\to} j$. The value function for the initial position $i \in S$ is thus given by

$$v_i^{(k)} \stackrel{\text{def}}{=} \max_{i_1 \ldots i_k} [A_{ii_1} + \cdots + A_{i_{k-1}i_k} + b(i_k)] \ . \qquad (4)$$

This is a particular case of Bellman chain (Akian, Quadrat and Viot [1]). As it is well known and obvious from (4), the value function is given by a product of matrices in the (max, +) algebra, that is, $v^{(k)} = A^k b$. Now, let us consider the vector of *generating series*

$$V_i \stackrel{\text{def}}{=} \bigoplus_k v_i^{(k)} X^k = \bigoplus_k (A^k b)_i X^k \in \mathbb{R}_{\max}[[X]] \ . \qquad (5)$$

An immediate comparison with (3) shows that *the generating series of an homogeneous Bellman chain with finite state are rational.*

---

[3] Given a semiring $\mathcal{S}$, $\mathcal{S}[[X]]$ denotes the semiring of formal series $H = \bigoplus_k H_k X^k$, equipped with componentwise sum and Cauchy product. $\mathcal{S}[X]$ denotes the subdioid of polynomials. We shall sometimes write $(H|X^k)$ instead of H_k, in line with the scalar product notation $(H|H') \stackrel{\text{def}}{=} \bigoplus_k H_k H'_k$.

[4] For a in a dioid $\mathcal{D}$, a^* is to be interpreted as the least upper bound of the set $\{a^0, a^1, a^2, \ldots\}$ with respect to the natural order $a \leq b \iff a \oplus b = b$. Here, $\mathcal{D} = (\mathcal{S}[[X]])^{n\times n}$ and the convergence of $(AX)^*$ is immediate, due to the fact that AX has no constant coefficient. In the scalar case ($a \in \mathcal{D} = \mathbb{R}_{\max}[[X]]$), a^* is well defined iff $a_0 = (a|X^0) \leq e$. See [12] for a more precise discussion.

2.3 Representation Theorems for Rational Series over Commutative Dioids

Definition 2 (Simple series). A series $s \in \mathcal{S}[[X]]$ is *simple* if it writes $s = aX^k(bX^p)^*$ with $a, b \in \mathcal{S}$, $k \in \mathbb{N}$, $p \in \mathbb{N}\backslash\{0\}$.

Theorem 3. *Let $\mathcal{S}$ be a commutative dioid. A series $s \in \mathcal{S}[[X]]$ is rational iff it is a sum of simple series.*

The proof consists in reducing an arbitrary rational expression to a sum of simple series via the following classical rational identities[5]

$$\begin{array}{rrl} (C) & (a \oplus b)^* & = a^* b^* \\ (SC) & (ab^*)^* & = e \oplus aa^* b^* \\ n \geq 1,\ (P(n)) & a^* & = (e \oplus a \oplus \cdots \oplus a^{n-1})(a^n)^* \ . \end{array}$$

Definition 4 (Weak and strong stabilization). The dioid $\mathcal{S}$ satisfies the weak stabilization condition[6] if

$$\forall a, b, \lambda, \mu \in \mathcal{S}, \quad \exists c, \nu \in \mathcal{S},\ \exists K \in \mathbb{N},\ \ \forall k \geq K, \qquad a\lambda^k \oplus b\mu^k = c\nu^k \ . \quad (6)$$

Strong stabilization holds if the value $\nu = \lambda \oplus \mu$ is allowed (whenever $a, b \neq \varepsilon$).

We also introduce the following central notion which already appears in the theory of nonnegative rational series [3].

Definition 5 (Merge of series). The merge of the series $s^{(0)}, \ldots, s^{(c-1)}$ is the series t such that $\forall k \in \mathbb{N}$, $\forall i \in \{0, \ldots, c-1\}$, $t_{i+kc} = s_k^{(i)}$.

For instance, the series $(X^2)^* \oplus 1X(1X^2)^* = 0 \oplus 1X \oplus 0X^2 \oplus 2X^2 \oplus 0X^3 \oplus 3X^4 \oplus \ldots$ is the merge of the series $X^* = 0 \oplus 0X \oplus 0X^2 \oplus \ldots$ and $1(1X)^* = 1 \oplus 2X \oplus 3X^2 \oplus \ldots$

Definition 6 (Ultimately geometric series). The series s is ultimately geometric if $\exists K \in \mathbb{N}, \exists \alpha \in \mathcal{S}, \forall k \geq K, s_{k+1} = \alpha s_k$.

We have the following central characterization first noted by Moller [19] for $\mathcal{S} = \mathbb{R}_{\max}$.

Theorem 7 [12]. *Let $\mathcal{S}$ be a commutative dioid satisfying the weak stabilization condition. Then, a series is rational iff it is a merge of ultimately geometric series.*

As a corollary, we obtain a version of the classical periodicity theorem [2, 6].

Corollary 8 (Cyclicity). *Assume that $\mathcal{S}$ is commutative without divisors of zero and satisfies the strong stabilization condition. Let $A \in \mathcal{S}^{n \times n}$ be an irreducible matrix. Then, $\exists \alpha \in \mathcal{S}$ and $k \geq 0$, $c \geq 1$ such that $A^{k+c} = \alpha A^k$.*

The proof [12] consists in showing that all the merged series which appear in $(AX)^*_{ij}$ have the same asymptotic rate — independent of ij — due to the irreducibility.

Example 1 (Computing the value function of a Bellman chain). Let us consider the Bellman Chain (4), whose transition rewards are described on Fig. 1. For instance,

---

[5] For a more complete study of rational identities, see Bonnier and Krob [16, 14]. The two labels (C) and (SC) recall that these identities are specific to commutative rational series.

[6] The strong stabilization is borrowed to Dudnikov and Samborkskiĭ [6, 18]. See [12] for a comparison of these properties.

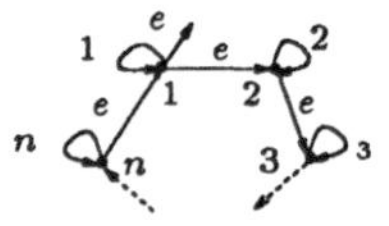

$$\begin{aligned} V_1 &= (1X)V_1 \oplus XV_2 \oplus e \\ V_2 &= (2X)V_2 \oplus XV_3 \\ &\cdots \\ V_n &= (nX)V_n \oplus XV_1 \ . \end{aligned}$$

Fig. 1. A simple bellman chain with its generating equations

the arc $n \xrightarrow{e} 1$ means that $A_{n1} = e$. We take as final reward a Dirac at node 1: $b_i = e$ if $i = 1$, $b_i = \varepsilon$ otherwise. Let v be the value function defined by (4) and consider the *generating series* V given by (5). V can be computed by performing a Gaussian elimination[7] on the system $V = AXV \oplus b$ displayed on Fig. 1. After some computations making an intensive use of the rational identities (C),(SC), we get eventually $V_1 = (1X)^* \oplus X^n(nX)^*$. Taking the coefficient of V_1 at X^p, we obtain the value function with initial state 1 and horizon p

$$v_1^{(p)} = \max(p, n(p-n)) \ .$$

3 Skew Rational Series

Let σ be an endomorphism of the semiring $\mathcal{S}$. The semiring of *skew series* over $\mathcal{S}$, denoted by $\mathcal{S}[[X;\sigma]]$ is by definition the set of formal sums $s = \bigoplus_{n\in\mathbb{N}} s_n X^n$ equipped with the usual componentwise sum and the skew product:

$$(s \otimes t)_n \stackrel{\text{def}}{=} \bigoplus_{p+q=n} s_p \otimes \sigma^p(t_q) \ .$$

This noncommutative product arises from the rule $Xa = \sigma(a)X, \forall a \in \mathcal{S}$.

3.1 Skew Series and Discounted Bellman Chains

Let $A \in \mathbb{R}_{\max}^{n\times n}$ and let us consider the following Markovian discounted optimization problem (discounted "Bellman chain") with initial state i, final state j and horizon k:

$$(A^{[k]})_{ij} \stackrel{\text{def}}{=} \max_{i_1,\ldots,i_{k-1}} \left(A_{ii_1} + \beta \times A_{i_1i_2} + \cdots + \beta^{k-1} \times A_{i_{k-1}j}\right) \ , \tag{7}$$

where $\beta \in]0,1]$ denotes the discount rate. We have the following Hamilton-Jacobi-Bellman equation

$$(A^{[k+1]})_{ij} = \max_q \left(A_{iq} + \beta \times (A^{[k]})_{qj}\right) \ . \tag{8}$$

After introducing the automorphism of $\mathbb{R}_{\max}$ and of $\mathbb{R}_{\max}^{n\times n}$:

$$x \in \mathbb{R}_{\max},\ \sigma(x) \stackrel{\text{def}}{=} \beta \times x \ , \qquad A \in \mathbb{R}_{\max}^{n\times n},\ \sigma(A)_{ij} \stackrel{\text{def}}{=} \sigma(A_{ij}) \ ,$$

we rewrite (8) as $A^{[k+1]} = A \otimes \sigma(A^{[k]})$, hence

$$A^{[k]} = A \otimes \sigma(A) \otimes \cdots \otimes \sigma^{k-1}(A) \ . \tag{9}$$

Since $A^{[k]} = ((AX)^*|X^k)$, the asymptotic study of the sequence of "skew powers" $A^{[k]}$ reduces to the evaluation of the matrix $(AX)^*$ whose entries are skew rational series.

[7] E.g. V_2 can be eliminated by noting that $V_2 = (2X)V_2 \oplus XV_3$ is equivalent to $V_2 = (2X)^*XV_3$.

3.2 Theorems of Representation for (max, +) Rational Skew Series

We again consider *simple* series which write $s = (aX^p)^*bX^n$ or equivalently $s = bX^n(fX^p)^*$ with $a, b, f \in \mathbb{R}_{\max}, p \geq 1, n \geq 0$.

Theorem 9. *A skew series $s \in \mathbb{R}_{\max}[[X; \sigma]]$ is rational iff it is a sum of simple series.*

This theorem is rather surprising because rational series usually cannot be expressed with a single level of star in noncommutative structures. The proof uses the machinery of noncommutative rational identities, such as

$$(S) \qquad (a \oplus b)^* = a^*(ba^*)^*$$

together with a few specific "commutative" identities, in particular

$$\forall a, b \in \mathbb{R}_{\max}, \quad p \geq 1, \quad ((a \oplus b)X^p)^* = (aX^p)^*(bX^p)^* \ . \tag{10}$$

Definition 10. The series $s \in \mathbb{R}_{\max}[[X; \sigma]]$ is *ultimately skew geometric* iff

$$\exists K \in \mathbb{N}, \alpha \in \mathbb{R}_{\max} \quad k \geq K \Rightarrow s_{k+1} = \alpha\sigma(s_k) \qquad (= \alpha + \beta \times s_k) \ . \tag{11}$$

Theorem 11. *A series $s \in \mathbb{R}_{\max}[[X; \sigma]]$ is rational iff it is a merge of ultimately skew geometric series.*

We obtain as a corollary the following remarkable periodicity theorem first proved by Braker and Resing [4] (for primitive matrices).

Theorem 12 (Skew cyclicity). *For an irreducible matrix $A \in \mathbb{R}_{\max}^{n\times n}$, $\exists c \geq 1$, $\forall ij$, $\exists \alpha_{ij}$ such that $A_{ij}^{[k+c]} = \alpha_{ij}\sigma^c(A_{ij}^{[k]})$ for k large enough.*

Example 2 (Value function of a discounted Bellman chain). Let us compute the value function v for the discounted version of the Bellman Chain described in Fig. 1. Using the identity (10), it is not too difficult to obtain $V_1 = (1X \oplus (\alpha X)^*X^n)^*$ with $\alpha = \bigoplus_{1\leq i\leq n-1} \sigma^i(i+1)$. An application of the identity (S) gives $V_1 = (1X)^* \oplus (\alpha X)^*X^n$, hence $v_1^{(p)} = 1^{[p]} \oplus \alpha^{[p-n]}$, which rewrites:

$$v_1^{(p)} = \max\left(\frac{1-\beta^p}{1-\beta}, \alpha\frac{1-\beta^{p-n}}{1-\beta}\right) \quad \text{with } \alpha = \max_{1\leq i\leq n-1}(\beta^i(i+1)) \ .$$

4 Rational Series in Several Commuting Indeterminates

4.1 Timed Event Graphs with Unknown Resources and Holding Times

As shown in [2, 5], the algebraic modelization of Timed Event Graphs uses the two operators[8]

$$\begin{aligned} \delta &: u \in \mathbb{R}_{\max}^{\mathbb{Z}} \mapsto y \in \mathbb{R}_{\max}^{\mathbb{Z}},\ y(k) = 1 + u(k) \ , \\ \gamma &: u \in \mathbb{R}_{\max}^{\mathbb{Z}} \mapsto y \in \mathbb{R}_{\max}^{\mathbb{Z}},\ y(k) = u(k-1) \ . \end{aligned}$$

When some task has an unknown duration τ_1, we get an equation of the form $y(k) = \tau_1 + u(k)$, which suggests to introduce a new operator $\delta_1 : \delta_1 u(k) \stackrel{\text{def}}{=} \tau_1 + u(k)$. In

[8] The signals u and y are *dater functions* [5, §4.1], δ and γ are the shifts in dating and in counting.

the same vein, an unknown initial marking q_1 (say an unknown number of parts, of machines) is represented by a new operator $\gamma_1 : \gamma_1 u(k) = u(k - q_1)$. Then, the dater functions of a TEG with unknown resources and holding times satisfy the following polynomial equations[9]:

$$x = Ax \oplus Bu, \;\; y = Cx, \quad A \in (\mathbb{B}[\gamma_i, \delta_i])^{n\times n}, B \in (\mathbb{B}[\gamma_i, \delta_i])^{n\times p}, C \in (\mathbb{B}[\gamma_i, \delta_i])^{r\times n}$$

(with essentially as many δ_i as unknown holding times and as many γ_i as unknown markings). The transfer $H = CA^*B$ is a rational series of $\mathbb{B}[[\gamma_i, \delta_i]]$.

Example 3 (Transfer of a machine with two part types). Consider a single machine producing 2 different parts with processing times t_1, t_2. Under a cyclic scheduling, we obtain the TEG shown on Fig. 2,(a). The input u_i represents the arrivals of row materials for

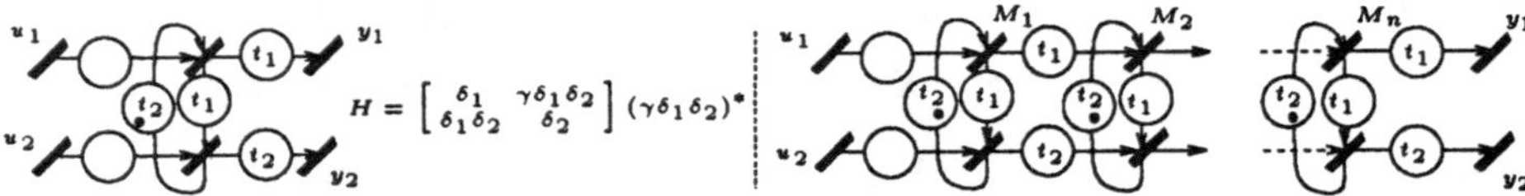

Fig. 2. (a): A Single Machine Producing 2 Types of Parts. (b): A Cascade of n Machines.

part i ($u_i(k)$ = date of k-th arrival), and the output y_i represents the dates of production of the same part. For instance, the expression[10] $H_{11} = \delta_1(\gamma\delta_1\delta_2)^*$ shows that for the autonomous regime, one part of type 1 exits every $t_1 + t_2$ units of time (after time t_1).

4.2 Some Algebraic Results

Let $\mathcal{S}$ denote a (commutative) dioid. It follows from the commutative rational identities (C),(SC) that a series $s \in \mathcal{S}[[X_1, \ldots, X_n]]$ is rational iff it can be written $s = \bigoplus_{i=1}^{k} u_i v_i^*$ where u_i, v_i are polynomials. However, we have a much more precise result inspired by the theory of rational subsets of $\mathbb{N}^k$ [8].

Definition 13. A *simple* series s writes $s = aX_1^{\beta_1} \ldots X_k^{\beta_k}(\bigoplus_{i=1}^{r} a_i X_1^{\alpha_{i1}} \ldots X_k^{\alpha_{ik}})^*$, where the vectors $(\alpha_{1,i})_{1\leq i\leq k}, \ldots, (\alpha_{r,i})_{1\leq i\leq k}$ form a free family of $\mathbb{N}^k$.

Theorem 14. *Assume that $\mathcal{S}$ is totally ordered. Then, a series $s \in \mathcal{S}[[X_1, \ldots, X_m]]$ is rational iff it is a sum of simple series.*

The proof uses essentially the following kind of rational identity:

$$(aX)^*(bY)^*(cXY)^* = (aX)^*(bY)^* \oplus (aX)^*(cXY)^* \oplus (bY)^*(cXY)^* \; . \qquad (12)$$

[9] $\mathbb{B} = \{\varepsilon, e\}$ denotes the boolean semiring.

[10] For the autonomous regime starting at time 0 (e.g. if an infinite quantity of inputs become available at time 0, see [2, §5.4.4.1]), the monomial transfer $\gamma^n \bigotimes_i \delta_i^{k_i}$ can be interpreted as "the event n occurs at the earliest at time $\sum_i k_i \times t_i$".

Example 4 (Transfer of a flowshop with n identical machines). Let us consider a cascade of n identical machines of type shown on Fig. 2,(b). The transfer matrix of the flowshop is equal to H^n. An easy induction gives

$$H^n = \begin{bmatrix} \delta_1^n \oplus \gamma\delta_1^2\delta_2^2(\delta_1 \oplus \delta_2)^{n-2} & \gamma\delta_1\delta_2(\delta_1 \oplus \delta_2)^{n-1} \\ \delta_1\delta_2(\delta_1 \oplus \delta_2)^{n-1} & \delta_2^n \oplus \gamma\delta_1^2\delta_2^2(\delta_1 \oplus \delta_2)^{n-2} \end{bmatrix} (\gamma\delta_1\delta_2)^*$$

(modulo the licit additional simplification rules $\delta_i^s \oplus \delta_i^t = \delta_i^{\max(s,t)}$). This expression specifies the input/output behavior in function of the unknown processing times t_1, t_2. In particular, the term[10] $\delta_1^n \oplus \gamma\delta_1^2\delta_2^2(\delta_1 \oplus \delta_2)^{n-2}$ in the expression of H_{11}^n means that, for the autonomous regime, the part of type 1 numbered 0 has a transfer time of $n \times t_1$, while the part of the same type numbered 1 has a transfer time of $2(t_2 + t_1) + (n-2)\max(t_1, t_2)$.

Example 5 (Heaviside calculus for some special variational inequalities). Let us search for the least solution u, v of the system[11]:

$$0 \geq \lambda - \frac{\partial v}{\partial x} - \frac{\partial v}{\partial t},\ 0 \geq \max(\alpha - \frac{\partial u}{\partial x}, \beta - \frac{\partial u}{\partial t}),\ u \geq \max(\gamma + v, f),\ v \geq \mu + u \ . \tag{13}$$

A discretization gives with the $\mathbb{R}_{\max}$ notation:

$$\begin{aligned} u_h(x,t) &\geq \alpha^h u_h(x-h,t) \oplus \beta^h u_h(x,t-h) \oplus \gamma v_h(x,t) \oplus f(x,t) \ , \\ v_h(x,t) &\geq \lambda^h v_h(x-h,t-h) \oplus \mu u_h(x,t) \ . \end{aligned} \tag{14}$$

Introducing the space and time shifts $Xu(x,t) \stackrel{\text{def}}{=} u(x-h,t)$ and $Tu(x,t) \stackrel{\text{def}}{=} u(x,t-h)$, we get

$$\begin{aligned} u_h &\geq (\alpha^h X \oplus \beta^h T)u_h \oplus \gamma v_h \oplus f \ , \\ v_h &\geq \lambda^h XT v_h \oplus \mu u_h \ . \end{aligned} \tag{15}$$

A Gaussian elimination gives the least solution $u = (\alpha^h X \oplus \beta^h T \oplus (\lambda^h XT)^* \gamma\mu)^* f$. The convergence of this star in $\mathbb{R}_{\max}[[X, T]]$ yields the compatibility condition $\gamma\mu \leq e$. After some computation involving the rational identities (C), (SC), (12), we obtain

$$u = \left\{(\alpha^h X)^*(\beta^h T)^* \oplus \gamma\mu(\alpha^h X)^*(\lambda^h XT)^* \oplus \gamma\mu(\beta^h T)^*(\lambda^h XT)^*\right\} f \ .$$

We have for the second term of this sum:

$$\begin{aligned} (\alpha^h X)^*(\lambda^h XT)^* f(x,t) &= \sup\nolimits_{p,n\geq 0}[\alpha hp + \lambda hn + f(x - hp - hn, t - hn)] \\ &= \sup\nolimits_{q\geq n\geq 0}[\alpha h(q-n) + \lambda hn + f(x - hq, t - hn)] \ . \end{aligned}$$

After an analogous argument for the two other terms, we introduce the three kernels

$$k_1(\xi,\tau) = \alpha\xi + \beta\tau,\ k_2(\xi,\tau) = \begin{cases} \alpha(\xi - \tau) + \lambda\tau + \gamma + \mu & \text{if } \xi \geq \tau, \\ -\infty & \text{otherwise,} \end{cases}$$

$$k_3(\xi,\tau) = \begin{cases} \beta(\tau - \xi) + \lambda\xi + \gamma + \mu & \text{if } \tau \geq \xi, \\ -\infty & \text{otherwise.} \end{cases}$$

Then, letting $h \to 0$, we get the explicit solution

$$u(x,t) = \sup_{\xi,\tau\geq 0}[\max(k_1, k_2, k_3)(\xi,\tau) + f(x - \xi, t - \tau)] \ .$$

[11] u, v, f are maps $\mathbb{R}^2 \to \mathbb{R}$. $\alpha, \beta, \lambda, \gamma, \mu$ are constant. We do not address the regularity issues here.

5 Rational Series in Non Commuting Indeterminates

5.1 Definition and Basic Examples

Let Σ^* denote the free monoid over a finite alphabet Σ. The series $y = \bigoplus_{w\in\Sigma^*}(y|w)w$ (in $\mathbb{R}_{\max}\langle\langle\Sigma\rangle\rangle$) is *recognizable* iff there exists $\alpha \in \mathbb{R}_{\max}^{1\times n}$, $\beta \in \mathbb{R}_{\max}^{n\times 1}$, and a morphism $\mu : \Sigma^* \to \mathbb{R}_{\max}^{n\times n}$ such that $(y|w) = \alpha\mu(w)\beta$. The general version of the Kleene-Schützenberger theorem states that recognizable and rational series coincide.

Example 6 (Deterministic cost). Let (Q, q_0, δ) denote a finite deterministic automaton, $\alpha : Q \to \mathbb{R}_{\max}$ a final cost and $\sigma : Q\times\Sigma \to \mathbb{R}_{\max}$ a transition cost. Let $w = w_k \dots w_1 \in \Sigma^k$ denote a sequence of decisions (w_i denotes the i-th letter of w, read read from right to left). We consider the cost

$$(c|w) \stackrel{\text{def}}{=} \sum_{n=1}^{k} \sigma(q_{n-1}, w_n) + \alpha(q_k), \quad \text{subject to } q_n = \delta(q_{n-1}, w_n) \tag{16}$$

(by convention, $(c|w) = \varepsilon$ if $\delta(q_0, w)$ is undefined). This is the discrete counterpart of the usual integral cost $\int_0^T \sigma(x(t), u(t))dt + \alpha(x(T))$ for the system $\dot{x} = f(x, u)$. The series $\bigoplus_w (c|w)w$ is recognizable (take β = Dirac at q_0 and $\forall a \in \Sigma, \forall p, q \in Q$, $\mu(a)_{pq} = \sigma(q, a)$ if $\delta(q, a) = p$ and ε otherwise).

Example 7 (A workshop with different schedules). We consider a workshop with 2 machines M_1, M_2 and 2 different regimes of production corresponding to the processing of 2 differents parts (a) and (b), as represented[12] by the TEGs (a) and (b) displayed on Fig. 3. We assume that the workshop can switch from a regime to the other, according

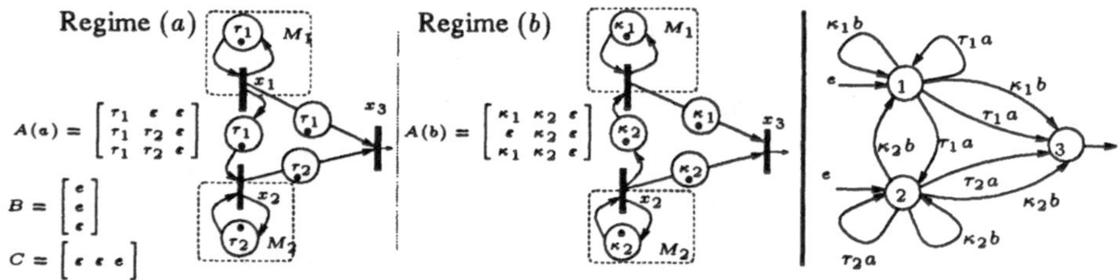

Fig. 3. A workshop with two production regimes and its equivalent (max, +) automaton

to an open loop schedule described by a word $w \in \{a, b\}^*$. E.g. the word baa means that the two machines first follow two tasks with times and precedence constraints determined by the TEG (a) and after that one task described by the TEG (b). The behavior of the workshop under the schedule $w = w_k \dots w_1$ is determined by the system of non stationary (max, +) linear equations

$$x_w(0) = B, \; x_w(n) = A(w_n)x_w(n-1), \; y_w(n) = Cx_w(n) \;, \quad 1 \le n \le k \;, \tag{17}$$

[12] The interpretation of regime (b) is the following. A part is processed by M_2 during κ_2 time units. Then it is sent to M_1 (processing time of κ_1). For simplicity, we have neglected the transportation times. A dual interpretation can be provided for regime (a). The initial condition $B = [e, e, \varepsilon]^T$ require M_1 and M_2 to begin to work at time $e = 0$.

where $A(a)$, B, C and $A(b)$, B, C stand for the linear representations of the TEGs (a) and (b) displayed on Fig. 3. The output $y_w(n)$ represents the date of completion of the latest task of the sequence $w_{n-1}\dots w_1$. The map $w \mapsto (y|w) \stackrel{\text{def}}{=} y_w(k)$ is recognizable (take $\alpha = C, \beta = B, \mu(a) = A(a), \mu(b) = A(b)$). The linear representation α, μ, β can be visualized by the automaton with multiplicities[13] in the (max, +) semiring displayed on Fig. 3.

5.2 Asymptotic Behavior of (max, +) Automata

The asymptotic analysis of a recognizable series y consists in estimating $(y|w)$ where w is a word of length $k \to \infty$. We consider the *worst case performance*:

$$r_k^{\text{worst}} \stackrel{\text{def}}{=} \bigoplus_{w\in\Sigma^k} (y|w) = \max_{w_k,\dots,w_1\in\Sigma} \max_{i_k,\dots,i_0} [\alpha_{i_k} + \mu(w_k)_{i_k i_{k-1}} + \dots + \mu(w_1)_{i_1 i_0} + \beta_{i_0}] \ .$$

Theorem 15 [11, 10]. *Let y be the series given by the linear representation α, μ, β. Let $M = \bigoplus_{a\in\Sigma} \mu(a)$. Then $r_k^{\text{worst}} = \alpha M^k \beta$.*

This is a *superposition principle* which states that the worst case behavior is obtained by mere superposition of the TEGs with transition matrices $\mu(a), a \in \Sigma$.

Example 8. For the automaton of Ex. 7, the asymptotic behavior is determined by the eigenvalue λ of $M = A(a) \oplus A(b)$, i.e. $\lim_k (r_k^{\text{worst}})^{\otimes\frac{1}{k}} = \lim_k r_k^{\text{worst}}/k = \lambda = \max_{i=1,2} \max(\tau_i, \kappa_i)$.

We next turn to the dual *optimal case* measure of performance[14]:

$$r_k^{\text{opt.}} \stackrel{\text{def}}{=} \min_{\substack{w\in\Sigma^k,\\ (y|w)\neq\varepsilon}} (y|w) = \min_{\substack{w_k,\dots,w_1\in\Sigma,\\ (y|w)\neq\varepsilon}} \max_{i_k,\dots,i_0} [\alpha_{i_k} + \mu(w_k)_{i_k i_{k-1}} + \dots + \mu(w_1)_{i_1 i_0} + \beta_{i_0}] \ .$$

This min-max problem consists in finding a schedule minimizing the transfer time of the k-th input. We remark that for the subclass of *deterministic*[15] series, $(y|w)$ does not involve maximization, so that the computation of $r^{\text{opt.}}$ and r^{worst} become dual. More formally, we introduce $y' = \bigoplus_{(y|w)\neq -\infty} (y|w) w \in \mathbb{R}_{\min}\langle\langle\Sigma\rangle\rangle$ (where $\mathbb{R}_{\min} \stackrel{\text{def}}{=} (\mathbb{R} \cup \{+\infty\}, \min, +)$) and we get:

Proposition 16. *For a deterministic series y, $r_k^{\text{opt.}} = \alpha'(M')^k\beta'$, where α', μ', β' is a $\mathbb{R}_{\min}$-linear representation of y', $M' = \bigoplus_{a\in\Sigma} \mu'(a)$ and all the operations are interpreted in $\mathbb{R}_{\min}$.*

Thus, we are reduced to characterize the series which admit deterministic representations. Under some finiteness and integrity conditions on α, μ, β, e.g. if $\forall a, i, j, \mu(a)_{ij} \in \mathbb{Z}$, the series y is deterministic [11, 10, 13]. However, the minimal dimension of a deterministic representation can be arbitrarily larger than the dimension of α, μ, β. Thus computing

[13] See [7]. The multiplicity of w – i.e. $(y|w)$ – is equal to the sum of the multiplicative weights of all the paths with label w from the input to the output. E.g. $(y|ba) = \tau_1\kappa_1 \oplus \tau_2\kappa_2 \oplus \tau_1\kappa_2$.

[14] The mean case performance $r_k^{\text{mean}} = \mathbb{E}(y|w^{(k)})$ – where $w^{(k)}$ is a random word of length k – is dealt with in the first order ergodic theory of TEGs [2, 17].

[15] A series is deterministic if it admits a representation of the form (16).

$r^{\text{opt.}}$ from Prop. 16 can be much more complex than computing r^{worst} from Th. 15. Some different techniques based on rational simplifications can sometimes be applied as in Ex. 9, but it is hopeless to obtain universal canonical forms since the equality of (max, +) rational series is undecidable [15]. It remains an open question whether the asymptotic behavior of $r^{\text{opt.}}$ can be exactly evaluated for some reasonable classes of series without appealing to the above determinization procedure.

Example 9. For the automaton of Ex. 7, $\lim_k r_k^{\text{opt.}}/k = \min(\max_i(\tau_i), \max_i(\kappa_i))$. When $\tau_2 \geq \tau_1$, $\kappa_2 \geq \kappa_1$, this follows for instance from the easily obtained expression $y = (\tau_2 a \oplus \kappa_2 b)(\tau_2 a \oplus \kappa_2 b)^*$.

References

1. M. Akian, J.P. Quadrat, and M. Viot. Bellman processes. *This conference*, 1994.
2. F. Baccelli, G. Cohen, G. Olsder, and J.P. Quadrat. *Synchronization and Linearity*. Wiley, 1992.
3. J. Berstel and C. Reutenauer. *Rational Series and their Languages*. Springer, 1988.
4. H. Braker. *Algorithms and Applications in Timed Discrete Event Systems*. PhD thesis, Delft University of Technology, Dec 1993.
5. G. Cohen. Dioids and Discrete Event Systems. *This conference*, 1994.
6. P. Dudnikov and S. Samborskiĭ. Endomorphisms of semimodules over semirings with an idempotent operation. *Math. in USSR, Izvestija*, 38(1), 1992.
7. S. Eilenberg. *Automata, Languages and Machines*, volume A. Acad. Press, 1974.
8. S. Eilenberg and M.P. Schützenberger. Rational sets in commutative monoids. *J. Algebra*, 13:173–191, 1969.
9. S. Gaubert. *Théorie des systèmes linéaires dans les dioïdes*. Thèse, École des Mines de Paris, July 1992.
10. S. Gaubert. Performance evaluation of timed automata. Research Report 1922, INRIA, May 1993.
11. S. Gaubert. Timed automata and discrete event systems. *Proceedings of the ECC93*, Groningen, July 1993.
12. S. Gaubert. On rational series in one variable over certain dioids. Research Report 2162, INRIA, Jan. 1994.
13. S. Gaubert. On semigroups of matrices in the (max, +) algebra. Research Report 2172, INRIA, Jan. 1994.
14. D. Krob. Complete systems of $\mathcal{B}$-rational identities. *Theor. Comp. Sci.*, 89, 1991.
15. D. Krob. The equality problem for rational series with multiplicities in the tropical semiring is undecidable. Research report, Université de Paris 7, LITP, 1992.
16. D. Krob and A. Bonnier-Rigny. A complete system of identities for one letter rational expressions with multiplicities in the tropical semiring. Research Report 93.07, Université de Paris 7, LITP, 1993.
17. J. Mairesse. Stochastic linear systems in the (max, +) algebra. *This conference*, 1994.
18. V. Maslov and S. Samborskiĭ, editors. *Idempotent analysis*, volume 13 of *Adv. in Sov. Math.* AMS, RI, 1992.
19. P. Moller. *Théorie algébrique des Systèmes à Événements Discrets*. Thèse, École des Mines de Paris, 1988.
20. I. Simon. The nondeterministic complexity of a finite automaton. M. Lothaire, editor, *Mots*. Hermes, 1990.

Stochastic Linear Systems in the $(\max, +)$ Algebra

*Jean Mairesse**

INRIA Sophia-Antipolis, B.P. 93, 06902 Sophia Antipolis Cedex, France

1 Introduction

A large class of computer or communication networks accept a description as Stochastic Discrete Events Systems and can be modelized by Stochastic Petri Networks (SPN). More precisely, a class of SPN, Stochastic Event Graphs (SEG), appears to be very efficient in describing models with synchronization, blocking and/or fork-join properties. These networks have an easy algebraic interpretation in a non conventional algebra. More precisely, it is possible to show that a SEG can be represented as a "linear" system of equations in the so-called (max, +) algebra.

In this review, our starting point are stochastic "linear" systems in the (max, +) algebra. The main domain of applications of the results presented here consists in SEG. However, our level of abstraction in the presentation will be that of "linear" algebra. We will not consider SPN which are not SEG as they do not provide a "linear" equation. We will present only qualitative results. For more insights on all modelling aspects and also for quantitative results, the reader is referred to [2].

There is an equivalent way of introducing our subject. It is a counterpart of the classical theory of products of random matrices but in another algebraic structure, the (max, +) algebra.

2 $(\max, +)$ Linear Systems

Definition 1 ((max, +) algebra). We consider the semiring $(\mathbb{R}^\star, \oplus, \otimes)$, where $\mathbb{R}^\star = \mathbb{R} \cup \{-\infty\}$. The law $\oplus$ is "max" and $\otimes$ is the usual addition. We set $\varepsilon = -\infty$ and $e = 0$. The element ε is neutral for the operation $\oplus$ and absorbing for $\otimes$. The element e is neutral for $\otimes$. The law $\oplus$ is idempotent, i.e. $a \oplus a = a$. $(\mathbb{R}^\star, \oplus, \otimes)$ is an example of idempotent semiring or dioid. We shall write it $\mathbb{R}_{\max}$.

In the rest of the paper, the notations "$+, \times$" will stand for the usual addition and multiplication. Nevertheless, we shall write ab for $a \otimes b$ whenever there is no possible

* Research supported by the Direction des Recherches Etudes et Techniques (DRET) under contract n° 91 815.

confusion. We define the product spaces $\mathbb{R}^J_{\max}$, $\mathbb{R}^{J\times J}_{\max}$. Matrix product is defined the usual way just by replacing $+$, $\times$ by $\oplus$, $\otimes$.

$$A, B \in \mathbb{R}^{J\times J}_{\max}, \quad (A \otimes B)_{ij} = \max_k (A_{ik} + B_{kj}) = \bigoplus_k A_{ik} \otimes B_{kj} .$$

Matrix-vector product is defined in a similar way.

We consider systems whose dynamic behavior is driven by a recursive equation of the form:

$$x_i(n+1) = \max_{1\le j\le J} (A_{ij}(n) + x_j(n)), \; i = 1, \dots, J . \tag{1}$$

We allow $A_{ij}(n)$ to be equal to $-\infty$. With the notations previously introduced, it can be rewritten as:

$$x(n+1) = A(n) \otimes x(n) . \tag{2}$$

Here $x(n) = (x_1(n), x_2(n), \dots, x_J(n))'$ and $A(n)$ is a $J \times J$ matrix. The passage from equation(1) to equation (2) is more than a simple notation game. Many ideas and techniques from classical linear algebra can be transferred to equations like (2) which are "linear" with respect to operations max and $+$.

Let us consider a probability space $(\Omega, \mathcal{F}, P, \theta)$. The probability P is stationary and ergodic with respect to the shift θ. We are interested in systems of the type:

$$x(n+1) = A(n) \otimes x(n) \;, \quad n \in \mathbb{N} \;, \quad x(0) = x_0 \;, \tag{3}$$

where $x(n)$ and $A(n)$ are finite random variables in $\mathbb{R}^J$ and $\mathbb{R}^{J\times J}$ respectively. We are sometimes going to use the notation $x(n, x_0)$ to emphasize the value of the initial condition. We will consider models where the sequence $\{A(n), n \in \mathbb{N}\}$ is respectively **i.i.d.** or **stationary** and **ergodic** ($A(n+1) = A(n) \circ \theta$).

For such systems, we will consider two kinds of asymptotic results.

- First order limits, on ratios $\lim_n \frac{\|x(n)\|}{n}$, $\lim_n \frac{x_i(n)}{n}$.
- Second order limits, on differences
  $\lim_n x_i(n+1) - x_i(n), \forall i, \quad \lim_n x_j(n) - x_i(n), \forall i \neq j.$

In the deterministic case, when $A(n) \equiv A$ is fixed, these quantities are closely related to the solutions of the following "eigenvalue" problem:

$$\max_{1\le j\le J} (A_{ij} + x_j) = \lambda + x_i, \; i = 1, \dots, J \text{ or } A \otimes y = \lambda \otimes y ,$$

where y is a column vector (the "eigenvector") and λ is a real constant (the "eigenvalue"). For a deterministic matrix A, first order limits as defined above are eigenvalues of A. Second order limits can be expressed in terms of the eigenvectors of A. Results for the deterministic case are presented in [3]. We present here only the results that we strictly need. For a complete study, see [2] and [4].

3 Deterministic Spectral Theory

The graph of a square matrix A is a directed graph having a number of nodes equal to the dimension of A. This graph contains an arc from i to j iff $A_{ji} \neq \varepsilon$. For each circuit $\zeta = \{t_1, t_2, \cdots, t_j, t_{j+1} = t_1\}$ in the graph, we define its average weight by $p(\zeta) = (A_{t_1 t_j} \otimes \cdots \otimes A_{t_3 t_2} \otimes A_{t_2 t_1})/j$.

Matrix $A = (A_{ij}, i, j = 1, \ldots, J)$ is irreducible if its graph is strongly connected, i.e. $\forall i, j\ \exists (i_1 = i, i_2, \ldots, i_n = j)$ s.t. $A_{ii_2} \otimes A_{i_2 i_3} \otimes \cdots \otimes A_{i_{n-1} j} > \varepsilon$.

Theorem 2. *If A is an irreducible matrix it admits a unique (non ε) eigenvalue, λ. It is equal to the maximal average weight of the circuits of A, $\lambda = \max_\zeta\ p(\zeta)$. We call also λ the* ***Lyapunov exponent*** *of A, by analogy with classical system theory.*

We normalize a matrix by subtracting its eigenvalue to each coordinate. A normalized matrix has an eigenvalue of e. For a normalized matrix A of size J, we define $A^+ = A \oplus A^2 \oplus \cdots \oplus A^J$. We check that $A^+ \oplus A^{J+1} = A^+$.

Theorem 3. *Let A be a normalized matrix. If i is such that $A^+_{ii} = e$ then column $A^+_{.i}$ is an eigenvector of A. Every eigenvector of A writes as a "linear" (in $\mathbb{R}_{\max}$) combination of columns of A^+.*

Definition 4. For an irreducible matrix A, with eigenvalue λ, we define:

Critical graph A circuit ζ of A is said to be critical if its average weight is maximal, i.e. if $p(\zeta) = \lambda$. The critical graph consists of the nodes and arcs of A which belong to a critical circuit.

Cyclicity The cyclicity of a strongly connected graph is the greatest common divisor of the lengths of all the circuits. The cyclicity of a general graph is the least common multiple of the cyclicities of its strongly connected subgraphs.

Definition 5. An irreducible matrix A is called **scs1-cyc1** if its critical graph has a unique strongly connected subgraph (scs1) and if the cyclicity of the critical graph is one (cyc1).

Proposition 6. *Let A be a* ***scs1-cyc1*** *matrix. Then A^n has a unique eigenvector $\forall n \geq 1$ and $\exists M$ such that $\forall m \geq M,\ A^{m+1} = \lambda \otimes A^m$ (λ is the unique eigenvalue of A).*

Definition 7. The projective space $\mathbb{PR}^J_{\max}$ is defined as the quotient of $\mathbb{R}^J_{\max}$ by the parallelism relation: $u, v \in \mathbb{R}^J_{\max}, u \simeq v \iff \exists a \in \mathbb{R}_{\max} \setminus \{\varepsilon\}$ s.t. $u = a \otimes v$.

Let π be the canonical projection of $\mathbb{R}^J_{\max}$ into $\mathbb{PR}^J_{\max}$. We define in a similar way $\mathbb{PR}^{J \times J}_{\max}$ and $\mathbb{PR}^J = \{\pi(u),\ u \in \mathbb{R}^J\}$.

For example $(e, -1)'$ and $(1, e)'$ are in the same parallelism class, i.e. are two representatives of the same vector of $\mathbb{PR}^2_{\max}$. It is easy to check that if $\alpha \in \mathbb{R}$ and if u is an eigenvector of a matrix A, then $\alpha \otimes u$ is also one. It is the motivation for the introduction of the projective space.

4 First Order Limits

First order's theorems have been treated by Baccelli ([1] or [2]) using Kingman's sub-additive ergodic theorem. Let us state the main results. We consider the usual $\mathcal{L}_\infty$ norm: for $A \in \mathbb{R}^{J \times J}_{\max}$, $\|A\| = \bigoplus_{i,j=1}^{J} A_{ij}$.

4.1 Closed Systems

Proposition 8. *Let $\{A(n)\}$ be a stationary and ergodic sequence of matrices. We suppose that $\forall i, j,\ P(A(0)_{ij} = \varepsilon) = 0$ and $+\infty > E(A(0)_{ij}) > \varepsilon$. Then there exists a constant $\lambda \in \mathbb{R}$ such that, for all initial condition x_0, and for every $i \in \{1, \ldots, J\}$,*

$$\lim_n \frac{x_i(n, x_0)}{n} = \lim_n E\left(\frac{x_i(n, x_0)}{n}\right) = \lambda, \quad P - a.s.$$

*The constant λ is called the **Lyapunov exponent** of the system.*

Remark. This definition of a Lyapunov exponent is coherent with the one of Th. 2. Indeed for every irreducible and deterministic matrix A, there exists d and M such that $\forall m \geq M,\ A^{m+d} = \lambda^d \otimes A^m$, where λ is the eigenvalue of A (see [2] for example). It is now clear that $\forall x_0 \in \mathbb{R}^J_{\max}$, $\lim_n A^n x_0 / n = \lambda$.

Remark. The assumptions can be weakened and replaced by

$$\lim_n P\left(A(n) \otimes A(n-1) \otimes A(0) \text{ irred.}\right) = 1,\ E(A(0)_{ij} / A(0)_{ij} \neq \varepsilon) > \varepsilon .$$

Proof. It is straightforward to check that $\forall A, B \in \mathbb{R}^{J\times J}_{\max}$, $\|A \otimes B\| \leq \|A\| \otimes \|B\|$. We define $Z_{m+k,m} = \|A(m+k-1) \otimes \cdots \otimes A(m)\|$. We have:

$$\begin{aligned}\|A(m+ k -1) \ldots A(m)\| &= \|A(0) \circ \theta^{k-1} \ldots A(0) \circ \theta^p \otimes A(0) \circ \theta^{p-1} \ldots A(0)\| \circ \theta^m \\ &\leq \|A(0) \circ \theta^{k-1} \ldots A(0) \circ \theta^p\| \circ \theta^m \otimes \|A(0) \circ \theta^{p-1} \ldots A(0)\| \circ \theta^m ,\end{aligned}$$

that is $Z_{m+k,m} \leq Z_{m+k,m+p} + Z_{m+p,m}$.

We have furthermore that $\forall i, j$:

$$(A(m) \ldots A(0))_{ij} \geq (A(m) \ldots A(p))_{ij} \otimes (A(p-1) \ldots A(0))_{ij} .$$

If we denote $K_{ij} = E(A(0)_{ij})$ and $K = \inf_{ij} K_{ij}$, we conclude that $E(Z_{m,0}) \geq K \times m$. We set $\mathrm{e} = (e, \ldots, e)'$. We apply Kingman's sub-additive ergodic theorem, see [5], to $Z_{m,o} = x(m, \mathrm{e})$. We conclude by remarking that for every finite initial condition x_0, we have $\|x(m, \mathrm{e})\| \otimes \min_i (x_0)_i \leq \|x(m, x_0)\| \leq \|x(m, \mathrm{e})\| \otimes \|x_0\|$. □

4.2 Open Systems

Definition 9. A stochastic matrix $\{A(\omega),\ \omega \in \Omega\}$ has a fixed structure if $P(A_{ij} = \varepsilon) = 1$ or $P(A_{ij} = \varepsilon) = 0,\ \forall i, j$.

We suppose that $A(0)$ has a fixed structure. We decompose the graph of $A(0)$ into its maximal strongly connected subgraphs (mscs). If we replace each mscs by one node, we obtain an associated reduced graph which is acyclic. We associate with each node $\tilde{u}$ of the reduced graph a constant $\lambda_{\tilde{u}}$ which is the Lyapunov exponent of the corresponding mscs in isolation (see Th. 8). We denote by $\pi^*[\tilde{u}]$ the set of predecessors of $\tilde{u}$ (including $\tilde{u}$) in the reduced graph. Then we have:

Theorem 10. *Let $\{A(n)\}$ be a stationary and ergodic sequence of matrices. We suppose that $A(0)$ has a fixed structure. We suppose also that $E(A(0)_{ij}) < +\infty, \forall i, j$. Let us consider $i \in \{1, \ldots, J\}$, i belongs to the mscs $\tilde{u}$.*

$$\lim_n \frac{x_i(n, x_0)}{k} = \lim_n E\left(\frac{x_i(n, x_0)}{n}\right) = \bigoplus_{\tilde{v} \in \pi^{\cdot}[\tilde{u}]} \lambda_{\tilde{v}}, \; P-a.s..$$

Intuitively, the Lyapunov exponent of a mscs has to be interpreted as the inverse of a throughput. The dynamic of the system is imposed by the mscs having the smallest throughput. In order to get a deeper intuition of this result, one can look at the example following Th. 13.

5 Second Order Limits

In the rest of the paper we will say that our model is **stable** if $\pi(x(n, x_0))$ converges to a unique stationary distribution, **uniformly** over $x_0 \in \mathbb{R}^J$. The type of convergence (weak convergence, total variation convergence) will be precised.

5.1 Closed Systems

Definition 11 pattern. Let $A(\omega)$ be a random matrix. We say that $\tilde{A}$ is a pattern of A if $\tilde{A}$ is a deterministic matrix which belongs to the support of the random matrix A. This definition includes the cases where $\tilde{A}$ is only accumulation point (A is a discrete r.v.) or boundary point (A is a continuous r.v.).

Theorem 12. *Let the sequence $\{A(n)\}$ be stationary and ergodic. We suppose there is an integer K such that among the patterns of $A(K-1) \otimes \cdots \otimes A(0)$, there exists an* ***scs1-cyc1*** *matrix C. Let M be such that: $\forall m \geq M$, $C^m = \lambda \otimes C^{m-1}$ (see Prop. 6). We suppose also that C^M is a pattern of $A(MK-1) \otimes \cdots \otimes A(0)$. Then the system is* ***stable.*** *If C is in the interior of the support, there is total variation convergence to the stationary distribution. Otherwise there is only weak convergence.*

Remark. When we have the stronger assumption that the sequence $\{A(n)\}$ is **i.i.d.**, the sufficient condition of the theorem becomes: there exists a **scs1-cyc1** matrix $C = C_{K-1} \otimes \cdots \otimes C_0$ where C_i is a pattern of $A(0)$, $\forall i \in \{0, \ldots, K-1\}$.

Proof. Under the **i.i.d.** assumption, the proof uses Prop. 6. Let C be a scs1-cyc1 pattern. We denote by c the unique eigenvector of C. There exists M such that $\forall m \geq M, \forall x_0 \in \mathbb{R}^J$, $\pi(C^m \otimes x_0) = \pi(c)$. If we suppose that $P(A(K-1) \ldots A(0) = C) > 0$ (discrete case), then it is straightforward to prove that $\pi(c)$ is a regeneration point for $\pi(x(n))$. We obtain in consequence that $\pi(x(n))$ is Harris ergodic. When we do not have $P(A(K-1) \ldots A(0) = C) > 0$, the proof is more technical. Under stationary and ergodic assumptions, we make use of Borovkov's theory of renovating events [6]. □

Remark. There exists a converse to this theorem. More precisely, if there is total variation convergence to a unique stationary regime then there exists, with positive probability, patterns of the type described in the previous theorem.

We deduce from this theorem that a good way to show the stability of a system "$x(n+1) = A(n)x(n)$" is to extract some deterministic matrices from the support of $A(0)$ and to build a scs1-cyc1 pattern with them. In most cases an extracted model with two matrices will be enough to conclude. We are going to illustrate in §6 the phenomena of uniqueness or multiplicity of stationary regimes with models of two matrices.

5.2 Open Systems

The sequence $\{A(n), n \in \mathbb{N}\}$ is stationary and ergodic. Matrices $A(n)$ have a fixed structure, $P - a.s.$ This structure is of the form:

$$A(n) = \begin{pmatrix} \tilde{U}(n) & \varepsilon \\ \tilde{B}(n) & \tilde{A}(n) \end{pmatrix} .$$

The block $\tilde{U}$ is a square matrix of size $I \times I$, irreducible. It is interpreted as the input of our system. The block $\tilde{A}$ is a square matrix of size $(J-I) \times (J-I)$, irreducible and aperiodic. The block $\tilde{B}$ is the matrix of the communication between the source $\tilde{U}$ and the system $\tilde{A}$. We suppose that the block $\tilde{U}$ in isolation has a unique stationary regime (see Th. 12). We have the following theorem.

Theorem 13. *Let u and a be the Lyapunov exponents of $\tilde{U}$ and $\tilde{A}$ respectively (see Prop. 8). If $a < u$, there is a unique stationary regime for $\pi(x(n))$, regardless of the initial condition. There is total variation convergence to the stationary regime. If $a > u$, then the increments of the form*

$$\delta_{ji}(n, x_0) = x_j(n, x_0) - x_i(n, x_0), \; i = 1, \ldots, I, \quad j = I+1, \ldots, J ,$$

tend to $+\infty$, $P - a.s.$, for all finite initial conditions.

For the proof the reader is referred to Baccelli [1]. A good way to get an intuition of this result is to consider deterministic matrices.

Example 1. We consider first a matrix for which $(u = 1) > (a = e)$.

$$A = \begin{pmatrix} 1 & \varepsilon \\ e & e \end{pmatrix} , \; A^n = \begin{pmatrix} n & \varepsilon \\ n-1 & e \end{pmatrix} .$$

We choose $x_0 = (u, v)'$. We have $A^n x_0 = (nu, (n-1)u \oplus v)'$. For n large enough, we have $A^n x_0 = nu \otimes (e, -1)' \implies \pi(A^n x_0) = \pi(e, -1)'$.

On the other hand, if we consider:

$$A = \begin{pmatrix} e & \varepsilon \\ e & 1 \end{pmatrix} , \; A^n = \begin{pmatrix} e & \varepsilon \\ n-1 & n \end{pmatrix} ,$$

we have $\pi(A^n x_0) = \pi(e, (n-1) \oplus (v-u)n)'$. We check that $x_2(n) - x_1(n) = (n-1) \oplus (v-u)n$ tends to $+\infty$ for all finite x_0.

For a stochastic model the idea remains the same. If $u > a$, the source which is slower imposes its dynamic. If $u < a$, everything happens asymptotically as if $\tilde{U}$ and $\tilde{A}$ were in isolation.

6 Illustration of Multiple Regimes

The canonical projection π was defined in Def. 7. It can be interpreted geometrically. It is nothing else than the orthogonal projection on the hyperspace orthogonal to the vector $\mathbf{1} = (1, \ldots, 1)'$. The projective space $\mathbb{PR}^J$ is isomorphic to $\mathbb{R}^{J-1}$. For irreducible matrices of size 3, a graphical representation of the set of eigenvectors is possible in $\mathbb{R}^2 \simeq \mathbb{PR}^3$, see [7].

In Fig. 1, we have represented some points in $\mathbb{PR}^3$. The three axes are the orthogonal

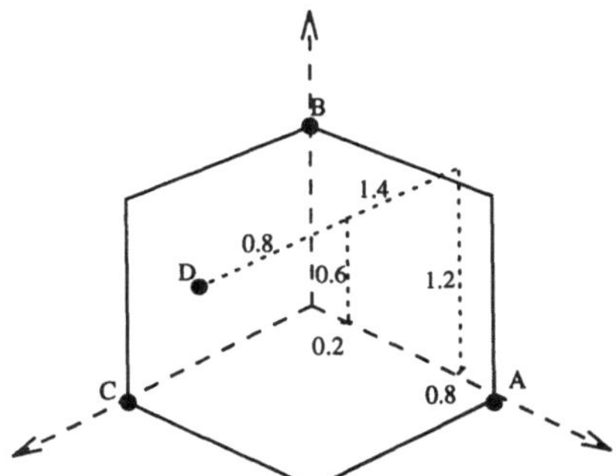

Fig. 1. Graphical representation in $\mathbb{PR}^3$

projection of the basis of $\mathbb{R}^3$. The represented points are $A = \pi(\mathbf{e_1}) = \pi(1, e, e)'$, $B = \pi(\mathbf{e_2}) = \pi(e, 1, e)'$, $C = \pi(\mathbf{e_3}) = \pi(e, e, 1)'$ and $D = \pi(0.2, 0.6, 0.8)'$.

The regular hexagon in Fig. 1 is the section of the unit square (i.e. the unit ball of $\mathbb{R}^3$ for the $\mathcal{L}_\infty$ norm) by the projection plane. One can also check that this hexagon is the convex hull in a (max, +) sense of the points A, B and C, i.e. the set of points:

$$\{\pi(\alpha \otimes \mathbf{e_1} \oplus \beta \otimes \mathbf{e_2} \oplus \gamma \otimes \mathbf{e_3}),\ \alpha, \beta, \gamma \in \mathbb{R}\}\ .$$

The practical way of representing a point X of $\mathbb{PR}^3$ is to choose a vector ($\in \mathbb{R}^3$) in the parallelism class of X and to draw it in the three axes obtained by projection of the orthonormal basis of $\mathbb{R}^3$ (it is easy to check that the point we obtain does not depend on the representative in the parallelism class). The point D of Fig. 1 illustrates this, we have drawn two constructions: one corresponding to (0.2, 0.6, 0.8) and the other one to $(0.8,\ 1.2,\ 1.4) = 0.6 \otimes (0.2,\ 0.6,\ 0.8)$.

Let us consider the matrix

$$M = \begin{pmatrix} e & -1 & -1 \\ -1 & e & -1 \\ -1 & -1 & e \end{pmatrix}\ .$$

Its eigenvalue is e. We check that this matrix verifies $M^+ = M$. By Th. 3, the three columns of M are the extremal eigenvectors of M. The complete set of eigenvectors of M is the convex hull of these three vectors. In the projective space $\mathbb{PR}^3$, it is precisely the hexagon of Fig. 1.

Let us consider now a matrix of the form $N = P^{-1} \otimes M \otimes P$. Matrix M is defined above, P is defined by $P_{ii} = u_i \neq \varepsilon,\ i \in \{1, 2, 3\},\ P_{ij} = \varepsilon,\ \forall i \neq j$ and P^{-1} is the inverse of P in $\mathbb{R}_{\max}$. It is possible to prove that the set of eigenvectors of N is obtained from the one of M by a translation of $\pi(u) = \pi(u_1, u_2, u_3)'$ in the projective space.

We consider stochastic models of type (3) with two matrices, i.e. models where $A(n) = A$ with probability $p > 0$ and $A(n) = B$ with probability $1 - p > 0$, A and B being irreducible deterministic matrices of dimension 3. We suppose that $\{A(n)\}$ is **i.i.d.**. More precisely, we consider two systems:

1. The first one with $A = M$ and $B = N$ for $u = (0.4, 1.1, e)'$.
2. The second one with $\tilde{A} = M$ and $\tilde{B} = N$ for $\tilde{u} = (0.8, .1.5, e)'$.

We have represented the set of eigenvectors of the matrices in Fig. 2.

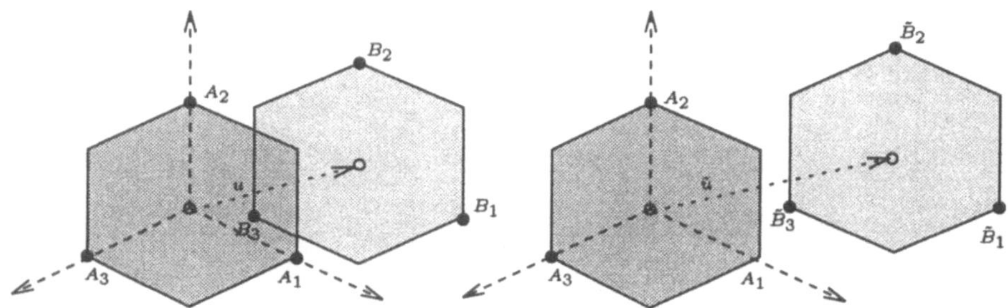

Fig. 2. Stochastic linear systems with two matrices

The intersection Π of the sets of eigenvectors of A and B is not empty. It implies a continuum of stationary regimes. Indeed if we consider an initial condition $x_0 \in \Pi$, we have $\pi(x(n, x_0)) = \pi(x_0), \forall n$.

On the other hand the sets of eigenvectors of $\tilde{A}$ and $\tilde{B}$ do not intersect. It is possible to prove that the product $\tilde{B}\tilde{A}\tilde{B}\tilde{A}$ is **scs1-cyc1**. Hence there is a unique stationary regime by Th. 12. It is in fact possible to prove the following result.

Theorem 14. *We consider A and B, two irreducible matrices of dimension 3. The corresponding stochastic system has a unique stationary regime (or equivalently there is a finite product of A and B which is scs1-cyc1) if and only if the sets of eigenvectors of A and B have an intersection which is empty or restricted to one point.*

Remark. The necessity of the condition is obvious as each common eigenvector is a stationary regime. For larger dimensions, this theorem is not true anymore and we need stronger assumptions on the structure of matrices A and B.

Here is another problem worth considering. We consider an **i.i.d.** model $x(n + 1) = A(n)x(n)$ but now $x_0 \in \mathbb{R}^J$ is fixed. Is it possible for the Markov Chain $\pi x(n, x_0)$ to have several classes of recurrence? The answer, illustrated in Fig. 3, is yes.

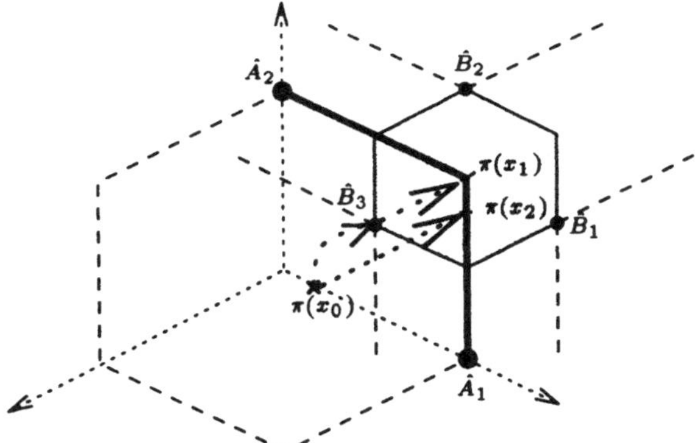

Fig. 3. A single initial condition and several stationary regimes

We consider a stochastic model with two matrices $\hat{A}$ and $\hat{B}$.

$$\hat{A} = \begin{pmatrix} e & -2 & -2 \\ -2 & e & -2 \\ -2 & -2 & -4 \end{pmatrix}, \quad \hat{B} = N \text{ with } \hat{u} = (2, e, e) .$$

It is easy to verify that $\hat{A}^+ = \hat{A}$. Hence by Th. 3, the first two columns of $\hat{A}$ are the extremal eigenvectors of $\hat{A}$. They are represented in Fig. 3 by the points $\hat{A}_1$ and $\hat{A}_2$. The broken segment between these two points is the set of eigenvectors of $\hat{A}$. There are multiple stationary regimes (Th. 14).

We fix now an initial condition $x_0 = (0.5, e, e)'$. By direct computation, we obtain that $\hat{A}x_0 = x_1 = (0.5, e, -1.5)'$ and $\hat{A}\hat{B}x_0 = \hat{A}\hat{B}^k x_0 = x_2 = (1, 1, -1)'$. Vectors x_1 and x_2 are common eigenvectors of $\hat{A}$ and $\hat{B}$. We conclude that $\lim_n P(x(n, x_0) = x_1) = p$ and $\lim_n P(x(n, x_0) = x_2) = 1 - p$.

References

1. Baccelli F., *Ergodic Theory of Stochastic Petri Networks*, Annals of Probability, Vol **20**, n° **1**, pp 375–396, 1992.
2. Baccelli F., Cohen G., Olsder G.J., Quadrat J.P., *Synchronization and Linearity*, John Wiley & Sons, 1993.
3. Cohen G., *Dioids and Discrete Event Systems*, Proc. 11th Int. Conf. on Analysis and Optim. of Systems, Sophia-Antipolis, Juin 1994.
4. Gaubert S., *Théorie des systèmes linéaires dans les dioïdes*, Thèse de doctorat, École Nationale Supérieure des Mines de Paris, 1992.
5. Kingman J., *Subadditive ergodic theory*, Annals of Probability, **1**, pp. 883–909, 1973.
6. Mairesse J., *Products of Irreducible Random Matrices in the* (max, +) *Algebra — Part I*, INRIA Report n° 1939, 1993.
7. Mairesse J., *A Graphical Representation for Matrices in the* (max, +) *Algebra*, INRIA Report n° 2078, 1993.

Cycle Times and Fixed Points of Min-Max Functions

*Jeremy Gunawardena**

Department of Computer Science,
Stanford University, Stanford, CA 94305, USA.
jeremy@cs.stanford.edu

1 Introduction

It will be convenient to use the infix operators $a \vee b$ and $a \wedge b$ to stand for maximum (least upper bound) and minimum (greatest lower bound) respectively: $a \vee b = \max(a, b)$ and $a \wedge b = \min(a, b)$. Note that addition distributes over both maximum and minimum:

$$h + (a \vee b) = h + a \vee h + b, \qquad h + (a \wedge b) = h + a \wedge h + b. \tag{1}$$

Definition 1. A min-max expression, f, is a term in the grammar:

$$f := x_1, x_2, \cdots \mid f + a \mid f \wedge f \mid f \vee f$$

where $x_1, x_2, \cdots$ are variables and $a \in \mathbb{R}$.

For example, $x_1 + x_2 \wedge x_3 + 2$ and $x_1 \vee 2$ are forbidden but $x_1 - 1 \vee x_2 + 1$ is allowed. (In expressions such as these $+$ has higher binding than $\wedge$ or $\vee$.)

Definition 2. A min-max function of dimension n is any function, $F : \mathbb{R}^n \to \mathbb{R}^n$, each of whose components, $F_i : \mathbb{R}^n \to \mathbb{R}$, is a min-max expression of n variables $x_1, \cdots, x_n$.

The theory of min-max functions is concerned with studying F as a dynamical system: with studying the behaviour of the sequence $\mathbf{x}$, $F(\mathbf{x})$, $F^2(\mathbf{x}), \cdots$, for varying $\mathbf{x} \in \mathbb{R}^n$. For a discussion of the origins of this problem see [3, §1] and for information on applications see [4].

There are three existing sources of information about the behaviour of such functions. The earliest is the work on the linear theory of max-plus algebra, [1, Chapter 3]. This work applies to max-only functions: those in which $\wedge$ does not appear in any F_i. (Dually for min-plus algebra and min-only functions. Because $-(a \wedge b) = (-a \vee -b)$, dual results do not require a separate proof and we leave it to the reader to formulate them

* Visiting Scholar at the Department of Computer Science from Hewlett-Packard's Stanford Science Centre.

where necessary.) It is easy to see that any max-only function can be written in canonical form:

$$F_i(x_1, \cdots, x_n) = (a_{i1} + x_1 \vee \cdots \vee a_{in} + x_n),$$

where $a_{ij} \in \mathbb{R} \cup \{-\infty\}$, and that this expression is unique, [3, Lemma 2.1]. If $A = (a_{ij})$ is the corresponding matrix in max-plus algebra then, using max-plus matrix notation, $F(\mathbf{x}) = A\mathbf{x}$ (considering vectors as column vectors). Hence we can expect the theory of min-max functions to include the linear theory as a special case.

Let $\mathbf{c}(h) = (h, h, \cdots, h)$ denote the vector each of whose components has the same value h.

Definition 3. F has a fixed point, $\mathbf{x} \in \mathbb{R}^n$, if, and only if, $F(\mathbf{x}) = \mathbf{x} + \mathbf{c}(h)$, for some $h \in \mathbb{R}$.

If F is a max-only function and A is the corresponding matrix then, for a fixed point $\mathbf{x}$ of F, $A\mathbf{x} = h\mathbf{x}$. That is, $\mathbf{x}$ is an eigenvector of A with eigenvalue h. This number, h, will turn out to be part of the cycle time vector of F which we shall study in §2.

The second source of work on min-max functions is due to Olsder, who undertook in [5] the first investigation of a system with mixed constraints. He considered separated functions $F : \mathbb{R}^n \to \mathbb{R}^n$ for which F_i is max-only for $1 \leq i \leq k$ and F_i is min-only for $k + 1 \leq i \leq n$. Olsder gave a necessary and sufficient condition for the existence of a fixed point, provided that F satisfied certain reasonable conditions, [5, Theorem 2.1], [1, Theorem 9.25].

The third and final work is [3]. This paper has several new results including a proof of eventual periodicity for min-max functions of dimension 2, [3, Theorem 3.3], and a formula for the cycle time of an arbitrary min-max function with a fixed point, [3, Theorem 5.1] (see (11) below).

The purpose of the present paper is to describe some new results and conjectures which provide a coherent explanation of this previous work. They capture some of the essential properties of min-max functions and provide a foundation for a deeper study of the remaining unsolved problems. They also shed a new light on aspects of the linear theory. For reasons of space, most proofs have been omitted; they can be found in the full version of this paper.

The author is very grateful to Professor Geert-Jan Olsder for his encouragement and hospitality during a visit to the University of Delft during which some of the ideas in this paper were developed. The work presented here was undertaken as part of project STETSON, a joint project between Hewlett-Packard Laboratories and Stanford University on asynchronous hardware design.

2 The cycle time vector

We shall frequently use numerical operations and relations and apply them to vectors. These should always be assumed to be applied to each component separately. Hence $\mathbf{u} \leq \mathbf{v}$ means $u_i \leq v_i$ for each i. Similarly, $(\bigwedge_l \mathbf{a}_l)_i = \bigwedge_l (\mathbf{a}_l)_i$. We begin by recalling some simple properties of a min-max function F. First, F is continuous. Second, F is monotone:

$$\mathbf{u} \leq \mathbf{v} \implies F(\mathbf{u}) \leq F(\mathbf{v}), \tag{2}$$

Third, F is homogeneous, in the sense that, for any $h \in \mathbb{R}$,

$$F(\mathbf{u} + \mathbf{c}(h)) = F(\mathbf{u}) + \mathbf{c}(h), \tag{3}$$

This follows easily from (1), [3, Lemma 2.3]. The next property is not quite so obvious. Let $|\mathbf{u}|$ denote the l^∞, or maximum, norm on $\mathbb{R}^n$: $|\mathbf{u}| = \bigvee_{1 \le i \le n} |u_i|$, where $|u_i|$ is the usual absolute value on real numbers.

Lemma 4 (Non-expansive property). *Let F be a min-max function of dimension n. If $\mathbf{u}, \mathbf{v} \in \mathbb{R}^n$ then $|F(\mathbf{u}) - F(\mathbf{v})| \le |\mathbf{u} - \mathbf{v}|$.*

F is not contractive. If it were, the Contraction Theorem, [2, Theorem 2.1], would imply that the dynamic behaviour of F was trivial. But suppose that at some point $\mathbf{x} \in \mathbb{R}^n$

$$\lim_{s \to \infty} \frac{F^s(\mathbf{x})}{s} \tag{4}$$

exists. (Note that this is a vector quantity in $\mathbb{R}^n$.) It then follows from Lemma 4 that this limit must exist at all points of $\mathbb{R}^n$ and must have the same value. In the applications of min-max functions, the state vector $\mathbf{x}$ is often interpreted as a vector of occurrence times of certain events and the vector $F(\mathbf{x})$ as the times of next occurrence. Hence the limit (4) can be thought of as the vector of asymptotic average times to the next occurrence of the events:

$$\frac{F^s(\mathbf{x}) - F^{s-1}(\mathbf{x}) + \cdots + F(\mathbf{x}) - \mathbf{x}}{s} = \frac{F^s(\mathbf{x}) - \mathbf{x}}{s},$$

which tends to (4) as $s \to \infty$, [3, §1]. This motivates the following definition.

Definition 5. Let F be a min-max function. If the limit (4) exists somewhere, it is called the cycle time vector of F and denoted by $\mathcal{X}(F) \in \mathbb{R}^n$.

This brings us to our first problem. When does the cycle time vector exist? Suppose first that F is a max-only function of dimension n and that A is the associated $n \times n$ matrix in max-plus algebra. We recall that the precedence graph of A, [1, Definition 2.8], denoted $\mathcal{G}(A)$, is the directed graph with annotated edges which has nodes $\{1, 2, \cdots, n\}$ and an edge from j to i if, and only if, $A_{ij} \neq -\infty$. The annotation on this edge is then the real number A_{ij}. A path in this graph has the usual meaning of a sequence of directed edges and a circuit is a path which starts and ends at the same node, [1, §2.2]. The weight of a path p, $|p|_{\mathrm{w}}$, is the sum of the annotations on the edges in the path. The length of a path, $|p|_\ell$, is the number of edges in the path. If g is a circuit, the ratio $|g|_{\mathrm{w}}/|g|_\ell$ is the cycle mean of the circuit, [1, Definition 2.18].

Definition 6. If A is an $n \times n$ matrix in max-plus algebra, let $\mu(A)$ be the vector $\mu_i(A) = \bigvee\{\, |g|_{\mathrm{w}}/|g|_\ell \mid g$ a circuit in $\mathcal{G}(A)$ upstream from node $i\}$. Dually, if B is a matrix in min-plus algebra, then $\eta(B)$ will denote the vector of minimum cycle means.

A circuit is upstream from node i if there is a path in $\mathcal{G}(A)$ from some node on the circuit to node i. Because of the conventions of canonical form, [3, §2], if A is a matrix associated to a max-only function then any node in $\mathcal{G}(A)$ has at least one upstream circuit and so $\mu(A) \in \mathbb{R}^n$.

Proposition 7. *Let F be a max-only function and A the associated matrix in max-plus algebra. The limit (4) always exists and the cycle time vector is given by $\chi(F) = \mu(A)$.*

We now return to the case of an arbitrary min-max function of dimension n. Each component of F can be placed in conjunctive normal form:

$$F_k(\mathbf{x}) = (A^k_{11} + x_1 \vee \cdots \vee A^k_{1n} + x_n) \wedge \cdots \wedge (A^k_{\ell(k)1} + x_1 \vee \cdots \vee A^k_{\ell(k)n} + x_n), \quad (5)$$

where $A^k_{ij} \in \mathbb{R} \cup \{-\infty\}$, [3, §2]. Here $\ell(k)$ is the number of conjunctions in the component F_k. It is an important property of min-max functions that conjunctive normal form is unique up to re-ordering of the conjunctions, [3, Theorem 2.1]. We can now associate a max-plus matrix A to F by choosing, for the k-th row of the matrix, one of the $\ell(k)$ conjunctions in (5): $A_{kj} = A^k_{i_k j}$ where $1 \leq i_k \leq \ell(k)$ specifies which conjunction is chosen in row k.

Definition 8. The matrix A constructed in this way is called a max-only projection of F. The set of all max-only projections is denoted $\mathsf{P}(F)$. Dually, the set of min-only projections of F, $\mathsf{Q}(F)$, is constructed from the disjunctive normal form of F.

If A is any max-only projection of F, it is clear from the construction above that $F(\mathbf{x}) \leq A\mathbf{x}$ for any $\mathbf{x} \in \mathbb{R}^n$. It follows from (2) that $F^s(\mathbf{x}) \leq A^s\mathbf{x}$ for all $s \geq 0$. Now choose $\epsilon > 0$. It then follows from Proposition 7 that, for all sufficiently large s, $F^s(\mathbf{x})/s \leq \mu(A) + \mathbf{c}(\epsilon)$. Since this holds for any max-only projection, and there are only finitely many such, we see that $F^s(\mathbf{x})/s \leq (\bigwedge_{A \in \mathsf{P}(F)} \mu(A)) + \mathbf{c}(\epsilon)$ for all sufficiently large s. By a dual argument applied to the min-only projections of F, we can conclude that

$$\left(\bigvee_{B \in \mathsf{Q}(F)} \eta(B)\right) - \mathbf{c}(\epsilon) \leq \frac{F^s(\mathbf{x})}{s} \leq \left(\bigwedge_{A \in \mathsf{P}(F)} \mu(A)\right) + \mathbf{c}(\epsilon). \quad (6)$$

for all sufficiently large s.

Conjecture 9 (The Duality Conjecture). *For any min-max function F,*

$$\bigvee_{B \in \mathsf{Q}(F)} \eta(B) = \bigwedge_{A \in \mathsf{P}(F)} \mu(A).$$

The significance of this should be clear. It implies that any min-max function has a cycle time and gives us a formula for computing it. Of course (4) could still exist without the Conjecture being true. We should note in passing that since ϵ was arbitrary, (6) already tells us that

$$\bigvee_{B \in \mathsf{Q}(F)} \eta(B) \leq \bigwedge_{A \in \mathsf{P}(F)} \mu(A). \quad (7)$$

What evidence is there to support the Duality Conjecture? It is easy to show that it holds for any max-only (dually, any min-only) function. In dimension 2 it follows from the geometric methods of [3] that (4) always exists. It can also be shown that the Duality Conjecture holds but this requires a complicated combinatorial argument (unpublished). Finally, we shall see in the next section that the Conjecture holds for any function with a fixed point.

3 The existence of fixed points

Lemma 10. *Let F be any min-max function of dimension n and let $\mathsf{P}(F) = \{A_1, \cdots, A_N\}$. The function $\{1, \cdots, n\} \times \{1, \cdots, N\} \to \mathbb{R}$ taking $(i, j) \to \mu_i(A_j)$ has a saddle point:*

$$\bigvee_{1\leq i\leq n} \bigwedge_{1\leq j\leq N} \mu_i(A_j) = \bigwedge_{1\leq j\leq N} \bigvee_{1\leq i\leq n} \mu_i(A_j).$$

If A is an $n \times n$ matrix in max-plus algebra let $\mu(A)$ denote the maximum cycle mean of A: $\mu(A) = \bigvee_{1\leq i\leq n} \mu_i(A)$. Similarly, let $\eta(B)$ denote the minimum cycle mean of a min-plus matrix B. It follows from Lemma 10 and (7) that

$$\mathbf{c}\left(\bigvee_{B\in\mathsf{Q}(F)} \eta(B)\right) \leq \bigvee_{B\in\mathsf{Q}(F)} \eta(B) \leq \bigwedge_{A\in\mathsf{P}(F)} \mu(A) \leq \mathbf{c}\left(\bigwedge_{A\in\mathsf{P}(F)} \mu(A)\right). \tag{8}$$

Now let F be any min-max function and suppose that $\mathbf{x}$ is a fixed point of F: $F(\mathbf{x}) = \mathbf{x} + \mathbf{c}(h)$. It follows from (3) that $F^s(\mathbf{x}) = \mathbf{x} + s\mathbf{c}(h)$. Hence the cycle time vector exists and is given by $\chi(F) = \mathbf{c}(h)$. All the components of the cycle time vector are equal.

What can we say about the cycle means? It was shown in [3, Theorem 5.1] that when F has a fixed point, $\bigvee_{B\in\mathsf{Q}(F)} \eta(B) = \bigwedge_{A\in\mathsf{P}(F)} \mu(A)$ and if h denotes this common value then $\chi(F) = \mathbf{c}(h)$. This should be considered as a generalization to min-max functions of the classical "Perron-Frobenius" theorem on the eigenvalues of an irreducible max-plus matrix, [1, Theorem 3.23]. It follows from (8) that in this case the Duality Conjecture is true for F. In fact we can distinguish three conditions which are implied by the existence of a fixed point:

$$\bigwedge_{A\in\mathsf{P}(F)} \mu(A) = \mathbf{c}(h), \tag{9}$$

$$\bigvee_{B\in\mathsf{Q}(F)} \eta(B) = \mathbf{c}(h), \tag{10}$$

$$\bigvee_{B\in\mathsf{Q}(F)} \eta(B) = h = \bigwedge_{A\in\mathsf{P}(F)} \mu(A). \tag{11}$$

Because of (8), (11) implies both (9) and (10). If the Duality Conjecture holds then Lemma 10 shows that either (9) or (10) separately imply (11) and all the conditions are equivalent.

The remainder of this section is concerned with determining when the necessary conditions above are also sufficient. As usual, we first consider the simple case.

Proposition 11. *Let F be a max-only function and A the associated matrix in max-plus algebra. F has a fixed point if, and only if, $\mu(A) = \mathbf{c}(h)$, for some $h \in \mathbb{R}$.*

Neither this result nor Proposition 7 have appeared before in the max-plus literature, to the best of our knowledge.

Theorem 12. *Let F be any min-max function. F has a fixed point if, and only if,*

$$\bigvee_{B \in \mathsf{Q}(F)} \eta(B) = \bigwedge_{A \in \mathsf{P}(F)} \mu(A).$$

Proof. If F has a fixed point then the result is just (11) above. So suppose that (11) holds. As usual, we may assume that $h = 0$. Let $X \in \mathsf{P}(F)$ be any max-only projection for which $\mu(X) = h$. Then, by (8),

$$\mathbf{c}(h) \geq \mu(X) \geq \bigwedge_{A \in \mathsf{P}(F)} \mu(A) = \mathbf{c}(h).$$

It follows that $\mu(X) = \mathbf{c}(h)$. Hence, by Proposition 11, X has a real eigenvector $\mathbf{v}$. By construction of the max-only projections, it follows that $F(\mathbf{v}) \leq X\mathbf{v} = \mathbf{v}$. (Recall that h was assumed to be 0.) Hence, by (2), $F^s(\mathbf{v})$ is a decreasing sequence. We can apply a similar argument to the min-only projections of B and find a vector $\mathbf{u}$ such that $\mathbf{u} \leq F(\mathbf{u})$. Again using (2), $F^s(\mathbf{u})$ is an increasing sequence. However, neither sequence can respectively decrease or increase too far because F is non-expansive. It is easy to see from Lemma 4 that $F^s(\mathbf{v})$ must be bounded below and $F^s(\mathbf{u})$ must be bounded above. But a bounded monotonic sequence must converge. Hence there exist vectors $\mathbf{u}_\infty$ and $\mathbf{v}_\infty$, not necessarily distinct, such that $\lim_{s\to\infty} F^s(\mathbf{v}) = \mathbf{v}_\infty$ and $\lim_{s\to\infty} F^s(\mathbf{u}) = \mathbf{u}_\infty$. Since F is continuous it follows that $F(\mathbf{v}_\infty) = \mathbf{v}_\infty$ and $F(\mathbf{u}_\infty) = \mathbf{u}_\infty$. This completes the proof. □

Theorem 12 is relatively weaker than Proposition 11 because the former uses (11) while the latter uses (9). (11) is inconvenient in practice because it requires information on both $\mathsf{P}(F)$ and $\mathsf{Q}(F)$. For example, if the Duality Conjecture were true then it could be shown that the conditions of Olsder's fixed point theorem for separated functions, [5, Theorem 2.1], imply (9) and we could deduce Olsder's theorem as a corollary of Theorem 12. In the absence of the Conjecture, this appears difficult. As another example suppose that F has a periodic point, $\mathbf{x}$, where $F^k(\mathbf{x}) = \mathbf{x} + \mathbf{c}(h)$, [3, Definition 2.3]. It follows easily from (3) that the cycle time vector exists and $\chi(F) = \mathbf{c}(h/k)$. If the Duality Conjecture were true we could deduce the strong result that a min-max function has a periodic point if, and only if, it has a fixed point. This, and stronger results, are already known to hold in dimension 2, [3, Theorem 3.3].

4 Conclusion

The main contribution of this paper is the identification of the Duality Conjecture and the demonstration of its significance for the deeper study of min-max functions.

References

1. F. Baccelli, G. Cohen, G. J. Olsder, and J.-P. Quadrat. *Synchronization and Linearity*. Wiley Series in Probability and Mathematical Statistics. John Wiley and Sons, 1992.

2. K. Goebel and W. A. Kirk. *Topics in Metric Fixed Point Theory*, volume 28 of *Cambridge Studies in Advanced Mathematics*. Cambride University Press, 1990.
3. J. Gunawardena. Min-max functions, Part I. Technical Report STAN-CS-93-1474, Department of Computer Science, Stanford University, May 1993. Submitted to *Discrete Event Dynamic Systems*.
4. J. Gunawardena. Timing analysis of digital circuits and the theory of min-max functions. In *TAU'93, ACM International Workshop on Timing Issues in the Specification and Synthesis of Digital Systems*, September 1993.
5. G. J. Olsder. Eigenvalues of dynamic max-min systems. *Discrete Event Dynamic Systems*, 1:177–207, 1991.

The Characteristic Equation and Minimal State Space Realization of SISO Systems in the Max Algebra

Bart De Schutter and Bart De Moor

ESAT - Katholieke Universiteit Leuven, Kardinaal Mercierlaan 94, 3001 Leuven, Belgium, email: *bart.deschutter@esat.kuleuven.ac.be*, *bart.demoor@esat.kuleuven.ac.be*.
Bart De Schutter is a research assistant and Bart De Moor is a research associate with the N.F.W.O. (Belgian National Fund for Scientific Research).

1 Introduction

1.1 Overview

There exist many modeling and analysis frameworks for discrete event systems: Petri nets, formal languages, generalized semi-Markov processes, perturbation analysis and so on. In this paper we consider systems that can be modeled using max algebra.

In the first part we study the characteristic equation of a matrix in the max algebra ($\mathbb{R}_{\max}$). We determine necessary and for some cases also sufficient conditions for a polynomial to be the characteristic polynomial of a matrix with elements in $\mathbb{R}_{\max}$. Then we indicate how to construct a matrix such that its characteristic polynomial is equal to a given monic polynomial in $\mathbb{S}_{\max}$, the extension of $\mathbb{R}_{\max}$.

In the second part of this paper we address the minimal state space realization problem. Based on the results of the first part we propose a procedure to find a minimal state space realization of a single input single output (SISO) discrete event system in the max algebra, given its Markov parameters. Finally we illustrate this procedure with an example.

1.2 Notations

One of the mathematical tools used in this paper is the max algebra. In this introduction we only explain the notations we use to represent the max-algebraic operations. A complete introduction to the max algebra can be found in [1].

In this paper we use the following notations: $a \oplus b = \max(a, b)$ and $a \otimes b = a + b$. $\epsilon = -\infty$ is the neutral element for $\oplus$ in $\mathbb{R}_{\max} = \mathbb{R} \cup \{\epsilon\}, \oplus, \otimes$. The inverse element of $a \neq \epsilon$ for $\otimes$ in $\mathbb{R}_{\max}$ is denoted by $a^{\otimes^{-1}}$. The division is defined as follows: $\frac{a}{b} = a \otimes b^{\otimes^{-1}}$ if $b \neq \epsilon$. E_n is the n by n identity matrix in $\mathbb{R}_{\max}$.

We also use the extension of the max algebra $\mathbb{S}_{\max}$ that was introduced in [1, 6]. $\mathbb{S}_{\max}$ is a kind of symmetrization of $\mathbb{R}_{\max}$. We shall restrict ourselves to the most important features of $\mathbb{S}_{\max}$. For a more formal derivation the interested reader is referred to [6].

There are three kinds of elements in $\mathbb{S}_{\max}$: the positive elements ($\mathbb{S}^{\oplus}_{\max} = \mathbb{R}_{\max}$), the negative elements ($\mathbb{S}^{\ominus}_{\max}$) and the balanced elements ($\mathbb{S}^{\bullet}_{\max}$). The positive and the negative elements are called signed ($\mathbb{S}_{\max}^{\vee} = \mathbb{S}_{\max}^{\oplus} \cup \mathbb{S}_{\max}^{\ominus}$). The $\ominus$ operation in $\mathbb{S}_{\max}$ is defined as follows: $a \ominus b = a \quad \text{if } a > b\,,$

$a \ominus b = \ominus b \quad \text{if } a < b\,,$

$a \ominus a = a^{\bullet}\ .$

If $a \in \mathbb{S}_{\max}$ then it can be written as $a = a^{+} \ominus a^{-}$ where a^{+} is the positive part of a and a^{-} is the negative part of a. If $a \in \mathbb{S}_{\max}^{\oplus}$ then $a^{+} = a$ and $a^{-} = \epsilon$, if $a \in \mathbb{S}_{\max}^{\ominus}$ then $a^{+} = \epsilon$ and $a^{-} = \ominus a$ and if $a \in \mathbb{S}_{\max}^{\bullet}$ then $a^{+} = a^{-}$.

In $\mathbb{S}_{\max}$ we have to use balances (∇) instead of equalities. Loosely speaking an $\ominus$ sign in a balance indicates that the element should be at the other side. If both sides of a balance are signed we can replace the balance by an equality. So $x \ominus 4 \ \nabla\ 2$ means $x \ \nabla\ 2 \oplus 4$ and if x is signed we get $x = 2 \oplus 4 = 4$ as a solution.

To select submatrices of a matrix we use the following notation: $A([i_1, \ldots, i_k], [j_1, \ldots, j_l])$ is the matrix resulting from A by eliminating all rows except for rows $i_1, \ldots, i_k$ and all columns except for columns $j_1, \ldots, j_l$.

1.3 Some Definitions and Properties

Definition 1 (Determinant). Consider a matrix $A \in \mathbb{S}_{\max}^{n \times n}$. The determinant of A is defined as

$$\det A = \bigoplus_{\sigma \in \mathcal{P}_n} \operatorname{sgn}(\sigma) \otimes \bigotimes_{i=1}^{n} a_{i\sigma(i)}$$

where $\mathcal{P}_n$ is the set of all permutations of $\{1, \ldots, n\}$, and $\operatorname{sgn}(\sigma) = 0$ if the permutation σ is even and $\operatorname{sgn}(\sigma) = \ominus 0$ if the permutation is odd.

Theorem 2. *Let $A \in \mathbb{S}_{\max}^{n \times n}$. The homogeneous linear balance $A \otimes x \ \nabla\ \epsilon$ has a non-trivial signed solution if and only if* $\det A \ \nabla\ \epsilon$.

Proof. See [6]. The proof given there is constructive so it can be used to find a solution. □

Definition 3 (Characteristic equation). The characteristic equation of a matrix $A \in \mathbb{S}_{\max}^{n \times n}$ is defined as $\det(A \ominus \lambda \otimes E_n) \ \nabla\ \epsilon$.

This leads to

$$\lambda^{\otimes^{n}} \oplus \bigoplus_{p=1}^{n} a_p \otimes \lambda^{\otimes^{n-p}} \ \nabla\ \epsilon \tag{1}$$

with

$$a_p = (\ominus 0)^{\otimes^{p}} \otimes \bigoplus_{\varphi \in C_p^n} \det A([i_1, i_2, \ldots, i_p], [i_1, i_2, \ldots, i_p]) \tag{2}$$

where C_p^n is the set of all combinations of p numbers out of $\{1, \ldots, n\}$ and $\varphi = \{i_1, i_2, \ldots, i_p\}$. Equation (1) will be called a *monic* balance, since the coefficient of $\lambda^{\otimes^{n}}$ equals 0 (i.e. the identity element for $\otimes$).

In $\mathbb{S}_{\max}$ every monic n-th order linear balance is the characteristic equation of an $n \times n$ matrix. However, this is not the case in $\mathbb{R}_{\max}$ as will be shown in the next section.

2 Necessary Conditions for a Polynomial to Be the Characteristic Polynomial of a Positive Matrix

A positive matrix is a matrix the elements of which lie in $\mathbb{R}_{\max}$. In this section we state necessary conditions for the coefficients of the characteristic polynomial of a positive matrix. These conditions will play an important role when one wants to determine the minimal order of a SISO system in the max algebra.

From now on we assume that $A \in \mathbb{R}_{\max}^{n \times n}$. If we define $\alpha_p = a_p^+$ and $\beta_p = a_p^-$ ($\alpha_p, \beta_p \in \mathbb{R}_{\max}$) and if we move all terms with negative coefficients to the right hand side (1) becomes

$$\lambda^{\otimes^n} \oplus \bigoplus_{i=2}^{n} \alpha_i \otimes \lambda^{\otimes^{n-i}} \;\nabla\; \beta_1 \otimes \lambda^{\otimes^{n-1}} \oplus \bigoplus_{j=2}^{n} \beta_j \otimes \lambda^{\otimes^{n-j}} . \tag{3}$$

There are three possible cases: $\alpha_p = \epsilon$, $\beta_p = \epsilon$ or $\alpha_p = \beta_p$. We already have omitted α_1, since we always have that $a_1 \in \mathbb{S}_{\max}^{\ominus}$ and thus $\alpha_1 = a_1^+ = \epsilon$.

The most stringent property for α_p and β_p that was proven in [3] is:

Property 4. $\forall i \in \{2, \ldots, n\}$ *at least one of the following statements is true:*

$$\alpha_i \leq \bigoplus_{r=1}^{\lfloor \frac{i}{2} \rfloor} \beta_r \otimes \beta_{i-r} \quad or \quad \alpha_i < \bigoplus_{r=2}^{\lfloor \frac{i}{2} \rfloor} \alpha_r \otimes \alpha_{i-r} \quad or \quad \alpha_i < \bigoplus_{r=2}^{i-1} \alpha_r \otimes \beta_{i-r} ,$$

where $\lfloor x \rfloor$ stands for the largest integer number less than or equal to x.

This property gives necessary conditions for the coefficients of an $\mathbb{S}_{\max}$ polynomial to be the characteristic polynomial of a positive matrix. For more properties and extensive proofs the reader is referred to [3].

3 Necessary and Sufficient Conditions for a Polynomial to Be the Characteristic Polynomial of a Positive Matrix

In the next subsections we determine case by case necessary and sufficient conditions for (3) to be the characteristic equation of a positive matrix and indicate how such a matrix can be found (see [3] for proofs). For the lower dimensional cases we can give an analytic description of the matrix we are looking for. For higher dimensional cases we shall first state a conjecture and then sketch a heuristic algorithm that will (in most cases) find a solution.

In all cases we have $\alpha_1 = \epsilon$ as a necessary condition. We also define $\kappa_{i,j} = \frac{\alpha_j}{\beta_i}$ if $\beta_i \neq \epsilon$ and $\kappa_{i,j} = \epsilon$ if $\beta_i = \epsilon$.

3.1 The 1×1 Case

There is no extra condition. The matrix $[\beta_1]$ has $\lambda \;\nabla\; \beta_1$ as its characteristic equation.

3.2 The 2 × 2 Case

The necessary and sufficient condition is: $\alpha_2 \leq \beta_1 \otimes \beta_1$. The characteristic equation of the matrix $\begin{bmatrix} \beta_1 & \beta_2 \\ 0 & \kappa_{1,2} \end{bmatrix}$ is $\lambda^{\otimes^2} \oplus \alpha_2 \nabla \beta_1 \otimes \lambda \oplus \beta_2$.

3.3 The 3 × 3 Case

The necessary and sufficient conditions are $\begin{cases} \alpha_2 \leq \beta_1 \otimes \beta_1 \\ \alpha_3 \leq \beta_1 \otimes \beta_2 \text{ or } \alpha_3 < \beta_1 \otimes \alpha_2 \end{cases}$.

The corresponding matrix is $\begin{bmatrix} \beta_1 & \beta_2 & \beta_3 \\ 0 & \kappa_{1,2} & \kappa_{1,3} \\ \epsilon & 0 & \epsilon \end{bmatrix}$.

3.4 The 4 × 4 Case

First we distinguish three possible cases:

Case A: $\alpha_4 \leq \beta_1 \otimes \beta_3$ or $\alpha_4 < \beta_1 \otimes \alpha_3$

Case B: $\alpha_4 > \beta_1 \otimes \beta_3$ and $\alpha_4 \geq \beta_1 \otimes \alpha_3$ and $\alpha_4 \leq \beta_2 \otimes \beta_2$ and ($\beta_1 = \epsilon$ or $\alpha_2 = \epsilon$ or $\beta_4 = \alpha_4$)

Case C: $\alpha_4 > \beta_1 \otimes \beta_3$ and $\alpha_4 \geq \beta_1 \otimes \alpha_3$ and $\alpha_4 \leq \beta_2 \otimes \beta_2$ and $\alpha_2 = \beta_2 \neq \epsilon$ and $\beta_4 = \epsilon$.

If the coefficients don't fall into exactly one of these three cases, they cannot correspond to a positive matrix.
The necessary and sufficient conditions are:

$$\begin{cases} \alpha_2 \leq \beta_1 \otimes \beta_1 \\ \alpha_3 \leq \beta_1 \otimes \beta_2 \text{ or } \alpha_3 < \beta_1 \otimes \alpha_2 \\ \text{for Case A: no extra conditions} \\ \text{for Case B: } \beta_1 \otimes \alpha_4 \leq \beta_2 \otimes \alpha_3 \text{ or } \beta_1 \otimes \alpha_4 < \beta_2 \otimes \beta_3 \\ \text{for Case C: } \beta_1 \otimes \alpha_3 = \beta_2 \otimes \alpha_2 \text{ and } \beta_1 \otimes \alpha_4 = \beta_2 \otimes \alpha_3 \end{cases} .$$

We find for Case A: $\begin{bmatrix} \beta_1 & \beta_2 & \beta_3 & \beta_4 \\ 0 & \kappa_{1,2} & \kappa_{1,3} & \kappa_{1,4} \\ \epsilon & 0 & \epsilon & \epsilon \\ \epsilon & \epsilon & 0 & \epsilon \end{bmatrix}$, for Case B: $\begin{bmatrix} \beta_1 & \beta_2 & \beta_3 & \beta_4 \\ 0 & \kappa_{1,2} & \kappa_{1,3} & \epsilon \\ \epsilon & 0 & \epsilon & \kappa_{2,4} \\ \epsilon & \epsilon & 0 & \epsilon \end{bmatrix}$ and for Case C: $\begin{bmatrix} \beta_1 & \beta_2 & \epsilon & \epsilon \\ 0 & \epsilon & \epsilon & \epsilon \\ \epsilon & 0 & \kappa_{2,3} & \kappa_{2,4} \\ \epsilon & \epsilon & 0 & \epsilon \end{bmatrix}$.

3.5 The General Case

Here we have not yet found sufficient conditions, but we shall outline a heuristic algorithm that will in most cases result in a positive matrix for which the characteristic polynomial will be equal to the given polynomial.

Extrapolating the results of the previous subsections and supported by many examples we state the following conjecture:

Conjecture 5. *If* $\lambda^{\otimes^n} \oplus \bigoplus_{i=2}^{n} \alpha_i \otimes \lambda^{\otimes^{n-i}} \nabla \beta_1 \otimes \lambda^{\otimes^{n-1}} \oplus \bigoplus_{j=2}^{n} \beta_j \otimes \lambda^{\otimes^{n-j}}$ *is the characteristic equation of a matrix* $A \in \mathbb{R}_{\max}^{n \times n}$ *then it is also the characteristic equation of an upper Hessenberg matrix of the form*

$$K = \begin{bmatrix} k_{0,1} & k_{0,2} & k_{0,3} & \dots & k_{0,n-1} & k_{0,n} \\ 0 & k_{1,2} & k_{1,3} & \dots & k_{1,n-1} & k_{1,n} \\ \epsilon & 0 & k_{2,3} & \dots & k_{2,n-1} & k_{2,n} \\ \vdots & \vdots & \vdots & \ddots & \vdots & \vdots \\ \epsilon & \epsilon & \epsilon & \dots & 0 & k_{n-1,n} \end{bmatrix} .$$

We shall use this conjecture in our heuristic algorithm to construct a matrix for which the characteristic polynomial will be equal to a given polynomial. However, in [5] we have presented a method to construct such a matrix that works even if Conjecture 5 would not be true. The major disadvantage of this method is its computational complexity. Therefore we now present a heuristic algorithm that will be much faster on the average. If a result is returned, it is right. But it could be possible that sometimes no result is returned although there is a solution (in which case we have to fall back on the method of [5]).

A heuristic algorithm:

First we check whether the coefficients of the given polynomial satisfy the conditions of Property 4. Then we reconstruct the a_p^-'s by setting $a_1^- = \beta_1$ and $a_p^- = \max(\alpha_p - \delta, \beta_p)$ for $p = 2, 3, \dots, n$ where δ is a small strictly positive real number.

Consider $K_1 = \begin{bmatrix} a_1^- & a_2^- & a_3^- & \dots & a_n^- \\ 0 & a_1^- & a_2^- & \dots & a_{n-1}^- \\ \epsilon & 0 & a_1^- & \dots & a_{n-2}^- \\ \vdots & \vdots & \vdots & \ddots & \vdots \\ \epsilon & \epsilon & \epsilon & \dots & a_1^- \end{bmatrix}$ and $K_2 = \begin{bmatrix} \epsilon & \epsilon & \epsilon & \dots & \epsilon \\ \epsilon & \kappa_{1,2} & \kappa_{1,3} & \dots & \kappa_{1,n} \\ \epsilon & \epsilon & \kappa_{2,3} & \dots & \kappa_{2,n} \\ \vdots & \vdots & \vdots & \ddots & \vdots \\ \epsilon & \epsilon & \epsilon & \dots & \kappa_{n-1,n} \end{bmatrix}$

where $\kappa_{i,j} = \dfrac{\alpha_j}{a_i^-}$ if $a_i^- \neq \epsilon$ and $\kappa_{i,j} = \epsilon$ if $a_i^- = \epsilon$.

We shall make a judicious choice out of the elements of K_1 and K_2 to compose a matrix for which the characteristic equation will coincide with (3).

We start with $A = \begin{bmatrix} a_1^- & a_2^- & \dots & a_n^- \\ 0 & \epsilon & \dots & \epsilon \\ \vdots & \vdots & \ddots & \vdots \\ \epsilon & \epsilon & \dots & \epsilon \end{bmatrix}$. Now we shall column by column transfer non-ϵ elements of K_2 to A (one element per column) such that the coefficients of the characteristic equation of A are less than or equal to those of (3). If this doesn't lead to a valid result we shift a_1^- along its diagonal and repeat the procedure. We keep shifting a_1^- until it reaches the n-th column. If this still doesn't yield a result we put a_1^- back in the first column and repeat the procedure but now with a_2^-, and so on. Finally, if we have found A we remove redundant entries: these are elements that can be removed without altering the characteristic equation.

4 Minimal State Space Realization

4.1 Realization and Minimal Realization

Suppose that we have a single input single output (SISO) discrete event system that can be described by an n-th order state space model

$$x[k+1] = A \otimes x[k] \;\oplus\; b \otimes u[k] \tag{4}$$

$$y[k] = c \otimes x[k] \tag{5}$$

with $A \in \mathbb{R}_{\max}^{n\times n}$, $b \in \mathbb{R}_{\max}^{n\times 1}$ and $c \in \mathbb{R}_{\max}^{1\times n}$. u is the input, y is the output and x is the state vector.

We define the unit impulse e as: $e[k] = 0$ if $k = 0$ and $e[k] = \epsilon$ otherwise. If we apply a unit impulse to the system and if we assume that the initial state $x[0]$ satisfies $x[0] = \epsilon$ or $A \otimes x[0] \leq b$, we get the impulse response as the output of the system:

$$x[1] = b\,,\; x[2] = A \otimes b\,,\; \ldots\,,\; x[k] = A^{\otimes^{k-1}} \otimes b \;\Rightarrow\; y[k] = c \otimes A^{\otimes^{k-1}} \otimes b\;. \tag{6}$$

Let $g_k = c \otimes A^{\otimes^k} \otimes b$. The g_k's are called the Markov parameters.

Let us now reverse the process: suppose that A, b and c are unknown, and that we only know the Markov parameters (e.g. from experiments – where we assume that the system is max-linear and time-invariant and that there is no noise present). How can we construct A, b and c from the g_k's? This process is called realization. If we make the dimension of A minimal, we have a minimal realization. Although there have been some attempts to solve this problem [2, 7, 8], this problem has at present – to the authors' knowledge – not been solved entirely.

4.2 A Lower Bound for the Minimal System Order

Property 6. *The Markov parameters of the system with system matrix $A \in \mathbb{S}_{\max}^{n\times n}$ satisfy the characteristic equation of A:*

$$\bigoplus_{p=0}^{n} a_p \otimes g_{k+n-p} \;\nabla\; \epsilon \quad \textit{for } k = 0, 1, 2, \ldots\;,$$

where $a_0 = 0$.

Suppose that we have a system that can be described by (4)–(5), with unknown system matrices. If we want to find a minimal realization of this system the first question that has to be answered is that of the minimal system order.

Consider the semi-infinite Hankel matrix $H = \begin{bmatrix} g_0 & g_1 & g_2 & \cdots \\ g_1 & g_2 & g_3 & \cdots \\ g_2 & g_3 & g_4 & \cdots \\ \vdots & \vdots & \vdots & \ddots \end{bmatrix}$. Let $H(:,i)$ be the i-th column of H. As a direct consequence of Property 6 we have that

$$\bigoplus_{p=0}^{n} a_p \otimes H(:, k+n-p) \nabla \epsilon \quad \text{for } k = 1, 2, \ldots \ . \tag{7}$$

Now we reverse this reasoning: first we construct a p by q Hankel matrix

$$H_{p,q} = \begin{bmatrix} g_0 & g_1 & \cdots & g_{q-1} \\ g_1 & g_2 & \cdots & g_q \\ \vdots & \vdots & \ddots & \vdots \\ g_{p-1} & g_p & \cdots & g_{p+q-2} \end{bmatrix}$$

with p and q large enough: $p, q \gg n$, where n is the real (but unknown) system order. Then we try to find n and $a_0, a_1, \ldots, a_n$ such that the columns of $H_{p,q}$ satisfy an equation of the form (7), which will lead to the characteristic equation of the unknown system matrix A.

We propose the following procedure:
First we look for the largest square submatrix of $H_{p,q}$ with consecutive column indices,

$$H_{\text{sub},r} = H_{p,q}([i_1, i_2, \ldots, i_r], [j+1, j+2, \ldots, j+r]) \ ,$$

the determinant of which is not balanced: $\det H_{\text{sub},r} \not\nabla \epsilon$. If we add one arbitrary row and the $j+r+1$-st column to $H_{\text{sub},r}$ we get an $r+1$ by $r+1$ matrix $H_{\text{sub},r+1}$ that has a balanced determinant. So according to Theorem 2 the set of linear balances $H_{\text{sub},r+1} \otimes a \nabla \epsilon$ has a signed solution $a = \begin{bmatrix} a_r & a_{r-1} & \ldots & a_0 \end{bmatrix}^t$. We now search a solution a that corresponds to the characteristic equation of a matrix with elements in $\mathbb{R}_{\max}$ (this should not necessarily be a signed solution). First of all we normalize a_0 to 0 and then we check if the necessary (and sufficient) conditions of §3 for α_p and β_p are satisfied, where $\alpha_p = a_p^+$ and $\beta_p = a_p^-$. If they are not satisfied we augment r and repeat the procedure.
We continue until we get the following stable relation among the columns of $H_{p,q}$:

$$H_{p,q}(:, k+r) \oplus a_1 \otimes H_{p,q}(:, k+r-1) \oplus \ldots \oplus a_r \otimes H_{p,q}(:, k) \nabla \epsilon \tag{8}$$

for $k \in \{1, \ldots, q-r\}$. Since we assumed that the system can be described by (4)–(5) and that $p, q \gg n$, we can always find such a stable relationship by gradually augmenting r. The r that results from this procedure is a lower bound for the minimal system order.

4.3 Determination of the System Matrices

In [5] we have described a method to find all solutions of a set of multivariate polynomial equalities in the max algebra. Now we can use this method to find the A, b and c matrices of an r-th order SISO system with Markov parameters $g_0, g_1, g_2, \ldots$. If the algorithm doesn't find any solutions, this means that the output behavior can't be described by an r-th order SISO system. In that case we have to augment our estimate of the system order and repeat the procedure. Since we assume that the system can be described by the state space model (4)–(5) we shall always get a minimal realization.

However, in many cases we can use the results of the previous section to find a minimal realization. Starting from the coefficients $a_1, a_2, \ldots, a_r$ of (8) we search a matrix A with elements in $\mathbb{R}_{\max}$ such that its characteristic equation is

$$\lambda^{\otimes^n} \oplus \bigoplus_{p=1}^{r} a_p \otimes \lambda^{\otimes^{r-p}} \nabla \epsilon \ . \tag{9}$$

Once we have found the A matrix, we have to find b and c with elements in $\mathbb{R}_{\max}$ such that

$$c \otimes A^{\otimes^k} \otimes b = g_k \qquad \text{for } k = 0, 1, 2, \ldots \ . \tag{10}$$

In practice it seems that we only have to take the transient behavior and the first cycles of the steady-state behavior into account. So we may limit ourselves to the first, say, N Markov parameters.

Let's take a closer look at equations of the form $c \otimes R \otimes b = s$ with $c \in \mathbb{R}_{\max}^{1 \times n}$, $R \in \mathbb{R}_{\max}^{n \times n}$, $b \in \mathbb{R}_{\max}^{n \times 1}$ and $s \in \mathbb{R}_{\max}$. This equation can be rewritten as

$$\bigoplus_{i=1}^{n} \bigoplus_{j=1}^{n} c_i \otimes r_{ij} \otimes b_i = s \ . \tag{11}$$

So if we take the first N Markov parameters into account, we get a set of N multivariate polynomial equations in the max algebra, with the elements of b and c as unknowns and $R = A^{\otimes^{k-1}}$ and $s = g_{k-1}$ in the k-th equation. This problem can also be solved using the algorithm described in [5].

However, one has to be careful since it is not always possible to find a b and a c for every matrix that has (9) as its characteristic equation (see [3] for an example). In that case we have to search another A matrix or we could fall back on the method described in [4, 5], which finds all possible minimal realizations.

5 Example

We now illustrate the procedure of the preceding section with an example.

Example 1. Here we reconsider the example of [2, 8]. We start from a system with system matrices

$$A = \begin{bmatrix} 1 & -1 & 2 \\ -1 & 2 & 0 \\ -3 & 1 & 2 \end{bmatrix}, \quad b = \begin{bmatrix} 0 \\ \epsilon \\ \epsilon \end{bmatrix} \quad \text{and} \quad c = \begin{bmatrix} 0 & \epsilon & \epsilon \end{bmatrix} \ .$$

Now we are going to construct the system matrices from the impulse response of the system. This impulse response is given by $\{g_k\} = 0, 1, 2, 3, 4, 5, 6, 8, 10, 12, 14, \ldots$. First we construct the Hankel matrix

$$H_{8,8} = \begin{bmatrix} 0 & 1 & 2 & 3 & 4 & 5 & 6 & 8 \\ 1 & 2 & 3 & 4 & 5 & 6 & 8 & 10 \\ 2 & 3 & 4 & 5 & 6 & 8 & 10 & 12 \\ 3 & 4 & 5 & 6 & 8 & 10 & 12 & 14 \\ 4 & 5 & 6 & 8 & 10 & 12 & 14 & 16 \\ 5 & 6 & 8 & 10 & 12 & 14 & 16 & 18 \\ 6 & 8 & 10 & 12 & 14 & 16 & 18 & 20 \\ 8 & 10 & 12 & 14 & 16 & 18 & 20 & 22 \end{bmatrix} .$$

The determinant of $H_{\text{sub},2} = H_{8,8}([1,7],[1,2]) = \begin{bmatrix} 0 & 1 \\ 6 & 8 \end{bmatrix}$ is not balanced. We add the second row and the third column and then we search a solution of the set of linear balances

$$\begin{bmatrix} 0 & 1 & 2 \\ 1 & 2 & 3 \\ 6 & 8 & 10 \end{bmatrix} \otimes \begin{bmatrix} a_2 \\ a_1 \\ a_0 \end{bmatrix} \nabla \epsilon .$$

The solution $a_0 = 0$, $a_1 = \ominus 2$, $a_2 = 3$ satisfies the necessary and sufficient conditions for the 2 by 2 case since $\alpha_1 = \epsilon$ and $\alpha_2 = 3 \leq 4 = 2 \otimes 2 = \beta_1 \otimes \beta_1$. This solution also corresponds to a stable relation among the columns of $H_{8,8}$:

$$H_{8,8}(:, k+2) \oplus 3 \otimes H_{8,8}(:, k) = 2 \otimes H_{8,8}(:, k+1) ,$$

for $k \in \{1, 2, \ldots, 6\}$, or to the following characteristic equation:

$$\lambda^{\otimes^2} \ominus 2 \otimes \lambda \oplus 3 \nabla \epsilon .$$

This leads to a second order system with $A = \begin{bmatrix} 2 & \epsilon \\ 0 & 1 \end{bmatrix}$. Using the technique of [5] we get a whole set of solutions for b and c. One of the solutions is $b = \begin{bmatrix} -4 \\ 0 \end{bmatrix}$ and $c = \begin{bmatrix} \epsilon & 0 \end{bmatrix}$. Apart from a permutation of the two state variables this result is the same as that of [8].

Another example, that doesn't satisfy the assumptions of [8] – where only impulse responses that exhibit a uniformly up-terrace behavior are considered –, can be found in [3].

6 Conclusions and Future Research

We have derived necessary and for some cases also sufficient conditions for an $\mathbb{S}_{\max}$ polynomial to be the characteristic polynomial of an $\mathbb{R}_{\max}$ matrix. So if we have a monic polynomial in $\mathbb{S}_{\max}$ these results allow us

1. to check whether the given polynomial can be the characteristic polynomial of a positive matrix and

2. to construct a matrix such that its characteristic polynomial is equal to the given polynomial.

Based on these results we have proposed a procedure to find a minimal state space realization of a SISO system, given its Markov parameters. This procedure is an alternative to the method of [4], which finds all possible minimal realizations but which has one disadvantage: its computational complexity. Since we allow a Hessenberg form for the system matrix A, our method incorporates both the companion form of [2] and the bidiagonal form of [8].

References

1. F. Baccelli, G. Cohen, G.J. Olsder, and J.P. Quadrat, *Synchronization and Linearity*. New York: John Wiley & Sons, 1992.
2. R.A. Cuninghame-Green, "Algebraic realization of discrete dynamic systems," in *Proceedings of the 1991 IFAC Workshop on Discrete Event System Theory and Applications in Manufacturing and Social Phenomena*, Shenyang, China, pp. 11–15, June 1991.
3. B. De Schutter and B. De Moor, "Minimal state space realization of SISO systems in the max algebra," Technical Report 93-57, ESAT/SISTA, Katholieke Universiteit Leuven, Kardinaal Mercierlaan 94, B-3001 Leuven, Belgium, 1993.
4. B. De Schutter and B. De Moor, "Minimal realization in the max algebra is an extended linear complementarity problem," Technical Report 93-70A, ESAT/SISTA, Katholieke Universiteit Leuven, Kardinaal Mercierlaan 94, B-3001 Leuven, Belgium, 1993. Submitted to Systems & Control Letters.
5. B. De Schutter and B. De Moor, "A method to find all solutions of a set of multivariate polynomial equalities and inequalities in the max algebra," Technical Report 93-71, ESAT/SISTA, Katholieke Universiteit Leuven, Kardinaal Mercierlaan 94, B-3001 Leuven, Belgium, 1993. Submitted to Discrete Event Dynamic Systems: Theory and Applications.
6. S. Gaubert, *Théorie des Systèmes Linéaires dans les Dioïdes*. PhD thesis, Ecole Nationale Supérieure des Mines de Paris, France, July 1992.
7. G.J. Olsder and R.E. de Vries, "On an analogy of minimal realizations in conventional and discrete-event dynamic systems," in *Proceedings of the CNRS/CNET/INRIA Seminar, Algèbres Exotiques et Systèmes à Evénements Discrets*, INRIA, Rocquencourt, France, June 1987.
8. L. Wang and X. Xu, "On minimal realization of SISO DEDS over max algebra," in *Proceedings of the 2nd European Control Conference*, Groningen, The Netherlands, pp. 535–540, June 1993.

A Max-Algebra Solution to the Supervisory Control Problem for Real-time Discrete Event Systems

Darren D. Cofer and Vijay K. Garg

Department of Electrical and Computer Engineering
University of Texas, Austin, TX 78712-1084, USA

1 Introduction

To date, work using max-algebra to model timed DES has concentrated on performance analysis rather than control. Indeed, this is true for much of the work on real-time systems and timed DES in general. By contrast, a well-developed theoretical framework (namely that of Ramadge and Wonham) already exists for controlling untimed DES. With this motivation, our work here shows how a similar approach to control may be applied to timed DES modelled in max-algebra.

Suppose that we are given a system having some events whose occurrence is controllable, along with a specification for the desired system performance. This specification is given as a range of acceptable execution times for events. Events are controllable by imposing delays on their execution. The problem is to determine whether the system can be controlled to meet the specification and if so, how the controllable events should be regulated to meet that specification.

2 Model and Motivation

DES which are subject to time synchronization constraints can be modelled by automata known as *timed event graphs*. A timed event graph is a Petri net in which a time delay is associated with each place and with forks and joins permitted only at its transitions. Each transition $t_i \in T$ in the graph corresponds to an event in the system. In [1] and [3] these systems are modelled using a *dioid* algebra.

The dynamic behavior of such systems can also be studied by a max-algebra structure called a *moduloid*. A moduloid over a dioid D is a commutative idempotent monoid M with binary operation $\oplus$ together with an operation called scalar multiplication mapping $D \times M \to M$. Scalar multiplication distributes over the sum ($\oplus$) in both D and M, is associative with the product in D, and has the same identity element as D. We let ε denote the null element for $\oplus$ in M. M is *complete* if it is closed for arbitrary $\oplus$ operations. Properties of moduloids are discussed in [6] and [1].

Definition 1. A function $f : M \to M$ is called *lower-semicontinuous (l.s.c.)* [1] if for any collection of elements $\{X_j \in M \mid j \in J\}$

$$\bigoplus_{j \in J} f(X_j) = f\left(\bigoplus_{j \in J} X_j\right).$$

Proposition 2. *Let F be the set of l.s.c. functions on M with the operations of function maximization and composition. Then F is a dioid. If M is complete, so is F. Furthermore, if scalar multiplication is defined to be the action of functions in F on M, then M is a moduloid over F [1], [2].*

Using the moduloid structure the behavior of a timed event graph is governed by the equation

$$X = AX \oplus B \tag{1}$$

where X is the sequence of firing time vectors for events, B is a sequence of initial firing time vectors, and A is a matrix of delay functions at places. The least solution of (1) is $X = A^*B$ where $A^* = \bigoplus_{i \geq 0} A^i$. The set of vector sequences forms an idempotent commutative monoid under pointwise maximization. The delay functions are of the form $a\gamma^m$, where γ is the index backshift function, which is easily shown to be l.s.c. It is this framework that we use throughout the remainder of the paper. This structure can also be used to model systems more general than timed event graphs, including some untimed DES [2].

To motivate our approach to controlling timed DES, we look to the control framework established by Ramadge and Wonham for untimed DES [5]. In untimed models of DES, system behavior may be described by the sequences of events performed by the system, normally taken to be a finite state machine. The *language* L of a system is the set of all event sequences it can generate. Events are classified as *controllable*, meaning that their occurrence may be prevented, or *uncontrollable*. Control is accomplished by disabling certain of the controllable events to avoid undesirable behaviors. Observe that such a controller or supervisor may restrict system behavior, but not introduce any new behaviors.

Definition 3. Language K is said to be *controllable* with respect to L and the uncontrollable events Σ_u if

$$\mathrm{pr}(K).\Sigma_u \cap L \subseteq \mathrm{pr}(K) \tag{2}$$

where $\mathrm{pr}(k)$ is the *prefix closure* of K.

If K is not controllable, we can optimize performance by finding the least restrictive controllable behavior which is a subset of K, called the *supremal controllable sublanguage*.

3 Supervisory Control of Timed DES

Now return to the max-algebra model of a timed event graph governed by (1). Suppose that some events $T_c \subseteq T$ are designated as controllable, meaning that their transitions may be delayed from firing until some arbitrary later time. This is similar in its mechanics to the controlled Petri net concept introduced in [4] for untimed DES. The delayed enabling times $u_i(k)$ for the controllable events are to be provided by a supervisor. Let U represent the sequence of transition enabling times provided by the supervisor, with $u_i(k) = \varepsilon$ for t_i uncontrollable. Then the supervised system is described by $X = AX \oplus B \oplus U$.

Notice that as in the untimed model no new behavior is introduced by the supervisor in the sense that system operation can never be accelerated — events can only be delayed. Suppose then that we wish to slow the system down as much as possible without causing any event to occur later than some sequence of execution times Y. Such a specification could be used to prevent buffer overflows, ensure the availability of sufficient processing time to accomplish a task, or to synchronize events in independent systems.

Consider the region of acceptable behaviors given by

$$\mathcal{Y} = \{Z \in M \mid Z \leq Y\}. \tag{3}$$

A natural supervisory action to consider is to delay the controllable events as much as allowed by Y. That is, let $U_i = Y_i$ for $t_i \in T_c$. However, even if all the transitions are controllable we cannot guarantee that the system will be confined to $\mathcal{Y}$. This is because the sequence Y does not necessarily account for the delays between successive events within the system.

To account for the interconnecting delays within the system, we define the *closure* of a sequence Y by A^*Y. Algebraically A^*Y is the solution to (1) with the restriction that no events occur earlier than permitted by Y. Clearly the uncontrolled behavior $X = A^*B$ of the system is closed.

Example 1. Consider the system

$$\begin{bmatrix} X_1 \\ X_2 \end{bmatrix} = \begin{bmatrix} 2\gamma & \varepsilon \\ 3 & \varepsilon \end{bmatrix} \begin{bmatrix} X_1 \\ X_2 \end{bmatrix} \oplus \begin{bmatrix} 0 \\ 0 \end{bmatrix}$$

with the specification sequence

$$\begin{bmatrix} Y_1 \\ Y_2 \end{bmatrix} = \left\{ \begin{bmatrix} 1 \\ 3 \end{bmatrix}, \begin{bmatrix} 3 \\ 5 \end{bmatrix}, \begin{bmatrix} 5 \\ 7 \end{bmatrix}, \ldots \right\}.$$

Even if event t_2 (corresponding to sequence X_2) is controllable it cannot fire for the kth time until $3y_2(k)$ which is always greater than $y_2(k)$. If we allow the system to go as slow as Y then we must be willing to accept it going as slow as

$$A^*Y = \left\{ \begin{bmatrix} 1 \\ 4 \end{bmatrix}, \begin{bmatrix} 3 \\ 6 \end{bmatrix}, \begin{bmatrix} 5 \\ 8 \end{bmatrix}, \ldots \right\}.$$

Let us now consider exactly what it means for a behavior to be controllable. To restrict the system to a given acceptable region, uncontrollable actions must not result in behaviors which lie outside of the acceptable region. Furthermore, we must permit the system to execute behaviors which lead to the acceptable region, even though they may not lie in that region themselves. Therefore, these potentially acceptable behaviors as well as the region of acceptable behavior should be invariant or closed under uncontrollable actions. For untimed DES the acceptable and potentially acceptable behaviors are given by $K \cup \mathrm{pr}(K) = \mathrm{pr}(K)$. Since uncontrollable events in the system act by concatenation we have the controllability condition stated in (2).

For timed DES with acceptable behavior $\mathcal{Y}$ as in (3), the potentially acceptable sequences are $\mathcal{Z} = \{Z \in M \mid A^*Z \in \mathcal{Y}\}$. In this case $\mathcal{Z} \subseteq \mathcal{Y}$, but this may not be true for different $\mathcal{Y}$. To compute the effect of uncontrollable events, let I_c denote the matrix having the identity function on diagonal elements i for which $t_i \in T_c$ and ε elsewhere. Then for any desired sequence Y the supervisor provides firing times $U = I_cY$ which results in $X = A^*(I_cY \oplus B)$.

Definition 4. A set of sequences $\mathcal{Y}$ is *controllable* with respect to A, B, and T_c if

$$A^*(I_c\mathcal{Y} \oplus B) \subseteq \mathcal{Y}$$

Proposition 5. *If $\mathcal{Y}$ specifies sequences no later than Y as in (3) then $\mathcal{Y}$ is controllable if and only if*

$$A^*I_cY \oplus X \leq Y \tag{4}$$

where $X = A^*B$ is the uncontrolled system behavior.

Example 2. For the system in Example 1 with t_1 controllable, suppose we are given as a specification sequences less than or equal to

$$Y = \left\{ \begin{bmatrix} 0 \\ 3 \end{bmatrix}, \begin{bmatrix} 3 \\ 5 \end{bmatrix}, \begin{bmatrix} 6 \\ 7 \end{bmatrix}, \ldots \right\}.$$

To determine if Y is controllable, we compute

$$\begin{aligned} A^*I_cY &= \begin{bmatrix} (2\gamma)^* & \varepsilon \\ 3(2\gamma)^* & 0 \end{bmatrix} \begin{bmatrix} Y_1 \\ \varepsilon \end{bmatrix} \\ &= \left\{ \begin{bmatrix} 0 \\ 3 \end{bmatrix}, \begin{bmatrix} 3 \\ 6 \end{bmatrix}, \begin{bmatrix} 6 \\ 9 \end{bmatrix}, \ldots \right\} > Y \end{aligned}$$

so Y is not controllable. □

Proposition 6. *If $Y \geq B$ then A^*Y is a controllable sequence.*

Example 3. Consider Y from Example 2. Then

$$A^*Y = \left\{ \begin{bmatrix} 0 \\ 3 \end{bmatrix}, \begin{bmatrix} 3 \\ 6 \end{bmatrix}, \begin{bmatrix} 6 \\ 9 \end{bmatrix}, \ldots \right\}$$

$$\text{and } A^*I_c(A^*Y) \oplus X = \left\{ \begin{bmatrix} 0 \\ 3 \end{bmatrix}, \begin{bmatrix} 3 \\ 6 \end{bmatrix}, \begin{bmatrix} 6 \\ 9 \end{bmatrix}, \ldots \right\}$$

so A^*Y is controllable. □

If specification Y is uncontrollable this means that some event t_i cannot be guaranteed to occur before Y_i given that all the other events could occur as late as allowed by Y. Following the pattern of the untimed model, we would like to find the analog of a supremal controllable sublanguage. In the context of timed DES this would be the greatest controllable firing sequence specification which is less than or equal to Y. Call this the *supremal controllable subsequence* of Y, denoted $Y^{\uparrow}$.

Proposition 7. *Given any firing sequence $Y \geq X$, $Y^{\uparrow}$ exists and is unique.*

Proof. Let $C(Y)$ be the set of all controllable sequences less than or equal to Y. It is easy to show that $C(Y)$ is nonempty and closed under $\oplus$. Since M is complete $C(Y)$ has unique supremal element $\bigoplus_{Z \in C(Y)} Z$. □

4 Example: Cat and Mouse

To illustrate the control concepts introduced here, we propose the following problem. Suppose a cat chases a mouse through the three-room house in Fig. 1. One of the doors can be held shut to prevent the cat from entering the next room. The mouse's movement may not be restricted. Initially the cat and mouse are at opposite ends of room 3. Our objective is to shut the controllable door at the proper times to keep the cat from catching the mouse.

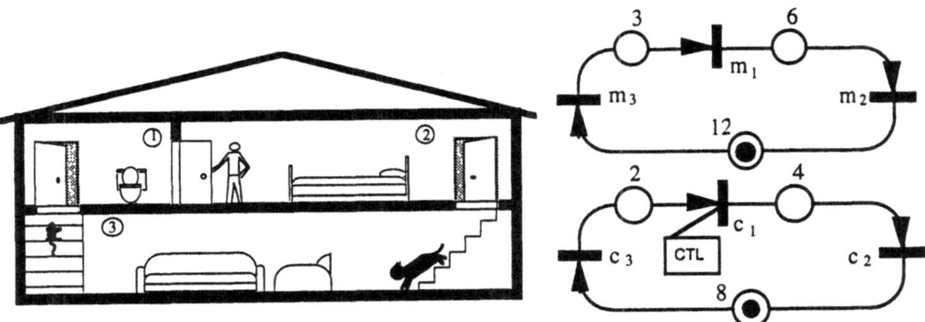

Fig. 1. Timed event graphs for cat and mouse

Timed event graphs for the cat and mouse are also shown in Fig. 1. Event m_1 is the mouse leaving room 1, and so on. The controllable event is c_1. The initial condition gives the mouse a headstart since it can exit room 1 immediately but the cat will take 8 seconds to cross the room and exit. The mouse is governed by

$$\begin{bmatrix} m_1 \\ m_2 \\ m_3 \end{bmatrix} = \begin{bmatrix} \varepsilon & \varepsilon & 3 \\ 6 & \varepsilon & \varepsilon \\ \varepsilon & 12\gamma & \varepsilon \end{bmatrix} \begin{bmatrix} m_1 \\ m_2 \\ m_3 \end{bmatrix} \oplus \begin{bmatrix} 0 \\ 0 \\ 0 \end{bmatrix} \tag{5}$$

and the cat by

$$\begin{bmatrix} c_1 \\ c_2 \\ c_3 \end{bmatrix} = \begin{bmatrix} \varepsilon & \varepsilon & 2 \\ 4 & \varepsilon & \varepsilon \\ \varepsilon & 8\gamma & \varepsilon \end{bmatrix} \begin{bmatrix} c_1 \\ c_2 \\ c_3 \end{bmatrix} \oplus \begin{bmatrix} 0 \\ 0 \\ 8 \end{bmatrix}.$$

Since these systems are not connected we have simply

$$\begin{bmatrix} X_m \\ X_c \end{bmatrix} = \begin{bmatrix} M & \varepsilon \\ \varepsilon & C \end{bmatrix} \begin{bmatrix} X_m \\ X_c \end{bmatrix} \oplus \begin{bmatrix} B_m \\ B_c \end{bmatrix}.$$

Our objective of sparing the mouse is equivalent to the following two specifications:

(1) The cat must leave each room after the mouse does so that the mouse stays ahead of the cat for all time. This is expressed in terms of the event occurrence times by $X_c > X_m$.

(2) The mouse must not enter a room until the cat has left. This means that the mouse must not "lap" the cat; otherwise, the cat could turn around (within the same room) and catch it. Then we must have

$$X_c \leq \begin{bmatrix} \varepsilon & \varepsilon & \gamma^{-1} \\ \gamma^{-1} & \varepsilon & \varepsilon \\ \varepsilon & 0 & \varepsilon \end{bmatrix} X_m = \begin{bmatrix} \gamma^{-1} m_3 \\ \gamma^{-1} m_1 \\ m_2 \end{bmatrix} = Y. \tag{6}$$

We would also like to let the mouse get as far ahead of the cat as possible consistent with the specifications. Our strategy will be to find the supremal controllable subsequence for specification (2) and then check if that sequence also satisfies specification (1).

Specification (2) can obviously be met if the vector sequence Y in (6) is controllable with respect to the cat system. We check the controllability of Y by computing

$$C^* I_c Y \oplus X_c = \begin{bmatrix} \gamma^{-1} m_3 \\ 4\gamma^{-1} m_3 \\ 12 m_3 \end{bmatrix} > \begin{bmatrix} \gamma^{-1} m_3 \\ 3\gamma^{-1} m_3 \\ 9 m_3 \end{bmatrix} = Y$$

where the last equality follows from the state equation of X_m (5). Therefore Y is not controllable. This implies that we cannot delay the cat as much as allowed by the specification. Therefore we must reduce the delay imposed on the cat at the controllable door by finding $Y^\dagger$.

Notice that the worst deviation from controllability occurs for event c_3 which needs to be less than or equal to $9m_3$. For this simple system it is fairly straightforward to determine from the C^* matrix that event c_3 is linearly related to event c_1. Consequently, a "gain" of $9m_3/12m_3 = -3$ applied at c_1 will attenuate c_3 by the same amount resulting in $c_3 = 9m_3$ as desired.

We must next check if this new specification is controllable. Let $Z = \begin{bmatrix} -3\gamma^{-1} m_3 & \gamma^{-1} m_1 & m_2 \end{bmatrix}^T$ and compute

$$C^* I_c Z \oplus X_c = \begin{bmatrix} -3\gamma^{-1} m_3 \\ 1\gamma^{-1} m_3 \\ 9 m_3 \end{bmatrix} < \begin{bmatrix} -3\gamma^{-1} m_3 \\ 3\gamma^{-1} m_3 \\ 9 m_3 \end{bmatrix} = Z.$$

Therefore Z is controllable. Note that Z differs from Y only in the first component. If this value were chosen any higher then we would have $C^* I_c Z \oplus X_c > Z$ in the third component and Z would not be controllable. Thus $Z = Y^{\uparrow}$. [1]

Finally, we see that this control delays the cat sufficiently to meet the first part of the specification since

$$X_c = \begin{bmatrix} -3\gamma^{-1}m_3 \\ 1\gamma^{-1}m_3 \\ 9m_3 \end{bmatrix} = \begin{bmatrix} 15m_1 \\ 13m_2 \\ 9m_3 \end{bmatrix} > X_m.$$

Therefore there exists a supervisor S for the cat and mouse system which meets both of our specifications which is given by $c_1 = 15m_1$, resulting in the controlled system

$$\begin{bmatrix} X_m \\ X_c \end{bmatrix} = \begin{bmatrix} M & \varepsilon \\ S & C \end{bmatrix} \begin{bmatrix} X_m \\ X_c \end{bmatrix} \oplus \begin{bmatrix} B_m \\ B_c \end{bmatrix}.$$

5 Conclusion

For timed systems the supervisory control problem is to impose delays on controllable events to modify system behavior to meet some specified performance goal. Using the tools of max-algebra it is possible to compute the uncontrolled behavior of a timed event graph, define a specification for some new desired behavior, and determine whether the specification can be realized by any supervisor given the set of controllable events. When the desired behavior cannot be realized (i.e, it is uncontrollable), a minimally restrictive behavior which meets the original specification always exists. All of these concepts have direct analogy in the control of untimed automata.

References

1. F. Baccelli, G. Cohen, G. J. Olsder, J. P. Quadrat, *Synchronization and Linearity*, Wiley, New York, 1992.
2. D. D. Cofer, V. K. Garg, "A Generalized Max-algebra Model for Timed and Untimed DES," T. R. SCC-93-02, Dept. of ECE, Univ. of Texas, Austin, 1993.
3. G. Cohen. P. Moller, J. P. Quadrat, M. Viot, "Algebraic tools for the performance evaluation of DES," *Proc. IEEE*, 77(1):39–58, 1989.
4. B. H. Krogh, "Controlled Petri Nets and Maximally Permissive Feedback Logic,"*Proc. 25th Allerton Conf. on Comm., Ctrl. and Comp.*, Urbana, IL, 1987.
5. P. J. Ramadge, W. M. Wonham, "The control of DES," *Proc. IEEE*, 77(1):81–98, 1989.
6. E. Wagneur, "Moduloids and pseudomodules," *Discrete Math.*, 98:57–73, 1991.

[1] In general, $Y^{\uparrow}$ can be found using a fixed point algorithm, similar to the approach in [5]. However, this is beyond the scope of the present paper.

Stable Earliest Starting Schedules for Periodic Job Shops: a Linear System Approach

Tae-Eog Lee

Department of Industrial Engineering
Korea Advanced Institute of Science and Technology (KAIST)
373-1 Gusung-Dong, Yusong-Gu, Taejon 305-701, Korea
E-mail: telee@sorak.kaist.ac.kr

1 Introduction

We consider a periodic job shop that repeatedly produces an identical, small set of items. The smallest such set that is taken to have the same proportion of items as the production requirement is called *minimal part set* (MPS). The operations of each MPS is assigned to the machines in an identical pattern. The processing order at each machine is kept same for each MPS.

A performance measure for periodic shops is the throughput rate or its reciprocal, cycle time. The *cycle time* is the average time for which each item (or equivalently MPS) is produced. Another important measure is the *MPS makespan* that is the makespan to complete an MPS. Minimizing the MPS makespan has each MPS kept in process for the minimum possible time and hence minimizes the WIP.

A schedule that repeats an identical timing pattern each MPS is called a *stable schedule* and leads to steady production. When each operation is often started as soon as all its preceding operations complete, the starting times of the operations need not to be computed prior to actual production and the deliberate timing control is not necessary. A schedule that is formed by such a starting policy is called an *earliest starting schedule*.

Therefore, the most desirable schedule would be a schedule that is both stable and earliest starting while the cycle time and the MPS makespan are kept as the minimum. Lee and Posner (1990) show that for a given processing order at each machine, under some conditions, there can exist a lot of stable earliest starting schedules (called *SESS*) with only a few beginning operations delayed appropriately and that such a SESS has the minimum cycle time over all the feasible schedules. Hence, we are interested in determining a schedule among all SESSs that minimizes the MPS makespan, denoted by SESS*.

In this paper, we characterize the set of all SESSs in a periodic job shop using the linear system model based on the minimax algebra and develop a polynomially bounded algorithm that constructs a SESS*. To do this, instead of using the deliberately derived linear system matrix or graph, we directly characterize the eigenvalue and eigenvectors using the original event modeling graph. To focus on the issue, we assume that the processing order of the operations at each machine is given. We further assume non-preemptive processing of operations, buffers with infinite capacity, known processing times, negligible setup and transportation times, and no disruptive events like machine breakdowns.

2 A Directed Graph Model of Periodic Job Shops

We introduce a directed graph model of periodic job shops. We let N denote the set of the processing operations for the items that comprise the MPS. The processing time of operation $i \in N$ is denoted by p_i. Let M be the set of machines. Each machine repeats an identical sequence of production operations. The immediate processing precedences between the operations of the MPS are represented by a set of ordered pairs of operations, E, where $(i, j) \in E$ if operation i precedes operation j. The precedence relations come from the technologically required processing precedence of the operations of each item and the presumably appropriately determined processing order of the operations assigned to each machine.

To represent the cyclic processing order at each machine, we introduce some notation. We denote by $\alpha(m)$ and $\beta(m)$ the first and the last operation of the MPS on machine m, respectively. For the determined machine processing order, we let R be the set of the ordered pair of the last operation and the first operation on each machine, $\{(\beta(m), \alpha(m)) \mid m \in M\}$. Each $(i, j) \in R$ implies that after the completion of the last operation i of an MPS on a given machine, the first operation j of the next MPS on the same machine starts.

To model the processing precedences, we construct a directed graph $G(N, E \cup R)$ with node set N and arc set $E \cup R$. Associated with each arc $(i, j) \in E \cup R$ are two weights. The first weight, p_i, represents the processing time of operation i. We sometimes refer to this weight as the "length" between i and j. The second weight is 1 if $(i, j) \in R$ and 0 otherwise. Since the second weights correspond to the advance of MPSs, we call the arcs corresponding to elements of R *recycling arcs*. The graph $G(N, E \cup R)$ with the two weights is called *precedence constraints graph* (PCG) by Lee and Posner (1990). We note that the PCG is regarded as an *event modeling graph* and sometimes called an *event graph*.

3 The Linear System Model

We now develop a linear system model. Let x_i^r denote the starting time of the r-th occurrence of operation i (that is, of the r-th MPS). A schedule is specified by the earliest starting times, $\{x_i^r \mid i \in N, r = 1, 2, \dots\}$. If $x_i^{r+d} - x_i^r = d\mu, i \in N$ for all $r \geq r_0$, some real constant μ, and a positive integer d, then the system is said to attain a *periodical steady state*. Especially, when $d = 1$, the system is said to attain a *stable steady state*. After the system attains a stable steady state, an identical MPS schedule pattern (that corresponds to the stable steady state) repeats every MPS. Once a stable steady state is known, we can construct a SESS by letting the first MPS schedule have the repeating MPS schedule pattern. Therefore, a stable steady state defines a SESS.

The earliest starting times $\{x_i^r \mid i \in N, r \geq 1\}$ satisfy the following dynamic programming equation;

$$x_j^{r+1} = \begin{cases} \max(\max_l(x_l^{r+1} + a_{lj}), x_i^r + p_i) & \text{if } (i,j) \in R, \\ \max_l(x_l^{r+1} + a_{lj}) & \text{otherwise,} \end{cases}$$

where $a_{ij} = p_i$ if $(i, j) \in E$ and $a_{ij} = -\infty$ otherwise.

Cunninghame-Green (1979) proposes the minimax algebra $(\overline{\mathbb{R}}, \oplus, \otimes)$ where $\oplus$ and $\otimes$ are algebraic operations on $\mathbb{R} \cup \{-\infty, \infty\} (\equiv \overline{\mathbb{R}})$ such that for $a, b \in \overline{\mathbb{R}}$, $a \otimes b \equiv a + b$ and $a \oplus b \equiv \max\{a, b\}$. Using the algebra, the dynamic programming equation is represented in a matrix form:

$$\mathbf{x}^{r+1} = \mathbf{x}^{r+1} \otimes \mathbf{A} \oplus \mathbf{x}^r \otimes \mathbf{B},$$

where $\mathbf{x}^r \equiv (x_1^r, x_2^r, \ldots, x_j^r, \ldots, x_{|N|}^r)$, $\mathbf{A} \equiv (a_{ij})_{|N|\times|N|}$, and $\mathbf{B} \equiv (b_{ij})_{|N|\times|N|}$ such that $b_{ij} = p_i$ if $(i, j) \in R$ and $b_{ij} = -\infty$ otherwise.

The matrix equation can be solved for $\mathbf{x}^{r+1}$;

$$\begin{aligned} \mathbf{x}^{r+1} &= \mathbf{x}^{r+1} \otimes \mathbf{A} \oplus \mathbf{x}^r \otimes \mathbf{B} \\ &= (\mathbf{x}^{r+1} \otimes \mathbf{A} \oplus \mathbf{x}^r \otimes \mathbf{B}) \otimes \mathbf{A} \oplus \mathbf{x}^r \otimes \mathbf{B} \\ &= (\ldots((\mathbf{x}^{r+1} \otimes \mathbf{A} \oplus \mathbf{x}^r \otimes \mathbf{B}) \otimes \mathbf{A} \oplus \mathbf{x}^r \otimes \mathbf{B})\ldots) \otimes \mathbf{A} \oplus \mathbf{x}^r \otimes \mathbf{B} \\ &= \mathbf{x}^{r+1} \otimes \mathbf{A}^n \oplus \mathbf{x}^r \otimes \mathbf{B} \otimes (\mathbf{A}^n \oplus \mathbf{A}^{n-1} \oplus \mathbf{A}^{n-2} \oplus \cdots \oplus \mathbf{I}), \end{aligned}$$

where $\mathbf{I}$ is the *identity matrix* such that all diagonal elements are 0 and all others are $-\infty$. Since $G(N, E)(= G(\mathbf{A}))$ does not have any circuit, every element of $\mathbf{A}^r$ is $-\infty$ for all $r > |N| - 1$ (Carré 1971). Therefore, $\bigoplus_{r=0,\ldots,\infty} \mathbf{A}^r$ converges to a matrix $\mathbf{A}^* \equiv \mathbf{I} \oplus \mathbf{A} \oplus \mathbf{A}^2 \oplus \cdots \oplus \mathbf{A}^{|N|-1}$. By letting $\widehat{\mathbf{A}} = \mathbf{B} \otimes \mathbf{A}^*$, we have a linear system for the starting times of operations for each MPS;

$$\mathbf{x}^{r+1} = \mathbf{x}^r \otimes \widehat{\mathbf{A}} \ .$$

The linear system model is considered as a simplified version of that of Cohen et al. (1985), where part release decisions are incorporated into the model.

4 Characterization of Stable Earliest Starting Schedules

To investigate SESSs, we characterize stable steady states of the linear system by identifying eigenvalues and eigenvectors $(\mathbf{x}, \lambda)$ such that $\mathbf{x} \otimes \widehat{\mathbf{A}} = \mathbf{x} \otimes \lambda$.

A circuit (cycle) that does not contain any smaller circuit is called an *elementary circuit*. For each circuit, the *circuit ratio* is the ratio of the sum of the first arc weights to the sum of the second arc weights. The *critical circuit ratio* of the graph is the maximum of the circuit ratios over all elementary circuits. We let λ denote the critical circuit ratio of a given PCG. An elementary circuit with the maximum circuit ratio is called a *critical circuit.*

Cunninghame-Green (1979) discusses eigenvalue properties of matrices on $(\overline{\mathbb{R}}, \oplus, \otimes)$. For a given matrix $\mathbf{H} \equiv (h_{ij})_{|N|\times|N|}$, an associated graph $G(\mathbf{H})$ is defined to have an arc (i, j) with the first weight h_{ij} and the second weight 1 when $h_{ij} \neq -\infty$. In view of the first weights of $G(\mathbf{H})$, matrix $\mathbf{H}$ is considered as the incidence matrix of $G(\mathbf{H})$. For a circuit in $G(\mathbf{H})$, the circuit ratio is defined to be the ratio of the sum of the lengths (the first weights) of the arcs in the circuit to the number of arcs (the sum of the second weights). Matrix $\Gamma(\mathbf{H})$ is defined as $\bigoplus_{r=1,\ldots,|N|} \mathbf{H}^r$. He also shows that for a finite eigenvalue h the row vector of $\Gamma(h^{-1} \otimes \mathbf{H})$ is an eigenvector of $\mathbf{H}$ and calls such an eigenvector a *fundamental* eigenvector.

We introduce a modification of the PCG that is proposed for some other purpose by Lee and Posner (1990). For the PCG with the critical circuit ratio λ, the first weight of each arc (i, j) in the PCG is replaced by c_{ij} where $c_{ij} = p_i$ if $(i, j) \in E'$ and $c_{ij} = p_i - \lambda$ if $(i, j) \in R$. This new graph is denoted by PCG$'$. It can be shown that the PCG$'$ does not have any circuit of positive length. The longest path from node i to j is denoted by γ_{ij}. Let N_α be $\{i \in N \mid i = \alpha(m), m \in M\}$, the set of the beginning operations and N_β be $\{i \in N \mid i = \beta(m), m \in M\}$, the set of last operations. A maximal strongly connected subgraph of the PCG is called a *component*. For the PCG, when a component contains a critical circuit, it is called *critical*. A component that has no preceding component is called a *root component*.

It can be shown that there is one-to-one correspondence between the circuits in the PCG and $G(\widehat{\mathbf{A}})$. Using the fact, we can establish that the linear system matrix has a finite eigenvalue and a finite corresponding eigenvector and apply the linear system theory of Cunninghame-Green (1979) to our model.

Theorem 1. *If every root component of the PCG is critical, then the associated linear system matrix $\widehat{\mathbf{A}}$ has the following properties;*

1. $\Gamma(\lambda^{-1} \otimes \widehat{\mathbf{A}})_{ij} = \gamma_{ij}$ *if* $i \in N_\beta$ *and* $-\infty$ *otherwise.*
2. $\widehat{\mathbf{A}}$ *has a finite eigenvalue* λ *and a finite corresponding eigenvector, where* λ *is the critical circuit ratio of the PCG.*
3. *Every finite eigenvector has the only possible corresponding finite eigenvalue* λ.
4. *For each critical circuit of the PCG with the set of recycling arcs* $\{(i_l, j_l) \mid l = 1, \ldots, q\}$, *each row vector of* $\Gamma(\lambda^{-1} \otimes \widehat{\mathbf{A}})$ *that corresponds to each one of* $i_1, i_2, \ldots, i_q$ *is a fundamental eigenvector of* $\widehat{\mathbf{A}}$ *with the corresponding eigenvalue* λ.
5. *The eigenvectors that correspond to the nodes of the same critical circuit of the PCG are all equivalent.*
6. *The set of all finite eigenvectors is the same as the set of all possible finite linear combinations of the fundamental eigenvectors,* $\{\bigoplus_{k=1,\ldots,p} w_k \otimes \mathbf{y}^k \mid -\infty < w_k < \infty, k = 1, \ldots, p\}$, *where* $\mathbf{y}^k, k = 1, \ldots, p$, *are the maximally non-equivalent fundamental eigenvectors.*

By Theorem 1, we can directly perform steady state analysis using the PCG rather than $G(\widehat{\mathbf{A}})$, where the computation of $\widehat{\mathbf{A}} = \mathbf{B} \otimes (\mathbf{I} \oplus \mathbf{A} \oplus \mathbf{A}^2 \oplus \ldots)$ requires a version of Gauss-Seidel procedure (Carré 1971). Thus, relationships between the structure of the PCG and the steady state behavior can be easily identified. These are illustrated in Example 1.

Example 1. . Consider the PCG given in Fig. 1(a). There are five machines. The critical circuit ratio is 5. All the strongly connected subgraphs(components) are identified and their precedence relations are illustrated in Fig. 1(b). The PCG has root components, *1* and *2*. Both are critical. The PCG$'$ is shown in Fig. 1(c). For vectors and matrices associated

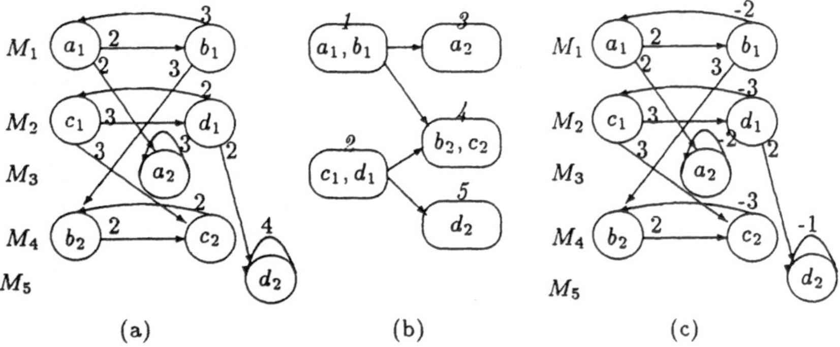

Fig. 1. The directed graphs for Example 1

with set N, we order the operations as $(a_1, b_1, c_1, d_1, a_2, b_2, c_2, d_2)$. By computing γ_{ij} as the longest path lengths of the PCG′, we have

$$\Gamma(\lambda^{-1} \otimes \widehat{\mathbf{A}}) = \begin{pmatrix} -\infty & -\infty & -\infty & -\infty & -\infty & -\infty & -\infty & -\infty \\ -2 & 0 & -\infty & -\infty & 0 & 3 & 5 & -\infty \\ -\infty & -\infty & -\infty & -\infty & -\infty & -\infty & -\infty & -\infty \\ -\infty & -\infty & -3 & 0 & -\infty & -3 & 0 & 2 \\ -\infty & -\infty & -\infty & -\infty & -2 & -\infty & -\infty & -\infty \\ -\infty & -\infty & -\infty & -\infty & -\infty & -\infty & -\infty & -\infty \\ -\infty & -\infty & -\infty & -\infty & -\infty & -3 & -1 & -\infty \\ -\infty & -\infty & -\infty & -\infty & -\infty & -\infty & -\infty & -1 \end{pmatrix}.$$

Two critical circuits π_1 and π_2 in the PCG are $(a_1 \to b_1 \to a_1)$ and $(c_1 \to d_1 \to c_1)$, respectively. Thus, the fundamental eigenvectors that correspond to π_1 and π_2 are $\mathbf{y}^1 = (-2, 0, -\infty, -\infty, 0, 3, 5, -\infty)$ and $\mathbf{y}^2 = (-\infty, -\infty, -3, 0, -\infty, -3, 0, 2)$, respectively. They are not finite. We may associate SESSs with $\mathbf{y}^1$ and $\mathbf{y}^2$. However, neither of them defines a complete schedule although each of them defines a stable steady state of the linear system. A complete SESS is then formed by appropriatetly combining the two partial schedules, as specified by Theorem 1. For instance, a SESS is defined by $2 \otimes \mathbf{y}^1 \oplus 1 \otimes \mathbf{y}^2 = (0, 2, -3, 0, 2, 5, 7, 2)$.

5 Minimization of the MPS Makespan

In this section, we develop a polynomial time algorithm to determine a SESS*. From the fact that the set of SESS is defined by the finite linear combinations of fundamental eigenvectors, we first present a linear programming model for SESS*.

Theorem 2. *Suppose that every root component of the PCG is critical. Then, the schedule that minimizes the MPS makespan over all the SESSs is defined by a finite vector* $\mathbf{x}$ *such that* $x_i = \max\{t_k^* + \gamma_{ki}\}$, $k = 1, \ldots, p$, $i \in N$, *and the minimum MPS makespan is* z^*, *where* $(\mathbf{t}^*, z^*)$ *is an optimal solution and the optimal solution value of the following* *linear program, respectively;*

$$(P1) \qquad \begin{array}{lll} z = \min v & & \\ \text{subject to} & v \geq t_k + \gamma_{ki}, & k = 1, \ldots, p, \quad i \in N, \\ & t_k + \gamma_{ki} \geq 0, & k = 1, \ldots, p, \quad i \in N\ . \end{array}$$

Instead of using the simplex algorithm to solve the linear program, we develop a polynomially bounded algorithm using a graph that is slightly modified from the PCG′. The graph is constructed as follow. For the PCG′, add dummy nodes, u and v, that represent the start and the completion of all the operations of an MPS, respectively. Add dummy arcs (u, j) for only each node $j \in N_\alpha$ that belongs to a critical circuit. Add dummy arc (j, v) for each $j \in N_\beta$. Observe that we do not add dummy arcs (u, j) for $j \in N_\alpha$ that does not belong to any critical circuit. Both of the first and second weights of each dummy arc are zero. The modified PCG′ is denoted by MPCG′. The longest path length from i to j in the MPCG′ is denoted by η_{ij} to avoid confusion with γ_{ij}.

We establish the following theorem using the properties of MPCG′.

Theorem 3. *When every root component of the PCG is critical, an earliest starting schedule with delays, $x_i^1 = \eta_{ui}, i \in N_\alpha$, is a SESS*.*

As a summary of the previous results, we present an algorithm to construct a SESS*.

Algorithm Φ*: Determine the starting times for the beginning operations, $\{x_i^1 \mid i \in N_\alpha\}$ of a SESS*.

1. Compute the critical circuit ratio λ of the PCG.
2. Identify G_C from the PCG and find all the root components.
3. Construct the PCG′ from the PCG.
4. Find all the critical circuits of the PCG by finding all circuits of length zero in the PCG′.
5. Check whether every root component has a critical circuit. If not, there is no SESS and stop the procedure.
6. Construct the MPCG′.
7. Compute η_{ui} and set $x_i^1 = \eta_{ui} - \min_{k \in N_\alpha} \eta_{uk}$ for all $i \in N_\alpha$.

Note that the first MPS schedule can be explicitly obatined beforehand by computing $x_i^1 = \eta_{ui} - \min_{k \in N_\alpha} \eta_{uk}$ for all $i \in N$. Step 1 can be performed in $O(|N|^3)$ or $O(|E \cup R||N| \log |N|)$ by using one of the two algorithms of Karp and Orlin (1981) because the second weights are 0 or 1. For Step 2, the algorithm of Tarjan (1972) is used. It takes $O(|E \cup R|)$ time. The Floyd-Warshall algorithm (Floyd 1962) performs Step 4 in $O(|N|^3)$ time. Step 7 can be completed in $O(|E \cup R||N|)$ time using the algorithm of Glover et al. (1985).

Example 1 (Continued). For the problem in Example 1, using Algorithm Φ*, the SESS is defined by $\mathbf{x}^1 = (0, 2, 0, 3, 2, 5, 7, 5)$. The MPS makespan is 9 while the MPS makespan of the previous SESS is 12.

6 Final Remarks

Since we characterized stable steady states of the linear system using the original event modeling graph (the PCG), rather than the deliberately derived linear system matrix or graph ($G(\widehat{\mathbf{A}})$), we could make use of the structural properties of the PCG to develop the efficient algorithm. It remains to investigate the dual system (in (min, +) algebra) for latest allowable schedules that meet a given deadline.

References

Ahuja, R.K., Magnanti, T.L., Orlin, J.B.: Network Flows. Prentice-Hall, NJ. (1993)

Carré, B.A.: An algebra for network routing problems. Journal of Institute of Mathematics and Applications **7** (1971) 273–294

Cohen, G., Dubois, D., Quadrat, J.P., Viot, M.: A linear–system–theoretic view of discrete–event processes and its use for performance evaluation in manufacturing. IEEE Transactions on Automatic Control **30** (1985) 210–220

Cunninghame-Green, R.A.: Minimax algebra. Springer-Verlag, NY. (1979)

Floyd, R.W.: Algorithm 97, shortest path. Communication in ACM **5** (1962) 345

Glover, F., Kingman, D., Phillips, N.: A new polynomially bounded shortest path algorithm. Operations Research **33** (1985) 65–73

Karp, R.M., Orlin, J.B.: Parametric shortest path algorithms with an application to cyclic staffing. Discrete Applied Mathematics **3** (1981) 37–45

Lee, T.E., Posner, M.E.: Performance measures and schedules in periodic job shops. The Ohio State University, Department of Industrial and Systems Engineering. Working Paper 1990-012. (1990)

Tarjan, R.: Depth-first search and linear graph algorithms. SIAM Journal of Computing **1** (1972) 146–160

Time Discrete and Continuous Control Problems Convergence of Value Functions

Sergei N. Samborski

Université de CAEN, Mathématiques, 14032 CAEN Cedex, FRANCE
e-mail : `Samborsk@UNIV-CAEN.fr`

In this paper we are concerned with properties of the so-called value functions associated with time continuous control problems and with their time discrete approximation. The question we wish to consider the convergence of corresponding value functions.

1 Introduction

Let $X \subset \mathbb{R}^n$ be the compact set in $\mathbb{R}^n$ which is the closure of an open domain. Let be a subset of $\mathbb{R}^n$, $\phi : X \longrightarrow \mathbb{R} \cup \{\infty\}$, and $L : X \times U \longrightarrow \mathbb{R} \cup \{\infty\}$ two given functions.

A. Let $F : X \times U \longrightarrow X$ be a continuous mapping. Let us determine the value function $S_n(x, \Phi)$ as Infimum of values of the sum :

$$\sum_{i=0}^{n} L(x_i, u_i) + \Phi(x_n)$$

with respect to all possible systems $\{(x_i, u_i)\}_{i=0}^{n}$ such, that $x_0 = x$ and $F(x_i, u_i) = x_{i+1}$.

The optimality principle gives immediately the formula :

$$S_n(x, \Phi) = (A^n \Phi)(x).$$

where $A^n = A \circ \ldots \circ A$ and the mapping A transferring functions from X to $\mathbb{R} \cup \{\infty\}$ is of the form :

$$(A\omega)(x) = \inf_{u \in U} (L(x, u) + \omega(F(x, u)). \tag{1}$$

B. Let $f : X \times U \longrightarrow X$ be a Lipschitz mapping. We determine the value function $S_t(x, \Phi)$ as the infimum of values of the expression :

$$\int_0^t L(x(\tau)), u(\tau))d\tau + \Phi(x(t))$$

over all possible curves $(x(\cdot), u(\cdot))$, where $u(\cdot)$ is a Lipschitzian function, $x(\cdot)$ a differentiable function and :

$$\dot{x}(t) = f(x(t), u(t)), \quad x(0) = x_0.$$

It follows from the optimality principle, that the mapping U_t, which transforms a function $\Phi : X \longrightarrow \mathbb{R} \cup \{\infty\}$ into a function $S_t(\cdot, \Phi)$ possess a semi-group property with respect to t

$$U_{t_1+t_2} = U_{t_1} \circ U_{t_2}.$$

One of the most popular methods of computation of a function $S_t(\cdot, \Phi) = U_t\Phi$ is the following time-discretization. Let t be fixed and n is natural. One considers the discrete problem A) with :

$$F_{t/n}(x, u) = \frac{t}{n} f(x, u) + x$$

and the mapping $A_{t/n}$ is determined by the formula (1) when $F = F_{t/n}$.
Then, one considers the function $(A^n_{t/n}(\Phi)$, when n is sufficiently large as a "good approximation" of the function $S_t(\cdot, \Phi) = U_t\Phi$. Below we investigate a new notion of convergence giving to this procedure a rigorous justification.

2 The Maslov Spaces (Functional Semi-Modules)

Let $\mathcal{R} = \mathbb{R} \cup \{\infty\}$ with operations $\oplus = \min$ and $\odot = +$. We denote neutral elements with respect to these operations by $\mathbf{0}$ (i.e. $+\infty$) and $\mathbf{1}$ (i.e. 0). It is convenient to consider the metric $\rho(a, b) = |\exp(-a) - \exp(-b)|$ in $\mathcal{R}$. Let $(\cdot, \cdot)$ be a scalar product :

$$(f, g) = \inf_x (f(x) + g(x)$$

and Φ — the set of test-functions — a subset in the set of continuous functions from X to $\mathcal{R}$. We will consider two cases : Φ is a set of all continuous functions, or $\Phi = \Phi_0$ is a set of "finite" functions, that is a set of continuous functions, which tend to $\mathbf{0}$ while their argument tends to the boundary ∂X of the set X.

In everyone of these cases we invent an equivalence relation on the set of lower bounded functions :

$$f \sim g \text{ if } (f, \varphi) = (g, \varphi), \quad \forall \varphi \in \Phi \ \text{ (or } \Phi_0).$$

We also introduce a convergence in the set of classes of equivalence :

$$f_n \longrightarrow f \text{ if } (f_n, \varphi) \longrightarrow (f, \varphi), \quad \forall \varphi \in \Phi \ \text{ (or } \Phi_0).$$

Denote these spaces by $\mathcal{P}(X)$, respectively $\mathcal{P}^o(X)$.

These spaces have a semi-module structure with the continuous operations $\oplus$: $\mathcal{P} \times \mathcal{P} \longrightarrow \mathcal{P}$ and $\odot : \mathcal{R} \times \mathcal{P} \longrightarrow \mathcal{P}$, which are determined, for instance, the following way : for $f, g \in \mathcal{P}$, $\lambda \in \mathcal{R}$: $f \oplus g$ and $\lambda \odot f$ are the elements of $\mathcal{P}$ such that $(f \oplus g, \varphi) = (f, \varphi) \oplus (g, \varphi)$ and $(\lambda \odot f, \varphi) = \lambda \odot (f, \varphi)$ for every $\varphi \in \Phi$(or Φ_0), [2].

The mappings $A_{t/n}$ and U_t, considered above, occur as endomorphisms of the semi-modules $\mathcal{P}(X)$.

Theorem 1 (V. Maslov, S. Samborski). *Let $\Phi \in \mathcal{P}(X)$, $t \geq 0$. Then, in the space $\mathcal{P}(X)$ one has*

$$\lim_{n \longrightarrow \infty} A^n_{t/n}\Phi = U_t\Phi.$$

3 Metrics in the Maslov Spaces

A notion of convergence on the equivalence classes is induced by each of the following metrics. Let $\varepsilon > 0$ and let χ_ε be a set of characteristic functions of balls of diameter ε which are intersected with X, $(X \subset \mathbb{R}^n)$: by definition the characteristic function of a ball B is the mapping $X \longrightarrow \mathcal{R}$ equal to $\mathbb{1}$ for all $X \in B \cap X$ and $\mathbb{0}$ for another $x \in X$. Let χ_ε^o be the set of characteristic functions of balls of diameter ε, belonging to X, which do not intersect the boundary ∂X of the set X. Let

$$r(f, g) = \inf\left\{\sqrt{\varepsilon^2 + \mu^2} \mid \rho((f, \varphi), (g, \varphi)) < \mu \text{ for } \varphi \in \chi_\varepsilon\right\},$$

$r^o(f, g)$ is determined by the same formula with χ_ε replaced by χ_ε^o.

Proposition 2. *The maps r and r^o are metrics on the set of all equivalence classes constructed above by the set of test functions Φ and Φ_0 respectively. The convergence, defined by these each of metrics coincides with the respective weak convergence, introduced above. The spaces $\mathcal{P}(X)$ and $\mathcal{P}^o(X)$ with these metrics are complete metric semi-modules.*

4 Spaces $\mathcal{P}_H(X)$

Let $H : X \times \mathbb{R}^n \longrightarrow \mathbb{R}$ be a continuous mapping (Hamiltonian). We will suppose that $H(x, \cdot)$ is convex (not strictly, in general) with respect to the second argument when the first one is fixed. The mapping H induce a differential operator, which acts from the set of continuously differentiable functions C^1 on X to the set C of continuous functions on X, and therefore, to the space $\mathcal{P}^o(X)$ because of the evident injection $i : C(X) \longrightarrow \mathcal{P}^o(X)$. This differential operator $\mathcal{H}$ is defined by the following rule :

$$(\mathcal{H}f)(x) = -H(x, Df(x)), \text{ where } Df \text{ is a gradient of } f.$$

Let us note $\mathcal{P}^o_{c^1}(X)$ the algebraic semi-module in $\mathcal{P}^o(X)$, generated by a subset iC^1, so $\mathcal{P}^o_{c^1}(X)$ consists of finite sums :

$$\bigoplus_i f_i \qquad (f_i \in C^1(X)).$$

Now we will extend the operator $\mathcal{H}$ to this semi-module. This can be done in the following way. (Evidently it is sufficient to define $\mathcal{H}(f_1 \oplus f_2)$, $f_i \in C^1(X)$, $i = 1, 2$).

Let $\mathcal{H}(f_1 \oplus f_2)$ be the equivalence class of $\mathcal{P}^o(X)$ which has the representative :

$$x \longrightarrow \begin{cases} \mathcal{H}(f_1)(x) & \text{if } f_1(x) < f_2(x), \\ \mathcal{H}(f_2)(x) & \text{if } f_1(x) > f_2(x), \\ \mathcal{H}(f_1)(x) \oplus \mathcal{H}(f_2)(x) & \text{if } f_1(x) = f_2(x). \end{cases}$$

It is easy to show the correctness of this definition. A simple example : Let $\mathcal{H} = \dfrac{\partial}{\partial x}$, $f_1(x) = x$, $f_2(x) = -x$.

Then $f_1 \oplus f_2) = -|x|$ and $\dfrac{\partial}{\partial x}(f_1 \oplus f_2)(x)$ is the element of $\mathcal{P}^o$ (that is, an equivalence class), which contains the function equals to 1 when $x < 0$ and -1 when $x > 0$. If $x = 0$, then this function can take an every value $a \geq -1$; this does not change the class.

Let us introduce the metric r_H on the semi-module $\mathcal{P}^o_{c^1}$ by the following rule : let Aff be the set of affine functions from X to $\mathbb{R}$, that is of functions of a type $x \longmapsto \langle a, x\rangle + b$, where $\langle \cdot, \cdot \rangle$ is a scalar product in $\mathbb{R}^n$, $a \in \mathbb{R}^n$, $b \in \mathbb{R}$.

Let $r_H(f, g) = \max\{r(f, g), \sup\limits_{\varphi \in \mathrm{Aff}} r^o(\mathcal{H}(f \oplus \varphi), \mathcal{H}(g \oplus \varphi))\}$.

Proposition 3. *The function r_H is a metric on the space $\mathcal{P}_{c^1}$.*

Definition 4. Let $\mathcal{P}_H$ be a completion of $\mathcal{P}_{c^1}$ by the metric r_H.

This completion naturally injects into $\mathcal{P}(X)$ and this injection is the identity on $\mathcal{P}^o_{c^1}$. We will consider elements of $\mathcal{P}_H$ belonging to $\mathcal{P}$.

Proposition 5. *The mapping $\mathcal{H}$ extends to a continuous mapping from $\mathcal{P}_H$ to $\mathcal{P}$. This means : if $f = \lim f_i$ in the metric r_H, $f_i \in \mathcal{P}_{c^1}$, then $\lim \mathcal{H}(f_i)$ exists in the metric of $\mathcal{P}$ and depends only on f, not on the sequence f_i. We will denote this limit by $\mathcal{H}(f)$.*

Theorem 6. *The set $\mathcal{P}_H$ is a metric semi-module (that is the operations $\oplus : \mathcal{P}_H \times \mathcal{P}_H \longrightarrow \mathcal{P}_H$ and $\odot : \mathcal{R} \times \mathcal{P}_H \longrightarrow \mathcal{P}_H$ are continuous) and the operator $\mathcal{H}$ acts continuously from $\mathcal{P}_H$ to $\mathcal{P}$. Let $\mathcal{A}$ be a subset of $\mathcal{P}$, provided with a metric, that converts $\mathcal{A}$ into a complete metric semi-module, $\mathcal{A}$ contains all C^1-functions, and the mapping $\mathcal{H}$ acts continuously from $\mathcal{A}$ to $\mathcal{P}$. Then $\mathcal{A} \subseteq \mathcal{P}_H$ and the topology, generated by the metric in $\mathcal{A}$ is stronger then or equal to that of $\mathcal{P}_H$.*

So $\mathcal{P}_H$ is the weakest complete metric sub-semi-module in $\mathcal{P}$ such that the mapping $\mathcal{H}$ extends continuously from $\mathcal{P}_H$ to $\mathcal{P}$.

Let X be the set $[0, 1] \times [0, 1] \subset \mathbb{R}^2$.

Example 1. Let $H(x, p) = p_1$, then $\mathcal{H}(f) = -\partial f/\partial x_1$. Every class of equivalence determined by a function semi-bounded from below of the type $(x_1, x_2) \longrightarrow f(x_2)$ belongs to $\mathcal{P}_H$ (f may be discontinuous or have infinite values).

Example 2. $H = |p_1| + |p_2|$. Then both functions $(x_1, x_2) \longrightarrow \max(x_1, x_2)$ and $(x_1, x_2) \longrightarrow \min(x_1, x_2)$ belong to $\mathcal{P}_H$.

Example 3. $H = p_1^2 + p_2^2$. The function $(x_1, x_2) \longrightarrow \min(x_1, x_2)$ belongs to $\mathcal{P}_H$ but $(x_1, x_2) \longrightarrow \max(x_1, x_2)$ does not belong to $\mathcal{P}_H$.

More generally : if the hamiltonian $H(x, \cdot) : \mathbb{R}^n \longrightarrow \mathbb{R}$ is strictly convex for every x and the sets $\{p \mid H(x, p) \leq \lambda\} \subseteq \mathbb{R}^n$ are bounded for $x \in X, \lambda \in \mathbb{R}$ than $\mathcal{P}_H$ consists of continuous functions possessing the following property : suppose that for some $x_0 \in X$, $x_0 \notin \partial X$ and some differentiable function φ the function $x \longrightarrow (f - \varphi)(x)$ has a minimum in x_0 ; then the function f is differentiable in x_0 [2]. This means that for these Hamiltonians the sets $\mathcal{P}_H$ coincide.

Example 4. $H = |p_1 + p_2|$. The discontinuous function

$$(x_1, x_2) \longrightarrow \begin{cases} x_1 & \text{for } x_1 \geq x_2, \\ x_2 + 1 & \text{for } x_1 < x_2, \end{cases}$$

determines a class of equivalence which belongs to $\mathcal{P}_H$.

Theorem 7. *Let us define*

$$H(x, p) = \min_{u \in U}(-\langle p, f(x, u)\rangle + L(x, u)) \tag{2}$$

for the control problem B and let Φ be a function belonging to the semi-module $\mathcal{P}_H$. Then for every $t \geq 0$, for the n functions $A_{t/n}\phi, A^2_{t/n}, \phi, \ldots, A^n_{t/n}\phi$ and $U_t\phi$ belong to $\mathcal{P}_H$ and in the metric r_H we have

$$\lim_{n \to \infty} A^n_{t/n}\phi = U_t\Phi.$$

Theorem 8. *Let $\phi \in \mathcal{P}(X)$. Suppose that for some $T \geq 0$ the function $(U_t\phi)(\cdot) : X \longrightarrow \mathcal{R}$ does not depend on t for $t \geq T$. Then $U_T\phi \in \mathcal{P}_H$ and $U_T\phi$ satisfies the equation $\mathcal{H}(y) = 0$ defined by H from Eq. (2).*

Definition 9. Let $y \in \mathcal{P}_H$ be a solution of $\mathcal{H}(y) = 0$, equal to g on the boundary ∂X of a domain X. We call this solution stable if for some $\lambda_0 < 0$ and for every $\lambda \in [\lambda_0, 0]$ the equation $\mathcal{H}(y) = \lambda$ has a solution $y_\lambda \in \mathcal{P}_H$ equal to g on ∂X and $y = \lim_{\lambda \to 0} y_\lambda$ in the space $\mathcal{P}(X)$.

Theorem 10. *Suppose, that under the conditions of Theorem 8, the restriction of the function $U_T\phi$ to the boundary ∂X equals g. Then $U_T\phi$ is the unique stable solution of $\mathcal{H}(y) = 0$ equal to g on ∂X.*

Remark. A non-stable solution might not be unique in $\mathcal{P}_H$; nevertheless the non-uniqueness does not represent the generic situation.

References

1. Maslov, V.: Asymptotic Methods for Solving Pseudodifferential Equations. Nauka, Moscow (1987) (in Russian).
2. Maslov, V., Samborski, S.: Stationary Hamilton-Jacobi and Bellman equations. Advances in Soviet Mathematics **13** AMS (1992) 119–133.

Bellman Processes

Marianne Akian[1], *Jean-Pierre Quadrat*[1] *and Michel Viot*[2]

[1] INRIA-Rocquencourt, B.P. 105, 78153 Le Chesnay Cedex, France
[2] CMA, École Polytechnique, 9128 Palaiseau Cedex, France

1 Introduction

The mapping $\lim_{\varepsilon\to 0}\log_\varepsilon$ defines a morphism of algebra between the asymptotics (around zero) of positive real functions of a real number and the real numbers endowed with the two operations min and plus, indeed:

$$\lim_{\varepsilon\to 0}\log_\varepsilon(\varepsilon^a+\varepsilon^b)=\min(a,b),\quad \log_\varepsilon(\varepsilon^a\varepsilon^b)=a+b.$$

This morphism send probability calculus into optimal control problems. Therefore almost of the concepts introduced in probability calculus have an optimization counterpart. The purpose of this paper is to make a presentation of known and new results of optimal control with this morphism in mind. The emphasis of this talk is i) on the trajectory point of view by opposition to the cost point of view and ii) on the optimization counterpart of processes with independent increments.

2 Inf-Convolution and Cramer Transform

Definition 1. Given two mappings f and g from $\mathbb{R}$ into $\overline{\mathbb{R}}\overset{\text{def}}{=}\mathbb{R}\cup\{+\infty\}$, the *inf-convolution of f and g* is the mapping $z\in\mathbb{R}\mapsto\inf_{x,y}[f(x)+g(y)\mid x+y=z]$. It is denoted $f\,\square\,g$. When f and g are lower bounded $f\,\square\,g$ is also lower bounded.

Example 1. For $m\in\mathbb{R}$ let us define the convex Dirac function:

$$\delta^c_m(x)=\begin{cases}+\infty & \text{for } x\neq m,\\ 0 & \text{for } x=m,\end{cases}$$

and consider the function $\mathcal{M}^p_{m,\sigma}(x)=\frac{1}{p}(|x-m|/\sigma)^p$ for $p\geq 1$ with $\mathcal{M}^p_{m,0}=\delta^c_m$. We have the formula

$$\mathcal{M}^p_{m,\sigma}\,\square\,\mathcal{M}^p_{\bar m,\bar\sigma}=\mathcal{M}^p_{m+\bar m,[\sigma^{p'}+\bar\sigma^{p'}]^{1/p'}}\text{ with } 1/p+1/p'=1\,.$$

This result is the analogue of $\mathcal{N}(m,\sigma)*\mathcal{N}(\bar m,\bar\sigma)=\mathcal{N}(m+\bar m,\sqrt{\sigma^2+\bar\sigma^2})$, in the particular case $p=2$, where $\mathcal{N}(m,\sigma)$ denotes the Gaussian law of mean m and standard deviation σ and $*$ the convolution operator.

Therefore there exists a morphism between the set of quadratic forms endowed with the inf-convolution operator and the set of exponentials of quadratic forms endowed with the convolution operator. This morphism is a particular case of the Cramer transform that we will define later. Let us first recall the definition of the Fenchel transform.

Definition 2. Let $c \in \mathcal{C}_x$, where $\mathcal{C}_x$ denotes the set of mappings from $\mathbb{R}$ into $\overline{\mathbb{R}}$ convex, l.s.c. and proper. Its Fenchel transform is the function from $\mathbb{R}$ into $\overline{\mathbb{R}}$ defined by $\hat{c}(\theta) \stackrel{\text{def}}{=} [\mathcal{F}(c)](\theta) \stackrel{\text{def}}{=} \sup_x [\theta x - c(x)]$.

Example 2. The Fenchel transform of $\mathcal{M}^p_{m,\sigma}$ is $[\mathcal{F}(\mathcal{M}^p_{m,\sigma})](\theta) = \frac{1}{p'}|\theta\sigma|^{p'} + m\theta$, with $1/p + 1/p' = 1$. The particular case $p = 2$ corresponds to the characteristic function of a Gaussian law.

Theorem 3. *For $f, g \in \mathcal{C}_x$ we have i) $\mathcal{F}(f) \in \mathcal{C}_x$, ii) $\mathcal{F}$ is an involution that is $\mathcal{F}(\mathcal{F}(f)) = f$, iii) $\mathcal{F}(f \,\square\, g) = \mathcal{F}(f) + \mathcal{F}(g)$, iv) $\mathcal{F}(f + g) = \mathcal{F}(f) \,\square\, \mathcal{F}(g)$.*

Definition 4. The Cramer transform $\mathcal{C}$ is a function from $\mathcal{M}$, the set of positive measures, into $\mathcal{C}_x$ defined by $\mathcal{C} \stackrel{\text{def}}{=} \mathcal{F} \circ \log \circ \mathcal{L}$, where $\mathcal{L}$ denotes the Laplace transform.

From the definition and the properties of the Laplace and Fenchel transform the following result is clear.

Theorem 5. *For $\mu, \nu \in \mathcal{M}$ we have $\mathcal{C}(\mu * \nu) = \mathcal{C}(\mu) \,\square\, \mathcal{C}(\nu)$.*

The Cramer transform changes the convolutions into inf-convolutions. In Table 1 we summarize the main properties and examples concerning the Cramer transform. The difficult results of this table can be found in Azencott [4]. In this table we have denoted by $\breve{A}$ the interior of the set A.

3 Decision Variables

The morphism between convolution and inf-convolution described in the previous section suggests the existence of a formalism adapted to optimization analogous to probability calculus. Some of the notions given here have been introduced in Bellalouna [5]. Another similar and independent work can be found in Del Moral [8]. We start by defining cost measures which can be seen as normalized idempotent measures of Maslov [13].

Definition 6. The triplet $(U, \mathcal{U}, \mathbb{K})$ where U is a topological space, $\mathcal{U}$ the set of the open sets of U and $\mathbb{K}$ a mapping from $\mathcal{U}$ into $\overline{\mathbb{R}}^+$ such that: *i*) $\mathbb{K}(U) = 0$, *ii*) $\mathbb{K}(\emptyset) = +\infty$, *iii*) $\mathbb{K}\left(\bigcup_n A_n\right) = \inf_n \mathbb{K}(A_n)$ for any $A_n \in \mathcal{U}$, is called a *decision space* and $\mathbb{K}$ is called a *cost measure*.

A map $c : u \in U \mapsto c(u) \in \overline{\mathbb{R}}^+$ such that $\mathbb{K}(A) = \inf_{u \in A} c(u)$, $\forall A \subset U$ is called a *cost density* of the cost measure $\mathbb{K}$.

The set $D_c \stackrel{\text{def}}{=} \{u \in U \mid c(u) \neq +\infty\}$ is called the *domain* of c.

Theorem 7. *Given a l.s.c. positive real valued function c such that $\inf_u c(u) = 0$, $\mathbb{K}(A) = \inf_{u \in A} c(u)$ for all A open set of U defines a cost measure. Conversely any cost measure defined on the open sets of a Polish space admits a unique minimal extension $\mathbb{K}_*$ to $\mathcal{P}(U)$ (the set of the parts of U) having a density c which is a l.s.c. function on U satisfying $\inf_u c(u) = 0$.*

Table 1. Properties of the Cramer transform.

$\mathcal{M}$	$\log(\mathcal{L}(\mathcal{M})) = \mathcal{F}(\mathcal{C}(\mathcal{M}))$	$\mathcal{C}(\mathcal{M})$
μ	$\hat{c}_\mu(\theta) = \log \int e^{\theta x} d\mu(x)$	$c_\mu(x) = \sup_\theta(\theta x - \hat{c}(\theta))$
0	$-\infty$	$+\infty$
δ	0	δ^c
δ_a	θa	δ_a^c
$e^{-H(x)}$ $H(x) \stackrel{\text{def}}{=} 0$ if $x \geq O$ $+\infty$ elsewhere	$H(-\theta) - \log(-\theta)$	$H(x) - 1 - \log(x)$
$\lambda e^{-\lambda x - H(x)}$	$H(\lambda - \theta) + \log(\lambda/(\lambda - \theta))$	$H(x) + \lambda x - 1 - \log(\lambda x)$
$p\delta + (1-p)\delta_1$	$\log(p + (1-p)e^\theta)$	$x \log(\frac{x}{1-p})$ $+(1-x)\log(\frac{1-x}{p})$ $+H(x) + H(1-x)$
$\frac{1}{\sigma\sqrt{2\pi}} e^{-\frac{1}{2}(x-m)^2/\sigma^2}$	$m\theta + \frac{1}{2}(\sigma\theta)^2$	$\mathcal{M}^2_{m,\sigma}$
Inf. divis. distrib. Feller [10]	$m\theta + \frac{1}{p'}\|\sigma\theta\|^{p'}$	$\mathcal{M}^p_{m,\sigma}$ with $p > 1, \ 1/p + 1/p' = 1$
	$m\theta + H(-\theta + 1/\sigma)$ $+H(\theta + 1/\sigma)$	$\mathcal{M}^1_{m,\sigma}$
	$a\theta \vee b\theta, \ a \leq b$	$H(x-a) + H(-x+b)$
$\mu * \nu$	$\hat{c}_\mu + \hat{c}_\nu$	$c_\mu \mathbin{\square} c_\nu$
$k\mu$	$\log(k) + \hat{c}$	$c - \log(k)$
$\mu \geq 0$	$\hat{c}$ convex l.s.c.	c convex l.s.c.
$m_0 \stackrel{\text{def}}{=} \int \mu$	$\hat{c}(0) = \log(m_0)$	$\inf_x c(x) = -\log(m_0)$
$m_0 = 1$	$\hat{c}(0) = 0$	$\inf_x c(x) = 0$
$\mathcal{P} \stackrel{\text{def}}{=} \{\mu \geq 0 \mid m_0 = 1\}$ $S_\mu \stackrel{\text{def}}{=} \overline{\text{cvx}(\text{supp}(\mu))}$ $\mu \in \mathcal{P}$	$\hat{c}$ strictly convex in $D_{\hat{c}}$ C^∞ in $\breve{D}_{\hat{c}}$	$D_c \stackrel{\text{def}}{=} \text{dom}(c)$ $\breve{D}_c = \breve{S}_\mu$ C^1 in $\breve{D}_c$
$m_0 = 1, \ m \stackrel{\text{def}}{=} \int x\mu$	$\hat{c}'(0) = m$	$c(m) = 0$
$m_0 = 1, \ m_2 \stackrel{\text{def}}{=} \int x^2\mu$	$\hat{c}''(0) = \sigma^2 \stackrel{\text{def}}{=} m_2 - m^2$	$c''(m) = 1/\sigma^2$
$m_0 = 1$ $\hat{c} = \|\sigma\theta\|^{p'}/p' + o(\|\theta\|^{p'})$	$\hat{c}^{(p')}(0^+) = \Gamma(p')\sigma^{p'}$	$c^{(p)}(0^+) = \Gamma(p)/\sigma^p$

Proof. This precise result is proved in Akian [1]. See Maslov [13] for the first result of this kind. See also Del Moral [8] for analogous results.

We have seen that the images by the Cramer transform of the probability measures are C^1 and convex cost density functions.

By analogy with the conditional probability we define now the conditional cost excess.

Definition 8. The *conditional cost excess* to take the best decision in A knowing that it must be taken in B is

$$\mathbb{K}(A|B) \stackrel{\text{def}}{=} \mathbb{K}(A \cap B) - \mathbb{K}(B) .$$

Definition 9. By analogy with random variables we define decision variables and related notions.

1. A *numerical decision vector* X on $(U, \mathcal{U}, \mathbb{K})$ is a mapping from U into $\mathbb{R}^n$. It induces $\mathbb{K}_X$ a cost measure on $(\mathbb{R}^n, \mathcal{B})$ ($\mathcal{B}$ denotes the set of open sets of $\mathbb{R}^n$) defined by $\mathbb{K}_X(A) = \mathbb{K}_*(X^{-1}(A)), \ \forall A \in \mathcal{B}$. The cost measure $\mathbb{K}_X$ has a l.s.c. density denoted c_X . When $n = 1$ we call it a *decision variable.*
2. A decision variable is said *regular* when its cost measure is regular.
3. Two decision variables X and Y are said *independent* when:
$$c_{X,Y}(x, y) = c_X(x) + c_Y(y).$$
4. The *conditional cost excess* of X knowing Y is defined by:
$$c_{X|Y}(x, y) \stackrel{\text{def}}{=} \mathbb{K}_*(X = x \mid Y = y) = c_{X,Y}(x, y) - c_Y(y).$$
5. The *optimum* of a decision variable is defined by $\mathbb{O}(X) \stackrel{\text{def}}{=} \arg\min_x c_X(x)$ when the minimum exists. When a decision variable X satisfies $\mathbb{O}(X) = 0$ we say that it is *centered.*
6. When the optimum of a decision variable X is unique and when near the optimum, we have:
$$c_X(x) = \frac{1}{p}\left|\frac{x - \mathbb{O}(x)}{\sigma}\right|^p + o(|x - \mathbb{O}(x)|^p) \ ,$$
we define the *sensitivity of order p* of $\mathbb{K}$ by $\mathbb{S}^p(X) \stackrel{\text{def}}{=} \sigma$. When a decision variable satisfies $\mathbb{S}^p(X) = 1$ we say that it is *of order p and normalized.* When we speak of sensitivity without making the order precise, we implicitly mean that this order is 2.
7. The numbers
$$|X|_p \stackrel{\text{def}}{=} \inf\left\{\sigma \mid c_X(x) \geq \frac{1}{p}|(x - \mathbb{O}(X))/\sigma|^p\right\} \text{ and } \|X\|_p \stackrel{\text{def}}{=} |X|_p + |\mathbb{O}(X)|$$
define respectively a seminorm and a norm on the set of decision variables having a unique optimum such that $\|X\|_p$ is bounded. The corresponding set of decision variables is called $\mathbb{D}^p$.
8. The *mean* of a decision variable X is $\mathbb{M}(X) \stackrel{\text{def}}{=} \inf_x(x + c_X(x))$, *the conditional mean* is $\mathbb{M}(X \mid Y = y) \stackrel{\text{def}}{=} \inf_x(x + c_{X|Y}(x, y))$.
9. The *characteristic function* of a decision variable is $\mathbb{F}(X) \stackrel{\text{def}}{=} \mathcal{F}(c_X)$ (clearly $\mathbb{F}$ characterizes only decision variables with cost in $\mathcal{C}_x$).

Example 3. For a decision variable X of cost $\mathcal{M}^p_{m,\sigma}$, $p > 1$, we have
$$\mathbb{O}(X) = m, \ \mathbb{S}^p(X) = |X|_p = \sigma, \ \mathbb{M}(X) = m - \frac{1}{p'}\sigma^{p'} \ .$$

The role of the Laplace or Fourier transform in the probability calculus is plaid by the Fenchel transform in the decision calculus.

Theorem 10. *If the cost density of a decision variable is convex, admits a unique minimum and is of order p, we have:*
$$\mathbb{F}(X)'(0) = \mathbb{O}(X), \ \ \mathbb{F}(X - \mathbb{O}(X))^{(p)}(0) = \Gamma(p')[\mathbb{S}^p(X)]^{p'} \ .$$

Theorem 11. *For two independent decision variables X and Y of order p and $k \in \mathbb{R}$ we have*

$$c_{X+Y} = c_X \,\square\, c_Y, \quad \mathbb{F}(X+Y) = \mathbb{F}(X) + \mathbb{F}(Y), \quad [\mathbb{F}(kX)](\theta) = [\mathbb{F}(X)](k\theta)\ ,$$
$$\mathbb{O}(X+Y) = \mathbb{O}(X) + \mathbb{O}(Y), \quad \mathbb{O}(kX) = k\mathbb{O}(X), \quad \mathbb{S}^p(kX) = |k|\mathbb{S}^p(X)\ ,$$
$$[\mathbb{S}^p(X+Y)]^{p'} = [\mathbb{S}^p(X)]^{p'} + [\mathbb{S}^p(Y)]^{p'}, \quad (|X+Y|_p)^{p'} \leq (|X|_p)^{p'} + (|Y|_p)^{p'}\ .$$

4 Independent Sequences of Decision Variables

We consider, in this section, sequences of independent decision variables and the analogues of the classical limit theorems of the probability calculus.

Definition 12. A *sequence of independent decision variables identically costed of cost* c on $(U, \mathcal{U}, \mathbb{K})$ (i.i.c.) is an application X from U into $\mathbb{R}^{\mathbb{N}}$ which induces a density cost satisfying

$$c_X(x) = \sum_{i=0}^{\infty} c(x_i), \quad \forall x = (x_0, x_1, \ldots) \in \mathbb{R}^{\mathbb{N}}\ .$$

Remark. 1. The cost density is finite only on minimizing sequences of c, elsewhere it is equal to $+\infty$.
2. We have defined a decision sequence by its density and not by its value on the open sets of $\mathbb{R}^{\mathbb{N}}$ because the density can be defined easily.

In order to state the limit theorems, we define several type of convergence of sequences of decision variables.

Definition 13. For the numerical decision sequence $\{X_n, n \in \mathbb{N}\}$ we say that

1. X_n*converges weakly* towards X, denoted $X_n \overset{w}{\to} X$, if for all f in $C_b(\mathbb{R})$ (where $C_b(\mathbb{R})$ denotes the set of uniformly continuous and lower bounded functions on $\mathbb{R}$) we have $\lim_n \mathbb{M}[f(X_n)] = \mathbb{M}[f(X)]$. When the test functions used are the set of affine functions we say that it converges *weakly*$*$ ($w*$);
2. $X_n \in \mathbb{D}^p$ *converges in p-sensitivity* towards $X \in \mathbb{D}^p$ denoted $X_n \overset{\mathbb{D}^p}{\longrightarrow} X$, if $\lim_n \|X_n - X\|_p = 0$;
3. X_n *converges in cost* towards X, denoted $X_n \overset{\mathbb{K}}{\longrightarrow} X$, if for all $\epsilon > 0$ we have $\lim_n \mathbb{K}\{u \mid |X_n(u) - X(u)| \geq \epsilon\} = +\infty$;
4. X_n *converges almost surely* towards X, denoted $X_n \overset{\text{a.s.}}{\longrightarrow} X$, if we have $\mathbb{K}\{u \mid \lim_n X_n(u) \neq X(u)\} = +\infty$.

Some relations between these different kinds of convergence are given in the following theorem.

Theorem 14. *1. Convergence in sensitivity implies convergence in cost but the converse is false.*
2. Convergence in cost implies almost sure convergence and the converse is false.
3. Almost sure convergence does not imply weak convergence.
4. Convergence in cost implies the weak convergence.

Proof. See Akian [2]. In Bellalouna [5] the point 3 has been proved previously.

We have the analogue of the law of large numbers.

Theorem 15. *Given a sequence of independent decision variables belonging to* $\mathbb{D}^p$, $p \geq 1$, *identically costed (i.i.c.)* $\{X_n,\ n \in \mathbb{N}\}$ *we have:*

$$\lim_{N\to\infty} Y_N \stackrel{\text{def}}{=} \frac{1}{N}\sum_{n=0}^{N-1} X_n = \mathbb{O}(X_0) ,$$

where the limit can be taken in the sense of the weak, almost sure, cost and p-sensitivity convergence.

Proof. We have only to estimate the convergence in sensitivity. The results follows from simple computation of the p-seminorm of Y_N. It satisfies $(|Y_N)|_p)^{p'} = N(|X_0|_p)^{p'}/N^{p'}$ thanks to theorem 11.

We have the analogue of the central limit theorem of the probability calculus.

Theorem 16. *Given an i.i.c. sequence* $\{X_n, n \in \mathbb{N}\}$ *centered of order p we have*

$$\text{weak} * \lim_N Z_N \stackrel{\text{def}}{=} \frac{1}{N^{1/p'}}\sum_{n=0}^{N-1} X_n = X ,$$

where X is a decision variable with cost equal to $\mathcal{M}^p_{0,\mathbb{S}^p(X_0)}$.

Proof. We have $\lim_N[\mathbb{F}(Z_N)](\theta) = \frac{1}{p'}[\mathbb{S}^p(X_0)\theta]^{p'}$.

5 Bellman Chains

We can generalize i.i.c. sequences to the analogue of Markov chains that we will call Bellman chains.

Definition 17. A finite valued Bellman chain (E, C, ϕ) with $i)$ E a finite set called the state space of $|E|$ elements, $ii)$ $C : E \times E \mapsto \overline{\mathbb{R}}$ satisfying $\inf_y C_{xy} = 0$ called the transition cost, $iii)$ ϕ is a cost measure on E called the initial cost, is the decision sequence $\{X_n\}$ on $(U, \mathcal{U}, \mathbb{K})$, taking its values in E, such that $c_X(x \stackrel{\text{def}}{=} (x_0, x_1, \ldots)) = \phi_{x_0} + \sum_{i=0}^{\infty} C_{x_i x_{i+1}}$, $\forall x \in E^{\mathbb{N}}$.

Theorem 18. *For any function f from E into* $\overline{\mathbb{R}}$, *a Bellman chain satisfies the Markov property* $\mathbb{M}\{f(X_n) \mid X_0, \ldots, X_{n-1}\} = \mathbb{M}\{f(X_n) \mid X_{n-1}\}$.

The analogue of the forward Kolmogorov equation giving a way to compute recursively the marginal probability to be in a state at a given time is the following Bellman equation.

Theorem 19. *The marginal cost* $w^n_x = \mathbb{K}(X_n = x)$ *of a Bellman chain is given by the recursive forward equation:* $w^{n+1} = w^n \otimes C \stackrel{\text{def}}{=} \min_{x\in E}(w^n_x + C_{x.})$ *with* $w^0 = \phi$.

The cost measure of a Bellman chain is normalized which means that its infimum on all the trajectories is 0. In some applications we would like to avoid this restriction. This can be done by introducing the analogue of the multiplicative functionals of the trajectories of a stochastic process.

Theorem 20. *The value*

$$v_x^n \stackrel{\text{def}}{=} \mathbb{M}\left\{\sum_{k=n}^{N-1} f(X_k) + \psi(X_N) \mid X_n = x\right\}$$

with $f, \psi \in \overline{\mathbb{R}}^{|E|}$ *can be computed recursively by*

$$v^n = F \otimes C \otimes v^{n+1} = f(\cdot) + \min_y (C_{.y} + v_y^{n+1}), \quad v^N = \psi ,$$

where F *is the* $(|E|, |E|)$ *matrix defined by* $F_{xy} \stackrel{\text{def}}{=} f_x$ *if* $x = y$ *and* $+\infty$ *elsewhere. The matrix* F *can be seen as a normalizing factor of the transition cost.*

6 Continuous-Time Bellman Processes

We can define easily continuous time decision processes which correspond to deterministic controlled processes with a cost associated to each trajectory. We discuss here only decision processes with continuous trajectories.

Definition 21. Associated to continuous time decision processes we have the following definitions.

1. A *continuous time Bellman process* X_t on $(U, \mathcal{U}, \mathbb{K})$, with continuous trajectories, is a function from U into $\mathcal{C}(\mathbb{R}^+)$ (where $\mathcal{C}(\mathbb{R}^+)$ denotes the set of continuous functions over $\mathbb{R}^+$ into $\mathbb{R}$) having the cost density

   $$c_X(x(\cdot)) \stackrel{\text{def}}{=} \phi(x(0)) + \int_0^\infty c(t, x(t), x'(t))dt ,$$

   with $c(t, \cdot, \cdot)$ a family of transition costs (that is a mapping c from $\mathbb{R}^3$ into $\overline{\mathbb{R}}^+$ such that $\inf_y c(t, x, y) = 0, \ \forall t, x)$ and ϕ a cost density on $\overline{\mathbb{R}}$. When the integral is not defined the cost is by definition equal to $+\infty$.
2. The Bellman process is said *homogeneous if* c *does not depend on the time* t.
3. The Bellman process is said *with independent increments* if c does not depend on the state x. Moreover if this process is homogeneous c is reduced to the cost density of a decision variable.

Example 4. The following processes are fundamental.

1. The *p-Brownian decision process*, denoted by B_t^p, is the process with independent increments and transition cost density $c(t, x, y) = \frac{1}{p} y^p$.
2. The *p-diffusion decision process*, denoted by $\mathcal{M}_{m,\sigma,t}^p$ will correspond to the transition cost density $c(t, x, y) = \mathcal{M}_{m(t,x),\sigma(t,x)}^p(y)$.

As in the discrete time case, the marginal cost to be in a state x at a time t can be computed recursively using a forward Bellman equation.

Theorem 22. *The marginal cost* $w(t, x) \stackrel{\text{def}}{=} \mathbb{K}(X_t = x)$ *is given by the Bellman equation:*

$$\partial_t w + \hat{c}(\partial_x w) = 0, \quad w(0, x) = \phi(x) , \tag{1}$$

where $\hat{c}$ *means here* $[\hat{c}(\partial_x w)](t, x) \stackrel{\text{def}}{=} \sup_y [y \partial_x w(t, x) - c(t, x, y)]$.

Example 5. 1. For the Brownian decision process B_t^p starting from 0, the marginal cost to be at time t in the state x satisfies the Bellman equation:

$$\partial_t w + (1/p')[\partial_x w]^{p'} = 0, \quad w(0, \cdot) = \delta^c .$$

Its solution can be computed explicitly, it is $w(t, x) = \mathcal{M}^p_{0,t^{1/p'}}(x)$. Therefore we have

$$\mathbb{M}[f(B_t^p)] = \inf_x \left[f(x) + \frac{x^p}{pt^{\frac{p}{p'}}} \right] .$$

2. The marginal cost of the p-diffusion decision process $\mathcal{M}^p_{m,\sigma,t}$ starting from O, satisfies the Bellman equation: $\partial_t w + m\partial_x w + \frac{1}{p'}[\sigma \partial_x w]^{p'} = 0$, $w(0, \cdot) = \delta^c$. Explicit solution is known only in particular cases for instance when $m(t, x) = -\alpha x$ and σ is constant. This case gives a generalization of the well known LQ problem. The solution is $w(t, x) = (1/p)q(t)(x/\sigma)^p$ where $q(t)$ satisfies the Bernouilli equation

$$\dot{q}/p - \alpha q + q^{p'}/p' = 0, \quad q(0) = +\infty .$$

The backward Bellman equation gives a mean to compute the analogue of the multiplicative functionals of a stochastic process.

Theorem 23. *The functional* $v(t, x) = \mathbb{M}\{\psi(X_T) \mid X_t = x\}$, *where* ψ *is a mapping from* $\mathbb{R}$ *into* $\overline{\mathbb{R}}$ *and* X_t *is a Bellman process of transition cost c and initial cost* ϕ, *can be computed recursively by the Bellman equation*

$$\partial_t v(t, x) + \inf_y [y\partial_x v(t, x) + c(t, x, y)] = 0, \quad v(T, x) = \psi(x) .$$

Moreover we have $\mathbb{M}[\psi(X_T)] = \inf_x [w(t, x) + v(t, x)]$, $\forall t$, *where* w *is the solution of (1).*

Example 6. In the case of the p-Brownian decision process $X_t = B_t^p$ we have an explicit formula dual of the formula given in the previous example:

$$v(t, x) \stackrel{\text{def}}{=} \mathbb{M}\{\psi(X_T) \mid X_t = x\} = \inf_y \left[\psi(y) + \frac{1}{p}(y - x)^p/(T - t)^{p/p'} \right] ,$$

moreover when the initial cost of the Brownian is ϕ we have:

$$\mathbb{M}\{\psi(X_T)\} = \inf_{x,y} \left[\psi(y) + \frac{1}{p}(y - x)^p/T^{p/p'} + \phi(x) \right] .$$

Other explicit formulas involving stopping time, in the particular case $p = 2$, are given in [15].

Theorem 24. *Given n an integer,* $h = T/n$, *the linear interpolation of the Bellman chain of transition cost* $c_h(x, y)$ *defines a decision process with trajectories in* $C[0, T]$. *This process converges weakly, when n goes to* ∞, *towards the order p process* $\mathcal{M}^p_{m,\sigma,t}$ *as soon as:*

$$[\mathcal{F}(c_h(x, x + \cdot)](\theta) = [m(x)\theta + (1/p')(\sigma(x)\theta)^{p'}]h + o(h) ,$$

or equivalently if

$$i)\ \mathbb{O}_{X_k=x}[X_{k+1} - X_k] = m(x)h + o(h), \quad ii) \mathbb{S}^p_{X_k=x}[X_{k+1} - X_k] = \sigma(x)h^{1/p'} + o(h^{1/p'}) .$$

This result is called min-plus invariance principle. See Samborski and Dudnikov in these proceedings for related results.

7 Conclusion

Let us conclude by summarizing the morphism between probability calculus and decision calculus in Table 2.

Table 2. Morphism between probability calculus and decision calculus.

Probability	Decision
$+$	min
$\times$	$+$
Measure: $\mathbb{P}$	Cost: $\mathbb{K}$
$\int d\mathbb{P}(x) \stackrel{\text{def}}{=} 1$	$\min_x \mathbb{K}(x) \stackrel{\text{def}}{=} 0$
$\mathcal{N}_{m,\sigma}$	$\mathcal{M}^2_{m,\sigma}$
Convolution	Inf-Convolution
Laplace	Fenchel
Random Variable: X	Decision Variable: X
$\mathbb{E}(X) \stackrel{\text{def}}{=} \int x dF(x)$	$\mathbb{M}(X) \stackrel{\text{def}}{=} \inf_x \{x + \mathbb{K}(x)\}$
	$\mathbb{O}(X) \stackrel{\text{def}}{=} \arg\min_x \mathbb{K}(x)$
$\sigma(X) \stackrel{\text{def}}{=} \sqrt{\mathbb{E}[(X - \mathbb{E}(X))^2]}$	$\mathbb{S}^2(X) \stackrel{\text{def}}{=} \sqrt{1/c''_X(\mathbb{O}(X))}$
$\Phi(X) \stackrel{\text{def}}{=} \mathbb{E}(e^{\theta X})$	$\mathbb{F}(X) \stackrel{\text{def}}{=} -\mathbb{M}(-\theta X)$
$\log \Phi(X)'(0) = \mathbb{E}(X)$	$\mathbb{F}(X)'(0) = \mathbb{O}(X)$
$\log \Phi(X)''(0) = (\sigma(X))^2$	$\mathbb{F}(X)''(0) = (\mathbb{S}^2(X))^2$
Markov Chain	Bellman Chain
Kolmogorov Eq.	Bellman Eq.
Stochastic Process	Decision Process
Brownian Motion	2-Brownian Decision Process
Diffusion Process	2-Diffusion Decision Process
Heat Equation	Quadratic HJB Equation
$\partial_t + m(x)\partial_x + \frac{1}{p'}(\sigma(x))^{p'}\partial_x^{(p')}$	$\partial_t v + m(x)\partial_x v - \frac{1}{p'}\lvert\sigma(x)\partial_x v\rvert^{p'}$
Gaussian Kernel	Quadratic Kernel
$\frac{1}{\sqrt{2\pi t}}e^{-(x-y)^2/2t}$	$(x-y)^2/2t$
Invariance Principle	(Min,+) Invariance Principle

Notes and Comments. Bellman [6] was aware of the interest of the Fenchel transform (which he calls max transform) for the analytic study of the dynamic programming equations. The Cramer transform is an important tool in large deviations literature [4], [12], [17]. Maslov has developed a theory of idempotent integration [13]. In [15] and [7] the law of large numbers and the central limit theorem for decision variables has been given in the particular case $p = 2$. In two independent works [5] and [8] the study of decision variables have been started. Some aspects of [18] are strongly related to this morphism between probability and decision calculus in particular the morphism between LQG and LEG problem and the link with H_∞ problem. In [14] idempotent Sobolev spaces have been introduced as a way to study HJB equation as a linear object.

We would also like to thank P.L. Lions and R. Azencott for some nice comments showing us the role of the Cramer transform in the morphism between probability and decision calculus. These comments have been an important step in the maturation process of this paper.

References

1. Akian, M.: Idempotent integration and cost measures. INRIA Report (1994), to appear.
2. Akian, M.: Theory of cost measures: convergence of decision variables. INRIA Report (1994), to appear.
3. Attouch, H., Wets, R.J.B.: Isometries of the Legendre-Fenchel transform. Transactions of the American Mathematical Society **296** (1986) 33–60.
4. Azencott, R., Guivarc'h, Y., Gundy, R.F.: Ecole d'été de Saint Flour 8. Lect. Notes in Math., Springer-Verlag, Berlin (1978).
5. Bellalouna, F.: Processus de décision min-markovien. Thesis dissertation, University of Paris-Dauphine (1992).
6. Bellman, R., Karush, W.: Mathematical programming and the maximum transform. SIAM Journal of Applied Mathematics **10** (1962).
7. Baccelli, F., Cohen, G., Olsder, G.J., Quadrat, J.P.: Synchronization and Linearity: an Algebra for Discrete Event Systems. John Wiley and Sons, New York (1992).
8. Del Moral, P.: Résolution particulaire des problèmes d'estimation et d'optimisation non-linéaires. Thesis dissertation, Toulouse, France (1994), to appear.
9. Del Moral, P., Thuillet, T., Rigal, G., Salut, G.: Optimal versus random processes: the non-linear case. LAAS Report, Toulouse, France (1990).
10. Feller, W.: An Introduction to Probability Theory and its Applications. John Wiley and Sons, New York (1966).
11. Fenchel, W.: On the conjugate convex functions. Canadian Journal of Mathematics **1** (1949) 73–77.
12. Freidlin, M.I., Wentzell, A.D.: Random Perturbations of Dynamical Systems. Springer-Verlag, Berlin (1979).
13. Maslov, V.: Méthodes Opératorielles. Éditions MIR, Moscou (1987).
14. Maslov, V., Samborski, S.N.: Idempotent Analysis. Advances In Soviet Mathematics **13** Amer. Math. Soc., Providence, (1992).
15. Quadrat, J.P.: Théorèmes asymptotiques en programmation dynamique. Note CRAS Paris **311** (1990) 745–748.
16. Rockafellar, R.T.: Convex Analysis. Princeton University Press Princeton, N.J. (1970).
17. Varadhan, S.R.S.: Large deviations and applications. CBMS-NSF Regional Conference Series in Applied Mathematics, N. 46, SIAM Philadelphia, Penn., (1984).
18. Whittle, P.: Risk Sensitive Optimal Control. John Wiley and Sons, New York (1990).

Maslov Optimisation Theory: Stochastic Interpretation, Particle Resolution

Pierre DEL MORAL, Jean-Charles NOYER, Gérard SALUT

Laboratoire d'Automatique et d'Analyse des Systèmes du CNRS
7, Avenue du Colonel Roche 31077 Toulouse Cédex, France

1 Optimisation Processes Theory

Let $\mathcal{A} \stackrel{\text{def}}{=} [0, +\infty[$ be the semiring of real numbers endowed with the commutative semigroup laws $\oplus = \sup$, $\odot = +$, the neutral elements $\mathbf{1}, \mathbf{0}$ and the exponential metric ρ. We denote $(E, \mathcal{E})$ an $\mathcal{A}$-normal and locally compact space. Let (Ω, σ) be a measurable space, μ a Maslov measure on (Ω, σ), $\mathcal{M}(\Omega, \sigma)$, respectively $\mathcal{E}(\Omega, \sigma)$, the semiring of measurable functions, respectively step functions. The Maslov integral of $f \in \mathcal{M}(\Omega, \sigma)$ relative to μ may be defined as a Lebesgue integral:

$$\int_{\Omega}^{\oplus} f \odot \mu \stackrel{\text{def}}{=} \sup\left\{\int_{\Omega}^{\oplus} f_e \odot \mu : 0 \le f_e \le f, \quad f_e \in \mathcal{E}(\Omega, \sigma)\right\} \tag{1}$$

where
$$\int_{\Omega}^{\oplus} f_e \odot \mu \stackrel{\text{def}}{=} \bigoplus_{d \in f_e(\Omega)} d \odot \mu\left(f_e^{-1}(\{d\})\right) \ . \tag{2}$$

We use the following concepts for set theory: A Maslov measure $\mathbb{P}$ on (Ω, σ) such that $\mathbb{P}(\Omega) = \mathbf{1}$ is called a performance measure and $(\Omega, \sigma, \mathbb{P})$ an optimisation basis. A measurable function X from (Ω, σ) into $(E, \mathcal{E})$ is now called an optimisation variable, we denote $\mathbb{P}^X$ its performance measure, p^X its density function. Then for every $A \in \mathcal{E}$

$$\mathbb{P}(\{\omega \in \Omega : X(\omega) \in A\}) = \mathbb{P}^X(A) = \int_A^{\oplus} \mathrm{p}^X(x) \odot dx \ .$$

Let $\mathcal{L}(\Omega, \sigma, \mathbb{P})$ be the semiring of $\mathcal{A}$-valued optimisation variables X such that $\int_{\Omega}^{\oplus} X \odot \mathbb{P} < +\infty$. For every $X \in \mathcal{L}(\Omega, \sigma, \mathbb{P})$ we define its *Maslov Expectation* by $\int_{\Omega}^{\oplus} X \odot \mathbb{P} \stackrel{\text{def}}{=} \mathbb{E}(X)$.

Definition 1. Let $(X_i)_{i \in I}$ be a non necessarily finite family of E-valued optimisation variables on the same optimisation basis $(\Omega, \sigma, \mathbb{P})$. We say they are $\mathbb{P}$-independent when, for every finite subset $J = \{t_1, \ldots, t_n\} \subset I, n \ge 1$,

$$\mathbb{P}_J^X = \bigodot_{j=1}^{n} \mathbb{P}_{t_j}^X \ ,$$

where $\mathbb{P}_J^X$, respectively $\mathbb{P}_{t_j}^X$, $1 \leq j \leq n$, is the performance measure of $X_J = (X_{t_1}, \ldots, X_{t_n})$, respectively X_{t_i}, $j \in [1, n]$.

Conditioning : Let A,B $\in \sigma$ with $\mathbb{P}(B) \neq 0$. We define the conditional performance of A relative to B and we denote $\mathbb{P}(A/B)$, by

$$\mathbb{P}(A/B) = \frac{\mathbb{P}(A \cap B)}{\mathbb{P}(B)}_{\ominus} ,$$

Where, for every $a, b \in \mathcal{A}, b \neq 0$, $\frac{a}{b}_{\ominus} \stackrel{\text{def}}{=} a \odot (-1)b$. This definition is the extension of Bayes formula to performance measures. Define $\mathbf{L}(\Omega, \sigma, \mathbb{P}) \stackrel{\text{def}}{=} \mathcal{L}(\Omega, \sigma, \mathbb{P}) /_{\mathbb{P}-\text{a.s.}}$

Theorem 2. *Let X, Y be two E-valued optimisation variables defined on the same optimisation basis $(\Omega, \sigma, \mathbb{P})$. We denote $\sigma(X, Y)$ the σ-algebra spanned by X and Y, $\sigma(Y)$ the σ-algebra spanned by Y. For every $\varphi \in \mathbf{L}(\Omega, \sigma(X, Y), \mathbb{P})$ there exists a unique function $\mathbb{E}(\varphi/\sigma(Y)) \in \mathbf{L}(\Omega, \sigma(Y), \mathbb{P})$ such that:*

$$\forall \psi \in \mathbf{L}(\Omega, \sigma(Y), \mathbb{P}) \quad , \quad \mathbb{E}(\varphi \odot \psi) = \mathbb{E}(\mathbb{E}(\varphi/\sigma(Y)) \odot \psi) \ .$$

To prove this, we use the conditional performance of X relative to Y, $\mathrm{p}^{X/Y}(x/y) \stackrel{\text{def}}{=} \frac{\mathrm{p}^{X,Y}(x,y)}{\mathrm{p}^{Y}(y)}_{\ominus}$ when $\mathrm{p}^Y(y) \neq 0$, 0 otherwise.

Moreover for every $a, b \in \mathcal{A}$, $\varphi, \phi \in \mathbf{L}(\Omega, \sigma(X, Y, Z), \mathbb{P})$, $\psi \in \mathbf{L}(\Omega, \sigma(Y), \mathbb{P})$

$$\mathbb{E}(a \odot \varphi \oplus b \odot \phi / Y) = a \odot \mathbb{E}(\varphi/Y) \oplus b \odot \mathbb{E}(\phi/Y), \quad \mathbb{E}(\psi/Y) = \psi \ ,$$

$$\mathbb{E}(\mathbb{E}(\phi/X, Y)/Y) = \mathbb{E}(\phi/Y) \ .$$

Maslov Processes: A E-valued optimisation process with time space I is a system $(\Omega, \sigma, \mathbb{P}, X = \{X_t\}_{t \in I})$ defined by an Optimisation Basis $(\Omega, \sigma, \mathbb{P})$ and a family of E-valued optimisation variables $(X = \{X_t\}_{t \in I})$ defined on $(\Omega, \sigma, \mathbb{P})$. The following result is the extension of the Markov causality principe to optimisation processes. It states that the Bellman optimality equation coincide with the Chapman-Kolmogorov transport equation of the associated Maslov process.

Proposition-Definition 3. *Let $(\Omega, \sigma, \mathbb{P}, X = \{X_t\}_{t \in I})$ a E-valued optimisation process. We say that X is a Maslov process when the future depends on the past only through the present time, in other words for every subdivision $s_1 \leq \ldots \leq s_m \leq t \leq t_1 \leq \ldots \leq t_n$, $n, m \geq 1$ of I, $\varphi \in \mathcal{M}(E^{m+1}, \otimes_{i=0}^m \mathcal{E})$, $\psi \in \mathcal{M}(E^{n+1}, \otimes_{i=0}^n \mathcal{E})$, $\mathbb{P}$-pp*

$$\mathbb{E}(\varphi(X_{\underline{s_m}}, X_t) \odot \psi(X_t, X_{\underline{t_n}}) / X_t) = \mathbb{E}(\varphi(X_{\underline{s_m}}, X_t) / X_t) \odot \mathbb{E}(\psi(X_t, X_{\underline{t_n}}) / X_t) \ ,$$

where $X_{\underline{t_n}} \stackrel{\text{def}}{=} (X_{t_1}, \ldots, X_{t_n})$ and $X_{\underline{s_m}} \stackrel{\text{def}}{=} (X_{s_1}, \ldots, X_{s_m})$. In that case, for every $0 \leq r \leq s \leq t$ in I,

$$\mathrm{p}_{t/r}^X(z/x) = \int_E^{\oplus} \mathrm{p}_{t/s}^X(z/y) \odot \mathrm{p}_{s/r}^X(y/x) \odot dy \quad \textit{(Bellman Optimality Equation)},$$

where $\mathrm{p}_{\tau_1/\tau_2}^X \stackrel{\text{def}}{=} \mathrm{p}^{X_{\tau_1}/X_{\tau_2}}$, $\tau_1 \leq \tau_2$.

From the above concepts we derive(see [13] or [7], [10]) a law of large numbers in the wide sense and a Central Limit theorem at the same level of generality as that of probability theory. These results have to be compared with the time asymptotic approach in [5] or [11] which can be viewed as ergodic properties of some optimisation processes in such a framework.

Among the most exciting developments in this theory are those concerning *the changes of reference performance measure*(see [13] or [7], [10]). As in stochastic theory, given a Maslov process X we may derive a change of reference performance which "*modifies the drift of X*", the induced Radon-Nykodim derivatives are the analogues of Girsanov exponentials in stochastic theory and they perform as $\mathbb{P}$-*martingales in such a framework.*

We conclude that Filtering and Optimisation problems can be embedded in a unique theory of normed measures (see [13], [7], [10] for details). Only the choice of the reference semiring indicates whether we assign a measure of probability or performance to an event.

2 Particle Optimisation.

The recently developed particle methods for nonlinear filtering problems (see [13], [12], [8], [9]) provide new approaches for above Filtering and Optimisation problems. They are indeed, efficient non linear stochastic procedures to obtain global solutions.

Log-Exp Transformation: We briefly recall the Log-Exp transformation(for details see [13]). This transformation leads to useful conclusions because it makes explicit the relationship between the performance and the probability measure of an event.

Let $\nu > 0, d \geq 1$, we denote D_d^ν the class of probability measures p on $\mathbb{R}^d$ such that $\int_{\mathbb{R}^d}^{\oplus} \log(p(x)^\nu) \odot dx \stackrel{\text{def}}{=} \mathbb{N}_\nu(p) > 0$ and, $\mathbb{D}_d^\nu$ the class of performance measures $\mathbb{p}$ on $\mathbb{R}^d$ such that $\int_{\mathbb{R}^d} \exp\left(\frac{\mathbb{p}(x)}{\nu}\right) dx \stackrel{\text{def}}{=} N_\nu(\mathbb{p}) > 0$. We use the conventions, when discrete events are embedded in a continuous fashion:

$$\begin{cases} \log\left(\sum_{n\geq 0} p_n \delta_{z_n}\right) & \stackrel{\text{def}}{=} \bigoplus_{n\geq 0} \log(p_n) \odot 1_{z_n} \ , \\ \exp\left(\bigoplus_{n\geq 0} \mathbb{p}_n \odot 1_{z_n}\right) & \stackrel{\text{def}}{=} \sum_{n\geq 0} \exp(\mathbb{p}_n) \delta_{z_n} \ . \end{cases} \tag{3}$$

These spaces are in a one to one correspondence by the following transformations:

$$\mathbb{E}\text{xp}_\nu(\mathbb{p}) \stackrel{\text{def}}{=} \frac{e^{\frac{1}{\nu}\mathbb{p}}}{N_\nu(\mathbb{p})} \ , \quad \mathbb{L}\text{og}_\nu(p) \stackrel{\text{def}}{=} \frac{\log p^\nu}{\mathbb{N}_\nu(p)}_{\oplus} \ , \quad \mathbb{E}\text{xp}_\nu = \mathbb{L}\text{og}_\nu^{-1} \ . \tag{4}$$

Let $(\Omega, \mathcal{F}, \mathbb{P})$ be an optimisation basis, $\mathbf{F} \stackrel{\text{def}}{=} (\mathcal{F}_k)_{k\geq 0}$ an increasing filtration of $\mathcal{F}$ on which are defined two optimisation processes U and V with values in $\mathbb{R}^{n_X}$ and $\mathbb{R}^{n_Y}$, $n_X, n_Y \geq 1$. We assume for every $k \geq 0$

$$\left(U_{\underline{k}}, V_{\underline{k}}\right) \stackrel{\text{def}}{=} (U_0, U_1, \ldots, U_k, V_0, V_1, \ldots, V_k) \text{ are } \mathbb{P} - \text{independent.}$$

Let us now define on $(\Omega, \mathcal{F}, \mathbb{P})$ the following optimisation processes, for every $k \geq 0$

$$(\mathcal{O}) \begin{cases} X_k = F(X_{k-1}, U_k) \ , & X_0 = U_0 \ , \\ Y_k = H(X_k) + V_k \ , \end{cases}$$

with F a measurable function from $\mathbb{R}^{n_X} \times \mathbb{R}^{n_X}$ into $\mathbb{R}^{n_X}$, H a measurable function from $\mathbb{R}^{n_X}$ into $\mathbb{R}^{n_Y}$. From the above

$$\mathrm{p}^{U_{\underline{k}},Y_{\underline{k}}}(u,y) = \mathrm{p}^{U_{\underline{k}}}(u) \odot \mathrm{p}^{V_{\underline{k}}}(y - H(\phi(u))) \ ,$$

where $\phi(u) = X_{\underline{k}}$ the state path associated to the value $U = u_{\underline{k}}$.

Assume $\mathcal{U}_{k,x} \stackrel{\text{def}}{=} \{u \in (\mathbb{R}^{n_X})^{[0,k]} : \phi_k(u) = x\}$. Because $\mathrm{p}^{X_k,Y_{\underline{k}}}$ is the performance measure of $(X_k, Y_{\underline{k}})$, for every $x \in \mathbb{R}^{n_X}$, $y \in \mathbb{R}^{n_Y[0,k]}$ it follows that $\mathrm{p}^{X_k,Y_{\underline{k}}}(x,y) = \sup_{u\in\mathcal{U}_{k,x}} \mathrm{p}^{U_{\underline{k}},Y_{\underline{k}}}(u,y)$. Define

1. $\mathrm{op}(X_k / y) \stackrel{\text{def}}{=} \arg\sup_{x\in\mathbb{R}^{n_X}} \mathrm{p}^{X_k,Y_{\underline{k}}}(x,y)$;
2. $\mathrm{op}(U_{\underline{k}} / y) \stackrel{\text{def}}{=} \arg\sup_{u\in(\mathbb{R}^{n_X})^{[0,k]}} \mathrm{p}^{U_{\underline{k}},Y_{\underline{k}}}(u,y)$.

In every optimisation problem we have to compute either 1 or 2. It is now straightforward to apply the Log-Exp transformation. For every $\nu > 0$ such that

$$\forall k \geq 0 \quad N_\nu(k) \stackrel{\text{def}}{=} \int_{(\mathbb{R}^{n_X}\times\mathbb{R}^{n_Y})^{[0,k]}} \exp\left(\frac{1}{\nu}\mathrm{p}^{U_{\underline{k}},Y_{\underline{k}}}(u,y)\right) du\, dy > 0 \ .$$

The measure $p^{W^\nu_{\underline{k}},Y^\nu_{\underline{k}}} \stackrel{\text{def}}{=} \mathbb{E}\mathrm{xp}_\nu\left(\mathrm{p}^{U_{\underline{k}},Y_{\underline{k}}}\right) = \frac{1}{N_\nu(k)}\exp\left(\mathrm{p}^{U_{\underline{k}},Y_{\underline{k}}}\right)$ is the probability measure associated to the filtering problem $\mathcal{F}^\nu$ defined for every $k \geq 0$, by:

$$(\mathcal{F}^\nu)\begin{cases} X^\nu_k = F\left(X^\nu_{k-1}, W^\nu_k\right) \ , & X^\nu_0 =, W^\nu_0 \ , \\ Y^\nu_k = H\left(X^\nu_k\right) + V^\nu_k \ , & \end{cases}$$

where W^ν, V^ν are two P-independent stochastic processes with probability measures $\mathbb{E}\mathrm{xp}_\nu\left(\mathrm{p}^U\right)$ and $\mathbb{E}\mathrm{xp}_\nu\left(\mathrm{p}^V\right)$.

Example 1. Let $n_X = n_Y = 1$, $H(x) = x$, $\tau \in [0,k]$, $0 < \lambda < 1$, c a fixed real value

1. $\mathrm{p}^{V_\tau}(u) = -\frac{1}{2}v^2 \Longrightarrow \mathrm{p}^{V_{\underline{k}}}(c - H(\phi(u))) = -\frac{1}{2}\sum_{\tau=0}^{k}(c - \phi_\tau(u))^2$,
2. $\mathrm{p}^{U_\tau}(u) = -\frac{1}{2}u^2 \Longrightarrow p^{W^\nu_\tau}(u) \stackrel{\text{def}}{=} \mathbb{E}\mathrm{xp}_\nu\left(\mathrm{p}^{U_\tau}\right)(u) = \frac{1}{\sqrt{2\nu\pi}}e^{-\frac{1}{2\nu}u^2}$,
3. $\mathrm{p}^{U_\tau}(u) = \log\left(\frac{\lambda}{\lambda\oplus(1-\lambda)}\right) \odot 1_1 \oplus \log\left(\frac{1-\lambda}{\lambda\oplus(1-\lambda)}\right) \odot 1_0 \Longrightarrow p^{W^\nu_\tau}(u) \stackrel{\text{def}}{=}$ $\mathbb{E}\mathrm{xp}_\nu\left(\mathrm{p}^{U_\tau}\right)(u) = \frac{\lambda^{\frac{1}{\nu}}}{\lambda^{\frac{1}{\nu}}+(1-\lambda)^{\frac{1}{\nu}}}\delta_1 + \left(1 - \frac{\lambda^{\frac{1}{\nu}}}{\lambda^{\frac{1}{\nu}}+(1-\lambda)^{\frac{1}{\nu}}}\right)\delta_0$,
4. initial constraint: $\mathrm{p}^{U_0} = 1_{x_0} \Longrightarrow p^{W^\nu_0} \stackrel{\text{def}}{=} \mathbb{E}\mathrm{xp}_\nu\left(\mathrm{p}^{U_0}\right) = \delta_{x_0}$,
5. Final constraint: $\mathrm{p}^{V_k} = 1_0 \Longrightarrow \mathrm{p}^{V_k}(c - H(\phi_k(u))) = 1_c(\phi_k(u)) \Longrightarrow p^{V^\nu_k} \stackrel{\text{def}}{=}$ $\mathbb{E}\mathrm{xp}_\nu\left(\mathrm{p}^{U_k}\right) = \delta_0$.

In other words one may regard the optimisation problem $\mathcal{O}$ as the maximum likelihood estimation problem associated to a filtering problem $\mathcal{F}^\nu$, $\nu > 0$. This last procedure computes the most likely trajectory in function space and then uses the final state of this trajectory as an estimate. We then have converted $\mathcal{O}$ to a form which can be treated by filtering theory. We now state a theorem which will serve as a tool for computing the maximum likelihood estimate by a sequential particle procedure similar to those developed for filtering problems.

Theorem 4. *Let X a $\mathbb{R}^n$-valued optimisation variable defined on a performance basis $(\Omega, \mathcal{F}, \mathbb{P})$ which has a unique optimal value* $\text{op}(X)$. *Let $N \geq 1$, $\left(\check{X}^i\right)_{1 \leq i \leq N}$ a sequence of N P-independent random variables defined on a stochastic basis $(\Omega, \mathcal{F}, P)$ with the same probability distribution. Define* $\text{op}_N(X) \stackrel{\text{def}}{=} \arg\sup_{1 \leq i \leq N} \mathbf{p}^X\left(\check{X}^i\right)$. *When the following conditions are satisfied*

1. $\mathcal{H}_1$ (Regularity):

$$\exists \epsilon_0 > 0, a > 0 : \forall 0 < \epsilon \leq \epsilon_0 \ ,$$
$$\rho\left(\mathbf{p}^X(\text{op}(X)), \mathbf{p}^X(x)\right) \leq \epsilon \Rightarrow \|x - \text{op}(X)\|_{\mathbb{R}^n} \leq a\epsilon \ , \quad (5)$$

2. $\mathcal{H}_2$ (Detectability):

$$\forall \epsilon > 0 \ , \quad P\left(\rho\left(\mathbf{p}^X(\text{op}(X)), \mathbf{p}^X\left(\check{X}^1\right)\right) \leq \epsilon\right) > 0 \ ,$$

we have

$$\forall \epsilon > 0 \quad \lim_{N \to +\infty} P\left(\|\text{op}_N(X) - \text{op}(X)\|_{\mathbb{R}^n} \leq \epsilon\right) = 1 \ . \quad (6)$$

The Log-Exp transformation makes explicit the relationship between the performance and the probability measure of an event. It provides good probability distribution candidates for the previous detectability assumption. We first recall some details about the decomposition of the probability measure $p^{W_{\underline{k}}^\nu, Y_{\underline{k}}^\nu} = \mathbb{E}\text{xp}_\nu\left(\mathbf{p}^{U_{\underline{k}}, Y_{\underline{k}}}\right)$. With obvious abusive notations:

1. $p\left(w_{\underline{k}}^\nu, y_{\underline{k}}^\nu\right) = p\left(y_{\underline{k}}^\nu / w_{\underline{k}}^\nu\right) p\left(w_{\underline{k}}^\nu\right) \quad p\left(w_{\underline{k}}^\nu\right) = \prod_{l=0}^k p\left(w_l^\nu\right)$. We define for every $k \geq 0$ $\check{W}_{\underline{k}}^{\nu,0}$ a stochastic process with probability distribution $p\left(w_{\underline{k}}^\nu\right)$
2. Let σ be a subdivision of $\mathbb{N}$, $k \geq 0$, $d \geq 0$, $\overline{\sigma} = (\overline{\sigma}_l)_{l \geq 0}$ the partition of $\mathbb{N}$ induced by σ and defined by $\overline{\sigma}_l =]\sigma_{l-1}, \sigma_l]$, $l \geq 0$, with convention $\sigma_{-1} = -1.\Sigma_d \stackrel{\text{def}}{=} \left\{\sigma^0, \ldots, \sigma^{d-1}\right\}$, the partition of $\mathbb{N}$ formed by the d complementary subdivisions $(\sigma^q)_{0 \leq q \leq (d-1)}$, defined for every $0 \leq q \leq (d-1)$ by $\sigma^q = \left(\sigma_k^q\right)_{k \geq 0}$ with $\sigma_k^q = kd + q$. For every $k \geq 0$,

$$p\left(w_{\overline{\sigma}_{\underline{k}}}^\nu, y_{\overline{\sigma}_{\underline{k}}}^\nu\right) = \prod_{l=0}^k p\left(y_{\overline{\sigma}_l}^\nu / w_{\overline{\sigma}_{\underline{(l-1)}}}^\nu\right) \prod_{l=0}^k p\left(w_{\overline{\sigma}_l}^\nu / w_{\overline{\sigma}_{\underline{l-1}}}^\nu, y_{\overline{\sigma}_{l-1}}^\nu\right) \ . \quad (7)$$

We define for every $k \geq 0$ $\check{W}_{\underline{k}}^{\nu,d}$ a stochastic process depending on Y^ν with a conditional probability measure version $\prod_{l=0}^k p\left(w_{\overline{\sigma}_l}^\nu / w_{\overline{\sigma}_{\underline{l-1}}}^\nu, y_{\overline{\sigma}_{l-1}}^\nu\right)$.
In [13] we give some procedures for sampling such trajectories. Moreover we introduce a new Sampling/Resampling procedure.

With the same lines of arguments as in the proof of the previous theorem:

$$\bigoplus_{i=1}^N \mathbf{p}^{U_{\underline{k}}, Y_{\underline{k}}}\left(\check{W}_{\underline{k}}^{\nu,q,i}, Y_{\underline{k}}^\nu\right) \xrightarrow[N \to +\infty]{} \mathbf{p}^{U_{\underline{k}}, Y_{\underline{k}}}\left(\text{op}\left(U_{\underline{k}} / Y_{\underline{k}}^\nu\right), Y_{\underline{k}}^\nu\right) \ ,$$

with additional regularity conditions:

$$\mathrm{op}^{N,q,\nu}\left(X_k \Big/ Y_{\underline{k}}^{\nu}\right) \xrightarrow[N \to +\infty]{} \mathrm{op}\left(X_k \Big/ Y_{\underline{k}}^{\nu}\right) ,$$

where $\mathrm{op}^{N,q,\nu}\left(X_k \Big/ Y_{\underline{k}}^{\nu}\right) \stackrel{\text{def}}{=} \phi_k\left(\mathrm{op}^{N,q,\nu}\left(U_{\underline{k}} \Big/ Y_{\underline{k}}^{\nu}\right)\right)$ denotes the final value of the trajectory controlled by

$$\mathrm{op}^{N,q,\nu}\left(U_{\underline{k}} \Big/ Y_{\underline{k}}^{\nu}\right) \stackrel{\text{def}}{=} \arg \sup_{u_{\underline{k}} \in \left\{\check{W}_{\underline{k}}^{\nu,q,1},\dots,\check{W}_{\underline{k}}^{\nu,q,N}\right\}} \mathrm{p}^{U_{\underline{k}},Y_{\underline{k}}}\left(u_{\underline{k}}, Y_{\underline{k}}^{\nu}\right) .$$

Moreover the weights in the algorithm represent the likelihood of each trajectory. They are time degenerative(see [13]). We then introduce a regularization kernel as it was shown in particle filtering(see [13], [7], [10], [12]). This kind of regularization is necessary to avoid degeneracy, while using the useful content of the path performance data.

Finally we prove the following theorem.

Theorem 5. *If the optimisation problem* $(\mathcal{O})$ *satisfies regularity and detectability conditions for some* $\nu > 0, d \geq 1$ *(see [13]), then for every* $\epsilon > 0$

$$\lim_{N\to+\infty} \sup_{k\geq 0} P\left(\left\|\mathrm{op}_{r(N)}^{N,\nu,d}\left(X_k \Big/ Y_{\underline{k}}^{\nu}\right) - \mathrm{op}\left(X_k \Big/ Y_{\underline{k}}^{\nu}\right)\right\| > \epsilon\right) = 0 ,$$

where

1. $r(N)_{N\geq 1}$ *denotes a sequence of optimal regularization kernels.*
2. $\mathrm{op}_{r(N)}^{N,q,\nu}\left(X_k \Big/ Y_{\underline{k}}^{\nu}\right) \stackrel{\text{def}}{=} \phi_k\left(\mathrm{op}_{r(N)}^{N,q,\nu}\left(U_{\underline{k}} \Big/ Y_{\underline{k}}^{\nu}\right)\right)$ *the final value of the trajectory controlled by* $\mathrm{op}_{r(N)}^{N,q,\nu}\left(U_{\underline{k}} \Big/ Y_{\underline{k}}^{\nu}\right) \stackrel{\text{def}}{=} \arg\sup_{u_{\underline{k}} \in \left\{\check{W}_{\underline{k}}^{\nu,q,1},\dots,\check{W}_{\underline{k}}^{\nu,q,N}\right\}} \mathrm{p}_{r(N)}^{U_{\underline{k}},Y_{\underline{k}}}\left(u_{\underline{k}}, Y_{\underline{k}}^{\nu}\right)$.
3. $\mathrm{p}_{r(N)}^{U_{\underline{k}},Y_{\underline{k}}}$ *the regularized performance induced by* $r(N)$.

Concluding Remark

As a matter of conclusion, it should be underlined that the above particle resolution has nothing to do with simulated annealing. The latter is a search technique concerning static optimisation problems, where the "search dynamics" has a degree of arbitrariness left to temperature. The former concerns dynamic optimisation problems and the trajectories that are generated describe the whole probability/performance space in such a way that arbitrariness is excluded, except as far as the number of particles is concerned.

References

1. V. Maslov : Méthodes opératorielles. Edition Mir, 1987.
2. T. Huillet, G. Salut, Etude de la dualité entre les processus stochastiques et les processus optimaux. Rapport LAAS N° 89025, Janvier 1989.

3. T. Huillet, G. Salut, Optimal versus Random processes: the regular case. Rapport LAAS $N°$ 89189, Juin 1989, 23p.
4. T. Huillet,G. Rigal, G. Salut, Optimal versus Random processes: a general framework Rapport LAAS $N°$ 89251, Juillet 1989, 6p.
5. J.P. Quadrat : Théorèmes asymptotiques en programmation dynamique. Comptes Rendus à l'Académie des Sciences, 311:745-748,1990.
6. P. Del Moral, T. Huillet, G. Rigal, G. Salut : Optimal versus random processes: the nonlinear case. Rapport LAAS $N°$ 91131, Avril 1991.
7. P. Del Moral, G. Rigal, G. Salut : Estimation et commande optimale non linéaire: un cadre unifié pour la résolution particulaire. Rapport LAAS $N°$ 91137, Rapport intermédiaire $N°$ 4 DIGILOG-DRET $N°$ 89.34.553.00.470.75.01, Avril 1991.
8. P. Del Moral, G. Rigal, G. Salut, "Estimation et commande optimale non linéaires: la résolution particulaire en estimation/filtrage-Résultats expérimentaux" Convention D.R.E.T., $N°$ 89.34.553.00.470.75.01, Rapport intermédiaire $N°$ 2, Janvier 1992
9. P. Del Moral, G. Rigal, G. Salut, "Filtrage non-linéaire non-gaussien appliqué au recalage de plates formes inertielles, Mise en ouevre par méthodes particulaires", Contrat S.T.C.A.N $N°$ A.91.77.013, Rapport final, Juin 1992.
10. P. Del Moral, G. Rigal, G. Salut : Estimation et commande optimale non linéaire: Méthodes de résolutions particulaires. Rapport final de contrat DIGILOG-DRET $N°$ 89.34.553.00.470.75.01, Octobre 1992.
11. F. Baccelli, G. Cohen, G.J. Olsder, J.P. Quadrat: Synchronization and linearity. An algebra for discrete event systems. John Wiley and sons, 1992.
12. P. Del Moral, J.C. Noyer, G. Rigal, G. Salut, "Traitement non-linéaire du signal par réseau particulaire: application Radar", $14^{\text{ième}}$ Colloque GRETSI, Juan-les-Pins, 13-16 Septembre 1993.
13. P. Del Moral : Résolution particulaire des problèmes d'estimation et d'optimisation non linéaires. Thèse Université Paul Sabatier, Toulouse, à paraître.

Networks Methods for Endomorphisms of Semimodules Over Min-Plus Algebras

Peter I. Dudnikov[1] *and Sergei N. Samborski*[2]

[1] 252510 Bojarka, Kiev obl., ul. Lenina 42-1, Ukraine
[2] Dept. of Maths., University of Caen, 14032 Caen Cedex, France.
e-mail: Samborsk@univ-caen.fr

1 Introduction

The usual problem of Dynamic Programming

$$\sum_{i=1}^{n} L(x_i, u_i) + \Phi(x_n) \to \inf, \tag{1}$$

where $x_i \in X$, $u_i \in \mathbb{R}^n$, $x_{i+1} = x_i + u_i$, $L : X \times \mathbb{R}^n \to \mathbb{R} \cup \{\infty\}$, $\Phi : X \to \mathbb{R} \cup \{\infty\}$, $L(x, u) = \infty$ for $x + u \notin X$, leads to the study of the mapping

$$(Af)(x) = \inf_{\xi \in X}(L(x, \xi - x) + f(\xi)).$$

According to the optimality principle, the *value function* $S_n(x)$ of the problem (1) (i.e. the value of inf over the set of admissible pairs $(x_1, u_1), \ldots, (x_n, u_n)$, where $x_1 = x$) is determined by iterations of the mapping A:

$$S_n(x) = (A^n \Phi)(x).$$

It is natural to consider the mapping A as an "integral" operator with the kernel $a(x, \xi) = L(x, \xi - x)$ in a functional semimodule over the semiring $\mathcal{R} = \mathbb{R} \cup (+\infty)$ with the operations $\oplus = \min$, $\odot = +$ [1, 2]. The usual method of computation of an integral operator consists in replacing it by a sum over the nodes of a network. Below we investigate convergence in terms of which this procedure applied to the operator A of the problem (1) is correct.

2 Weak Convergence

In the semimodule $\mathcal{R} = (\mathbb{R} \cup (+\infty), \oplus, \odot)$ we introduce the metric $\rho(a, b) = |\exp(-a) - \exp(-b)|$ and denote $+\infty$ (the neutral element with respect to $\oplus$) by $\mathbf{0}$. Let $X \subset \mathbb{R}^n$ be the closure of a bounded open domain, Ψ be the set of functions $\varphi : X \to \mathcal{R}$ that are differentiable at the points of X,where their values are finite. Let $(\cdot, \cdot)$ be the following scalar product

$$(f, g) = \oint_X f \odot g \stackrel{\text{def}}{=} \inf_{x \in X}(f(x) + g(x)).$$

The set of equivalence classes with respect to the equivalence relation $f \sim g \Leftrightarrow (f, \varphi) = (g, \varphi), \forall \varphi \in \Psi$ [2] forms a topological semimodule with the convergence

$$f_i \to f \text{ whenever } \rho((f_i, \varphi), (f, \varphi)) \to 0, \forall \varphi \in \Psi$$

(see [2]). This semimodule, called the Maslov space, is denoted $\mathcal{P}(X)$. The kernel $a : X \times X \to \mathcal{R}$ determines an endomorphism $f \to Af$,

$$(AF)(x) = \oint_X a(x, \xi) \odot f(\xi) \stackrel{\text{def}}{=} \inf_{\xi \in X}(a(x, \xi) + f(\xi))$$

of the semimodule $\mathcal{P}(X)$.

For every $i \in \mathbb{N}$, let S_i be a finite subset of X (a network).

Definition 1. We shall say that a sequence $\{f_i, f_i \in \mathcal{P}(S_i)\}_{i \in \mathbb{N}}$ weakly converges to an element $f \in \mathcal{P}(X)$, if for every function $\varphi \in \Psi$ one has

$$\rho((f_i, \varphi|_{S_i})_i, f(\varphi)) \to 0,$$

where $(\cdot, \cdot)_i$ is the scalar product in $\mathcal{P}(S_i)$, i.e. $(\alpha, \beta)_i = \bigoplus_{x \in S_i} \alpha(x) \odot \beta(x)$.

Theorem 2. *Let $\{S_i\}_{i \in \mathbb{N}}$ be a sequence of finite subsets in X, $\{a_i, a_i \in \mathcal{P}(S_i \times S_i)\}_{i \in \mathbb{N}}$ a sequence of matrices which weakly converges to a kernel $a \in \mathcal{P}(X \times X)$, $\{\Phi_i, \Phi_i \in \mathcal{P}(S_i)\}_{i \in \mathbb{N}}$ a sequence which weakly converges to $\Phi \in \mathcal{P}(X)$. Then for the endomorphisms A_i of $\mathcal{P}(S_i)$ determined by the matrix a_i, namely $(A_i\alpha)(x) = \bigoplus_{\xi \in S_i} a_i(x, \xi) \odot \alpha(\xi)$, the sequence $\{A_i \Phi_i\}_{i \in \mathbb{N}}$ weakly converges to $A\Phi$.*

3 Approximation

In fact there exists a notion of strong convergence in a functional semimodule with a topology defined by the Hamiltonian that corresponds to the initial Lagrangian from (1). In order to define this convergence we introduce a finite difference analogue of the $\mathcal{P}_H$-convergence [3].

Let $H(x, \cdot)$ be the Legendre transform (the Fourier transform in terms of the operations in $\mathcal{R}$) of a function $(x, \eta) \to L(x, \eta)$ with respect to the second argument. We shall suppose that $H(x, p)$ is finite for $x \in X$, $p \in \mathbb{R}^n$ and $H : X \times \mathbb{R}^n \to \mathbb{R}$ is continuous. The following proposition was obtained in [2] in the case of strictly convex Hamiltonians that tend to ∞ as $|p| \to \infty$ and in [3] in the general situation.

Proposition 3. *There exists a closed topological semimodule $\mathcal{P}_H$ which continuous operations $\oplus : \mathcal{P}_H \times \mathcal{P}_H \to \mathcal{P}_H$ and $\odot : \mathcal{P}_H \times \mathcal{P}_H \to \mathcal{P}_H$ that is an algebraic sub-semimodule in $\mathcal{P}(X)$ such that the set of differentiable functions is dense in $\mathcal{P}_H$ and the differential expression $y \to \mathcal{H}(y)$, $(\mathcal{H}(y))(x) = H(x, \mathcal{D}y)$ — $\mathcal{D}$ is the gradient — is extended from the set of differentiable functions to a continuous mapping from $\mathcal{P}_H$ to $\mathcal{P}$. $\mathcal{P}_H$ possesses the property of maximality (with respect to inclusion), and is therefore unique.*

In the case of a non-strictly convex Hamiltonian $p \to H(x, p)$ the semimodule $\mathcal{P}_H$ may contain some discontinuous functions.

Let $S_i X \subset X$ be a finite subset (a network) with an adjacency relation for its elements such that for every $x \in S_i$ the set of nodes, adjacent to x, is determined. Let $f_i : S_i \to \mathcal{R}$ and $f_i \to \mathcal{D}_{(i)} f_i$ be the difference gradient:

$$(\mathcal{D}_{(i)} f_i)(s) = \sum_j \frac{f_i(s_j) - f_i(s)}{d(s_j, s)} \overrightarrow{(S_j, s)},$$

where the sum is taken over the nodes s_j adjacent to s. We suppose that the sequence of networks $\{S_i\}_{i=1}^{\infty}$ is such that for every differentiable function φ we have: $\lim \mathcal{D}_{(i)}\varphi = \mathcal{D}\varphi$ (in $\mathbb{R}^n$). This gives rise to the mappings $\mathcal{H}_i$ from $\mathcal{P}(S_i)$ to $\mathcal{P}(S_i)$ defined by

$$(\mathcal{H}_i f_i)(x) = H(x, (\mathcal{D}_{(i)} f_i)(x)) \quad , \quad x \in S_i .$$

Let Aff be the set of affine mappings $X \to \mathbb{R}$.

Definition 4. We shall say that a sequence $\{f_i \mid f_i \in \mathcal{P}(S_i)\}_{i=1}^{\infty}$ H-converges to $f \in \mathcal{P}_H$, if f_i weakly converges to f and $\mathcal{H}_i(f_i \oplus \Psi|_{S_i})$ weakly converge to $\mathcal{H}(f \oplus \Psi)$, $\forall \Psi \in$ Aff uniformly with respect to $\Psi \in$ Aff.

Theorem 5. *Let A be an integral endomorphism with the kernel $a(x, \xi) = L(x, \xi - x)$, $H(x, \cdot)$ be Legendre transform of the function $z \to L(x, z)$. Suppose that a sequence $\{a_i, a_i \in \mathcal{P}(S_i \times S_i)\}_{i=1}^{\infty}$ weakly converges to $a \in \mathcal{P}(X \times X)$, a sequence $\{\Phi_i, \Phi_i \in \mathcal{P}(S_i)\}_{i=1}^{\infty}$ H-converges to $\Phi \in \mathcal{P}_H(X)$. Then $A\Phi \in \mathcal{P}_H(X)$ and the sequence $\{A_i \Phi_i\}$ of the elements of $\mathcal{P}(S_i)$ H-converges to $A\Phi$, where A_i are the endomorphisms of $\mathcal{P}(S_i)$ defined by the matrix a_i.*

References

1. Maslov, V.: On a new superposition principle for optimization problems. Russian Math. Surveys **42**, (1987).
2. Maslov, V. P. Samborski, S.N. Idempotent Analysis. Advances in Soviet Mathematics **13**, Amer. Math. Soc., Providence, (1992).
3. Samborski, S.N : Time discrete and continuous control problems. Convergence of value functions. Paper presented at this conference.

Subdirect Sum Decomposition of Finite Dimensional Pseudomodules

Edouard Wagneur[1,2]

[1] LAN-URA823 (CNRS), École des Mines de Nantes,
1, rue de la Noë, F-44072, Nantes Cedex 03, France
[2] GERAD, Montréal, Canada

1 Introduction

A pseudoring $(P, +, \cdot)$ is an idempotent, completely ordered, and conditionnally complete semiring with minimal element 0, such that $(P \setminus \{0\}, \cdot)$ is a commutative group. For simplicity, the reader may think of P as $\mathbb{R} \cup \{-\infty\}$, with $+$ the max operator, and $\cdot$ the usual addition. The neutral element of $\cdot$ will be written 1 (this corresponds to the real number zero), and the 'multiplication' symbol $\cdot$ will usually be omitted. The structure of pseudomodule over a pseudo-ring is defined in a similar way as module over a ring ([CM1], [Wa1], see also [Cu1] and [Ga1], where slightly more general structures are studied). Additon in a pseudomodule will also be written additively, and 0 will also stand for its neutral element. The context will usually eliminate all risks of ambiguity. Independence of a system of generators X of a pseudomodule M is defined by the condition that for any $x \in X$, we have $x \notin M_{X \setminus \{x\}}$, where M_Y stands for the pseudomodule generated by Y (i.e. the set of finite linear combinations $\sum_Y \lambda_i y_i$, $\lambda_i \in P$).

Also, M is partially ordered by "$\leq$', where $x \leq y$ iff $x + y = y$. It follows that $(M, \leq)$ is a (sup) semilattice with least element 0, the neutral element of $+$. For $X \subset M$, the set X^+ of finite sums $\{\sum x_i \mid x_i \in X\}$ is a subsemilattice of M_X. Moreover, the set of (sup)-irreducible elements of X generates M_X. A *basis* of M_X is then extracted from the set of irreducible elements of X by keeping one and only one of the elements of the form λx, $\lambda \in P$ $(\lambda \neq 0)$. Morphisms of pseudomodules have been introduced in [Wa1]. It is a particular case of the idempotent semimodule morphisms studied in [MS1].

In the theory of modules, a module of finite type M over a principal ideal domain, is isomorphic to the biproduct $M_1 \oplus M_2$, where M_1 is free, and M_2 is a torsion module. Hence the classification problem reduces to that of torsion modules. The aim of this paper is to approach the classification problem of a finite dimensional pseudomodule over P in a similar way, although, as illustrated by the examples of §3, the pseudomodule structure is fairly more sophisticated. We have the following first classification result.

Theorem 1. *The free pseudomodule with n generators is isomorphic to* P^n.

The free pseudomodule over n generators is generated by $E_n = (e_i)_{i=1}^n$, with $e_i = (\delta_{i1}, \ldots, \delta_{in})$, where δ_{ij} is the Kronecker symbol. The free semilattice E_n^+ is isomorphic to $\mathbf{2}^n \setminus \{\varnothing\}$.

Except for the case $n = 1$, there are infinitely many nonisomorphic pseudomodules of dimension n ([Wa1]). Let X be a finite basis of M. The relation $x \prec y \iff \exists\lambda \in P$ such that $x \leq \lambda y$. defines a quasi-order on M, hence also on X and X^+. It is easy to see that $x \sim y \iff x \prec y$ and $y \prec x$ is a congruence relation with respect to the semilattice structure of M. For every $x \in M$, let $\mathbf{c}_x$ stand for the equivalence class of x. Clearly $\prec$ induces an order relation on the set of equivalence classes (since the context will usually limit the risks of ambiguity, will we will also write $\leq$ for this order relation). Let $S(X^+) = X^+|_\sim$. Our second classification result states that $S(X^+)$ is an intrinsic invariant of M.

Theorem 2. *For any basis X, the quotient map* $S_X\colon X^+ \to S(X^+), x \mapsto \mathbf{c}_x$ *is an epimorphism of (join) semilattices, and* $S(X^+)$ *is independent of the particular choice of basis X of M.*

From Theorem 2, for a given finite basis X of M, there are two types of elements in X^+, according to whether, for $\mathbf{c} \in S(X^+)$, $S_X^{-1}(\mathbf{c})$ is a singleton or not. This leads to the concepts of semiboolean element in the first case, and of torsion cycle in the second case. Our third classification result may be seen as the conterpart to the decomposition theorem for modules over a principal ideal domain, and may be stated as follows, where the subdirect sum $M_1 \check{\oplus} M_2$ of two pseudomodules will be defined in §2 below.

Theorem 3. *For every finite dimensional pseudomodule over P, there is a unique semiboolean pseudomodule* M_1*, and a unique torsion pseudomodule* M_2*, such that* $M \cong M_1 \check{\oplus} M_2$ *(where* $M \cong N$ *means that M and N are isomorphic).*

In §2, we define the concepts needed, and prove our classification results. As a consequence of Theorem 3, the classification of pseudomodules reduces to simpler classification problems, which may be dealt with separately. In §3, some insight into this problem is given through the study of various examples.

2 Semiboolean and Torsion Pseudomodules

Proof of Theorem 2: Let $\mathbf{c}_x, x \in S^+$ as in §1. It is easy to see that $\{\mathbf{c}_x \mid x \in X\}$ is the set of irreducible elements of the semilattice it generates, and that this semilattice coincides with $S(X^+)$.

Theorem 2 states that the semilattice $S(X^+)$ is an intrinsic invariant of M_X, we will write $S(M)$ for this semilattice. Examples 3 and 7 below show that it is not sufficient for the characterization of the isomorphy class of M.

Definition 4. The semilattice $S(M)$ is called the *structural* semilattice of M.

It is easy to see that, for every basis X the map $S_X\colon X^+ \to S(M)$ is a fibration, i.e. $\forall \mathbf{c} \in S(M)$, $S_X^{-1}(\mathbf{c})$ is a semilattice. We call it the *fiber* over $\mathbf{c}$.

Definition 5. We say that M is *semiboolean* if there is a basis X such that S is an isomorphism of semilattices.

Let $\mathbf{c}_1, \ldots, \mathbf{c}_m$ be the set of irreducible elements of $S(M)$. For $x \in S(M)$, we say that that the fiber $S_X^{-1}(x)$ is a *torsion cycle* if it does not reduce to a singleton, and $x \in S(M)$ is called a *torsion point*. Irreducible torsion points are called torsion points of order 0. W.l.o.g. we may assume that $\mathbf{c}_{k_1+1}, \ldots, \mathbf{c}_m$ is the set of torsion points of order 0. The set $\{\mathbf{c}_1, \ldots, \mathbf{c}_{k_1}\}$ generates a semilattice $S^1(M)$, which may contain torsion points. Such a point x is called a torsion point of order 1 if $\exists \mathbf{c}_i, \mathbf{c}_j$ $(1 \leq i, j \leq k_1)$ such that $x = \mathbf{c}_i + \mathbf{c}_j$. W.l.o.g. let $\mathbf{c}_{k_2+1}, \ldots, \mathbf{c}_{k_1}$ be the subset of $\{\mathbf{c}_1, \ldots, \mathbf{c}_{k_1}\}$ which generates all torsion points of order 1, and let $S^2(M)$ stand for the semilattice generated by $\{\mathbf{c}_1, \ldots, \mathbf{c}_{k_2}\}$. Then, by construction, $S^2(M)$ contains no torsion point of order 0 or 1. After ℓ similar steps, we obtain a set $\{\mathbf{c}_1, \ldots, \mathbf{c}_{k_\ell}\}$ of irreducible elements which generate a semilattice $S^\ell(M)$, containing no torsion point of order less than ℓ. The torsion points x of $S^\ell(M)$ (if any) which can be written as a nonredundant sum of the form $x = \sum_{j=1}^{\ell} \mathbf{c}_{i_j}$ $(1 \leq i_1, \ldots, i_\ell \leq k_\ell$, where ℓ is minimal, are called torsion points of order ℓ, and (w.l.o.g. we may assume that) $\mathbf{c}_1, \ldots, \mathbf{c}_{k_{\ell+1}}$ generate a semilattice $S^{\ell+1}(M)$, which contains no torsion point of order less than or equal to ℓ.

After a finite number of steps, we get a set of irreducible elements $\{\mathbf{c}_1, \ldots, \mathbf{c}_k\}$ which generate a semilattice $S_1(M)$, containing no torsion point. The complement $\{\mathbf{c}_{k+1}, \ldots, \mathbf{c}_m\}$ generates a semilattices $S_2(M)$ containing all the generators of the torsion points of $S(M)$. Since $\{\mathbf{c}_1, \ldots, \mathbf{c}_m\}$ is a poset, then $S(M)$ cannot be written as a product (or coproduct) of $S_1(M)$ and $S_2(M)$ in general. However, since $S_1(M) \cup S_2(M)$ is a poset, such that the insertions $s_i \colon S_i(M) \to S_1(M) \cup S_2(M)$ are monomorphisms of posets, $i = 1, 2$, we can define the *subdirect* sum $S_1(M) \check{\oplus} S_2(M)$ to be the intersection of all semilattices S which contain $S_1(M) \cup S_2(M)$ as a subset (i.e. $\imath \colon S_1(M) \cup S_2(M) \to S$ is a monomorphism of posets).

Lemma 6. *Let $S(M)$ be the structural semilattice of M, then $S(M) = S_1(M) \check{\oplus} S_2(M)$, where $S_1(M)$ contains no torsion point.*

Definition 7. We say that $S_1(M)$ (resp. $S_2(M)$) is the semiboolean (resp. torsion) component of $S(M)$.

Let $X = \{x_1, \ldots, x_n\}$ be an arbitrary basis of M. The subset $X_1 \subset X$ of generators which is mapped onto the set $\{\mathbf{c}_1, \ldots, \mathbf{c}_k\}$ of irreducible elements of $S_1(M)$ is well-defined and, up to a rescaling $x_i \mapsto \lambda_i x_i$, generates a torsion free pseudomodule M_1. Then $X_2 = X \setminus X_1$ generates a torsion pseudomodule M_2. Clearly, M_1 and M_2 are independent of the choice of X.

The direct sum (or biproduct) of two pseudomodules may be introduced in a similar way as in the theory of modules over a ring, However, as suggested by Lemma 6, the analogue of the classical decomposition of a module into the direct sum of a free module and and a torsion module will not hold in general. But the subdirect sum of two pseudomodules M_1, M_2 such that $M_1 \cup M_2$ is a poset while the insertions $m_i \colon M_i \to M_1 \cup M_2$ are monomorphisms of posets, $i = 1, 2$, may me defined in a similar way as that of semilattices by requiring that $M_1 \check{\oplus} M_2$ stands for the intersection of all pseudomodules M containing $M_1 \cup M_2$, (with the insertion $\imath \colon M_1 \cup M_2 \to M$ a monomorphism of posets).

Proof of Theorem 3. This follows directly from our definition of the subdirect sum of pseudomodules, together with Lemma 6.

Definition 8. We say that M_1 (resp. M_2) is the semiboolean (resp. torsion) part of M.

The next section is devoted to the analysis of various examples, and will also serve as an informal introduction to the concept of *bending* in a pseudomodule.

3 Examples

In this section, we assume that $P = \mathbb{R}_{\max}$, i.e. $(\mathbb{R} \cup -\infty, \max, +)$, and we consider subpseudomodules of P^m only. Our examples show that the intrinsic invariant given by the structural semilattice is not sufficient for the characterisation of the isomorphy class of a pseudomodule. Thus we have to define another construction, which depends on the choice of the basis. This construction (an oriented weighted graph) will arise from the equivalence class of matrices, as suggested by the commutative diagram below, and leads to the concept of *standard* basis. A formal definition of the concept is beyond the scope of this paper. The general idea may be described as follows. Let X, Y be two bases of M and $\Phi\colon M \to M$ an automorphism. It is well-known that Φ is isotone, i.e. $\forall x_i, x_j \in X$, $x_i \le x_j \Rightarrow \Phi(x_i) \le \Phi(x_2)$. Since a change of basis has the form $x_i \mapsto \lambda_i x_i$, this means that, except in the case of the free pseudomodule, the "extension by linearity" of a change of basis **is not an automorphism in general** (for example, for $X = \{(0, 1), (1, 1)\}$, and $\lambda > 1$, the change of basis of M_X defined by $\varphi(x_1) = \lambda x_1$, $\varphi(x_2) = x_2$ is not a morphism, since $x_1 \le x_2$, while $\varphi(x_1) \not\le \varphi(x_2)$). An n-dimensional sub-pseudomodule of P^m can always be represented by a morphism $\varphi \in \mathrm{Hom}(P^n, P^m)$ with 'source rank' n, i.e. by an $m \times n$ matrix of column rank n. An equivalence class will then correspond to a factorisation of the form

$$\begin{array}{ccc} P^n & \xrightarrow{\Phi_1} & P^n \\ \varphi \downarrow & & \tilde{\varphi} \downarrow \\ P^m & \xleftarrow{\Phi_2} & P^n \end{array}$$

where Φ_i, $i = 1, 2$, is the product of a permutation matrix and a diagonal matrix, and $\tilde{\varphi}$ some "canonical" matrix whose structure has to be specified.

This type of equivalence between maps (or matrices), which will be written $\varphi \simeq \tilde{\varphi}$, can be decomposed into elementary steps as follows :

- multiplication of a row (column) by a scalar,
- permutation of two rows (columns).

This reduction of matrices leads to a "standard" form (which has to be defined), which yields an easy way to recognize the type of pseudomodule defined by a given map. Another problem is to be able to determine completely the isomorphy class of a given pseudomodule (e.g. Example 6).

For a pseudomodule $M \subset P^m$ with (finite) basis (alternatively, for a matrix A with independent columns), we seek for a weighted oriented graph $\Gamma_X(M)$, with vertices $x_i \in X^+$, where $X = \{x_1, x_2, \ldots, x_n\}$, is a standard basis (or the x_i are the columns of a matrix $B \simeq A$ in standard form). There is an arc from x_i to x_j iff $x_i \prec x_j$, and its weight w_{ij} is defined by $w_{ij} = \inf\{\lambda \in P \mid x_i \le \lambda x_j\}$. Since P is conditionnally complete with least element 0, these weights are well-defined.

For $\dim M = 2$, let $A = \begin{pmatrix} a & c \\ b & d \end{pmatrix}$. W.l.o.g., we may assume that $a = 1$. Assume first that $b = 0$. Since $x_1 = \begin{pmatrix} 1 \\ 0 \end{pmatrix}$, and $x_2 = \begin{pmatrix} c \\ d \end{pmatrix}$ are independent, we necessarily have $d \ne 0$. Thus, either $c = 0$, or $c \ne 0$.

Example 1. If $c = 0$, then, by an elementary column transformation (multiply x_2 by d^{-1}), we get a matrix $B = \begin{pmatrix} 1 & 0 \\ 0 & 1 \end{pmatrix} \simeq A$. Therefore, $\Gamma_X(M)$ coincides with the weighted oriented graph of $(X^+, \leq)$, with $x_3 = x_1 + x_2$, and $w_{13} = w_{23} = 1$.

Example 2. If $c \neq 0$, then, by elementary column and row transformations (multiply x_2 by c^{-1}, and row 2 by cd^{-1}), we get a matrix $B = \begin{pmatrix} 1 & 0 \\ 1 & 1 \end{pmatrix} \simeq A$. Again, $\Gamma_X(M)$ coincides with the weighted oriented graph of $(X^+, \leq)$, with $x_1 + x_2 = x_2$, and $w_{12} = 1$.

Example 3. If $b \neq 0$, then ($c \neq 0$, $d \neq 0$, and) by elementary column and row transformations we get a matrix $B = \begin{pmatrix} 1 & 1 \\ 1 & \lambda \end{pmatrix} \simeq A$, where we may assume w.l.o.g. that $\lambda > 1$ (for if $\lambda < 1$, just multiply the second row by λ^{-1}, and interchange the two columns of the matrix). Clearly, if $\mu > 1$, and $C = \begin{pmatrix} 1 & 1 \\ 1 & \mu \end{pmatrix} \simeq B$, then we must have $\mu = \lambda$.

To B, we associate a graph $\Gamma_X(M)$ with vertices x_1, x_2, and two arcs a_{12}, a_{21} with respective weights $w_{12} = 1$, $w_{21} = \lambda > 1$. Clearly, if $N \cong M$, then $\Gamma_X(M) \cong \Gamma_Y(N)$ (in the graph theoretic sense). Conversely, given $\Gamma = (x_1, x_2; a_{12}, a_{21}; 1, \lambda)$, we readily construct a matrix B as above, hence (a basis X and) a sub-pseudomodule $M_X \subset P^2$ with $\Gamma_X(M) = \Gamma$.

In a geometrical representation, where P^2 is identified with the positive cone of the Euclidian plane, we have $\lambda = (1 + \mathrm{tg}(\alpha))/(1 - \mathrm{tg}(\alpha))$, with α the angle defined by the generators $(1, 1)$ and $(\lambda, 1)$. We may say that λ is a measure for this angle, and conclude that, geometrically, sub-pseudomodules of P^2 with this type of graph are classified by the angle of the cone defined by their generators.

The case $n = 3$ is already quite complex. Recall the 7 nonisomorphic structures whose graph $\Gamma(M)$ are subsemillattices of $\mathbf{2}^3$ [Wa1]. Our next two examples show that (at least) two 3-dimensional semiboolean pseudomodules cannot be embeded in P^m for $m < 4$.

Example 4. Let $A = \begin{pmatrix} 1 & 1 & 0 \\ 1 & 0 & 1 \\ 0 & 1 & 1 \\ 0 & 0 & 1 \end{pmatrix}$, and let $x_1 \in P^4$ be given by the i-th column of A ($i = 1, 2, 3$). We have $x_1 + x_2 < x_1 + x_3 = x_2 + x_3$, which is impossible in E_3^+ when $\{x_1, x_2, x_3\}$ is an antichain. Note that $\Gamma_X(M) = X^+$.

Example 5. Let $A = \begin{pmatrix} 1 & 0 & 0 \\ 1 & 1 & 0 \\ 0 & 1 & 1 \\ 0 & 0 & 1 \end{pmatrix}$. With x_i, $i = 1, 2, 3$ as above. We have $x_1 + x_2 < x_1 + x_3$, and $x_2 + x_3 < x_1 + x_3$, which is also impossible for an antichain $\{x_1, x_2, x_3\}$ in E_3^+. Also $\Gamma_X(M) = X^+$, as above.

Example 6. Let $A = \begin{pmatrix} 1 & 1 & \xi^{-1} \\ 0 & 1 & 1 \\ 0 & 0 & 1 \end{pmatrix}$, where $\xi < 1$. With the same notation as above, we have $x_1 \leq x_2 \leq x_3$, and $x_1 \leq \xi x_3$, i.e. the triangle inequality holds strictly. There are two paths between x_1 and x_3. The "long" one, through x_2, has weight $w_{12}w_{23} = 1$, while the 'short" one has weight $w_{13} = \xi < 1$, and "bends" the graph $\Gamma_X(M)$. Clearly M is semiboolean, and it is easy to see that, as long as we constrain $\Gamma_X(M)$ to have $w_{12} = w_{23} = 1$, we cannot eliminate the bending coefficient ξ from $\Gamma_X(M)$ (alternatively, there is no elementary columns and row transformations leading to a matrix $B \simeq A$ with the property that $\inf\{\lambda \in P \mid y_1 \leq \lambda y_2\} \inf\{\lambda \in P \mid y_2 \leq \lambda y_3\} = 1$, and $\inf\{\lambda \in P \mid y_1 \leq \lambda y_3\} \neq \xi$, where the y_i correspond to the columns of B).

Hence every pseudomodule defined by a matrix such as $B = \begin{pmatrix} 1 & 1 & \nu \\ 0 & \lambda & \lambda \\ 0 & 0 & \mu \end{pmatrix}$ is isomorphic to M iff $\nu = \xi^{-1}$. The isomorphy classes of sub-pseudomodules of P^n $(n > 1)$ with the same graph structure $\Gamma_X(M)$ are completely determined by the value $\xi \leq 1$ of the short path (for $\xi = 1$, we have the standard chain $(1, 0, 0)$, $(1, 1, 0)$, $(1, 1, 1)$).

Note also that $B \not\simeq A$, for $\mu \neq 1$. This means that the equivalence class of a matrix A does not capture all the isomorphy class of $M = \mathrm{Im}A$.

Example 7. Let $A = \begin{pmatrix} \lambda & 1 & 1 \\ 1 & \lambda & 1 \\ 1 & 1 & \lambda \end{pmatrix}$, where $\lambda > 1$. The graph $\Gamma_X(M)$ is complete, with arc weights $w_{12} = w_{23} = 1$, and $w_{21} = w_{32} = \lambda^2$, with bending $w_{13} = \lambda^{-1} < w_{12}w_{23} = 1$, and $w_{31} = \lambda^3 < w_{21}w_{32} = \lambda^4$. By symmetry, it is easy to see that these torsion and bending coefficients are minimal, provided $w_{12} = w_{23} = 1$ (i.e. for every permutation of the three generators satisfying this condition, we have the same torsion and bending coefficients).

The following example shows that there exists a torsion pseudomodule $M \subset P^3$ of arbitrary dimension n $(1 < n \leq \infty)$.

Example 8. Let $x_i = (1, i, i^2)'$, $i = 1, 2, \ldots,$. The x_i are independent. Indeed, if x_j were dependent for some j, then it could be expressed as a nonredundant linear combination of three of the other x_i's only (since there are only 3 components) : $x_j = \lambda_i x_i + \lambda_k x_k + \lambda_\ell x_\ell$, say. Then we would have

$$\lambda_i + \lambda_k + \lambda_\ell = 1, \tag{1}$$

$$i\lambda_i + k\lambda_k + \ell\lambda_\ell = j, \tag{2}$$

$$i^2\lambda_i + k^2\lambda_k + \ell^2\lambda_\ell = j^2. \tag{3}$$

From (1), we must have $\lambda_p \leq 1$, $p = i, k, \ell$, with equality in at least one case. W.l.o.g. we may assume $\lambda_i = 1$, therefore $i < j$, and the system becomes :

$$k\lambda_k + \ell\lambda_\ell = j, \tag{2'}$$

$$k^2\lambda_k + \ell^2\lambda_\ell = j^2. \tag{3'}$$

Since $\lambda_p \leq 1$, for $p = k, \ell$, then $k < j$ yields $\ell\lambda_\ell = j$, and $\ell^2\lambda_\ell = j^2$, which is impossible. Hence $k > j$. Similarly $\ell > j$, therefore $\lambda_p < 1$, $p = k, \ell$. W.l.o.g., we may assume that $k\lambda_k = j$. Since $\lambda < 1 \Rightarrow \lambda^2 < \lambda$, we get $j^2 = k^2\lambda_k^2 < k^2\lambda_k j^2 \leq j$, which yields the desired contradiction.

To get a geometric insight into such pseudomodules, just take a polygon Q with vertices $x_1, \ldots, x_n$ in general position in P^3. Then M is generated by the vertices of Q.

Our last two examples show that a torsion pseudomodule may not have its generators belong to a torsion cycle, and show some insight into our algorithm of § 2 for the selection of the torsion-free subpseudomodule of a general pseudomodule.

Example 9. Let M be given by par the matrix $A = \begin{pmatrix} 1 & 1 & 0 \\ 1 & 0 & 1 \\ 0 & 1 & \lambda \end{pmatrix}$, with $\lambda > 1$. The graph $\Gamma_X(M)$ has 5 vertices : $x_1 = (1, 1, 0), x_2 = (1, 0, 1), x_3 = (0, 1, \lambda), x_4 = x_1 + x_2, x_5 = x_1 + x_3 = x_2 + x_3$, with arc weights $w_{14} = w_{15} = w_{24} = w_{25} = w_{35} = w_{45} = 1$, and $w_{54} = \lambda > 1$. An order one torsion cycle binds the vertices x_4 and x_5 (Fig. 1).

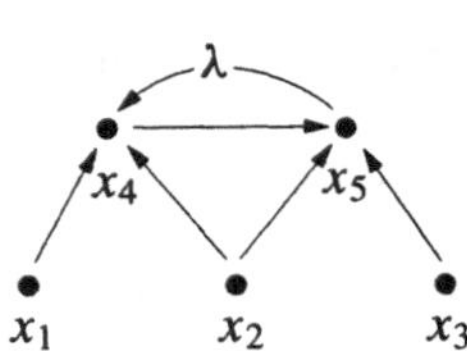

Fig. 1. Order one torsion of Example 9

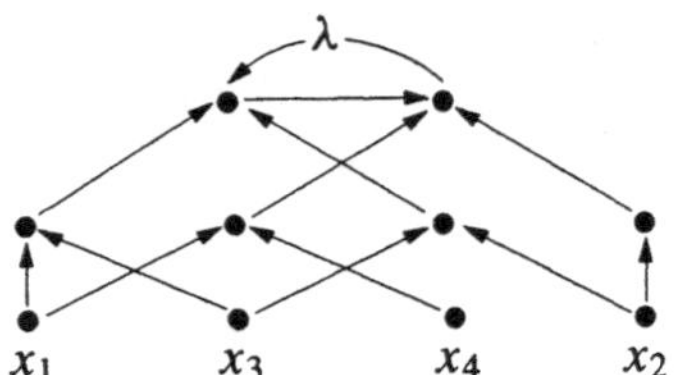

Fig. 2. Order one torsion of Example 10

Example 10. Let $A = \begin{pmatrix} 1 & 0 & 0 & 1 \\ 0 & 1 & 1 & 0 \\ 1 & 0 & 1 & 0 \\ 0 & 1 & 0 & \lambda \end{pmatrix}$, with $\lambda > 1$. The generators are an antichain, as well as $x_1+x_3 = \begin{pmatrix}1 & 1 & 1 & 0\end{pmatrix}'$, $x_2+x_3 = \begin{pmatrix}0 & 1 & 1 & 1\end{pmatrix}'$, $x_1+x_4 = \begin{pmatrix}1 & 0 & 1 & \lambda\end{pmatrix}'$ and $x_2+x_4 = \begin{pmatrix}1 & 1 & 0 & \lambda\end{pmatrix}'$ while $x_1 + x_2 = \begin{pmatrix}1 & 1 & 1 & 1\end{pmatrix}'$ and $x_3 + x_4 = \begin{pmatrix}1 & 1 & 1 & \lambda\end{pmatrix}'$ belong to an order one torsion cycle, i.e. $x_1 + x_2 < x_3 + x_4 < \lambda(x_1 + x_2)$ (Fig. 2).

Acknowledgement. The author wishes to thank Stéphane Gaubert of INRIA for very helpful comments on earlier versions of the paper.

References

[CM1] Cohen, G., Moller, P., Quadrat, J.P., Viot,M.: Une théorie linéaire des systèmes à évènements discrets. Rapport de recherche INRIA # 362. Le Chesnay, 1985.

[Cu1] Cunninghame-Green, R. A.: Minimax Algebra. Lecture Notes in Economics and Mathematical Systems, 83, SpringerVerlag, 1979.

[Ga1] Gaubert, S.: Théorie des Systèmes linéaires dans les Dioïdes. Thèse. École des Mines de Paris, juillet 1992.

[MS1] Maslov, V.P. and Sasmborski, S.N., Eds.: Idempotent Analysis. Advances in Soviet Mathematics, American Mathematical Society, Vol. 13, 1991.

[Wa1] Wagneur, E.: Moduloïds and Pseudomodules. 1. Dimension Theory. Discr. Math., 98, 57–73, 1991.

Hybrid Systems

The Algorithmic Analysis of Hybrid Systems*

R. Alur[1], *C. Courcoubetis*[2], *T. Henzinger*[3], *P. Ho*[3], *X. Nicollin*[4], *A. Olivero*[4], *J. Sifakis*[4] *and S. Yovine*[4]

[1] AT&T Bell Laboratories, Murray Hill, USA
[2] University of Crete, Heraklion, Greece
[3] Cornell University, Ithaca, USA
[4] VERIMAG, Grenoble, France

We extend the timed-automaton model for real-time systems [AD90] to a formal model for hybrid systems: while the continuous variables of a timed automaton are clocks that measure time, the continuous variables of a hybrid system are governed by arbitrary differential equations. We then adopt the verification methodology for timed automata [ACD90, ACD+92, HNSY92] to analyze hybrid systems: while the verification problem is decidable for timed automata, we obtain semidecision procedures for the class of hybrid systems whose continuous variables change in a piecewise linear fashion.

1 A Model for Hybrid Systems

We specify hybrid systems by graphs whose edges represent discrete transitions and whose vertices represent continuous activities [ACHH93, NOSY93].

A *hybrid system* $H = (\mathrm{Loc}, \mathrm{Var}, \mathrm{Lab}, \mathrm{Edg}, \mathrm{Act}, \mathrm{Inv})$ consists of six components:

- A finite set Loc of vertices called *locations*.
- A finite set Var of real-valued *variables*. A *valuation* ν for the variables is a function that assigns a real-value $\nu(x) \in \mathbb{R}$ to each variable $x \in \mathrm{Var}$. We write V for the set of valuations.
  A *state* is a pair (ℓ, ν) consisting of a location $\ell \in \mathrm{Loc}$ and a valuation $\nu \in V$. We write Σ for the set of states.
- A finite set Lab of *synchronization labels* that contains the *stutter label* $\tau \in \mathrm{Lab}$.
- A finite set Edg of edges called *transitions*. Each transition $e = (\ell, a, \mu, \ell')$ consists of a *source* location $\ell \in \mathrm{Loc}$, a *target* location $\ell' \in \mathrm{Loc}$, a synchronization label $a \in \mathrm{Lab}$, and a *transition relation* $\mu \subseteq V^2$. We require that for each location $\ell \in \mathrm{Loc}$, there is a *stutter transition* of the form $(\ell, \tau, \mathrm{Id}, \ell)$ where $\mathrm{Id} = \{(\nu, \nu) \mid \nu \in V\}$.
  The transition e is *enabled* in a state (ℓ, ν) if for some valuation $\nu' \in V$, $(\nu, \nu') \in \mu$. The state (ℓ', ν'), then, is a *transition successor* of the state (ℓ, ν).
- A labeling function Act that assigns to each location $\ell \in \mathrm{Loc}$ a set of *activities*. Each activity is a function from the nonnegative reals $\mathbb{R}^+$ to V. We require that the activities of each location are *time-invariant*: for all locations $\ell \in \mathrm{Loc}$, activities $f \in \mathrm{Act}(\ell)$,

* An extended version of this paper has been submitted for publication to the special issue of Theoretical Computer Science on Hybrid Systems to appear in December 1994.

and nonnegative reals $t \in \mathbb{R}^+$, also $(f+t) \in \mathrm{Act}(\ell)$, where $(f+t)(t') = f(t+t')$ for all $t' \in \mathbb{R}^+$.
For all locations $\ell \in \mathrm{Loc}$, activities $f \in \mathrm{Act}(\ell)$, and variables $x \in \mathrm{Var}$, we write f^x the function from $\mathbb{R}^+$ to $\mathbb{R}$ such that $f^x(t) = f(t)(x)$.
- A labeling function Inv that assigns to each location $\ell \in \mathrm{Loc}$ an *invariant* $\mathrm{Inv}(\ell) \subseteq V$.

The hybrid system H is *time-deterministic* if for every location $\ell \in \mathrm{Loc}$ and every valuation $\nu \in V$, there is at most one activity $f \in \mathrm{Act}(\ell)$ with $f(0) = \nu$. The activity f, then, is denoted by $\varphi_\ell[\nu]$.

The Runs of a Hybrid System At any time instant, the state of a hybrid system is given by a control location and values for all variables. The state can change in two ways,

- by a *discrete* and *instantaneous* transition that can change both the control location and the values of the variables according to the transition relation, or
- by a *time delay* that changes only the values of the variables according to the activities of the current location.

The system may stay at a location only if the location invariant is true; that is, some discrete transition must be taken before the invariant becomes false.

A *run* of the hybrid system H, then, is a finite or infinite sequence

$$\rho: \quad \sigma_0 \mapsto^{t_0}_{f_0} \sigma_1 \mapsto^{t_1}_{f_1} \sigma_2 \mapsto^{t_2}_{f_2} \cdots$$

of states $\sigma_i = (\ell_i, \nu_i) \in \Sigma$, nonnegative reals $t_i \in \mathbb{R}^+$, and activities $f_i \in \mathrm{Act}(\ell_i)$, such that for all $i \geq 0$,

1. $f_i(0) = \nu_i$,
2. for all $0 \leq t \leq t_i$, $f_i(t) \in \mathrm{Inv}(\ell_i)$,
3. the state σ_{i+1} is a transition successor of the state $\sigma_i' = (\ell_i, f_i(t_i))$.

The state σ_i' is called a *time successor* of the state σ_i; the state σ_{i+1}, a *successor* of σ_i. We write $[H]$ for the set of runs of the hybrid system H.

Notice that if we require all activities to be smooth functions, then the run ρ can be described by a piecewise smooth function whose values at the points of higher-order discontinuity are finite sequences of discrete state changes. Also notice that for time-deterministic systems, we can omit the subscripts f_i from the *next relation* $\mapsto$.

The run ρ *diverges* if ρ is infinite and the infinite sum $\sum_{i \geq 0} t_i$ diverges. The hybrid system H is *nonzeno* if every finite run of H is a prefix of some divergent run of H.

Hybrid Systems as Transition Systems With the hybrid system H, we associate the labeled transition system $\mathcal{T}_H = (\Sigma, \mathrm{Lab} \cup \mathbb{R}^+, \to)$, where the *step relation* $\to$ is the union of the *transition-step relation* $\to^a$, for $a \in \mathrm{Lab}$,

$$\frac{(\ell, a, \mu, \ell') \in \mathrm{Edg} \qquad (\nu, \nu') \in \mu \qquad \nu, \nu' \in \mathrm{Inv}(\ell)}{(\ell, \nu) \to^a (\ell', \nu')}$$

and the *time-step relation* $\rightarrow^t$, for $t \in \mathbb{R}^+$,

$$\frac{f \in \mathrm{Act}(\ell) \qquad f(0) = \nu \qquad \forall 0 \leq t' \leq t.\ f(t') \in \mathrm{Inv}(\ell)}{(\ell, \nu) \rightarrow^t (\ell, f(t))} .$$

Notice that the stutter transitions ensure that the transition system $\mathcal{T}_H$ is reflexive.

There is a natural correspondence between the runs of the hybrid system H and the paths through the transition system $\mathcal{T}_H$: for all states $\sigma, \sigma' \in \Sigma$, where $\sigma = (\ell, \nu)$,

$$\exists f \in \mathrm{Act}(\ell), \sigma \mapsto^t_f \sigma' \text{ iff } \exists \sigma'' \in \Sigma, a \in \mathrm{Lab}.\ \sigma \rightarrow^t \sigma'' \rightarrow^a \sigma'.$$

It follows that for every hybrid system, the set of runs is closed under prefixes, suffixes, stuttering, and fusion.

For time-deterministic hybrid systems, the rule for the time-step relation can be simplified. *Time can progress* by the amount $t \in \mathbb{R}^+$ from the state (ℓ, ν) if this is permitted by the invariant of location ℓ; that is,

$$\mathsf{tcp}_\ell[\nu](t) \text{ iff } \forall 0 \leq t' \leq t.\ \varphi_\ell[\nu](t') \in \mathrm{Inv}(\ell).$$

Now we can rewrite the time-step rule for time-deterministic systems as

$$\frac{\mathsf{tcp}_\ell[\nu](t)}{(\ell, \nu) \rightarrow^t (\ell, \varphi_\ell[\nu](t))} .$$

Example: Thermostat The temperature of a room is controlled through a thermostat, which continuously senses the temperature and turns a heater on and off. The temperature is governed by differential equations. When the heater is off, the temperature, denoted by the variable x, decreases according to the exponential function $x(t) = \theta e^{-Kt}$, where t is the time, θ is the initial temperature, and K is a constant determined by the room; when the heater is on, the temperature follows the function $x(t) = \theta e^{-Kt} + h(1 - e^{-Kt})$, where h is a constant that depends on the power of the heater. We wish to keep the temperature between m and M degrees and turn the heater on and off accordingly.

The resulting time-deterministic hybrid system is shown in Fig. 1. The system has two locations: in location 0, the heater is turned off; in location 1, the heater is on. The transition relations are specified by guarded commands; the activities, by differential equations; and the location invariants, by logical formulas.

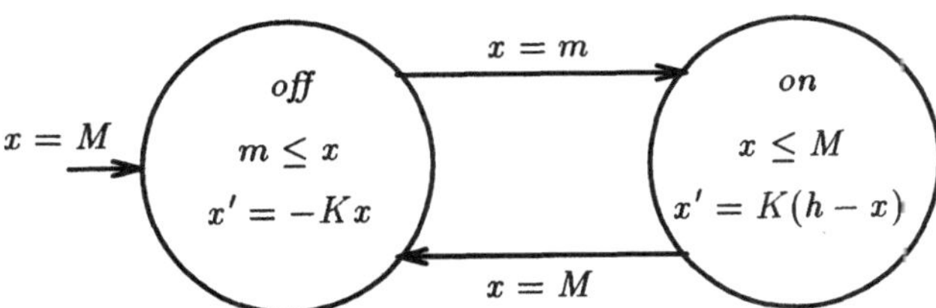

Fig. 1. Thermostat

The Parallel Composition of Hybrid Systems Let $H_1 = (\text{Loc}_1, \text{Var}, \text{Lab}_1, \text{Edg}_1, \text{Act}_1, \text{Inv}_1)$ and $H_2 = (\text{Loc}_2, \text{Var}, \text{Lab}_2, \text{Edg}_2, \text{Act}_2, \text{Inv}_2)$ be two hybrid systems over a common set Var of variables. The two hybrid systems synchronize on the common set $\text{Lab}_1 \cap \text{Lab}_2$ of synchronization labels; that is, whenever H_1 performs a discrete transition with the synchronization label $a \in \text{Lab}_1 \cap \text{Lab}_2$, then so does H_2.

The *product* $H_1 \times H_2$ is the hybrid system $(\text{Loc}_1 \times \text{Loc}_2, \text{Var}, \text{Lab}_1 \cup \text{Lab}_2, \text{Edg}, \text{Act}, \text{Inv})$ such that

- $((\ell_1, \ell_2), a, \mu, (\ell_1', \ell_2')) \in \text{Edg}$ iff
  (1) $(\ell_1, a_1, \mu_1, \ell_1') \in \text{Edg}_1$ and $(\ell_2, a_2, \mu_2, \ell_2') \in \text{Edg}_2$,
  (2) either $a_1 = a_2 = a$, or $a_1 = a$ and $a_2 = \tau$, or $a_1 = \tau$ and $a_2 = a$,
  (3) $\mu = \mu_1 \cap \mu_2$;
- $\text{Act}(\ell_1, \ell_2) = \text{Act}_1(\ell_1) \cap \text{Act}_2(\ell_2)$;
- $\text{Inv}(\ell_1, \ell_2) = \text{Inv}_1(\ell_1) \cap \text{Inv}_2(\ell_2)$.

It follows that all runs of the product system are runs of both component systems:

$$[H_1 \times H_2]_{\text{Loc}_1} \subseteq [H_1] \quad \text{and} \quad [H_1 \times H_2]_{\text{Loc}_2} \subseteq [H_2]$$

where $[H_1 \times H_2]_{\text{Loc}_i}$ is the projection of $[H_1 \times H_2]$ on Loc_i.

Notice also that the product of two time-deterministic hybrid systems is again time-deterministic.

2 Linear Hybrid Systems

A *linear term* over the set Var of variables is a linear combination of the variables in Var with integer coefficients. A *linear formula* over Var is a boolean combination of inequalities between linear terms over Var.

The time-deterministic hybrid system $H = (\text{Loc}, \text{Var}, \text{Edg}, \text{Act}, \text{Inv})$ is *linear* if its activities, invariants, and transition relations can be defined by linear expressions over the set Var of variables:

1. For all locations $\ell \in \text{Loc}$, the activities $\text{Act}(\ell)$ are defined by a set of differential equations of the form $\dot{x} = k_x$, one for each variable $x \in \text{Var}$, where $k_x \in \mathbb{Z}$ is an integer constant: for all valuations $\nu \in V$, variables $x \in \text{Var}$, and nonnegative reals $t \in \mathbb{R}^+$,
$$\varphi_\ell^x[\nu](t) = \nu(x) + k_x \cdot t.$$
We write $\text{Act}(\ell, x) = k_x$ to refer to the *rate* of the variable x at location ℓ.
2. For all locations $\ell \in \text{Loc}$, the invariant $\text{Inv}(\ell)$ is defined by a linear formula ψ over Var:
$$\nu \in \text{Inv}(\ell) \text{ iff } \nu(\psi).$$
3. For all transitions $e \in \text{Edg}$, the transition relation μ is defined by a guarded set of nondeterministic assignments
$$\psi \Rightarrow \{x := [\alpha_x, \beta_x] \mid x \in \text{Var}\},$$

where the guard ψ is a linear formula and for each variable $x \in \text{Var}$, both interval boundaries α_x and β_x are linear terms:

$$(\nu, \nu') \in \mu \text{ iff } \nu(\psi) \wedge \forall x \in \text{Var}.\ \nu(\alpha_x) \leq \nu'(x) \leq \nu(\beta_x).$$

If $\alpha_x = \beta_x$, we write $\mu(e, x) = \alpha_x$ to refer to the updated value of the variable x after the transition e.

Notice that every run of a linear hybrid system can be described by a piecewise linear function whose values at the points of first-order discontinuity are finite sequences of discrete state changes.

Special Cases of Linear Hybrid Systems Various special cases of linear hybrid systems are of particular interest:

- If $\text{Act}(\ell, x) = 0$ for each location $\ell \in \text{Loc}$, then x is a *discrete variable*. Thus, a discrete variable changes only when the control location changes. A *discrete system* is a linear hybrid system all of whose variables are discrete.
- A discrete variable x is a *proposition* if $\mu(e, x) \in \{0, 1\}$ for each transition $e \in \text{Edg}$. A *finite-state system* is a linear hybrid system all of whose variables are propositions.
- If $\text{Act}(\ell, x) = 1$ for each location ℓ and $\mu(e, x) \in \{0, x\}$ for each transition e, then x is a *clock*. Thus, (1) the value of a clock increases uniformly with time, and (2) a discrete transition either resets a clock to 0, or leaves it unchanged. A *timed automaton* [AD90] is a linear hybrid system all of whose variables are propositions or clocks.
- If there is a nonzero integer constant $k \in \mathbb{Z}$ such that $\text{Act}(\ell, x) = k$ for each location ℓ and $\mu(e, x) \in \{0, x\}$ for each transition e, then x is a *skewed clock*. Thus, a skewed clock is similar to a clock except that it changes with time at some fixed rate different from 1. A *multirate timed system* is a linear hybrid system all of whose variables are propositions and skewed clocks. An *n-rate timed system* is a multirate timed system whose skewed clocks proceed at n different rates.
- If $\text{Act}(\ell, x) \in \{0, 1\}$ for each location ℓ and $\mu(e, x) \in \{0, x\}$ for each transition e, then x is an *integrator*. Thus, an integrator is a clock that can be stopped and restarted; it is typically used to measure accumulated durations. An *integrator system* is a linear hybrid system all of whose variables are propositions and integrators.
- A discrete variable x is a *parameter* if $\mu(e, x) = x$ for each transition $e \in \text{Edg}$. Thus, a parameter is a symbolic constant. For each of the subclasses of linear hybrid systems listed above, we obtain *parameterized* versions by admitting parameters.

Notice that linear hybrid systems, and all of the subclasses of linear hybrid systems listed above, are closed under parallel composition.

2.1 Examples of Linear Hybrid Systems

A Water-Level Monitor The water level in a tank is controlled through a monitor, which continuously senses the water level and turns a pump on and off. The water level changes as a piecewise-linear function over time. When the pump is off, the water level, denoted by the variable y, falls by 2 inches per second; when the pump is on, the water level rises by 1 inch per second. Suppose that initially the water level is 1 inch and the pump is turned on. We wish to keep the water level between 1 and 12 inches. But from the time that the monitor signals to change the status of the pump to the time that the change becomes effective, there is a delay of 2 seconds. Thus the monitor must signal to turn the pump on before the water level falls to 1 inch, and it must signal to turn the pump off before the water level reaches 12 inches.

The linear hybrid system of Fig. 2 describes a water level monitor that signals whenever the water level passes 5 and 10 inches, respectively. The system has four locations: in locations 0 and 1, the pump is turned on; in locations 2 and 3, the pump is off. The clock x is used to specify the delays: whenever the control is in location 1 or 3, the signal to switch the pump off or on, respectively, was sent x seconds ago. In the next section, we will prove that the monitor indeed keeps the water level between 1 and 12 inches.

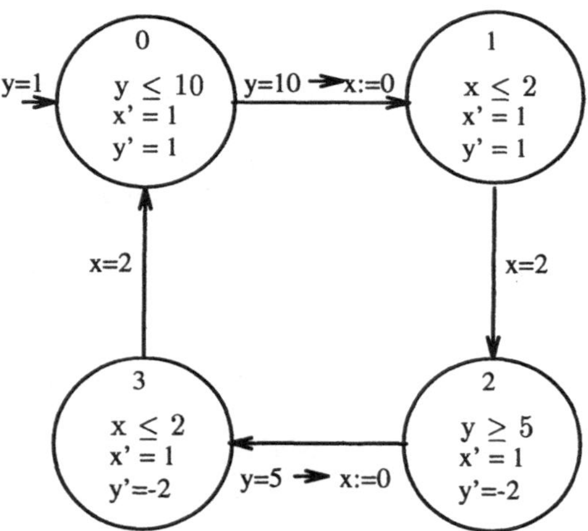

Fig. 2. Water-level monitor

A Leaking Gas Burner Now we consider an integrator system. In [CHR92], the duration calculus is used to prove that a gas burner does not leak excessively. It is assumed that (1) any leakage can be detected and stopped within 1 second and (2) the gas burner will not leak for 30 seconds after a leakage has been stopped. We wish to prove that the accumulated time of leakage is at most one twentieth of the time in any interval of at least 60 seconds. The system is modeled by the hybrid system of Fig. 3. The system has two locations: in location 1, the gas burner leaks; location 2 is the nonleaking location. The integrator z records the cumulative leakage time; that is, the accumulated amount of time that the system has spent in location 1. The clock x records the time the system has spent in the current location; it is used to specify the properties (1) and (2). The clock y records the total elapsed time. In the next section, we will prove that $y \geq 60 \Rightarrow 20z \leq y$

is an invariant of the system.

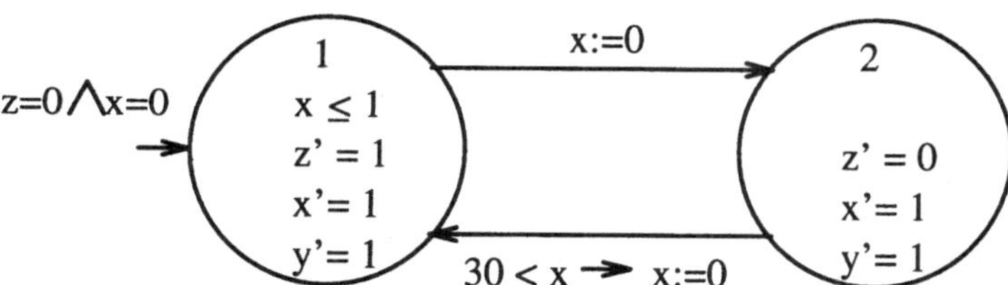

Fig. 3. Leaking gas burner

A Temperature-Control System This example appears in [JLHM91]. The system controls the coolant temperature in a reactor tank by moving two independent control rods. The goal is to maintain the coolant between the temperatures θ_m and θ_M. When the temperature reaches its maximum value θ_M, the tank must be refrigerated with one of the rods. A rod can be moved again only if T time units have elapsed since the end of its previous movement. If the temperature of the coolant cannot decrease because there is no available rod, a complete shutdown is required. Fig. 4 shows the hybrid system of this example. The values of clocks x_1 and x_2 represent the times elapsed since the last use of rod 1 and rod 2, respectively.

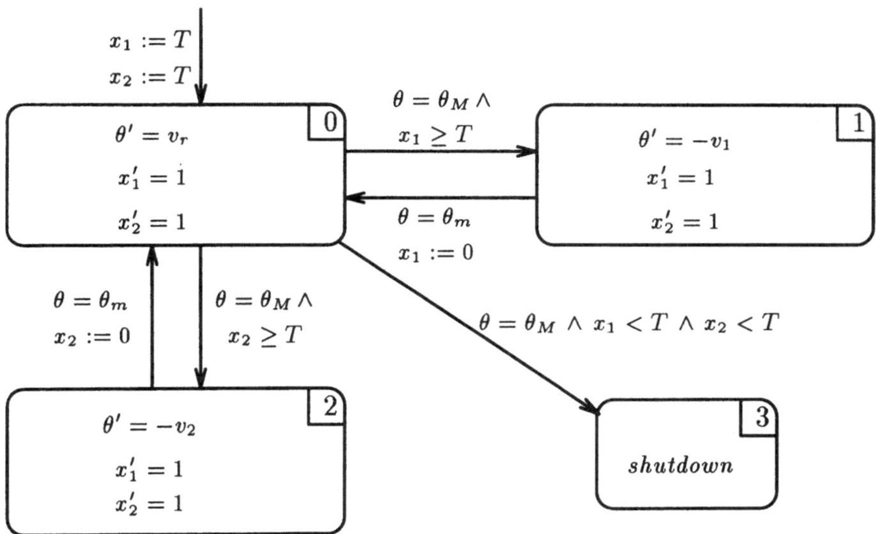

Fig. 4. Temperature control system

2.2 The Reachability Problem for Linear Hybrid Systems

Let σ and σ' be two states of a hybrid system H. The state σ' is *reachable* from the state σ, written $\sigma \mapsto^* \sigma'$, if there a run of H that starts in σ and ends in σ'. The *reachability question* asks, then, if $\sigma \mapsto^* \sigma'$ for two given states σ and σ' of a hybrid system H.

The reachability problem is central to the verification of hybrid systems. In particular, the verification of invariance properties is equivalent to the reachability question: a set $R \subseteq \Sigma$ of states is an invariant of the hybrid system H iff no state in $\Sigma - R$ is reachable from an initial state of H.

A linear hybrid system is *simple* if all linear atoms in location invariants and transition guards are of the form $x \leq k$ or $k \leq x$, for a variable $x \in \text{Var}$ and an integer constant $k \in \mathbb{Z}$. In particular, for multirate timed systems the simplicity condition prohibits the comparison of skewed clocks with different rates.

Theorem 1. *The reachability problem is decidable for simple multirate timed systems.*

Theorem 2. *The reachability problem is undecidable for 2-rate timed systems.*

Theorem 3. *The reachability problem is undecidable for simple integrator systems.*

3 The Verification of Linear Hybrid Systems

We present a methodology for analyzing linear hybrid systems that is based on predicate transformers for computing the step predecessors and the step successors of a given set of states. Throughout this section, let $H = (\text{Loc}, \text{Var}, \text{Lab}, \text{Edg}, \text{Act}, \text{Inv})$ be a linear hybrid system.

3.1 Forward Analysis

Given a location $\ell \in \text{Loc}$ and a set of valuations $P \subseteq V$, the *forward time closure* $\langle P \rangle_\ell^{\nearrow}$ of P at ℓ is the set of valuations that are reachable from some valuation $\nu \in P$ by letting time progress:

$$\nu' \in \langle P \rangle_\ell^{\nearrow} \text{ iff } \exists \nu \in V, t \in \mathbb{R}^+ .\ \nu \in P \wedge \mathsf{tcp}_\ell[\nu](t) \wedge \nu' = \varphi_\ell[\nu](t).$$

Thus, for all valuations $\nu' \in \langle P \rangle_\ell^{\nearrow}$, there exist a valuation $\nu \in P$ and a nonnegative real $t \in \mathbb{R}^+$ such that $(\ell, \nu) \rightarrow^t (\ell, \nu')$.

Given a transition $e = (\ell, a, \mu, \ell')$ and a set of valuations $P \subseteq V$, the *postcondition* $\mathsf{post}_e[P]$ of P with respect to e is the set of valuations that are reachable from some valuation $\nu \in P$ by executing the transition e:

$$\nu' \in \mathsf{post}_e[P] \text{ iff } \exists \nu \in V .\ \nu \in P \wedge (\nu, \nu') \in \mu.$$

Thus, for all valuations $\nu' \in \mathsf{post}_e[P]$, there exists a valuation $\nu \in P$ such that $(\ell, \nu) \rightarrow^a (\ell', \nu')$.

A set of states is called a *region*. Given a set $P \subseteq V$ of valuations, by (ℓ, P) we denote the region $\{(\ell, \nu) \mid \nu \in P\}$; we write $(\ell, \nu) \in (\ell, P)$ iff $\nu \in P$. The forward time closure and the postcondition can be naturally extended to regions: for $R = \bigcup_{\ell \in \mathrm{Loc}} (\ell, R_\ell)$,

$$\langle R \rangle^{\nearrow} = \bigcup_{\ell \in \mathrm{Loc}} (\ell, \langle R_\ell \rangle_\ell^{\nearrow}),$$

$$\mathsf{post}[R] = \bigcup_{e=(\ell,\ell') \in \mathrm{Edg}} (\ell', \mathsf{post}_e[R_\ell]).$$

A *symbolic run* of the linear hybrid system H is a finite or infinite sequence

$$\varrho: \quad (\ell_0, P_0)\,(\ell_1, P_1) \ldots (\ell_i, P_i) \ldots$$

of regions such that for all $i \geq 0$, there exists a transition e_i from ℓ_i to ℓ_{i+1} and

$$P_{i+1} \;=\; \mathsf{post}_{e_i}[\langle P_i \rangle_{\ell_i}^{\nearrow}];$$

that is, the region (ℓ_{i+1}, P_{i+1}) is the set of states that are reachable from a state $(\ell_0, \nu_0) \in (\ell_0, P_0)$ after executing the sequence $e_0, \ldots, e_i$ of transitions. There is a natural correspondence between the runs and the symbolic runs of the linear hybrid system H. The symbolic run ϱ represents the set of all runs of the form

$$(\ell_0, \nu_0) \mapsto^{t_0} (\ell_1, \nu_1) \mapsto^{t_1} \cdots$$

such that $(\ell_i, \nu_i) \in (\ell_i, P_i)$ for all $i \geq 0$. Besides, every run of H is represented by some symbolic run of H.

Given a region $I \subseteq \Sigma$, the *reachable region* $(I \mapsto^*) \subseteq \Sigma$ of I is the set of all states that are reachable from states in I:

$$\sigma \in (I \mapsto^*) \text{ iff } \exists \sigma' \in I.\, \sigma' \mapsto^* \sigma.$$

Notice that $I \subseteq (I \mapsto^*)$.

The following proposition suggests a method for computing the reachable region $(I \mapsto^*)$ of I.

Proposition 4. *Let $I = \bigcup_{\ell \in \mathrm{Loc}} (\ell, I_\ell)$ be a region of the linear hybrid system H. The reachable region $(I \mapsto^*) = \bigcup_{\ell \in \mathrm{Loc}} (\ell, R_\ell)$ is the least fixpoint of the equation*

$$X = \langle I \cup \mathsf{post}[X] \rangle^{\nearrow}$$

or, equivalently, for all locations $\ell \in \mathrm{Loc}$, the set R_ℓ of valuations is the least fixpoint of the set of equations

$$X_\ell = \langle I_\ell \cup \bigcup_{e=(\ell',\ell) \in \mathrm{Edg}} \mathsf{post}_e[X_{\ell'}] \rangle_\ell^{\nearrow}\;.$$

Let ψ be a linear formula over Var. By $[\![\psi]\!]$ we denote the set of valuations that satisfy ψ. A set $P \subseteq$ of valuations is *linear* if P is definable by a linear formula; that is, $P = [\![\psi]\!]$ for some linear formula ψ. If Var contains n variables, then a linear set of valuations can be thought of as a union of polyhedra in n-dimensional space.

Lemma 5. *For all linear hybrid systems H, if $P \subseteq V$ is a linear set of valuations, then for all locations $\ell \in$ Loc and transitions $e \in$ Edg, both $\langle P \rangle_\ell^\nearrow$ and $\mathsf{post}_e[P]$ are linear sets of valuations.*

Given a linear formula ψ, we write $\langle \psi \rangle_\ell^\nearrow$ and $\mathsf{post}_e[\psi]$ for the linear formulas that define the sets of valuations $\langle [\![\psi]\!] \rangle_\ell^\nearrow$ and $\mathsf{post}_e[[\![\psi]\!]]$, respectively.

Let $l \notin$ Var be a *control variable* that ranges over the set Loc of locations and let $R = \bigcup_{\ell \in \mathrm{Loc}} (\ell, R_\ell)$ be a region. The region R is *linear* if for every location $\ell \in$ Loc, the set R_ℓ of valuations is linear. If the sets R_ℓ are defined by the linear formulas ψ_ℓ, then the region R is defined by the linear formula

$$\psi = \bigvee_{\ell \in \mathrm{Loc}} (l = \ell \wedge \psi_\ell);$$

that is, $[\![\psi]\!] = R$. Hence, by Lemma 5, for all linear hybrid systems, if R is a linear region, then so are both $\langle R \rangle^\nearrow$ and $\mathsf{post}[R]$.

Using Proposition 4, we compute the reachable region $(I \mapsto^*)$ of a region I by successive approximation. Lemma 5 ensures that all regions computed in the process are linear. Since the reachability problem for linear hybrid systems is undecidable, the successive-approximation procedure does not terminate in general. The procedure does terminate for simple multirate timed systems (Theorem 1).

3.2 Backward Analysis

The forward time closure and the postcondition define the successor of a region R. Dually, we can compute the predecessor of R.

Given a location $\ell \in$ Loc and a set of valuations $P \subseteq V$, the *backward time closure* of P at ℓ is the set of valuations from which it is possible to reach some valuation $\nu \in P$ by letting time progress:

$$\nu' \in \langle P \rangle_\ell^\swarrow \text{ iff } \exists \nu \in V, t \in \mathbb{R}^+.\ \nu = \varphi_\ell[\nu'](t) \wedge \nu \in P \wedge \mathsf{tcp}_\ell[\nu'](t).$$

Thus, for all valuations $\nu' \in \langle P \rangle_\ell^\swarrow$, there exist a valuation $\nu \in P$ and a nonnegative real $t \in \mathbb{R}^+$ such that $(\ell, \nu') \rightarrow^t (\ell, \nu)$.

Given a transition $e = (\ell, a, \mu, \ell')$ and a set of valuations $P \subseteq V$, the *precondition* $\mathsf{pre}_e[P]$ of P with respect to e is the set of valuations from which it is possible to reach a valuation $\nu \in P$ by executing the transition e:

$$\nu' \in \mathsf{pre}_e[P] \text{ iff } \exists \nu \in V.\ \nu(P) \wedge (\nu', \nu) \in \mu.$$

Thus, for all valuations $\nu' \in \mathsf{pre}_e[P]$, there exists a valuation $\nu \in P$ such that $(\ell, \nu') \rightarrow^a (\ell', \nu)$.

The backward time closure and the precondition can be naturally extended to regions: for $R = \bigcup_{\ell \in \mathrm{Loc}} (\ell, R_\ell)$,

$$\langle R \rangle^\swarrow = \bigcup_{\ell \in \mathrm{Loc}} (\ell, \langle R_\ell \rangle_\ell^\swarrow) \ ,$$

$$\mathsf{pre}[R] = \bigcup_{e=(\ell',\ell) \in \mathrm{Edg}} (\ell', \mathsf{pre}_e[R_\ell]) \ .$$

Given a region $R \subseteq \Sigma$, the *initial region* $(\mapsto^* R) \subseteq \Sigma$ of R is the set of all states from which a state in R is reachable:

$$\sigma \in (\mapsto^* R) \text{ iff } \exists \sigma' \in R.\, \sigma \mapsto^* \sigma'.$$

Notice that $R \subseteq (\mapsto^* R)$.

The following proposition suggests a method for computing the initial region $(\mapsto^* R)$ of R.

Proposition 6. *Let $R = \bigcup_{\ell \in \mathrm{Loc}}(\ell, R_\ell)$ be a region of the linear hybrid system H. The initial region $I = \bigcup_{\ell \in \mathrm{Loc}}(\ell, I_\ell)$ is the least fixpoint of the equation*

$$X = \langle R \cup \mathsf{pre}[X]\rangle^{\swarrow}$$

or, equivalently, for all locations $\ell \in$ Loc, the set I_ℓ of valuations is the least fixpoint of the set

$$X_\ell = \langle R_\ell \cup \bigcup_{e=(\ell,\ell')\in \mathrm{Edg}} \mathsf{pre}_e[X_{\ell'}]\rangle_\ell^{\swarrow}$$

of equations.

Lemma 7. *For all linear hybrid systems H, if $P \subseteq V$ is a linear set of valuations, then for all locations $\ell \in$ Loc and transitions $e \in$ Edg, both $\langle P\rangle_\ell^{\swarrow}$ and $\mathsf{pre}_e[P]$ are linear sets of valuations.*

It follows that for all linear hybrid systems, if R is a linear region, then so are both $\langle R\rangle^{\swarrow}$ and $\mathsf{pre}[R]$. Given a linear formula ψ, we write $\langle \psi\rangle_\ell^{\swarrow}$ and $\mathsf{pre}_e[\psi]$ for the linear formulas that define the sets of valuations $\langle [\![\psi]\!]\rangle_\ell^{\swarrow}$ and $\mathsf{pre}_e[[\![\psi]\!]]$, respectively.

Example: The Gas Burner We apply backward analysis to prove that the design requirement

$$\psi_R \;=\; (y \geq 60 \;\Rightarrow\; 20z \leq y)$$

is an invariant of the gas burner system; that is, the region R defined by ψ_R is not reachable from the set I of initial states defined by the linear formula

$$\psi_I \;=\; (l = 1 \;\wedge\; x = y = z = 0).$$

The set $(\mapsto^* R)$ of states from which it is possible to reach a state in R is characterized by the least fixpoint of the two equations

$$\psi_1 = \langle (l = 1 \;\wedge\; (y \geq 60 \;\wedge\; 20z > y) \vee \mathsf{pre}_{(1,2)}[\psi_2]\rangle_1^{\swarrow}$$
$$\psi_2 = \langle (l = 2 \;\wedge\; (y \geq 60 \;\wedge\; 20z > y) \vee \mathsf{pre}_{(2,1)}[\psi_1]\rangle_2^{\swarrow}$$

which can be iteratively computed as

$$\psi_{1,i} = \psi_{1,i-1} \vee \langle \mathsf{pre}_{(1,2)}[\psi_{2,i-1}]\rangle_1^{\swarrow}$$
$$\psi_{2,i} = \psi_{2,i-1} \vee \langle \mathsf{pre}_{(2,1)}[\psi_{1,i-1}]\rangle_2^{\swarrow}$$

where $\psi_{1,0} = \langle(y \geq 60 \wedge 20z > y)\rangle_1^{\checkmark}$ and $\psi_{2,0} = \langle(y \geq 60 \wedge 20z > y)\rangle_2^{\checkmark}$. Then,

$$\begin{aligned}
\psi_{1,1} &= (-19 < 20z - 19x - y \wedge 2 < z - x \wedge -1 \leq -x \wedge 0 \leq x)\\
\psi_{1,2} &= (-19 < 20z - y \wedge 11 < 20z + x - y \wedge 2 < z \wedge 0 \leq z \wedge 0 \leq x)\\
\psi_{2,1} &= (-8 < 20z - 19x - y \wedge 1 < z - x \wedge -1 \leq -x \wedge 0 \leq x)\\
\psi_{2,2} &= (-19 < 20z - y \wedge 2 < z \wedge 11 < 20z + x - y \wedge 0 \leq x)\\
\psi_{3,1} &= (-8 < 20z - 19x - y \wedge 1 < z - x \wedge -1 \leq -x \wedge 0 \leq x)\\
\psi_{3,2} &= (-8 < 20z - y \wedge 1 < z \wedge 22 < 20z + x - y \wedge 0 \leq x)\\
\psi_{4,1} &= (0 < z - x \wedge 3 < 20z - 19x - y \wedge -1 \leq -x \wedge 0 \leq x)\\
\psi_{4,2} &= (-8 < 20z - y \wedge 1 < z \wedge 22 < 20z + x - y \wedge 0 \leq x)\\
\psi_{5,1} &= (0 < z - x \wedge 3 < 20z - 19x - y \wedge -1 \leq -x \wedge 0 \leq x)\\
\psi_{5,2} &= (0 < z \wedge 3 < 20z - y \wedge 33 < 20z + x - y \wedge 0 \leq x)\\
\psi_{6,1} &= (-1 < z - x \wedge 14 < 20z - 19x - y \wedge -1 \leq -x \wedge 0 \leq x)\\
\psi_{6,2} &= (0 < z \wedge 3 < 20z - y \wedge 33 < 20z + x - y \wedge 0 \leq x)\\
\psi_{7,1} &= (-1 < z - x \wedge 14 < 20z - 19x - y \wedge -1 \leq -x \wedge 0 \leq x)\\
\psi_{7,2} &= (14 < 20z - y \wedge 44 < 20z + x - y \wedge 0 \leq z \wedge 0 \leq x)\\
\psi_{8,1} &= (25 < 20z - 19x - y \wedge -1 \leq z - x \wedge -1 \leq -x \wedge 0 \leq x)\\
\psi_{8,2} &= (14 < 20z - y \wedge 44 < 20z + x - y \wedge 0 \leq z \wedge 0 \leq x)\\
\psi_{9,1} &= (25 < 20z - 19x - y \wedge -1 \leq z - x \wedge -1 \leq -x \wedge 0 \leq x)\\
\psi_{9,2} &= (25 < 20z - y \wedge 55 < 20z + x - y \wedge 0 \leq z \wedge 0 \leq x)
\end{aligned}$$

Since $\psi_9 \Rightarrow \psi_8$, the solution ψ is

$$\bigvee_{0 \leq i \leq 8} (l = 1 \wedge \psi_{i,1}) \vee (l = 2 \wedge \psi_{i,2}),$$

which contains no initial states; that is, $\psi_I \wedge \psi = \text{false}$. It follows that the design requirement ψ_R is an invariant.

3.3 Minimization

We extend the next relation $\mapsto$ to regions: for all regions R and R', we write $R \mapsto R'$ if some state $\sigma' \in R'$ is a successor of some state $\sigma \in R$, that is

$$R \mapsto R' \text{ iff } \exists \sigma \in R, \sigma' \in R'.\, \sigma \mapsto \sigma'.$$

We write $\mapsto^*$ for the reflexive-transitive closure of $\mapsto$.

Let π be a partition of the state space Σ. A region $R \in \pi$ is *stable* if for all $R' \in \pi$,

$$R \mapsto R' \text{ implies } \forall \sigma \in R.\, \{\sigma\} \mapsto R'$$

or, equivalently,

$$R \cap \mathsf{pre}[\langle R' \rangle^{\checkmark}] \neq \varnothing \text{ implies } R \subseteq \mathsf{pre}[\langle R' \rangle^{\checkmark}].$$

The partition π is a *bisimulation* if every region $R \in \pi$ is stable. The partition π *respects* the region R_f if for every region $R \in \pi$, either $R \subseteq R_f$ or $R \cap R_f = \emptyset$.

If a partition π that respects the region R_f is a bisimulation, then it can be used to compute the initial region $(\mapsto^* R_f)$: for all regions $R \in \pi$, if $R \mapsto^* R_f$ then $R \subseteq (\mapsto^* R_f)$, otherwise $R \cap (\mapsto^* R_f) = \emptyset$. Thus, our objective is to construct the coarsest bisimulation that respects a given region R_f, provided there is a finite bisimulation that respects R_f.

If we are given, in addition to R_f, an initial region I that restricts our interest to the reachable region $(I \mapsto^*)$, then it is best to use an algorithm that performs a simultaneous reachability and minimization analysis of transition systems [BFH90, LY92].

The minimization procedure of [BFH90] is given below. Starting from the initial partition $\{R_f, \Sigma - R_f\}$ that respects R_f, the procedure selects a region R and checks if R is stable with respect to the current partition; if not, then R is split into smaller sets. Additional book-keeping is needed to record which regions are reachable from the initial region I. In the following procedure, π is the current partition, $\alpha \subseteq \pi$ contains the regions R that have been found reachable from I, and $\beta \subseteq \pi$ contains the regions R that have been found stable with respect to π. The function $\mathsf{split}[\pi](R)$ splits the region $R \in \pi$ into subsets that are "more" stable with respect to π:

$$\mathsf{split}[\pi](R) := \begin{cases} \{R', R - R'\} \text{ if } & \exists R'' \in \pi.\ R' = \mathsf{pre}[\langle R'' \rangle^{\swarrow}] \cap R \wedge R' \subset R, \\ \{R\} & \text{otherwise.} \end{cases}$$

The minimization procedure returns YES iff $I \mapsto^* R_f$.

State space minimization:

$\pi := \{R_f, \Sigma - R_f\}$; $\alpha := \{R \mid R \cap I \neq \emptyset\}$; $\beta := \emptyset$
while $\alpha \neq \beta$ **do**
  choose $R \in (\alpha - \beta)$
  let $\alpha' := \mathsf{split}[\pi](R)$
  if $\alpha' = \{R\}$ **then**
    $\beta := \beta \cup \{R\}$
    $\alpha := \alpha \cup \{R' \in \pi \mid R \mapsto R'\}$
  else
    $\alpha := \alpha - \{R\}$
    if $\exists R' \in \alpha'$ such that $R' \cap I \neq \emptyset$ **then** $\alpha := \alpha \cup \{R'\}$ **fi**
    $\beta := \beta - \{R' \in \pi \mid R' \mapsto R\}$
    $\pi := (\pi - \{R\}) \cup \alpha'$
  fi
od
return there is $R \in \alpha$ such that $R \subseteq R_f$.

If the regions R_f and I are linear, from Lemma 7 it follows that all regions that are constructed by the minimization procedure are linear. The minimization procedure terminates if the coarsest bisimulation has only a finite number of equivalence classes. An alternative minimization procedure is presented in [LY92], which can also be implemented using the primitives $\langle\rangle^{\checkmark}$ and pre.

Example: The Water-Level Monitor Let H be the hybrid automaton defined in Fig. 2. We use the minimization procedure to prove that the formula $1 \leq y \leq 12$ is an invariant of H. It follows that the water-level monitor keeps the water level between 1 and 12 inches.

Let the set I of initial states be so defined by the linear formula

$$\psi_I = (l = 0 \wedge x = 0 \wedge y = 1)$$

and let the set R_f of "bad" states be defined by the linear formula

$$\psi_f = (y < 1 \vee y > 12).$$

The initial partition is $\pi_1 = \{$

$$\begin{array}{ll} \psi_{00} = (l = 0 \wedge 1 \leq y \leq 12), & \psi_{01} = (l = 0 \wedge (y < 1 \vee y > 12)), \\ \psi_{10} = (l = 1 \wedge 1 \leq y \leq 12), & \psi_{11} = (l = 1 \wedge (y < 1 \vee y > 12)), \\ \psi_{20} = (l = 2 \wedge 1 \leq y \leq 12), & \psi_{21} = (l = 2 \wedge (y < 1 \vee y > 12)), \\ \psi_{30} = (l = 3 \wedge 1 \leq y \leq 12), & \psi_{31} = (l = 3 \wedge (y < 1 \vee y > 12))\}. \end{array}$$

The bad states are represented by ψ_{i1}, for $i \in \{0, 1, 2, 3\}$. Since the set I of initial states is contained in ψ_{00}, that is $\psi_I \Rightarrow \psi_{00}$, let $\alpha = \{\psi_{00}\}$. Considering $\psi = \psi_{00} \in \alpha$, we find that $\text{split}[\pi_1](\psi_{00}) = \{$

$$\begin{array}{l} \psi_{000} = (l = 0 \wedge 1 \leq y < 10), \\ \psi_{001} = (l = 0 \wedge 10 \leq y \leq 12)\}. \end{array}$$

Therefore, $\pi_2 = \{\psi_{000}, \psi_{001}, \psi_{01}, \psi_{10}, \psi_{11}, \psi_{20}, \psi_{21}, \psi_{30}, \psi_{31}\}$. Now $\psi_I \Rightarrow \psi_{000}$, so take $\alpha = \{\psi_{000}\}$ and $\beta = \emptyset$. Considering $\psi = \psi_{000}$, we find that it is stable with respect to π_2. Thus $\alpha = \alpha \cup \{R' \in \pi \mid R \mapsto R'\} = \{\psi_{000}, \psi_{001}, \psi_{10}\}$ and $\beta = \{\psi_{000}\}$. Since $\psi = \psi_{001}$ is also stable in π_2 and is not reaching any new states not in α, α remains the same and $\beta = \{\psi_{000}, \psi_{001}\}$. However, considering $\psi = \psi_{10}$, we obtain $\text{split}[\pi_2](\psi_{10}) = \{$

$$\begin{array}{l} \psi_{100} = (l = 1 \wedge 0 \leq x < 2 \wedge 1 \leq y \leq 12), \\ \psi_{101} = (l = 1 \wedge x \geq 2 \wedge 1 \leq y \leq 12)\}. \end{array}$$

Now, ψ_{100} and ψ_{101} together with π_2, except for ψ_{10}, constitute π_3. The new β is obtained by removing $\{R' \in \pi \mid R' \mapsto R\} = \psi_{000}$ from the old β. The new α becomes $\{\psi_{000}, \psi_{001}\}$. Now $\psi = \psi_{000}$ is stable in π_3.

Hence $\alpha = \{\psi_{000}, \psi_{001}, \psi_{100}\}$ and $\beta = \{\psi_{000}, \psi_{001}\}$. Since $\psi = \psi_{100}$ is stable in π_3, we have $\alpha = \{\psi_{000}, \psi_{001}, \psi_{100}, \psi_{101}, \psi_{20}\}$ and $\beta = \{\psi_{000}, \psi_{001}, \psi_{100}\}$. $\psi = \psi_{101}$ is also stable in π_3, so $\beta = \{\psi_{000}, \psi_{001}, \psi_{100}, \psi_{101}\}$ and α remains unchanged. Considering $\psi = \psi_{20}$, we obtain $\text{split}[\pi_3](\psi_{20}) = \{$

$$\begin{array}{l} \psi_{200} = (l = 2 \wedge 5 < y \leq 12), \\ \psi_{201} = (l = 2 \wedge 1 \leq y \leq 5)\}. \end{array}$$

Now π_4 contains ψ_{200} and ψ_{201}, and thus ψ_{100} must be reconsidered. It is split into split$[\pi_4](\psi_{100}) = \{$

$$\begin{aligned}\psi_{1000} &= (l = 1 \wedge 0 \leq x < 2 \wedge 3 < y \leq 12 \wedge 3 < y - x \leq 12),\\ \psi_{1001} &= (l = 1 \wedge 0 \leq x < 2 \wedge 1 \leq y \leq 5 \wedge 1 \leq y - x \leq 3)\}.\end{aligned}$$

Thus π_5 contains ψ_{1000} and ψ_{1001}. After finding that ψ_{000}, ψ_{1000} and ψ_{200} all are stable, we finally have $\alpha = \{\psi_{000}, \psi_{001}, \psi_{1000}, \psi_{200}, \psi_{201}, \psi_{30}\}$ and $\beta = \{\psi_{000}, \psi_{001}, \psi_{1000}, \psi_{200}\}$. So let $\psi = \psi_{201}$. It is stable, so $\beta = \beta \cup \{\psi_{200}\}$ and α does not change. Then $\psi = \psi_{30}$ is partitioned into $\{$

$$\begin{aligned}\psi_{300} &= (l = 3 \wedge 0 \leq x < 2 \wedge 1 \leq y \leq 12),\\ \psi_{301} &= (l = 3 \wedge x \geq 2 \wedge 1 \leq y \leq 12)\}.\end{aligned}$$

ψ_{200} has to be considered again. It is stable with respect to the current partition. Then $\psi = \psi_{300}$ is considered and split$[\pi_6](\psi_{300}) = \{$

$$\begin{aligned}\psi_{3000} &= (l = 3 \wedge 0 \leq x < 2 \wedge 1 \leq y \leq 12 \wedge 5 \leq y + 2x < 14),\\ \psi_{3001} &= (l = 3 \wedge 0 \leq x < 2 \wedge 1 \leq y < 5 \wedge 1 \leq y + 2x < 5)\}.\end{aligned}$$

We must consider ψ_{200} again. It turns out that it is still stable. After considering $\psi = \psi_{3000}$, we have $\beta = \{\psi_{000}, \psi_{001}, \psi_{1000}, \psi_{200}, \psi_{201}, \psi_{3000}\}$ and $\alpha = \alpha \cup \{\psi_{000}\}$. Now the partition is

$$\pi_7 = \{\psi_{000}, \psi_{001}, \psi_{01}, \psi_{1000}, \psi_{1001}, \psi_{101}, \psi_{11}, \psi_{200}, \psi_{201}, \psi_{21}, \psi_{3000}, \psi_{3001}, \psi_{301}, \psi_{31}\}.$$

Since ψ_{000} is stable in π_7, we have $\alpha = \beta = \{\psi_{000}, \psi_{001}, \psi_{1000}, \psi_{200}, \psi_{201}, \psi_{3000}\}$. Notice that α contains no bad states from R_f, that is $\psi \wedge \psi_f =$ false for all $\psi \in \alpha$. Therefore, the invariant property has been verified.

3.4 Model Checking

Previously, we presented three semidecision procedures for the reachability problem of linear hybrid systems. Now we address the more general problem of whether the given linear hybrid system H satisfies a requirement that is expressed in the real-time temporal logic TCTL [ACD90].

Timed Computation Tree Logic Let C be a set of clocks not in Var; that is, $C \cap \text{Var} = \emptyset$. A *state predicate* is a linear formula over the set $\text{Var} \cup C$ of variables.

The formulas of TCTL are built from the state predicates by boolean connectives, the two temporal operators $\exists\mathcal{U}$ and $\forall\mathcal{U}$, and the reset quantifier for the clocks in C. The formulas of TCTL, then, are defined by the grammar

$$\phi ::= \psi \mid \neg\phi \mid \phi_1 \vee \phi_2 \mid z.\ \phi \mid \phi_1 \exists\mathcal{U} \phi_2 \mid \phi_1 \forall\mathcal{U} \phi_2$$

where ψ is a state predicate and $z \in C$. The formula ϕ is *closed* if all occurrences of a clock $z \in C$ are within the scope of a reset quantifier z.

The closed formulas of TCTL are interpreted over the state space Σ of the linear hybrid system H. Intuitively, a state σ satisfies the TCTL-formula $\phi_1 \exists\mathcal{U}\phi_2$ if there exists a run of H from σ to a state σ' satisfying ϕ_2 such that $\phi_1 \vee \phi_2$ continuously holds along the run. Dually, the state σ satisfies the TCTL-formula $\phi_1 \forall\mathcal{U}\phi_2$ if every divergent run from σ leads to a state σ' satisfying ϕ_2 such that $\phi_1 \vee \phi_2$ continuously holds along from σ to σ'. Clocks can be used to express timing constraints. For instance, the TCTL-formula $z.\ (\text{true}\exists\mathcal{U}(\phi \wedge z \leq 5))$ asserts that there is a run on which ϕ is satisfied within 5 time units.

We use the standard abbreviations such as $\forall \Diamond \phi$ for $\text{true}\forall\mathcal{U}\phi$, $\exists \Diamond \phi$ for $\text{true}\exists\mathcal{U}\phi$, $\exists \Box \phi$ for $\neg\forall \Diamond \neg\phi$, and $\forall \Box \phi$ for $\neg\exists \Diamond \neg\phi$. We also put timing constraints as subscripts on the temporal operators. For example, the formula $z.\ \exists \Diamond (\phi \wedge z < 5)$ is abbreviated to $\exists \Diamond_{<5} \phi$.

Let $\rho = \sigma_0 \mapsto^{t_0} \sigma_1 \mapsto^{t_1} \ldots$ be a run of the linear hybrid system H, with $\sigma_i = (\ell_i, \nu_i)$ for all $i \geq 0$. A *position* π of ρ is a pair (i, t) consisting of a nonnegative integer i and a nonnegative real $t \leq t_i$. The positions of ρ are ordered lexicographically; that is, $(i, t) \leq (j, t')$ iff $i < j$, or $i = j$ and $t < t'$. For all positions $\pi = (i, t)$ of ρ,

- the state $\rho(\pi)$ at the position π of ρ is $(\ell_i, \varphi_{\ell_i}[\nu_i](t))$, and
- the time $\delta_\rho(\pi)$ at the position π of ρ is $t + \sum_{j<i} t_j$.

A *clock valuation* ξ is a function from C to $\mathbb{R}^+$. For any nonnegative real $t \in \mathbb{R}^+$, by $\xi + t$ we denote the clock valuation ξ' such that $\xi'(z) = \xi(z) + t$ for all clocks $z \in C$. For any clock $z \in C$, by $\xi[z := 0]$ we denote the valuation ξ' such that $\xi'(z) = 0$ and $\xi'(z') = \xi(z')$ for all clocks $z' \neq z$.

An *extended state* (σ, ξ) consists of a state $\sigma \in \Sigma$ and a clock valuation ξ. The extended state (σ, ξ) *satisfies* the TCTL-formula ϕ, denoted $(\sigma, \xi) \models \phi$, if

$(\sigma, \xi) \models \psi$ iff $(\sigma, \xi)(\psi)$;
$(\sigma, \xi) \models \neg\phi$ iff $(\sigma, \xi) \not\models \phi$;
$(\sigma, \xi) \models \phi_1 \vee \phi_2$ iff $(\sigma, \xi) \models \phi_1$ or $(\sigma, \xi) \models \phi_2$;
$(\sigma, \xi) \models z.\ \phi_1$ iff $(\sigma, \xi[z := 0]) \models \phi_1$;
$(\sigma, \xi) \models \phi_1 \exists\mathcal{U}\phi_2$ iff there is a run ρ of H and a position π of ρ such that (1) $\rho(0, 0) = \sigma$, (2) $(\rho(\pi), \xi + \delta_\rho(\pi)) \models \phi_2$, and (3) for all positions $\pi' \leq \pi$ of ρ, $(\rho(\pi'), \xi + \delta_\rho(\pi')) \models \phi_1 \vee \phi_2$;
$(\sigma, \xi) \models \phi_1 \forall\mathcal{U}\phi_2$ iff for all divergent runs ρ of H with $\rho(0, 0) = \sigma$ there is a position π of ρ such that $(\rho(\pi), \xi + \delta_\rho(\pi)) \models \phi_2$ and for all positions $\pi' \leq \pi$ of ρ, $(\rho(\pi'), \xi + \delta_\rho(\pi')) \models \phi_1 \vee \phi_2$.

Let ϕ be a closed formula of TCTL. A state $\sigma \in \Sigma$ satisfies ϕ, denoted $\sigma \models \phi$, if $(\sigma, \xi) \models \phi$ for all clock valuations ξ. The linear hybrid system H satisfies ϕ, denoted $H \models \phi$, if all states of H satisfy ϕ. The *characteristic set* $[\![\phi]\!] \subseteq \Sigma$ of ϕ is the set of states that satisfy ϕ.

The Model-Checking Algorithm Given a closed TCTL-formula ϕ, a model-checking algorithm computes the characteristic set $[\![\phi]\!]$. We present the symbolic model-checking algorithm for timed automata [HNSY92], which is a semidecision procedure for model checking TCTL-formulas over linear hybrid systems.

The procedure is based on fixpoint characterizations of the TCTL-modalities in terms of a binary next operator $\triangleright$. Given two regions $R, R' \subseteq V$, the region $R \triangleright R'$ is the set of states σ that have a successor $\sigma' \in R'$ such that all states between σ and σ' are contained in $R \cup R'$: $(\ell, \nu) \in (R \triangleright R')$ iff

$$\exists(\ell', \nu') \in R', t \in \mathbb{R}^+. \ ((\ell, \nu) \mapsto^t (\ell', \nu') \ \wedge \ \forall 0 \leq t' \leq t. \ (\ell, \nu + t') \in (R \cup R'));$$

that is, the $\triangleright$ operator is a "single-step until" operator.

To define the $\triangleright$ operator syntactically, we introduce some notation. For a linear formula ψ, we extend the tcp operator such that

$$\mathsf{tcp}_\ell[\psi][\nu](t) \text{ iff } \forall 0 \leq t' \leq t. \ \varphi_\ell[\nu](t') \in (\mathrm{Inv}(\ell) \cap [\![\psi]\!]);$$

that is, all valuations along the evolution by time t from the state (ℓ, ν) satisfy not only the invariant of location ℓ but also ψ. For a state $\sigma = (\ell, \nu) \in \Sigma$ we write $\varphi[\sigma]$ for the function $\varphi_\ell[\nu]$, and for a region $R = \bigcup_{\ell \in \mathrm{Loc}} (\ell, R_\ell)$ we write

$$\mathsf{tcp}[R][\sigma](t) \text{ iff } \mathsf{tcp}_\ell[R_\ell][\nu](t) \text{ for all locations } \ell.$$

Now, for two regions $R, R' \subseteq \Sigma$, we define the region $R \triangleright R'$ as

$$\sigma \in (R \triangleright R') \text{ iff } \exists t \in \mathbb{R}^+. \ (\varphi[\sigma](t) \in (\mathsf{pre}[R'] \cup R') \ \wedge \ \mathsf{tcp}[R \cup R'][\sigma](t)).$$

Lemma 8. *For all linear hybrid systems H, if R and R' are two linear regions of H, then so is $R \triangleright R'$.*

In [HNSY92] it is shown that for nonzeno systems, the meaning of both TCTL-modalities $\exists\mathcal{U}$ and $\forall\mathcal{U}$ can be computed iteratively as fixpoints, using the $\triangleright$ operator. While for simple multirate timed systems, the iterative fixpoint computation always terminates, this is no longer the case for linear hybrid systems in general. Lemma 8, however, ensures that all regions that are computed by the process are linear and each step of the procedure is, therefore, effective.

Here, we present the method for some important classes of TCTL-formulas:

- Let R and R' be the characteristic sets of the two TCTL-formulas ϕ and ϕ', respectively. The characteristic set of the formula $\phi\exists\mathcal{U}\phi'$ can be iteratively computed as $\bigcup_i R_i$ with
  - $R_0 = R'$, and
  - for all $i \geq 0$, $R_{i+1} = R_i \cup (R \triangleright R_i)$.
- To check if the TCTL-formula ϕ is an invariant of H, we check if the set of initial states is contained in the characteristic set of the formula $\forall\Box\phi$. This characteristic set can be iteratively computed as $\bigcap_i R_i$ with
  - $R_0 = [\![\phi]\!]$, and
  - for all $i \geq 0$, $R_{i+1} = R_i \cap \neg(\mathrm{true} \triangleright \neg R_i)$.
- The real-time response property asserting that a given event occurs within a certain time bound is expressed in TCTL by a formula of the form $\forall\Diamond_{\leq c}\phi$, whose characteristic set can be iteratively computed as $\neg\bigcup_i R_i[z := 0]$ with
  - $R_0 = z > c$, and
  - for all $i \geq 0$, $R_{i+1} = R_i \cup ((\neg R) \triangleright R_i)$,

  where $R = [\![\phi]\!]$ and $z \in C$.

Example: The Temperature-Control System The goal is to maintain the temperature of the coolant between lower and upper bounds θ_m and θ_M. If the temperature rises to its maximum θ_M and it cannot decrease because no rod is available, a complete shutdown is required.

Now, let $\Delta\theta = \theta_M - \theta_m$. Clearly, the time the coolant needs to increase its temperature from θ_m to θ_M is $\tau_r = \frac{\Delta\theta}{v_r}$, and the refrigeration times for rod 1 and rod 2 are $\tau_1 = \frac{\Delta\theta}{v_1}$ and $\tau_2 = \frac{\Delta\theta}{v_2}$, respectively.

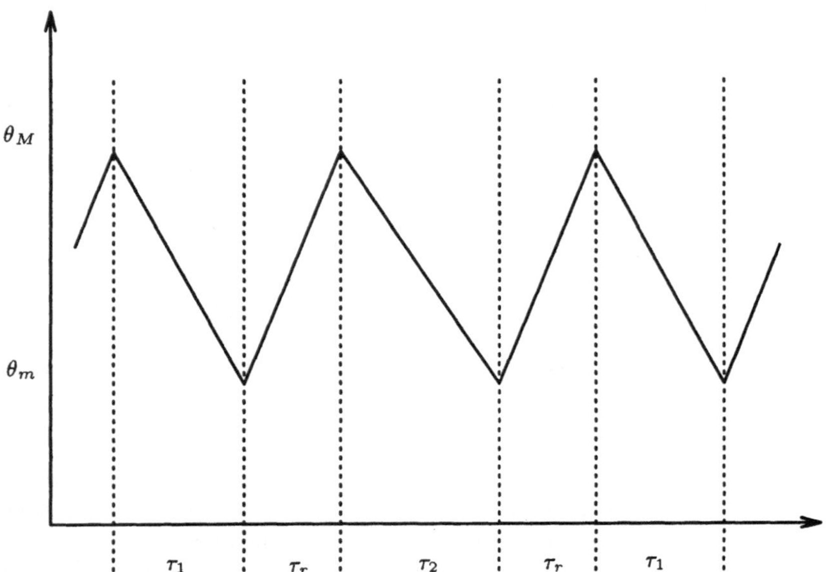

Fig. 5. Refrigeration times

The question is whether the system will ever reach the shutdown state. Clearly, if temperature rises at a rate slower than the time of recovery for the rods, i.e., $\tau_r \geq T$, shutdown is unreachable. Moreover, it can be seen that $2\tau_r + \tau_1 \geq T \wedge 2\tau_r + \tau_2 \geq T$ is a necessary and sufficient condition for never reaching the shutdown state (see Fig. 5).

The property stating that state 3 (shutdown) is always unreachable corresponds to the following TCTL formula:

$$l = 0 \wedge \theta \leq \theta_M \wedge x_1 \geq T \wedge x_2 \geq T \Rightarrow \forall\Box\neg(l = 3)$$

or equivalently,

$$l = 0 \wedge \theta \leq \theta_M \wedge x_1 \geq T \wedge x_2 \geq T \Rightarrow \neg\exists\Diamond(l = 3)$$

- Let $v_r = 6$, $v_1 = 4$, $v_2 = 3$, $\theta_m = 3$, $\theta_M = 15$ and $T = 6$. In this case the condition $2\tau_r + \tau_1 \geq T \wedge 2\tau_r + \tau_2 \geq T$ holds. We compute using KRONOS [OY92] the characteristic set of $\exists \Diamond l = 3$. The results obtained at each iteration are shown below, where each ψ_i has been computed according to the method described in §3.

$$\begin{aligned}
\psi_0 &= l = 3 \\
\psi_1 &= (l = 0 \wedge \theta \leq 15 \wedge 6x_1 < \theta + 21 \wedge 6x_2 < \theta + 21) \vee l = 3 \\
\psi_2 &= (l = 0 \wedge \theta \leq 15 \wedge 6x_1 < \theta + 21 \wedge 6x_2 < \theta + 21) \vee \\
&\quad (l = 1 \wedge 3 \leq \theta \leq 15 \wedge 4x_2 + \theta < 19) \vee \\
&\quad (l = 2 \wedge 3 \leq \theta \leq 15 \wedge 3x_1 + \theta < 15) \vee l = 3 \\
\psi_3 &= (l = 0 \wedge \theta \leq 15 \wedge (6x_1 < \theta + 21 \wedge 6x_2 < \theta + 21 \vee 6x_2 + 3 < \theta)) \vee \\
&\quad (l = 1 \wedge 3 \leq \theta \leq 15 \wedge 4x_2 + \theta < 19) \vee \\
&\quad (l = 2 \wedge 3 \leq \theta \leq 15 \wedge 3x_1 + \theta < 15) \vee l = 3 \\
\psi_4 &= \psi_3
\end{aligned}$$

The state predicate $\neg \bigvee_{i=0}^{3} \psi_i[z := 0]$ representing the meaning of $\neg\exists \Diamond (l = 3)$ is

$$\begin{aligned}
& l = 0 \wedge \theta \leq 15 \wedge (\theta + 21 \leq 6x_1 \wedge \theta \leq 6x_2 + 3 \vee \theta + 21 \leq 6x_2) \vee \\
& l = 1 \wedge 3 \leq \theta \leq 15 \wedge 19 \leq 4x_2 + \theta \vee \\
& l = 2 \wedge 3 \leq \theta \leq 15 \wedge 15 \leq 3x_1 + \theta
\end{aligned}$$

Since the state predicate $l = 0 \wedge \theta \leq 15 \wedge x_1 \geq 6 \wedge x_2 \geq 6$ characterizing the set of initial states implies the above predicate, the system satisfies the invariant as required.

- Suppose that we change the time of recovery to $T = 8$. Now, the condition $2\tau_r + \tau_1 \geq T \wedge 2\tau_r + \tau_2 \geq T$ is no longer satisfied. Again, we compute using Kronos the characteristic set of $\exists\Diamond l = 3$. The results obtained at each iteration are shown below.

$$\begin{aligned}
\psi_0 &= l = 3 \\
\psi_1 &= (l = 0 \wedge \theta \leq 15 \wedge 6x_1 < \theta + 33 \wedge 6x_2 < \theta + 33) \vee l = 3 \\
\psi_2 &= (l = 0 \wedge \theta \leq 15 \wedge 6x_1 < \theta + 33 \wedge 6x_2 < \theta + 33) \vee \\
&\quad (l = 1 \wedge 3 \leq \theta \leq 15 \wedge 4x_2 + \theta < 27) \vee \\
&\quad (l = 2 \wedge 3 \leq \theta \leq 15 \wedge 3x_1 + \theta < 21) \vee l = 3 \\
\psi_3 &= (l = 0 \wedge \theta \leq 15 \wedge (6x_1 + 3 < \theta \vee 6x_2 < \theta + 3 \vee \\
&\quad (6x_1 < \theta + 33 \wedge 6x_2 < \theta + 33))) \vee \\
&\quad (l = 1 \wedge 3 \leq \theta \leq 15 \wedge 4x_2 + \theta < 27) \vee \\
&\quad (l = 2 \wedge 3 \leq \theta \leq 15 \wedge 3x_1 + \theta < 21) \vee l = 3 \\
\psi_4 &= (l = 0 \wedge \theta \leq 15 \wedge (6x_1 + 3 < \theta \vee 6x_2 < \theta + 3 \vee \\
&\quad (6x_1 < \theta + 33 \wedge 6x_2 < \theta + 33))) \vee \\
&\quad (l = 1 \wedge 3 \leq \theta \leq 15 \wedge 4x_2 + \theta < 27) \vee \\
&\quad (l = 2 \wedge 3 \leq \theta \leq 15) \vee l = 3
\end{aligned}$$

$$\psi_5 = (l = 0 \wedge \theta \leq 15 \wedge (\theta + 33 \leq 6x_2 \vee 6x_1 < \theta + 33 \vee 6x_2 < \theta + 3)) \vee$$
$$(l = 1 \wedge 3 \leq \theta \leq 15 \wedge 4x_2 + \theta < 27) \vee (l = 2 \wedge 3 \leq \theta \leq 15) \vee$$
$$l = 3$$
$$\psi_6 = (l = 0 \wedge \theta \leq 15 \wedge (\theta + 33 \leq 6x_2 \vee 6x_1 < \theta + 33 \vee 6x_2 < \theta + 3)) \vee$$
$$(l = 1 \wedge 3 \leq \theta \leq 15) \vee (l = 2 \wedge 3 \leq \theta \leq 15) \vee l = 3$$
$$\psi_7 = (l = 0 \wedge \theta \leq 15) \vee (l = 1 \wedge 3 \leq \theta \leq 15) \vee (l = 2 \wedge 3 \leq \theta \leq 15) \vee$$
$$l = 3$$
$$\psi_8 = \psi_7$$

The state predicate $\neg \bigvee_{i=0}^{7} \psi_i[z := 0]$ representing the meaning of $\neg\exists \Diamond (l = 3)$ is

$$l = 0 \wedge \theta > 15 \vee$$
$$l = 1 \wedge (\theta < 3 \vee \theta > 15) \vee$$
$$l = 2 \wedge (\theta < 3 \vee \theta > 15)$$

and since the state predicate $l = 0 \wedge \theta \leq 15 \wedge x_1 \geq 6 \wedge x_2 \geq 6$ characterizing the set of initial states does not imply the above predicate we have that shutdown is reachable.

Table 1 shows the number of iterations and the running times (measured in seconds) obtained with KRONOS on a SUN 4 Sparc Station for verifying the above formula on the Temperature Control System with different values of the parameters.

Table 1. Performances for the Temperature Control System

parameters						number of	running
θ_m	θ_M	v_r	v_1	v_2	T	iterations	times
3	15	6	4	3	6	4	0.033
3	15	6	4	3	8	4	0.033
10	190	45	30	18	20	6	0.083
250	1100	34	25	10	80	4	0.033

References

[ACD90] R. Alur, C. Courcoubetis, and D. Dill. Model-checking for real-time systems. In *Proc. 5th Symp. on Logics in Computer Science*, pages 414–425. IEEE Computer Society Press, 1990.

[ACD+92] A. Alur, C. Courcoubetis, D. Dill, N. Halbwachs, and H. Wong-Toi. Minimization of timed transition systems. In *CONCUR 92: Theories of Concurrency*. Lecture Notes in Computer Science 630, Springer-Verlag, 1992.

[ACHH93] R. Alur, C. Courcoubetis, T. A. Henzinger, and Pei-Hsin Ho. Hybrid automata: an algorithmic approach to the specification and analysis of hybrid systems. In *Workshop on Theory of Hybrid Systems*, Lyngby, Denmark, June 1993. Lecture Notes in Computer Science 736, Springer-Verlag.

[AD90] R. Alur and D. Dill. Automata for modeling real-time systems. In *Proc. 17th ICALP*, pages 322–335. Lecture Notes in Computer Science 443, Springer-Verlag, 1990.

[BFH90] A. Bouajjani, J.Cl. Fernandez, and N. Halbwachs. Minimal model generation. In E.M. Clarke and R.P. Kurshan, editors, *Proc. 2nd Workshop on Computer-Aided Verification*, Rutgers University, June 1990. Lecture Notes in Computer Science 531, Springer-Verlag.

[CHR92] Z. Chaochen, C. A. R. Hoare, and A. P. Ravn. A calculus of durations. *Information Processing Letters*, 40(5):269–276, 1992.

[HNSY92] T.A. Henzinger, X. Nicollin, J. Sifakis, and S. Yovine. Symbolic model-checking for real-time systems. In *Proc. 7th Symp. on Logics in Computer Science*. IEEE Computer Society Press, 1992.

[JLHM91] M. Jaffe, N. Leveson, M. Heimdahl, and B. Melhart. Software requirements analysis for real-time process-control systems. *IEEE Transactions on Software Engineering*, 17(3):241–258, 1991.

[LY92] D. Lee and M. Yannakakis. Online minimization of transition systems. In *ACM Symp. on Theory of Computing*. ACM Press, 1992.

[NOSY93] X. Nicollin, A. Olivero, J. Sifakis, and S. Yovine. An approach to the description and analysis of hybrid systems. In *Workshop on Theory of Hybrid Systems*, Lyngby, Denmark, June 1993. Lecture Notes in Computer Science 736, Springer-Verlag.

[OY92] A. Olivero and S. Yovine. *Kronos: a Tool for Verifying Real-time Systems. User's Guide and Reference Manual*. VERIMAG, Grenoble, France, 1992.

A Unified Framework for Hybrid Control

Michael S. Branicky[1], *Vivek S. Borkar*[2] *and Sanjoy K. Mitter*[1]

[1] LIDS and EECS Department, M.I.T., Cambridge, MA 02139-7205,
email: branicky@mit.edu

[2] Department of Electrical Engineering, Indian Institute of Science, Bangalore 560012, India

1 Introduction

Hybrid control systems are control systems that involve both continuous dynamics and discrete phenomena. Some examples include computer disk drives [7], transmissions and stepper motors [5], constrained robotic systems [1], and intelligent vehicle/highway systems [11]. The continuous dynamics is usually modelled by a differential equation

$$\dot{x}(t) = \xi(t) \ , \quad t \geq 0 \ . \tag{1}$$

Here, $x(t)$ is the *continuous component* of the state taking values in some subset of a Euclidean space. $\xi(t)$ is a *controlled vector field* which generally depends on $x(t)$, the *continuous component* $u(t)$ of the control policy, and the aforementioned discrete phenomena. (We shall make this more precise later on.) The discrete phenomena generally considered are of four types:

1. Autonomous switching: Here the vector field $\xi(\cdot)$ changes discontinuously when the state $x(\cdot)$ hits certain "boundaries" [9, 10]. The simplest example of this is when it changes depending on a "clock" which may be modelled as a supplementary state variable [5].
2. Autonomous jumps: Here $x(\cdot)$ jumps discontinuously on hitting prescribed regions of the state space [1].
3. Controlled switching: Here $\xi(\cdot)$ changes abruptly in response to a control command with an associated cost. This can be interpreted as switching between different vector fields [13].
4. Controlled jumps: Here $x(\cdot)$ changes discontinuously in response to a control command, with an associated cost [2].

In this paper, we study a model that subsumes all these phenomena and study an associated control problem. The next section reviews models of hybrid systems from the control and dynamical systems literature. Section 3 develops the framework for hybrid control and defines a control problem. The existence of an optimal control for this problem is established in §4. Section 5 gives a formal derivation of the associated *generalized quasi-variational inequalities*. To conserve space in this extended abstract, we have deleted the proofs of the theorems.

2 Review of Models of Hybrid Dynamical Systems

One can find the following four explicit models of hybrid systems in the control and dynamical systems literature: Brockett model (B, [5, Type A and B models]), Tavernini model (T, [10]), Back et al. model (BGM, [1]), Nerode-Kohn model (NK, [9]). For sure, there are many others and no review is attempted here (for this see the collection [8]). In the paper, we briefly review and compare these four models.

Our comparison results may be summarized by the following inclusion diagram: $B \subset BGM \supset NK \supset T$. B is distinct from the T and NK since it allows the possibility of a (countably) infinite number of vector fields. We also show that autonomous switching can be viewed as a special case of autonomous jumps.

3 The Control Problem

Our state space for $x(\cdot)$ will be $S = \bigcup_{i=1}^{\infty} \overline{S_i}$ where each S_i is a connected open subset of some Euclidean space $\mathbb{R}^{d_i}$, $d_i \in \mathbb{N}$, with a Lipschitz boundary ∂S_i.[3] We also specify *a priori* regions $A_i, B_i, C_i \subset S_i$, $i \in \mathbb{N}$, such that the following hold. For each i, C_i is closed with empty interior and $\partial C_i = C_i$ is Lipschitz and contains ∂S_i. $A_i, B_i \subset S_i$ are closed. Let $B^{\hat{i}} = \bigcup_{j \neq i} B_j$, $i \in \mathbb{N}$. Let U, V be compact metric "action" spaces. The following bounded uniformly continuous maps are assumed to be known:

1. *vector fields* $f_i : \overline{S_i} \times \overline{S_i} \times U \to \mathbb{R}^{d_i}$, $i \in \mathbb{N}$, which are bounded (uniformly in i), uniformly Lipschitz continuous in the first argument, uniformly equicontinuous with respect to the rest.
2. *switching map* $G : \left(\bigcup_i C_i\right) \times V \to \bigcup_i B^{\hat{i}}$ such that $G(C_i \times V) \subset B^{\hat{i}}$, for all i.
3. *switching delay* $\Delta_1 : \left(\bigcup_i C_i\right) \times V \to \mathbb{R}^+$.
4. *impulse delay* $\Delta_2 : \bigcup_i \left(A_i \times B^{\hat{i}}\right) \to \mathbb{R}^+$.
5. *running cost* $k : \overline{S_i} \times \overline{S_i} \times U \to \mathbb{R}^+$.
6. *switching cost* $c_1 : \left(\bigcup_i C_i\right) \times V \to \mathbb{R}^+$.
7. *impulse cost* $c_2 : \bigcup_i \left(A_i \times B^{\hat{i}}\right) \to \mathbb{R}^+$, satisfying the conditions

$$c_2(x, y) \geq c_0 > 0, \qquad \forall x \in A_i, y \in B^{\hat{i}}, \tag{2}$$

$$c_2(x, y) < c_2(x, z) + e^{-a\Delta_2(x,z)} c_2(z, y), \quad \forall x \in A_i, z \in B^{\hat{i}} \cap A_j, y \in B^{\hat{i}} \cap B^{\hat{j}}, \tag{3}$$

for all $i \in \mathbb{N}$, $i \neq j \in \mathbb{N}$, resp., a a positive discount factor.

The dynamics of the control system can now be described as follows. There is a sequence of *pre-jump times* $\{\tau_i\}$ and another sequence of *post-jump times* $\{\Gamma_i\}$ satisfying $0 = \Gamma_0 \leq \tau_1 < \Gamma_1 < \tau_2 < \Gamma_2 < \cdots \leq \infty$, such that on each interval $[\Gamma_{j-1}, \tau_j)$ with non-empty interior, $x(\cdot)$ evolves according to (1) in some $S_i \backslash C_i$, $i \in \mathbb{N}$. At the next pre-jump time (say, τ_j) it jumps to some $S_k \backslash C_k$, $k \neq i$ according to one of the following two possibilities:

[3] The state dimension may change to take into account component failures or change in dynamical description based on discrete events—controlled or autonomous—which change it, e.g., the collision of two inelastic particles.

1. $x(\tau_j) \in C_i$, in which case it must jump to $x(\Gamma_j) = G(x(\tau_j), v_j) \in B^{\hat{i}}$ at time $\Gamma_j = \tau_j + \Delta_1(x(\tau_j), v_j)$, $v_j \in V$ being a control input. This is done at a cost of $c_1(x(\tau_j), v_j)$ paid at time τ_j. We call this phenomenon an autonomous jump.
2. $x(\tau_j) \in A_i$ and the controller chooses to[4] move the trajectory discontinuously to $x(\Gamma_j) \in B^{\hat{i}}$ at time $\Gamma_j = \tau_j + \Delta_2(x(\tau_j), x(\Gamma_j))$ at a cost of $c_2(x(\tau_j), x(\Gamma_j))$ paid at time τ_j. We call this an impulsive jump.

For $t \in [0, \infty)$, let $[t] = \max_j\{\Gamma_j \mid \Gamma_j \leq t\}$. The vector field $\xi(t)$ of (1) is given by

$$\xi(t) = f_i(x(t), x([t]), u(t)) \tag{4}$$

where i is such that $x(t), x([t]) \in \overline{S_i}$ and $u(\cdot)$ is a U-valued control process, assumed to be measurable. In addition to the costs associated with the jumps as above, the controller also incurs a *running cost* of $k(x(t), x([t]), u(t))$ per unit time during the intervals $[\Gamma_{j-1}, \tau_j)$, $j \in \mathbb{N}$. Let $a > 0$ be a *discount factor*. The total discounted cost is defined as

$$\int_T e^{-at} k(x(t), x([t]), u(t))\, dt + \sum_i e^{-a\sigma_i} c_1(x(\sigma_i), v_i) + \sum_i e^{-a\zeta_i} c_2(x(\zeta_i), x(\zeta_i')) \tag{5}$$

where $T = \mathbb{R}^+ \backslash \left(\bigcup_i [\tau_i, \Gamma_i)\right)$, $\{\sigma_i\}$ (resp. $\{\zeta_i\}$) are the successive pre-jump times for autonomous (resp. impulsive) jumps and ζ_j' is the post-jump time for the jth impulsive jump. The *decision* or *control* variables over which (5) is to be minimized are the continuous control $u(\cdot)$, the discrete control $\{v_i\}$ exercised at the pre-jump times of autonomous jumps, the pre-jump times $\{\zeta_i\}$ of impulsive jumps, and the associated *destinations* $\{x(\zeta_i')\}$. As for the periods $[\tau_j, \Gamma_j)$, we shall follow the convention that the system remains frozen during these intervals. Note that (2) rules out from consideration infinitely many impulsive jumps in a finite interval and (3) rules out the merging of post-jump time of an impulsive jump with the pre-jump time of the next impulsive jump.

Our framework clearly includes conventional impulse control [2] and hybrid systems with purely autonomous jumps [1]. As discussed in the previous section, the latter includes autonomous switching as a special case. Likewise, we show in the paper that our framework includes setting of parameters or timers upon hitting boundaries and controlled switching.

4 Existence of Optimal Controls

Let $J(x)$ denote the infimum of (5) over all choices of $(u(\cdot), \{v_i\}, \{\zeta_i\}, \{x(\zeta_i')\})$ when $x(0) = x$. We need one more assumption on our model above:

(A0) $d(A_i, C_i) > 0$ and $\inf_{i \in \mathbb{N}} d(B_i, C_i) > 0$, d being the appropriate Euclidean distance.

With this assumption we have

Theorem 1. *A finite optimal cost exists for any initial condition. There are only finitely many autonomous jumps in finite time.*

[4] He does not have to

It is easy to construct examples that show that an assumption like the second part of (A0) is necessary for the above results.

Next, we show that $J(x)$ will be attained for all x if we extend the class of admissible $u(\cdot)$ to "relaxed" controls, under the following additional assumptions:

(A1) Each B_i is bounded and for each i, there exists an integer $N(i) < \infty$ such that for $x \in A_i$, $y \in B_j$, $j > N(i)$, $c_2(x, y) > \sup_z J(z)$.

(A2) For each i, C_i is an oriented C^1-manifold without boundary and at each point x on C_i, $f_i(x, z, u)$ is "transversal" to C_i for all choices of z, u. By this we require that (1) the flow lines be transversal in the normal sense[5] and (2) the vector field does not vanish on C_i.

The "relaxed" control framework [12] is as follows: We suppose that $U = \mathcal{P}(U')$, defined as the space of probability measures on a compact space U' with the topology of weak convergence [3]. Also

$$f_i(x, z, u) = \int f_i'(x, z, u)u(dy) \ , \quad i \in \mathbb{N} \ ,$$

$$k(x, z, u) = \int k'(x, z, u)u(dy) \ ,$$

for suitable $\{f_i'\}$, k' satisfying the appropriate continuity/Lipschitz continuity requirements. The relaxed control framework and its implications in control theory are well known and the reader is referred to [12] for details.

Theorem 2. *An optimal trajectory exists for any initial condition.*

It is easy to see why Theorem 2 may fail in absence of (A1). Suppose, for example, $k(x, z, u) \equiv \alpha_i$ and $c_1(x, v) \equiv \beta_i$ when $x \in \overline{S_i}$, $c_2(x, y) \equiv \gamma_{i,j}$ when $x \in \overline{S_i}$, $y \in \overline{S_j}$, with α_i, β_i, $\gamma_{i,j}$ strictly decreasing with i, j. It is easy to conceive of a situation where the optimal choice would be to "jump to infinity" as fast as you can. In the paper we give examples which show that the theorem may fail in the absence of (A2).

Coming back to the relaxed control framework, say that $u(\cdot)$ is a *precise* control if $u(\cdot) = \delta_{q(\cdot)}(dy)$ for a measurable $q : [0, \infty) \to U'$ where δ_z denotes the Dirac measure at $z \in U'$. Let D denote the set of measures on $[0, T] \times U'$ of the form $dt \ u(t, dy)$ where $u(\cdot)$ is a relaxed control, and D_0 its subset corresponding to precise controls. It is known that D_0 is dense in D with respect to the topology of weak convergence [12]. In conjunction with the additional assumption (A3) below, this allows us to deduce the existence of ϵ-optimal control policies using precise $u(\cdot)$, for every $\epsilon > 0$.

(A3) Same as (A2) but with ∂A_i (the boundary of A_i) replacing C_i.

Theorem 3. *Under* (A1)–(A3), *for every* $\epsilon > 0$ *an* ϵ*-optimal control policy exists wherein* $u(\cdot)$ *is precise.*

[5] Transversality implies that C_i is $(d_i - 1)$-dimensional.

Remark. If $\{\overline{f_i}(x, z, y) \mid y \in U'\}$ are convex for each x, z, a standard selection theorem [12] allows us to replace $u^\infty(\cdot)$ by a precise control which will then be optimal. Even otherwise, using Caratheodory's theorem (viz. Each point in a compact subset of $\mathbb{R}^n$ is expressible as a convex combination of at most $n+1$ of its extreme points) and the aforementioned selection theorem, one may suppose that for $t \geq 0$, the support of $u^\infty(t)$ consists of at most $d_i + 1$ points when $x(t) \in \overline{S_i}$.

5 The Value Function

In the foregoing, we had set $[0] = 0$ and thus $x([0]) = x(0) = x_0$. More generally, for $x(0) = x_0 \in \overline{S_{i_0}}$, we may consider $x([0]) = y$ for some $y \in \overline{S_{i_0}}$, making negligible difference in the foregoing analysis. Let $V(x, y)$ denote the optimal cost corresponding to this initial data. Then in dynamic programming parlance, $(x, y) \mapsto V(x, y)$ defines the "value function" for our control problem.

In view of (A2), we can speak of the *right side* of C_i as the side on which $f_i(\cdot, \cdot, \cdot)$ is directed towards C_i, $i \in \mathbb{N}$. A similar definition is possible for the right side of ∂A_i (in light of (A3)). Say that $(x_n, y_n) \to (x_\infty, y_\infty)$ *from the right* in $\bigcup_i \left(\overline{S_i} \times \overline{S_i}\right)$ if $y_n \to y_\infty$ and either $x_n \to x_\infty \notin \bigcup_i (C_i \cup \partial A_i)$ or $x_n \to x_\infty \in \bigcup_i (C_i \cup \partial A_i)$ from the right side. V is said to be *continuous from the right* if $(x_n, y_n) \to (x_\infty, y_\infty)$ from the right implies $V(x_n, y_n) \to V(x_\infty, y_\infty)$.

Theorem 4. *V is continuous from the right.*

We can now formally derive the *generalized quasi-variational inequalities* $V(\cdot, \cdot)$ is expected to satisfy [Derivation deleted.]: For $(x, y) \in \overline{S_i} \times \overline{S_i}$,

$$V(x, y) \leq \min_{z \in \widehat{B^i}} \left(c_2(x, z) + e^{-a\Delta_2(x,z)} V(z, z)\right) \quad \text{on } A \tag{6}$$

$$V(x, y) \leq \min_{v} \left(c_1(x, v) + e^{-a\Delta_1(x,v)} V(G(x, v), G(x, v))\right) \quad \text{on } \bigcup_i \left(C_i \times \overline{S_i}\right) \tag{7}$$

$$\min_u \left(\langle \nabla_x V(x, y), f_i(x, y, u)\rangle - aV(x, y) + k(x, y, u)\right) \leq 0 \tag{8}$$

and

$$\left(V(x, y) - \min_{z \in \widehat{B^i}} \left(c_2(x, z) + e^{-a\Delta_2(x,z)} V(z, z)\right)\right) \cdot \left(\min_u \left(\langle \nabla_x V(x, y), f_i(x, y, u)\rangle - aV(x, y) + k(x, y, u)\right)\right) = 0 \quad \text{on } A \tag{9}$$

((9) states that at least one of (6), (8) must be an equality on A.) (6)–(9) generalize the traditional *quasi-variational inequalities* encountered in impulse control [2]. We do not address the issue of well-posedness of (6)–(9). The following "verification theorem", however, can be proved by routine arguments.

Theorem 5. *Suppose (6)–(9) has a "classical" solution V which is continuously differentiable "from the right" in the first argument and continuous in the second. Suppose $x(\cdot)$ is an admissible trajectory of our control system with initial data (x_0, y_0) and $u(\cdot)$, $\{v_i\}$, $\{\sigma_i\}$, $\{\zeta_i\}$, $\{\tau_i\}$, $\{\Gamma_i\}$ the associated controls and jump times, such that the following hold: For a.e. $t \notin T$,*

$$\langle \nabla_x V(x(t), x([t])), f_i(x(t), x([t]), u(t))\rangle + k(x(t), x([t]), u(t)) = \min_u (\langle \nabla_x V(x(t), x([t])), f_i(x(t), x([t]), u)\rangle + k(x(t), x([t]), u))$$

where i is such that $x(t) \in \overline{S_i}$.

$$V(x(\sigma_i), x([\sigma_i])) = c_1(x(\sigma_i), v_i) + \exp\{-a\Delta_1(x(\sigma_i), v_i)\}V(G(x(\sigma_i), v_i), G(x(\sigma_i), v_i))$$

for all i.

$$V(x(\zeta_i), x([\zeta_i])) = c_2(x(\zeta_i), x(\zeta_i')) + \exp\{-a\Delta_2(x, x(\zeta_i'))\}V(x(\zeta_i'), x(\zeta_i'))$$

for all i. Then $x(\cdot)$ is an optimal trajectory.

6 Conclusions

The foregoing presents some initial steps towards developing a unified "state space" paradigm for hybrid control. Several open issues suggest themselves:

1. A daunting problem is to characterize the value function as the unique viscosity solution of the generalized quasi-variational inequalities (6)–(9).
2. Many of our assumptions can possibly be relaxed at the expense of additional technicalities or traded off for alternative sets of assumptions that have the same effect. For example, the condition $d(A_i, C_i) > 0$ could be dropped by having c_2 penalize highly the impulsive jumps that take place too close to C_i. (In this case, (A3) has to be appropriately reformulated.)
3. An important issue here is to develop good computational schemes to compute near-optimal controls. See [6] for some related work.
4. Another possible extension is in the direction of replacing $\overline{S_{i_0}}$ by smooth manifolds with boundary embedded in a Euclidean space. See [4] for some related work.

References

1. A. Back, J. Guckenheimer, and M. Myers. A dynamical simulation facility for hybrid systems. TR 92-6, Mathematical Sciences Institute, Cornell Univ., April 1992. Revised.
2. A. Bensoussan and J.-L. Lions. *Impulse Control and Quasi-Variational Inequalities*. Gauthier-Villars, Paris, 1984.
3. P. Billingsley. *Convergence of Probability Measures*. Wiley, New York, NY, 1968.
4. R. W. Brockett. Smooth multimode control systems. In Clyde Martin, ed., *Berkeley-Ames Conf. Nonlinear Problems in Control and Fluid Dynamics*, pp. 103–110. Math-Sci Press, Brookline, MA, 1983.

5. R. W. Brockett. Hybrid models for motion control systems. CICS-P-364, Center for Intelligent Control Systems, Massachusetts Institute of Technology, March 1993.
6. O. L. V. Costa and M. H. A. Davis. Impulse control of piecewise deterministic processes. *Math. Control Signals Syst.*, 2:187–206, 1989.
7. A. Gollu and P. Varaiya. Hybrid dynamical systems. In *Proc. 28th IEEE Conference on Decision and Control*, pages 2708–2712, Tampa, FL, December 1989.
8. Grossman, R. L. et al., eds. *Hybrid Systems*, vol. 736 of *Lecture Notes in Computer Science*. Springer-Verlag, New York, 1993.
9. A. Nerode and W. Kohn. Models for hybrid systems: Automata, topologies, stability. TR 93-11, Mathematical Sciences Institute, Cornell Univ., March 1993. Revised.
10. L. Tavernini. Differential automata and their discrete simulators. *Nonlinear Analysis, Theory, Methods, and Applications*, 11(6):665–683, 1987.
11. P. Varaiya. Smart cars on smart roads: Problems of control. *IEEE Transactions on Automatic Control*, 38(2):195–207, February 1993.
12. L. C. Young. *Lectures on the Calculus of Variations and Optimal Control Theory*. Chelsea, New York, 2nd edition, 1980.
13. J. Zabczyk. Optimal control by means of switching. *Studia Mathematica*, 65:161–171, 1973.

Reasoning About Hybrid Systems With Symbolic Simulation

Sanjai Narain

Bellcore, 2E-260, 445 South Street, Morristown, NJ 07962, USA
email: narain@thumper.bellcore.com

1 Introduction

The discrete-event simulation technique e.g. [Sur], [Eva], [Zei] is widely used to model and simulate dynamic systems in diverse areas such as telecommunications, manufacturing, military or transportation. This technique is based on the assumption that system behavior can be represented by its history i.e., a sequence of discrete events which occur in that system. From a history, system state can be computed at any point of time. The technique defines a method for specifying how events are caused, and a method for computing histories from that specification.

The discrete-event technique is powerful enough to model and simulate hybrid systems. The behavior of a hybrid system over time is viewed as a sequence of phases. In each phase, the behavior is described by differential equations. However, at phase boundaries these equations change. Phase boundaries are regarded as discrete events, so event causation is specified using the discrete-event technique. This specification is then used to compute the sequence of phase boundaries. From this sequence, the differential equations governing system behavior at any point of time are computed.

In [NCC1] is presented DMOD (Declarative MODeling), a formalization of the discrete-event technique. In the context of an industrial example of fiber-optics networks, it is shown how DMOD models are substantially simpler to develop. In particular, it is substantially simpler to model the phenomenon of *event preemption*. This phenomenon occurs especially in hybrid systems in which continuous events occur. See Section 3 for details.

This paper shows how one can reason with DMOD models to prove temporal properties of modeled systems. The paper shows this in the context of a (toy) railroad crossing, a hybrid system with discrete events and continuous time and state. Reasoning is done by means of a technique called *symbolic simulation*. Whereas in (traditional) simulation input parameters are fixed, in symbolic simulation input parameters can be constrained variables. Symbolic simulation computes histories which are parameterized with these variables. The histories for any particular values of these variables are obtained by instantiation. Simple manipulation of symbolic histories allows one to prove safety, liveness and possibility properties as well as to do sensitivity analysis. An application of symbolic DMOD to verification of a network protocol is described in [Seb].

DMOD models of systems must satisfy strong requirements before symbolic simulation can be performed. These restrictions allow the symbolic simulation procedure to be significantly simpler and more efficient than procedures for more obvious forms of symbolic evaluation. However, as illustrated in this paper, models of non-trivial systems satisfying these requirements can still be developed. See Section 4.4 for more details.

DMOD has been implemented in the commercially available Quintus Prolog. Prolog [Kow] is a powerful language based upon symbolic logic in which one can specify not only relations but also procedures for computing those relations. It is used simultaneously as a language for specifying DMOD models, for specifying temporal properties, for manipulating differential equations and for computing performance metrics.

An experimental version of *symbolic* DMOD, however, has been implemented in CLP(R) [JL1]. CLP(R) is an extension of Prolog [Kow] in which one can reason with linear constraints on real numbers. However, the abstract symbolic simulation framework is general and can be used to reason about hybrid systems with non-linearities as well.

Section 2 provides an overview of DMOD. Section 3 presents a DMOD model of a railroad crossing, and examples of simulations of it. Section 4 shows examples of reasoning with this model using symbolic simulation and presents the symbolic simulation procedure. Section 5 contains a summary and conclusions. A complete, formal treatment of symbolic simulation is presented in a forthcoming paper [Nar].

2 Overview of DMOD

DMOD is based upon the assumption that the behavior of a system can be modeled by its history, i.e. the sequence of the *discrete* events which occur in it. Thus, simulation of a system can be regarded as computing its history. DMOD defines a formal framework for modeling systems and computing their histories. A DMOD *structure* is a tuple *{Events, time, causes}* where:

- *Events* is an enumerably infinite set of objects called *events*.
- *time(E)* is a function mapping an event *E* to a non-negative real number called the timestamp of *E*.
- *causes(E,HE,F)* is a relation between events *E* and *F*, and a finite sequence *HE* of events in *Events* satisfying:
  - $\forall E \forall HE.\{F \mid causes(E,HE,F)\}$ is finite
  - $\forall E \forall HE \forall F. causes(E,HE,F) \supset time(E) \leq time(F)$

An event is assumed to be discrete and instantaneous. Note that information about an event's timestamp is contained within the event. *causes(E,HE,F)* means *E* causes *F* given that *HE* is the sequence of all the events which have occurred before *E*.

Let *P* be a DMOD structure and *S=[E0,E1,..]* be a finite, or enumerably infinite sequence of events. Then *S* is said to be *temporally ordered* if for each i, $i \geq 0 \supset time(Ei) \leq time(Ei+1)$ whenever *Ei+1* exists.

Let *P={Events, time, causes}* be a DMOD structure. Let *S=[E0,E1,E2,..]* be a finite or enumerably infinite sequence of events. Then *S* is said to be *causally-sound* if every event in *S* has a cause in *S*, i.e.:

$$\forall j. j>0 \supset \exists i. i<j \wedge causes(Ei,[E0,..,Ei-1],Ej).$$

Note that the singleton sequence *[E0]* is trivially causally sound.

Let *P={Events, time, causes}* be a DMOD structure and *S=[E0,E1,E2,..]* be a finite or enumerably infinite sequence of events. Then *S* is said to be *causally-complete* iff it contains all caused events, i.e.

$$\forall G. \forall i. causes(Ei,[E0,..,Ei-1],G) \supset \exists j. j>i \wedge Ej=G$$

Let *P* be a DMOD structure and *init* be an event. A history *H* for *P* beginning with *init* is a finite or or enumerably infinite sequence of events such that:

- *H* begins with *init*
- *H* is temporally ordered
- An event occurs at most once in *H*
- *H* is causally-sound
- *H* is causally-complete

Let *S* be a set of events. Then $E \in S$ is said to be an *earliest event in S* if the timestamp of *E* is less than or equal to that of every other event in *S*. Note that if *S* contains concurrent events (i.e. with the same timestamp), there can be more than one earliest event in *S*. Also, if *S* is finite, it always contains an earliest event.

Let *Seq=[E0,E1,..]* be a temporally ordered sequence of events. Then *Seq* is said to be *strictly progressive* iff either *Seq* is finite, or for each real number *t* there is an *i* such that *t<time(Ei)*. In particular, a sequence of events is not strictly progressive if its timestamps converge.

Let *E0,E1,..,Ek* be a sequence of events. The set of all effects of *Ek* w.r.t. *E0,..,Ek-1* is the set *{F | causes(Ei,[E0,..,Ei-1],F)}*. By the restriction on *causes*, this set is finite.

Procedure I. Let *P={Events, time, causes}* be a DMOD structure and *init* be an event. The procedure computes histories for *P* beginning with *init*. Each such history is of the form *E0,E1,..* where *E0=init*. The procedure also computes an auxiliary sequence of "queues" of events *Q0,Q1,..* where for each *i*, *Qi* denotes the set of events "waiting to occur" just before *Ei* occurs. In particular, *Q0=[init]*. Suppose that for some $k \geq 0$ the sequences of events and queues have been computed, respectively, as *E0,..,Ek* and *Q0,..,Qk*. Let *Effects* be the set of effects of *Ek* w.r.t. *E0,..,Ek-1*. Form the union of *Effects* and *Qk* and delete all events in this union already occurring in *E0,..,Ek*. Let the result be *Qk+1*. If *Qk+1* is empty, halt. Otherwise, let *Ek+1* be an earliest event in *Qk+1* □

Note that *Qk+1* can contain more than one earliest event (i.e. concurrent events) so the procedure is non-deterministic. A different history would be computed for each choice of *Em+1*. We can now prove:

Theorem 1. Soundness and Completeness of Procedure I. Let *P = {Events, time, causes}* be a DMOD structure and *init* be an event. Let *E0=init*. A strictly progressive sequence of events *E0,E1,..* is computed by the above procedure if and only if it is a history.

The proof is straightforward.

The main task in defining a DMOD structure is defining the causality relation. As Prolog is a relational language, it is an excellent candidate for defining causality. This paper assumes familiarity with Prolog.

3 Example: A Railroad Crossing

We now show, by an example of a (toy) railroad crossing, how DMOD can be used to model hybrid systems.

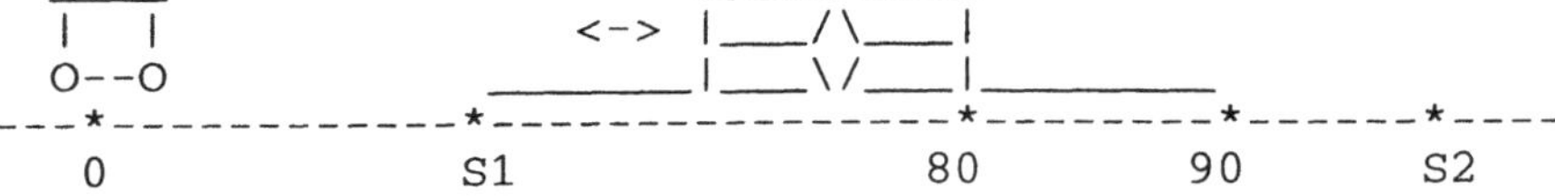

In the above figure, an engine moves on a track from left to right with a fixed speed, *40*. The track can be crossed between *80* and *90*. A barrier slides back and forth to close or open the crossing. When the engine reaches *S1*, *sensor(1)* senses the engine and activates closing of the barrier. When the engine reaches *S2*, *sensor(2)* senses the engine and activates opening of the barrier. The barrier opens and closes with a fixed speed, *10*.

The DMOD model of the above crossing is given below. We assume that the initial event is always of the form *start([S1, S2], 0)* where *S1, S2* are real values denoting, respectively, positions of *sensor(1)* and *sensor(2)*. In our implementation we have defined a Prolog procedure called *simulate(Init,H)*, which prints out histories *H* of the crossing starting with the initial event *Init*. If we type *simulate(start([40, 100],0), H)*, we obtain:

start([40, 100], 0)
begin_j(e(1), 40, 0, 0)
sensed(e(1), sensor(1), 1)
start(barrier, close, 1)
check(start(barrier, close, 1), sensed(e(1), sensor(2)), end(barrier, close, 2), 2)
end(barrier, close, 2)
sensed(e(1), sensor(2), 2.5)
start(barrier, open, 2.5)
end(barrier, open, 3.5)

From this history many types of information can be computed. For example, to compute velocity of barrier immediately after *start(barrier,close,1)* let the initial segment of history upto this event be *Hist*. Now, typing *velocity(barrier, V, Hist)* yields *V = 10*. Typing *position(barrier, P, Hist)* yields *P = 80*. Similarly, for velocity and position of the engine.

We now present the formal DMOD model and definitions of procedures *velocity* and *position*. Each event is of the form *f(U1,..,Un,T)* where each *Ui* is a term and *T* is the event's timestamp. In particular, *time(f(U1,..,Un,T)) = T*. Events are of the forms *start([S1,S2], 0)*, *begin_j(e(K), V, P, T)*, *sensed(e(K), sensor(M), T)*, *starts(barrier, OpenOrClose, T)*, *end(barrier, OpenOrClose, T)* and *check(Cause, PreemptionEventType, Effect, T)*. Their meanings become clear from discussions of causality rules which link these.

Note that continuous events, such as the closing of a barrier are modeled using two discrete events *start(barrier, close, T)* and *end(barrier, close, T)*. However, we now need to model event preemption, the phenomenon that even if a continuous event has begun, its ending may be precluded by the satisfaction of a preemption condition before the ending. Event preemption is simple to model in DMOD. A rule of the form *E causes F provided preemption condition P is not satisfied in between E and F*, is modeled as follows: *E* is made to cause an auxiliary event *check(E,P,F,TF)* where *TF* is the timestamp of *F*. This event definitely occurs. When it does, it extracts the segment of history between *E* and itself and checks whether *P* ever holds over this segment. If *P* does not hold, *F* is caused, otherwise no event is caused. The rules for checking whether *P* ever holds can be easily encoded in Prolog.

We now define the causality relation using Prolog. In the following, histories are maintained in reverse temporal order. Thus, in *causes(E, HE, F)*, the first event in *HE* has the highest timestamp of any event in *HE*. This convention yields more efficient programs. In the rule below, the initial event causes an event of engine *e(1)* to begin journey to the right, from position *0*, with velocity *40*, at time *0*.

```
causes(E,HE,F) if
    E=start([S1, S2], 0),
    F=begin_j(e(1), 40, 0, 0).
```

If *Engine* begins journey at time *T* with position *P* and velocity *V*, it is sensed by *sensor(1)* after time taken to travel *(S1-P)/V* where *S1* is the position of *sensor(1)*. *S1* is obtained by retrieving the initial event in *HE*. A similar rule is written for *sensor(2)*.

```
causes(E,HE,F) if
    E=begin_j(Engine,V,P,T),
    F=sensed(Engine,sensor(1),T+Delay),
    member(start([S1,S2], 0), HE),
    Delay = (S1-P)/V.
```

If *Engine* reaches *sensor(1)*, the barrier starts to close immediately. Similarly, *sensor(2)* activates barrier opening.

```
causes(E,HE,F) if
    E=sensed(Engine,sensor(1),T),
    F=start(barrier,close,T).
```

If barrier starts to close, then after *Delay* check that the barrier has fully closed. Check this by showing that no event of *sensor(2)* sensing *e(1)* occurs in between. *Delay* is the time taken by barrier to reach fully closed position, *90*. It is equal to *(90-P)/V* where *P,V* are current position and velocity of barrier. Similarly, for beginning and ending of barrier opening, except that the opening is not preempted.

```
causes(E,HE,F) if
    E=start(barrier,close,T),
    F=check(E, sensed(e(1),sensor(2)), end(barrier,close,FT),FT),
    position(barrier,P,[E|HE]),
    velocity(barrier,V,[E|HE]),
    Delay = (90-P)/V,
    FT = T+Delay.
```

The following rule states that *check(Cause, PreemptionEventType, Effect, T)* causes *Effect* provided in *EventsAfterCause*, the sequence of events between *Cause* and the checking event, an event of *PreemptionEventType* does not occur.

```
causes(E,HE,F) if
    E=check(Cause, PreemptionEventType, Effect,T),
    F=Effect,
    events_after(Cause, HE, EventsAfterCause),
    absent(PreemptionEventType, EventsAfterCause).
```

We now outline how velocity and position of the barrier are computed. Velocity and position of engine are computed similarly. The following rules state that the velocity of the barrier immediately after the initial event is *0*, immediately after it starts closing is *10*, immediately after it starts opening is *-10* and immediately after it ends opening or closing is *0*. An underscore _ represents a variable whose value is irrelevant to the present computation. Implicit is another rule that if *E* is none of these four forms, then velocity of barrier immediately after *E* is the same as that after *E*'s predecessor.

```
velocity(barrier,0,Hist) if Hist=[start(_,0)].
velocity(barrier,10,Hist) if Hist=[start(barrier,close,_)|_].
velocity(barrier,-10,Hist) if Hist=[start(barrier,open,_)|].
velocity(barrier,0,Hist) if Hist=[end(barrier,_,_)|_].
```

The first rule below states that the position of the barrier immediately after the initial event is *80*. The next rule states that the position of any object immediately after *E* is the position immediately after the predecessor of *E* plus the incremental displacement *V*(TE - THE)*. Note that the displacement is a linear function of time and velocity.

```
position(barrier,80,H) if H=[start(_,0)].
position(Object,NewPos,Hist) if
    Hist=[E|HE],
    position(Object,OldPos,HE),
    velocity(Object,V,HE),
    time(E,TE),
    time(HE,THE),
    NewPos = OldPos+V*(TE-THE).
```

4 Symbolic DMOD

Up to now we computed histories for initial events of the form *start([S1, S2], 0)* where *S1, S2* were real numbers. We now present a general framework for computing a set of histories even when *S1, S2* are variables constrained by the constraint *InitCon*. Each such *symbolic* history *H* contains occurrences of *S1, S2*. Associated with *H* is a constraint *histcon(H)* which is the conjunction of *InitCon* and the constraint that the symbolic history is temporally ordered.

The set of symbolic histories satisfies two main properties. First, for any specific value of *S1, S2* satisfying *InitCon*, the corresponding history is obtained by instantiating one of the symbolic histories with these values. Second, where *H* is a symbolic history, for every value of *S1,S2* satisfying *histcon(H)*, the corresponding instance of *H* is a history computed for that value of *S1, S2*.

These two properties of symbolic histories allow us to prove a wide range of temporal properties. To show that a property *P* holds for each value of *S1, S2* satisfying *InitCon*, show that *P* holds for each symbolic history. If each symbolic history is finite, and the number of symbolic histories is finite, then one can finitely reason about an infinite number of (instantiated) histories. *Safety* and *liveness* properties can be proved in this manner. To derive conditions under which it is *possible* for property *P* to be satisfied, check whether a symbolic history *H* satisfies *P*, and if so, read off *histcon(H)*. Finally, we can write algorithms which would take symbolic histories as inputs and compute performance metrics symbolically, i.e. as symbolic functions of *S1, S2*. These functions can be used to do sensitivity analysis.

We now illustrate the above ideas in the context of the railroad crossing. We then present the symbolic simulation procedure and discuss its design. As mentioned previously, symbolic DMOD has been implemented in CLP(R).

4.1 Symbolic Histories

The procedure *simulate(start([S1, S2],0),H)* prints the symbolic history for *S1, S2* as well as the associated constraints. For example, to check what symbolic histories are possible if *S1>0, S1<60, S2=100*, we type *S1>0, S1<60, S2=100, simulate(start([S1,S2],0),H)*. We obtain a single symbolic history *H* with *histcon(H) = (S1>0, S1<60, S2=100)*.

start([S1, 100], 0)
begin_j(e(1), 40, 0, 0)
*sensed(e(1), sensor(1), 0.025 * S1)*
*start(barrier, close, 0.025 * S1)*
*[check(start(barrier, close, 0.025 * S1),*
sensed(e(1), sensor(2)),
*end(barrier, close, 0.025*S1 + 1), 0.025*S1 + 1)]*
*end(barrier, close, 0.025 * S1 + 1)*
sensed(e(1), sensor(2), 2.5)
start(barrier, open, 2.5)
end(barrier, open, 3.5)

It is simple to verify that the instance of this history for *S1=40* is precisely that obtained by conventional simulation in Section 3. This symbolic history also shows that the behavior of the crossing is qualitatively the same for all values of *S1* satisfying *0<S1<60*. In particular, the barrier always ends closing. On the other hand, typing: *S1>60, S1<80, S2=100, simulate(start([S1,S2],0),H)* produces a single history but in which the barrier does not end closing. The engine reaches *sensor(2)* too quickly and prematurely activates opening of the barrier.

4.2 Possibility & Safety Properties

Let *unsafe(H)* hold if in the history *H* there are events *E* and *F* such that at some point of time between *E* and *F*, positions of both the barrier and engine are between *80* and *90*. The definition of *unsafe(H)* is easily encoded in *CLP(R)*. If we now type:

S1>0, S1<80, S2=100, simulate(start([S1, S2], 0),H), unsafe(H)

we obtain three values of *H* for which the associated constraints are: *40<S1<=60, S1=60* and *60<=S1<80*. It is easy to see that for any value of *S1* in the range *(40,80)*, the barrier does not have enough time to fully close before the engine enters the crossing. To give enough time for the barrier to fully close, we need to restrict *S1* to be less than *40*. To verify this we type:

S1>0, S1<40, S2=100, simulate(start([S1, S2], 0), H), unsafe(H)

CLP(R) responds with 'no', i.e. it was unable to compute any unsafe symbolic history if *S1,S2* are constrained as above.

4.3 Sensitivity Analysis

Let *distance_barrier(H,D)* mean that *D* is the distance traveled by the barrier (both forward and backward) given history *H*. If we now type *S1=60, S2>90, S2<100, simulate(start([S1, S2],0), H), distance_barrier(H, D)*, we obtain a single *H*, where *histcon(H)* is *S2 = 2*D + 60*. Thus we know that *D* is a linear function of *S2*. It can easily be verified by hand that this dependence is, in fact, correct.

4.4 A Symbolic Simulation Procedure

Let *P={Events, causes, time}* be a DMOD structure. Let *X1,..,Xn* be a finite, fixed set of variables, ranging over domains, respectively, *D1,..,Dn*. Let *Constraint* be a constraint on *X1,..,Xn*, i.e. an expression containing zero or more occurrences of *X1,..,Xn* and denoting a relation over the cross product of *D1,..,Dn*. An example of *Constraint* is *X1 + X2 > 3*. A (ground) substitution σ defines bindings of *X1,..,Xn* to objects in, respectively, *D1,..,Dn*. Let *E* be an expression containing zero or more occurrences of *X1,..,Xn*. Let σ bind *Xi, i≤n* to *ti*. Then *E*σ is the application of σ to *E*, i.e. the result of replacing, for each *i≤n*, *Xi* by *ti* in *E*. A substitution σ is said to be admissible w.r.t. a constraint *C*, *ad*(σ, *C*), if *C*σ is true. A constraint *C* is said to be consistent if there exists a substitution σ such that *ad*(σ, *C*).

An expression *E* containing zero or more occurrences of *X1,..,Xn* is said to be a symbolic event if there exists a substitution σ such that *E*σ $\in$ *Events*. A symbolic timestamp of a symbolic event *E*, *stime(E)*, is an expression *T* such that for every substitution σ and timestamp *X*, *time(E*σ*) = X iff X = T*σ. We assume that a symbolic timestamp of every symbolic event can be computed.

Let *E0,..,Ek* be a sequence of symbolic events and *C* a constraint. A set of symbolic events *S* is said to be a set of symbolic effects of *Ek* w.r.t. *C* and *E0,..,Ek-1* if:

- *S* is finite
- Let *F* be a symbolic event in *S*. For every substitution σ such that *ad*(σ, *C*) we have *causes(Ek, (E0,..,Ek-1), F)*σ.
- For any substitution σ such that *ad*(σ, *C*) let there be an event *G* such that *causes(Ek*σ*, (E0,..,Ek-1)*σ*, G)*. Then, there exists a symbolic event *F* in *S* such that *F*σ*=G*.

For example, let *HE = [start([S1, S2], 0)], E = begin_j(e(1), 40, 0, 0)* and *C = (S1>0, S1<60, S2=100)*. Then, the set of symbolic effects of *E*, w.r.t. *C* and *HE* is *{sensed(e(1), sensor(1), S1/40), sensed(e(1), sensor(2), 2.5)}*.

Suppose that *causes* is defined by the rules *(causes(start([D],0), _, g(D)) if D>4)* and *(causes(start([D],0), _, h(D)) if D>4,D<6)*. Let *InitCon* be *D>4,D<6*. Then, the set of symbolic effects of *start([D],0)* is *{g(D), h(D)}*. On the other hand if *InitCon* is *D>4*, then the set of (all) symbolic effects, in our sense, does not exist. In general, we require that the set of symbolic effects of *E* w.r.t. *H* and *C* be computed without strengthening *C*. This requirement is strong, but its advantage is that it eliminates the need to reconcile different constraints on different effects of the same event, e.g. *D>4* with *g(D)* and *(D>4, D<6)* with *h(D)*.

As outlined below, CLP(R) provides a simple way of checking whether constraints are strengthened. However, models can also be written in such a way that input parameters are never constrained in the computation of causality. Then, computation of symbolic effects can become more efficient because the check for constraint strengthening can be eliminated. For example, in the railroad crossing model, causality rules never constrain *S1, S2*.

The symbolic simulation procedure can now be formulated. It is closely analogous to the procedure for the non-symbolic case.

Procedure II. Let *{Events, time, causes}* be a DMOD structure. Let an initial symbolic event be *start([X1,..,Xn], 0)*, each *Xi* a variable, and *InitCon* a constraint upon *X1,..,Xn*. Each symbolic history for this structure will be of the form *E0,E1,..* where *E0=start([X1,..,Xn], 0)*. Also computed will be an auxiliary sequence of queues of symbolic events *Q0,Q1,..* where *Q0=[start([X1,..,Xn], 0)]*. Suppose that for some *k≥0* the sequences of symbolic events and queues of symbolic events have been computed, respectively, as *E0,..,Ek* and *Q0,..,Qk*. Let *Ck* be the constraint:

$$stime(E0) \leq stime(E1) \leq ... \leq stime(Ek)$$

Let *Effects* be the set of symbolic effects of *Ek* w.r.t. *E0,..,Ek-1* and *(InitCon* $\wedge$ *Ck)*. Form the union of *Effects* and *Qk* and delete from this set all symbolic events in *E0,..,Ek*. Let the result be *Qk+1 = {F1, .., Fj}, j≥0*. If *Qk+1* is empty, print *E0,..,Ek* as a symbolic history and let *histcon([E0,..,Ek]) = (InitCon* $\wedge$ *Ck)*. Otherwise, let *G* be a symbolic event in *Qk+1* and let *QueueCon* be the conjunction of the constraints:

$$stime(G) \leq stime(F0), ..., stime(G) \leq stime(Fj)$$

Let *Con* = *(InitCon* ∧ *Ck* ∧ *QueueCon)* be consistent. Check whether there is an event *Ej*, $0 \leq j \leq k$ and a substitution α admissible w.r.t. *Con* such that $Ej\alpha = G\alpha$. If so, halt with failure. Otherwise, let *Ek+1* = *G* □

The requirement in the last paragraph is to ensure that we do not generate symbolic events such as *f(X), f(Y)* because instances of these could lead to duplicate events which are disallowed in DMOD histories.

The two main operations in the above procedure are computing the set of symbolic effects of an event, and checking consistency of constraints. Both of these operations are easy to implement in CLP(R). For the first, let *E* be a symbolic event, *HE* a sequence of symbolic events, and *F* a new variable. Suppose typing *C*, *causes(E,HE,F)* returns a value for *F* and a constraint *C1*. If *C1=C* then the semantics of *CLP(R)* ensures that *F* is a symbolic effect of *E* by the causality rule involved. Repeat this step for all causality rules. Implementation of the second operation is based upon the fact that CLP(R) collects sets of constraints as it computes and automatically checks whether they are consistent.

5 Summary and Conclusions

This paper presented DMOD, a formalization of the popular discrete-event modeling and simulation technique. The paper showed how DMOD could be used to model and simulate a simple hybrid system. In particular, discrete-events, continuous time, and continuous state are all conveniently modeled. The paper also outlined a technique called symbolic simulation and showed how one could perform limited, yet useful forms of reasoning with this technique. Thus, symbolic DMOD represents a significant step beyond the discrete-event technique. The main limitation of DMOD is that one cannot reason in any straightforward way when symbolic histories are infinite. These arise in oscillating systems e.g. a room thermostat which switches heat on and off infinitely often e.g.[ACH+]. However, the building blocks for doing such reasoning are present in symbolic simulation and Prolog, and we expect to develop appropriate techniques in the near future.

Acknowledgement. I am grateful to Ritu Chadha, Ernest Cohen and Yow-Jian Lin for helpful comments.

References

[ACH+] Alur, R., Courcoubetis, C., Henzinger, T., Ho, P.-H.: Hybrid automata: An algorithmic approach to the specification and verification of hybrid systems. Hybrid Systems, R. Grossman, A. Nerode, A. Ravn, H. Rischel (eds.), Lecture Notes in Computer Science **736** (1993) 209-229.

[CA1] Chadha, R., Alt, D.: Analysis of ripple effect in fiber optics rings: a symbolic simulation study. Submitted for publication (1993)

[Eva] Evans, J.B.: Structures of discrete-event simulation. An introduction to the engagement strategy. Ellis Horwood, New York (1988)

[Hal] Halbwachs, N.: Delay analysis in synchronous programs. Proceedings of Fifth Conference on Computer-Aided Verification, Lecture Notes in Computer Science **697**, Springer Verlag (1993)

[JL1] Jaffar, J., Lassez, J.-L.: Constraint logic programming. Proceedings of ACM Symposium on Principles of Programming Languages, Munich, Germany (1987).

[Kow] Kowalski, R.: Logic for problem solving. Elsevier North Holland, New York (1979).

[LW1] Lin, Y.-J., Wuu, G.: A constrained approach for temporal intervals in the analysis of timed transitions. Proceedings of Protocol Specification, Testing and Verification Conference, Stockholm, Sweden (1991)

[NCC1] Narain, S., Cockings, O., Chadha, R.: A formal model of SONET's alarm surveillance procedures and their simulation. Proceedings of FORTE: Formal Description Techniques, Boston, Massachusetts (1993).

[Nar] Narain, S.: Symbolic Discrete-Event Simulation. Proceedings of Discrete-Event Systems Workshop, Institute of Mathematics and its Applications, University of Minnesota, Minneapolis, Minnesota (1993).

[Ost] Ostroff, J.: Constraint logic programming for reasoning about discrete-event systems. Journal of Logic Programming 11 (1991)

[Seb] Sebuktekin, I.: A protocol modeling and validation exercise using DMOD and symbolic simulation. Submitted for publication (1993)

[SLN] Sekar, R., Lin, Y.-J., Narain, S.: On modeling and reasoning about hybrid systems. Proceedings of Protocol Specification, Testing and Verification Conference (1992).

[Sur] Suri, R.: Infinitesimal perturbation analysis for general discrete-event dynamic systems. Journal of the ACM, July, (1987).

[Var] Varaiya, P: Smart cars on smart roads: Problems of control. IEEE Transactions on Automatic Control, **38(2)** (1993).

[Zei] Zeigler, B.: Multifacetted modeling and discrete-event simulation. Academic Press, New York (1984).

Simple Hybrid Control Systems – Continuous FDLTI Plants with Quantized Control Inputs and Symbolic Measurements

Jörg Raisch

ISR, Universität Stuttgart, Pfaffenwaldring 9, D-70550 Stuttgart, FRG *and*
Systems Control Group, Dep. of Electrical and Computer Engineering, University of Toronto, Toronto, Ont., Canada M5S 1A4
email: raisch@isr.uni-stuttgart.d400.de *or* raisch@control.toronto.edu

1 Introduction

Hybrid control systems involve both continuous-valued and discrete-valued, or quantized, signals. The dynamic behaviour of a hybrid system can therefore be interpreted as the interaction of a continuous and a discrete-event component. In many areas of application, the number of hybrid problems is abundant. Here, it suffices to mention process control systems where actuators can often take only a finite number of positions (such as in on/off heating and cooling systems) and sensors often give only crudely quantized symbolic measurement information (such as "liquid level in tank A is above/below value B or C"). Many process control problems share the following features: 1. Specifications are "finer" than the measurement quantization levels (such as "on average, the liquid level in tank A should be as close as possible to some value D between B and C") and therefore have to be formulated in terms of continuous plant variables. 2. Plant time constants are extremely large compared to the sampling intervals of the actual control system, and time can therefore be considered to be continuous in both plant and controller.

We will restrict ourselves to the simplest control set-up which reflects these features. It is of some practical importance and at the same time allows for a number of analysis and design problems to be solved by applying well-known control-theoretic tools. We will deal with a finite-dimensional linear time-invariant (FDLTI) system with input variable $u(t) \in \mathbb{R}^q$ and output variable $y(t) \in \mathbb{R}^p$ evolving in continuous time (Fig. 1). The system has minimal realization

$$\dot{x}(t) = Ax(t) + Bu(t) \tag{1}$$

$$y(t) = Cx(t) \tag{2}$$

with state variable $x(t) \in \mathbb{R}^n$. Unlike in a conventional scenario, the controller has access only to a quantized, or symbolic, version $y_d(t)$ of the continuous-valued signal $y(t)$. It generates a piecewise constant signal $u_d(t)$ with finite range, i.e. a discrete-valued, or quantized, signal, which is then inserted into $u(t)$. In Fig. 1, discrete-valued signals are represented by dashed lines whereas continuous-valued signals are shown as solid lines. The quantization level in both the controller input and output can be arbitrarily coarse; our set-up therefore covers extreme cases involving binary signals.

We will first give details on the interface between the continuous-valued world of the plant and the discrete-valued world of the controller. We will then point out that both controller input and output signals can be interpreted as sequences of timed (measurement and control) events. The sequence of measurement events contains *precise* (albeit incomplete) information on y and therefore the state x. We will discuss which part of the state space is controllable to, and reachable from, the origin by an admissible sequence of control events. With these results, we can easily give conditions for observability of the state variable from measurement events.

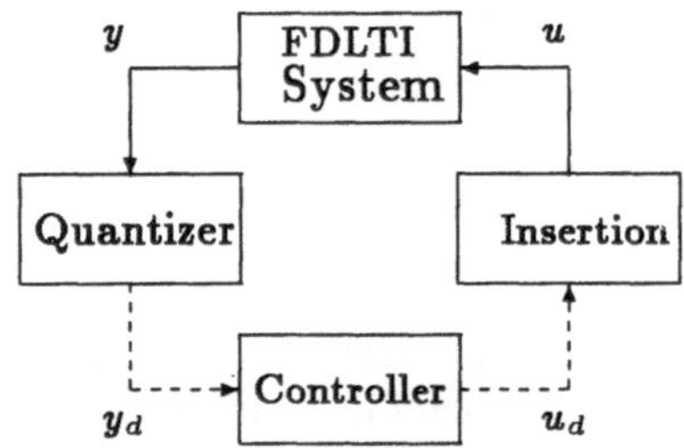

Fig. 1. System under consideration

2 Interface Between Continuous- and Discrete-Valued Signals

Only a quantized version $y_d(t)$ of the output signal $y(t)$ is available to the controller: Its measurement information is limited to

$$y_{d_i}(t) = Q_i(y_i(t)), \quad i = 1, \dots, p, \tag{3}$$

where the quantizer for the i-th component of the measurement signal, $Q_i : \mathbb{R} \to \{\alpha_i^1, \dots, \alpha_i^{M_i+1}\}$, assigns the same symbol α_i^r to every y_i within a given interval:

$$Q_i(y_i) = \begin{cases} \alpha_i^1 & \text{for } -\infty < y_i \leq y_i^1 \\ \alpha_i^r & \text{for } y_i^{r-1} < y_i \leq y_i^r,\ r = 2, \dots, M_i \\ \alpha_i^{M_i+1} & \text{for } y_i^{M_i} < y_i < \infty\ . \end{cases} \tag{4}$$

This quantization procedure partitions $\mathbb{R}^p$ into $\prod_{i=1}^{p}(M_i + 1)$ rectilinear "quantization boxes" with edges parallel to the coordinate axes. It is illustrated in Fig. 2 for $M_i = 4$. As time is continuous, knowledge of the symbolic measurement signal $\{y_d(t) \mid t \geq 0\}$ means that we know exactly when any component of $y_d(t)$ changes its value. This, in turn, implies *precise* (albeit incomplete) information on the underlying continuous variable y. Suppose, for example, that y_{d_i} changes from α_i^r to α_i^{r+1} (or from α_i^{r+1} to α_i^r) at time t_k^y. Then, at time t_k^y, y_i takes its threshold value y_i^r.

$$y_{d_i}(t_k^y) = \alpha_i^r \quad \& \quad \lim_{(\epsilon>0)\to 0} y_{d_i}(t_k^y \pm \epsilon) = \alpha_i^{r+1} \quad \Longrightarrow \quad y_i(t_k^y) = y_i^r\ .$$

Each such change, i.e. each new piece of precise information, can be interpreted as a *timed measurement event*:

$$\langle t = t_k^y,\ y_i(t_k^y) = y_i^r \rangle, \quad k = 1, 2, \dots \tag{5}$$

(or $\langle t_k^y,\ y_i^r \rangle$, for short). Conversely, from the initial information $y_d(0)$ and the sequence of measurement events (5), one can immediately tell the quantization box the signal $y(t)$ is currently in and therefore reconstruct its symbolic version $y_d(t)$. Hence, given $y_d(0)$, all incoming measurement information is captured in the sequence of events (5). Recently, it has come to the author's attention that Stiver and Antsaklis (1993) suggest a similar characterization of measurement events.

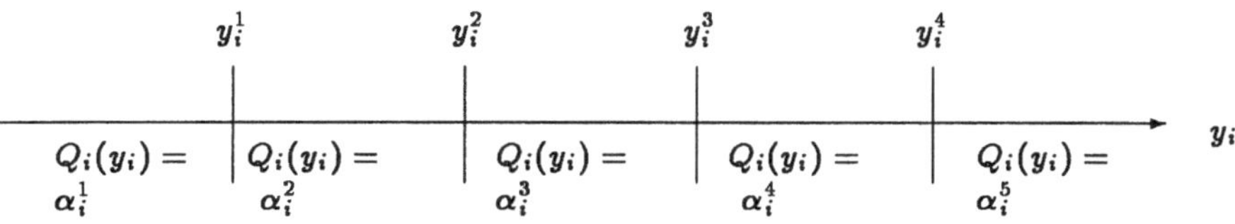

Fig. 2. Quantization of i-th component of measurement signal

The controller output $u_d(t)$ is a piecewise constant signal with finite range, which is simply inserted[1] into the continuous-valued variable u:

$$\begin{aligned} \text{Ins} : \{u_1^1, \dots, u_1^{N_1}\} \times \dots \times \{u_q^1, \dots, u_q^{N_q}\} \to \mathbb{R}^q \\ u(t) = \text{Ins}(u_d(t)) = u_d(t) \ . \end{aligned} \tag{6}$$

The controller output $u_d(t)$ (and hence the plant input $u(t)$) can also be interpreted as a sequence of events: Denote time instants where any (say, the j-th,) component of the signal changes by $t_k^u, k = 1, 2, \dots$ and let the "new" signal value be u_j^s, i.e. $u_{d_j}(t_k^u) = u_j^s$ and $\lim_{(\epsilon>0)\to 0} u_{d_j}(t_k^u - \epsilon) \neq u_j^s$. Each control strategy is then simply a sequence of timed control events

$$\langle t = t_k^u,\ u_j(t_k^u) = u_{d_j}(t_k^u) = u_j^s \rangle, \quad k = 1, 2, \dots \tag{7}$$

(or $\langle t_k^u,\ u_j^s \rangle$, for short), which is generated by the controller in response to the sequence of measurement events.

3 Controllability/reachability with discrete-valued inputs

In this section, we will investigate, which part of the state space $\mathbb{R}^n$ is controllable to (reachable from) the origin by a piecewise constant quantized control signal $u_d(t)$, i.e. by a sequence of control events (7). More precisely, we will show that under intuitively appealing assumptions, an element in $\mathbb{R}^n$ is controllable to (reachable from) the origin by discrete-valued inputs if and only if it is controllable (reachable) by a certain class of constrained continuous-valued inputs.

[1] if the "real" controller output is a non-numeric (symbolic) signal $\tilde{u}_d$, we add a one-to-one map $\tilde{u}_d \to u_d$, which is subsequently thought of as part of the controller.

Definition 1 (C-controllability and C-reachability). A control input $u(t)$ for system (1) is *C-admissible* on the interval $[0, T]$ if

1. $u_i(t)$ is piecewise continuous for $t \in [0, T]$, $i = 1, \dots, q$;
2. $\underline{u}_i \le u_i(t) \le \overline{u}_i$ for $t \in [0, T]$, $i = 1, \dots, q$ (i.e. the control variable "lives" inside a q-dimensional parallelepiped U).

Denote the set of all inputs that are C-admissible on $[0, T]$ by $\mathcal{U}_c^T$. The set of all states that can be controlled to (reached from) the origin by C-admissible inputs in given time T is denoted by $\mathcal{C}_c^T$ and $\mathcal{R}_c^T$, respectively.

Definition 2 (D-controllability and D-reachability). A control input $u(t)$ for system (1) is *D-admissible* on the interval $[0, T]$ if

1. $u_i(t)$ is piecewise constant for $t \in [0, T]$, $i = 1, \dots, q$;
2. $u_i(t) \in \{u_i^1, \dots, u_i^{N_i}\}$ for $t \in [0, T]$, $i = 1, \dots, q$,

(i.e. if for every $t \in [0, T]$, u(t) is the result of the insertion (6)). For convenience, the discrete values of u_i are assumed to be ordered ($u_i^s < u_i^{s+1}$, $s = 1, \dots, N_i - 1$). Denote the set of all control inputs that are D-admissible on $[0, T]$ by $\mathcal{U}_d^T$. The set of all states that can be controlled to (reached from) the origin by D-admissible inputs in given time T is denoted by $\mathcal{C}_d^T$ and $\mathcal{R}_d^T$, respectively. Note that the requirement of $u_i(t)$ being piecewise constant implies that for any finite T, every $x \in \mathcal{R}_d^T$ can be reached by applying a *finite* sequence of control events (7).

Theorem 3 (**Equivalence of C-controllability/C-reachability and D-controllability/D-reachability**). *Assume that*

1. *rank* $[b_i, Ab_i, \dots, A^{n-1}b_i] = n$ *for* $i = 1, \dots, q$, *where* b_i *is the* i*-th column of* B *(the system (1) is controllable in the "usual sense" from each component of the input vector* $u(t)$*),*
2. *for all finite* T, $u(t) \equiv 0 \in \mathcal{U}_c^T$ *and* $u(t) \equiv 0 \in \mathcal{U}_d^T$ *(zero is both a C-admissible and a D-admissible control input), and*
3. $\underline{u}_i = u_i^1$ *and* $\overline{u}_i = u_i^{N_i}$, $i = 1, \dots, q$ *(maximum and minimum values of C-admissible and D-admissible control inputs coincide);* $\underline{u}_i < 0 < \overline{u}_i$ *for at least one* $i \in \{1, \dots, q\}$ *("doing nothing" is not an extreme control policy for all components of* u*).*

Then,

$$x \in \mathcal{C}_c^T \text{ iff } x \in \mathcal{C}_d^T \tag{8}$$

$$x \in \mathcal{R}_c^T \text{ iff } x \in \mathcal{R}_d^T, \tag{9}$$

i.e. C-controllability/C-reachability in time T *implies and is implied by D-controllability/D-reachability in time* T.

Proof. The "if" part in (8) is trivial, as $\mathcal{U}_d^T \subset \mathcal{U}_c^T$. The "only if" part follows from an application of Pontryagin's maximum principle (e.g. Boltyanski (1971)). Details of the proof can be found in Raisch (1993).

The "if" part in (9) is also trivial. To prove the "only if" part of (9), define "controllability operators" $\mathbf{C}_c^T : \mathcal{U}_c^T \to \mathbb{R}^n$ and $\mathbf{C}_d^T : \mathcal{U}_d^T \to \mathbb{R}^n$ with action

$$\mathbf{C}_c^T(u(t)) := -\int_0^T e^{-At} Bu(t)dt \qquad \text{and} \qquad \mathbf{C}_d^T(u(t)) := -\int_0^T e^{-At} Bu(t)dt \ .$$

The ranges of these operators are just the sets of states that can be controlled to the origin in time T by C-admissible and D-admissible inputs, respectively. Because of (8), it is obvious that

$$\text{Range}\mathbf{C}_c^T = \mathcal{C}_c^T = \mathcal{C}_d^T = \text{Range}\mathbf{C}_d^T \ . \tag{10}$$

Analogously, we define "reachability operators" $\mathbf{R}_c^T : \mathcal{U}_c^T \to \mathbb{R}^n$ and $\mathbf{R}_d^T : \mathcal{U}_d^T \to \mathbb{R}^n$ with action

$$\mathbf{R}_c^T(u(t)) := \int_0^T e^{A(T-t)} Bu(t)dt = -e^{AT}\mathbf{C}_c^T(u(t)) \tag{11}$$

$$\mathbf{R}_d^T(u(t)) := \int_0^T e^{A(T-t)} Bu(t)dt = -e^{AT}\mathbf{C}_d^T(u(t)) \ . \tag{12}$$

Clearly, the range of $\mathbf{R}_c^T$ ($\mathbf{R}_d^T$) determines the set of states that can be reached from the origin in time T by applying C-admissible (D-admissible) control inputs. From (10)-(12), it follows immediately that the range of $\mathbf{R}_c^T$ and the range of $\mathbf{R}_d^T$ coincide, and therefore $\mathcal{R}_c^T = \text{Range}\mathbf{R}_c^T = \text{Range}\mathbf{R}_d^T = \mathcal{R}_d^T$. □

How to determine the discrete-time version of $\mathcal{R}_c^T$ is discussed, for example, in Lasserre (1993) and Lin (1970). Their results can be used to approximate $\mathcal{R}_c^T$ and, because of (9) and (12), $\mathcal{R}_d^T$ and $\mathcal{C}_d^T$.

4 Observability with discrete-valued measurements

Adapting the terminology in Sontag (1990), we say that a given D-admissible control input $u(t)$ (i.e. a given sequence of control events $\langle t_k^u, u_j^s \rangle$, $k = 1, 2, \dots$) *distinguishes* between two initial states $x(0)$ and $\tilde{x}(0)$, if it produces different sequences of measurement events for both initial conditions. $x(0)$ and $\tilde{x}(0)$ are called *distinguishable*, if there exists at least one sequence $\langle t_k^u, u_j^s \rangle$, $k = 1, 2, \dots$ that distinguishes between them. Finally, the hybrid system consisting of plant (1)-(2), quantizers (3)-(4), and insertion (6) is said to be *observable* on a set $\mathcal{O}_d$, if every pair of initial states in $\mathcal{O}_d$ is distinguishable.

Theorem 4 (Observability of hybrid systems). *Denote the i-th row of C by c_i. For any $T \in \mathbb{R}^+$, the hybrid system (1)-(2), (3)-(4), (6) is observable on $\mathcal{C}_d^T$ if for at least one $i \in \{1, \dots, p\}$*

1. *there exists an $r \in \{1, \dots, M_i\}$ such that $c_i x = y_i^r$ intersects $\mathcal{C}_d^T \cap \mathcal{R}_d^T$ and*
2. *$\text{rank}[c_i', A'c_i', \dots, (A')^{n-1}c_i'] = n$.*

Proof. Suppose first that we have been able to find a sequence of control events that generates n measurement events $\langle t_k^y, y_i^r \rangle$, $k = 1, \ldots, n$. Then

$$y_i^r = c_i e^{At_k^y} x(0) + h_k(u(t)), \quad k = 1, \ldots, n,$$

where $h_k(u(t))$ is known. $x(0)$ is the *unique* initial condition that results in this particular sequence of measurement events if and only if the row vectors $c_i e^{At_k^y}$, $k = 1, \ldots, n$, span $\mathbb{R}^n$. For the contrary to be true (i.e. for these rows to be linearly dependent), it is necessary (and sufficient) that the homogeneous equations $0 = c_i e^{At_k^y} x(0)$, $k = 1, \ldots, n$, have a common non-trivial solution (i.e. we can start the continuous system (1),(2) with zero-input in $x(0) \neq 0$ and get $y_i(t) = c_i x(t) = 0$ at n different instants of time $t_1^y, \ldots, t_n^y$). Because of the rank condition in the theorem, $y_i(t) \equiv 0 \quad \forall t \geq 0$ is not possible. Therefore, the only way how linear dependency between the $c_i e^{At_k^y}$ can occur is that the event times t_k^y are "pathological", i.e. they coincide with the points where the i-th component of the autonomous solution of (1),(2) intersects the t-axis. Clearly, these points form a subset of measure zero of $\mathbb{R}^+$. Now, it remains to show that there exists a control signal that generates measurement events with "nonpathological" timing. This is indeed the case for all initial states in $\mathcal{C}_d^T$: As the "event-generating" hyperplane $c_i x = y_i^r$ intersects $\mathcal{C}_d^T \cap \mathcal{R}_d^T$, there certainly exists a sequence of control events that drives $x(0)$ onto the hyperplane in any desired time $t_a \geq 2T$ (for example by forcing x_0 into the origin first, see Fig. 3). This can be repeated to generate other measurement events at any time t_b with $t_b - t_a \geq 2T$, t_c with $t_c - t_b \geq 2T$ and so on. In fact, any randomly chosen control signal that makes the state-trajectory cross the "event-generating" hyperplane often enough, will do this in a "nonpathological way", enabling us to distinguish between initial conditions and thus to reconstruct $x(0)$, with probability 1. □

Remark. In general, the "pathological" distances between measurement event times depend on *all* the eigenvalues of the matrix A. For asymptotically stable A and time going to infinity, they approach the inverse of the well known pathological sampling frequencies $(\lambda - \mu)/(2m\pi\sqrt{-1})$, $\quad m = \pm 1, \pm 2, \ldots$, where λ and μ are any two eigenvalues of A.

Remark. If hyperplanes $c_i x = y_i^r$ intersect $\mathcal{C}_d^T \cap \mathcal{R}_d^T$ for more than one i, the rank condition in the theorem can obviously be relaxed. If this is true for all $i \in \{1, \ldots, p\}$, then the rank condition reduces to $\operatorname{rank}[C', A'C', \ldots, (A')^{n-1}C'] = n$.

Remark. Note that the above notion of observability only guarantees the *existence* of control sequences that distinguish between any two initial states. Whether we will be able to find such a sequence depends on the system under consideration.

5 Example

The following simple example serves to illustrate the concepts described in the previous sections (actually, it is so simple that everything becomes almost trivial): Assume that a cart of mass $M = 1$ moves without friction on a plane. The control input is the horizontal force u applied to the cart, the states are defined as position (x_1) and velocity (x_2). An idealized system model is given by

$$\dot{x} = \begin{bmatrix} 0 & 1 \\ 0 & 0 \end{bmatrix} x + \begin{bmatrix} 0 \\ u \end{bmatrix}.$$

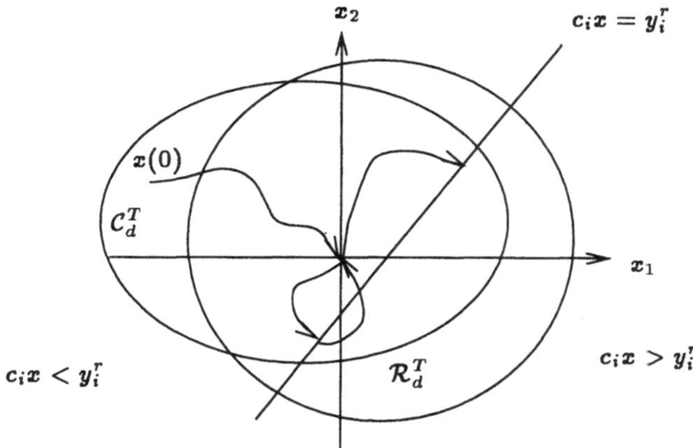

Fig. 3. Example for generating "nonpathologically" spaced measurement events

Assume that u can only take three distinct values: $u(t) = u_d(t) \in \{-1, 0, 1\}$. Suppose furthermore that velocity cannot be measured and position only in a highly quantized manner:

$$y_d(t) = Q(\underbrace{[\,1\;0\,]\,x(t)}_{y(t)}) = \begin{cases} \alpha^1 \text{ if } y \leq y^1 \\ \alpha^2 \text{ if } y^1 < y \leq y^2 \\ \alpha^3 \text{ if } y > y^2 \end{cases},$$

where $y^1 < 0$ and $y^2 > 0$ are given threshold values. Thus, we have a continuous system ($x \in \mathbb{R}^2$) and discrete-valued input and measurement sets. The state trajectories for the three possible input values are shown on the left of Fig. 4: For $u_d = 1$, the state trajectories are depicted as solid lines, for $u_d = 0$ as dotted and for $u_d = -1$ as dashed lines. For $u_d = 0$, all points on the x_1 axis (x-marked) are stationary. Obviously, the whole state space is D-controllable and D-reachable.

If we want to drive the system state into the origin, we can use the strategy shown in the right part of Fig. 4 (there, $y^1 = -2$, $y^2 = 1$, an "x" marks the occurence of a measurement event and an "o" the occurrence of a control event): Use the initial information $y_d(0)$ (i.e. the knowledge that at time $t = 0$, the system is in the left/middle/right "quantization box") to generate a sequence of control events (a control signal) that creates two measurement events: For $y_d(0) = \alpha^1$, we select $\langle t_1^u = 0,\ 1\rangle$, i.e. $u_d(t) = 1; t \geq 0$. This will eventually generate two measurement events $\langle t_1^y, y(t_1^y) = y^1\rangle$ – the state crosses into the middle quantization box – and $\langle t_2^y, y(t_2^y) = y^2\rangle$ – the state leaves the middle quantization box. Reconstruct the current state and drive it into the origin – e.g. in the time-optimal way shown in Fig. 4 by applying the sequence of control events $\langle t_2^u = t_2^y,\ -1\rangle$ and $\langle t_3^u,\ 1\rangle$; then switch the control to zero: $\langle t_4^u,\ 0\rangle$. t_3^u and t_4^u can be computed from $x(t_2^y)$.

In theory (no modelling errors, no disturbances), we can do this exactly. In practice, we will not be *exactly* in the origin (nor in any other point on the x_1-axis), when we switch the control to zero. This will cause the state to drift slowly away from its desired position (the origin). As this does not show in the quantized measurement until $y(t) = y^1$ or $y(t) = y^2$, it will be hidden from the "outside world" and there is nothing we can do about it. Once a new measurement event occurs, we can update our state estimate and apply the same policy again. This heuristic approach is reminiscent of *Receding horizon control*. It seems likely that aspects of receding horizon control theory (as discussed, e.g., in Michalska and Mayne (1993)) will prove helpful in extending the approach to deal with performance and robustness issues.

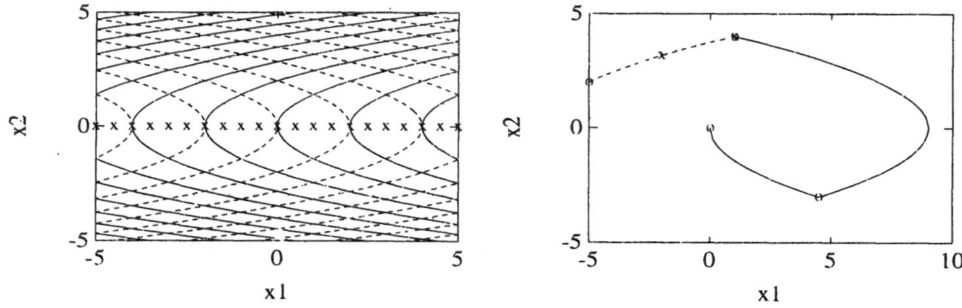

Fig. 4. All possible state trajectories (left), state trajectory for a candidate control strategy (right)

References

J. A. Stiver and P. J. Antsaklis. On the controllability of hybrid control systems. In *32th IEEE Conference on Decision and Control*, 1993.

V.G. Boltyanski. *Mathematical Methods of Optimal Control*. Holt, Rinehart and Winston, New York, 1971.

J. Raisch. Simple hybrid control systems – continuous FDLTI plants with quantized control inputs and measurements. Systems Control Group Report 9305, Department of Electrical and Computer Engineering , University of Toronto, May 1993.

J. B. Lasserre. Reachable, controllable sets and stabilizing control of constrained linear systems. *Automatica*, 29:531–536, 1993.

J. N. Lin. Determination of reachable set for a linear discrete system. *IEEE Transactions on Automatic Control*, 15:339–342, 1970.

E. D. Sontag. *Mathematical Control Theory*. Springer-Verlag, Berlin, 1990.

H. Michalska and D. Q. Mayne. Robust receding horizon control of constrained nonlinear systems. *IEEE Transactions on Automatic Control*, 38:1623-1633, 1993.

Controllability and Control-Law Synthesis of Linear Hybrid Systems

Michael Tittus and Bo Egardt

Control Engineering Lab, Chalmers University of Technology, S-412 91 Gothenburg, Sweden
e-mail: mt@contrl.chalmers.se, be@contrl.chalmers.se

1 Introduction

Systems that are described by continuously changing variables alongside with discrete variables are called hybrid systems. A number of different philosophies have been pursued in order to establish a mathematical framework that is able to handle hybrid systems. These range from the construction of hybrid automata [1] and structures [2], to the definition of interfaces between the continuous and the discrete part [3]. Other approaches use Petri Nets [4], bond-graphs [5] or a synchronuous approach [6], just to name a few.

The present work is based on the hybrid automata model introduced by [1]. The system's dynamics are described by sequences of phases in which the system's variables change continuously. During the transitions between phases—which happen in zero-time—discrete state changes can take place. This framework has been developed with the purpose to describe linear hybrid systems and to verify that control-laws and specifications are in harmony with each other, i.e. the system satisfies all invariants when running.

In this paper we have reformulated the problem from the verification problem 'Does the system have a run with the given specifications and control-laws?' discussed in [1], to a control problem 'With the given specifications, does there exist a control-law that ensures a run of the hybrid system?'. In other words, we are asking if the specifications and restrictions are reasonable or not, and if they are, a possible path (i.e. sequence of phases) is produced that leads the system from a specified initial state to a final state and a set of corresponding control-laws is generated.

A geometrical approach is used to calculate an acceptable run. Starting with the final state, the coaccessible state-space (i.e. state-space from which the final state can be reached) along each path is calculated. If the initial state is included in the coaccessible state-space of a path, an acceptable run can be generated. Based on this calculation a set of control-laws can be synthesized.

In the following sections we shall briefly summarize and extend the hybrid automata introduced in [1] to suit our purposes, discuss the notion of controllability for hybrid systems and describe the generation of correct control-laws. We shall also briefly describe the algorithm used.

2 The Hybrid Framework

The framework used here is a derivation of the one introduced in [1] and largely follows the same nomenclature.

2.1 Some Definitions

In accordance with the two types of variables (discrete and continuous) we encounter two kinds of state changes—continuous and discrete ones. Since discrete state changes happen at discrete points in time we divide the time axis into *phases*, with each change in a discrete variable separating two adjacent phases. The transition between two phases we shall refer to as a *phase-transition*. Within each phase the system's state transformation can be described by a set of differential equations.

In our model, phases correspond to different settings of discrete control-variables and denote the discrete state of the system. *Control location* is thus used as a synonym for phase. A system state, therefore, is defined as a pair consisting of the *data-state* (i.e. an interpretation of all continuous or data variables) and the *control location* (i.e. an interpretation of the discrete control-variables).

2.2 Hybrid Automata

We define a hybrid automata as a seven-tuple $\mathcal{A} = (V_D, Q, s_1, s_2, \mu_1, \mu_2, \mu_3)$.

- A finite set V_D of real-valued data variables. Σ_D stands for the set of all data states.
- A finite set Q of control locations. Each location $l \in Q$ is encoded as an integer value of an index variable i, which therefore ranges over the set Q. We write $Q = \{l_1, l_2, \ldots, l_n\}$. Thus, the system state is a pair $\Sigma = Q \times \Sigma_D$, as defined above.
- An initial state s_1 and a final state s_2 with $s_1, s_2 \in Q$. Each can either be a system state, i.e. a singular point in the state-space, or a limited sub-space in the system's data state-space.
- A labeling function μ_1 that is used to assign to each phase l a function describing its data-state transformation $\mu_1(l)$ in the form of first-order differential equations. $\mu_1(l)$ is also called the activity in l.
- A labeling function μ_2 that assigns to each location $l \in Q$ a phase-invariant $\mu_2(l) \subseteq \Sigma_D$. As long as the system resides in a phase l it has to satisfy the invariant $\mu_2(l)$. Otherwise an exception is triggered. $\mu_2(l)$ is also used to denote the sub-space in the system's state-space that is defined by the invariant $\mu_2(l)$. The correct interpretation is usually clear from the context.
- A labeling function μ_3 that assigns to each pair $e \in Q^2$ of locations a phase-transition relation $\mu_3(e) \subseteq \Sigma_D^2$. A state (σ', l') is called a successor of the state (σ, l) iff $(\sigma, \sigma') \in \mu_3(l, l')$. In general, μ_3 is defined as

$$\mu_3(l_i, l_j) = \begin{cases} \mathtt{x := } f(x) & \text{transition} \\ \mathtt{undefined} & \text{no transition} \end{cases}$$

so that it also determines the connectivity between phases.

Restrictions Made in this Paper. In this paper we will restrict ourselves to the following cases:

- The activities $\mu_1(l)$ within a phase l are restricted to linear functions defined by a set of first-order differential equations of the form $\dot{\boldsymbol{x}} = \boldsymbol{k}_l$ with $\boldsymbol{k}_l$ a constant vector.
- The phase invariant $\mu_2(l)$ is defined by means of a set of linear inequalities over Σ_D and has the form $\boldsymbol{A} \cdot \boldsymbol{x} + \boldsymbol{b} \leq 0$ with $\boldsymbol{A}$ and $\boldsymbol{b}$ being constant matrices and $\boldsymbol{x} \in \Sigma_D$.
- The phase-transition relation $\mu_3(e)$ is defined by an assignment $\boldsymbol{x} := \alpha_x$ for $\boldsymbol{x} \in \Sigma_D$, where α_x is a linear term over Σ_D of the form $\boldsymbol{R} \cdot \boldsymbol{x} + \boldsymbol{c}$, with $\boldsymbol{R}$ and $\boldsymbol{c}$ being again constant matrices.

Formulation of Specifications and Other Restrictions. The guiding factor in the design of control laws are specifications and restrictions imposed on the system by the laws of physics, and the intentions of system engineers. Both are formulated as invariants. Only the "acceptable" is defined and thus everything that is not explicitly denoted as acceptable is by definition an exception.

Invariants can be described on two different levels:

- *System-invariants*, which are required to hold independent of the location or state the system is in. They apply automatically if not overwritten by stricter local specifications. Examples for this kind of invariants might be physical restrictions as for example maximum levels or pressures.
- On the lower level the more general system-invariants can be strengthened for certain phases and certain data variables by means of *phase-invariants*.

3 Controllability of Hybrid Systems

Hybrid systems are assumed to be controlled by control-signals that make them change their phase or control location. All control-variables can be manipulated. Hence, the system can be controlled to run through any possible sequence of phases, the only restriction being the invariants of the different phases. We can thus define controllability of a hybrid system as follows:

Definition 1. A hybrid system $\mathcal{A}$ is said to be controllable—with respect to its state-transition $\Phi(s_1, s_2)$—if there exists at least one sequence of discrete control-inputs that give $\mathcal{A}$ an acceptable run $\pi(s_1, \ldots, s_2)$, i.e. they *(1)* take $\mathcal{A}$ from s_1 to s_2, and *(2)* all invariants applying to $\Phi(s_1, s_2)$ are continuously satisfied.

There are no requirements that all available phases have to be visited during a run and no restrictions are made that one phase or even data-state cannot be revisited.

The main focus in this section is on the problem of deciding whether a hybrid system is controllable or not. This question is also of importance to the system designer since an uncontrollable system means that the system's specifications are not compatible with the control objective to take the system from s_1 to s_2. Once an acceptable run is found, corresponding control laws can be generated.

3.1 Some More Definitions

An important requirement when it comes to a phase transition between two phases, l_1 and l_2, is that such a discrete state change or "jump" will not lead the system into forbidden areas in location l_2, i.e. trigger an exception. Hence, we define for each possible transition from a location l_1 to a location l_2 a so-called *jump-set* from l_1 to l_2.

Definition 2. The *jump-set* $\lambda(l_1 \rightarrow l_2)$ is the set of all points of the system's data-state space Σ_D that satisfy $\mu_2(l_1)$ and from which $\mu_2(l_2)$ is reachable by means of the phase-transition relation $\mu_3(l_1, l_2)$.

$\lambda(l_1 \rightarrow l_2)$ will even denote the set of linear inequalities used as a mathematical description of this sub-set—the so-called jump-conditions.

The jump-conditions therefore make sure that the system will always "jump" into the invariant space of the destination phase and can thus give guidelines for the design of correct control-laws, as will be seen later. Since both $\mu_2(l)$ and $\mu_3(l_1, l_2)$ are given by linear expressions, the jump-conditions will of necessity be of the same form.

Definition 3. The *extended jump-set* $\lambda_{\text{ext}}(l_1 \rightarrow l_2)$ is the union of $\lambda(l_1 \rightarrow l_2)$ and all states $x \in \mu_2(l_1)$ such that $\lambda(l_1 \rightarrow l_2)$ is reachable from x via the data-state transformation $\mu_1(l_1)$.

Thus, the extended jump-set $\lambda_{\text{ext}}(l_1 \rightarrow l_2)$ guarantees, when satisfied, that a correct transition $e = (l_1, l_2)$ sooner or later will be possible. That is, once the system has entered phase l_1 and is within $\lambda_{\text{ext}}(l_1 \rightarrow l_2)$ the system will "drift" into the jump-set $\lambda(l_1 \rightarrow l_2)$ because of the phase activity $\mu_1(l_1)$.

3.2 Deciding Controllability

The basic idea in determining whether or not a hybrid system $\mathcal{A}$ is controllable is fairly simple. As we mentioned above, a transition from a location l_i to another location l_{i+1} is or becomes possible only if the system constantly satisfies the conditions describing the extended jump-set $\lambda_{\text{ext}}(l_i \rightarrow l_{i+1})$ while in phase l_i. If we continue this line of thought, then, in order to have a controllable path $\pi(l_{i-1}, l_i, l_{i+1})$ we have to make sure that the transition $e = (l_{i-1}, l_i)$ transfers the system into the extended jump-set $\lambda_{\text{ext}}(l_i \rightarrow l_{i+1})$, so as to guarantee a continuation of the path. Hence, in order to check the controllability of path π the allowed space for phase l_i is $\lambda_{\text{ext}}(l_i \rightarrow l_{i+1})$. We say that the invariant state-space of location l_i is reduced to $\lambda_{\text{ext}}(l_i \rightarrow l_{i+1})$ for this specific path.

In Fig. 1(a) this principle is illustrated. Interesting sub-spaces of the state-space are graphically represented as two-dimensional areas within the nodes of the respective location. In order to test if a certain path is controllable it is tested if the final state can be reached from the initial state. This is achieved by generating the extended jump-spaces for each step in the path starting with the final state. The invariant $\mu_2(l)$ of each phase in the path is reduced to the extended jump-set before considering the next transition along the path. If $\lambda(s_1 \rightarrow l_1)$ turns out not to be void, then we can say that the system is controllable along this specific path. Otherwise, a different path has to be tried.

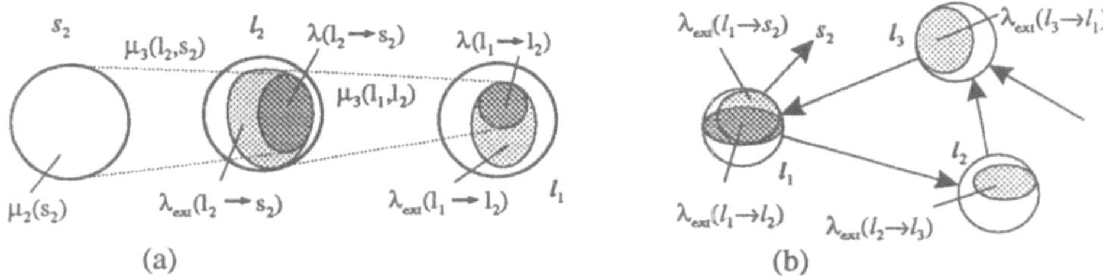

Fig. 1. (a) Deciding controllability along a path (b) Loop reduction

In the case of s_1 and s_2 consisting not only of a single point in the state-space, but a sub-set thereof we note that an acceptable run $\pi(s_1, l_1, \ldots, l_n, s_2)$ does not guarantee that each point in s_2 can be reached from every point $x \in s_1$. For a path π to qualify, we merely demand that there exists at least one state $x \in s_1$ from which some state $y \in s_2$ can be reached along π.

A path is said to contain a possible loop if one and the same location can appear more than once in a path. Sometimes it is necessary to include a loop in a path in order to find an acceptable run of the system since a loop may increase the extended jump-set of a phase. As can be seen in Fig. 1(b), a new extended jump-set, $\lambda_{ext,new}(l_1 \rightarrow s_2)$, can be formed as the union of the original $\lambda_{ext}(l_1 \rightarrow s_2)$ and $\lambda_{ext}(l_1 \rightarrow l_2)$, since from all states within that union the system can reach the jump-set $\lambda(l_1 \rightarrow s_2)$ and from there s_2. The jump-set is reached either via the data-state transformation $\mu_1(l_1)$ or via the loop and $\mu_1(l_1)$.

The number of loops to be included in the path depends on whether another loop increases the jump-set or not, that is, whether $\lambda_{ext,new}(l_1 \rightarrow s_2) \supset \lambda_{ext}(l_1 \rightarrow s_2)$ or not. Once a fixpoint $\lambda_{max}(l_1 \rightarrow s_2)$ in the computation of $\lambda_{ext,new}(l_1 \rightarrow s_2)$ is reached, the loop can be eliminated from the path and $\lambda_{max}(l_1 \rightarrow s_2)$ serves as phase invariant for the further iteration of the path.

Unfortunately, the algorithm in its present form can only describe convex sub-sets of the state-space. Since $\lambda_{max}(l_1 \rightarrow s_2)$ can become non-convex (see Fig. 1(b)) it is never explicitly computed in the algorithm.

3.3 Example and Algorithm

An algorithm [7] to decide the controllability of a hybrid system $\mathcal{A}$ has been implemented in MATLAB using a geometrical approach to compute the different sub-spaces of the state-space by means of hyperplanes. We shall use a simple example with two data-variables to illustrate the workings of the algorithm.

Example. The considered hybrid system is shown in Fig. 2(a). The nodes of the directed graph represent the different phases, and the vertices the possible phase transitions. Labels s_1 and s_2 denote the initial and final location, i.e. l_0 and l_4.

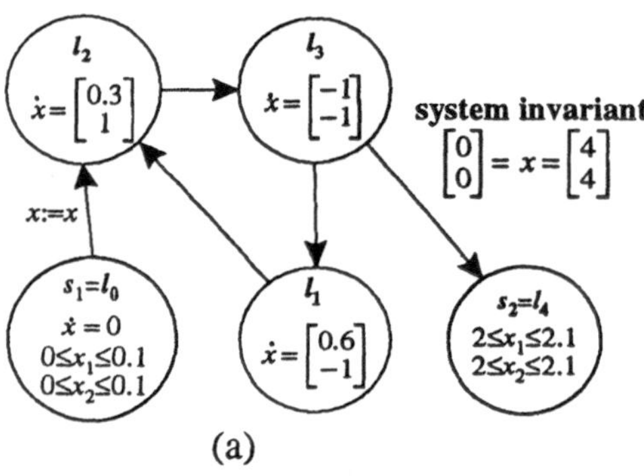

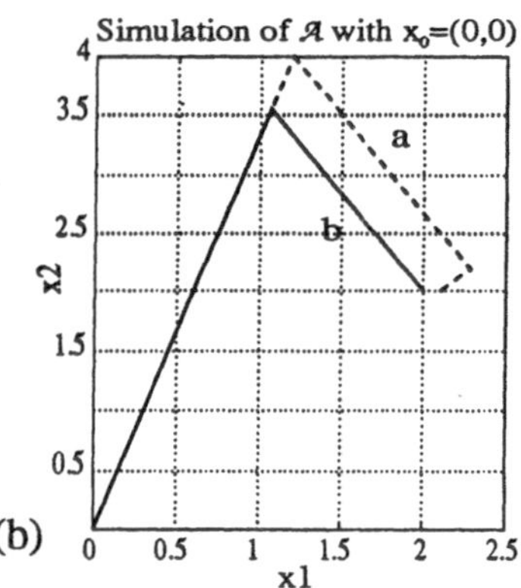

Fig. 2. (a) Hybrid system $\mathcal{A} = (\{x_1, x_2\}, \{l_0, l_1, l_2, l_3, l_4\}, l_0, l_4, \mu_1, \mu_2, \mu_3)$ (b) Simulation of run π of system $\mathcal{A}$

Each location l contains besides its label two expressions: its activity $\mu_1(l)$, and its invariant $\mu_2(l)$ expressed as a set of inequalities. Phase invariants that are valid for each phase are written seperately as system invariant. Furthermore, each phase transition from i to j is labeled with its phase-transition relation $\mu_3(l_i, l_j)$ in the form of a mapping $\boldsymbol{x} := \boldsymbol{A} \cdot \boldsymbol{x} + \boldsymbol{b}$. Phase-transition relations of the form $\boldsymbol{x} := \boldsymbol{x}$ and data-state transformations of the form $\dot{\boldsymbol{x}} = \boldsymbol{0}$ are omitted.

As a first step, all elementary paths, that is, paths without loops, leading from s_1 to s_2 are computed and tested for controllability. Next paths with loops are checked. As soon as an acceptable run has been found the algorithm terminates.

The system $\mathcal{A}$ is found to be controllable and a path $\pi(s_1, l_2, l_3, l_1, l_2, l_3, s_2)$ is found to satisfy all invariants. When reaching phase s_1 we find that $\mu_2(s_1) \supseteq \lambda_{\text{ext}}(s_1 \rightarrow l_2)$, that is, that the run is acceptable.

We also want to note that all sub-sets of the state-space need to be limited by hyperplanes in all directions in order for the algorithm to work.

4 Generation of a Control-Law

A control-law is a sequence of discrete inputs that cause the system to transfer from the initial state s_1 to the final state s_2 along a specific path π. An acceptable run can only be guaranteed iff the system transfers from a location l_i to the next location l_j in the path while in the jump-space $\lambda(l_i \rightarrow l_j)$. Hence, these jump-sets control as transition guards the execution of the input-sequence.

Each jump-set is implemented as a set of linear inequalities in the system's continuous variables. As soon as all inequalities belonging to a certain transition evaluate to TRUE the corresponding guard is changed from FALSE to TRUE. It is changed back to FALSE as soon as at least one of these inequalities evaluates to FALSE. The system *has* to change location while the guard is TRUE.

The two extreme cases of control-laws are to change phase *as soon as* the jump-conditions are satisfied, or *just before* the system leaves the jump-set. In order to accomplish the second control strategy, knowledge about the system's dynamical behavior has to be used to predict the data-state since the transition has to take place *before* the guard becomes FALSE.

Two simulations of $\mathcal{A}$ are shown in Fig. 2(b). Simulation a is done with a control-law that changes location as late as possible whereas in simulation b locations are changed as soon as possible. Both final points lie within the specification set up for s_2.

5 Conclusions

The concept of controllability for hybrid systems has been defined and an algorithm has been presented that is able to decide the controllability for linear hybrid systems. Besides the computation of an acceptable run, the control-law for executing this acceptable run is synthesized as a sequence of phase-transitions with the computed jump-spaces acting as control-law. This control-law guarantees a correct transition from the initial to the final state of the system following a certain path.

The main problem is that the complexity and, as a result, the execution time of the algorithm increases exponentially with the number of data-variables, that is, the dimension of the state-space.

References

1. R. Alur, C. Courcoubetis, T.A. Henzinger, and P.-H. Ho. Hybrid automata: An algorithmic approach to the specification and verification of hybrid systems. In *Workshop on Theory of Hybrid Systems*, Lecture Notes in Computer Science 736, pages 209–229, Lyngby, Denmark, Oct. 1992.
2. X. Nicollin, A. Olivero, J. Sifakis, and S. Yovine. An approach to the description and analysis of hybrid systems. In *Workshop on Theory of Hybrid Systems*, Lyngby, Denmark, Oct. 1992.
3. J.A. Stiver and P.J. Antsaklis. Modeling and analysis of hybrid control systems. In *Proceedings of the 31st CDC*, pages 3748–3751, Tucson, Arizona, 1992.
4. J. Le Bail, H. Alla, and R. David. Hybrid petri nets. In *European Control Conference*, pages 1472–1477, Grenoble, France, July 1991.
5. J.-E. Strömberg, J. Top, and U. Söderman. Modelling mode switching in dynamical systems. In J.W. Nieuwenhuis, C. Praagman, and H.L. Trentelman, editors, *Proc. of the 2nd ECC '93*, pages 848–853, Groningen, The Netherlands, June 1993.
6. A. Benveniste and G. Berry. The synchronous approach to reactive and real-time systems. *Proceedings of the IEEE*, 79(9), Sep. 1991.
7. M. Tittus. An algorithm for the generation of control laws for linear hybrid systems. Technical Report CTH/RT/R-93/0013, Control Engineering Lab, Chalmers University of Technology, S-412 96 Gothenburg, Sweden, 1993.

Simulation and Perturbation Analysis

Sample-Path-Based Continuous and Discrete Optimization of Discrete Event Systems: From Gradient Estimation to "Rapid Learning"

Christos G. Cassandras

Department of Electrical and Computer Engineering
University of Massachusetts
Amherst, MA 01003, USA

1 Introduction

In the performance evaluation, control, and optimization of stochastic Discrete Event Systems (DES), one is faced with two basic problems:

$$\textbf{(P1)}\ \text{evaluate } J(\theta) = E[L(\theta)] \text{ over all } \theta \in \Theta$$
$$\textbf{(P2)}\ \text{find } \theta \in \Theta \text{ to minimize } J(\theta) = E[L(\theta)]$$

where θ is some parameter to be chosen from some set Θ, and $J(\theta)$ is the performance measure of interest. Note that we are careful to distinguish between $L(\theta)$, the performance obtained over a specific sample path of the system (e.g., a single simulation run or an observation of the real system over some period of time), and $J(\theta)$, the expectation over all possible sample paths. In **(P1)**, our goal is simply to evaluate the response surface $J(\theta)$ over the entire parameter range Θ. On the other hand, it is sometimes possible to formulate a well-defined optimization problem **(P2)**, which we then seek to solve. There are two key difficulties associated with these problems. First, the function $J(\theta)$ is often simply unknown due to the complexity of the DES. Second, even if one succeeds in obtaining an analytical model for the DES, it is generally difficult to obtain accurate values for the parameters on which it depends. As an example, models for packetized voice and video traffic needed in the analysis of high-speed communication networks involve a number of parameters that are hard to accurately estimate. In short, because of the lack of analytical expressions for $J(\theta)$, the solution of problems **(P1)** and **(P2)** generally requires repetitive simulation in order to estimate $J(\theta)$ over the set Θ.

When the parameter θ takes real values and $J(\theta)$ is differentiable, estimating derivatives of the form $\partial J/\partial\theta$ is a process that allows us to invoke a number of standard gradient-based optimization techniques that can be used to solve **(P2)**. Moreover, derivative estimates can be used to approximate $J(\theta)$ in trying to obtain the global response surface required in **(P1)**. An obvious way to estimate $\partial J/\partial\theta$ is to estimate finite-difference approximations of the form $\Delta J/\Delta\theta$ through either simulation or direct observation as follows: observe a sample path under θ and obtain $L(\theta)$. Next, perturb θ by $\Delta\theta$ and observe $L(\theta+\Delta\theta)$. Finally, $[L(\theta+\Delta\theta)-L(\theta)]/\Delta\theta$ is an estimate of $\partial J/\partial\theta$. However, this approach involves $(n+1)$ sample paths if there are n parameters of interest, and it quickly becomes prohibitively time-consuming. In addition, accurate gradient estimation requires "small" $\Delta\theta$; however, division by a small number in $\Delta J/\Delta\theta$ leads to a host of numerical problems. This has motivated the effort towards derivative estimation based on information from a single observed sample path, and has led to Perturbation Analysis (PA) and Likelihood Ratio (LR) techniques.

Turning our attention to discrete parameters, suppose that the decision space is some finite set $\Theta = \{\theta_1, \ldots, \theta_m\}$. Examples of such parameters are "thresholds" based on which it can be shown that many optimal control policies in queueing systems can be constructed, buffer capacities in queueing models, numbers of "kanban" in so-called Just-In-Time (JIT) production control policies, etc. More generally, we can view Θ as a set of (potentially very large) "control actions" that affect the performance of a DES. In the spirit of extracting information from a single sample path, we can pose the following question: by observing a sample path under θ_1, is it possible to infer the performance of the system under all (or at least many) other parameters $\theta_2, \ldots, \theta_m$? More precisely, can we construct sample paths under $\theta_2, \ldots, \theta_m$ from the information contained in the observed sample path under θ_1? This is referred to as the "constructability" problem. When this problem can be solved, it is tantamount to generating m simulation runs in parallel.

The area of gradient estimation for DES is well-studied and well-documented (see [13], [19], [4], [3], [22]). In this paper, we limit ourselves to a brief survey of these techniques to bring the reader up to speed with the most recent developments (§3). The main objective of the paper is to concentrate on the sample path constructability problem, describe it in a formal setting, and then review informally, in tutorial fashion, the two main approaches that can be used to solve problems **(P1)** and **(P2)** over a discrete set Θ (§4). In §5, we also present a recently proposed approach for solving a class of problems of the form **(P2)** with discrete Θ. The interesting feature of this approach is that it combines gradient estimation with techniques used to solve the constructability problem. We begin, however, in §2 with a modeling framework for DES that we will adopt for our purposes.

2 Stochastic Timed Automata

We choose to describe a DES by means of a *state automaton*, i.e. a five-tuple (E, X, Γ, f, x_0), where E is a countable *event set*, X is a countable *state space*, $\Gamma(x)$ is a set of *feasible* or *enabled* events, defined for all $x \in X$ with $\Gamma(x) \subseteq E$, f is a *state transition function*, $f : X \times E \to X$, defined only for events $i \in \Gamma(x)$ when the state is x and not defined for $i \notin \Gamma(x)$, and x_0 is an *initial state*, $x \in X$. The feasible event set $\Gamma(x)$ is needed to reflect the fact that it is not always physically possible for some events to occur. For every $i \in \Gamma(x)$, there is an associated *clock* value (or *residual lifetime*) y_i, which represents the amount of time required until event i occurs, with the clock running down at unit rate. The clock value of event i always starts with a lifetime, which is an element of an externally provided clock sequence $\mathbf{v}_i = \{v_{i,1}, v_{i,2}, \ldots\}$, which we view as the input to the automaton: the model is endowed with a clock structure, $V = \{\mathbf{v}_i, i \in E\}$, giving rise to a *timed automaton* $(E, X, \Gamma, f, x_0, V)$.

Let us informally describe how a timed automaton operates (for details, see [19] or [4]). If the current state is x, we look at all clock values y_i, $i \in \Gamma(x)$. The triggering event e' is the event which occurs next at that state, i.e., the event with the smallest clock value:

$$e' = \arg\min \{y_i,\ i \in \Gamma(x)\}. \tag{1}$$

Once this event is determined, the next state, x', is specified by $x' = f(x, e')$, where f is the given state transition function. More generally, f can be replaced by transition probabilities $p(x'; x, i)$ where $i \in \Gamma(x)$. The amount of time spent at state x defines the *interevent time* (between the event that caused a transition into x and the event e'):

$$y^* = \min \{y_i,\ i \in \Gamma(x)\}. \tag{2}$$

Thus, time is updated through $t' = t + y^*$, and clock values through $y_i' = y_i - y^*$, except for e' and any other events which were not feasible in x, but become feasible in x'. For any such event, the clock value is set to a new lifetime obtained from the next available element in the event's clock sequence.

Finally, in a *stochastic* timed automaton $(E, X, \Gamma, f, x_0, G)$, the clock structure is replaced by a set of probability distribution functions $G = \{G_i,\ i \in E\}$. In this case, whenever a lifetime for event i is needed, we obtain a sample from G_i. The state sequence generated through this mechanism is a stochastic process known as a *Generalized Semi-Markov Process* (GSMP) (see also [13], [19]). This provides the framework for generating sample paths of stochastic DES. A sample path is defined by a sequence $\{e_k, t_k\},\ k = 1, 2, \ldots,$ where e_k is the kth event taking values from E, and t_k is its occurrence time; or, equivalently, by $\{x_k, t_k\},\ k = 1, 2, \ldots,$ where x_k is the kth state entered when an event occurs at time t_k.

3 A Brief Review of Gradient Estimation Techniques

Perturbation Analysis (PA) and Likelihood Ratio (LR) techniques both attempt to estimate performance derivatives of the form $\partial J/\partial\theta$ from information extracted from a single observed sample path of a DES. The two techniques, however, are radically different. We begin by briefly describing the simplest form of PA, Infinitesimal Perturbation Analysis (IPA), and contrast it to the LR approach.

3.1 Infinitesimal Perturbation Analysis (IPA)

The basic idea of IPA is to use the sample derivative $\partial L/\partial\theta$ as an estimate of $\partial J/\partial\theta$. In this approach, the underlying sample space is taken to be $[0, 1]^\infty$. Thus, an event lifetime $v_{i,k}$ corresponding to the kth occurrence of event $i \in E$ with lifetime distribution G_i is viewed as the result of a transformation $v_{i,k} = G_i^{-1}(u)$, where u is a sample from the uniform distribution over [0, 1]. Accordingly, if G_i is parameterized by θ, then $v_{i,k}(\theta) = G_i^{-1}(u, \theta)$. As θ varies, the underlying random numbers are taken to be fixed, and it can be shown that the derivative $\partial v_{i,k}(\theta)/\partial\theta$ is obtained through

$$\frac{\partial v_{i,k}}{\partial\theta} = -\frac{[\partial G_i(x;\theta)/\partial\theta]_{(v_{i,k},\theta)}}{[\partial G_i(x;\theta)/\partial x]_{(v_{i,k},\theta)}} \tag{3}$$

where the notation $[\cdot]_{(v_{i,k},\theta)}$ means that the term in brackets is evaluated at the point $(v_{i,k},\theta)$. For many performance measures of interest, $L(\theta)$ can be expressed as a sum of event lifetimes and shown to be a continuous and piecewise differentiable function of θ. In such cases, it is quite simple to evaluate $\partial L/\partial\theta$ as a function of lifetime derivatives $\partial v_{i,k}/\partial\theta$. IPA provides unbiased and strongly consistent estimates for a large class of DES, including Jackson-like queueing networks, as long as the sample functions do not exhibit discontinuities in θ. Moreover, a structural condition, known as the *commuting condition* in its simplest form, can be used as an unbiasedness test for IPA. The commuting condition (in its simplest form) requires the following: if an event α occurs at state x followed by β and the final state is y, then reversing the event order leads to the same final state y. For full details on IPA see [13], [19], [3] and Chapter 9 of [4].

3.2 The Likelihood Ratio (LR) approach

In this approach (also known as the "score function" methodology), a realization is viewed as remaining fixed when θ varies; the effect of θ is to change the probability measure that characterizes the realization. The LR approach is attractive because, compared to IPA, it yields unbiased estimates for a wider class of problems. Unfortunately, however, the variance of the LR derivative estimates increases with the length of a sample path. Thus, except for regenerative DES with reasonably short regenerative periods, this approach is limited in practice. In addition, to apply this technique, one needs to know precisely what the probability density functions of the event lifetimes with a dependence on θ are. This is not so in IPA, where, for instance, any $G_i(x,\theta)$ for which θ is a scale parameter gives $\partial v_{i,k}/\partial\theta = v_{i,k}/\theta$ in (3). Further details on the LR approach may be found in [21], [15], [22].

The limitations of IPA have motivated the effort to extend it beyond the confines of the commuting condition. In what follows, we provide a very brief summary of these extensions.

3.3 IPA for sample functions with discontinuities

In general, when $L(\theta)$ is discontinuous in θ, IPA gives biased estimates. It is possible, however, to define constructions which provide sample paths equivalent to the original one while eliminating the discontinuities at the same time. When this is possible, IPA can be directly applied. A collection of such cases is provided in Chapter 9 of [4].

3.4 Finite and Extended Perturbation Analysis (FPA and EPA)

In cases where IPA fails, it is possible to use a finite difference approximation to $\partial L/\partial\theta$, of the form $\Delta L/\Delta\theta$, where ΔL is still obtained from a single sample path observed under θ. This is referred to as Finite PA (FPA), and actually preceded IPA in the development of this field [17]. It should be clear, however, that in this case we can only obtain an approximation to the derivative required. In some cases, IPA can be used up to some event on the observed sample path, at which point one has to shift to FPA for some period of time. This is referred to as Extended PA [18].

3.5 Smoothed Perturbation Analysis (SPA)

The most general way to solve the problem of sample path discontinuities that limits IPA is to replace $L(\theta)$ by a conditional expectation of the form $E[L(\theta)|z]$, where z is a "characterization" of the observed sample path [16], [14], [11]. Because of the smoothing property of the conditional expectation, $E[L(\theta)|z]$ is generally continuous in θ. The price to pay, however, is in determining the appropriate characterization for the problem of interest and extracting additional information from the sample path. A number of special forms of SPA have emerged that solve various classes of interesting problems, including Rare Perturbation Analysis (RPA) [2], several so called "marking" and "phantomizing" techniques [8], and Structural PA [10] to name a few. It is also noteworthy that SPA may be used to derive second derivative estimates for a wide class of DES, including Jackson-like queueing networks [1].

4 The Constructability Problem and "Rapid Learning"

Let us now concentrate on problems **(P1)** and **(P2)** for a *discrete* parameter set $\Theta = \{\theta_1, \ldots, \theta_m\}$. Let some value from this set, θ_1, be fixed. A sample path of a DES depends on θ_1 and on ω, an element of the underlying sample space Ω, which (as in PA) is taken to be $[0, 1]^\infty$. Let such a sample path be $\{e_k^1, t_k^1\}, k = 1, 2, \ldots$. Then, assuming all events and event times $e_k^1, t_k^1, k = 1, 2, \ldots$, are directly observable, the problem is to construct a sample path $\{e_k^j, t_k^j\}, k = 1, 2, \ldots$, for any θ_j, $j = 2, \ldots, m$, as shown in Fig. 1. We refer to this as the *constructability problem*. Ideally, we would like this construction to take place on line, i.e., while the observed sample path evolves. Moreover, we would like the construction of all $(m-1)$ sample paths for $j = 2, \ldots, m$ to be done in parallel.

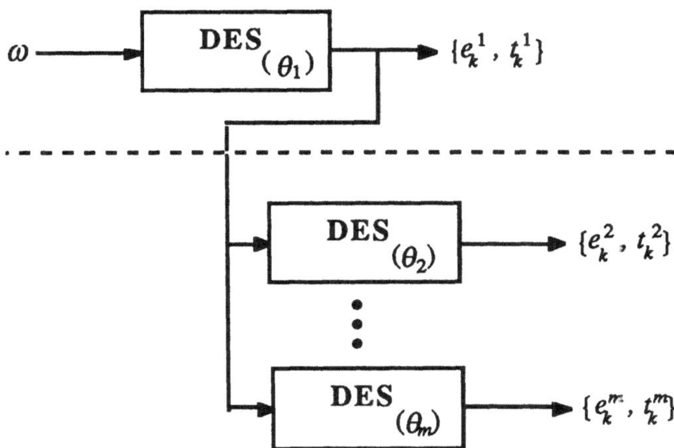

Fig. 1. The sample path constructability problem

The solution to this problem gives rise to what we call "rapid learning", since it enables us to learn about the behavior of a DES under all possible parameter values in Θ from a single "trial", i.e., a sample path under one parameter value. Returning to problem **(P2)**, note that any sample performance metric $L(\theta)$ is obtained as a function of $\{e_k, t_k\}$, $k = 1, 2, \ldots$. Thus, from an optimization standpoint, if performance estimates $L(\theta_1), \ldots, L(\theta_m)$ are all available at the conclusion of one sample path, we can immediately select a candidate optimal parameter $\theta^* = \arg\min \{L(\theta),\ \theta \in \Theta\}$. This is potentially the true optimal choice, depending on the statistical accuracy of the estimates $L(\theta_1), \ldots, L(\theta_m)$ of $J(\theta_1), \ldots, J(\theta_m)$. If the accuracy at the end of one trial is not sufficient, the process may be repeated several more times, each time observing the sample path with the optimal parameter to date. It is, however, important to point out that what is of interest in this process is accuracy in the *order* of the estimates $L(\theta_1), \ldots, L(\theta_m)$, not their actual *cardinal* value. This idea can be further exploited to develop a new setting for solving optimization problems as recently described in [20]. It is also interesting to point out that this approach is inherently based on processing system data in parallel, as clearly illustrated in Fig. 1. Therefore, as a "parallel simulation" technique it is ideally suited for emerging massively parallel and distributed processing computer environments (see also [12] for a different perspective on parallel simulation).

A natural question that now arises is "under what conditions can one solve the constructability problem?" This question was addressed in [5] and [6], and the constructability condition presented in [6] is briefly reviewed next. Suppose we are constructing a sample path $\{e_k^j, t_k^j\}$, $k = 1, 2, \ldots$, and the current state is x_k^j. In view of (1), to proceed with the construction, we need knowledge of (*i*) the feasible events defining $\Gamma(x_k^j)$, and (*ii*) the corresponding clock values to be compared, $y_{i,k}$, $i \in \Gamma(x_k^j)$. Observability is a property that addresses part (*i*) of this requirement. It applies to two state sequences, $\{x_k\}$, and $\{x_k^j\}$, $k = 0, 1, \ldots$, resulting from two sample paths:

Definition 1. A sample path $\{e_k^j, t_k^j\}$, $k = 1, 2, \ldots$, is *observable* with respect to $\{e_k, t_k\}$ if $\Gamma(x_k^j) \subseteq \Gamma(x_k)$ for all $k = 0, 1, \ldots$.

Intuitively, this condition guarantees that all feasible events required at state x_k^j to proceed with the sample path construction are "observable", since they are a subset of the feasible events at the observed state x_k. Unfortunately, this condition alone does not guarantee that part (*ii*) of the requirement above is satisfied, since the clock values in the constructed sample path are generally not identical to those of the observed sample path, not even in distribution. To address this issue, let v_i be the lifetime of some event i currently feasible, and let z_i be its age, i.e., if y_i is its clock value, we have $z_i = v_i - y_i$. We can then define the conditional probability distribution of the kth clock value $y_{i,k}$ in a sample path given the event age $z_{i,k}$ (an observable quantity), and denote it by $H(\cdot, z_{i,k})$. Using similar notation for the constructed sample path $\{e_k^j, t_k^j\}$, $k = 1, 2, \ldots$, the constructability condition presented in [6] is as follows.

Definition 2. A sample path $\{e_k^j, t_k^j\}$, $k = 1, 2, \ldots$, is *constructable* with respect to $\{e_k, t_k\}$ if:

$$\Gamma(x_k^j) \subseteq \Gamma(x_k) \text{ for all } k = 0, 1, \ldots \tag{4}$$

$$\text{and } H^j(\cdot, z_{i,k}^j) = H(\cdot, z_{i,k}) \text{ for all } i \in \Gamma(x_k^j) \tag{5}$$

One can then easily establish the following two results:

Theorem 3. *If all event processes are Markovian (generated by Poisson processes), then observability implies constructability.*

Theorem 4. *If* $\Gamma(x_k^j) = \Gamma(x_k)$ *for all* $k = 0, 1, \ldots,$ *then constructability is satisfied.*

The first result follows from the memoryless property: condition (5) is satisfied regardless of the event ages of the feasible events at state x_k^j. The second result holds because (5) is now satisfied due to the fact that $z_{i,k}^j = z_{i,k}$ for all $k = 0, 1, \ldots$.

The constructability condition (4)-(5) is naturally satisfied for some simple DES. For example, consider an $M/M/1/K$ queueing system and let the parameter of interest be the queueing capacity K. It can be easily verified that all sample paths of an $M/M/1/L$ system are constructable with respect to a sample path of an $M/M/1/K$ system for all L such that $1 \leq L < K$ (e.g., see Chapter 9 of [4]). However, constructability is not satisfied for $L > K$, because (4) is violated for at least one state. To solve the constructability problem when one or both of (4)-(5) are violated, two techniques have been proposed, the Standard Clock approach and Augmented System Analysis, which are overviewed next.

4.1 The Standard Clock (SC) Approach

In the Standard Clock approach, introduced in [23], we limit ourselves to Markovian DES, i.e., all event lifetime distributions G_i in our model are assumed exponential with corresponding parameter λ_i. By Theorem 1, the problem of constructability is then reduced to that of observability. This, in turn, is resolved by adopting the well-known uniformization approach which exploits the properties of Markov chains. In particular, the way that the SC approach gets around the observability problem is by generating a sample path of a "master" system with $\Gamma(x) = E$ for all states x, and then constructing sample paths of various systems of interest for which $\Gamma(x^j) \subseteq E$. Trivially, then, each such sample path satisfies the observability condition (4). A sample path of this "master" system is constructed as follows. Since at any state the total event rate is $\Lambda = \sum_{i \in E} \lambda_i$ (the maximal event rate possible), all interevent times are exponentially distributed with parameter Λ. Thus, the first step is to generate an i.i.d time sequence $\{t_k\}$, $k = 1, 2, \ldots,$ with $(t_{k+1} - t_k)$ sampled from an exponential distribution with parameter Λ. This sequence is called a Standard Clock (SC). Note that it depends only on Λ. Next, since at any state event i occurs with probability λ_i/Λ, to determine the triggering event at any state we generate another i.i.d event sequence $\{e_k\}$, $k = 1, 2, \ldots,$ with e_k determined through $P[e_k = i] = \lambda_i/\Lambda$. Note that this sequence is generated completely independently from the SC itself.

We now wish to construct a sample path of some system corresponding to a parameter θ_j for which the event set is E but where it is generally the case that $\Gamma(x_k^j) \subseteq E$. To do so, we simply use the interevent times $(t_{k+1} - t_k)$ from the SC and the event sequence $\{e_k\}$ generated above. Whenever $e_{k+1} \in \Gamma(x_k^j)$, the state is updated using the state transition function f^j for this system. Whenever $e_{k+1} \notin \Gamma(x_k^j)$, this event is simply ignored; e_{k+1} is interpreted as a fictitious event that leaves the state unaffected, i.e., $f^j(x_k^j, e_{k+1}) = x_k^j$, and preserves the statistical validity of the sample path due to the memoryless property. This approach is well-suited for a simulation environment, where the SC sequence can be generated in advance, and then multiple sample paths can be constructed in parallel. If, on the other hand, an actual sample path $\{e_k, t_k\}$ of a DES is observed, then one observes only feasible events for the DES in question and must resort to a modified scheme described in [7].

Example 1. To construct multiple sample paths of an $M/M/1/K$ queueing system for different values of $K = 1, 2, \ldots$, using the SC approach, we first generate a SC with rate $\lambda + \mu$, where λ is the arrival rate and μ is the service rate. Note that this is done independently of K. The event set for this DES is $E = \{a, d\}$, where a is an arrival and d is a departure. An event sequence $\{e_k\}$ is then generated as follows: use a random number u_k and set $e_k = a$ if $u_k \leq \lambda/(\lambda + \mu)$ and $e_k = d$ otherwise. Then, if a state sequence of the system corresponding to a value K is denoted by $\{x_k^K\}$, a sample path of this system is obtained by setting

$$x_{k+1}^K = x_k^K + 1 \text{ if } e_{k+1} = a \text{ and } x_k^K < K$$
$$x_{k+1}^K = x_k^K - 1 \text{ if } e_{k+1} = d \text{ and } x_k^K > 0$$

We can then see that arrivals when the queue is full and departures when the queue is empty are simply ignored and leave the state unaffected. However, the same SC and event sequence are used to construct as many sample paths in parallel as desired corresponding to $K = 1, 2, \ldots$.

4.2 Augmented System Analysis

Augmented System Analysis (ASA) was introduced in [5] for purely Markovian DES, and was subsequently extended in [6]. The scope of ASA is similar to that of the SC approach, with three basic differences: (*i*) While the SC technique applies to discrete or continuous parameters, ASA is specifically geared towards discrete parameters, (*ii*) ASA is always driven by *actually observed* event and time sequences, i.e., there is no "master" system and hence fictitious events as in the SC construction, and (*iii*) It is applicable to more general models, beyond Markovian ones.

Let us begin by limiting ourselves once again to purely Markovian DES. The main contribution of ASA is in providing a formal technique for overcoming the problem of "unobservable" events, i.e., instances along the construction of a sample path $\{e_k^j, t_k^j\}$, $k = 1, 2, \ldots$, where $\Gamma(x_k^j) \supset \Gamma(x_k)$, violating (4). The key idea of this technique is to "suspend" the sample path construction for the jth system until a state x_k where observability is satisfied is next encountered. The resulting procedure is known as *event matching*. In order to describe this procedure, let us define a variable associated with every constructed sample path indexed by j (corresponding to parameter value θ_j in problems **(P1)** and **(P2)**) called the "mode" of the jth system and denoted by $\mathcal{M}_k^j$, $k = 0, 1, \ldots$. This variable is always initialized to the value "active" and remains unaffected unless a state x_k is entered such that $\Gamma(x_k^j) \supset \Gamma(x_k)$. At this point, we set $\mathcal{M}_k^j = x_k^j$, i.e., store the constructed sample path's state. The construction is subsequently suspended until a new state x_k is entered such that $\Gamma(\mathcal{M}_k^j) \subseteq \Gamma(x_k)$. At this point, we set $\mathcal{M}_k^j =$ active, and continue the process. This procedure is summarized in the following *Event Matching Algorithm*:

- *Step 1*: Initialize: $x_0^j = x_0$ and $\mathcal{M}_0^j =$ active for all j

With every observed event e_{k+1}, $k = 0, 1, \ldots$, for every sample path j:

- *Step 2*:
  1. If $\mathcal{M}_k^j =$ active, update state: $x_{k+1}^j = f^j(x_k^j, e_{k+1})$
  2. If $\Gamma(x_{k+1}^j) \supset \Gamma(x_{k+1})$, set: $\mathcal{M}_{k+1}^j = x_{k+1}^j$
- *Step 3*: If $\mathcal{M}_k^j \neq$ active, and $\Gamma(\mathcal{M}_k^j) \subseteq \Gamma(x_k)$, set: $\mathcal{M}_{k+1}^j =$ active.
- *Step 4*: If $\mathcal{M}_k^j \neq$ active, and $\Gamma(\mathcal{M}_k^j) \supset \Gamma(x_k)$, continue

It is worth pointing out the inherent parallelism of this procedure: whenever an event is observed, all sample path states are simultaneously updated, with the exception of those with $\mathcal{M}_k^j \neq$ active. Obviously, if observability is always satisfied for some j, then $\mathcal{M}_k^j =$ active and no suspension is needed. If, on the other hand, observability is violated at *Step 2.2* above, then at *Step 3*, when it is once again satisfied, we must in addition check that condition (5) is satisfied. This is always true for Markovian events, or it may be true in special cases where one can exploit the structure of the system, as in Example 3.

Example 2. Consider an $M/M/1/K$ queueing system for different values of $K = 1, 2, \ldots$. In contrast to the SC approach in Example 1, we must now base our analysis on an actual sample path observed under some value of K. Let us first assume that the nominal sample path corresponds to $K = 3$ and we are interested in constructing a sample path for the $M/M/1/2$ system. It is not difficult to verify that in this case the constructability condition (4)-(5) is satisfied. To see this, we construct what is referred to as an "augmented system", i.e., a hypothetical system whose state is of the form (x_k^2, x_k^3), where x_k^K denotes the queue length of the $M/M/1/K$ system, $K = 2, 3$. This hypothetical system is driven by the exact same event sequence as the observed sample path. In Fig. 2 we show the state transition diagrams of the $M/M/1/2$ and $M/M/1/3$ systems along with the state transition diagram of the augmented system. It is now easy to see that $\Gamma(x_k^2) \subseteq \Gamma(x_k^3)$ for all k. In fact, $\Gamma(x_k^2) = \Gamma(x_k^3)$ for all states with one exception: at the augmented system state (0, 1) we have $\Gamma(0) = \{a\} \subset \Gamma(1) = \{a, d\}$. In this case, however, observability is still satisfied. Thus, the mode of the $M/M/1/2$ system whose sample path we wish to construct is always "active", and steps 3 and 4 in the Event Matching Algorithm above are never invoked.

Example 3. Consider an $M/G/1/K$ (i.e., allow the service time distribution to be arbitrary). Let us now reverse the role of the two systems in Example 2: we observe a sample path of the $M/G/1/2$ system and wish to construct a sample path of the $M/G/1/3$ system. Interestingly, this role reversal causes the observability condition to be violated at the augmented system state (0, 1), since $\Gamma(1) = \{a, d\} \supset \Gamma(0) = \{a\}$. Intuitively, when the observed system is empty, there is no feasible d event available to be used in

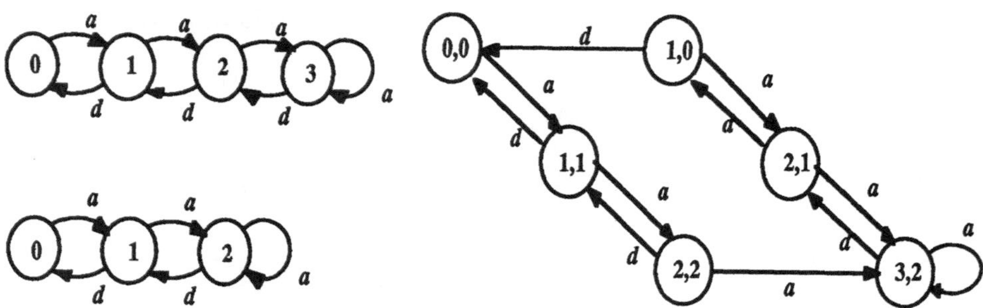

Fig. 2. State transition diagrams of $M/M/1/2$, $M/M/1/3$, and augmented system

the construction of the sample path of the $M/G/1/3$ system beyond this point. Therefore, the construction is suspended and we set $\mathcal{M}_k^2 = 1$. However, since the only event feasible is a, the next transition leads the observed system to $x_{k+1}^2 = 1$ and observability is satisfied. Moreover, condition (5) is also satisfied regardless of the service time distribution, since the age of event d for both systems is always zero when a occurs at state $x_k^2 = 0$.

It should be emphasized that the augmented system is useful only in checking whether observability is satisfied or not. The Event Matching Algorithm, however, does not require constructing such a system.

It is possible to extend ASA to DES with at most one non-Markovian event. This requires the *Age Matching Algorithm* described in [6] and Chapter 9 of [4]. The key idea is to now store not only the state where a sample path construction is suspended, but also the age of the non-Markovian event when this happens (if such an event is feasible at that state). Then, to reset the mode to "active" requires waiting until a point in time where x_k is such that observability is satisfied and the age of the observed non-Markovian event matches that stored at the point of suspension. Extensions of ASA to DES with completely arbitrary event lifetime distributions are possible through techniques such as the "time warping" scheme described in Chapter 9 of [4]. However, additional storage and computational costs need to be imposed. The general sample path constructability problem remains open. In addition, it is necessary to develop systematic frameworks for comparing various schemes, such as the SC and ASA approaches, in terms of factors such as computational cost, variance properties, and suitability for various parallel processing architectures.

5 Discrete Stochastic Optimization Problems

In this section, we describe an approach recently proposed in [9] which combines gradient techniques with "rapid learning" ideas from §4 to solve a class of discrete stochastic optimization problems of the form **(P2)**. Let us consider a stochastic resource allocation problem where K identical resources are to be allocated to N user classes. It is assumed that a queue is associated with the ith class, $i = 1, \ldots, N$, and users in that class form an arrival stream to that queue with arbitrary interarrival time distribution. When a user accesses a resource, he may hold it for some random period of time whose distribution is also arbitrary. The problem then is to determine the number of resources to be allocated to the ith class so that some particular cost criterion is minimized (e.g., the average waiting time of users who must wait in queue before accessing a resource).

To formally state this problem, let the resources be sequentially indexed so that the "state" or "allocation" $\mathbf{s}$ is represented by the K-dimensional vector $[s_1, \ldots, s_K]^T$ where $s_j \in \{1, \ldots, N\}$ is the user class index assigned to resource j. Let S be the set of feasible resource allocations $S = \{[s_1, \ldots, s_K] : s_j \in \{1, \ldots, N\}\}$. By "feasible" we mean that the allocation may have to be chosen to satisfy some basic requirements, such as 'stability' or 'fairness'. For any $\mathbf{s} \in S$, let $J_i(\mathbf{s}) = E[L_i(\mathbf{s})]$ be the class i expected cost function. Then, our objective is to select $\mathbf{s} \in S$ so as to solve the following discrete optimization problem:

$$\textbf{(P3)} \quad \min_{\mathbf{s} \in S} \sum_{i=1}^{N} \beta_i E[L_i(\mathbf{s})]$$

where β_i is a weight associated with user class i. This is generally a difficult problem, as it involves both optimization over a discrete set and estimation of the cost under all possible allocations $\mathbf{s} \in S$. This fits in the type of problems we can address using the parallel sample path construction techniques discussed earlier. Note, however, that the resource allocation problem here is combinatorially explosive. The approach we will follow tries to transform **(P3)** into a continuous optimization problem so as to take advantage of gradient-based techniques.

Let $\theta_i = P[s_j = i]$, independent of the resource index $j = 1, \ldots, K$. In other words, we assume that any one of the identical K resources is allocated to user class i with probability θ_i. In this setting, the probability vector $\boldsymbol{\theta} = [\theta_1, \ldots, \theta_N]$, with $\sum_{i=1}^{N} \theta_i = 1$, represents new control parameters which can be optimally selected so as to minimize the desired cost function. In particular, we replace problem **(P3)** by the following constrained continuous parameter optimization problem:

$$\textbf{(P4)} \quad \min_{\theta} \sum_{i=1}^{N} \beta_i E[L_i(\theta)] \; s.t. \; \sum_{i=1}^{N} \theta_i = 1$$

The relationship between the resource assignment probability vector $\boldsymbol{\theta}$ and the actual resource allocation $\mathbf{s}$ can be made precise by specifying the randomization mechanism used. Thus, let $\mathbf{r} = [r_1, \ldots, r_K]$ be a K-dimensional vector of independent random variables. Although there may be flexibility in choosing the distribution of r_j, for simplicity we assume that r_j is uniformly distributed in [0, 1]. Then, suppose that $\mathbf{r}$ is fixed by obtaining K independent random numbers. An allocation $\mathbf{s}(\boldsymbol{\theta}, \mathbf{r}) = [s_1, \ldots, s_K]$ is obtained using the standard inverse transform technique for generating random variates $s_1, \ldots, s_K$ as follows:

$$s_j = \min\{i : r_j \leq \sum_{k=1}^{i} \theta_k, \; i = 1, \ldots, N\}. \tag{6}$$

It is then clear that, for a given $\mathbf{r}$, by adjusting the parameters θ_i, $i = 1, \ldots, N$, we can affect s_j, $j = 1, \ldots, K$, and, consequently, control the allocation of the resources.

The next step in our approach is to seek a solution of **(P4)**. This is still a hard problem, since no closed-form expression is generally available for $E[L_i(\boldsymbol{\theta})]$. Here, however, we can at least resort to the gradient estimation techniques reviewed in §3. In particular, as shown in [9], it is possible to derive unbiased estimators of the derivatives $\partial E[L_i(\theta)]/\partial\theta_i$ using an approach similar to that of [8].

Since our goal then is to estimate $\partial E[L_i(\theta)]/\partial\theta_i$ from a single sample path, we need to generate such a sample path of the DES of interest (i.e., the process of arriving users who queue up and then hold a resource for some period of time). To do so, we need to specify an allocation $\mathbf{s}(\boldsymbol{\theta}, \mathbf{r})$. Assuming that a control vector $\boldsymbol{\theta}$ is specified, then a sample path is characterized by the vector $\mathbf{r}$ and we write in **(P4)** $E[L_i(\boldsymbol{\theta})] = E[L_i(\mathbf{s}(\boldsymbol{\theta}, \mathbf{r}))]$. Let us now assume that the derivatives $\partial E[L_i(\mathbf{s}(\boldsymbol{\theta}, \mathbf{r}))]/\partial\theta_i$ can be estimated from a sample path generated under a specific $\mathbf{r}$ and use D_i to denote the ith derivative estimate. We are then in a position to summarize our approach by means of the following optimization scheme:

1. Initially select some $\mathbf{r}$ and $\boldsymbol{\theta}^{(0)}$. Hence, determine an initial allocation $\mathbf{s}^{(0)} = \mathbf{s}(\boldsymbol{\theta}^{(0)}, \mathbf{r})$ through (6).
2. Observe a sample path under $\mathbf{s}(\boldsymbol{\theta}^{(0)}, \mathbf{r})$ and estimate the derivatives $D_i^{(0)}$ for all $i = 1, \ldots, N$.
3. For any $m = 0, 1, \ldots,$ iterate as follows:
$$\theta_i^{(m+1)} = \theta_i^{(m)} + \eta_m D_i^{(m)} \tag{7}$$
$$s_j^{(m+1)} = \min\{i : r_j \leq \sum_{k=1}^{i} \theta_k^{(m+1)},\ i = 1, \ldots, N\} \tag{8}$$
with some provisions to ensure that $\sum_{i=1}^{N} \theta_i^{(m)} = 1$ for all $m = 0, 1, \ldots$.
4. Observe a sample path under $\mathbf{s}(\boldsymbol{\theta}^{(m)}, \mathbf{r})$ to estimate the derivatives $D_i^{(m)}$ for all $i = 1, \ldots, N$, and repeat step 3 until the sequence $\boldsymbol{\theta}^{(m)}$, $m = 0, 1, \ldots,$ converges to some $\boldsymbol{\theta}^*(\mathbf{r})$, and hence an allocation $\mathbf{s}^* = \mathbf{s}(\boldsymbol{\theta}^*, \mathbf{r})$.

Of course, (7) is a standard stochastic approximation scheme driven by the derivative estimates D_i^m with $\{\eta_m\}$ an appropriately selected step size sequence. The important observation here, however, is that after an update in the auxiliary vector $\boldsymbol{\theta}^{(m)}$, *the allocation is also updated* using the transformation defined in (6). Moreover, note that $\mathbf{r}$ remains fixed throughout the process.

There are several crucial issues related to this scheme. First, we must see how to obtain derivative estimates $D_i^{(m)}$ from a single sample path. What is interesting here is that such derivative estimates are given by expressions dependent on $L_i(\mathbf{s}) - L_i(\mathbf{s}^j)$ where $\mathbf{s}^j$ is an allocation identical to $\mathbf{s}$ except that the jth resource which was allocated to user i is now removed. However, obtaining $L_i(\mathbf{s}^j)$ for multiple j simultaneously from a sample path under $\mathbf{s}$ is precisely the idea of "rapid learning". In short, this approach integrates both gradient estimation and parallel sample path construction techniques.

Second, in dealing with iterative schemes of the type (7), one is usually concerned with the conditions under which $\{\boldsymbol{\theta}^{(m)}\}$, $m = 0, 1, \ldots,$ indeed converges to the optimal probability vector. In our approach, however, it is the iterative scheme (8) that is of interest; therefore, the main issue is to identify a "stopping condition" in (7) such that the resulting $\boldsymbol{\theta}^*(\mathbf{r})$ yields an optimal allocation $\mathbf{s}(\boldsymbol{\theta}^*, \mathbf{r})$. Some results related to this issue and some applications of interest may be found in [9]. However, the full potential of the approach and a number of technical issues remain unresolved and are subjects of ongoing research.

Acknowledgements. This work is supported in part by the National Science Foundation under Grants EID-9212122 and ECS-9311776, and by a Grant from United Technologies/Otis Elevator Co.

References

1. Bao, G., and Cassandras, C.G., "First and Second Derivative Estimators for Closed Jackson-Like Queueing Networks Using Perturbation Analysis Techniques", *Proceedings 31st IEEE Conf. on Decision and Control*, pp. 698-703, 1993 (also submitted for publication to *Journal of Discrete Event Dynamic Systems*).
2. Bremaud, P., and Vazquez-Abad, F.J., "On the Pathwise Computation of Derivatives with Respect to the Rate of a Point Process: The Phantom RPA Method", *Queueing Systems*, 10, pp. 249-270, 1992.
3. Cao X.R., *Realization Probabilities - The Dynamics of Queueing Systems*, Springer-Verlag, 1994.
4. Cassandras, C.G., *Discrete Event Systems: Modeling and Performance Analysis*, Irwin, 1993.
5. Cassandras, C.G., and Strickland, S.G., "On-Line Sensitivity Analysis of Markov Chains", *IEEE Trans. on Automatic Control*, AC-34, 1, pp. 76-86, 1989.
6. Cassandras, C.G., and Strickland, S.G., "Observable Augmented Systems for Sensitivity Analysis of Markov and Semi-Markov Processes", *IEEE Trans. on Automatic Control*, AC-34, 10, pp. 1026-1037, 1989.
7. Cassandras, C.G., Lee, J.I., and Ho, Y.C., "Efficient Parametric Analysis of Performance Measures for Communications Networks", *IEEE Journal on Selected Areas in Communication*, Vol. 8, 9, pp. 1709-1722, 1990.
8. Cassandras, C.G., and Julka, V., "Marked/Phantom Algorithms for a Class of Scheduling Problems", *Proceedings 31st IEEE Conf. on Decision and Control*, pp. 3215-3220, 1992 (also submitted for publication to *Queueing Systems*).
9. Cassandras, C.G., and Julka, V., "A New Approach for Some Combinatorially Hard Stochastic Optimization Problems", *Proceedings 1994 Allerton Conf.*, 1993.
10. Dai, L., and Ho, Y.C., "Structural Infinitesimal Perturbation Analysis (SIPA) for Derivative Estimation in Discrete Event Dynamic Systems", to appear in *IEEE Trans. on Automatic Control.*
11. Fu, M., and Hu, J.-Q., "Extensions and Generalizations of Smoothed Perturbation Analysis in a Generalized Semi-Markov Process Framework", *IEEE Trans. on Automatic Control*, AC-37, pp. 1483-1500, 1992.
12. Fujimoto, R., "Parallel Discrete Event Simulation", *Communcations of ACM*, Vol. 33, 10, pp. 31-53, 1990.
13. Glasserman, P., *Gradient Estimation Via Perturbation Analysis*, Kluwer, 1991.
14. Glasserman, P., and Gong, W.B., "Smoothed Perturbation Analysis for a Class of Discrete Event Systems", *IEEE Trans. on Automatic Control*, AC-35, pp. 1218-1230, 1990.

15. Glynn, P., "Likelihood Ratio Gradient Estimation: An Overview", *Proceedings of 1987 Winter Simulation Conference*, pp. 366-375, 1987.
16. Gong, W.B., and Ho, Y.C., "Smoothed Perturbation Analysis of Discrete Event Systems", *IEEE Trans. on Automatic Control*, AC-32, pp. 858-856, 1987.
17. Ho, Y.C., Cao, X.R., and Cassandras, C.G., "Infinitesimal and Finite Perturbation Analysis for Queueing Networks", *Automatica*, 19, pp. 439-445, 1983
18. Ho, Y.C., and Li, S., "Extensions of Perturbation Analysis of Discrete Event Dynamic Systems", *IEEE Trans. on Automatic Control*, AC-37, pp. 258-262, 1988.
19. Ho, Y.C., and Cao X.R., *Perturbation Analysis of Discrete Event Dynamic Systems*, Kluwer, 1991.
20. Ho, Y.C., Sreenivas, R., and Vakili, P., "Ordinal Optimization of Discrete Event Dynamic Systems", *Journal of Discrete Event Dynamic Systems*, pp. 61-88, 1992.
21. Reiman, M.I., and Weiss, A., "Sensitivity Analysis for Simulations via Likelihood Ratios", *Operations Research*, 37, pp. 285-289, 1989.
22. Rubinstein, R., *Monte Carlo Optimization, Simulation and Sensitivity of Queueing Networks*, Wiley, 1986.
23. Vakili, P., "A Standard Clock Technique for Efficient Simulation", *Operations Research Letters*, 10, pp. 445-452, 1991.

Infinitesimal Perturbation Analysis of Generalized Semi-Markov Processes: A Tutorial

Xi-Ren Cao

The Hong Kong University of Science and Technology

1 Introduction

In this tutorial paper, we introduce the readers to some fundamental aspects of perturbation analysis. Perturbation analysis is a method that provides performance sensitivities by analyzing a single sample path of a stochastic discrete system (Glasserman [1990] and Ho and Cao [1990]). There exist many works in this field; it is impossible to give an survey of all these works in this short paper. We choose to present the topic by using the framework of Generalized Semi-Markov Process (GSMP) and a fundamental concept called *realization probability*. The material presented here can be considered as a short summary of a part of a recent book (Cao [1994]).

The main idea of this approach is that a performance sensitivity can be decomposed into a sum of the final effects of many small pertubrations, which can be measured by realization probabilties. Thus, realization probability can be viewed as a fundamental "building block" of sensitivity analysis. The main results are as follows: a small change of a parameter of an event lifetime distribution will induce a perturbation in the event termination time, the size of the perturbation can be determined by a perturbation generation function; a perturbation of an event will be propagated to other events along a sample path, the final effect of this perturbation to a performance can be determined by a quantity called realization probability, which is specified by a set of linear differential equations; the performance sensitivity obtained by analyzing a single sample path converges with probability one to the sensitivity of the steady-state performance, which equals the expected value of the generation function and the realization probability. We will review the main concepts and results and will not provide proofs.

2 The Generalized Semi-Markov Process (GSMP)

Let Φ and Ψ be subsets of positive integers. The events that may occur in state $\mathbf{x} \in \Phi$ are denoted by the set $B(\mathbf{x}) \subseteq \Psi$. A GSMP stays in a state $\mathbf{x}$ until an event $\alpha \in B(\mathbf{x})$ triggers a transition. Let $p(\mathbf{x}'; \mathbf{x}, \alpha)$ be the probability that the new state is $\mathbf{x}'$ after event α triggers a transition from state $\mathbf{x}$. Each event in $B(\mathbf{x})$ has a lifetime, and an event can trigger a transition only at the end of its lifetime. Associated with each event α, there is a clock whose reading is denoted as d_α. The clock runs at a speed $\sigma(\mathbf{x}, \alpha)$. If at time 0 the clock is set to d_α then at time τ the reading of the clock will be $d'_\alpha = d_\alpha - \sigma(\mathbf{x}, \alpha) \times \tau$. For simplicity, we assume $\sigma(\mathbf{x}, \alpha) = 1$ in this paper. The lifetime of an event ends when the associated clock reading reaches zero. d_α is also called the *residual lifetime* of event α.

The clock readings after the transition are determined as follows. New clock readings are independently generated for each $\beta \in N(\mathbf{x}', \mathbf{x}, \alpha) \equiv B(\mathbf{x}') - (B(\mathbf{x}) - \{\alpha\})$. The new clock reading for event $\beta \in N(\mathbf{x}', \mathbf{x}, \alpha)$ has a cumulative probability distribution (c.p.d.) $F(s; \mathbf{x}', \beta, \mathbf{x}, \alpha)$. For events in both $B(\mathbf{x})$ and $B(\mathbf{x}')$ except for event α, the old clock readings are kept after the transition, i.e., for $\beta \in O(\mathbf{x}', \mathbf{x}, \alpha) \equiv B(\mathbf{x}') \cap (B(\mathbf{x}) - \{\alpha\})$, d_β does not change at the transition. For events in $B(\mathbf{x})$, but not in $B(\mathbf{x}')$, the clocks are set to quit, i.e., if $\beta \in (B(\mathbf{x}) - \{\alpha\}) - B(\mathbf{x}')$, then $d_\beta = \infty$ after the transition.

Let T_i, $\mathbf{x}_i$, and β_i be the ith transition time, the ith state, and the ith triggering event, respectively. Let $s_{\alpha,j}$ be the jth lifetime of event α. At T_i (after the transition), the completion time for event $\alpha \in B(\mathbf{x}_i)$ is $\tau_{\alpha,i} := T_i + d_{\alpha,i}$, where $d_{\alpha,i}$ is the clocking reading (residual lifetime) of event α at T_i. We have $T_{i+1} = \tau_{\beta_{i+1},i}$. According to the clock setting rule, for $\alpha \in B(\mathbf{x}_{i+1})$ we have

$$\tau_{\alpha,i+1} = \begin{cases} \tau_{\alpha,i} & \text{if } \alpha \in O(\mathbf{x}_{i+1}; \mathbf{x}_i, \beta_i) \ , \\ \tau_{\beta_{i+1},i} + s_{\alpha,j} & \text{if } \alpha \in N(\mathbf{x}_{i+1}; \mathbf{x}_i, \beta_i) \ , \end{cases} \tag{1}$$

where $\tau_{\beta_{i+1},i} = T_{i+1}$ and $s_{\alpha,j}$ is the new lifetime of event α generated at T_{i+1}. The subscript j indicates that there are $j-1$ completions of event α before T_{i+1}. Equation (1) and the state sequence $\mathbf{x}_1, \mathbf{x}_2, \ldots$ completely determine the evolution of the GSMP. Equation (1) describes the coupling between the event completion times.

3 Event Coupling in a GSMP

Thoughout this paper, we make the following two assumptions:

Assumption 1. Every event in $B(\mathbf{x})$ may be a triggering event with a positive probability.

Assumption 2 (Noninterruptive). For all $\mathbf{x} \in \Phi$, if $\alpha, \beta \in B(\mathbf{x})$, $\alpha \neq \beta$, and $p(\mathbf{x}'; \mathbf{x}, \alpha) > 0$, then $\beta \in B(\mathbf{x}')$.

Let

$$\begin{pmatrix} \mathbf{x}_0 \\ \beta_0 \end{pmatrix}, \begin{pmatrix} \mathbf{x}_1 \\ \beta_1 \end{pmatrix}, \ldots, \begin{pmatrix} \mathbf{x}_n \\ \beta_n \end{pmatrix} \tag{2}$$

represent a sequence of the state and triggering event pairs of a GSMP, with β_i, $i = 0, 1, \ldots, n$, denoting the triggering event from state $\mathbf{x}_i$ to $\mathbf{x}_{i+1}$. Consider an event $\alpha_0 \in B(\mathbf{x}_0)$ and let $\Psi_0 = \alpha_0$. We define

$$\Psi_{i+1} = \begin{cases} \Psi_i & \text{if } \beta_i \notin \Psi_i \ , \\ \{\Psi_i - \beta_i\} \cup N(\mathbf{x}_{i+1}; \mathbf{x}_i, \beta_i) & \text{if } \beta_i \in \Psi_i \ , \end{cases} \qquad i = 0, 1, \ldots, n-1 \ . \tag{3}$$

The physical meaning of (3) is as follows. Suppose that the initial event completion time of α_0 at state $\mathbf{x}_0$ is somehow delayed by a small amount Δ. We say that $(\alpha_0, \mathbf{x}_0)$ obtains a *perturbation* Δ. $(\Psi_0, \mathbf{x}_0)$ denotes the set of events associated with $\mathbf{x}_0$ that have the perturbation. Because of event coupling described in (1), after a transition triggered by event β_0, the changes in the completion time of α_0 will affect those of the events in

$$\Psi_1 = \begin{cases} \alpha_0 & \text{if } \beta_0 \neq \alpha_0 \ , \\ N(\mathbf{x}_1; \mathbf{x}_0, \alpha_0) & \text{if } \beta_0 = \alpha_0 \ . \end{cases}$$

More precisely, the completion times of the events in Ψ_1 will be delayed by the same amount Δ, i.e., the perturbation of $(\Psi_0, \mathbf{x}_0)$ will be *propagated* to $(\Psi_1, \mathbf{x}_1)$. The same explanation applies to Ψ_i, $i = 1, 2, \ldots, n$. Ψ_i are called *perturbation sets,* and (3) is called the *perturbation propogation rule.* (3) can be viewed as a qualitative abstraction of (1). In (3), the actual values of the event lifetimes are ignored, and only the logical relation between event termination times are maintained.

The sequence $(\Psi_0, \mathbf{x}_0), (\Psi_1, \mathbf{x}_1), \ldots, (\Psi_n, \mathbf{x}_n)$, $\Psi_0 = \alpha_0$, is called an *event-coupling tree* with a root $(\alpha_0, \mathbf{x}_0)$. We can also define in a similar way an event-coupling tree with a root $(\Psi_0, \mathbf{x}_0)$ for any subset $\Psi_0 \subseteq B(\mathbf{x}_0)$. Based on any sample path of the GSMP and the definition in (3), a stochastic process $(\Psi_i, \mathbf{x}_i)$, called the *perturbation set process,* can be constructed. The state space of this process is $2^{\Psi} \times \Phi$.

Lemma 1. *$\{(B(\mathbf{x}), \mathbf{x}),$ all $\mathbf{x}\}$ and $\{(\varnothing, \mathbf{x}),$ all $\mathbf{x}\}$ are two absorbing sets of the perturbation-set process.*

Theorem 2. *Superposition and exclusion properties hold for event-coupling trees. More precisely, let $(\Psi_0, \mathbf{x}_0)$, $(\Psi_1, \mathbf{x}_1)$, $\ldots$, $(\Psi_n, \mathbf{x}_n)$; $(\Psi_{1,0}, \mathbf{x}_0)$, $(\Psi_{1,1}, \mathbf{x}_1)$, $\ldots$, $(\Psi_{1,n}, \mathbf{x}_n)$; and $(\Psi_{2,0}, \mathbf{x}_0)$, $(\Psi_{2,1}, \mathbf{x}_1)$, $\ldots$, $(\Psi_{2,n}, \mathbf{x}_n)$ be the event-coupling trees on the same sample path $\mathbf{x}_0, \mathbf{x}_1, \ldots, \mathbf{x}_n$ with roots $(\Psi_0, \mathbf{x}_0)$, $(\Psi_{1,0}, \mathbf{x}_0)$, and $(\Psi_{2,0}, \mathbf{x}_0)$, respectively. We have*

1. *Superposition: If $\Psi_0 = \Psi_{1,0} \cup \Psi_{2,0}$, then $\Psi_i = \Psi_{1,i} \cup \Psi_{2,i}$ for all $i \geq 0$; and*
2. *Exclusion: If $\Psi_{1,0} \cap \Psi_{2,0} = \varnothing$, then $\Psi_{1,i} \cap \Psi_{2,i} = \varnothing$ for all $i \geq 0$.*

From Theorem 2, trees with different roots do not overlap. Also, a tree that grows up on a set of different events is the union of the trees that grow up on each individual event in the set.

Corollary 3. *If $\Psi_{1,0} \subseteq \Psi_{2,0}$, then $\Psi_{1,i} \subseteq \Psi_{2,i}$ for all $i \geq 0$.*

Definition 4. If on a sample path of a GSMP, the perturbation set eventually reaches the absorbing set $\{(B(\mathbf{x}), \mathbf{x}),$ all $x\}$ (or $\{(\varnothing, x),$ all $x\}$)

$$\lim_{i\to\infty} \{\Psi_i - B(\mathbf{x}_i)\} = \varnothing \text{ (or } \lim_{i\to\infty} \Psi_i = \varnothing) \ ,$$

then we say that the event-coupling tree *prevails* (or *perishes*), or the perturbation in $(\Psi_0, \mathbf{x}_0)$ is *realized* (or *lost*), on the sample path.

Since Ψ is a finite space, the definition implies that Ψ_i will reach its limit, either $B(\mathbf{x}_i)$ or $\varnothing$, in a finite number of steps.

Corollary 5. *If $\Psi_{0,1} \subseteq \Psi_{0,2}$ and the event-coupling tree rooted on $(\Psi_{0,1}, \mathbf{x}_0)$ prevails on a sample path, then the event-coupling tree rooted on $(\Psi_{0,2}, \mathbf{x}_0)$ prevails on the same sample path.*

Corollary 6. *If on a sample path a tree with a root* $(\alpha, \mathbf{x}_0)$, $\alpha \in B(\mathbf{x}_0)$, *prevails, then all the other trees with* $(\alpha', \mathbf{x}_0)$, $\alpha' \in B(\mathbf{x}_0)$, $\alpha' \neq \alpha$ *must perish.*

The event-coupling tree provides a vital picture of the evolution of a perturbation. The perturbation realization, on the other hand, provides physical meanings to the concept. We will use both terminologies in the rest of this paper.

Definition 7. A GSMP is said to be *stable*, if on a sample path, any perturbation $(\Psi_0, \mathbf{x}_0)$ will be either realized or lost with probability one. That is, if $(\Psi_0, \mathbf{x}_0), (\Psi_1, \mathbf{x}_1), \ldots$, is a tree rooted on $(\Psi_0, \mathbf{x}_0)$, then

$$P\left\{\lim_{i\to\infty}[\Psi_i - B(\mathbf{x}_i)] = \varnothing\right\} + P\left\{\lim_{i\to\infty} \Psi_i = \varnothing\right\} = 1 \ .$$

In a stable GSMP, if any set of events obtain a small perturbation, then eventually every active event will obtain the same perturbation, or every active event will lost the perturbation. In the former case, all the event completion times are perturbed by the same amount of time. In the latter, all the event completion times are not changed at all. In both cases, the relative positions of the event completion times are not changed.

Let us develop some sufficient conditions for the stable GSMPs. If $B(\mathbf{x})$ contains only one event, then $\mathbf{x}$ is called a *one-event state*.

Lemma 8. *If an irreducible GSMP has a one-event state* $\mathbf{x}$, *then the GSMP is stable.*

This simple lemma covers a considerably wide range of systems, including queueing networks in which every customer has a positive probability of visiting a common server. Besides, if an irreducible GSMP has a one-event state, the GSMP possesses a regenerative structure and is hence ergodic [Cinlar 1975]. For a GSMP which does not have a one-event state, the following theorem helps to determine whether the GSMP is stable.

Theorem 9. *An irreducible GSMP is stable, if and only if there exists a sequence represented as (2), and an event* $\alpha_0 \in B(\mathbf{x}_0)$ *such that the corresponding event-coupling tree* $(\alpha_0, \mathbf{x}_0), (\Psi_1, \mathbf{x}_1), \ldots, (\Psi_n, \mathbf{x}_n)$ *prevails.*

4 The Effect of a Single Perturbation: Realization

Let $\kappa_{\alpha,i}$ and $\nu_{\alpha,i}$ be the ith starting time and the ith completion time, respectively, of event α. At time $t \in [\kappa_{\alpha,i}, \nu_{\alpha,i})$, the residual time (or the clock reading) of α is $d_\alpha = \nu_{\alpha,i} - t$ and the elapsed time is $r_\alpha = t - \kappa_{\alpha,i}$. Let $N = |\Psi|$ be finite, then each event $\alpha \in \Psi$ may be considered as an integer from 1 to N. Let $\mathbf{r}$ be an N-dimensional vector whose αth component is r_α. The system state can be denoted as $W(t) = (X(t), R(t))$, where $X(t)$ is the physical state of the GSMP, and $R(t)$ is the vector of the elapsed lifetimes at time t. The state space of this stochastic process is $\Phi \times \mathbb{R}_+^N$.

Suppose at time $t = 0$, the system is at state $\mathbf{w}_0 = (\mathbf{x}_0, \mathbf{r}_0)$ and the events in Ψ_0 obtain a perturbation. We denote it as $(\Psi_0, \mathbf{w}_0) = (\Psi_0, \mathbf{x}_0, \mathbf{r}_0)$. This means that at $t = 0$ the completion times of the events in Ψ_0 are delayed by the same small amount.

The perturbation will be propagated along a coupling tree on a sample path $W(t)$ with $W(0) = \mathbf{w}_0$. Let $V(t)$ be the perturbation set at time t. By definition, $V(t) = \Psi_i$ for $t \in [T_i, T_{i+1})$. Thus, each $\{V(t), X(t)\}$ produces an event-coupling tree defined in §3.

Definition 10. In a stable GSMP, $c(\Psi_0, \mathbf{w}_0) = P\{\omega : \lim_{t\to\infty}[V(t,\omega) - B(X(t,\omega))] = \varnothing \mid V(0) = \Psi_0, W(0) = \mathbf{w}_0\}$ is called the *realization probability* of the perturbation $(\Psi_0, \mathbf{w}_0)$.

By Definition 7, in a stable GSMP, the probability that the perturbation $(\Psi_0, \mathbf{w}_0)$ is lost is $1 - c(\Psi_0, \mathbf{w}_0)$. Since the realization probability is defined only for stable GSMPs, all the results related to realization probabilities hold only for stable GSMPs. We will not repeat this point in all our theorems.

Realization probabilities satisfy the following properties.

Theorem 11.

(i)

$$c(B(\mathbf{x}), \mathbf{x}, \mathbf{r}) = 1 \ . \tag{4}$$

(ii) If $\Psi_1 \cap \Psi_2 = \varnothing$ *and* $\Psi_1 \cup \Psi_2 = \Psi_3$, $\Psi_i \subseteq B(\mathbf{x})$, $i = 1, 2, 3$, *then*

$$c(\Psi_1, \mathbf{x}, \mathbf{r}) + c(\Psi_2, \mathbf{x}, \mathbf{r}) = c(\Psi_3, \mathbf{x}, \mathbf{r}) \ . \tag{5}$$

(iii)

$$\sum_{\beta \in B(\mathbf{x})} c(\beta, \mathbf{x}, \mathbf{r}) = 1 \ . \tag{6}$$

For further study, we require that all $F_\alpha(s)$ are absolutely continuous. Thus, the probability density functions $f_\alpha(s)$, defined by $F_\alpha(s) = \int_0^s f_\alpha(t)dt$, $\alpha \in \Psi$, exist and $|f_\alpha(s)| < \infty$, $s < \infty$. The hazard rate of the lifetime of event α is

$$g_\alpha(r_\alpha) = \frac{f_\alpha(r_\alpha)}{1 - F_\alpha(r_\alpha)} \ .$$

Assumption 3. $F(s; \mathbf{x}', \beta, \mathbf{x}, \alpha) = F_\beta(s)$ does not depend on α, $\mathbf{x}$, and $\mathbf{x}'$; all $F_\beta(s) < \infty$ for $s < \infty$ and are absolutely continuous.

For any N-dimensional vector $\mathbf{r}$ and any set $\Psi_0 \subseteq \Psi$, we define a vector $\mathbf{r}_{-\Psi_0}$ with components

$$\{\mathbf{r}_{-\Psi_0}\}_\alpha = \begin{cases} \{\mathbf{r}\}_\alpha & \text{if } \alpha \notin \Psi_0 \ , \\ 0 & \text{if } \alpha \in \Psi_0 \ . \end{cases}$$

We let $\Delta\mathbf{t}$ be a vector whose components are all Δt's.

Theorem 12. *Under Assumptions 2–3, we have*

$$\sum_{\beta \in B(\mathbf{x})} \frac{\partial c(\alpha, \mathbf{x}, \mathbf{r})}{\partial r_\beta} = c(\alpha, \mathbf{x}, \mathbf{r}) \sum_{\beta \in B(\mathbf{x})} g_\beta(r_\beta)$$

$$- \sum_{\mathbf{x}' \in \Phi} \sum_{\beta \in B(\mathbf{x}) - \alpha} g_\beta(r_\beta) p(\mathbf{x}'; \mathbf{x}, \beta) c(\alpha, \mathbf{x}', \mathbf{r}_{-N(\mathbf{x}'; \mathbf{x}, \beta)})$$

$$- \sum_{\mathbf{x}' \in \Phi} g_\alpha(r_\alpha) p(\mathbf{x}'; \mathbf{x}, \alpha) \left\{ \sum_{\gamma \in N(\mathbf{x}'; \mathbf{x}, \alpha)} c(\gamma, \mathbf{x}', \mathbf{r}_{-N(\mathbf{x}'; \mathbf{x}, \alpha)}) \right\} \quad (7)$$

for any $\alpha \in \Psi$, $\mathbf{x} \in \Phi$, *and* $\mathbf{r}$.

5 Parametric Sample-Path Sensitivity

Recall that the lifetime of event α, s, is determined by the distribution function $F_\alpha(s, \theta)$, where we explicitly indicate that θ is a parameter of interest. Let $\xi_\alpha = F_\alpha(s, \theta)$. Then ξ_α is a random varaible uniformly distributed in [0,1) and $s = F_\alpha^{-1}(\xi_\alpha, \theta)$. Suppose that θ changes to $\theta + \Delta\theta$. Then s will change to $s + \Delta s$ with

$$\Delta s = \frac{\partial F_\alpha^{-1}(\xi_\alpha, \theta)}{\partial \theta} \Delta\theta \ . \quad (8)$$

Let r, $0 \le r \le s$ be the elapsed service time and set $\rho = F_\alpha(r, \theta)$. As r increases from 0 to s, ρ increases from 0 to $\xi_\alpha = F_\alpha(s, \theta)$. Again, if θ changes to $\theta + \Delta\theta$, the elapsed time r changes to $r + \Delta r$ (corresponding to the same ρ) with

$$\Delta r = \left. \frac{\partial F_\alpha^{-1}(\rho, \theta)}{\partial \theta} \right|_{\rho = F(r, \theta)} \Delta\theta \ .$$

Taking the derivative with respect to r, we have

$$d(\Delta r) = G_\alpha(r, \theta) \Delta\theta dr \ ,$$

where

$$G_\alpha(r, \theta) = \frac{\partial}{\partial r} \left\{ \left. \frac{\partial F_\alpha^{-1}(\rho, \theta)}{\partial \theta} \right|_{\rho = F_\alpha(r, \theta)} \right\} \ .$$

From this, we have

$$\Delta s = \int_0^s d(\Delta r) = \left\{ \int_0^s G_\alpha(r, \theta)\, dr \right\} \Delta\theta \ .$$

$G_\alpha(r, \theta)$ is called the *perturbation generation function* [Cao 1991]. $d(\Delta r)$ is the perturbation generated in interval $[r, r + dr)$, $0 \le r < s$, due to $\Delta\theta$. In (8), we have

$$\frac{\partial F_\alpha^{-1}(\xi_\alpha, \theta)}{\partial \theta} = \lim_{\Delta\theta \to 0} \frac{F_\alpha^{-1}(\xi_\alpha, \theta + \Delta\theta) - F_\alpha^{-1}(\xi_\alpha, \theta)}{\Delta\theta} \ .$$

Note that the same random variable ξ_α is used for $\theta + \Delta\theta$ and θ.

In general, let $s_{\alpha,i}$ be the ith lifetime of α and $\xi_{\alpha,i}$ be the corresponding uniformly distributed random variables. Then $(\xi_{\alpha,i},\ all\ \alpha \in \Psi, i = 1, 2, \dots)$ represents all the randomness involved in the event lifetimes. Let $\mathbf{w}_0$ be the initial state and ς be a random vector representing all the other random phenomenon (i.e., the state transitions) in the system. Then the vector $\xi = (\xi_{\alpha,i}, \varsigma, \mathbf{w}_0,\ \text{all}\ \alpha \in \Psi, i = 1, 2, \dots)$ represents all the randomness involved in the GSMP. ξ is a random vector defined on Ω, and each $\xi(\omega)$, $\omega \in \Omega$, specifies a sample path of the GSMP.

With ξ, every performance function defined on a sample path can be expressed as a function of (ξ, θ). For example, the Lth transition time, T_L, on a sample path can be written as $T_L(\xi, \theta)$. We define

$$\frac{\partial T_L(\xi,\theta)}{\partial\theta} = \lim_{\Delta\theta\to 0} \frac{T_L^{(f)}(\xi,\theta+\Delta\theta) - T_L^{(f)}(\xi,\theta)}{\Delta\theta} \quad . \tag{9}$$

The same ξ is used for both θ and $\theta + \Delta\theta$ in the definition. For this reason, (9) is called a *sample derivative*, and $T_L(\xi, \theta)$, a sample function.

Let

$$\eta_L(\xi,\theta) = \frac{1}{T_L(\xi,\theta)}$$

be the *throughtput* defined on the sample path in the period of $[0, T_L]$. Then we have

$$\frac{1}{\eta_L(\xi,\theta)} \frac{\partial\eta_L(\xi,\theta)}{\partial\theta} = -\frac{1}{T_L(\xi,\theta)} \frac{\partial T_L(\xi,\theta)}{\partial\theta} \quad .$$

To continue the study, we need the following bounded assumption.

Assumption 4. The derivatives in the definition of $G_\alpha(r, \theta)$ exist, and there is a positive number $K > \infty$ such that in a neighborhood of θ, $\theta \in [\theta_c, \theta_b]$, $|G_\alpha(r, \theta)| < K$ holds.

Assumption 5. The GSMP is ergodic with a unique invariant measure $\pi(\mathbf{w}) = \pi(\mathbf{x}, \mathbf{r})$.

Theorem 13. *Under Assumptions 3–5, the normalized sample derivative the system throughput, $\frac{1}{\eta_L}\frac{\partial\eta_L(\xi,\theta)}{\partial\theta}$, converges with probability one to the negative expected value of the product of the perturbation generation function and the realization probability. That is,*

$$\begin{aligned}\lim_{L\to\infty} \frac{1}{\eta_L}\frac{\partial\eta_L(\xi,\theta)}{\partial\theta} &= -E_\pi\{G_\alpha[r_\alpha(t),\theta]c(\alpha,\mathbf{x},\mathbf{r})\} \\ &= -\sum_{\text{all }\mathbf{x}} \int_{\mathbb{R}_+^N} G_\alpha[r_\alpha(t),\theta]c(\alpha,\mathbf{x},\mathbf{r})\pi(\mathbf{x},\mathbf{r})\,d\mathbf{r}, \quad \text{w.p. } 1 \ .\end{aligned} \tag{10}$$

6 Sensitivity of Steady-State Performance

Assumption 6 (Glasserman's commuting condition). For any $\mathbf{x}_1$, $\mathbf{x}_2$, and $\mathbf{x}_3$, if $\{\alpha, \beta\} \in B(\mathbf{x}_1)$ and $p(\mathbf{x}_2; \mathbf{x}_1, \alpha)p(\mathbf{x}_3; \mathbf{x}_2, \beta) > 0$, then there is an $\mathbf{x}_4$ such that $p(\mathbf{x}_4; \mathbf{x}_1, \beta) = p(\mathbf{x}_3; \mathbf{x}_2, \beta)$ and $p(\mathbf{x}_3; \mathbf{x}_4, \alpha) = p(\mathbf{x}_2; \mathbf{x}_1, \alpha)$.

Theorem 14. *Under Assumptions 3 and 6, we have (Glasserman [1990])*

$$E\left\{\frac{\partial}{\partial\theta}\eta_L(\xi,\theta) \mid \mathbf{x}_0,\mathbf{r}_0\right\} = \frac{\partial}{\partial\theta}E\left\{\eta_L(\xi,\theta) \mid \mathbf{x}_0,\mathbf{r}_0\right\} \ .$$

Next, by definition, we have

$$\sum_{\text{all } \mathbf{x}_0}\left\{\int_{\mathbb{R}_r^N}\pi(\mathbf{x}_0,\mathbf{r}_0)\,d\mathbf{r}_0\right\} = 1 \ . \tag{11}$$

For any real number $r > 0$, let $\mathbb{R}_r^N = [0, r)^N$ be a hyper-cube in $\mathbb{R}_+^N$ and $\mathbb{R}_{r-}^N = \mathbb{R}_+^N - \mathbb{R}_r^N$ be the complementary set of $\mathbb{R}_r^N$ in $\mathbb{R}_+^N$. To prove the next theorem, we need the following technical assumption.

Assumption 7. There exist two real numbers $r^* > 0$ and $K^* > 0$ such that $|\frac{\partial\pi(\mathbf{x},\mathbf{r})}{\partial\theta}| < K^*\pi(\mathbf{x},\mathbf{r})$ for all $\mathbf{r} \in \mathbb{R}_{r^*-}^M$, and the integration in (11) uniformly converges in a neighborhood of θ.

Theorem 15. *Under Assumptions 3–7, the sample derivative of the system throughput with respect to a service distribution parameter, based on a single sample path in* $[0, T_L]$, *converges to the derivative of the steady-state throughput with probability one as the length of the sample path goes to infinity; moreover, the normalized derivative of the steady-state throughput,* $\frac{1}{\eta}\frac{\partial\eta}{\partial\theta}$, *equals the negative expected value of the product of the perturbation generation function and the realization probability. That is,*

$$\begin{aligned}
\frac{1}{\eta}\frac{\partial\eta}{\partial\theta} &= \lim_{L\to\infty}\frac{1}{\eta_L}\frac{\partial\eta_L(\xi,\theta)}{\partial\theta} = -E_\pi[G_\alpha(r_\alpha,\theta)c(\alpha,\mathbf{x},\mathbf{r})] \\
&= \sum_{\text{all } \mathbf{x}}\int_{\mathbb{R}_+^N} G_\alpha(r_\alpha,\theta)c(\alpha,\mathbf{x},\mathbf{r})\pi(\mathbf{x},\mathbf{r})d\mathbf{r} \ , \quad \text{w.p. } 1.
\end{aligned}$$

The realization probability can be determined by Eq. (4)–(7). Finally, the results in this paper have been extended to general performance measures.

References

1. X. R. Cao, *Realization Probabilities — the Dynamics of Queuing Systems,* Springer-Verlag, 1994.
2. E. Cinlar, *Introduction to Stochastic Processes,* Prentice Hall, Inc., Englewood Cliff, NJ, 1975.
3. P. Glasserman, *Gradient Estimation Via Perturbation Analysis,* Kluwer Academic, Norwell, MA, 1990.
4. Y. C. Ho and X. R. Cao, *Perturbation Analysis of Discrete Event Dynamic Systems,* Kluwer Academic Publishers, 1990.
5. W. Whitt, "Continuity of Generalized Semi-Markov Processes," *Mathematics of Operations Research,* Vol. 5, 494–501, 1980.

A Tutorial Overview of Optimization via Discrete-Event Simulation

Michael C. Fu

University of Maryland, College Park, MD 20742, USA

1 Introduction

In this tutorial, we survey several major methodologies for the optimization of discrete-event systems via simulation. We discuss both the discrete parameter case and the continuous parameter case. For the discrete parameter case, we focus on techniques for optimization from a finite set: multiple-comparison procedures and ranking-and-selection procedures. For the continuous parameter case, we discuss sequential response surface methodology procedures and stochastic approximation gradient-based procedures. Due to space limitations, no attempt is made here to be comprehensive; Fu (1994) expands upon the topics touched on here and provides a more extensive bibliography.

2 The Problem Setting

The general problem setting is the parametric optimization problem:

$$\min_{\theta \in \Theta} J(\theta), \tag{1}$$

where $J(\theta) = E[L(\theta, \omega)]$ is the performance measure of interest, $L(\theta, \omega)$ will be called the *sample* performance, ω represents the stochastic effects of the system, θ is a controllable vector of p parameters, and Θ is the constraint set on θ. Let us also define the optimum by $\theta_* = \arg\min_{\theta \in \Theta} J(\theta)$. In the experimental design literature, the performance measure is usually referred to as the *response* and the parameters as *factors*. Here, we will consider only the single response problem. For expository purposes, we will often discuss application of the various techniques to the following two discrete-event system simulation models.

Example 1. For a GI/G/1 queue, find the mean service time of the server that minimizes a cost function which trades off the expected mean time in system and the server speed:

$$\min_{\theta \in \Theta} c_0 E\left[\frac{1}{N}\sum_{i=1}^{N} T_i\right] + c_1/\theta, \tag{2}$$

where T_i is the time in system for the ith customer (T denoting the steady-state version), N is the number of customers served, c_0 and c_1 are given costs, θ is the mean service time, λ is the arrival rate, and $\Theta = [\delta, 1/\lambda - \delta]$, for some $\delta \approx 0$. For the $M/M/1$ queue in steady state, the optimum can be determined analytically as $\theta_* = (\lambda + \sqrt{c_0/c_1})^{-1}$.

Example 2. For an (s, S) inventory control system, find the values of s and $q = S - s$ to minimize a cost function on holding, ordering, and backlogging. The (s, S) ordering policy is to order up to S if the inventory level falls below s; otherwise, no order is placed. We will consider the zero order lead time, periodic review case:

$$\min_{\theta \in \Theta} \frac{1}{N} E\left[\sum_{i=1}^{N} C(X_i)\right], \tag{3}$$

where X_i is the inventory level (and position) at review epoch i, N is the number of periods in the horizon, $C(x) = hx^+ + px^- + I\{x < s\}[K + c(S - x)]$, h, p, K, and c are the holding, backlogging, order set-up, and order per-unit costs, respectively, $\theta = (s, q)$, and $\Theta = \mathbb{R} \times \mathbb{R}^+$.

The best possible convergence rate with "pure" stochastic optimization algorithms is generally of the order $n^{-1/2}$, where n represents (roughly) the computational effort. However, since it is an *asymptotic* convergence rate, the performance of the algorithm in the beginning may be better (or worse), i.e., we may have a "jump start" effect. In addition, order $n^{-1/2}$ is also the best convergence rate obtainable for simulation *estimation* (vs. optimization) of any (non-trivial) output random variable.

3 Optimization Over a Finite Set

Oftentimes, the number of choices in the parameter set is finite. This may be due to the nature of the problem itself, or it may be due to a reduction through other analyses, or it may be a simplifying step due to practical considerations. If the number of choices is not too large, then statistical procedures based on ranking and selection or multiple comparisons can be applied. Roughly speaking, ranking-and-selection procedures specify some criterion, such as choosing the best with some pre-specified confidence level, and then derive a statistical procedure, usually sequential, that meets the criterion. Multiple-comparisons procedures, on the other hand, specify the use of certain pairwise comparisons to make inferences in the form of confidence intervals; they are not inherently sequential procedures.

Again, we wish to solve the parametric optimization problem (1), where now the parameter set is finite: $\Theta = \{\lambda_1, \lambda_2, \ldots, \lambda_K\}$, i.e., we wish to find λ_i s.t. $\lambda_i = \theta_*$. Let us denote the estimate of performance from the jth sample path (replication) at λ_i by L_{ij}. Thus, our estimate of $J(\lambda_i)$ over n sample paths (replications) is simply the sample mean: $\widehat{J}_i = \bar{L}_i = \frac{1}{n}\sum_{j=1}^{n} L_{ij}$.

Procedures based on multiple comparisons (cf. Hochberg and Tamhane 1987) are of very basic importance in statistical inference, since applications inevitably require comparisons. Like most statistical techniques, the two major assumptions underlying the procedures are *independence* and *normality*. The former directly conflicts with some of the advantages of discrete-event simulation, e.g., the implementation of powerful variance reduction techniques such as common random numbers (CRN) and control variates. We will sketch the main ideas of three multiple-comparisons procedures:

- The "brute force" paired-t, Bonferroni, all-pairwise comparisons approach that works particularly well when CRN apply.
- An all-pairwise multiple comparisons (MCA) approach.
- A multiple comparisons with the best (MCB) approach more tailored to optimization purposes than the previous two approaches, and requiring far fewer comparisons.

The idea of the "brute force" approach is quite simple:

1. Calculate a difference estimate for each possible pair of replications.
2. Form the usual $(1-\alpha)100\%$ confidence intervals for each difference.
3. Apply the Bonferroni inequality to arrive at a lower bound on the *overall* confidence level.

After forming all $K(K-1)/2$ confidence intervals, one would simply look to see if there is a "clear winner," i.e., a λ_i such that the confidence interval for the difference with all other pairs is strictly negative. If not, one can crudely eliminate some candidates, estimate the number of additional replications needed to make conclusive inference, and repeat the process with the smaller set.

MCA works in principle similar to the above, except that instead of constructing separate confidence intervals and using Bonferroni to determine an overall confidence bound, a *simultaneous* set of confidence intervals at an overall $(1-\alpha)100\%$ level is formed:

$$(\widehat{J}_i - \widehat{J}_j) \pm r^{\alpha}_{K,K(n-1)} s/\sqrt{n}, \quad i < j, \quad s^2 = \frac{1}{K(n-1)} \sum_{i=1}^{K} \sum_{j=1}^{n} (L_{ij} - \bar{L}_i)^2,$$

where s^2 is the *pooled* sample variance and $r^{\alpha}_{K,K(n-1)}$ can be found in tables of Hochberg and Tamhane.

Thus, the difference between brute force and MCA is that MCA obtains an overall simultaneous confidence level, with the same confidence half-widths for each pairwise difference, whereas the brute force approach obtains a different confidence half-width for each pairwise difference and uses Bonferroni to obtain a bound on the overall confidence. Although the original version of MCA requires independence between the replications, a control-variate variation allowing the use of CRN was derived in Yang and Nelson (1991).

The intent of the multiple comparisons with the best (MCB) procedure is to reduce the number of comparisons, since we are interested in the optimization goal of picking only the best. The procedure is as follows (Hsu and Nelson 1988):

1. Form K confidence intervals for each choice with the *best of the rest*:

$$\left[-\left(\widehat{J}_i - \min_{j\neq i} \widehat{J}_j - d^{\alpha}_{K-1,K(n-1)} s\sqrt{\frac{2}{n}} \right)^{-}, \left(\widehat{J}_i - \min_{j\neq i} \widehat{J}_j + d^{\alpha}_{K-1,K(n-1)} s\sqrt{\frac{2}{n}} \right)^{+} \right]$$

where $d^{\alpha}_{K-1,K(n-1)}$ can be found in tables of Hochberg and Tamhane.

2. If only one of the confidence intervals falls on the negative side of 0, then the λ_i corresponding to that interval would be declared the optimum; otherwise, all of the λ_i with intervals having some part on the negative side of 0 could potentially be the optimum.

In practice, the pooled variance could be used to estimate the additional number of replications needed to make a more conclusive determination. Hsu and Nelson (1988) demonstrate the procedure for Example 2, the (s, S) inventory system.

We turn now to the ranking-and-selection procedures. In general, two approaches have been taken: indifference zone and subset selection. The method of Dudewicz and Dalal (1975) falls into the indifference-zone approach. It has two strong points that make it particularly suitable for optimization of discrete-event simulations: the variances do not have to be equal and they do not have to be known. However, independence must be maintained, thus precluding the use of CRN. The procedure guarantees that with user-specified probability at least P^* the selected λ_i will guarantee that $J(\lambda_i)$ is within δ of the optimal value $J(\theta_*)$, where δ represents the "indifference zone," i.e., $P\{J(\lambda_i) - J(\theta_*) < \delta\} \geq P^*$, including the possibility that $\lambda_i = \theta_*$.

The basic idea of the procedure is the following:

1. Take a first-stage set of replications for each of the K different parameter settings to calculate the first-stage sample means and sample variances.
2. Use the first-stage sample variances to determine the number of second-stage replications needed for each of the K different parameter settings.
3. Use the second-stage replications to get the second-stage sample means.
4. Take a weighted average of the first-stage and second-stage sample means.
5. Choose the λ_i with the smallest weighted average estimate of $J(\lambda_i)$.

A subset-selection procedure would be algorithmically similar, with the notable exception being the last step, where instead of selecting a single λ_i, a *subset* of all λ_i having the weighted average estimate of $J(\lambda_i)$ within some preselected distance is selected, up to some maximum number. Although very powerful tools, at present the major disadvantage of ranking-and-selection procedures for simulation optimization is the requirement of independence over competing designs, which precludes the use of most variance reduction techniques such as CRN.

4 Response Surface Methodology

Broadly speaking, response surface methodology (RSM) attempts to fit a polynomial (possibly after some initial transformation on the variables) of appropriate degree to the response of the system of interest. The application of RSM to simulation optimization falls into two main categories: metamodels and sequential procedures. In the context of optimization, it usually takes the form of the latter, whereby through successive experimental stages, one attempts to "home in" on the optimal region where a "final" (usually quadratic) polynomial is fitted and the optimum determined through the usual deterministic means. We will briefly outline the general approach in the context of discrete-event simulation.

Instead of exploring the entire feasible region, which may be impractical or computationally prohibitive, small subregions are explored in succession, where successive subregions are selected for their potential improvement. A point, e.g., the center of the subregion currently being explored, "represents" the current "best" θ value. The basic algorithm consists of two phases:

- **Phase I**
  In this phase, first-order experimental designs are used to get a least-squares fit. Then, a steepest descent direction is estimated from the model, and a new subregion chosen to explore via

$$\theta_{n+1} = \theta_n - a_n \nabla J_n, \tag{4}$$

  where θ_n is the representative point of the nth explored subregion, ∇J_n is the estimated (from the fitted linear response) gradient direction, and a_n is a the step size determined by a line search or some other means. This is repeated until the linear response surface becomes inadequate, which is indicated when the slope is "approximately" zero, by which the interaction effects become larger than the main effects.
- **Phase II**
  A quadratic response surface is fitted using more detailed second-order experimental designs; then the optimum determined analytically from this fit.

From the algorithm, one can see that Phase II is done just once, whereas Phase I is iterated a number of times. Thus, for each iteration of Phase I, one should strive to expend fewer replications, whereas in Phase II, the region should be explored quite thoroughly by using a large number of replications.

RSM sequential procedures provide a very general methodology for optimization via simulation. RSM's biggest advantage is its generality, but its biggest drawback if applied blindly is its computational requirements. Other techniques or analyses based on the nature of the system of interest which can be used to improve the efficiency of RSM are crucial. For example, efficient gradient estimation techniques may be used to *complement* the sequential aspects of RSM by reducing the number of simulation points needed.

5 Gradient-Based Algorithms

In this section, we consider gradient-based stochastic optimization algorithms, where the "best guess" of the optimal parameter is updated iteratively based on an estimate of the gradient of the performance measure with respect to the parameter. The basic underlying assumption of stochastic approximation is that the original problem given by (1) can be solved by finding the zero of the gradient, i.e., by solving $\nabla J(\theta) = 0$. Of course, in practice, this may lead only to local optimality. The general form of the stochastic algorithm takes the following form:

$$\theta_{n+1} = \Pi_\Theta \left(\theta_n - a_n \widehat{\nabla} J_n\right), \quad \widehat{\nabla} J_n = [\widehat{\nabla_1} J_n \dots \widehat{\nabla_p} J_n]^T, \tag{5}$$

where θ_n is the parameter value at the beginning of iteration n, $\widehat{\nabla} J_n$ is an estimate of $\nabla J(\theta_n)$ from iteration n, a_n is a (positive) sequence of step sizes, and Π_Θ is a projection onto Θ. When finite differences are used to estimate $\nabla J(\theta_n)$, (5) is called a Kiefer-Wolfowitz algorithm; when a direct (possibly unbiased) estimator is used for $\nabla J(\theta_n)$, (5) is called a Robbins-Monro-like algorithm (cf. Kushner and Clark 1978). The usual requirements needed for the convergence of (5) to the optimum are that (i) the step size go to zero at a rate not too fast to lead to convergence to the wrong value and not too slow to avoid convergence to a value at all, and (ii) that the bias of the gradient estimate go to zero. One set of common assumptions on the step sizes is $\sum_n a_n = \infty$, $\sum_n a_n^2 < \infty$, which for example the harmonic series $a_n = a/n$ (for some constant a) satisfies. In terms of practical implementation for discrete-event simulation, one must select various parameters in the algorithm such as the initial step size a and the observation horizon, as well as a projection rule and a stopping rule for the algorithm.

We discuss three gradient estimation techniques: finite differences (FD), perturbation analysis (PA), and the likelihood ratio (LR) method. Finite difference techniques alter the input and analyze the resulting input, whereas PA and LR involve an "add-on" to the simulator itself, which involves *additional* accumulations and calculations. However, the underlying simulator (by which we mean the event-generation scheme) is *not* altered, and as a result both LR and PA can also be implemented for on-line gradient estimation and optimization.

The most obvious way to estimate the gradient is to run multiple simulations to estimate some secant as an approximation to the tangent. We call this the finite difference (FD) estimate. The symmetric difference version is given by

$$\widehat{\nabla}_i J_n = \frac{\widehat{J}(\theta_n + c_n e_i) - \widehat{J}(\theta_n - c_n e_i)}{2c_n}, \tag{6}$$

where e_i denotes the ith unit vector. Note that this estimate requires $2p$ simulations. The forward difference would simply replace $\widehat{J}(\theta_n - c_n e_i)$ with $\widehat{J}(\theta_n)$ and hence would require only $p+1$ simulations; however, the convergence rate when used in a stochastic approximation algorithm is worse.

A potentially more efficient version of finite differences is the simultaneous perturbation (SP) finite difference estimate proposed by Spall (1992). Let $\{\Delta_1, \ldots, \Delta_p\}$ be a set of i.i.d. perturbations satisfying the conditions given in Spall (1992), and define the vector $\Delta = [\Delta_1 \ldots \Delta_p]$. Then, the SP estimator is given by

$$\widehat{\nabla}_i J_n = \frac{\widehat{J}(\theta_n + c_n \Delta) - \widehat{J}(\theta_n - c_n \Delta)}{2c_n \Delta_i}. \tag{7}$$

Note that whereas in the finite-difference estimators, there is a pair of numerators for each parameter, thus requiring $2p$ simulations, here the *same* pair is used in the numerator for all parameters, and the denominator changes; thus, only two simulations are required.

Finite difference-based (Kiefer-Wolfowitz) algorithms require $c_n \to 0$ (at an appropriate rate) for convergence, and generally the best asymptotic convergence rate achievable is $O(n^{-1/3})$, versus $O(n^{-1/2})$ when an unbiased estimate is used. Although this procedure has the dual disadvantages of being computationally more intensive and having a slower convergence rate, it is straightforward to implement and the most generally applicable.

Perturbation analysis is one technique for obtaining unbiased gradient estimates efficiently. The books by Ho and Cao (1991) and Glasserman (1991) cover perturbation analysis up to 1990 (with an extensive bibliography), concentrating on the two most developed forms: infinitesimal perturbation analysis (IPA) and smoothed perturbation analysis (SPA). Recently developed "extensions" include discontinuous perturbation analysis, augmented perturbation analysis, and rare perturbation analysis (Fu 1994 contains references for these).

We illustrate the estimators for our two examples. For the $GI/G/1$ queue, an unbiased gradient estimator for mean steady-state system time T is given by the IPA estimator

$$\left(\frac{dT}{d\theta}\right)_{\text{IPA}} = \frac{1}{N}\sum_{m=1}^{M}\sum_{i=1}^{n_m}\sum_{j=1}^{i}\frac{dX_{(j,m)}}{d\theta}, \tag{8}$$

where n_m is the number of customers served in the mth busy period, M is the number of busy periods, $N = \sum_{m=1}^{M} n_m$ is the total number of customers served, and $X_{(j,m)}$ is the service time of the jth customer in the mth busy period. An unbiased estimator for the second derivative is given by the following SPA estimator:

$$\left(\frac{d^2T}{d\theta^2}\right)_{\text{SPA}} = \frac{1}{N}\sum_{m=1}^{M}\sum_{i=1}^{n_m}\sum_{j=1}^{i}\frac{d^2X_{(j,m)}}{d\theta^2} + \frac{1}{M}\sum_{m=1}^{M}\frac{f_1(z_m)}{1-F_1(z_m)}\left(\sum_{i=1}^{n_m}\frac{dX_{(i,m)}}{d\theta}\right)^2, \tag{9}$$

where z_m is the age of the interarrival time at the end of the mth busy period, and f and F are the interarrival time p.d.f. and c.d.f., respectively.

For the (s, S) inventory system example, SPA can be used to derive the following consistent estimator (Fu 1994; see also Fu and Healy 1992):

$$\left(\frac{\partial J(s,q)}{\partial s}\right)_{\text{PA}} = \frac{1}{N}\left[\sum_{i:X_i>0} h - \sum_{i:X_i<0} p\right], \tag{10}$$

$$\left(\frac{\partial J(s,q)}{\partial q}\right)_{\text{PA}} = \frac{1}{N}\left[\sum_{i:X_i>0} h - \sum_{i:X_i<0} p\right] + \frac{1}{N+1}\sum_{j:X_j<s}\frac{g(Z_j)}{1-G(Z_j)} \cdot \left[cE[D] + hE[s-D]^+ + pE[D-s]^+ - \frac{\sum_{i=1}^{N} C(X_i)}{N}\right], \tag{11}$$

where $J(s, q)$ is the long-run average cost per period, $Z_j = X_{j-1} - s$, and D is a single period demand with p.d.f. and c.d.f. $g(\cdot)$ and $G(\cdot)$, respectively.

Another technique for obtaining unbiased gradient estimates is the likelihood ratio (LR) method, also known as the score function (SF) method (cf. Rubinstein and Shapiro 1993). The basic idea of the method is to differentiate the underlying probability measure of the system, but it can more generally be viewed as a special case of importance sampling. The LR estimator is as easily implementable as IPA, and it often works for systems where IPA fails. However, the resulting estimator may have variance problems for some systems; when IPA works, it usually has much lower variance. Variance comparisons between SPA and LR, on the other hand, seem to be quite problem dependent. Also, because the LR method requires the differentiation of a probability measure, the technique is not usually applicable to structural parameters such as s and S in the (s, S) inventory system.

We present a brief overview of the LR technique, and derive estimators for Example 1. We assume that the dependence on θ enters only through a random vector X with joint cumulative distribution function $F(\theta, \cdot)$ and density $f(\theta, \cdot)$ depending on a parameter (or vector of parameters) θ:

$$E[L(X)] = \int L(x)dF(\theta, x). \tag{12}$$

Differentiating (12), we have (assuming certain interchanges are valid)

$$\begin{aligned} \frac{\partial E[L]}{\partial \theta} &= \frac{\partial}{\partial \theta} \int L(x) f(\theta, x)dx = \int L(x) \frac{\partial f(\theta, x)}{\partial \theta} dx \\ &= \int L(x) \frac{\partial \ln f(\theta, x)}{\partial \theta} f(\theta, x)dx = E\left[L(X) \frac{\partial \ln f(\theta, X)}{\partial \theta}\right]. \end{aligned} \tag{13}$$

Thus, in a single simulation, one can estimate the derivative of the performance measure along with the performance measure itself. Higher derivatives can be handled in a similar manner. However, the "naive" estimator for (13) leads to unbounded variance for steady-state performance measures. For example, for the GI/G/1 queue, where the interarrival times and the service times comprise the random vector, the natural estimator would be given by

$$\left(\frac{dT}{d\theta}\right)_{\text{LR}} = \frac{1}{N} \sum_{i=1}^{N} T_i \frac{\partial \ln f}{\partial \theta} \ , \tag{14}$$

where X_i is the ith service time. For example, for exponential service times, $\frac{\partial \ln f}{\partial \theta} = \sum_{i=1}^{N} \left(\frac{X_i}{\theta^2} - \frac{1}{\theta}\right)$. The problem with these estimators is that if they are used to estimate *steady state* quantities by increasing the horizon length N, then it is obvious that the variance of the estimator will *increase* linearly. On the other hand, a regenerative estimator does not suffer from this problem:

$$\left(\frac{dT}{d\theta}\right)_{\text{LR}} = \frac{1}{N} \sum_{m=1}^{M} \left\{ \sum_{i=1}^{n_m} T_{(i,m)} \frac{\partial \ln f}{\partial \theta} \right\} - \frac{1}{N} \sum_{m=1}^{M} \left\{ n_m \frac{\partial \ln f}{\partial \theta} \right\} \frac{1}{N} \sum_{j=1}^{N} T_j, \tag{15}$$

where the subscript (i, m) denotes the ith customer in the mth busy period. The summations in these estimators are bounded by the length of the busy periods.

We now briefly describe the application of SA to simulation optimization. One of the earliest applications was the work by Azadivar and Talmage (1980), who implemented a version utilizing FD estimates with a number of "practical" heuristics to improve its performance. They empirically compared the performance of their algorithm with an RSM sequential procedure for a number of simple polynomial functions with additive noise and a single discrete-event system. According to their simulation results, for a given computational budget, their algorithm dominated the RSM procedure for every example.

The first application of PA to optimization was contained in the paper by Ho and Cao (1983). An IPA gradient estimate for throughput of a queueing network was incorporated into a simple stochastic approximation algorithm on an objective function with Lagrangian multipliers. The approach was to use long simulation runs to get a good estimate of the gradient; thus, the number of iterations was relatively small. In contrast, the work of Suri and Zazanis (1988) introduced the idea of "single-run" optimization using IPA. Instead of completing a long simulation run before updating the parameter, and repeating the procedure for just a few iterations, the parameter was updated after a very short observation horizon, and the simulation *continued*; between iterations the simulation mechanism was not reinitialized and restarted. The single simulation run was terminated when it was determined that the gradient was "close enough" to zero according to a given stopping criterion. Thus, a *single* run of approximately the same length it would take to estimate the performance itself also yielded an estimate of the *optimal* value of the parameter, providing significant computational savings over the previous implementation. The procedure was applied to the steady-state version of Example 1 for various interarrival time and service time distributions, and empirically, the algorithm worked quite well.

Examples incorporating LR estimators in SA algorithms are presented in Rubinstein and Shapiro (1993). In L'Ecuyer et al. (1994), a very comprehensive set of numerical experiments on the M/M/1 queue example are reported. Various algorithms utilizing IPA, LR, and FD estimates with CRN are considered and compared, with the IPA-based algorithms clearly superior. The numerous simulation results also show the obvious effect of step size selection.

6 Other Methods

Here, we list a few other approaches, some proposed recently:

- Utilizing non-gradient-based algorithms such as pattern search methods and random search methods, see, e.g., Jacobson and Schruben (1989).
- Using *each sample* to derive an entire performance curve "sample" and optimize the resulting curve using deterministic methods. Rubinstein (1991) and Healy and Schruben (1991) provide numerical results for many examples, including the M/M/1 queue of Example 1.
- Replacing cardinal optimization with ordinal optimization (Ho et al. 1992), i.e., instead of trying to find the best in a possibly uncountable infinite state space, just try to find better "satisficing" solutions; key components are the use of the standard clock methodology of Vakili (1991) and the exploitation of massively parallel simulation.
- Combining techniques, e.g., the proposed method by Ho et al. (1992) called the Gradient Surface Method (GSM), which combines the approaches of RSM and SA by utilizing gradient estimates for two different purposes in the GSM procedure: in the first phase providing the "point estimates" in a *gradient* surface least-squares fit, and in the second phase, providing the search direction in the gradient-based stochastic approximation algorithm.

- Employing massively parallel simulation for exploring a response surface in parallel, e.g., in the ordinal optimization approach mentioned earlier, or to provide multiple starting points for SA algorithms.

References

F. Azadivar and J.J. Talavage, Optimization of stochastic simulation-models, *Mathematics and Computers in Simulation* **22** (1980) 231–241.

E. J. Dudewicz and S. R. Dalal, Allocation of measurements in ranking and selection with unequal variances, *Sankhya* **B37** (1975) 28–78.

M. C. Fu, Sample path derivatives for (s, S) inventory systems, to appear in *Operations Research* **42**, No.2 (1994a).

M. C. Fu, Optimization via simulation: a review, to appear in *Annals of Operations Research* (1994b).

M. C. Fu and K. Healy, Simulation optimization of (s,S) inventory systems, *Proceedings of the Winter Simulation Conference* (1992) 506–514.

P. Glasserman, *Gradient Estimation Via Perturbation Analysis*, Kluwer (1991).

K. Healy and L.W. Schruben, Retrospective simulation response optimization, *Proceedings of the 1991 Winter Simulation Conference* (1991) 901–906.

Y. C. Ho and X. R. Cao, *Discrete Event Dynamic Systems and Perturbation Analysis,* Kluwer Academic (1991).

Y. C. Ho, L. Shi, L. Dai and W. B. Gong, Optimizing discrete event systems via the gradient surface method, *Discrete-Event Dynamic Systems* **2** (1992) 99–120.

Y. C. Ho, R. Sreenevas and P. Vakili, Ordinal optimization of DEDS, *Discrete-Event Dynamic Systems: Theory and Applications* **2** (1992) 61–88.

Y. Hochberg and A. C. Tamhane, *Multiple Comparison Procedures*, Wiley (1987).

J. C. Hsu and B.L. Nelson, Optimization over a finite number of system designs with one-stage sampling and multiple comparisons with the best, *Proceedings of the Winter Simulation Conference* (1988) 451–457.

S. H. Jacobson and L.W. Schruben, A review of techniques for simulation optimization, *Operations Research Letters* **8** (1989) 1–9.

H. J. Kushner and D. C. Clark, *Stochastic Approximation Methods for Constrained and Unconstrained Systems,* Springer-Verlag, New York (1978).

P. L'Ecuyer, N. Giroux, N., and P. W. Glynn, Stochastic optimization by simulation: numerical experiments with a simple queue in steady-state, to appear in *Management Science* (1994).

R. Y. Rubinstein, How to optimize discrete-event systems from a single sample path by the score function method, *Annals of Oper. Res. 27* (1991) 175–212.

R. Y. Rubinstein and A. Shapiro, *Discrete Event Systems: Sensitivity Analysis and Stochastic Optimization by the Score Function Method*, Wiley (1993).

J. C. Spall, Multivariate stochastic approximation using a simultaneous perturbation gradient approximation, *IEEE Trans. on Aut. Con.* **37** (1992) 332–341.

R. Suri and M. Zazanis, Perturbation analysis gives strongly consistent sensitivity estimates for the M/G/1 queue, *Management Science* **34** (1988) 39–64.

P. Vakili, Using a standard clock technique for efficient simulation, *Operations Research Letters* (1991) 445–452.

W.N. Yang and B.L. Nelson, Using common random numbers and control variates in multiple-comparison procedures, *Operations Research,* **39** (1991) 583–591.

Parallel Simulation of Discrete Event Systems

Richard M. Fujimoto[1]

College of Computing, Georgia Institute of Technology,
Atlanta GA 30332-0280, USA

1 Introduction

It is an unfortunate fact of life that all too often, a simulation study fails to achieve its full potential because the execution of the simulator requires too much time. In some cases, this occurs because a detailed simulation of a large, complex system is required. For example, a simulation of a computer system may require hours, or even days to simulate only a few seconds of the system's operation. In other cases, the simulation model may not be overly complex, but long runs are required to capture enough occurrences of certain rare events to obtain statistically significant results. A simulation of a multiplexer (a circuit that combines several incoming streams of traffic into a single output stream) in a telecommunication network may have to simulate 10^9 successful data transmissions before a buffer overflow occurs [Rob92]. Whatever the reason, prohibitively long execution times limit the effectiveness of simulation techniques, potentially leading to systems that are overly expensive, unreliable, or do not perform up to expectations. In interactive applications, e.g., simulators used for training purposes, long execution times may necessitate the use of overly simplified models that compromise the realism and validity of the exercise.

One approach to reducing model execution time is to utilize many processors that can simultaneously work on different portions of the simulation computation. Commercial multiprocessors containing hundreds or thousands of powerful microprocessors offer the potential of reducing month-long simulation runs to hours, day-long runs to minutes. Here, we focus attention on multiprocessor computers, however, the same principals are also applicable to distributed computing platforms, e.g., networks of high performance workstations.

For the purposes of this article, it is convenient to view a simulation as a computation that must compute the values of certain *state variables* across simulated time. The state variables capture the state of the system, e.g., the number of customers waiting for service in a simulation of a queue. Changes in the state of the system occur at discrete points in simulated time, e.g., when a new customer enters the system. This so-called space-time view of the simulation [BLC91] is depicted in Figure 1 using a two-dimensional graph, where the vertical axis represents the state variables, and the horizontal axis represents simulated time. The goal of the simulation program is to "fill in" the graph by computing the values of each of the state variables across simulated time. A parallel simulator attempts to use multiple processors that simultaneously "fill in" different portions of the space-time graph.

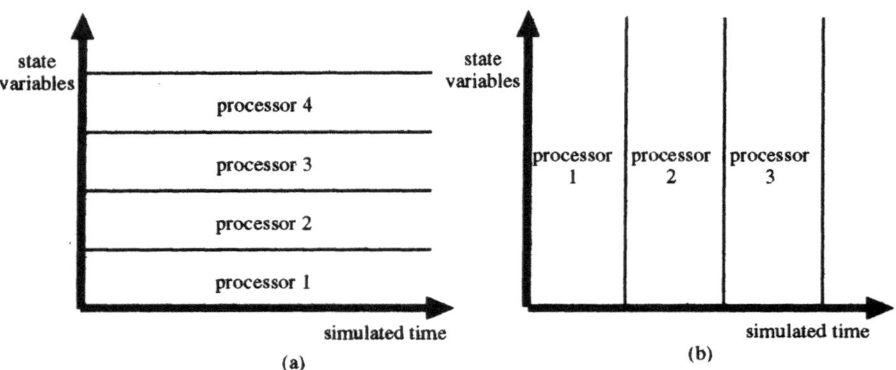

Fig. 1. Space-time diagram depicting simulated time on the horizontal axis, and state variables on the vertical axis. (a) approach exploiting space-parallelism. (b) approach exploiting time-parallelism.

In Figure 1(a) the graph is divided into horizontal "strips," with a *logical process* (or *LP*) responsible for the computation within each strip. This *space-parallel* approach can be viewed as partitioning the system being modeled into a collection of subsystems, and assigning a logical process to simulate each one. Each LP is a sequential, event-driven simulator. Different LPs may execute concurrently on different processors, and communicate by exchanging timestamped *events* or *messages* (we use the terms event and message synonymously here). For instance, in a simulation of a queueing network, each LP might model a single queue. A "customer departure" from one queue signals the arrival of the customer at another. This can be modeled by having the first LP send a message to the second with timestamp equal to the customer's arrival time at the second queue.

An alternative approach called the *time-parallel* method partitions the space-time graph into vertical strips as shown in Figure 1(b) and assigns a separate processor to each strip. The simulated time axis is divided into intervals $[T_1, T_2]$, $[T_2, T_3]$, ..., $[T_i, T_{i+1}]$, ... with processor i assigned the task of computing the portion of the space-time graph within the interval $[T_i, T_{i+1}]$.

The remainder of this article describes underlying principles and techniques behind the space- and time- parallel simulation approaches. This article is only concerned with speeding up the execution of a *single* execution of the simulation program. Another useful form of parallel execution is to perform multiple, independent executions of the simulator on different processors, e.g., to reduce variance of output statistics or to examine different parameter settings. Such approaches offer clear benefits when they can be applied, but are beyond the scope of the present discussion.

2 The Space-Parallel Approach

A central issue that must be addressed in space-parallel simulations is the so-called synchronization problem. Each logical process must process incoming events in timestamp order, or events in the simulated future might affect those in the past. The constraint that each LP must process events in timestamp order is referred to as the *local causality constraint*. Ensuring that each LP obeys this constraint, coupled with the assumptions that (1) LPs only interact by exchanging timestamped event messages (specifically, an LP may not directly access the state variables of another LP) and (2) each event has a unique timestamp (this can be guaranteed by appending "hidden" fields to the timestamp) is sufficient to ensure that the parallel execution will yield exactly the same results as a sequential execution of the simulator where events across all of the LPs are processed in timestamp order. It should be noted that the parallel synchronization mechanism does not guarantee that the simulation model is correct, but only that it produces the same results as the sequential simulator. Traditional validation procedures are required to ensure correctness.

Thus, the goal of the synchronization algorithm is to ensure that each LP processes events in timestamp order. Two classes of synchronization algorithms have been proposed to accomplish this. *Conservative* algorithms guarantee that events within an LP are never processed out of timestamp order, i.e., synchronization errors never occur. *Optimistic* algorithms allow synchronization errors to occur, but provide a mechanism to recover.

2.1 Conservative Algorithms

Historically, the first parallel simulation algorithms were based on conservative approaches. The principal task of any conservative protocol is to determine when it is "safe" to process an event, i.e., when have all events containing a smaller timestamp been received and processed by the LP. Chandy and Misra [CM79], and Bryant [Bry77] developed some of the first algorithms. They assume the topology indicating which LPs send messages to which others is fixed and known prior to execution, and messages arrive on each incoming link in timestamp order. This guarantees that the timestamp of the last message received on a link is a lower bound on the timestamp of any subsequent message that will later arrive on that link.

Messages arriving on each incoming link are stored in first-in-first-out order, which is also timestamp order because of the above restriction. Each link has a clock associated with it that is equal to the timestamp of the message at the front of that link's queue if the queue contains a message, or the timestamp of the last received message if the queue is empty. The process repeatedly selects the link with the smallest clock and, if there is a message in that link's queue, processes it. If the selected queue is empty, the process blocks. This protocol guarantees that each process will only process events in non-decreasing timestamp order, thereby ensuring no violations of the local causality constraint occur.

Although this approach avoids out-of-order event executions, it is prone to deadlock. A cycle of empty links with small link clock values (e.g., smaller than any unprocessed message in the simulator) can occur, resulting in each process waiting for the next process in the cycle. If there are relatively few unprocessed event messages compared to the number of links in the network, or if the unprocessed events become clustered in one portion of the network, deadlock may occur very frequently.

Null messages may be used to avoid deadlock. A null message with timestamp T_{null} sent from LP_A to LP_B is a promise by LP_A that it will not later send a message to LP_B carrying a timestamp smaller than T_{null}. Null messages do not correspond to any activity in the simulated system. Processes send null messages on each outgoing link after processing each event. A null message provides the receiver with additional information that may be used to determine that other events are safe to process.

How does a process determine the timestamps of the null messages it sends? The clock value of each *incoming* link provides a lower bound on the timestamp of the next event that will be removed from that link's buffer. When coupled with knowledge of the simulation performed by the process, this bound can be used to determine a lower bound on the timestamp of the next *outgoing* message on each output link. For example, if a queue server has a minimum service time of T, then the timestamp of any future departure event must be at least T units of simulated time larger than any arrival event that will be received in the future. Whenever a process finishes processing an event, it sends a null message on each of its output links indicating the best lower bound it can compute; the receiver of the null message can then compute new bounds on its outgoing links, send this information on to its neighbors, and so on. It can be shown that this algorithm avoids deadlock under some mild constraints [CM79].

This algorithm may generate many null messages, however. Chandy and Misra [CM81] later developed another approach that allows the computation to deadlock, but then detects and breaks it. The deadlock can be broken by observing that the message(s) containing the smallest timestamp is (are) always safe to process. Alternatively, one may use a distributed computation to compute lower bound information (not unlike the distributed computation using null messages described above) to enlarge the set of safe messages.

Numerous variations on these approaches have been developed, as well as others. Some protocols use a synchronous execution where the computation cycles between (i) determining which events are "safe" to process, and (ii) processing those events. To determine which events are safe, the *distance between LPs* is sometimes used. This "distance" is the minimum amount of simulation time that must elapse for an event in one LP to directly or indirectly affect another LP, and can be used by an LP to determine bounds on the timestamp of future events it might receive from other LPs. Space does not permit full elaboration of all of the techniques that have been proposed, however, these techniques are reviewed in [Fuj90, FN92].

2.2 Optimistic Algorithms

In contrast to conservative approaches that avoid violations of the local causality constraint, optimistic methods allow violations to occur, but are able to detect and recover from them. Optimistic approaches offer two important advantages over conservative techniques. First, they can exploit greater degrees of parallelism. If two events *might* affect each other, but the computations are such that they actually don't, optimistic mechanisms can process the events concurrently, while conservative methods must sequentialize execution. Second, conservative mechanism generally rely very heavily on application specific information (e.g., distance between objects) in order to determine which events are safe to process. While optimistic mechanisms can execute more efficiently if they exploit such information, they are less reliant on such information for correct execution. This allows the synchronization mechanism to be more transparent to the application program than conservative approaches, thereby simplifying software development. On the other hand, optimistic methods may require more overhead computations than conservative approaches, leading to certain performance degradations.

The Time Warp mechanism [Jef85] is the most well known optimistic method. When an LP receives an event with timestamp smaller than one or more events it has already processed, it rolls back and reprocesses those events in timestamp order. Rolling back an event involves restoring the state of the LP to that which existed prior to processing the event (checkpoints are taken for this purpose), and "unsending" messages sent by the rolled back events. An elegant mechanism called anti-messages is provided to "unsend" messages.

An anti-message is a duplicate copy of a previously sent message. Whenever an anti-message and its matching (positive) message are both stored in the same queue, the two are deleted (annihilated). To "unsend" a message, a process need only send the corresponding anti-message. If the matching positive message has already been processed, the receiver process is rolled back, possibly producing additional anti-messages. Using this recursive procedure all effects of the erroneous message will eventually be erased.

Two problems remain to be solved before the above approach can be viewed as a viable synchronization mechanism. First, certain computations, e.g., I/O operations, cannot be rolled back. Second, the computation will continually consume more and more memory resources because a history (e.g., checkpoints) must be retained, even if no rollbacks occur; some mechanism is required to reclaim the memory used for this history information. Both problems are solved by *global virtual time (GVT)*. GVT is a lower bound on the timestamp of any future rollback. GVT is computed by observing that rollbacks are caused by messages arriving "in the past." Therefore, the smallest timestamp among unprocessed and partially processed messages gives a value for GVT. Once GVT has been computed, I/O operations occurring at simulated times older than GVT can be committed, and storage older than GVT (except one state vector for each LP) can be reclaimed.

A number of other optimistic algorithms have been proposed [Fuj90, FN92]. Most attempt to limit the amount of optimistic computation. For instance, one approach limits optimistic execution to a sliding window of simulated time [SBW88]. Other proposals include introducing additional rollbacks to "pull back" LPs that have advanced too far ahead of others (e.g., see [MH92], among others), or delaying message sends until it is guaranteed that the send will not be later rolled back, thereby eliminating the need for anti-messages (e.g., see [Rey88]).

2.3 Performance

A large body of work has been concerned with evaluating the performance of parallel simulation mechanisms using analytic models and experimentation. For instance, Lin and Lazowska derive a condition for Time Warp to yield performance no worse than the critical path through the computation [LL90]. Lipton and Mizell show that Time Warp can outperform the Chandy/Misra protocols by an amount proportional to the number of processors, but Chandy/Misra can only outperform Time Warp by a constant factor [LM90]. Nicol compares a conservative protocol with Time Warp in [Nic91], highlighting the importance of overhead computations in comparing the relative merits of different synchronization protocols. Felderman and Kleinrock also compare the performance of conservative and optimistic assumptions but under a different set of distributional assumptions [FK91a]. Other analyses of synchronization algorithms using simulation time windows are reported in [Nic93, Dic92, Ste91].

Several analyses of Time Warp using Markov chains have recently appeared. An analysis of N processor Time Warp is described in [GAF91], and performance under limited memory conditions is described in [ACD+93]. Rollback behavior is studied in [LL91a, LWS91]. Earlier analyses of two processor Time Warp are described in [LMS83, MM84, FK91b].

Numerous results from experimental studies have reported that one to two orders of magnitude reduction in execution time can be obtained across a variety of simulation applications, e.g., digital logic circuits, combat models, telecommunication networks, queueing networks, Petri networks, and transportation systems [Fuj93]. These results demonstrate that parallel simulation techniques can achieve significant performance improvement for many applications.

3 The Time-Parallel Approach

Recently, time-parallel simulation methods have received considerable attention for attacking specific simulation problems with well-defined objectives, e.g., measuring the loss rate of a finite capacity queue. Recall that time-parallel algorithms divide the simulated time axis into intervals, and assign each interval to a different processor (see Figure 1(b)). This allows for massively parallel execution because simulations often span long periods of simulated time.

A central question that must be addressed by time-parallel simulators is ensuring the states computed at the "boundaries" of the time intervals "match." Specifically, it is clear that the state computed at the end of the interval $[T_{i-1}, T_i]$ must match the state at the beginning of interval $[T_i, T_{i+1}]$. Thus, this approach relies on being able to perform the simulation corresponding to the ith interval without first completing the simulations of the preceding $(i-1, i-2, \ldots, 1)$ intervals.

Because of the "state-matching" problem, time-parallel simulation is really more of a methodology for developing massively parallel algorithms for specific simulation problems than a general approach for executing arbitrary discrete-event simulation models on parallel computers. Time-parallel algorithms are currently not as robust as space-parallel approaches because they rely on specific properties of the system being modeled, e.g., specification of the system's behavior as recurrence equations and/or a relatively simple state descriptor. This approach is currently limited to a handful of important applications, e.g., queueing networks, Petri nets, cache memories, and statistical multiplexers in telecommunication networks. Space-parallel simulations offer greater flexibility and wider applicability, but concurrency is limited to the number of logical processes. In some cases, both time and space-parallelism can be used [GGN93].

One approach to solving the state matching problem is to have each processor "guess" the initial state of its simulation, and then simulate the system based on this guessed initial state [LL91b]. In general, the initial state will not match the final state of the previous interval. After the interval simulators have completed, a "fix-up" computation is performed to account for the fact that the wrong initial state was used. This might be performed, for instance, by simply repeating the simulation, using the final state computed in the previous interval as the new initial state. This "fix-up" process is repeated until the initial state of each interval matches the final state of the previous interval. In the worst case, N such iterations are required when there are N simulators. However, if the final state of each interval simulator is seldom dependent on the initial state, far fewer iterations will be needed.

In [HS90], the above approach is proposed to simulate cache memories using a least-recently-used replacement policy. This approach is effective for this application because the final state of the cache is not heavily dependent on the cache's initial state. A variation on this approach devised in the context of simulating statistical multiplexers for asynchronous transfer mode (ATM) switches precomputes certain points in time where one can guarantee that a buffer overflow (full queue) or underflow (empty queue) will occur [NFC93]. Because the state of the system, namely, the number of occupied buffers in the queue, is known at these points, independent simulations can be begun at these points in simulated time, thereby eliminating the need for a fix-up computation. Ammar and Deng also use a related approach for simulating Petri networks [AD92].

Another approach to time-parallel simulation is described in [GLM91]. Here, a queueing network simulation is expressed as a set of recurrence equations that are then solved using well-known parallel prefix algorithms. The parallel prefix computation enables the state of the system at various points in simulated time to be computed concurrently. Another approach also based on recurrence equations is described in [BC93] for simulating timed Petri nets. New massively parallel algorithms for less tractable recurrence equations were developed for trace-driven cache simulations [NGLR92], and for circuit-switched communication networks [EGLW93, GGN93].

4 Future Directions

Space-parallel simulation algorithms provide "general purpose" methods for executing general discrete event simulation problems on parallel computers. They offer great robustness and flexibility, but are limited in the degree of concurrency that is obtained by the number of logical processes making up the simulation. Current "hot" topics in space-parallel simulation techniques include memory management, load balancing, hardware support, and performance evaluation of parallel discrete event simulation techniques. One of the principal challenges in the parallel simulation field is in simplifying the development of parallel simulation models. Currently, this requires a great deal of expertise in parallel computation, as well as simulation modeling. The impact of this technology on the general simulation community is likely to be limited until this obstacle is overcome.

Time-parallel simulation algorithms exploit specific properties of the simulation problem at hand to gain the ability of achieving truly massively parallel execution. These methods often offer much greater parallelism than space-parallel methods, at the cost of reduced robustness and flexibility, e.g., to accommodate changes in the simulation model. An important challenge that remains for these methods is determining to what extent these techniques can be generalized.

Both time and space-parallel simulation techniques have enjoyed a significant amount of success in speeding up the execution of specific simulation applications, however, their use has been largely limited to a few research laboratories and universities. Major obstacles must be overcome to make the technology more readily usable by the general simulation community before these techniques will gain widespread acceptance.

References

[ACD+93] I. F. Akyildiz, L. Chen, S. Das, R. M. Fujimoto, and R. F. Serfozo. The effect of memory capacity on time warp performance. *Journal of Parallel and Distributed Computing*, 18(4):411–422, August 1993.

[AD92] H. Ammar and S. Deng. Time warp simulation using time scale decomposition. *ACM Transactions on Modeling and Computer Simulation*, 2(2):158–177, April 1992.

[BC93] F. Baccelli and M. Canales. Parallel simulation of stochastic petri nets using receurrence equations. *ACM Transactions on Modeling and Computer Simulation*, 3(1):20–41, January 1993.

[BLC91] R. Bagrodia, W.-T. Liao, and K.M. Chandy. A unifying framework for distributed simulation. *ACM Transactions on Modeling and Computer Simulation*, 1(4), October 1991.

[Bry77] R. E. Bryant. Simulation of packet communication architecture computer systems. MIT-LCS-TR-188, Massachusetts Institute of Technology, Cambridge, Massachusetts, 1977.

[CM79] K. M. Chandy and J. Misra. Distributed simulation: A case study in design and verification of distributed programs. *IEEE Transactions on Software Engineering*, SE-5(5):440–452, September 1979.

[CM81] K. M. Chandy and J. Misra. Asynchronous distributed simulation via a sequence of parallel computations. *Communications of the ACM*, 24(4):198–205, April 1981.

[Dic92] P.M. Dickens. *Performance Analysis of Parallel Simulations*. PhD thesis, University of Virginia, December 1992.

[EGLW93] S. G. Eick, A. G. Greenberg, B. D. Lubachevsky, and A. Weiss. Synchronous relaxation for parallel simulations with applications to circuit switched networks. *ACM Transactions on Modeling and Computer Simulation*, 3(4):287–314, October 1993.

[FK91a] R. Felderman and L. Kleinrock. Bounds and approximations for self-initiating distributed simulation without lookhead. *ACM Transactions on Modeling and Computer Simulation*, 1(4):386–406, October 1991.

[FK91b] R. Felderman and L. Kleinrock. Two processor time warp analysis: Some results on a unifying approach. In *Advances in Parallel and Distributed Simulation*, volume 23, pages 3–10. SCS Simulation Series, Jan. 1991.

[FN92] R. M. Fujimoto and D. M. Nicol. State of the art in parallel simulation. In *1992 Winter Simulation Conference Proceedings*, pages 122–127, December 1992.

[Fuj90] R. M. Fujimoto. Parallel discrete event simulation. *Communications of the ACM*, 33(10):30–53, October 1990.

[Fuj93] R. M. Fujimoto. Parallel and distributed discrete event simulation: Algorithms and applications. In *1993 Winter Simulation Conference Proceedings*, pages 106–114, December 1993.

[GAF91] A. Gupta, I. F. Akyildiz, and R. M. Fujimoto. Performance analysis of Time Warp with multiple homogenous processors. *IEEE Transactions on Software Engineering*, 17(10):1013–1027, October 1991.

[GGN93] B. Gaujal, A. G. Greenberg, and D. M. Nicol. A sweep algorithm for massively parallel simulation of circuit-switched networks. *Journal of Parallel and Distributed Computing*, 18(4):484–500, August 1993.

[GLM91] A.G. Greenberg, B.D. Lubachevsky, and I. Mitrani. Algorithms for unboundedly parallel simulations. *ACM Trans. on Comp. Systems*, 9(3):201–221, 1991.

[HS90] P. Heidelberger and H. Stone. Parallel trace-driven cache simulation by time partitioning. In *Proceedings of the 1990 Winter Simulation Conference*, pages 734–737, New Orleans, December 1990.

[Jef85] D. R. Jefferson. Virtual time. *ACM Transactions on Programming Languages and Systems*, 7(3):404–425, July 1985.

[LL90] Y.-B. Lin and E. Lazowska. Optimality considerations of "time warp" parallel simulation. In *Distributed Simulation*, volume 22, pages 29–34. SCS Simulation Series, Jan. 1990.

[LL91a] Y.-B. Lin and E.D. Lazowska. A study of Time Warp rollback mechanisms. *ACM Transactions on Modeling and Computer Simulation*, 1(1):51–72, January 1991.

[LL91b] Y.-B. Lin and E.D. Lazowska. A time-division algorithm for parallel simualtion. *ACM Transactions on Modeling and Computer Simulation*, 1(1):73–83, January 1991.

[LM90] R. Lipton and D. Mizell. Time Warp vs. Chandy-Misra: A worst-case comparison. In *Distributed Simulation*, volume 22, pages 137–143. SCS Simulation Series, Jan. 1990.

[LMS83] S. Lavenberg, R. Muntz, and B. Samadi. Performance analysis of a rollback method for distributed simulation. In *Performance '83*, pages 117–132, Elsevier Science Pub., (North Holland), 1983.

[LWS91] B. Lubachevsky, A. Weiss, and A. Shwartz. An analysis of rollback-based simulation. *ACM Transactions on Modeling and Computer Simulation*, 1(2):154–192, April 1991.

[MH92] V.K. Madisetti and D. Hardaker. Synchronization mechanisms for distributed event-driven computation. *ACM Transactions on Modeling and Computer Simulation*, 2(1), January 1992.

[MM84] D. Mitra and I. Mitrani. Analysis and optimum performance of two message passing parallel processors synchronized by rollback. In *Performance '84*, pages 35–50, Elsevier Science Pub., (North Holland), 1984.

[NFC93] I. Nikolaidis, R. M. Fujimoto, and A. Cooper. Parallel simulation of high-speed network multiplexers. In *Proceedings of the IEEE Conference on Decision and Control*, December 1993.

[NGLR92] D. Nicol, A. Greenberg, B. Lubachevsky, and S. Roy. Massively parallel algorithms for trace-driven cache simulation. In 6th *Workshop on Parallel and Distributed Simulation*, volume 24, pages 3–11. SCS Simulation Series, Jan. 1992.

[Nic91] D.M. Nicol. Performance bounds on parallel self-initiating discrete-event simulations. *ACM Transactions on Modeling and Computer Simulation*, 1(1):24–50, January 1991.

[Nic93] D.M. Nicol. The cost of conservative synchronization in parallel discrete-event simulations. *Journal of the ACM*, April 1993. To appear. Available as technical report 90-20 from ICASE, Mail Stop 132C, NASA Langley Research Center, Hampton, VA 23665.

[Rey88] P. F. Reynolds, Jr. A spectrum of options for parallel simulation. In *1988 Winter Simulation Conference Proceedings*, pages 325–332, December 1988.

[Rob92] J. W. Roberts, editor. *Performance Evaluation and Design of Multiservice Networks*. Commission of the European Communities, Luxembourg, 1992.

[SBW88] L. M. Sokol, D. P. Briscoe, and A. P. Wieland. MTW: a strategy for scheduling discrete simulation events for concurrent execution. In *Proceedings of the SCS Multiconference on Distributed Simulation*, volume 19, pages 34–42. SCS Simulation Series, July 1988.

[Ste91] J. Steinman. Speedes:synchronous parallel environment for emulation and discrete event simulation. In *Advances in Parallel and Distributed Simulation*, volume 23, pages 95–103. SCS Simulation Series, Jan. 1991.

On the Existence and Estimation of Performance Measure Derivatives for Stochastic Recursions *

Peter W. Glynn[1] *and Pierre L'Ecuyer*[2]

[1] Department of Operations Research, Stanford University
Stanford, CA 94305-4022
[2] Département d'IRO, Université de Montréal
C.P. 6128, Succ. A, Montréal, H3C 3J7, CANADA

1 Introduction

In this paper, we study a class of Markov chains $X = (X_n : n \geq 0)$ that arise as solutions to stochastic recursions of the form

$$X_{n+1} = f(X_n, Z_{n+1}), \tag{1}$$

where $Z = (Z_n : n \geq 1)$ is an i.i.d. sequence, called the *innovations* sequence. Our motivation stems from the fact that discrete-event simulation are typically formulated and implemented as stochastic recursions of the form (1). Our aim is to study the behavior of X under perturbations of the distribution that governs Z. Suppose that θ is a real-valued parameter under which the Z_n's have common distribution K_θ (say). We will provide conditions under which

(i) the expectation of a r.v. (random variable) defined over a randomized time-horizon is differentiable in θ; and
(ii) the stationary probability measure of X is differentiable in θ (in a sense to be made more precise in §4).

Our conditions are based on stochastic Lyapunov functions and can be expressed in terms of K_θ and the one-step transition function of X. In addition to giving conditions for model "smoothness", our approach also provides derivative estimators which can be computed via simulation. In particular, we develop a likelihood ratio (LR) derivative estimator for the derivative of the steady-state expectation of a functional defined on a Harris recurrent Markov chain.

In §2, we consider a finite horizon model where the horizon is a randomized stopping time and provide sufficient conditions under which the expected performance measure is differentiable. We also construct LR derivative estimators where the LR can be based on either the filtration associated with the "innovation process" or that associated with the

* The research of the first author was supported by the U. S. Army Research Office under Contract No. DAAL03-91-G-0101 and by the NSF under Contract No. DDM-9101580. The second author's research was supported by NSERC-Canada grant No. OGP0110050 and FCAR-Québec grant No. 93ER1654.

Markov chain itself. In §3, we construct LRs for Harris recurrent Markov chains, while in §4, we study the derivative of such LRs and construct LR derivative estimators for the steady-state performance measures. The main results are stated here without proof, due to space limitations. A complete version of the paper, including all the proofs and some examples, will appear elsewhere.

2 LRs for Finite-Horizon Stochastic Recursions

We consider the recursion (1), where for each n, X_n and Z_n take values in separable metric spaces S_1 and S_2, respectively, and $f : S_1 \times S_2 \to S_1$ is jointly measurable. For each $\theta \in \Lambda$, we assume that there exists a family of measurable functions $\{r_n(\theta) : S_1^{n+1} \to [0,1],\ n \geq 0\}$ and a r.v. T such that $P_\theta[T = n \mid X] = r_n(\theta, X_0, \ldots, X_n)$. The random stopping time T is the time horizon, and is determined as $T = \inf\{j : \sum_{n=1}^{j} r_n(\theta, X_0, \ldots, X_n) \geq U\}$, where U is a uniform r.v. over $(0, 1)$. For each $\theta \in \Lambda = (a, b)$, K_θ and μ_θ are probability measures on S_2 and S_1, that act as the respective distribution of Z_n and X_0 under θ. Let $P_{\theta,x}$ denote the probability law of the process, conditional on $X_0 = x$, and $E_{\theta,x}$ be the corresponding expectation. The sequence X is a time-homogeneous Markov chain having the one-step transition function $P(\theta, x, dy) = P_{\theta,x}[X_1 \in dy]$ for $x, y \in S_1$. Define $P_\theta(d\omega) = \int_{S_1} \mu_\theta(dx) P_{\theta,x}(d\omega)$.

Assumption 1. *There exists $\epsilon > 0$ such that for each $\theta \in \Lambda_\epsilon = (\theta_0 - \epsilon, \theta_0 + \epsilon)$,*

(i) K_θ is absolutely continuous with respect to K_{θ_0};
(ii) μ_θ is absolutely continuous with respect to μ_{θ_0};
(iii) $r_n(\theta, x_0, \ldots, x_n) > 0$ implies $r_n(\theta_0, x_0, \ldots, x_n) > 0$ for all $n \geq 0$ and all $(x_0, \ldots, x_n) \in S_1^{n+1}$.

Let $k(\theta, z)$ and $u(\theta, x)$ be densities such that $K_\theta(dz) = k(\theta, z) K_{\theta_0}(dz)$ and $\mu_\theta(dx) = u(\theta, x)\mu_{\theta_0}(dx)$. Let $\rho(\theta)$ denote $r_T(\theta, X_0, \ldots, X_T)/r_T(\theta_0, X_0, \ldots, X_T)$ and let $\mathcal{G}_n = \sigma(U, X_0, Z_1, \ldots, Z_n)$ for each n.

Theorem 1. *Let Y be a non-negative $\mathcal{G}_T$-measurable r.v. and let Assumption 1 be in force. Then, there exists $\epsilon > 0$ such that*

$$E_\theta[Y\, I(T < \infty)] = E_{\theta_0}[Y \tilde{L}(\theta) I(T < \infty)] \tag{2}$$

for $\theta \in \Lambda_\epsilon$, where I denotes the indicator function and

$$\tilde{L}(\theta) = u(\theta, X_0)\rho(\theta) \prod_{i=1}^{T} k(\theta, Z_i). \tag{3}$$

Let $p(\theta, x, y)$ be the density of $P(\theta, x, \cdot)$ with respect to $P(\theta_0, x, \cdot)$. Set $\mathcal{F}_n = \sigma(U, X_0, \ldots, X_n)$. One obtains an alternative LR representation by conditioning: if Y is a non-negative $\mathcal{F}_T$-measurable r.v. then

$$E_\theta[Y I(T < \infty)] = E_{\theta_0}[Y L(\theta) I(T < \infty)] \tag{4}$$

where

$$L(\theta) = u(\theta, X_0)\rho(\theta)\prod_{i=1}^{T} p(\theta, X_{i-1}, X_i). \tag{5}$$

Applying (4) to $L(\theta)$ yields $L(\theta) = E_{\theta_0}[\tilde{L}(\theta) \mid \mathcal{F}_T]$ on the set $\{T < \infty\}$. One can use (2) or (4) to estimate functionals of the measure P_θ, while simulating X under θ_0. Since $L(\theta)$ is a conditional expectation of $\tilde{L}(\theta)$, the estimation based on (4) is statistically more efficient (or at least as efficient) in terms of variance; see Fox and Glynn (1986) for a similar argument. Generally speaking, the more information Z contains relative to X, the greater the gain in statistical efficiency should be. However, (2) could be much easier to implement, because the densities $p(\theta, \cdot, \cdot)$ are often rather complicated functions in practice. Therefore, there is typically a trade-off between variance reduction on the one side and ease of implementation and computational costs on the other.

The stopping time T is *non-randomized* if each $r_n(\theta, X_1, \ldots, X_n)$ is either 0 or 1. In that case, $\rho(\theta) = 1$ P_{θ_0}-a.s. and the LRs simplify accordingly.

To derive a LR representation for the derivative of P_θ, we shall require that K_θ be suitably smooth in θ. To simplify the notation, we denote P_{θ_0} and E_{θ_0} by P and E, respectively. A "prime" denotes the derivative with respect to θ.

Assumption 2. *(i) There exists $\epsilon > 0$ such that for all $\theta \in \Lambda_\epsilon$, $P_\theta[T < \infty] = 1$;*

(ii) There exists $\epsilon > 0$ such that for each $x \in S_1$ and $z \in S_2$, $u(\cdot, x_0)$ and $k(\cdot, z)$ are continuously differentiable over Λ_ϵ. Also, $\rho'(\theta_0) = \lim_{h\to 0}(\rho(\theta_0 + h) - \rho(\theta_0))/h$ exists P-a.s.;

(iii) For each $p > 0$, there exists $\epsilon = \epsilon(p)$ such that

$$E\left[\sup_{\theta\in\Lambda_\epsilon} |u'(\theta, X_0)|^p + \sup_{\theta\in\Lambda_\epsilon} |k'(\theta, Z_1)|^p\right] < \infty \text{ and } \sup_{\theta\in\Lambda_\epsilon} E\left[\left|\frac{\rho(\theta) - 1}{\theta - \theta_0}\right|^p\right] < \infty.$$

To differentiate (2) or (4), and pass the derivative inside the expectation operator, we need to control the behavior of $\tilde{L}(\theta)$ and $L(\theta)$. For that, we will make the following assumption, to control the r.v. T.

Assumption 3. *There exists $z > 1$ such that $E[z^T] < \infty$.*

Theorem 2. *Let Y be a $\mathcal{F}_T$-measurable r.v. and assume that there exists $\delta > 0$ such that $E[|Y|^{1+\delta}] < \infty$. If Assumptions 1–3 hold, then $E_\theta[Y]$ is differentiable at θ_0 and the derivative is given by $E[Y\tilde{L}'(\theta_0)] = E[YL'(\theta_0)]$, where*

$$\tilde{L}'(\theta) = \tilde{L}(\theta)\left[\frac{u'(\theta, X_0)}{u(\theta, X_0)} + \frac{\rho'(\theta)}{\rho(\theta)} + \sum_{i=1}^{T}\frac{k'(\theta, Z_i)}{k(\theta, Z_i)}\right];$$

$$L'(\theta) = L(\theta)\left[\frac{u'(\theta, X_0)}{u(\theta, X_0)} + \frac{\rho'(\theta)}{\rho(\theta)} + \sum_{i=1}^{T}\frac{p'(\theta, X_{i-1}, X_i)}{p(\theta, X_{i-1}, X_i)}\right].$$

3 LRs for Harris-Recurrent Stochastic Recursions

We now turn our attention to the construction of LRs and derivative estimators for infinite-horizon (steady-state) systems.

Assumption 4. *There exists $\epsilon > 0$, an integer $m \geq 0$, a (measurable) subset $A \subseteq S_1$, a probability φ on S_1, and a measurable function $\lambda \geq 0$ for which*

i) For each $\theta \in \Lambda_\epsilon$, there exists a measurable function $g(\theta, \cdot) \geq 0$ and a constant $\epsilon(\theta) > 0$ such that: (a) $E_{\theta,x}[g(\theta, X_1)] \leq g(\theta, x) - \epsilon(\theta)$ for $x \notin A$; and (b) $\sup_{x \in A} E_{\theta,x}[g(\theta, X_1)] < \infty$.
ii) $P_{\theta,x}[X_m \in dy] \geq \lambda(x)\, \varphi(dy)$ for $x, y \in S_1,\ \theta \in \Lambda_\epsilon$;
iii) $\inf\{\lambda(x) : x \in A\} \stackrel{\Delta}{=} \lambda_* > 0$.

In most applications, A is a compact set, and (ii–iii) follow via a continuity argument. Let $T(A) = \inf\{n \geq 1 : X_n \in A\}$. Then, (i) ensures that $\sup_{x \in A} E_{\theta,x}[T(A)] < \infty$ and $P[X_n \in A \text{ i.o.}] = 1$ (see Nummelin 1984). The function $g(\theta, \cdot)$ is called a "test function" or "stochastic Lyapunov function".

Remark. Allowing $m = 0$ in Assumption 4 is non-standard, but permits us to simplify the estimators nicely for systems which have a regenerative state. To be more precise, suppose that there is a specific state $x_* \in S_1$ that is hit a.s. in finite time from any other state. Define $A = \{x_*\}$ and $\varphi(dy) = I[x_* \in dy]$. Then, Assumptions 4(ii-iii) hold with $m = 0$, $\lambda(x) = I[x = x_*]$, and $\lambda_* = 1$. In fact, this degenerate case is the only case where Assumption 4 can hold for $m = 0$.

We can now use the so-called "splitting method" (Athreya and Ney 1978, and Nummelin 1978), to show that for each $\theta \in \Lambda_\epsilon$, X possesses a unique σ-finite stationary measure $\pi(\theta)$ having a regenerative representation. Assumption 4 (ii) ensures the existence of a family of transition functions $Q(\theta)$ such that

$$P_{\theta,x}[X_m \in dy] = \lambda(x)\, \varphi(dy) + (1 - \lambda(x))\, Q(\theta, x, dy) \tag{6}$$

for $\theta \in \Lambda_\epsilon$, $x, y \in S_1$. Roughly speaking, if X currently occupies state $x \in S$, then m time units later, with probability $\lambda(x)$, the chain will be distributed according to φ. There is then a random time τ for which X_τ is distributed independently of $X_{\tau-m}$, and the stationary distribution $\pi(\theta)$ can be represented in terms of a ratio formula expressed over the time interval $[0, \tau]$. If $m = 0$ and $\lambda_* = 1$, then φ is concentrated on a single state x_* and τ is the first hitting time of x_*.

For $m \geq 1$, we will shrink $\lambda(x)$ to make sure that the measures $Q_\beta(\theta, x, \cdot)$ and $P^m(\theta, x, \cdot) \equiv P_{\theta,x}[X_m \in \cdot\,]$ are equivalent. Fix $\beta \in (0, 1)$ and let

$$\begin{aligned} \varphi_\beta(x, dy) &= \beta\lambda(x)\, \varphi(dy), \\ Q_\beta(\theta, x, dy) &= (1 - \beta)\lambda(x)\, \varphi(dy) + (1 - \lambda(x))\, Q(\theta, x, dy) \\ &= P_{\theta,x}[X_m \in dy] - \beta\lambda(x)\varphi(dy). \end{aligned} \tag{7}$$

Then, there exist densities $w_i(\theta, x, y)$, $i = 0, 1$, such that

$$\begin{aligned} Q_\beta(\theta, x, dy) &= w_0(\theta, x, y)\, P^m(\theta, x, dy); \\ \varphi_\beta(x, dy) &= w_1(\theta, x, y) P^m(\theta, x, dy), \end{aligned} \tag{8}$$

and these densities satisfy $w_0(\theta, x, y) + w_1(\theta, x, y) = 1$.

Let $S_0 = -m$ and $S_j = \inf\{n \geq S_{j-1} + m : X_n \in A\}$ for $j \geq 1$ be a sequence of hitting times of A, spaced at least m steps apart. Define a r.v. τ such that

$$P_\theta[\tau = S_n + m \mid Z] = w_1(\theta, X_{S_n}, X_{S_n+m}) \prod_{j=1}^{n-1} w_0(\theta, X_{S_j}, X_{S_j+m})$$

for each $n \geq 1$, and let γ be such that $S_\gamma + m = \tau$. On can identify τ with the finite horizon T of the previous section. Its distribution is concentrated on $\{S_n + m,\ n \geq 1\}$, and one has

$$r_\tau(\theta, X_0, \ldots, X_\tau) = w_1(\theta, X_{S_\gamma}, X_{S_\gamma+m}) \prod_{j=1}^{\gamma-1} w_0(\theta, X_{S_j}, X_{S_j+m})$$

and $\mu_\theta \equiv \varphi$. This τ is a time at which the distribution of X is independent of its position at time $\tau - m$. It is the desired "regeneration time" for X under P_θ, and under Assumption 4, there exists $\epsilon > 0$ such that for $\theta \in \Lambda_\epsilon$,

$$\pi_\theta(dx) = \frac{E_\theta\left[\sum_{j=0}^{\tau-1} I(X_j \in dx)\right]}{E_\theta[\tau]}. \tag{9}$$

To construct a LR representation of π_θ in terms of π_{θ_c}, let Assumption 1 be in force and let $p_n(\theta, x, y)$ be the density of $P^n(\theta, x, \cdot)$ with respect to $P^n(\theta_0, x, \cdot)$ for $n \geq 1$ and $\theta \in \Lambda_\epsilon$. Then, it is possible to show that

$$\rho(\theta) = \left(\prod_{j=1}^{\gamma-1} \frac{q(\theta, X_{S_j}, X_{S_j+m})}{p_m(\theta, X_{S_j}, X_{S_j+m})}\right) \frac{1}{p_m(\theta, X_{\tau-m}, X_\tau)},$$

$$\tilde{L}(\theta) = \left(\prod_{i=1}^{\tau} k(\theta, Z_i)\right) \rho(\theta), \quad \text{and} \quad L(\theta) = \left(\prod_{i=1}^{\tau} p(\theta, X_{i-1}, X_i)\right) \rho(\theta).$$

Combining this with (9) and Theorem 1, we obtain:

Corollary 3. *Under Assumptions 4 and 1(i), there exists $\epsilon > 0$ such that for all $\theta \in \Lambda_\epsilon$,*

$$\pi_\theta(dx) = \frac{E_{\theta_0}\left[\sum_{j=0}^{\tau-1} I(X_j \in dx)\, \tilde{L}(\theta)\right]}{E_{\theta_0}\left[\tau \tilde{L}(\theta)\right]} = \frac{E_{\theta_0}\left[\sum_{j=0}^{\tau-1} I(X_j \in dx)\, L(\theta)\right]}{E_{\theta_0}[\tau L(\theta)]}.$$

In the degenerate case where $m = 0$, one can take $\beta = 1$, so $\rho(\theta) \equiv 1$.

4 A LR Representation for the Derivative of the Stationary Distribution

A glance at $\tilde{L}(\theta)$ and $L(\theta)$ suggests that a LR derivative formula for the stationary distribution should require differentiability of $p_m(\cdot)$ and $q(\cdot)$ with respect to θ. By recognizing that the derivative of $p_m(\cdot)$ can be defined in terms of the conditional expectation of the derivative of $k(\cdot)$, the required differentiability can be proved by imposing appropriate regularity conditions on $k(\cdot)$. These conditions force $p_m(\cdot, x, y)$ and $q(\cdot, x, y)$ to be well-behaved.

Assumption 5. *(i) There exists $\epsilon > 0$ such that for each $z \in S_2$, $k(\cdot, z)$ is continuously differentiable over Λ_ϵ;*
(ii) For each $p > 0$, there exists $\epsilon = \epsilon(p)$ such that $E\left[\sup_{\theta\in\Lambda_\epsilon} |k'(\theta, Z_1)|^p\right] < \infty$;
(iii) For each $r \in \mathbb{R}$, $\lim_{\epsilon\to 0} E\left[\sup_{\theta\in\Lambda_\epsilon} |k(\theta, Z_1)|^r\right] = 1$.

To show that Assumption 2 (iii) and 3 hold, and apply the results of § 2, we need to control the behavior of τ and γ. Under suitable hypotheses (below), it can be shown that τ and γ have a geometrically dominated tail. We shall assume the existence of the following Lyapunov function:

Assumption 6. *There exists a function $g \geq 0$ defined on S_1, $\epsilon > 0$, and $r < 1$ such that $E_{\theta_0,x}[g(X_1)] \leq r\, g(x) - \epsilon$ for $x \notin A$ and $\sup_{x\in A} E_{\theta_0,x}[g(X_1)] < \infty$.*

The set A is then a Kendall set for the Markov chain having transition function $P(\theta_0)$, i.e., if $T(A) = \inf\{n \geq 1 : X_n \in A\}$, then there exists $z > 1$ such that $\sup_{x\in A} E_{\theta_0,x}\left[z^{T(A)}\right] < \infty$. (See Nummelin (1984), pp. 90–91 and Chap. 16 of Meyn and Tweedie (1993).)

Proposition 4. *Under Assumptions 1(i), 4, and 5, $P[\gamma > k] \leq (1 - \beta\lambda_*)^k$ for $k \geq 0$. If, in addition, Assumption 6 is in force, then there exists $z > 1$ such that $E[z^\tau] < \infty$.*

It is then possible to show that

$$\begin{aligned}\frac{\rho'(\theta)}{\rho(\theta)} &= \sum_{i=1}^{\gamma-1} \frac{q'(\theta, X_{S_i}, X_{S_i+m})}{q(\theta, X_{S_i}, X_{S_i+m})} - \sum_{i=1}^{\gamma} \frac{p'_m(\theta, X_{S_i}, X_{S_i+m})}{p_m(\theta, X_{S_i}, X_{S_i+m})} \\ &= \sum_{i=1}^{\gamma-1} \frac{p'_m(\theta, X_{S_i}, X_{S_i+m})}{p_m(\theta, X_{S_i}, X_{S_i+m})} \frac{w_1(\theta, X_{S_i}, X_{S_i+m})}{w_0(\theta, X_{S_i}, X_{S_i+m})} - \frac{p'_m(\theta, X_{S_\gamma}, X_{S_\gamma+m})}{p_m(\theta, X_{S_\gamma}, X_{S_\gamma+m})}\end{aligned}$$

when these expressions exist. Under Assumptions 1(i) and 4–6, one can prove that $\rho'(\theta_0)$ exists a.s.. Verifying that condition does not seem direct, because there is no a priori reason to expect almost sure differentiability, or even continuity over $\theta \in \Lambda_\epsilon$, of the r.v.'s $p_m(\theta, X_{S_i}, X_{S_{i+m}})$ and $q(\theta, X_{S_i}, X_{S_{i+m}})$. The proof also turns out to be quite involved.

Then, Assumptions 1–3 hold for our Harris-recurrent setup and Theorem 5 below shows that the stationary distributions π_θ are in fact differentiable in a very strong sense, namely in an extended version of the total variation norm. (For $f \equiv 1$, the notion of convergence in Theorem 5 is precisely that of total variation.) Recent work of Vásquez-Abad and Kushner (1992) also addresses this question. The hypotheses given there are quite different and, in particular, are not given in terms of conditions that can be checked directly from the transition function of the chain. For a measure μ on S_1 and a S_1-measurable function f, we adopt the notation $\mu f = \int_{S_1} f(y)\,\mu(dy)$. Let $f \geq 0$ be S_1-measurable. This and our previous assumptions will imply that $\pi_{\theta_0} f^{1+\delta}$ is finite (see Tweedie 1983).

Assumption 7. *There exists a function $g \geq 0$ defined on S_1 and $\epsilon > 0$ such that $E_{\theta_0,x}[g(X_1)] \leq g(x) - \epsilon\, f(x)^{1+\delta}$ for $x \notin A$, and $\sup_{x\in A} E_{\theta_0,x}[g(X_1)] < \infty$.*

Theorem 5. *Let Assumptions 1(i) and 4–7 hold. Then, there exists a finite signed measure π' such that*

$$\lim_{h\to 0} \sup_{|g|\leq f} \left| \frac{\pi_{\theta_0+h}\, g \; - \; \pi_{\theta_0}\, g}{h} \; - \; \pi' g \right| = 0$$

and

$$\pi'(\cdot) = \frac{E\left[\sum_{n=0}^{\tau-1} \left[I(X_n \in \cdot) \; - \; \pi_{\theta_0}(\cdot)\right] \tilde{L}'(\theta_0)\right]}{E[\tau]} .$$

Noting that $L'(\theta_0) = E[\tilde{L}'(\theta_0) \mid \mathcal{F}_\tau]$ and that $Y(g)$ and τ are both $\mathcal{F}_\tau$-measurable, we obtain the following corollary to Theorem 5.

Corollary 6. *Under the assumption of Theorem 5, $\pi_\theta g$ is differentiable at $\theta = \theta_0$ for any g satisfying $|g| \leq f$, and*

$$\frac{d}{d\theta}\, \pi_\theta g \,\bigg|_{\theta=\theta_0} = \frac{E\left[(Y(g) - (\pi_{\theta_0} g)\tau)\tilde{L}'(\theta_0)\right]}{E[\tau]} = \frac{E\left[(Y(g) - (\pi_{\theta_0} g)\,\tau)L'(\theta_0)\right]}{E[\tau]} .$$

References

Athneya, K. B. and P. Ney (1978). A new approach to the limit theory of recurrent Markov Chains. *Trans. Amer. Math. Soc.* **245**, 493–501.

Fox, B. L. and P. W. Glynn (1986). Discrete-time conversion for simulating semi-Markov processes. *Operations Research Letters* **5**, 191–196.

Meyn, S. P. and R. L. Tweedie (1993). *Markov Chains and Stochastic Stability*. Springer-Verlag, New York.

Nummelin, E. (1978). A splitting technique for Harris recurrent Markov chains. *Z. Wahrscheinlichkeitstheorie* **43**, 309–318.

Nummelin, E. (1984). *General Irreducible Markov Chains and Non-negative Operators*, Cambridge University Press, New York.

Tweedie, R. L. (1983). The existence of moments for stationary Markov chains. *J. Appl. Prob.* **20**, 191–196.

Vásquez-Abad, F. and H. Kushner (1992). Estimation of the derivative of a stationary measure with respect to a control parameter. *J. Appl. Prob.* **29**, 343–352.

Perturbation Analysis for the $GI/G/1$ Queue with Two Priority Classes

Naoto Miyoshi[1] *and Toshiharu Hasegawa*[2]

[1] Division of Applied Systems Science, Faculty of Engineering, Kyoto University, Kyoto 606-01, JAPAN
e-mail: miyoshi@kuamp.kyoto-u.ac.jp

[2] Department of Applied Mathematics and Physics, Faculty of Engineering, Kyoto University, Kyoto 606-01, JAPAN
e-mail: hasegawa@kuamp.kyoto-u.ac.jp

1 Introduction

Perturbation analysis (PA) is a technique for estimating the derivatives of a performance measure from a single sample path of a stochastic discrete event system and the extensive bibliography on it can be found in Ho and Cao (1991). In particular, *infinitesimal perturbation analysis* (IPA) is an approach which takes the sample derivative as the estimator we want and is the easiest and most efficient means of calculating the derivative estimators. However, as shown by Cao (1985) and Glasserman (1991), IPA is not applicable to performance measures which are discontinuous in the parameter of interest and, in that case, the method called *smoothed perturbation analysis* (SPA), introduced by Gong and Ho (1987) and extended and generalized by Fu and Hu (1992), can often be applied to smooth the discontinuities using conditional expectation. However, the form of the SPA estimator derived by Fu and Hu (1992) essentially requires the need for the additional simulation, except a few special cases.

In this paper, we consider the $GI/G/1$ queue with two priority classes. After each service, the server chooses a customer with high priority whenever at least one is waiting in the queue. A low priority customer is served only if there is no high priority customer in the queue when a service is completed. We assume that the service discipline is *nonpreemptive*, that is, each service is never interrupted once it starts. For this model, we intend to derive the estimators for the derivatives of the mean system time of the customer in the high priority class with respect to parameters of the service time distributions for the both classes. Our approach is mainly based on that developed by Fu and Hu (1992). Although the estimator they derive requires the additional simulation except a few special cases, we show that, using the technique of sample path analysis, the estimators can be derived directly from a single sample path without an additional simulation.

2 Model Description and Problem Formulation

We consider the $GI/G/1$ queue with two priority classes. Customers arrive at the queue with an unlimited buffer and are served by a single server. The interarrival times between customers are independent and identically distributed (i.i.d.) with distribution function F and density f. Each arriving customer is from class 1 with probability p or from class 2 with $1-p$, independently of the arrival process. Each service is nonpreemptive and class 1 customers are served with higher priority than the class 2 customers, that is, after each service completion, a class 1 customer, if any, is taken for service prior to class 2 customers. The service time distribution function is denoted by G_i, $i = 1, 2$, for class i customers and the service discipline is assumed to be FCFS within each class. It is assumed that the system starts empty at time 0 and empties often.

Our model has the two types of *events*: Events a and d represent, respectively, an arrival to and a departure from the system, where event d is *active* if and only if the system is nonempty. Event a is classified into two types, that is, a_1 with probability p and a_2 with $1-p$, where a_i, $i = 1, 2$ corresponds to a class i arrival.

Our performance measure of interest is mean system time (time in queue plus in service) of a class 1 customer, denoted by $\mathrm{E}[T]$, and we are interested in estimating $\partial\mathrm{E}[T]/\partial\theta_i$, $i = 1, 2$, where $\theta_i \in \Theta_i$ is a parameter of the service time distribution for the class i customers and each Θ_i is an open interval. We call $\theta = (\theta_1, \theta_2)$ the *parameter vector*.

Let $T_{kj}(\theta)$ be the system time of the jth class 1 customer in the kth busy period. Suppose that N class 1 customers are observed in the given sample path and K busy periods are included in the sample path of N class 1 customers. Then, our *sample performance* is given by

$$T(\theta) = \frac{1}{N}\sum_{k=1}^{K}\sum_{j=1}^{N_k} T_{kj}(\theta) \ , \tag{1}$$

where N_k denotes the number of the class 1 customers in the kth busy period. Note that the given sample path may include the incomplete last busy period in it.

Let A_{ki} represent the interarrival time between the $(i-1)$st and ith class 1 customers in the kth busy period, where A_{k1} is the time from the beginning of the kth busy period to the arrival epoch of the first class 1 customer in this busy period. (If the kth busy period starts with an arrival of a class 1 customer, $A_{k1} = 0$.) Let $X_{ki}(\theta_1)$ and $Y_{ki}(\theta_2)$ represent the service times of the ith class 1 customer and the ith class 2 customer in the kth busy period, respectively. Then, $T_{kj}(\theta)$ in (1) is written as

$$T_{kj} = \sum_{i=1}^{j} X_{ki} + \sum_{i=1}^{m_{kj}} Y_{ki} - \sum_{i=1}^{j} A_{ki} \ , \tag{2}$$

where m_{kj} means the number of the class 2 customers in the kth busy period who are served before the jth class 1 customer in this busy period. In the next section, we derive the unbiased estimator for $\partial\mathrm{E}[T]/\partial\theta_i$, $i = 1, 2$, using the expressions (1) and (2).

3 Derivations of the Estimator

We first assume that $X_{ki}(\theta_1)$ and $Y_{ki}(\theta_2)$ are a.s. continuously differentiable on Θ_1 and Θ_2, respectively, and also $\mathrm{d}X_{ki}/\mathrm{d}\theta_1$ ($\mathrm{d}Y_{ki}/\mathrm{d}\theta_2$) can be determined by X_{ki} (Y_{ki}) and θ_1 (θ_2). These assumptions are valid for a broad class of random variables (see Suri (1987) and Cassandras *et al.* (1991) for discussions of these assumptions).

According to the result by Fu and Hu (1992), our estimator is expressed as the sum of two parts: One is called the *IPA contribution* and the other the *SPA contribution*. The IPA contribution estimates the effect in case that the sufficiently small perturbation does not change the event occurrence order in the sample path. Then, the IPA contribution is the mere sample derivative with respect to the parameter of interest. From (1) and (2), the IPA contributions with respect to θ_1 and θ_2 are, respectively, given by

$$\frac{\partial T(\theta)}{\partial \theta_1} = \frac{1}{N}\sum_{k=1}^{K}\sum_{j=1}^{N_k}\sum_{i=1}^{j}\frac{\mathrm{d}X_{ki}}{\mathrm{d}\theta_1} \quad \text{and} \quad \frac{\partial T(\theta)}{\partial \theta_2} = \frac{1}{N}\sum_{k=1}^{K}\sum_{j=1}^{N_k}\sum_{i=1}^{m_{kj}}\frac{\mathrm{d}Y_{ki}}{\mathrm{d}\theta_2}\,. \tag{3}$$

Now, we derive the SPA contribution which estimates the expected changes in the performance when any changes in the event occurrence order take place due to the perturbation in θ_i. If no other event occurs between two events along the sample path, we say these two events are *adjacent*. Since we can check, similarly to Fu and Hu (1992), that the probability of the order change in nonadjacent events due to the introduction of the perturbation of size $\Delta\theta_i$ is of $o(\Delta\theta_i)$ under some mild condition, it suffices to consider only adjacent event order changes.

Suppose that $\Delta\theta_i > 0$ for $i = 1, 2$, that is, we consider the right-hand derivative as $\Delta\theta_i \downarrow 0$. Moreover, we assume that each service time is nondecreasing in the parameter of interest. In what follows, the original (unperturbed) path at parameter vector θ is referred to as the *nominal* path and the path due to the introduction of perturbation in θ_i, $i = 1, 2$, is referred to as the *perturbed* path. In our system, under positive perturbations in the service times, the type of adjacent event pair in which the order changes can occur is only d/a pair. Furthermore, we know, from Fu and Hu (1992), that if an adjacent event pair satisfies the *local commuting condition*, that is, if a system state through the occurrences of these two possible events is independent of their order, there is no contribution from the interchange in their occurrence order (see Glasserman (1991) for a more detailed discussion of the commuting condition). Therefore, we see that the only adjacent event pair which yields nonzero SPA contribution is the d/a_1 pair at the state such that the server is busy and no class 1 customer is waiting in the queue but at least one class 2 customer is waiting. Indeed, in such a situation, after the occurrences of these two events, a class 2 customer is in service and a class 1 customer is waiting in the queue in the nominal path but, when the occurrence order changes in the perturbed path, the class 1 customer is in service and only class 2 customers are in the queue and, afterwards, the behavior of the system differs. We call the adjacent event pair *critical* if it is d/a_1 at the state mentioned above.

Consider the jth critical event pair in the nominal sample path and the situation where the departure occurs at t_j, followed subsequently by the class 1 arrival at $t_j + \tau$. Suppose that the class 1 customer who arrives at $t_j + \tau$ is the lth one in the kth busy period. When the perturbation of size $\Delta\theta_i$ is introduced, the accumulated perturbation $\Delta_i t_j$, for the departure epoch t_j, is given by $\Delta_1 t_j = \sum_{i=1}^{l-1}(\mathrm{d}X_{ki}(\theta_1)/\mathrm{d}\theta_1)\Delta\theta_1$ or $\Delta_2 t_j = \sum_{i=1}^{m_{kl}-1}(\mathrm{d}Y_{ki}(\theta_2)/\mathrm{d}\theta_2)\Delta\theta_2$. If the accumulated perturbation $\Delta_i t_j$ becomes equal to or greater than τ, we will see that the jth critical event pair changes that order. Thus, the

SPA contribution is expressed as

$$\sum_{j=1}^{C} \lim_{\Delta\theta_i \downarrow 0} \frac{\mathrm{E}[\Delta_i T(\theta)\, \mathbf{1}\{\tau \leq \Delta_i t_j\} \mid z_j]}{\Delta\theta_i} , \tag{4}$$

where $\mathbf{1}\{\cdot\}$ is the indicator function. In (4), C denotes the number of the critical event pairs and $\Delta_1 T(\theta) = T(\theta_1+\Delta\theta_1, \theta_2) - T(\theta_1, \theta_2)$ ($\Delta_2 T(\theta)$ similarly). Condition z_j represents a set of some quantities available from the observed nominal path, called *characterization* by Gong and Ho (1987) and must be selected. The numerator $\mathrm{E}[\Delta_i T(\theta)\, \mathbf{1}\{\tau \leq \Delta_i t_j\} \mid z_j]$ in (4) represents the conditional expectation of the effect which the order change in the jth critical event pair gives to the sample performance. Characterization z_j should be chosen such that the conditional density of τ can be calculated at t_j. Since τ is the residual interarrival time at t_j, we can condition on that age ξ_j. Moreover, τ must be equal to or less than the class 2 service time scheduled at t_j in the nominal path, denoted by $\eta_j (= Y_{k,m_{kl}})$. Therefore, characterization z_j we use contains ξ_j and η_j (and eventually, the sample path quantities such that it determines the sample path completely if τ is given). The conditional density for τ at t_j given z_j is expressed by

$$f_\tau(x \mid z_j) = \begin{cases} \dfrac{f(\xi_j + x)}{F(\xi_j + \eta_j) - F(\xi_j)} & \text{if } x \leq \eta_j , \\ 0 & \text{if } x > \eta_j . \end{cases}$$

Integrating with respect to $f_\tau(x \mid z_j)$, we obtain

$$\begin{aligned} &\mathrm{E}[\Delta_i T(\theta)\, \mathbf{1}\{\tau \leq \Delta_i t_j\} \mid z_j] \\ &= \int_0^{\Delta_i t_j} \mathrm{E}[\Delta_i T(\theta,\ \tau = x) \mid \tau \leq \Delta_i t_j,\ z_j]\, f_\tau(x \mid z_j)\, \mathrm{d}x + o(\Delta\theta_i) \\ &= \mathrm{E}[\Delta_i T(\theta,\ \tau = \sigma) \mid \tau \leq \Delta_i t_j,\ z_j]\, f_\tau(\sigma \mid z_j) \Delta_i t_j + o(\Delta\theta_i) \quad \exists \sigma \in [0,\ \Delta_i t_j] , \end{aligned}$$

where we use, in the second equality, the mean value theorem under the assumption that f is continuous. Dividing by $\Delta\theta_i$ and taking $\Delta\theta_i \downarrow 0$, we have

$$\lim_{\Delta\theta_i \downarrow 0} \frac{\mathrm{E}[\Delta_i T(\theta) \mathbf{1}\{\tau \leq \Delta_i t_j\} \mid z_j]}{\Delta\theta_i} = \\ f_\tau(0 \mid z_j) \frac{\partial t_j}{\partial \theta_i} \lim_{\Delta\theta_i \downarrow 0} \mathrm{E}[\Delta_i T(\theta,\ \tau = 0) \mid \tau \leq \Delta_i t_j,\ z_j] . \tag{5}$$

In (5) above, $\lim_{\Delta\theta_i \downarrow 0} \mathrm{E}[\Delta_i T(\theta,\ \tau = 0) \mid \tau \leq \Delta_i t_j,\ z_j]$ represents the expected change in the sample performance between that in the nominal path with the additional condition $\tau = 0^+$ and in the perturbed path with $\tau = 0$, under $\tau \leq \Delta_i t_j$ with $\Delta\theta_i \downarrow 0$. Although the derivation of this term requires, in general, an additional simulation, we show below that the sample path analysis makes possible to obtain it without the additional simulation.

Let z_j contain the sample path quantities enough to determine the sample path completely if τ is given. Under condition $\tau = 0^+$ in the nominal path, the class 1 customer who arrives at $t_j + \tau$ and is assumed to be the lth one in the kth busy period has the system time $\eta_j + X_{kl}$. While, under condition $\tau = 0$ in the perturbed path, the same customer becomes to have the system time X_{kl}. In other words, the system time of that customer becomes shortened by η_j due to the order change in the jth critical event pair with $\Delta\theta_i \downarrow 0$. Furthermore, if, under condition $\tau = 0$ in the perturbed path, each of the following class 1 customers arrives before the departure of the preceding class 1 customer, he has similarly the negative additional wait of size $-\eta_j$. Thus, the number of the class 1 customers who become to have that negative additional wait due to the jth critical event order change under $\Delta\theta_i \downarrow 0$ is given by

$$n_j = \min\left\{n \geq 1\colon \sum_{i=1}^{n} A_{k,l+i} > \sum_{i=0}^{n-1} X_{k,l+i}\right\} \ . \tag{6}$$

Note that n_j is not greater than the number of the class 1 customers included in the class 1's *local* busy period that starts at $t_j + \tau$. Whereas, if any of the following class 1 customers arrives after the departure of the preceding class 1 customer under $\tau = 0$ in the perturbed path, then the service of the class 2 customer with length η_j is scheduled and, afterwards, each of the system times of the following class 1 customers is identical to that under $\tau = 0^+$ in the nominal path. Therefore, using n_j in (6), we have

$$\lim_{\Delta\theta_i \downarrow 0} \mathrm{E}[\Delta_i T(\theta,\ \tau = 0) \mid \tau \leq \Delta_i t_j,\ z_j] = -\frac{n_j\,\eta_j}{N} \ . \tag{7}$$

This can be obtained through the observed sample path only. Combining (3) and (4) with (5), (7), we have the expression as the unbiased estimator for $\partial\mathrm{E}[T(\theta)]/\partial\theta_i$, for $i = 1$, 2:

$$\left(\frac{\partial\mathrm{E}[T]}{\partial\theta_i}\right)_{PA} = \frac{\partial T(\theta)}{\partial\theta_i} - \frac{1}{N}\sum_{j=1}^{C} \frac{f(\xi_j)}{F(\xi_j + \eta_j) - F(\xi_j)}\frac{\partial t_j}{\partial\theta_i} \cdot n_j\eta_j \ . \tag{8}$$

4 Simulation Examples

This section contains the simulation results of some examples for validating our estimators. For the comparison purpose, the analytical results in the steady-state are used, where all analytical values are calculated by differentiating the formulas given in Takagi (1991). When the analytical results are not available, the simulation results are compared with the finite difference (FD) estimates, whose formulas are given by

$$\left(\frac{\partial\mathrm{E}[T]}{\partial\theta_1}\right)_{FD} := \frac{T(\theta_1 + \Delta\theta_1,\ \theta_2) - T(\theta_1 - \Delta\theta_1,\ \theta_2)}{2\Delta\theta_1} \ , \tag{9}$$

and $(\partial\mathrm{E}[T]/\partial\theta_2)_{FD}$ similarly. The length of every sample path is in total 100,000 class 1 customers. The table entries other than analytical values are given in the form mean $\pm$ standard deviation, taken from 30 independent replications. Parameters θ_1 and θ_2 are the mean service times and are fixed at $\theta_1 = 1$ and $\theta_2 = 2$, respectively. The results are presented for three different values of ρ ($:= \lambda_1\theta_1 + \lambda_2\theta_2$), that is, $\rho = 0.2$, 0.5 and 0.8, where $\lambda_1 = p\lambda$, $\lambda_2 = (1 - p)\lambda$ and p is fixed at 0.75.

Example 1 ($M/M/1$). The case of an $M/M/1$ queue was simulated with mean interarrival time $1/\lambda$ and mean service times θ_i for the class i customers. The experimental results are given in Table 1 with the analytical values and show good agreement.

Table 1. Estimates for the $M/M/1$ priority queue

ρ	$(E[T])_{est}$	$E[T]$	$(\partial E[T]/\partial\theta_1)_{PA}$	$\partial E[T]/\partial\theta_1$	$(\partial E[T]/\partial\theta_2)_{PA}$	$\partial E[T]/\partial\theta_2$
0.2	1.319±0.005	1.318	1.316±0.006	1.316	0.182±0.004	0.182
0.5	1.998±0.013	2.0	2.284±0.024	2.285	0.569±0.013	0.571
0.8	3.155±0.025	3.154	4.757±0.300	4.834	1.206±0.098	1.231

Example 2 ($M/G/1$ with different service distributions). An $M/G/1$ queue was simulated with mean interarrival time $1/\lambda$, where the service time distribution for the class 1 customers is uniform on $[0,\ 2\theta_1]$ and that for the class 2 customers is the second-order Erlang with mean θ_2. The experimental results are given in Table 2 with the analytical values and, again, good agreement is indicated.

Table 2. Estimates for the $M/(U,\ E_2)/1$ priority queue

ρ	$(E[T])_{est}$	$E[T]$	$(\partial E[T]/\partial\theta_1)_{PA}$	$\partial E[T]/\partial\theta_1$	$(\partial E[T]/\partial\theta_2)_{PA}$	$\partial E[T]/\partial\theta_2$
0.2	1.228±0.004	1.227	1.213±0.004	1.213	0.137±0.002	0.136
0.5	1.713±0.008	1.714	1.879±0.014	1.878	0.430±0.007	0.429
0.8	2.537±0.017	2.538	3.656±0.140	3.651	0.925±0.049	0.923

Example 3 ($U/E_2/1$). The case of a $U/E_2/1$ queue was simulated with interarrival times uniform on $[0,\ 2/\lambda]$ and the service times of the second-order Erlang distribution with means θ_1 and θ_2. For this case, no analytical results exist with which to compare the numerical results. Thus, for comparison purpose, the results from the FD given by (9) were used, where $\Delta\theta_i = 0.05$, for $i = 1, 2$. The experimental results are given in Table 3 and the PA estimates show smaller variances than the FD's.

Table 3. Estimates for the $U/E_2/1$ priority queue

ρ	$(E[T])_{est}$	$(\partial E[T]/\partial\theta_1)_{PA}$	$(\partial E[T]/\partial\theta_1)_{FD}$	$(\partial E[T]/\partial\theta_2)_{PA}$	$(\partial E[T]/\partial\theta_2)_{FD}$
0.2	1.124±0.004	1.117±0.004	1.116±0.057	0.078±0.001	0.072±0.043
0.5	1.450±0.006	1.550±0.012	1.560±0.091	0.340±0.005	0.345±0.100
0.8	2.171±0.015	3.184±0.147	3.225±0.214	0.949±0.072	0.869±0.230

5 Concluding Remarks

We have considered the $GI/G/1$ queue with two priority classes and have derived the unbiased SPA estimator for the derivative of the mean system time of high priority class customers with respect to each parameter of the service time distributions for the both classes. Using the technique of sample path analysis in Sect. 3, we can derive the estimator corresponding to the low class customers at the same time. In Miyoshi and Hasegawa (1994), moreover, the multiclass priority queue is dealt with and the unbiased derivative estimator for the system time of any class is derived, where every class has its own arrival source differently from the model in this paper.

Acknowledgement

The authors wish to express their sincere gratitude to the anonymous reviewer for the helpful comments and suggestions.

References

Cao, X. R., *IEEE Trans. Automat. Contr.*, **30**, pp. 845–853 (1985).

Cassandras, C. G., W. B. Gong and J. I. Lee, *J. Optimiz. Theory and Applicat.*, **70**, pp. 491–519 (1991).

Fu, M. C. and J. Q. Hu, *IEEE Trans. Automat. Contr.*, **37**, pp. 1483–1500 (1992).

Glasserman, P., *Operat. Res.*, **39**, pp. 724–738 (1991).

Gong, W. B. and Y. C. Ho, *IEEE Trans. Automat. Contr.* **32**, pp. 858-867 (1987).

Ho, Y. C. and X. R. Cao, *Discrete Event Dynamic Systems and Perturbation Analysis*, Kluwer Academic, Boston (1991).

Miyoshi, N. and T. Hasegawa, Preprint (1994).

Suri, R., *J. ACM*, **34**, pp. 686–717 (1987).

Takagi, H., *Queueing Analysis: A Foundation of Performance Evaluation Volume 1, Vacation and Priority Systems Part 1*, North-Holland, Amsterdam (1991).

Supply Management in Assembly Systems: The Case of Random Lead Times

Chengbin CHU, Jean-Marie PROTH*, Yorai WARDI** and Xiaolan XIE**

* INRIA-Lorraine, Technopôle Metz 2000, 4 rue Marconi, 57070 Metz - FRANCE

** School of Electrical and Computer Technology, Georgia Institute of Technology, Atlanta, GA 30332 - USA

1 Introduction

Large companies, for example the automotive, aircraft and computer industries, often order many components from subcontractors on a just-in-time basis. There typically is some uncertainty in the yields of the subcontractors, for instance randomness in the delivery times. Since this can have a significant adverse effect on the delivery schedules of finished products, the determination of adequate supply management policies can be quite important.

The works reported in the literature deal with supplier selection strategies [CL1], [MK1] and [HR1], with the supplier diversification when the quantities are uncertain, or with restricted situations such as uniformly or exponentially distributed lead times, or no conflict in the use of the components [CP1]. But we observe that no satisfactory solution exists in the general case. The reason is that two closely related decisions have to be made at each decision point, namely:

(i) the choice of the products to be assembled, taking into account the components available; note that conflicts may arise when the inventory levels of the components are insufficient to meet the product backloggings and demands since different types of products may require the same types of components,

(ii) the choice of the number of components of each type to be ordered, taking into account the random yields of the subcontractors.

In this work, we assume that a component is ordered to a single subcontractor. As a consequence, the lead time of a given component is related to a single random variable, but no restrictive hypothesis is made on such a random variable.

The criterion to be optimized is the sum of the backlogging costs (which concern the products) and the inventory costs (which concern the components).

The remainder of the paper is organised as follows. In section 2, we set the problem and introduce the notations. Section 3 is devoted to the basics of sensitivity analysis via Finite Perturbation Analysis (FPA) and Infinitesimal Perturbation Analysis (IPA). In section 4, we propose a parametrized policy based on common sense and show how to use the sensitivity analysis to improve iteratively the parameters' values. We give some examples in section 5 and propose some concluding remarks in section 6

2 Problem Setting and Notations

We use the following notations:

d_i $i = 1,2,\dots,n$, is the demand of product i at the end of each elementary period; note that the demand does not depend on time, but the approach can be easily extended to stationary random demand,

x_j^k $j = 1,2,\dots,m$, is the number of components of type j ordered at the beginning of the k-th period,

y_i^k $i = 1,2,\dots,n$, is the number of products of type i assembled during the k-th period,

$l_{i,j}$ is the number of j-type components required to assemble one unit of product i,

b_i is the backlogging cost of one unit of product i during one elementary period,

a_j is the cost for keeping in stock one unit of component j during one elementary period.

For each component j, we denote by p_j^q, $q = 1,2,\dots,n_j$, the probability for a set of components ordered at the beginning of the k-th period to arrive at the end of the (k+q-1)-th period, and thus to be available for the products assembled during the (k+q)-th period. Note that $\sum_{q=1}^{n_j} p_k^q = 1, \forall j$.

For k = 1,2,..., we denote by:

s_j^k the inventory level of component j during the k-th period, after removing the components required for the products which are assembled during this period,

v_i^k the backlogging of product i during the k-th period.

The following state equations hold:

$$s_j^k = s_j^{k-1} - \sum_{i=1}^{n} y_i^{k-1} . l_{i,j} + Z_j^{k-1}$$

where Z_j^{k-1} is the number of j-type components arriving at the end of the (k-1)-th period. This value may be the sum of several quantities ordered before the (k-1)-th period.

$$v_i^k = v_i^{k-1} + d_i - y_i^{k-1} \text{ and } s_j^k \geq 0 \text{ whatever } j, k, \text{ and } v_i^k \geq 0 \text{ whatever } i, k$$

The goal is to define the x_j^k's and y_i^k's values which minimize:

$$G(X,Y) = \lim_{K \to +\infty} \left[\frac{1}{K} \sum_{k=1}^{K} \left(a_j\, s_j^k + b_i\, v_i^k \right) \right]$$

where X and Y are the vectors whose components are respectively the x_j^k's and y_i^k's values.

3 Sensitivity Analysis via FPA and IPA

Let θ be the vector whose discrete components are the parameters to be adjusted and $C(\theta)$ the value of the criterion to be optimized when θ holds. Given an increment $\Delta\theta$, FPA approach consists of evaluating the finite difference $\Delta C = C(\theta+\Delta\theta) - C(\theta)$ in a single simulation run (i.e. the same as the one used to evaluate $C(\theta)$) without performing an additional simulation at $\theta+\Delta\theta$.
The general formulation of the algorithm is then:

$$\theta_{k+1} = \mathcal{P}_\theta\,(\theta_k - z_k\,\Delta C_k)$$

where:

θ_{k+1} is the value of the parameters at the beginning of the (k+1)-th iteration,

$\mathcal{P}_\theta$ is the projection of the parameters' values on the set of feasible values,

z_k, $k = 1,2,\ldots$, is a serie of parameters such that

$$\sum_{k=1}^{\infty} z_k = +\infty \text{ and } \sum_{k=1}^{\infty} z_k^2 < +\infty$$

ΔC_k is the finite difference ΔC when only the k-th component of $\Delta\theta$ is different from 0 and equal to $\tau > 0$. τ is a positive value which should be selected by the user.

FPA is used for discrete parameters. In this paper, parameters h_j are concerned by this type of approach.

Parameters α_j, which are continuous, are subject to IPA. When continuous parameters are concerned, the objective is to solve $\nabla C(\theta) = 0$. The general formulation of the algorithm is then:

$$\theta_{k+1} = \mathcal{P}_\theta\left(\theta_k - z_k \hat{\nabla} C(\theta_k)\right)$$

where $\hat{\nabla} C(\theta_k)$ is an evaluation of $\nabla C(\theta_k)$.

In practice, we decrease the value of z_k only when the sign of ΔC_k changes.

4 Design of an Assembly Policy based on Sensitivity Analysis

This section is the key of the paper. We propose a parametrized policy based on common sense and a fast algorithm to change the parameters' values in order to improve the value of the criterion.

4.1 Parametrized Policy

The proposed policy is twofold:

(i) The choice of the product to be assembled (i.e. the choice of the y_i^k values) is made in the decreasing order of a cost which is the sum of the backlogging cost of the product and the inventory costs of its components, taking into account the number of components needed to assemble one unit of product. We observe that the first step of the strategy does not depend on parameters;

(ii) The x_j^k's values are given by:

$$x_j^k = \left[h_j - \left(s_j^k + \sum_{q=1}^{n_j} (\alpha_j)^q U_{j,q}^k \right) \right]^+ , \forall j, k \qquad (1)$$

where:

the threshold value $h_j \geq 0$ and the discount factor $0 < \alpha_j < 1$ are the parameters to be adjusted,

$a^+ = \text{Max}\ (a, 0)$,

$U_{j,q}^k$ is the expected value of the quantity of components j arriving at the end of period (k+q-1).

4.2 Sensitivity Analysis

The goal of this subsection is to evaluate the sensitivity of G(X,Y) with regard to the values of the parameters.

a. The first step of the process consists of evaluating the increase of the inventory level of component j on period k if s_j^{k-1} increases by one. Assuming the continuity of the $l_{i,j}^{'}$'s value, we obtain:

$$w_j^k = \begin{cases} -\sum_{i \in E_k} \left[\sum_{j_1 \neq j} \left(a_{j_1} l_{i,j_1} / l_{i,j} + b_i / l_{i,j} \right) \right] + \text{card}(E_k).a_j & \text{if } s_j^{k-1} < h_j \\ 0 \quad \text{otherwise} \end{cases} \qquad (2)$$

where:

$$E_k = \left\{ i / i \in \{1, \ldots, n\},\ v_i^{k-1} > 0 \text{ and } l_{i,j_1} \leq s_{j_1}^{k-1}, \forall\, j_1 \neq j \right\}$$

$$\overline{E}_k = \{1, \ldots, n\} \setminus E_k$$

Similarly, when s_j^{k-1} decreases by one:

$$w_j^k = \begin{cases} +\sum_{i \in E_k} \left[\sum_{j_1 \neq j} \left(a_{j_1} l_{i,j_1} / l_{i,j} + b_i / l_{i,j} \right) \right] - \text{card}(E_k).a_j & \text{if } s_j^{k-1} > 0 \\ 0 \quad \text{otherwise} \end{cases} \qquad (2')$$

b. The second step consists of evaluating the effect of the parameters' perturbation of the inventory level of j.

b_1. Taking into account relation (1) it is easy to show that, if h_j increases by $\Delta(h_j)$, then a reasonable evaluation of the change of the inventory level of component j is also $\Delta(h_j)$ on the whole simulation period. As a consequence, the change in the value of the cost function during period k can be evaluated by (see (2) and (2')):

$$\Delta_j^k(G) = \Delta(h_j)w_j^k \tag{3}$$

b_2. Let us denote by $\hat{\nabla}_j(\theta)$ an evaluation of the partial derivate of G(X,Y) with regard to α_j, and by $\hat{\nabla}_j^k(\theta)$ the restriction of $\hat{\nabla}_j(G)$ to the period at which the decision made at the beginning of period k applies. Then, its steady state:

$$\hat{\nabla}_j^k(G) = -\left(\sum_{q=1}^{n_j} q(\alpha_j)^{q-1}.U_{j,q}^k\right)w_j^k \tag{4}$$

4.3 Algorithm

The algorithm is explained in section 3. The following aspects are specific to this problem.

(i) The initial h_j's values are equal to the number of components required at each period, i.e. $h_j^{\circ} = \sum_{i=1}^{n} l_{ij}$.

(ii) The initial α_j's values are generated at random between 0 and 1.

(iii) The simulation is limited to 400 periods for each set of parameters' values.

(iv) The value of a parameter z_k (see section 3) changes only when the sign of the related cost variation changes.

(v) The algorithm stops either when the z_k's values are less than a given threshold, or the mean quantity of each component ordered for each period is close enough to the quantity required (a threshold of 0.1% was used).

5 Some Numerical Results

More than one hundred problems were generated at random as follows:

* the number of products and components are generated between 2 and 10,
* the inventory costs of the components are generated between 1 and 500,
* the backlogging costs of the products take their values between 1 and 40,
* a component ordered at the beginning of period k may arrive between the end of period k and the end of period k+9 according to probabilities generated at random,
* the components required to assemble a product, as well as the number of each component, are generated at random.

Some examples are given in table 1.

Table 1: Some numerical examples

Example	n	m	SR	EM	TINV	BPR
1	2	8	13	1.69	6.29	0.69
2	6	10	9	1.49	3.88	0.83
3	8	10	13	2.29	3.25	0.95
4	5	7	611	1.15	8.08	1.12
5	6	2	10	0.66	1.34	0.45
6	3	8	249	0.80	2.30	0.06
7	4	4	48	1.17	2.64	0.43

SR: Number of simulations to reach the result

EM: Maximal difference between the number of components required at each period and the mean number of components ordered (in %)

TINV: Maximal ratio (mean inventory level per period / number of components required per period)

BPR: Maximal ratio (mean backlogging per period / demand per period)

The efficiency of the algorithm can be evaluated by considering columns TINV and BPR (taking into account inventory and backlogging costs).

6 Concluding Remarks

The algorithm proposed in this paper has proved its efficiency on a large number of examples. Similar approaches should be applied to assembly systems subject to other types of randomness as, for instance, randomness in the quantities delivered. Furthermore, the problem of defining an efficient low bound to this problem remains open.

Bibliography

[CP1] Chu C., Proth J-M. and Xie X, "Supply management in assembly systems: the basic problem", R.R. INRIA #1624 (1992).

[CL1] Cohen M. and Lee H., "Resource development analysis of global manufacturing and distribution networks", *Journal of Manufacturing and Operations Management*, Vol. 2, No. 2, pp. 81-104 (1989).

[MK1] Hahn C.K., Kim K.H. and Kim J.S., "Costs of competition: implications for purchasing strategy", *Journal of Purchasing and Materials Management*, Vol. 22, No.3, pp. 2-7 (1986).

[HR1] Hendrick T.E. and Ruch W.A., "Determining performance appraisal criteria for the buyer", *Journal of Purchasing and Materials Management*, Vol. 24, No.2, pp. 18-26 (1988).

[HC1] Ho Y.C and Cao X.R., "Perturbation analysis of discrete event dynamic systems", Klewer Academic Publishers, Boston MA (1991).

[HC2] Ho Y.C., Cao X.R. and Cassandras C.G., "Infinitesimal and finite perturbation analysis for queueing networks" *Automatica*, Vol. 4, pp. 439-445 (1983).

Simulation Trees for Functional Estimation via the Phantom Method *

Felisa Vázquez-Abad and Pierre L'Ecuyer[1]

Département d'IRO
Université de Montréal, C.P. 6128, Succ. A
Montréal, H3C 3J7, CANADA

1 Functional Estimation

We consider a class of stochastic models for which the performance measure is defined as a mathematical expectation that depends on a parameter θ, say $\alpha(\theta)$, and we are interested in constructing estimators of α and of its derivative α' in functional form (i.e., entire functions of θ), which can be computed from a single simulation experiment. One approach for doing that is based on the use of likelihood ratios and score functions. In this paper, we consider a different approach, based on the phantom method, which can be used in the case where the parameter of interest is related to the sampling rate of a point process. This approach can be viewed in some sense as an efficient implementation of an infinite number of parallel simulations with common random numbers. We illustrate the approach with some examples.

Let $\{(\Omega, \Sigma, P_\theta),\ \theta \in \Theta\}$ be a family of probability spaces defined over the same measurable space, where Θ is a bounded subset of $\mathbb{R}^d$. The probability law P_θ depends on a parameter θ. Consider a finite-horizon discrete event model defined over that family of probability spaces and let $h(\theta, \omega)$ be some random variable of interest (e.g. total sojourn time of all the customers served during a given day in a queueing system, or the total number of rejected customers in a finite-buffer system, etc). Suppose that we are interested in the function

$$\alpha(\theta) = E_\theta[h(\theta, \omega)] = \int_\Omega h(\theta, \omega) P_\theta(d\omega).$$

Normally, a simulation performed at $\theta = \theta_0$ permits one to estimate $\alpha(\theta_0)$ and perhaps $\alpha'(\theta_0)$ or higher order derivatives. Techniques for doing that include the likelihood ratio or score function method, as well as perturbation analysis and its numerous variants (Glasserman 1991, Glynn 1990, Ho and Cao 1991, L'Ecuyer 1990, 1991, Rubinstein and Shapiro 1993). To obtain estimations at different values of θ, one would usually perform different simulations (perhaps with common random numbers) at all of those values of interest, which may become costly.

* This work has been partially supported by NSERC-Canada grants # WFA0139015 and # OGP0110050, as well as FCAR-Québec grant # 93-ER-1654.

One approach for estimating α in a functional form, i.e., for estimating $\alpha(\theta)$ for all $\theta \in \Theta$ from a single simulation run, is based on the "change of measure" idea, sometimes called *importance sampling* (Glynn and Iglehart 1989). To summarize the idea in a simplified form, suppose that $h(\theta, \omega) = h(\omega)$ does not depend (directly) on θ and that P_θ has a corresponding density f_θ. Then, assuming that the *likelihood ratio* $L(\theta_0, \theta, \omega) = (f_\theta / f_{\theta_0})(\omega)$ exists, α can be rewritten as

$$\alpha(\theta) = \int_\Omega h(\omega) f_\theta(\omega) d\omega = \int_\Omega [h(\omega) L(\theta_0, \theta, \omega)] f_{\theta_0}(\omega) d\omega.$$

So, if the simulation is performed at θ_0, then $h(\omega)L(\theta_0, \cdot, \omega)$ provides an unbiased functional estimator of α. The random variable $h(\omega)$ is computed only once, and the likelihood ratio is calculated at any value of θ of interest to compute the estimator. Under appropriate regularity conditions, an unbiased estimator of the derivative $\alpha'(\theta)$ is $h(\omega)S(\theta, \omega)L(\theta_0, \theta, \omega)$, where $S(\theta, \omega) = \frac{\partial}{\partial \theta} \ln L(\theta_0, \theta, \omega)$ is called the *score function*.

Several examples of this approach are given in Rubinstein and Shapiro (1993). L'Ecuyer (1993) also gives numerical illustrations for simple queueing systems. His examples show how dramatically the variance of the functional estimators for α and α' can increase (often with vertical asymptotes) as θ gets away from θ_0. It also increases exponentially fast (typically) as a function of the length of the simulation horizon. As a result, those functional estimators are typically useful only in a small neighborhood of θ_0 and only for short horizon models, or "steady-state" models with short regenerative cycles.

In this paper, we examine an alternative method for estimating the derivative in functional form, for the following situation. We suppose that during the simulation, there is sequence of independent binary decisions to be taken. Each decision is 1 with probability p and zero with probability $1 - p$, independently of previous history. The parameter of interest here is $\theta = p \in [0, 1] = \Theta$. This covers a rather large number of situations, including admission control, scheduling problems (Cassandras and Julka 1992) and the problems in our examples.

The phantom method introduced in Suri and Cao (1986) computes a perturbed sample path if a particular decision was reversed. This method was later used by Brémaud and Vázquez-Abad (1992) for derivative estimation (RPA) with respect to θ. In this paper, the perturbed paths may correspond to parameter values far from the nominal value of θ. Since no limit is involved in functional estimation, the applicability of the method is much wider than the applicability of RPA.

Strictly speaking, our estimators are not really computed in a "single" simulation run, because the simulation run will be "split" along the way into several parallel runs. This approach can be viewed in fact as an efficient way of implementing parallel simulations for several values of p, with common random numbers for the binary decisions. More specifically, from the sample path of a simulation performed at $p = 1$, plus a sequence of i.i.d. uniform random variates used for the binary decisions, we can recover all the information required to compute what would happen at all values of p in $[0, 1]$. It turns out that the sample path changes as a function of p only at a finite number of values of p. More importantly, we show that with our representation, the number of such values increases only linearly with the number of binary decisions that are made, and we show how functional estimation can be performed efficiently.

We define our setup more precisely and explain the main ideas in §2. In §3, we illustrate the approach with a few variants of a single-queue example. In a more elaborate version of the paper, we shall give further examples and numerical illustrations.

2 Simulation Trees for Functional Estimation

We consider a finite-horizon simulation, parameterized by $p \in \Theta = [0, 1]$. We suppose that in the model, the dependence on the parameter p is only via a sequence of i.i.d. Bernoulli(p) random decision variables $\eta_1(p), \ldots, \eta_\tau(p)$, where τ is a (a.s. finite) random variable whose distribution does not depend on p. We shall assume that a typical sample point $\omega \in \Omega$ takes the form $\omega = (\tilde{\omega}, u_1, u_2, \ldots)$, where $u_i \in [0, 1]$ for each i, and that the probability distribution $P_\theta \equiv P_p \equiv P$ does not depend on p. We also assume that the random variables defined by $U_i(\omega) = u_i$, for $i \geq 1$, are i.i.d. and follow the $U(0, 1)$ distribution. We define our binary (random) decision variables by $\eta_i(p) = I_{[u_i \leq p]}$, where I denotes the indicator function.

For a given ω, let $v_1 \leq \cdots \leq v_\tau$, denote the values of $u_1, \ldots, u_\tau$ sorted by increasing order, and let $r(i)$ denote the index of the u_j that corresponds to v_i, that is, such that $v_i = u_{r(i)}$. Let also $v_0 = 0$.

Proposition 1. *Let $h(p, \omega)$ be any measurable random variable, under the above assumptions. Then, $h(\cdot, \omega)$ is piecewise constant over* $[0, 1]$, *with at most $\tau + 1$ pieces, and possible jumps only at $v_1, \ldots, v_\tau$.*

Proof. For each $i \in \{0, 1, \ldots, \tau - 1\}$ and $p \in [v_i, v_{i+1})$, one has $\eta_{r(1)}(p) = \cdots \eta_{r(i)}(p) = 1$ and $\eta_{r(i+1)}(p) = \cdots \eta_{r(\tau)}(p) = 0$. For $p \in [v_\tau, 1]$, one has $\eta_j(p) = 1$ for all j. But since $h(p, \omega)$ depends on p only through the $\eta_i(p)$'s, this yields the desired result.

Proposition 1 can be used to estimate $\alpha(p)$ simultaneously for all $p \in [0, 1]$ as follows. Perform the simulation with $p = 1$, generating the u_i's along the way. Sort the u_i's dynamically while they are generated (e.g., using a heap, as suggested in Problem 1.9.7 of Bratley, Fox, and Schrage 1987). Let I_i denote the interval $I_i = [v_i, v_{i+1})$ for $i = 0, \ldots, \tau - 1$ and $I_\tau = [v_\tau, 1]$. The value of $h(p, \omega)$ is constant (and the sample path is the same) within each of those intervals, but may differ considerably between the intervals. To obtain a functional estimator one must generate the sample path and compute h within each interval. Formally, this is equivalent to performing τ simulation runs in parallel. However, there are many cases where a large fraction of the required computations is common to all intervals and does not have to be repeated. Think for example of the situation where $\eta_i(p)$ represents the decision of accepting customer i into a queueing system. While a different value of $\eta_i(p)$ changes the evolution of the system, many of the other random variables to generate could be the same in both cases. Furthermore, the sample paths associated with two neighboring intervals I_{i-1} and I_i potentially differ only from the time where $u_{r(i)}$ is generated. Before that time, there is not need to duplicate the simulation in order to distinguish those two intervals.

So, the simulation will be run by constructing a tree of sample paths as follows. Start by simulating the nominal sample path, say at $p = 1$. When u_1 is generated, split this sample path in two: one for the case $p < u_1$ (with $\eta_1(p) = 0$) and one for the case $p \geq u_1$ (with $\eta_1(p) = 1$). When u_2 is generated, one of these two sample paths is again split in two, which now yields three sample paths evolving in parallel, associated with the intervals $[0, \min(u_1, u_2))$, $[\min(u_1, u_2), \max(u_1, u_2))$, and $[\max(u_1, u_2), 1]$. Again, one of these sample paths is divided in two when u_3 is generated, and so on. Note that the size of the tree grows only linearly as a function of τ. At the end of those simulations, one obtains the value of $h(p, \omega)$ for each interval and, from that, constructs the piecewise constant function $h(\cdot, \omega)$. Of course, this can be repeated, say, N times, and the functional estimator $\bar{h}(\cdot, \omega)$ of α can be constructed by averaging out those N piecewise constant functions.

3 A GI/G/1 Queue with Mixed Service Time Distribution

Consider a single queue with arbitrary interarrival and service time distributions. Assume that the interarrival times are i.i.d. and that each arrival is of type I (or II) with probability p (respectively, $1 - p$). For each $i \geq 1$, let $W_i(p)$, $S_i(p)$, and $X_i(p) = W_i(p) + S_i(p)$ denote the *waiting* time, *service* time, and *sojourn* time for customer i, respectively, and A_i be the time between arrivals of customers i and $i + 1$. Define $\eta_i(p) = 1$ if arrival i is of type I, $\eta_i(p) = 0$ otherwise. We suppose that the two types of customers have different service time distributions, say B_{I} for type I and B_{II} for type II. For each i, let $S_{i,1}$ and $S_{i,2}$ be two random variables with respective distributions B_{I} and B_{II} (independent of the interarrival and other service times) and let

$$S_i(p) = \eta_i(p)S_{i,1} + (1 - \eta_i(p))S_{i,2}.$$

Assuming that the queue has infinite capacity, the evolution of this system can be described as usual by Lindley's equation:

$$W_{i+1}(p) = (W_i(p) + S_i(p) - A_i)^+, \tag{1}$$

for $i \geq 1$, where x^+ means $\max(x, 0)$, and $W_1(p) = s \geq 0$ is the initial state of the queue ($W_1(p) = 0$ corresponds to an initially empty system).

Let $\tau = \tau(\omega)$ be a (possibly random) stopping time for this system and $h(p, \omega) = f(W_1(p), S_1, \eta_1(p), \ldots, W_\tau(p), S_\tau, \eta_\tau(p))$, for some (measurable) function f. Observe that τ is assumed not to depend on p, but can be defined as a function of the system's evolution for a *fixed* value of p. For example, τ can be the number of customers in the first busy cycle when $p = 1$.

To see how functional estimation works in this case, observe that when p goes down from I_i to I_{i-1}, the only changes in the system's evolution are due to the fact that $\eta_{r(i)}(p)$ goes from 1 to 0, i.e., $S_i(p)$ changes from $S_{i,1}$ to $S_{i,2}$. In other words, the Lindley equations associated with adjacent intervals evolve in almost exactly the same way. The generation of the random variables A_i, $S_{i,1}$, $S_{i,2}$ and U_i need not be duplicated. During the simulation, a value of $W_i(p)$ must be maintained, using Eq. (1), for each interval I_j. When $W_{i+1}(p)$ is to be computed, i customers have arrived, and so the interval $[0, 1]$ is divided at that moment into $i + 1$ subintervals (i.e., the "simulation tree" has $i + 1$ nodes).

Of course, it may happen occasionally that the values of $W_i(p)$ become the same for two or more neighboring intervals I_j, which means that *coupling* occurs between the trajectories associated with those intervals. This can be exploited to reduce computations: whenever coupling occurs, merge the intervals involved for the computation of $W_j(p)$, $j > i$, by Lindley's equation. On the other hand, the different values of $W_j(p)$ over those different intervals for $j < i$ may have to be taken into account in the cost function $h(p, \omega)$.

As a special case, suppose that $S_{i,2} = 0$ a.s. for all i. This represents a system with only one type of customers, with service time distribution B_{I}, but where each customer is rejected with probability $1 - p$. Here, $\eta_i(p)$ represents the decision of accepting the i-th arrival into the system (admission control), and the service times are modeled as client attributes. For the rejected customers, $W_i(p)$ and $X_i(p)$ are "virtual" waiting times and sojourn times, but are nevertheless well defined, and $S_i(p) = 0$. In that situation, whenever $W_i(p_0)$ becomes zero for some p_0, then one also has $W_i(p) = 0$ for all $p \leq p_0$ and coupling is achieved. In other words, the systems with $p < p_0$ are *dominated* by the system with $p = p_0$, that is, the sample paths are *monotone* in p. A particular case of such coupling is when the system empties out in the nominal sample path, i.e., when $W_i(1)$ becomes zero. Then, the simulation tree shrinks back to a single node. If the cost function $h(p, \omega)$ is additive between busy cycles, with is typically the case, then one only needs to memorize the piecewise constant (sample) cost functions associated with the different busy cycles. The number of pieces in each such piecewise constant function will be equal to the number of customers in the cycle, plus one.

We will now examine a few more specific variants of that single-queue example and see how our approach works for each of them. Of course, many more variants could also be considered.

Example 1. Consider an $M/G/1$ queue with arrival rate λ, and suppose that we want to estimate some (expected) performance measure as a function of λ, for $0 \leq \lambda \leq \lambda_0$, where λ_0 is a positive constant smaller than the inverse of the mean service time (so that the system is stable). Define $p = \lambda/\lambda_0$. Then, simulating that system at λ is equivalent to simulating it at λ_0 and rejecting each arrival with probability $1 - p$. This "thinning" idea is well-known in the area of stochastic processes and is often used for simulating non-homogeneous Poisson processes; see Lewis and Shedler (1979), Bratley, Fox, and Schrage (1987), and Brémaud and Vázquez-Abad (1992). Estimating a performance measure as a function of λ is equivalent to estimating it as a function of p, which we have just seen how to do.

Of course, the same approach can also be used if the queue has finite capacity K, althought $S_i(p)$ would have to be defined differently than in the previous setup: let $S_i(p) = S_{i,1}$ if $\eta_i(p) = 1$ *and the queue is not full when customer i arrives* (which also depends on p); $S_i(p) = 0$ otherwise. In that case, the sample paths would no longer be monotone in p, i.e., we would loose the domination property of the systems with $p < p_0$ by the system with $p = p_0$. Indeed, when lowering p slightly, we might cut out an arrival that has an extremely long service time, and as a result, accept more customers later on. Nevertheless, coupling will occur and can be exploited as well. It is possible to recover domination by associating the service times with the server instead of with the arriving customers (i.e., by using $S_{i,1}$ as the service time of the ith customer that is *served* for any given value of p, as suggested for example in Problems 2.1.1 et 2.1.8 of Bratley, Fox, et Schrage 1987), but implementing functional estimation using that approach seems messy and not so much useful.

Example 2. Consider again an $M/G/1$ queue with arrival rate λ, and let θ be a scale parameter of the service time distribution B_θ. In other words, each service time S_i can be generated by generating a random variable Z_i from distribution B_1, and defining $S_i = \theta Z_i$. Alternatively, one can use surrogate estimation via time rescaling (see Vázquez-Abad 1993): generate the service times $S_i = Z_i$ using distribution B_1, the arrivals at rate $\lambda\theta_0$ (where θ_0 is the largest parameter value of interest), and accept any given arrival with probability $p = \theta/\theta_0$. Here, estimating a performance measure as a function of θ is clearly equivalent to estimating it as a function of p. The time units have been rescaled by the factor $p\theta_0$, which should be taken into account in the evaluation of $h(p, \omega)$.

Example 3. Consider again the admission control setup. Let $\tau = T$, a fixed constant, and

$$h(p, \omega) = \sum_{i=1}^{T} \eta_i(p) X_i(p),$$

the sum of sojourn times of all the customers that are accepted in the system. We have seen how to obtain a functional estimator of $\alpha(p)$, $0 \leq p \leq 1$, but now, suppose that we want to estimate the derivative of $\alpha(p)$ with respect to some parameter θ of the service time distribution $B_\text{I} \equiv B_\theta$, and we want to estimate it everywhere as a function of p. Let $\alpha'(\theta, p)$ denote that derivative. To estimate it, one can use under appropriate conditions (see Glasserman 1991 or L'Ecuyer 1990) the following infinitesimal perturbation analysis (IPA) estimator:

$$h'(\theta, p, \omega) = \sum_{i=1}^{T} \sum_{j=\phi_i(p)}^{i} \eta_j(p) S_j',$$

where $S_j' = (\partial B_\theta^{-1}/\partial\theta)(B_\theta(S_{j,1}))$, and $\phi_i(p) = \max\{j \mid 1 \leq j \leq i \text{ and } W_j(p) = 0\}$ if that set is non-empty, $\phi_i(p) = 1$ otherwise. In other words, $\phi_i(p)$ is the first customer with index ≥ 1 in the busy period to which customer i belongs, for that value of p. To estimate $\alpha'(\theta, \cdot)$ in functional form, just apply the same technique as above, with h replaced by h'. Note that this can also be accomplished with other kinds of derivative estimators.

References

Brémaud, P. and Vázquez-Abad, F. J., 1992. On the Pathwise Computation of Derivatives with Respect to the Rate of a Point Process: The Phantom RPA Method, *Queueing Systems*, **10**, 249–270.

Cassandras, C. G. and Julka, V., 1992. Marked/Phantom Slot Algorithms for a Class of Scheduling Problems, *Proceedings of the 31th IEEE Conf. on Decision and Control*, 3215–3220.

Glasserman, P., 1991. *Gradient Estimation via Perturbation Analysis*, Kluwer Academic.

Glynn, P. W., 1990. Likelihood Ratio Gradient Estimation for Stochastic Systems, *Communications of the ACM*, **33**, 75–84.

Glynn, P. W. and Iglehart, D. L., 1989. Importance Sampling for Stochastic Simulations. *Management Science* **35**, 1367–1392.

Ho, Y.-C. and Cao, X.-R., 1991. Discrete-Event Dynamic Systems and Perturbation Analysis. Kluwer Academic.

L'Ecuyer, P., 1990. A Unified View of the IPA, SF, and LR Gradient Estimation Techniques, *Management Science*, **36**, 1364–1383.

L'Ecuyer, P., 1991. An Overview of Derivative Estimation. *Proceedings of the 1991 Winter Simulation Conference*, 207–217.

L'Ecuyer, P., 1993. Two Approaches for Estimating the Gradient in Functional Form. *Proceedings of the 1993 Winter Simulation Conference*, to appear.

Lewis, P. A. W., and Shedler, G. S., 1979. Simulation of Nonhomogeneous Poisson Processes by Thinning, *Naval Research Logistics Quarterly*, **26**, 403–414.

Rubinstein, R. Y. and Shapiro, A., 1993. *Discrete Event Systems: Sensitivity Analysis and Stochastic Optimization by the Score Function Method*, Wiley.

Suri, R. and Cao, X.-R. 1986. The Phantom and Marked Customer Methods for Optimization of Closed Queueing Networks with Blocking and General Service Times. *ACM Performance Evaluation Review*, bf 12, 3, 243–256.

Vázquez-Abad, F. J. and L'Ecuyer, P., 1991. Comparing Alternative Methods for Derivative Estimation when IPA Does Not Apply Directly. *Proceedings of the 1991 Winter Simulation Conference*, 1004-1011.

Vázquez-Abad, F., 1993. Generalizations of the surrogate estimation approach for sensitivity analysis via simple examples. In preparation.

Infinitesimal Perturbation Analysis for Discrete Event Systems with Discrete Lifetime Distributions

Bernd Heidergott[1]

University of Hamburg, Institute for Mathematical Stochastics,Bundesstraße 55
D-20146 Hamburg, Germany

1 Introduction

Infinitesimal Perturbation Analysis (IPA) is a well established method for sensitivity analysis of Discrete Event Dynamic Systems (DEDS), see [1, 7]. During the last decade unbiasedness and strong consistency of IPA for performance measures like throughput, mean queue length or waiting times were proved for a broad class of systems, see [2, 3, 4].

Usually, these results are obtained under the assumption that the distributions of all lifetimes, such as service or interarrival times, are continuous. In our paper, we drop this assumption and establish unbiasedness and strong consistency of IPA for a broad class of timed performance measures for DEDSs having discrete lifetime distributions. The reason for this approach is twofold:

- For many real world systems, the assumption of continuous distributions fails to be true. For example, in manufacturing systems processing times are often constant or varying only over a finite range of values, that is, the lifetime distributions are discrete.
- In computer simulation, continuous distributions are only given via an approximation by discrete distributions. Therefore, applying IPA to computer simulation means to use IPA in an environment where all lifetime distributions are discrete.

The main difficulty caused by discrete lifetime distributions, is the positive probability of an simultaneous occurring of events, which may lead to non-differentiability of sample performance measures, see [8, 9].

The paper is organized as follows: In the subsequent section we state the mathematical framework needed for our analysis. Section 3 discusses the impact of discrete lifetime distributions on the pathwise construction of sample trajectories of a DEDS. In §4 we derive a modified perturbation propagation rule for IPA, such that we are able to prove unbiasedness and strong consistency of (modified) IPA for DEDSs with discrete lifetime distributions in §5.

2 The GSMS Framework

In this section we give a short description of the mathematical framework needed for our analysis. We use the notation introduced by Glasserman and refer to [2] for details.

For our purpose we define a Generalized Semi-Markov Scheme (GSMS) as a tuple (S, A, E, Φ, h, F, Θ), where S is the set of states and A is the set of possible events. By E we denote the "event list mapping", which for every state $s \in S$ yields a set of feasible events $E(s)$ associated with s. Φ is a mapping that specifies the new state $\Phi(s, e)$ reached from state s due to the occurrence of event e for all s and e, and h is the "selection rule" which selects an event $h(E(s))$ out of $E(s)$ (the meaning of h is described below). By $F = \{F_e : e \in A\}$ we denote the family of event lifetime distributions and we assume that the lifetime distribution of a particular event e^* depends on a parameter θ, with $\theta \in \Theta = (a, b) \subset \mathbb{R}$ for $a < b$. We denote by $X(e, n)$ the n-th lifetime of event e having c.p.f. F_e, i. e. $P(X(e, n) \leq x) = F_e(x)$, and we assume that the $X(e^*, n)$s are a.s. differentiable with respect to θ on whole Θ and satisfy some simple Lipschitz conditions, see [2, 3, 4] for details.

We assume that the GSMS is non-interruptive:

$$\forall e \in E(s): \qquad E(s) - \{e\} \subset E(\Phi(s, e))$$

which means, that all events in an old state s remain feasible in a new state $\Phi(s, e)$, except possibly for the event that activated the transition.

The $(n + 1)$ – th event, denoted by e_{n+1}, which causes the transition from state s_n to $s_{n+1} = \Phi(s_n, e_{n+1})$ is determined by random "clocks" $C_n(e)$ associated with every $e \in E(s_n)$. The event with the smallest remaining clock-time is the next to occur. However, if some lifetime distributions are discrete, e_{n+1} is not uniquely defined by the minimum of the clocks $C_n(e)$. Therefore we denote by H_{n+1} the set of all events with minimal lifetimes in s_n, i. e.

$$H_{n+1} := \{e \in E(s_n) \mid C_n(e) = c\}$$

with

$$c := \min\{C_n(e) \mid e \in E(s_n)\}$$

and choose a particular event out of H_{n+1} according to a "selection rule" h,

$$e' = h(H_{n+1}),$$

and let e' trigger the transition from s_n to s^h_{n+1} (the superscript denotes the dependence on h), thus we set $e^h_{n+1} = e'$. For $|H_{n+1}| \geq 2$, we call H_{n+1} a *synchronous* event.

Let $Z = \{Z_\theta : \theta \in \Theta\}$, with $Z_\theta = Z^h_\theta = (Z^h_t(\theta) : t \geq 0), \theta \in \Theta$, denote the family of stochastic processes given by the GSMS. With the help of h we are able to construct a unique sample path Z^h_θ by setting $Z^h_t(\theta) = s^h_n(\theta)$, for $t \in [\tau^h_n(\theta), \tau^h_{n+1}(\theta))$ where $\tau^h_n(\theta)$ denotes the jump epoch out of state $s^h_n(\theta)$. [1]

Remark. If s_n is determined by a synchronous event H_{n+1}, with $|H_{n+1}| = r > 1$, then $\tau_{n+1} = \cdots = \tau_{n+r}$, and an observer will see that the "event" H_{n+1} causes the process to jump from s_n to s_{n+r} at time τ_{n+1}.

[1] In the following, we will simplify the notation by omitting the θ and h indices where this causes no confusion.

We consider a class of timed performance measures L of Z_θ given by

$$L(\theta, n) = \frac{1}{\tau_n(\theta)} \sum_{i=0}^{n-1} f(s_i(\theta))(\tau_{i+1}(\theta) - \tau_i(\theta)) \tag{1}$$

where f is a bounded real valued function. For example, considering a single server station and setting f equal to one, if the server is busy and zero if not, $L(\theta, n)$ will denote the (transient) mean utilization of server j.

By (1) it is clear, that results on continuity and differentiability of the jump epochs are easily extended to $L(\theta, n)$. Therefore, we will restrict our further analysis on the jump epochs τ_n, $n \in \mathbb{N}$.

3 The Commuting Condition (Revisited)

The commuting condition, as introduced by Glasserman in [2], is a structural condition assuming

$$\forall s \in S : \forall e, e' \in E(s) : \quad \Phi(\Phi(s, e), e') = \Phi(\Phi(s, e'), e) \tag{c}$$

which simply means that the state reached is independent of the order of the events. If all lifetimes have continuous distributions, (c) implies a.s. continuity of the τ_n–s on Θ, see [2]. However, in presence of discrete distributions (c) has a different interpretation, see [5], which is given in the following lemma:

Lemma 1. *If (c) is satisfied, then Z_θ^h is independent of h.*

Proof. [2] Repeated application of h to an synchronous event H can be viewed as generating a permutation of the events in H. Since each permutation can be decomposed into a sequence of transpositions of adjacent events, (c) can be applied to show that Z_θ^h is independent of h. The complete proof follows by induction over all synchronous events. □

The meaning of Lemma 1 is, that a DEDS with discrete lifetime distributions, which does not satisfy the commuting condition, is not well defined, unless a selection rule is explicitly given. Typically, we simulate such a system without explicit definition of a selection rule and therefore apply the selection rule *implicitly* defined in the code of the simulation program. Unfortunately, we are not aware of this (hidden) selection rule and have no idea how it affects the results.

On the other hand, if the commuting condition is satisfied, we are free to choose h. We will use this freedom in the following section to construct selection rules h^+ and h^-, respectively, which will constitute continuity as well as one–sided differentiability, respectively, of τ_n.

---

[2] We give only sketches of the proofs, for details see [6].

4 Continuity and Differentiability

For the following, we need some additional notation. Let occ(e, k) denote the index of the transition triggered by the k-th occurrence of event e and let $gen(e, m)$ denote the index of the transition which caused the generation of the m-th lifetime of event e. Let further $N_e(n)$ denote the index of the current lifetime of e in τ_n, for $e \in E(s_n)$, i.e. $N_e(n) = m$ means, that at τ_n the (m)-th lifetime of e is running. With this notation, we rewrite τ_n as follows:

$$\tau_n = \tau_{\text{occ}(e_n, N_{e_n}(n))} = X(e_n, N_{e_n}(n)) + \tau_{gen(e_n, N_{e_n}(n))} \ . \tag{2}$$

From [2] we know the existence of a $\Delta_n > 0$, such that τ_m is a.s. differentiable on $(\theta - \Delta_n, \theta + \Delta_n)$ for all $m \leq n$, given no synchronous event occurred up to τ_n, i.e. $\tau_m < \tau_{m+1}$ for $m = 1 \cdots n - 1$. Suppose at time τ_{n+1} for the first time a synchronous event occurs and assume $H_{n+1} = \{\alpha, \beta\}$. How does H_{n+1} affect the differentiability of τ_{n+1}?

Suppose, we have

$$0 < \frac{d}{d\theta} \tau_{\text{occ}(\alpha, N_\alpha(n))} < \frac{d}{d\theta} \tau_{\text{occ}(\beta, N_\beta(n))} \ .$$

Note, that $d/d\theta \ \tau_{\text{occ}}$ is well defined by (2), since $gen(e, N_e(n)) < n+1$ for $e = \alpha, \beta$ and hence the differentiability of τ_{gen} follows from Glasserman [2], while the lifetimes are differentiable by assumption. α and β are unequally sensitive with respect to θ. Hence, H_{n+1} will split up for any change in θ. We have

$$\tau_{\text{occ}(\alpha, N_\alpha(n))}(\theta') \leq \tau_{\text{occ}(\beta, N_\beta(n))}(\theta') \quad \text{for } \theta' \in [\theta, \theta + \Delta_n)$$

and

$$\tau_{\text{occ}(\beta, N_\beta(n))}(\theta') \leq \tau_{\text{occ}(\alpha, N_\alpha(n))}(\theta') \quad \text{for } \theta' \in (\theta - \Delta_n, \theta]$$

Hence, τ_{n+1} behaves on $[\theta, \theta + \Delta_n)$ like $\tau_{\text{occ}(\alpha, N_\alpha(n))}$ and on $(\theta - \Delta_n, \theta]$ like $\tau_{\text{occ}(\beta, N_\beta(n))}$. An immediate consequence is, that τ_{n+1} is left differentiable on $(\theta - \Delta_n, \theta]$ with

$$\frac{d}{d\theta}^{-} \tau_{n+1} = \frac{d}{d\theta} \tau_{\text{occ}(\beta, N_\beta(n))} \quad ,$$

right differentiable on $[\theta, \theta + \Delta_n)$ with

$$\frac{d}{d\theta}^{+} \tau_{n+1} = \frac{d}{d\theta} \tau_{\text{occ}(\alpha, N_\alpha(n))}$$

and non-differentiable at θ, since

$$\frac{d}{d\theta}^{+} \tau_{n+1} \neq \frac{d}{d\theta}^{-} \tau_{n+1} \ .$$

The key observation is, that even though differentiability fails in the presence of a synchronous event, one–sided derivatives exist. To achieve one–sided differentiability, we have to order the elements of H_{n+1} according to their sensitivity with respect to θ, such that the event order is preserved for changes in θ. Therefore, we define selection rules h^+ and h^- by

$$h^+(H_{n+1}) = \arg_e\min\left\{\frac{d}{d\theta}^+ \tau_{\mathrm{occ}(e,N_e(n))} : e \in H_{n+1}\right\}$$

and

$$h^-(H_{n+1}) = \arg_e\max\left\{\frac{d}{d\theta}^- \tau_{\mathrm{occ}(e,N_e(n))} : e \in H_{n+1}\right\}$$

which simply means, that h^+ selects the event out of H_{n+1} which is least sensitive with respect to θ and h^- takes the event which is most sensitive with respect to θ. Now we are able to state the main result of our paper:

Lemma 2. *Suppose (c) is satisfied, then there exist selection rules h^+ and h^- such that:*

(i) $\tau_n^{h^+}$ *is right differentiable w.r.t. θ for all $n \in \mathbb{N}$*
(ii) $\tau_n^{h^-}$ *is left differentiable w.r.t. θ for all $n \in \mathbb{N}$*
(iii) τ_n *is continuous at θ for all $n \in \mathbb{N}$*

Proof. The arguments in the example are easily extended to the case of synchronous events with $|H_{n+1}| > 2$. Induction over all synchronous events using h^+ and h^- as defined above gives (i) and (ii). For the proof of (iii), observe that (i) and (ii) imply one–sided continuity of τ_n at θ. Hence, we have for all $n \in \mathbb{N}$

$$\lim_{\Delta\downarrow 0} \tau_n(\theta + \Delta) = \lim_{\Delta\downarrow 0} \tau_n^{h^+}(\theta + \Delta) = \tau_n^{h^+}(\theta)$$

and

$$\lim_{\Delta\downarrow 0} \tau_n(\theta - \Delta) = \lim_{\Delta\downarrow 0} \tau_n^{h^-}(\theta - \Delta) = \tau_n^{h^-}(\theta) \quad .$$

The proof follows by applying Lemma 2 to the equations above. □

5 Unbiasedness and Strong Consistency

The arguments of [2, 4] can be applied in the context of the results of §4. This enables us to obtain conditions for the unbiasedness and strong consistency of IPA for one–sided derivatives of the system performance.

Theorem 1 gives conditions for unbiasedness of IPA for one–sided derivatives of the expected performance over a finite time–horizon [3].

Theorem 1. *If (c) is satisfied and $E[\sup_{\theta\in\Theta} \tau_n] < \infty$ holds, then*

$$\frac{d}{d\theta}^{+/-} E[L^{+/-}(\theta, n)] = E\left[\frac{d}{d\theta}^{+/-} L^{+/-}(\theta, n)\right]$$

Proof. Using Lemma 2, the arguments given in [2] can be extended to the case of right and left derivatives. □

[3] The superscript $+/-$ refers to h^+ and h^-, respectively.

With Theorem 2 we give conditions for the strong consistency of IPA. We refer the reader to [3] for a detailed discussion on the "regenerative" condition in the theorem.

Theorem 2.
Suppose (c) is satisfied and let $\left\{(s_n, \tau_{n+1} - \tau_n, \frac{d}{d\theta}^{+/-}\tau_{n+1} - \frac{d}{d\theta}^{+/-}\tau_n) : n \in \mathbb{N}\right\}$ *be a regenerative process with regeneration points* $\{N_k : k \in \mathbb{N}\}$. *If in addition* $E[\tau_{N_1}] < \infty$, *then*

$$\lim_{n\to\infty} \frac{d}{d\theta}^{+/-} L^{+/-}(\theta, n) = \frac{d}{d\theta}^{+/-} E_\pi[f(s_0^{+/-}(\theta))] \quad a.e.\ on\ \Theta$$

holds, where π *denotes the stationary distribution of* Z_θ^h *and "a.e. on* Θ*" refers to the Lebesgue measure on the interval* Θ.

Proof. By lemma 2 and condition (i), π is well defined. Theorem 1 together with theorem 4 of [4] completes the proof. □

Theorems 1 and 2 suggest, that we have to simulate Z_θ^h either with h^+ or h^- to obtain estimators for right and left derivatives, respectively. But by Lemma 1, Z_θ^h is independent of h under the commuting condition. Thus, if (c) is satisfied, h^+ and h^- are not needed for the generation of the sample trajectories of Z_θ^h, but only for the correct "perturbation propagation". Therefore, we are able to estimate left and right derivatives simultaneously within a single simulation experiment by using h^+ and h^- as perturbation propagation rules for IPA.

6 Conclusion

We developed a modified perturbation propagation rule for IPA. We proved unbiasedness and strong consistency of the (modified) IPA estimator for the one–sided derivatives of performance measures for DEDSs having discrete lifetime distributions, provided that the DEDS satisfies the commuting condition.

It turned out, that the commuting condition can be interpreted as a condition for the completeness of the simulation model of a DEDS. This indicates, that the commuting condition is rather a restriction to the naive way of building a simulation model than to the applicability of IPA.

References

1. Glasserman, P.: Gradient Estimation via Perturbation Analysis. Kluwer Academic (1991)
2. Glasserman, P.: Structural conditions for perturbation analysis derivative estimation: finite-time performance indices. Operations Research **39** (1991) 724–738
3. Glasserman, P.: Regenerative derivatives of regenerative sequences. Adv. Appl. Prob. (1993) 116–139;
4. Glasserman,P. ,Hu, J.-Q. ,Strickland, S.: Strongly consistent steady-state derivative estimates. Prob. in the Eng. and Inf. Science **5** (1991) 391–413
5. Haas, P.J. ,Shedler G.S.: Simulations with simultaneous events. IBM Research Report RJ 5158(53498) (1986)

6. Heidergott, B.: Simultaneous events in queueing systems and their impact on perturbation analysis derivative estimation. Tech. Rep. No. 92-4 University of Hamburg (1992)
7. Ho, Y.-C., X.-R.: Perturbation Analysis of Discrete Event Dynamic Systems; Kluwer Academic (1991)
8. Li, S.,Ho, Y.-C.: Sample path and performance homogeneity of Discrete Event Dynamic Systems. JAC (1989) 907–915
9. Wardi, Y., McKinnon, W., Schuckle,R.: On perturbation analysis of queueing networks with finitely supported service time distributions. IEEE Trans. Automatic Control (1991) 863–867

Large Discrete Event Systems and Network Stability

Loss Networks in Thermodynamic Limit

Dimitri Botvich[1], *Guy Fayolle*[2] *and Vadim Malyshev*[2]

[1] Telecommunications, School of Electronic Engineering, Dublin City University, Glasnevin, Dublin 9, Ireland and
Laboratory of Large Random Systems, Faculty of Mechanics and Mathematics, Moscow State University, 119899, Moscow, Russia

[2] INRIA-Domaine de Voluceau, Rocquencourt, B.P. 105, 78153 Le Chesnay Cedex, France

1 Introduction

1.1 Abstract

The study deals with circuit switched loss networks (LN) in the *thermodynamic* limit. The model is the so-called *perturbed free loss network.* It consists of the superposition of a free LN and of a second LN, called the *perturbation.* In the free loss network a *fixed routing* is used and each route consists of only one link and different routes do not intersect. On the other hand, the perturbation can have a fixed, alternative or adaptive routing. Moreover, all (arrival) perturbation rates are proportional to some *perturbation parameter* $\epsilon > 0$. For sufficiently small $\epsilon > 0$, we prove that the limits

$$\lim_{\Lambda \nearrow \mathbb{R}^\nu} \lim_{t \to \infty} \mathbb{P}_t^{\Lambda,\epsilon}, \qquad \lim_{t \to \infty} \lim_{\Lambda \nearrow \mathbb{R}^\nu} \mathbb{P}_t^{\Lambda,\epsilon}$$

of the *time correlation functions* exist and are equal. When the perturbations involve only fixed routing, the limiting measure is the Gibbs one. In general the situation is more complicated since LNs with alternative or adaptive routing are no longer reversible. To cope with these more complicated situations, new cluster expansions for the limiting measure are obtained. In particular, limiting semi-groups are shown to depend analytically on ϵ. Our approach to these problems is based on a diagram estimation technique, previously applied to handle the dynamics of some quantum systems.

1.2 General Presentation

The aim of this study is to analyze circuit switched loss networks (LN) with fixed, alternative and adaptive routing in the *thermodynamic limit*, i.e. when these systems become infinitely large. Loss networks with fixed routing are suitable models of cellular radio systems, while LN with adaptive routing give a fair representation of large telecommunications systems. We note that both fixed and alternative routings are particular cases of *adaptive routings*, defined in §6. On the other hand, LN provide a special class of non space homogeneous interacting particle systems (e.g., see [12]). In particular, LN can be considered as *spin systems*, with a spin taking usually more than two values. The interaction in these systems is a kind of *hard core interaction* in the terminology of statistical mechanics.

In finite *volumes* (or areas), LN with fixed routing are *reversible*. This means, in particular, that their stationary measure has a *product form*, which can be written explicitly. In fact, this measure is a Gibbs one and the study of the thermodynamic limit has an equivalent formulation in terms of Gibbs fields. But, in general, LN operating with alternative or adaptive routing are no longer reversible, so that their stationary measures do not have any tractable analytic expression.

In the thermodynamic limit, the stationary measure of loss networks (as well as the Gibbs measure) can be non-unique. This case corresponds to *phase transition*. An example of translation invariant LN in dimension two is described in [13] (see also the review [10] for further references). In some cases in dimension one, translation invariant LN with fixed routing have no phase transition [6]; when they are not translation invariant, a phase transition phenomenon can exist, and then there are exactly two extreme stationary measures [6]. At this moment, it is worth remarking that the question of uniqueness of the stationary measure for one-dimensional LN with adaptive routing is still open. This problem does also arise in similar translation invariant models encountered in statistical mechanics, where the Gibbs measure in dimension one is unique, but in dimension more than one there exists a phase transition at sufficiently low temperature (in the context of LN, the role of the temperature is played by the arrival rates). In [10], a one-dimensional LN with fixed routing has been explicitly solved .

Our contribution here is to construct a new *cluster* expansion allowing us to analyze, in the thermodynamic limit, the dynamics and the stationary measure of a large variety of LN with adaptive routing and sufficiently small arrival intensities. This expansion proves to be extremely useful to deal with this class of problems and relies on an original approach, which consists in a diagram representation of measures. Similar methods have been applied to the dynamics of some quantum dynamical systems [2, 3, 5, 7].

The paper is organized as follows: In the next four sections, LN with fixed routing are considered; we prove several results and estimates (concerning in particular perturbed Markov chains), which lead to the fundamental Theorem 3 of §4. Then in §6 we introduce LN with alternative and adaptive routing and we explain why the proposed method works also in this more complicated situation.

2 Loss Networks and Thermodynamic Approximation

Following [11], in order to describe loss networks, we consider $G = (V, L)$ a non-oriented connected graph, where V is the *set of vertices* and L is the *set of links*. In the sequel, V and L are assumed to be countable. We suppose also that each link $g = (v, v') \in L$ comprises $c(g) \in \mathbb{Z}_+$ circuits, i.e. each link g has the *capacity* $c(g)$. Let $\mathcal{R}$ be a *set of routes*, where each *route* is just a finite subset of L. A call on route R uses $n_R(g) \in \mathbb{Z}_+$ circuits from link g. Calls requesting route R arrive according to a Poisson stream with intensity $\lambda_R \geq 0$. These Poisson streams are assumed to be mutually independent. A call of route $R = \{g_1, \dots, g_{|R|}\}$ is blocked and lost if, on some link $g_i \in R$, $i = 1, \dots, |R|$, there are less than $n_R(g_i)$ free circuits on link g_i. Otherwise the call is connected and occupies simultaneously $n_R(g_i)$ free circuits on g_i, for a duration exponentially distributed, with rate $\mu_R > 0$. Holding and arrival processes are supposed to be independent, as well as the successive holding periods. Hence a loss network $\mathcal{N}$ can be described formally by the array

$$\mathcal{N} = \{G, \mathcal{R};\ \lambda, \mu;\ c, n\}.$$

Remark. As well as for interacting particle systems (see [12]), the question of existence and uniqueness of this stochastic process is not straightforward. This problem will be considered in Section 4 and we assume for the moment that this process is well defined.

Thermodynamic approximation. It is natural to ask the question of how to approximate (in some sense) this process by a simpler one, for example, by a finite Markov chain. To this end, we introduce a formal scheme of *thermodynamic limit* or *thermodynamic approximation* of infinite loss networks.

Let be given a sequence of loss networks $\mathcal{N}^{(i)}$, $i \geq 1$, where $G^{(i)} = (V^{(i)}, L^{(i)})$ and (formally)

$$\mathcal{N}^{(i)} = \{G^{(i)}, \mathcal{R}^{(i)};\ \lambda^{(i)}, \mu^{(i)};\ c^{(i)}, n^{(i)}\},$$

so that all sets $V^{(i)}$, $L^{(i)}$, $G^{(i)}$, $\mathcal{R}^{(i)}$ are finite with

$$V^{(i)} \subset V^{(i+1)}, \quad \bigcup_i V^{(i)} = V, \qquad L^{(i)} \subset L^{(i+1)}, \quad \bigcup_i L^{(i)} = L,$$

$$\mathcal{R}^{(i)} \subset \mathcal{R}^{(i+1)}, \qquad \bigcup_i \mathcal{R}^{(i)} = \mathcal{R},$$

and, for all $j > i$,

$$\lambda^{(j)}{}_{|\mathcal{R}^{(i)}} = \lambda^{(i)}, \qquad \mu^{(j)}{}_{|\mathcal{R}^{(i)}} = \mu^{(i)},$$

$$c^{(j)}{}_{|\mathcal{R}^{(i)}} = c^{(i)}, \qquad n^{(j)}{}_{|\mathcal{R}^{(i)}} = n^{(i)}.$$

Then $\mathcal{N}$ will be referred to as the *thermodynamic limit* of the sequence of loss networks $\{\mathcal{N}^{(i)}, i \geq 1\}$. Whenever it is possible to associate each i with a volume Λ_i and to consider $\mathcal{N}^{(i)}$ as a localization of $\mathcal{N}$ in Λ_i (written later $\mathcal{N}_{\Lambda_i}$), one comes up with the standard notion of thermodynamic limit.

Markovian description of loss networks. An *admissible route configuration* for a loss network $\mathcal{N}$ is a function $\eta : \mathcal{R} \to \mathbb{Z}_+$, such that

$$\sum_{R \in \mathcal{R}: g \in R} \eta(R) n_R(g) \leq c(g), \tag{1}$$

for all $g \in G$, where $\eta(R) \geq 0$ denotes the number of calls requiring route $R \in \mathcal{R}$. By definition, $\eta(R) = 0$ means that there are no calls needing route R for the configuration η. Correspondingly, an *admissible route configuration in volume* Λ in a loss network $\mathcal{N}_\Lambda$ is a function $\eta^\Lambda : cal R_\Lambda \to \mathbb{Z}_+$ such that

$$\sum_{R \in \mathcal{R}_\Lambda : g \in R} \eta^\Lambda(R) n_R(g) \leq c(g), \quad \forall g \in G_\Lambda . \tag{2}$$

Denote by $\mathcal{A}$ (resp. $\mathcal{A}_\Lambda$) the *set of all admissible route configurations* (resp. in volume Λ). Let

$$\eta_t = \{\eta_t(R),\ R \in \mathcal{R}\}, \quad t \in \mathbb{R}_+ ,$$

respectively

$$\eta_t^\Lambda = \{\eta_t^\Lambda(R),\ R \in \mathcal{R}_\Lambda\}, \quad t \in \mathbb{R}_+$$

be the Markov process describing $\mathcal{N}$ (resp. $\mathcal{N}_\Lambda$). Thus, for any finite volume Λ, η_t^Λ is a continuous time homogeneous Markov chain with finite state space $\mathcal{A}_\Lambda$.

Generators of finite loss networks. Let H_Λ be the generator of the Markov chain η_t^Λ. Then the probability $\mathbb{P}_t^\Lambda(\eta_\Lambda)$ of η_t^Λ of being in state η_Λ at time t is given by

$$\mathbb{P}_t^\Lambda(\eta_\Lambda) = \mathbb{P}_0^\Lambda \ \exp(t \ H_\Lambda)(\eta_\Lambda)\,,$$

for any initial distribution $\mathbb{P}_0^\Lambda$.

Let $\mathcal{B}(\mathcal{A}_\Lambda)$ denote the finite dimensional space $\mathcal{B}(\mathcal{A}_\Lambda)$ of real functions $f : \mathcal{A} \to \mathbb{R}$. The generator H_Λ has the following local structure:

$$H_\Lambda = \sum_{R \in \mathcal{R}_\Lambda} (V_R^{\text{in}} + V_R^{\text{out}}), \tag{3}$$

where V_R^{in}, V_R^{out} are bounded linear operators : $\mathcal{B}(\mathcal{A}_\Lambda) \to \mathcal{B}(\mathcal{A}_\Lambda)$, corresponding respectively to arrivals and services of calls on route R. For each admissible route configuration η^Λ,the operator V_R^{in} transforms the δ-measure

$$\delta_{\eta^\Lambda}(\cdot) \equiv \delta(\cdot - \eta^\Lambda)$$

into the measure

$$\begin{cases} \lambda_R(\delta_{\eta^\Lambda+\delta_R} - \delta_{\eta^\Lambda}) & \text{if } \eta^\Lambda + \delta_R \in \mathcal{A}_\Lambda\,, \\ 0 & \text{otherwise,} \end{cases} \tag{4}$$

and all other δ-measures into 0, where, by definition, we put

$$\delta_R(R') = \begin{cases} 1, & R = R'\,, \\ 0, & R \neq R'\,, \end{cases}$$

and

$$(\eta^\Lambda + \delta_R)(R') = \begin{cases} \eta^\Lambda(R') + 1 & \text{if } R = R'\,, \\ \eta^\Lambda(R') & \text{otherwise.} \end{cases} \tag{5}$$

Analogously, V_R^{out} transforms $\delta_{\eta^\Lambda}(\cdot)$ into the measure

$$\begin{cases} \eta^\Lambda(R)\ \mu_R\ (\delta_{\eta^\Lambda-\delta_R} - \delta_{\eta^\Lambda}) & \text{if } \eta^\Lambda(R) > 0\,, \\ 0 & \text{otherwise,} \end{cases} \tag{6}$$

and all other δ-measures into 0.

In other words, we have

$$V_R^{\text{in}} f(\eta^\Lambda) = \begin{cases} \lambda_R(f(\eta^\Lambda + \delta_R) - f(\eta^\Lambda)) & \text{if } \eta^\Lambda + \delta_R \in \mathcal{A}_\Lambda\,, \\ 0 & \text{otherwise,} \end{cases} \tag{7}$$

and

$$V_R^{\text{out}} f(\eta^\Lambda) = \begin{cases} \eta^\Lambda(R)\mu_R(f(\eta^\Lambda - \delta_R) - f(\eta^\Lambda)) & \text{if } \eta^\Lambda(R) > 0\,, \\ 0 & \text{otherwise.} \end{cases} \tag{8}$$

Time correlation functions. For any finite set $A \subset \mathcal{R}$ and any admissible route configuration $\eta \in \mathcal{A}$, the quantity $s_A = (s_R \equiv \eta(R) \geq 0,\ R \in A)$ is called a *(finite) route subconfiguration* in η. For a given s_A, we introduce the *time correlation function* (or simply *correlation function*) given by

$$\mathbb{P}_t(s_A) \equiv \mathbb{P}_t^{\mathcal{N}}(s_A) = \mathbb{P}(\eta_t(R) = s_R,\ R \in A), \quad t \in \mathbb{R}_+\,, \tag{9}$$

for some initial distribution $\mathbb{P}_0^{\mathcal{N}}$. Accordingly, the *time correlation function in volume* Λ is given by

$$\mathbb{P}_t^\Lambda(s_A) \equiv \mathbb{P}_t^{\Lambda,\mathcal{N}}(s_A) = \mathbb{P}(\eta_t^\Lambda(R) = s_R,\ R \in A)\,, \quad t \in \mathbb{R}_+\,, \tag{10}$$

for some initial distribution $\mathbb{P}_0^{\Lambda,\mathcal{N}}$ and $A \subset \mathcal{R}_\Lambda$.

Remark. The Markov process η_t^Λ describing a loss network with fixed routing in volume Λ is reversible and its stationary measure is given by, after setting $\rho_R \equiv \frac{\lambda_R}{\mu_R}$,

$$\pi^\Lambda(\eta^\Lambda) \equiv \mathbb{P}_\infty^\Lambda(\eta^\Lambda) = \lim_{t\to\infty} \mathbb{P}_t^\Lambda(\eta^\Lambda) = Z_\Lambda^{-1} \prod_{R\in\mathcal{R}_\Lambda} \frac{(\rho_R)^{\eta^\Lambda(R)}}{(\eta^\Lambda(R))!}\,, \tag{11}$$

for each $\eta^\Lambda \in \mathcal{A}_\Lambda$, where Z_Λ is the normalizing constant (or *partition function*)

$$Z_\Lambda = \sum_{\eta^\Lambda\in\mathcal{A}_\Lambda} \prod_{R\in\mathcal{R}_\Lambda} \frac{(\rho_R)^{\eta^\Lambda(R)}}{(\eta^\Lambda(R))!}\,. \tag{12}$$

Therefore, when $t \to \infty$, the correlation function has a limit given by

$$\pi^\Lambda(s_A) \equiv \lim_{t\to\infty} \mathbb{P}_t^\Lambda(s_A) = Z_\Lambda^{-1} \left(\sum_{\substack{\eta^\Lambda\in\mathcal{A}_\Lambda,\\ \eta^\Lambda(R)=s_R,\, R\in A}} \prod_{R\in\mathcal{R}_\Lambda} \frac{(\rho_R)^{\eta^\Lambda(R)}}{(\eta^\Lambda(R))!} \right), \tag{13}$$

where $\pi^\Lambda(s_A)$ is the stationary probability of having a finite route subconfiguration s_A in a route configuration.

3 Loss Networks in $\mathbb{R}^\nu$

Here we introduce a class of loss networks which can be naturally localized in $\mathbb{R}^\nu$.

Let $\nu, d \in \mathbb{Z}_+$ be fixed. Denote by Γ_d the set of non-oriented connected graphs $G=(V, L)$ with a finite or countable number of vertices such that each vertex $v \in V$ has at most d adjacent links $\in L$. For fixed parameters $0 < D_1 < D_2 < \infty$, let $\Gamma_d^\nu(D_1, D_2)$ be a subset of Γ_d such that, for each $G = (V, L) \in \Gamma_d^\nu(D_1, D_2)$, there is a function $X : V \to \mathbb{R}^\nu$ satisfying the following properties: $(x_v \equiv X(v),\ v \in V)$

1. $\|x_v - x_{v'}\| \geq D_1$, if $v \neq v'$.
2. if $\|x_v - x_{v'}\| \geq D_2$ then $(v, v') \notin L$,

where $x_v \equiv X(v), \forall v \in V$ and, $\forall x = (x_1, \ldots, x_\nu) \in \mathbb{R}^\nu$,

$$\|x\| = \max_i \ |x_i|\,.$$

Thus $\Gamma_d^\nu(D_1, D_2)$ consists of graphs from Γ_d, for which there is a function X such that the set of points $\{x_v, v \in V\}$ has the *hard core* $D_1 > 0$ and that there are no links longer than D_2.

Example 1. The regular lattice $\mathbb{Z}^\nu$ belongs to $\Gamma_d^\nu(D_1, D_2)$, with $d = 2\nu$ and D_1, D_2 are constants subject to the inequalities $0 < D_1 < 1$, $D_2 > 1$.

Next we fix a graph $G = (V, L) \in \Gamma_d^\nu(D_1, D_2)$ together with a function X satisfying the above conditions. To define the thermodynamic limit in a convenient way, we also fix some vertex $v_0 \in V$, remarking that the limiting LN will not depend on the choice of v_0.

Graphs and boundary conditions in finite volumes. For a finite *volume* (an open bounded set) $\Lambda \subset \mathbb{R}^\nu$, we denote by $G_\Lambda = (V_\Lambda, L_\Lambda)$ the maximal connected component of the restriction of the graph G to the volume Λ containing x_{v_0}. Next, we chose the set of routes $\mathcal{R}_\Lambda$ of $\mathcal{N}$ as follows: $\mathcal{R}_\Lambda$ consists of routes $R = \{g_1, \ldots, g_{|R|}\}$ such that $g_i \in L_\Lambda$, for each $i = 1, \ldots, |R|$. In fact, this choice, for each Λ, of the set $\mathcal{R}_\Lambda$ corresponds to the choice of a boundary condition for a LN in a finite volume. Other (more general) boundary conditions will not be considered in this study, although the same methods could also be applied.

Thermodynamic limit of loss networks. A loss network $\mathcal{N}_\Lambda$ in volume Λ can be described by

$$\mathcal{N}_\Lambda = \{G_\Lambda, \mathcal{R}_\Lambda;\ \lambda^\Lambda, \mu^\Lambda;\ c^\Lambda, n^\Lambda\ \},$$

where λ^Λ, μ^Λ and c^Λ and n^Λ are given by the corresponding restrictions of λ, μ, c, n. Given a sequence of volumes $\Lambda_1 \subset \Lambda_2 \subset \cdots \subset \Lambda_i \subset \ldots$, such that $x_{v_0} \in \Lambda_i, \forall i \geq 1$ and

$$\bigcup_{i=1}^{\infty} \Lambda_i = \mathbb{R}^\nu\,,$$

we say that $\mathcal{N}$ is the *thermodynamic limit* of the sequence $\{\mathcal{N}_{\Lambda_i}, i \geq 1\}$.

In the sequel, we shall impose some technical conditions on these loss networks.

Condition 1. The function $c : L \to \mathbb{Z}_+$ is uniformly bounded:

$$\|c\|_\infty \equiv \sup_{g \in L} c(g) < \infty. \tag{14}$$

Condition 2. The arrival intensities are *uniformly bounded*:

$$\sup_{R \in \mathcal{R}} \lambda_R\ < \infty.$$

Condition 3. The routes are *uniformly bounded*: there exists a constant $D > 0$, such that $|R| \leq D$, for all $R \in \mathcal{R}$.

Condition 3*. The arrival intensities have an *exponential decay* with parameter $r > 0$:

$$\|\lambda\|_r \equiv \sup_{g \in L} \left(\sum_{k=1}^{\infty} e^{rk} \sum_{\substack{R: g \in R \\ |R|=k}} \lambda_R \right) < \infty. \tag{15}$$

Condition 4. The output intensities are *uniformly positive and bounded*:

$$\sup_{R \in \mathcal{R}} \mu_R < \infty, \qquad \mu_b \equiv \inf_{R \in \mathcal{R}} \mu_R > 0. \tag{16}$$

Clearly, Condition 3* follows from Conditions 2-3.

Routes. Formally, no further conditions need to be imposed on the routes $\mathcal{R}$, except Conditions 3 and 3*. But in realistic models it is natural to consider the following classes of routes: *connected* routes and *connected self-avoiding* routes. Connected route R has the form $\{g_1, \ldots, g_{|R|}\}$, $g_i \in L$, $i = 1, \ldots, |R|$, where the graph with links $g_i = (v_i, v_{i+1})$, $i = 1, \ldots, |R|$ and with vertices $\cup_{i=1}^{|R|+1} v_i$ is connected. We denote by $\mathcal{R}^c$ the set of connected routes in the LN and by $\mathcal{R}^c_g$, $g \in L$, the set of routes containing the link g. Connected self-avoiding route R have the form $\{g_1, \ldots, g_{|R|}\}$, $g_i = (v_i, v_{i+1}) \in L$, $i = 1, \ldots, |R|$, where the vertices $v_1, \ldots, v_{|R|}, v_{|R|+1}$ are distinct. Analogously, $\mathcal{R}^{sa}$ will denote the set of connected self-avoiding routes. Clearly, $\mathcal{R}^{sa} \subset \mathcal{R}^c$. Routings from $\mathcal{R}^c$ are used in models of cellular radio networks, while routes from $\mathcal{R}^{sa}$ rather apply to telephone networks.

4 Main Results

First we will consider the question of existence and uniqueness of the dynamics of a loss network in an infinite volume. Some general results about existence and uniqueness of the dynamics of interacting particle systems [12] will be used.

Let $\mathcal{N} = \{G, \mathcal{R};\ \lambda, \mu;\ c, n\}$. The state space of the Markov process η_t is the set $\mathcal{A}$ of all admissible route configurations. Condition 1 is supposed to hold. Let $N = \|c\|_\infty < \infty$ and let $S = \{0, 1, \ldots, N\}$ be a topological space endowed with the discrete topology. Then $S^{\mathcal{R}}$ is a compact metric space with the product topology. Since $\mathcal{A}$ is closed in $S^{\mathcal{R}}$, $\mathcal{A} \subset S^{\mathcal{R}}$ is also a compact metric space with the induced topology. Denote by $C(\mathcal{A})$ the set of continuous functions on $\mathcal{A}$, regarded as a Banach space with the norm

$$\|f\| = \sup_{\eta \in \mathcal{A}} \|f(\eta)\|.$$

For $f \in C(\mathcal{A})$ and $R \in \mathcal{R}$, let

$$\Delta_f(R) = \sup_{\substack{\eta, \eta' \in \mathcal{A}: \\ \eta(R') = \eta'(R'), \forall R' \neq R}} \left| f(\eta) - f(\eta') \right|.$$

This should be thought of as a measure of the extent the function f depends on the number of calls on route R. Define the set of functions

$$D(\mathcal{A}) = \left\{ f \in C(\mathcal{A}) : \sum_{R \in \mathcal{R}} \Delta_f(R) < \infty \right\}$$

which will play the role of a core for the generator of the Markov process η_t, $t \in \mathbb{R}_+$. Clearly, $D(\mathcal{A})$ is dense in $C(\mathcal{A})$.

For $\eta \in \mathcal{A}$, $R \in \mathcal{R}$, $m \in \mathbb{N}$, denote by

$$c_R(\eta, m) = \begin{cases} \lambda_R & \text{if } \eta(R) = m - 1 \text{ and } \eta^{(m,R)} \in \mathcal{A}, \\ (m+1)\mu_R & \text{if } \eta(R) = m + 1, \\ 0 & \text{otherwise,} \end{cases}$$

where $\eta^{(m,R)}(R) = m$ and $\eta^{(m,R)}(R') = \eta(R')$, for all $R' \neq R$.

Theorem 1. *Let $G \in \Gamma_d^{\nu}(D_1, D_2)$, $0 < D_1 < D_2 < \infty$. Assume condition 1 is satisfied for the loss network $\mathcal{N}$ and that*

$$\sup_{R \in \mathcal{R}} \max(\lambda_R, \mu_R) < \infty. \tag{17}$$

Then, for $f \in D(\mathcal{A})$, the series

$$Hf(\eta) = \sum_{R \in \mathcal{R}} \sum_{n \in \mathbb{N}} c_R(\eta, n) \left(f(\eta^{(n,R)}) - f(\eta) \right) \tag{18}$$

converges uniformly and defines a function in $C(\mathcal{A})$. Moreover, H is a Markov pregenerator and its closure $\overline{H}$ is a Markov generator of a Markov semigroup in $C(\mathcal{A})$ with the core $D(\mathcal{A})$.

The proof of Theorem 1 is identical to the proof of Proposition 3.2 in Chapter 1 of [12].

Corollary 2. *Under the conditions of Theorem 1, the dynamics of $\mathcal{N}$ (i.e. the Markov process η_t, $t \in \mathbb{R}_+$) is well defined and unique.*

Let

$$d_R = \sup_{\eta \in \mathcal{A}} \eta(R) \quad \forall R \in \mathcal{R}.$$

We say that two distinct routes $R, R' \in \mathcal{R}$ are *independent* if there is a route configuration $\eta \in \mathcal{A}$ such that $\eta(R') = d_R$ and $\eta(R) > 0$. For $R \in \mathcal{R}$, we denote by I_R the number of routes $R' \neq R \in \mathcal{R}$ which are not independent of R.

Let also

$$M = \sup_{R} \sum_{R' \neq R} c_R(R'), \tag{19}$$

where

$$c_R(R') = \begin{cases} \sup\limits_{\substack{\eta_1, \eta_2 \in \mathcal{A}: \\ \eta_1(\tilde{R}) = \eta_2(\tilde{R}), \tilde{R} \neq R'}} \sup\limits_{n \in \mathbb{N}} |c_R(\eta_1, n) - c_R(\eta_2, n)| & \text{if } R \neq R', \\ 0 & \text{otherwise.} \end{cases}$$

Clearly, $c_R(R')$ can take only the values λ_R or 0. More precisely, $c_R(R') = 0$ if R is independent of R' or if $\text{supp}(R) \cap \text{supp}(R') = \varnothing$; in all other cases, $c_R(R')$ is equal to λ_R, otherwise. Thus

$$M = \sup_{R \in \mathcal{R}} I_R \lambda_R .$$

Let

$$\varepsilon = \inf_{R \in \mathcal{R}} \inf_{\substack{\eta_1, \eta_2 \in \mathcal{A}: \eta_1(R) \neq \eta_2(R), \\ \eta_1(R') = \eta_2(R'), R' \neq R}} \{c_R(\eta_1, \eta_2(R)) + c_R(\eta_2, \eta_1(R))\}.$$

It is easy to see that

$$\epsilon \geq \inf_{R \in \mathcal{R}} (\lambda_R + \mu_R).$$

Theorem 3. *Assume all conditions of Theorem 1 are satisfied (including inequality (17)) for $\mathcal{N}$. If*

$$\sup_{R \in \mathcal{R}} I_R \lambda_R < \epsilon , \tag{20}$$

then the process η_t is ergodic.

The proof of Theorem 3 mimics the proof of Theorem 4.1 in Chapter 1 of [12], since (20) is equivalent to the condition $M < \epsilon$.

Corollary 4. *Under the conditions of Theorem 3, the loss network $\mathcal{N}$ is ergodic.*

Corollary 5. *If the conditions of Theorem 3 hold and (17) is replaced by the inequality*

$$\sup_{R \in \mathcal{R}} I_R < \inf_{R \in \mathcal{R}} (\lambda_R + \mu_R) , \tag{21}$$

then the process η_t is ergodic.

Remark. From corollary 5, it follows that, if conditions 1-4 are satisfied then, for λ_R, $R \in \mathcal{R}$ sufficiently small, $\mathcal{N}$ is ergodic.

4.1 Perturbing a Free Loss Network

A loss network

$$\mathcal{N}^f = \{G, \mathcal{R}^f;\ \lambda^f, \mu^f;\ c^f, n^f\} \tag{22}$$

is said to be *free* if each route consists of only one link, i.e. $R \in \mathcal{R}^f$ if $R = \{g\}$, for some $g \in L$. Moreover, in the sequel, we shall always suppose that different links correspond to different routes.

For two loss networks $\mathcal{N}_1, \mathcal{N}_2$ having the same graphs capacities, such that $\mathcal{R}_1 \cap \mathcal{R}_2 = \varnothing$, the sum $\mathcal{N} = \mathcal{N}_1 + \mathcal{N}_2$ can be defined in a obvious way. Let

$$\mathcal{N}(\epsilon) = \{G, \mathcal{R};\ \epsilon_{\text{in}}\lambda, \epsilon_{\text{out}}\mu;\ c, n\} , \tag{23}$$

where $\epsilon = (\epsilon_{\text{in}}, \epsilon_{\text{out}}) \in \mathbb{R}_+^2$.

Definition 6. Let $\mathcal{N}^f$ and $\mathcal{N}^p$ denote respectively a free loss network and a loss network which has no one-link route. The loss network

$$\mathcal{N}_\epsilon^{pf} = \mathcal{N}^f + \mathcal{N}^p(\epsilon), \tag{24}$$

will be referred to as a *perturbed free loss network* with the *perturbation* $\mathcal{N}^p(\epsilon)$, where ϵ is the *perturbation parameter*, $\epsilon \equiv (\epsilon_{\text{in}}, \epsilon_{\text{out}}) \in \mathbb{R}_+^2$. The operator

$$H_\Lambda^f = \sum_{R \in \mathcal{R}_\Lambda^f} \underbrace{1 \otimes \cdots \otimes h_R \otimes \cdots \otimes 1} \tag{25}$$

is the generator of $\mathcal{N}_\Lambda^f$ in volume Λ, where h_R ($R = \{g\}$) denotes the generator of a finite Markov chain with state space

$$\left\{0, n_R(g), \ldots, \left[\frac{c(g)}{n_R(g)}\right] n_R(g)\right\}.$$

Also let us denote by $H_{\Lambda,\epsilon} \equiv H_{\Lambda,\epsilon}^{pf}$ the generator of a perturbed free loss network in volume Λ. Then

$$H_{\Lambda,\epsilon} = H_\Lambda^f + \epsilon_{\text{out}} V_\Lambda^{\text{out}} + \epsilon_{\text{in}} V_\Lambda^{\text{in}}, \tag{26}$$

where the generator $\epsilon_{\text{out}} V_\Lambda^{\text{out}}$ (resp. $\epsilon_{\text{in}} V_\Lambda^{\text{in}}$) describes the service mechanism (resp. the arrival process) of calls in $\mathcal{N}^p(\epsilon)$. Let

$$H_{\Lambda,0} \stackrel{\text{def}}{\equiv} H_\Lambda^f + \epsilon_{\text{out}} V_\Lambda^{\text{out}} \tag{27}$$

be the generator of the Markov process where the arrivals in $\mathcal{N}^p(\epsilon)$ are cut off. Hence

$$H_{\Lambda,\epsilon} = H_{\Lambda,0} + \epsilon_{\text{in}} V_\Lambda^{\text{in}}. \tag{28}$$

For a finite route subconfiguration s_A, $|A| < \infty$, we denote by $\mathbb{P}_t^{\Lambda,\epsilon}(s_A)$ (resp. $\mathbb{P}_t^{\Lambda,0}(s_A)$) the correlation functions of the Markov process with generator $H_{\Lambda,\epsilon}$ (resp. $H_{\Lambda,0}$) and by

$$\pi_\epsilon^\Lambda(s_A) \equiv \mathbb{P}_\infty^{\Lambda,\epsilon}(s_A) = \lim_{t\to\infty} \mathbb{P}_t^{\Lambda,\epsilon}(s_A) \tag{29}$$

the corresponding stationary probabilities.

In order to formulate our main result, we will introduce now the notion of *A-connected diagrams*. Let $R_i = (g_1^{(i)}, \ldots, g_{l_i}^{(i)})$, $1 \le i \le k$, $g_j^{(i)} \in L_\Lambda$, $1 \le j \le l_i$ be a given route and denote by

$$\tilde{R}_i = \bigcup_j g_j^{(i)}$$

the *support* of this route.

Definition 7 (*A*-**connected diagrams**) . A *diagram* D is a sequence $((R_1, t_1), \ldots, (R_k, t_k))$, where $R_i \in \mathcal{R}$, $t_i \in \mathbb{R}$, for $1 \le i \le k$. Let A be a set of routes. The diagram D is said to be *A-connected* if

$$\tilde{R}_i \cap \sigma_{i-1}(A) \neq \varnothing\,, \quad \forall i = 1, \ldots, k\,,$$

where

$$\sigma_{i-1}(A) = \sigma_0(A) \cup \tilde{R}_1 \cup \cdots \cup \tilde{R}_{i-1}, \quad \sigma_0(A) = \tilde{A} \stackrel{\text{def}}{\equiv} \bigcup_{R\in A} \tilde{R}.$$

Notation. Hereafter multiple integrals and summations will frequently occur. To render formulas of a reasonable size, we shall write typically

$$\mathbf{ds}_k = ds_1\, ds_2 \dots ds_k\,,$$

and

$$\int_{[0,\infty]^n} F\, \mathbf{dt}_n = \int_0^\infty \cdots \int_0^\infty F dt_{\_} \dots dt_n$$

and

$$\int_{\Delta_n^t} F(t, s_1 s_2 \dots s_n)\, \mathbf{ds}_n = \int_0^t \int_0^{s_1} \cdots \int_0^{s_n} F(t, s_1 \dots s_n)\, ds_1\, ds_2 \dots ds_n$$

where, for example, $0 \le s_n \le \cdots \le s_1 \le t$

Theorem 8. *Let $G \in \Gamma_d^\nu(D_1, D_2)$, $0 < D_1 < D_2 < \infty$ and $\epsilon_{\text{out}} > 0$ be fixed. Suppose conditions 1, 2, 3 (or 3* with a parameter $r > 0$ sufficiently large) and 4 are satisfied for the networks $\mathcal{N}^f$ and $\mathcal{N}^p$. Then there are constants $\epsilon_0 > 0$, $C > 0$, such that, for all $\epsilon \equiv \epsilon_{\text{in}} \in [0, \epsilon_0]$, the Markov process η_t^ϵ, which describes $\mathcal{N}_\epsilon^{pf}$, is ergodic and, for each finite route subconfiguration $s_A = \{s_{R_1}, \dots, s_{R_k}\}$, $R_i \in \mathcal{R}$, $1 \le i \le k < \infty$, the following limits exist and are equal:*

$$\lim_{t\to\infty} \lim_{\Lambda \nearrow \mathbb{R}^\nu} \mathbb{P}_t^{\Lambda,\epsilon}(s_A) = \lim_{\Lambda \nearrow \mathbb{R}^\nu} \lim_{t\to\infty} \mathbb{P}_t^{\Lambda,\epsilon}(s_A) = \pi_\epsilon(s_A). \tag{30}$$

Moreover,

$$\pi_\epsilon(s_A) = \sum_{k=0}^\infty \epsilon^n C_n(s_A), \tag{31}$$

$$C_n(s_A) = \sum_{\substack{R_i \in \mathcal{R},\\ 1\le i\le n}} \int_{[0,\infty]^n} \pi_0 V_{R_n}^{\text{in}} \exp(t_n H_0) \dots V_{R_1}^{\text{in}} \exp(t_1 H_0)(s_A)\, \mathbf{dt}_n \tag{32}$$

where the summations and integrals are taken over all A-connected diagrams

$$D = ((R_1, t_1), \dots, (R_n, t_n))\ .$$

Moreover the constants $C_n(s_A)$ are independent of ϵ and there exists a constant $C(s_A) > 0$, independent of n, such that

$$|C_n(s_A)| \le C^n C(s_A). \tag{33}$$

Thus the series (31) is convergent for $|\epsilon| \le \frac{1}{C}$ and $\pi_\epsilon(s_A)$ depends analytically on ϵ.

The proof is given in the next section.

Remark. The existence of the first limit in (30) follows from Theorem 3.

Remark. When the loss network has a fixed routing, the Markov process $\eta_t^{\Lambda,\epsilon}$ is reversible and

$$\lim_{t\to\infty} \mathbb{P}_t^{\Lambda,\epsilon}$$

is the Gibbs measure in Λ. Therefore, studying the second limit in (30) is equivalent to characterizing the corresponding Gibbs measure in the thermodynamic limit, so that one can use powerful tools of statistical mechanics. In particular, for the measure π_ϵ, one can construct various cluster expansions (e.g. see [16]). On the other hand, the study of the first limit in (30) is usually more difficult. In fact, we will construct a new cluster expansion, allowing us to control both limits simultaneously and the method works also in non-reversible situations (e.g. for LNs with alternative or adaptive routing), as will show the generalization of Theorem 8 in §6.

5 Proof of Theorem 8

We prove this theorem under condition 3, i.e. when the routes and the input intensities are uniformly bounded. The case where the input intensities have an exponential decrease (see condition 3*) needs some slight modifications of the proof which will be omitted.

One shall proceed along the following lines. First, using perturbation series (see (60), Appendix B) for each finite volume Λ and route subconfiguration $s_A \subset \mathcal{R}_\Lambda$, we represent $\mathbb{P}_t^{\Lambda,\epsilon}(s_A)$ as a sum of A-connected diagrams in Λ. Next, an exponential bound will be proved for each diagram, which in turn will yield a cluster bound for the sum of all A-connected diagrams. Here a crucial role will be played by the so-called universal cluster bound (see Appendix A). Similar cluster bounds have been used to handle the dynamics of some infinite quantum systems (see e.g. [4]). In particular, this will show that the limits

$$\lim_{\Lambda\nearrow\mathbb{R}^\nu}\lim_{t\to\infty}\mathbb{P}_t^{\Lambda,\epsilon}(s_A),\qquad \lim_{t\to\infty}\lim_{\Lambda\nearrow\mathbb{R}^\nu}\mathbb{P}_t^{\Lambda,\epsilon}(s_A)$$

exist and are equal, for any sufficiently small $\epsilon \geq 0$.

It will be convenient to write, up to a slight distortion in the notation, $V_\Lambda \equiv V_\Lambda^{\text{in}}$, $V_R \equiv V_R^{\text{in}}$ for all Λ, R. Now, let a finite route subconfiguration s_A and the initial measure P_0^Λ be fixed. Then (see (60), Appendix B)

$$\mathbb{P}_t^{\Lambda,\epsilon}(s_A) = \sum_{k=0}^{\infty} \epsilon^k \int_{\Delta_k^t} \mathcal{K}(t,k;\mathbf{s}_k)\,\mathbf{ds}_k\,, \tag{34}$$

where we have set

$$\mathcal{K}(t,k;\mathbf{s}_k) = P_0^\Lambda \exp\left(s_k H_{\Lambda,0}\right)V_\Lambda \exp\left((s_{k-1}-s_k)H_{\Lambda,0}\right)\ldots V_\Lambda \exp\left((t-s_1)H_{\Lambda,0}\right)(s_A)\,.$$

Hereafter, a *diagram* D will often mean a time-dependent k-uple, still denoted by

$$D = ((R_1,s_1),\ldots,(R_k,s_k))\,,\quad R_i \in \mathcal{R}\,,\quad \text{with } 0 \leq s_k \leq \cdots \leq s_1 \leq t\,,$$

and its cardinality $|D|$ is equal to k. The k-th term of series in (34) is equal to

$$\begin{aligned}
&\int_{\Delta_k^t} P_0^\Lambda \exp\left(s_k H_{\Lambda,0}\right)V_\Lambda \exp\left((s_{k-1}-s_k)H_{\Lambda,0}\right)\ldots V_\Lambda \exp\left((t-s_1)H_{\Lambda,0}\right)(s_A)\,\mathbf{ds}_k\\
&\qquad = \sum_{R_1\in\mathcal{R}_\Lambda}\cdots\sum_{R_k\in\mathcal{R}_\Lambda}\int_{\Delta_k^t} I(D,t)(s_A)\,\mathbf{ds}_k\\
&\qquad = \sum_{R_1\in\mathcal{R}_\Lambda}\cdots\sum_{R_k\in\mathcal{R}_\Lambda}\mathcal{I}(\hat{D},t)(s_A)\,,
\end{aligned} \tag{35}$$

where, setting $\widehat{D} = (R_1, \ldots, R_k)$, with $|\widehat{D}| \equiv k$,

$$I(D,t)(s_A) \stackrel{\text{def}}{=} \tag{36}$$
$$P_0^\Lambda \exp\left(s_k H_{\Lambda,0}\right) V_{R_k} \exp\left((s_{k-1}-s_k)H_{\Lambda,0}\right) \ldots V_{R_1} \exp\left((t-s_1)H_{\Lambda,0}\right)(s_A),$$

is called the *contribution of the diagram D* and

$$\mathcal{I}(\widehat{D},t)(s_A) \stackrel{\text{def}}{=} \int_{\Delta_k^t} I(D,t)(s_A)\,\mathbf{d}\mathbf{s}_k\,.$$

For the sake of brevity, we shall often write

$$\sum_{|\widehat{D}|=k} \mathcal{I}(\widehat{D},t) = \sum_{R_1\in\mathcal{R}_\Lambda} \cdots \sum_{R_k\in\mathcal{R}_\Lambda} \mathcal{I}(\widehat{D},t). \tag{37}$$

where the $\sum_{D\subset\Lambda}$ is taken over all $\widehat{D} = (R_1, \ldots, R_k)$, such that $R_i \in L_\Lambda$, for $1 \le i \le k$. Using (34) one has

$$\mathbb{P}_t^{\Lambda,\epsilon}(s_A) = P_0^\Lambda \left(1 + \sum_{k=1}^{\infty} \epsilon^k \sum_{\substack{\widehat{D}\subset\Lambda:\\ |\widehat{D}|=k}} \mathcal{I}(\widehat{D},t)\right)(s_A). \tag{38}$$

To find $\mathbb{P}_t^{\Lambda,\epsilon}(s_A)$ one has to sum over all admissible route configurations $\eta^\Lambda \in \mathcal{A}_\Lambda$ containing the subconfiguration s_A. In other words,

$$\mathbb{P}_t^{\Lambda,\epsilon}(s_A) = \sum_{\substack{\eta^\Lambda\in\mathcal{A}_\Lambda:\\ \eta^\Lambda(R)=s_R,\ \forall R\in A}} \mathbb{P}_t^{\Lambda,\epsilon}(\eta^\Lambda). \tag{39}$$

Lemma 9. *Let a finite set $A \subseteq \mathcal{R}_\Lambda$ and a finite route subconfiguration s_A be fixed. If $D=((R_1,s_1),\ldots,(R_k,s_k))$, $\widetilde{R}_i \in \mathcal{R}_\Lambda$, is not an A-connected diagram, then*

$$\sum_{\substack{\eta^\Lambda\in\mathcal{A}_\Lambda:\\ \eta^\Lambda(R)=s_R,R\in A}} \mathbb{P}_0^\Lambda I(D,t)(\eta^\Lambda) = 0\,, \tag{40}$$

for all $0 \le s_k \le \cdots \le s_1 \le t$ and for each initial distribution $\mathbb{P}_0^\Lambda$.

Proof. From the definition of $I(D,t)(\eta^\Lambda)$ given in (36), two types of operators arise: V_R and $\exp\left(sH_{\Lambda,0}\right)$. Fist, we have

$$\delta_\eta \exp\left(sH_{\Lambda,0}\right) = \sum_{\eta'\in\mathcal{A}_\Lambda} T_s(\eta,\eta')\delta_{\eta'}\,,$$

where $T_s(\eta,\eta')$ is the s-time-transition probability to go from state η to state η'. Secondly, for each δ-measure δ_η, $\eta \in \mathcal{A}_\Lambda$, the operator V_R produces a positive measure $\lambda_R\delta_{\eta+\delta_R}$ and a negative measure $\lambda_R\delta_\eta$ on $\mathcal{A}_\Lambda$ (see Eq. (4) and (6)). These measures are concentrated on route configurations η and $\eta+\delta_R$, respectively, which differ only on route R, and $(\eta+\delta_R)(R) = \eta(R)+1$. Formally,

$$\delta_\eta V_R = \begin{cases} \lambda_R(\delta_{\eta^\Lambda+\delta_R} - \delta_{\eta^\Lambda}) & \text{if } \eta^\Lambda + \delta_R \in \mathcal{A}_\Lambda\,, \\ 0 & \text{otherwise.} \end{cases} \tag{41}$$

Both measures are conserved by the dynamics of $\exp\left(sH_{\Lambda,0}\right)$, from the very definition of $H_{\Lambda,0}$.

For $R \in \mathcal{R}_\Lambda$, let

$$\mathcal{R}^p_{\partial R} = \left\{R' \in \mathcal{R}^p_\Lambda : \mathrm{supp}(R') \cap \mathrm{supp}(R) \neq \varnothing,\ (L_\Lambda \setminus \mathrm{supp}(R)) \cap \mathrm{supp}(R') \neq \varnothing\right\}.$$

We first prove the following

Lemma 10. *Let $\Lambda \subset \mathbb{R}^\nu$ and some route $R \in \mathcal{R}_\Lambda$ be fixed. Then, for each admissible configuration $\eta = \eta_f + \eta_p \in \mathcal{A}_\Lambda$, $\eta_f \in \mathcal{A}^f_\Lambda$, $\eta_p \in \mathcal{A}^p_\Lambda$,*

$$\delta_\eta V_R \exp(sH_{\Lambda,0}) = \lambda_R \sum_{k\geq 0} \sum_{\eta_p^1} \cdots \sum_{\eta_p^k} \int_{\Delta_k^s} P(\eta_{p,R}, \overline{L}_k, \eta'_{p,R}) \Phi(\eta_{p,R}; \eta'_{p,R}, s|\overline{L}_k) \tag{42}$$

with

$$(\nu^{R,+}(\eta_{p,R}; \eta'_{p,R}, s|\overline{L}_k) - \nu^{R,-}(\eta_{p,R}; \eta'_{p,R}, s|\overline{L}_k)) \otimes \nu^R(\eta_{p,R}; \eta'_{p,R}, s|\overline{L}_k),$$

for all $s \geq 0$, where:

(i) $\eta_{p,R} \equiv \eta_{|\mathcal{R}^p_{\partial R}}$, $\sum_{\eta_p^1} \cdots \sum_{\eta_p^k}$ is taken over all sequences of finite subconfigurations $\eta_p^0 = \eta_{p,R}, \eta_p^1, \ldots, \eta_p^k = \eta'_{p,R}$ of routes in $\mathcal{R}^p_{\partial R}$, such that $\eta_p^{i+1} = \eta_p^i - \delta_{R(i)}$, where $R(i)$ is a route in $\mathcal{R}^p_{\partial R}$. The subconfiguration $\eta_{p,R}$ is a localization of η_p in $\mathcal{R}^p_{\partial R}$ and

$$\overline{L}_k = (\eta_p^0, r_0; \eta_p^1, r_1; \ldots; \eta_p^k, r_k), \quad r_0 \equiv 0.$$

(ii) The conditional probability measures

$$\nu^{R,+}(\eta_{p,R}; \eta'_{p,R}, s|\overline{L}_k), \quad \text{and} \quad \nu^{R,-}(\eta_{p,R}; \eta'_{p,R}, s|\overline{L}_k)$$

(for the process evolving with the generator $H_{\Lambda,0}$) are defined on route configurations in $\mathcal{R}_{\mathrm{supp}(R)}$, under the condition that $\eta_{p,R}(r)$, $0 \leq r \leq s$, satisfies

$$\eta_{p,R}(r) = \eta_p^i \quad \text{on the time interval } [r_i, r_{i+1}),\ \forall i \leq k-1,$$

the jumps taking place at the instants r_i, $0 \leq i \leq k-1$ and the initial measures being respectively $\delta_{\eta+\delta_R}$ and δ_η.

(iii) A similar definition holds for the measure $\nu^R(\eta_{p,R}; \eta'_{p,R}, s|\overline{L}_k)$, which is defined for the route configurations in $\mathcal{R}_{L_\Lambda \setminus \mathrm{supp}(R)}$ under the same conditions.

(iv) $P(\eta_{p,R}, \overline{L}_k, \eta'_{p,R})$ is the probability density function of the occurrence

$$\eta_{p,R}(r) = \eta_p^i, \quad \forall 0 \leq r \leq s \text{ and } 0 \leq i \leq k-1,$$

according to the definition given above in (ii).

The proof follows directly from (41) and from the definition of $H_{\Lambda,0}$.

Remark. The following properties take place:

(i) The measures $\nu^{R,+}(\eta_{p,R}; \eta'_{p,R}, s|\overline{L}_k)$ and $\nu^{R,-}(\eta_{p,R}; \eta'_{p,R}, s|\overline{L}_k)$ on one hand, $\nu^R(\eta_{p,R}; \eta'_{p,R}, s|\overline{L}_k)$ on the other hand, have non-intersecting supports. Moreover,

(ii) $\nu^{R,+}(\eta_{p,R}; \eta'_{p,R}, s|\overline{L}_k)$ and $\nu^{R,-}(\eta_{p,R}; \eta'_{p,R}, s|\overline{L}_k)$ have equal masses. The number of summands in 42 is uniformly bounded by a constant $C < \infty$ since, conditions 1 and 3, the number of different sequences of route subconfigurations $\eta_p^0, \dots, \eta_p^k$ in $\mathcal{R}_{\partial R}^p$ is uniformly bounded.

Continuing with the proof of Lemma 9, we note that, if the diagram D is not A-connected, then there exists R_i such that $\widetilde{R}_i \cap \sigma_{i-1}(A) = \varnothing$, so that, by (41) and Lemma 10, it is easy to establish by induction that the sum

$$\sum_{\substack{\eta^\Lambda \in \mathcal{A}_\Lambda: \\ \eta^\Lambda(R)=s_R, R\in A}}$$

gives the variation of the measures

$$\nu^{R_i,+}(\eta_{p,R_i}; \eta'_{p,R_i}, s_{i-1} - s_i|\overline{L}_k) \quad \text{and} \quad \nu^{R_i,-}(\eta_{p,R_i}; \eta'_{p,R_i}, s_{i-1} - s_i|\overline{L}_k),$$

respectively positive and negative, which both are defined on $\mathcal{R}_{\mathrm{supp} R_i}$ and have the same variation. As the stochastic process with generator $H_{\Lambda,0}$ conserves the mass of these measures, we obtain Lemma 9. □

5.1 Edges and Exponential Contribution of A-Connected Diagrams

We define the edges of an A-connected diagram $D = ((R_1, s_1), \dots, (R_k, s_k))$ in the following way. For each R_i, $1 \le i \le k$, let $j < i$ be the smallest integer such that

$$\gamma(R_i) \cap \sigma_j(A) \neq \varnothing.$$

Then we connect the levels s_i and s_j with a vertical line. By definition, A corresponds to the level $s_0 \equiv 0$. If there is no such $j \ge 1$, then $\gamma(R_i) \cap A \neq \varnothing$ and we connect the level s_i to the level s_0, i.e. to A. Next we fix constants $\gamma > 0$, $K > 0$, and define the *exponential contribution of the edge* $\psi_i = (s_i, s_j)$ by

$$\varphi_{K,\gamma}(\psi_i) = K \exp\{-\gamma(s_j - s_i)\}. \tag{43}$$

Then the *exponential contribution* $\varphi_{K,\gamma}(D)$ of the diagram D is defined by

$$\varphi_{K,\gamma}(D) = \prod_{i=1}^{k} \varphi_{K,\gamma}(\psi_i). \tag{44}$$

Lemma 11. *Let a volume $\Lambda \subset \mathbb{R}^\nu$ be fixed. Then there exist constants $K > 0$, $\gamma' > 0$ such that*

$$\left\| \nu^{R,+}(\eta_{p,R}; \eta'_{p,R}, s|\overline{L}_k) - \nu^{R,-}(\eta_{p,R}; \eta'_{p,R}, s|\overline{L}_k) \right\|_1 \le K' \exp(-\gamma' s), \tag{45}$$

for all $R \in \mathcal{R}$, $s \ge 0$, $r_0 = 0 \le r_1 \le \dots \le r_k \le s$ and for each sequence of admissible configurations $\eta_p^0 = \eta_{p,R}, \eta_p^1, \dots, \eta_p^k = \eta'_{p,R}$ of routes in $\mathcal{R}_{\partial R}^p$ such that $\eta_p^{i+1} = \eta_p^i + \delta_{R'(i)}$, for some route $R(i)$ from $\mathcal{R}_{\partial R}^p$, $i = 1, \dots, k$, where $\|\cdot\|_1$ denotes the variation norm.

Proof. Let the route R be fixed. On the interval $[r_i, r_{i+1})$, the measures $\nu^{R,+}(\eta_{p,R}; \eta'_{p,R}, s|\overline{L}_k)$ and $\nu^{R,-}(\eta_{p,R}; \eta'_{p,R}, s|\overline{L}_k)$ evolve according to the same finite irreducible Markov chain. The number of different Markov chains and the number of states for each of them are uniformly bounded for all R (see conditions 1,3). Thus, the bound (45) follows from standard results on the exponential convergence to steady state for an irreducible Markov chain, during each time interval $[r_i, r_{i+1})$. The parameters can be uniformly chosen, since the number of different sequences of route configurations $\eta_p^0, \ldots, \eta_p^k$ is uniformly bounded and conditions 2,4 are satisfied. □

Lemma 12. *There exist positive constants K' and γ' such that, for each A-connected diagram $D = ((R_1, s_1), \ldots, (R_k, s_k))$, $\widetilde{R}_i \in \mathcal{R}_\Lambda$, $I(D,t)$ can be estimated by*

$$|I(D,t)| \leq C(A) \prod_{i=1}^{k} \varphi_{K',\gamma'}(\psi_i)\,, \tag{46}$$

for all $\Lambda \subset \mathbb{R}^\nu$ and $t \geq 0$, where the constant $C(A) > 0$ is independent of Λ, k, and t.

Proof. This lemma can easily be proved along the same lines as Lemma 9, by using formula (42) and the probabilistic interpretation of the measures $\nu^{R,+}(\cdot)$, $\nu^{R,-}(\cdot)$ and $\nu^{R}(\cdot)$. □

The following lemma plays a crucial role.

Lemma 13. *For any $\gamma > 0$, $K > 0$,*

$$\sum_{\substack{D \text{ is } A\text{-connected} \\ |D|=k}} \int_{\Delta_k^t} \varphi_{K,\gamma}(D)\, \mathrm{d}\mathbf{s}_k \leq C^k\,, \tag{47}$$

for all $k \geq 1$, where $C = \frac{4K}{\gamma}$.

Proof. It is an immediate consequence of Lemma 17 in Appendix A , after choosing the function $g(t) = K \exp(-\gamma |t|)$. □

Lemma 14. *Let π_0, π_0^Λ be the stationary measures of the loss networks $\mathcal{N}^f$, $\mathcal{N}^f_\Lambda$, respectively. Then there exist positive constants K', γ', such that, for each A-connected diagram*

$$D = ((R_1, s_1), \ldots, (R_k, s_k)),\ \widetilde{R}_i \in \mathcal{R}_\Lambda\,,$$

1.

$$|\pi_0 V_{R_n} \exp(t_n H_0) \ldots V_{R_1} \exp(t_1 H_0)(s_A)| \leq C(A) \prod_{i=1}^{k} \varphi_{K',\gamma'}(\psi_i)\,, \tag{48}$$

for all $t \geq 0$, where the constant $C_1(A) > 0$ is independent of k, t and D.

2.

$$\left|\pi_0^\Lambda V_{R_n} \exp(t_n H_0^\Lambda) \ldots V_{R_1} \exp(t_1 H_0^\Lambda)(s_A)\right| \leq C_2(A) \prod_{i=1}^{k} \varphi_{K',\gamma'}(\psi_i) \tag{49}$$

for all $\Lambda \subset \mathbb{R}^\nu$ and $t \geq 0$, where the constant $C_2(A) > 0$ is independent of Λ, k, t and D.

3.

$$\left|\sum_{\substack{D \text{ is } A_{co} \\ |D|=k}} \int_{[0,\infty]^k} \pi_0 V_{R_k} \exp(t_k H_0) \ldots V_{R_1} \exp(t_1 H_0)(s_A)\,\mathrm{d}t_k - \int_{\Delta_k^t} I(D,t)(s_A)\,\mathrm{d}s_k\right|$$

$$\leq C_3(A) K' \exp(-\gamma' t)\,, \quad (50)$$

for all $\Lambda \subset \mathbb{R}^\nu$ and $t \geq 0$, where A_{co} in the sum stands for A-connected and the constant $C_3(A) > 0$ is independent of Λ, k, t and D.

Proof. The bounds (48), (49) are particular cases of Lemma 12 and the bound (50) can be easily proved by using Lemma 13. See also the proof of (65) in Appendix B. □

The proof of theorem 8 now follows directly from Lemma 14.

6 Loss Networks with Adaptive Routing

Here we introduce loss networks operating with adaptive (in particular alternative) routing. It seems useful to give a general formal definition of these LNs, since (see e.g. [9] and references therein) there are of practical importance. In this section, we slightly modify the notation.

Let $\mathcal{R}$ be a set of routes, where a route is a nonordered pair of vertices $\{v, v'\}$. This means we do not distinguish the routes from v to v' and from v' to v. Each route $R = \{v, v'\}$ has a finite set of subroutes S_R, where any subroute $r \in S_R$ is a connected subgraph $G_R^r = (V_R^r, L_R^r) \subset G$, such that $v, v' \in V_R^r$, where V_R^r, L_R^r are respectively sets of vertices and links. The support of a route R is defined as by

$$\mathrm{supp}(R) \equiv \bigcup_{r \in S_R} \mathrm{supp}(r).$$

Upon arrival at time t, a call of route R chooses a subroute from $r \in S_R$ and will use $n_R^r(g) \in \mathbb{Z}_+$ circuits on link g. Calls requesting route $R = \{v, v'\}$ form a Poisson stream of intensity $\lambda_R \geq 0$ and all these Poisson streams are independent.
The sets $\mathcal{A}$ (resp. $\mathcal{A}_\Lambda$) still denote all *admissible route configurations* (resp. in volume Λ). Now, an admissible route configuration for $\mathcal{N}$ with adapting routing is a function

$$\eta : \mathcal{R} \times \mathcal{R} \to \mathbb{Z}_+\,, \quad \text{such that} \quad \sum_{\substack{R, r \in \mathcal{R}: \\ g \in \mathrm{supp}(R), r \in S_R}} \eta(R, r) n_R^r(g) \leq c(g)\,, \quad (51)$$

for all $g \in G$, where $\eta(R, r) \geq 0$ denotes the number of calls requesting route $R \in \mathcal{R}$ and choosing subroute $r \in S_R$.
Similarly an admissible route configuration in volume Λ is a function

$$\eta_\Lambda : \mathcal{R}_\Lambda \times \mathcal{R}_\Lambda \to \mathbb{Z}_+ \quad , \text{ such that} \quad \sum_{\substack{R, r \in \mathcal{R}_\Lambda: \\ g \in \mathrm{supp}(R), r \in S_R}} \eta_\Lambda(R, r) n_R^r(g) \leq c(g) \quad (52)$$

for all $g \in G_\Lambda$, where $\eta_\Lambda(R, r) \geq 0$ denotes the number of calls requesting route $R \in \mathcal{R}_\Lambda$ and choosing subroute $r \in S_R$. By definition, $R \in \mathcal{R}_\Lambda$ means that $\mathrm{supp}(R) \subset G_\Lambda$.

A subroute is chosen according to the following procedure called *adaptive routing*. On the set $\mathcal{R} \times \mathcal{A}$, one defines two functions $\mathcal{K}_{\cdot}(\cdot)$, $\mathcal{E}_{\cdot}(\cdot)$. For each route R and each configuration η, $\mathcal{K}_R(\eta)$ is a finite sequence of subroutes from S_R and $\mathcal{E}_R(\eta)$ is the length of this sequence. The function $\mathcal{K}_R(\eta)$ will describe the set of possible alternative subroutes (the number of which is $\mathcal{E}_R(\eta)$) to connect an arriving call of route R at time t, when the current route configuration is η.

The mechanism is as follows: At time t, for a configuration η, an arriving call of route $R = \{v, v'\}$ tries to use subroutes from $\mathcal{K}_R(\eta)$, by turns, in the order specified by $\mathcal{K}_R(\eta)$. A call of route $R = \{v, v'\}$ can not use a subroute r if, on some link $g \in L_R^r$, there are less than $n_R^r(g)$ circuits free from this link. Otherwise, the call is connected and simultaneously holds $n_R^r(g)$ free circuits on link $g \in G_R^r$, and the holding time is exponentially distributed with rate $\mu_R^r > 0$ (function of R and r). A call is blocked and lost if all subroutes from $\mathcal{K}_R(\eta)$ are unworkable. As in the preceding sections, all random variables describing the arrival and service processes are supposed to be independent.

Remark. If $|S_R| \equiv 1$ and $\mathcal{E}_R \equiv 1$, for each $R \in \mathcal{R}$, then we get a LN with fixed routing. If the functions $\mathcal{E}_R(\eta)$ and $\mathcal{K}_R(\eta)$ do not depend on η, then we have the *alternative routing*. Thus fixed and alternative routings can be considered as particular cases of adaptive routing.

In the case of adaptive routing, the following conditions 3a, 3a*, 4a will replace conditions 3, 3*, 4 earlier introduced in §3.

Condition 3a. The routes are *uniformly bounded*: there exists a constant $D < \infty$ such that $|\text{supp}(R)| \leq D$ for all $R \in \mathcal{R}$.

Condition 3a*. The input intensities have an *exponential decay*, with parameter $d > 0$:

$$\|\lambda\|_r \equiv \sup_{g \in L}\Big(\sum_{k=1}^{\infty} e^{dk} \sum_{\substack{R: g \in R \\ |R|=k}} \lambda_R\Big) < \infty. \tag{53}$$

Condition 4a. The output intensities are *uniformly bounded*:

$$\sup_{R \in \mathcal{R}} \mu_R < \infty, \qquad \mu_b \equiv \inf_{R \in \mathcal{R}} \inf_{r \in S_R} \mu_R^r > 0.$$

Definition 15. The function $\mathcal{K}_{\cdot}(\cdot)$ is said to be *local* (with parameter $D > 0$), if it depends only on links belonging to a D-neighbourhood $L_{N(R)}$ of $\text{supp}(R)$, where

$$L_{N(R)} \stackrel{\text{def}}{=} \{g \in L : \text{dist}(g, \text{supp}(R)) \leq D\}.$$

Condition 5. The function $\mathcal{K}_{\cdot}(\cdot)$ of a loss network with adaptive routing is *local* with some finite parameter D.

Let us consider a perturbed free loss network $\mathcal{N}_\epsilon = \mathcal{N}^f + \mathcal{N}^p(\epsilon)$, $\epsilon = (\epsilon_{\text{in}}, \epsilon_{\text{out}})$,

$$\mathcal{N}^p(\epsilon) = \{G, \mathcal{R};\ \epsilon_{\text{in}}\lambda, \epsilon_{\text{out}}\mu;\ \mathcal{K}_{\cdot}(\cdot), \mathcal{E}_{\cdot}(\cdot);\ c, n\},$$

where $\mathcal{N}^f$ is a free loss network with fixed routing and the perturbation $\mathcal{N}^p$ is a loss network with adaptive routing. Before stating a variant of theorem 8, we have to define, as in §4.1, the operators V_R, $R \in \mathcal{A}$ and H_0.

Let Λ and $\eta \in \mathcal{A}_\Lambda$ be fixed. The operator $V_R \stackrel{\text{def}}{=} V_R^{\text{in},p}$, $\text{supp}(R) \subset \mathcal{R}_\Lambda$, corresponds to call arrivals of route R. Suppose that $\mathcal{K}_R(\eta) = (r_1, \ldots, r_m)$, where $m = \mathcal{E}_R(\eta)$, $r_i \in S_R$ for $i = 1, \ldots, m$. Let $N_R(\eta)$ be the serial number of the subroute chosen to connect a call of route R. (In case of blocking, we put $N_R(\eta) = 0$). Then for each admissible route configuration

$$\eta : \mathcal{R}_\Lambda \times \mathcal{R}_\Lambda \to \mathbb{Z}_+,$$

V_R transforms the δ-measure

$$\delta_\eta(\cdot, \cdot) \equiv \delta((\cdot, \cdot) - \eta)$$

into the measure

$$\begin{cases} \lambda_R(\delta_{\eta+\delta_{R,r}} - \delta_\eta) & \text{if } r = r_i \text{ for } i = N_R(\eta)\,, \\ 0 & \text{otherwise,} \end{cases}$$

and all other δ-measures into 0, where, by definition, we have put

$$(\eta + \delta_{R,r})(R', r') = \begin{cases} \eta(R', r') + 1 & \text{if } R = R' \text{ and } r = r'\,, \\ \eta(R', r') & \text{otherwise.} \end{cases}$$

The generator H_0 could be defined in a similar manner.

Theorem 16. *Let $G \in \Gamma_d^\nu(D_1, D_2)$, $0 < D_1 < D_2 < \infty$ and ϵ_{out} is fixed. Suppose conditions 1, 2, 3a (or 3a* with a parameter $d > 0$ sufficiently large), 4a and 5 are satisfied for the networks $\mathcal{N}^f$ and $\mathcal{N}^p$. Then there are constants $\epsilon_0 > 0$, $C > 0$, such that, for all $\epsilon \equiv \epsilon_{\text{in}} \in [0, \epsilon_0]$, the Markov process η_t^ϵ, which describes $\mathcal{N}_\epsilon^{pf}$ is ergodic and, for each finite route subconfiguration $s_A = \{s_{R_1}, \ldots, s_{R_k}\}$, $R_i \in \mathcal{R}$, $1 \le i \le k < \infty$, the limits in (30) exist and are equal. Moreover, (31)and (32) hold, with the bound (33), for some constant $C(s_A) > 0$ independent of n.*

Proof. It mimics the proof of Theorem 8, since the the main arguments relied on the fact that the operator V_R was depending only on the support of the route R. But our definition of $\text{supp}(R)$, as well as conditions 3a, 3a*, 4a and 5, ensure that the corresponding lemmas in the proof of Theorem 8 are still also valid in this case. □

Appendix A: a Universal Cluster Bound

Consider the class C_n of non-oriented graphs, such that each member G of the class be a $n - tree$, i.e. it is connected, has n edges and $n + 1$ vertices, denoted by $v \in V = \{0, 1, \ldots, n\}$. Moreover, for all $t \ge 0$, we associate to any vertex $v \in V$, a real number $t_v \ge 0$ (called a *v-time*), where

$$t_0 = 0 \le t_1 \le \cdots \le t_n \le t\,.$$

Setting then $\bar{t} = (t_0, t_1, \dots, t_n)$, the pair $(G, \bar{t})$ will be called a *t-diagram*. We always assume that the coordinate t_0 of vertex 0 is equal to 0. Unless otherwise mentioned, a n-uple explicitly written $\bar{t}$ will be subject to the above constraints and his components will be also called *coordinates*.

Let us fix some function $g : \mathbb{R} \to \mathbb{R}$. For a graph $G = (V, L) \in \mathcal{C}_n$, the quantity

$$\prod_{\substack{l \in L, \\ l=(v,v')}} g(t_v - t_{v'})$$

is called the *contribution* of the diagram $(G, \bar{t})$.

Lemma 17. *Let $g : \mathbb{R} \to \mathbb{R}$ be an even function, Riemann-integrable on any compact set $K \subset \mathbb{R}$. Then*

$$\sum_{G \in \mathcal{C}_n} \int_{\Delta_n^t} dt_1 \dots dt_n \prod_{\substack{l \in G, \\ l=(v,v')}} g(t_v - t_{v'}) \le 8^n \left(\|g\|_1^t\right)^n , \tag{54}$$

where $\|g\|_1^t \stackrel{\text{def}}{=} \int_0^t |g(s)| \, ds$.

Proof. It suffices to prove (54) for non-negative functions. Clearly, (54) is equivalent to

$$\sum_{G \in \mathcal{C}_n} \int_{\Delta_n^t} dt_1 \dots dt_n \prod_{\substack{l \in G, \\ l=(v,v')}} g(t_v - t_{v'}) \le 4^n \{\int_{-t}^t g(r) \, dr\}^n . \tag{55}$$

In fact, we will prove the following inequality:

$$\delta^n \sum_{\substack{t_i \in \mathbb{Z}_\delta, \\ 0<t_1<\dots<t_n<t}} \sum_{G \in \mathcal{C}_n} \prod_{\substack{l \in G, \\ l=(v,v')}} g(t_v - t_{v'}) \le 4^n \delta^n \sum_{\substack{r_1 \in \mathbb{Z}_\delta, \\ |r_1| \le t, \ r_1 \ne 0}} \cdots \sum_{\substack{r_n \in \mathbb{Z}_\delta, \\ |r_n| \le t, \ r_n \ne 0}} \prod_{i=1}^n g(r_i) , \tag{56}$$

for all $\delta > 0$, $n \ge 1$, where $\mathbb{Z}_\delta$ the one-dimensional δ-lattice, $\mathbb{Z}_\delta \stackrel{\text{def}}{=} \delta \cdot \mathbb{Z}$. Since both sides of (56) are approximations by Riemann sums of both sides of (55), it suffices to let $\delta \to 0+$ to obtain (55).

Notation. It will be convenient to introduce two types of vectors:

(i)

$$\bar{r} = (r_1, \dots, r_n), \quad \text{with } r_i \in \mathbb{Z}_\delta, \ \ 0 < |r_n| \le t, \ \ 1 \le i \le n , \tag{57}$$

noting that $\bar{r}$ is not necessarily positive;

(ii)

$$\bar{q} = (q_0, \dots, q_n), \quad \text{with } q_i \ge 0, \ q_i \in \mathbb{Z}_+ , \forall 1 \le i \le n \ \text{ and } \sum_{i=0}^n q_i = n . \tag{58}$$

When the components of a vector belong to $\mathbb{Z}_\delta$, it will be said to satisfy the δ-*condition*.

Sketch of proof. For each vector $\overline{r}$, an algorithm will be given, allowing us to produce at most 4^n different diagrams $(G, \overline{t})$, $G \in \mathcal{C}_n$ and $t_i \in \mathbb{Z}_\delta$, for $i \geq 1$, each diagram having a contribution equal to $\prod_{i=1}^n g(r_i)$. More exactly, it will be shown that each diagram $(G, \overline{t})$ can be constructed from a vector $\overline{r}$ (see the above definition) and, hence, 55 will follow.

Let us fix $t > 0$, $\overline{r}$ and $\overline{q}$ satisfying respectively (57) and (58). The algorithm $A(t, \overline{r}, \overline{q})$ presented hereafter constructs recursively at most one diagram $(G, \overline{t})$, which will be denoted by $I(t, \overline{r}, \overline{q})$ with $n + 1$ vertices and n edges, such that $\overline{t}$ satisfies the δ-condition and the contribution of $I(t, \overline{r}, \overline{q})$ is equal to $\prod_{i=1}^n g(r_i)$. Moreover, it will appear that

$$\bigcup_{G \in \mathcal{C}_n} \bigcup_{\substack{\overline{t}: t_i \in \mathbb{Z}_\delta, \\ t_0 \equiv 0 < t_1 < \cdots < t_n \leq t}} (G, \overline{t}) = \bigcup_{\overline{r}, \overline{q}} I(t, \overline{r}, \overline{q}) . \tag{59}$$

Since the number of distinct vectors $\overline{q}$, with $q_0 > 0$ does not exceed 4^n, inequality (56) follows immediately from (59).

The algorithm consists of at most $2(n + 1)$ steps enumerated by the sequence

$$(0, 1), \ldots, (0, q_0 + 1), \ldots, (n, 1), \ldots, (n, q_n + 1) .$$

If $q_0 = 0$, then the algorithm stops at step (0, 1) without constructing any diagram. Otherwise, it starts with a vertex called vertex 0 and $t_0 = 0$. At step (0, 1), the vertex 0 is referred to as being *current*. Then, choosing the number r_1, one constructs, from the current vertex 0, an edge leading to a new vertex r_1. At this moment, one will say that the number r_1 was *used* at step (0, 1). At the final stage of the algorithm, the new vertex will be labelled as vertex i, for some i, with $t_i = r_1$, and we proceed by induction.

Let $r_1, \ldots, r_a$ be the numbers already used and assume we are at step (i, j). Then the algorithm works as follows:

(1) At step (i, j), take the number r_{a+1} and, if $q_i \geq j$, construct an edge length r_{a+1} from the current vertex v; otherwise, change the current vertex from which to construct an edge of length r_{a+1} (see stages 2 and 3 of the algorithm).
(2) At step $(i, 1)$, choose one of the vertices produced beforehand, say v, call v the current vertex for steps $(i, 1), \ldots, (i, q_i + 1)$ and draw edges from v at each of these steps. The vertex v is said to have been *used* at step $(i, 1)$.
(3) The choice of v is rendered unique according to the following rule: Among all previously constructed vertices, not used at previous steps, v is the vertex having the smallest coordinate. If such vertex does not exist, then our algorithm stops and does not produce any diagram.
(4) If a vertex is constructed with a coordinate either negative, or strictly greater than t, or if it coincides with a previously constructed vertex, then the algorithm stops and does not produce any diagram.
(5) The algorithm *normally* stops at step $(n, q_n + 1)$.

When the algorithm stops *normally*, it produces a graph. Enumerating its vertices by increasing order of their coordinates, it is easy to see that we have in fact constructed a diagram $(G, \overline{t})$, $G \in \mathcal{C}_n$, such that $\overline{t}$ satisfies the δ-condition and the contribution is equal to $\prod_{i=1}^n g(r_i)$.

So it has just been shown that the algorithm constructs at most 4^n different diagrams $(G, \overline{t})$, each giving a contribution $\prod_{i=1}^n g(r_i)$.

It remains to prove that each diagram $(G, \bar{t})$ can be constructed by the algorithm for some $(\bar{r}, \bar{q})$, and (59) will be established. To that end, we slightly modify the above construction.

First, set $q_i = 0, \forall 0 \le i \le n$, and call the vertex 0 with coordinate t_0 *current*. Then consider all the links coming out from this vertex and let q_0 be their number. Next, order these links and let r_i be the (signed) length of the ith one. These links are said to be *are used* and the corresponding vertices to be *are constructed*, except vertex 0, which is also said to be *used*. Then, among all the already constructed (but not used) vertices, choose the one having the smallest coordinate, call it *current*, consider all the links coming out from the current vertex and let q_1 be the number of these links. Next order these links and let r_{i+q_0} be the length of the ith link $l = (v, v')$, counted positively if $t_v < t_{v'}$ and negatively otherwise. Say again that these links *are used* and the corresponding vertices are *constructed*, except the current vertex which is *used*, and proceed by induction. The process stops when all the links of diagram $(G, \bar{t})$ are *used*.

It is easy to see that if one takes the vectors $\bar{r}, \bar{q}$ obtained by the above procedure, then the algorithm will construct exactly the diagram $I(t, \bar{r}, \bar{q})$.

Lemma 17 is proved. □

Appendix B: Perturbation Theory for Stationary Probabilities of a Markov Chain

Here we derive a useful formula for stationary probabilities of a perturbed Markov chain.

Let $\mathcal{L}_0, \mathcal{L}_\epsilon, \epsilon \in \mathbb{R}$ be continuous-time-homogeneous Markov chains, with the same countable state space $\mathcal{A}$, defined by their respective generators H_0 and $H_\epsilon = H_0 + \epsilon V, \ \epsilon \in \mathbb{R}$. We assume that the operators H_0, V are bounded in $l_1(\mathcal{A})$. Then the probability $\mathbb{P}_t^\epsilon(\alpha)$, for the Markov chain $\mathcal{L}_\epsilon$, to be in α at time t can be expressed by the following formula:

$$\mathbb{P}_t^\epsilon(\alpha) =$$
$$\sum_{k=0}^{\infty} \epsilon^k \int_{\Delta_k^t} \mathbb{P}_0 \exp(s_k H_0) V \exp((s_{k-1} - s_k) H_0) \dots V \exp((t - s_1) H_0)(\alpha)\, \mathrm{d}s_k \,, \quad (60)$$

where P_0 is the initial distribution of $\mathcal{L}_\epsilon$ and

$$\Delta_k^t = \{(s_1, \dots, s_k) \in \mathbb{R}^k;\ 0 \le s_k \le \dots \le s_1 \le t\}.$$

This is a well known formula for the perturbation of a semi-group in a Banach space (see e.g. [6, Theorem 3.3.33]), which is easy to derive when H_0 and V are bounded. Indeed, we have

$$\mathbb{P}_t^\epsilon = P_0 \exp(t(H_0 + \epsilon V)) \,,$$

and it is possible to define, for all t, the following operator, bounded in $l_1(\mathcal{A})$,

$$W(t) = \exp(t(H_0 + \epsilon V)) \exp(-t H_0) \,.$$

Then

$$\frac{dW(t)}{dt} = \epsilon W(t) \exp(t H_0) V \exp(-t H_0) \,,$$

whence

$$W(t) = W(0) + \epsilon \int_0^t W(s)\exp(sH_0)V\exp(-sH_0)\,ds$$

or, equivalently,

$$\exp(tH_\epsilon) = \exp(tH_0) + \epsilon \int_0^t \exp(sH_\epsilon)V\exp((-s+t)H_0)\,ds\,. \tag{61}$$

Now (60) is obtained by iterating (61). The series (60) converges in l_1, since

$$\sum_{k=0}^{\infty} |\epsilon|^k \int_{\Delta_k^t} \|\exp(s_k H_0)V\exp((s_{k-1}-s_k)H_0)\dots V\exp((t-s_1)H_0)\|_1\,\mathbf{ds}_k$$

$$\leq \sum_{k=0}^{\infty} |\epsilon|^k \int_{\Delta_k^t} \|V\|_1{}^k\,\mathbf{ds}_k \;\leq\; \exp(t|\epsilon|\|V\|_1), \quad \forall t \in \mathbb{R}_+\,.$$

Theorem 18. *Let $\mathcal{L}_0$ and $\mathcal{L}_\epsilon \epsilon \in \mathbb{R}$ be irreducible time-continuous-Markov chains, with countable state space $\mathcal{A}$, defined by their respective generators H_0 and $H_\epsilon = H_0 + \epsilon V$, $\epsilon \in \mathbb{R}_+$. Assume also $\mathcal{L}_0$ is ergodic and that:*

(A) There exist positive constants C and δ such that, for a given initial distribution P_0,

$$\|P_0\exp(sH_0) - \pi_0\|_1 \leq C\exp(-\delta s)\,; \tag{62}$$

(B) The operator V is bounded in $l_1(\mathcal{A})$.

Then there exist positive constants ϵ_0, C_1, , δ_1 such that, for each $\epsilon \in [0, \epsilon_0]$, the Markov chain $\mathcal{L}_\epsilon$ is ergodic and its stationary distribution is given by the convergent series

$$\pi_\epsilon(\alpha) = \sum_{k=0}^{\infty} \epsilon^k \mathcal{E}(k)(\alpha) = \lim_{t\to\infty} \mathbb{P}_t^\epsilon(\alpha) = \lim_{t\to\infty} \sum_{k=0}^{\infty} \epsilon^k \mathcal{F}(t,k)(\alpha)\,, \tag{63}$$

for any $\alpha \in \mathcal{A}$ and any initial distribution P_0 of $\mathcal{L}_\epsilon$, where we set

$$\mathcal{F}(t,k)(\alpha) \stackrel{\text{def}}{=} \int_{\Delta_k^t} \mathbb{P}_0\exp(s_kH_0)V\exp((s_{k-1}-s_k)H_0)\dots V\exp((t-s_1)H_0)\,\mathbf{ds}_k(\alpha)\,,$$

and

$$\mathcal{E}(k)(\alpha) \stackrel{\text{def}}{=} \int_{[0,\infty]^k} \pi_0 V\exp(t_kH_0)\dots V\exp(t_1H_0)\,\mathbf{dt}_k(\alpha) \leq C_1^k\,.$$

Moreover,

$$\|\mathcal{E}(k)\|_1 \leq C_1^k\,, \tag{64}$$

and

$$\|\mathcal{E}(k) - \mathcal{F}(t,k)\| \leq C_1^k\exp(-\delta_1 t)\,, \quad \forall k \in \mathbb{Z}_+,\; t \in \mathbb{R}_+\,. \tag{65}$$

Remark. Conditions A and B in the above Theorem 18 are rather strong. It can be shown that they hold for Markov chains which are either finite or countable and satisfying Doeblin's condition. For our purpose, it suffices to consider *finite* Markov chains. It might be worth noting that, for countable Markov chains, more general results on the analyticity of stationary distributions can be proved by means of Lyapounov functions (see [8, 14, 17]).

Proof. First one shows that (62) yields

$$\|V\exp(sH_0)\|_1 \le C\|V\|_1 \exp(-\delta s)\,, \quad \forall s \in \mathbb{R}_+\,. \tag{66}$$

For each measure $x \in l_1(\mathcal{A})$, one can represent xV as $(xV)^+ - (xV)^-$, where

$$(xV)^+(\alpha) = \max\,(0, xV) \quad \text{and} \quad (xV)^-(\alpha) = \max\,(0, -xV).$$

Then

$$\|xV\|_1 = \|(xV)^+\|_1 + \|(xV)^-\|_1\,, \|(xV)^+\|_1 = \|(xV)^-\|_1\,.$$

The last equality proceeds from the fact that V is a difference of generators. Also, as $\exp(H_0 s)$ conserves positive measures, we get

$$\begin{aligned}(xV)^+\exp\,(H_0 s)(\beta) = \textstyle\sum_\alpha (xV)^+(\alpha)p^0_{\alpha\beta}(s) = \\ \|(xV)^+\|_1\pi^0_\beta + \textstyle\sum_\alpha (xV)^+(\alpha)(p^0_{\alpha\beta}(s) - \pi^0_\beta)\,,\end{aligned} \tag{67}$$

where $p^0_{\alpha\beta}(s)$ is the probability to be in state β at time s, when the initial distribution P_0 is δ_α. Now (66) follows from (67). Indeed,

$$\begin{aligned}&\|((xV)^+ - (xV)^-)\exp(H_0 s)\|_1 = \\ &\sum_\beta \left|\sum_\alpha \big((xV)^+(\alpha) - (xV)^-(\alpha)\big)\,(p^0_{\alpha\beta}(s) - \pi^0_\beta)\right| \le C\|V\|_1\exp(-\delta s)\,,\end{aligned} \tag{68}$$

for some constant $C > 0$. Then (62) and (66) yield

$$\|\pi_0 V\,\exp(t_k H_0)\ldots V\,\exp(t_1 H_0)\|_1 \le C^k\|V\|_1{}^k\exp\,(-(t_1 + \cdots + t_k)\delta) \tag{69}$$

and

$$\begin{aligned}\|(P_0\exp(s_k H_0) - \pi_0)V\,\exp((s_{k-1} - s_k)H_0)\ldots V\,\exp((t - s_1)H_0)\|_1 \\ \le C^k\|V\|_1{}^k\exp\,(-t\delta)\,,\end{aligned} \tag{70}$$

since $t = t - s_1 + \cdots + (s_{k-1} - s_k) + s_k$. Hence, we obtain

$$\|\mathcal{E}(k)\| \le \frac{C^k\|V\|_1{}^k}{\delta^k} \tag{71}$$

and

$$\begin{aligned}\|\mathcal{E}(k) - \mathcal{F}(t,k)\|_1 &\le \int_{\Delta^t_k} \mathcal{G}(t,k;\mathbf{s}_k)\,\mathbf{ds}_k + \int_{\widetilde{\Delta}^t_k} C^k\exp(-(t_1 + \cdots + t_k)\delta)\,\mathbf{dt}_k \\ &\le (C\|V\|_1)^k\exp\,(-\delta t)\frac{t^k}{k!} + \left(\frac{C\|V\|_1}{\delta - \delta_1}\right)^k\exp\,(-\delta_1 t)\,,\end{aligned} \tag{72}$$

for all $t \geq 0$, where δ_1 is fixed, $0 < \delta_1 < \delta, \mathcal{G}(t, k; \mathbf{s}_k)$ denotes the left-hand side member of the inequality (70) and

$$\widetilde{\Delta}_k^t = \left\{ (t_1, \ldots, t_k), t_i \geq 0, 1 \leq i \leq k \; ; \sum_i t_i \geq t \right\}.$$

Now choosing

$$0 < \epsilon_0 < \frac{\delta - \delta_1}{C \| V \|_1},$$

(63) and (65) follow from (71) and (72), for positive constants C_1 and δ_1, which can easily be found. Theorem 18 is proved. □

References

1. Blockmeyer, E., Halstrom, H.L., Jensen, A.: The life and works of A.K.Erlang. *Academy of Technical Sciences, Copenhagen* (1948)
2. Botvich, D.D., Malyshev, V.A.: Unitary equivalence of temperature dynamics of ideal and locally perturbed Fermi-gas. *Commun.Math.Phys.* (1983) **91** 301–312.
3. Botvich, D.D., Malyshev, V.A.: The Proof of Asymptotic Completeness Uniformly on the Number of Particle. *Izv.Akad.Nauk SSSR. Ser.Math.* (1990) **54** 132–145 (in Russian)
4. Botvich, D.D., Malyshev, V.A.: Asymptotic Completeness and All That for an Infinite Number of Fermions. *Advances in Soviet Mathematics*, Ed. Minlos R.A., **5**(1991) 39–98. Providence, Rhode Island, AMS Publications.
5. Botvich, D.D., Domnenkov, A.Sh.: Malyshev V.A. Examples of Asymptotic Completeness in Translation Invariant Systems with a non-bounded Number of Particles. *Acta Applicandae Mathematicae* **22** (1991) 117–137.
6. Botvich, D.D., Gajrat, A.: Stationary reversible measures of loss networks with fixed routing. (in preparation)
7. Botvich, D.D., Malyshev, V.A., Manita, A.D.: Translation Invariant Quantum Master Equation. *Helvetica Physica Acta* **64** (1991) 1072–1092
8. Fayolle, G., Menshikov, M.V., Malyshev, V.A.:*Topics in the Constructive Theory of Countable Markov Chains.* Cambridge University Press (1994)
9. Girard, A.: *Routing and dimensioning in circuit-switched networks.* Addison-Wesley (1990)
10. Kelly, F.P.: One-dimensional circuit switching networks. *Ann. Prob.* (1987) **15** 1166–1179.
11. Kelly, F.P.: Loss networks. *Annals of Applied Probability* **1** (1991) (3) 319–378
12. Ligget, T.M.: *Interacting Particle Systems.* Springer-Verlag, New York (1985)
13. Louth, G.M.: Stochastic networks: Dependence and routing. Ph.D.thesis, University of Cambridge (1990)
14. Malyshev, V.A., Menshikov, M.V.: Ergodicity, continuity and analyticity of countable Markov chains. *Trans. Moscow Math. Soc. (in Russian)* **39** (1979) 3–48. (Translation 1981, Issue I)
15. Malyshev, V.A.: Networks and Dynamical Systems. *Advances in Applied Probability* **25** (1993) 140–175
16. Malyshev, V.A., Minlos, R.A.: *Gibbs random fields.* Moscow Nauka (in russian) (1985). English translation Kluwer Academic Publishers (1991) Amsterdam
17. Ignatyuk, I.A., Malyshev, V.A.,Turova, T.S.: VINITI. Stability of infinite systems of stochastic equations. *Prob. Theory, Math. Stat., Theoretical cybernetics* Itogi nauki i techniki, Akad. Nauk SSSR, Vsesoyuz. Inst. Naucn. i Techn.Informacii, Moscow **27** (1990) 3–78. English translation in Journal of Soviet Mathematics (1992)

A Survey of Markovian Methods for Stability of Networks*

D. Down and S. Meyn

University of Illinois and the Coordinated Science Laboratory

1 Introduction

It has generally been taken for granted in queueing theory that stability of a network is guaranteed so long as the overall traffic intensity is less than unity, and in recent years there has been much analysis which supports this belief for special classes of systems, such as single class queueing networks (see Borovkov [2], Sigman [39], Meyn and Down [30], and Foss and Baccelli [1]). This intuition was shown to be false in general by Lu and Kumar [25], where it was demonstrated that for a multiclass system, with an unintelligently chosen buffer priority scheduling policy, instability will occur even for loads less than unity. Similar counterexamples were reported in [35]. The question of whether or not more natural policies would always be stable, such as the standard FIFO policy, remained open until the recent work of Bramson [4] and Seidman [38]. The re-entrant structure of these and all other counterexamples so far discovered is similar to the topology of many semiconductor manufacturing plants.

In this paper we describe recent methods for determining whether a given network or scheduling policy is stable. All of these methods are based on a single technique: *Foster's criterion*, which is a particular case of the stochastic Lyapunov function method. This technique has been successfully used to investigate the behavior of queues and queueing networks which admit a Markovian realization since [17] in a wide variety of contexts. In particular, Botvich and Zamyatin [3], Fayolle et. al. [15, 16], Malyshev et. al. [26, 28, 27], Men'sikov [29], and others have obtained a series of results which have shed much light on the stability of random walks and queueing networks on low dimensional spaces, and much of this work is based upon stochastic Lyapunov techniques for Markov chains, in the spirit of Kingman [21].

For a ψ-irreducible Markov chain, it is known [31, Chapters 10-13] that positive recurrence is equivalent to the existence of a Lyapunov function, which can always be taken as the mean time to reach some specific set of states. For queueing networks, it is convenient to focus on the empty state, or a compact set of states, and recently this has given rise to three seemingly different methods for addressing stability:

* Send all correspondence to the second author at: Coordinated Science Laboratory, 1308 W. Main, Urbana, IL, 61801 (meyn@decision.csl.uiuc.edu).

(i) *finding approximations to the mean emptying time for the network.* For certain networks, especially those with a monotone structure, this can be a straightforward approach to stability.
(ii) *construction of a linear program.* The LP approach gives an algorithmic approach to the construction of a Lyapunov function which is widely applicable, and can be used to establish stability or instability.
(iii) *construction of a fluid model.* If an associated fluid model is stable, then again one obtains a Lyapunov function for the system under general assumptions.

We shall show how the Lyapunov approach provides the tools for all of these, and unifies them into a coherent set of approaches to stability for complex queueing networks. The strength and diversity of these recent consequents of Foster's criterion are sometimes surprising, and we will show that they are applicable to a far wider class of systems than was originally thought possible.

Once stability is deduced, several methods have been recently devised to compare the performance of one policy over another [36, 8, 23]. We will not discuss this aspect of the theory in the present paper. We will also not attempt to give a complete bibliography of the stability theory literature.

We begin with a description of the models which we treat.

2 Network Models

2.1 The Re-entrant Line

For simplicity, in this paper we consider exclusively *re-entrant lines*, which are a particular example of the multiclass networks with feedback described by Harrison and Nguyen in [19]. A simple example of a re-entrant line is the network described in Figure 1 consisting of two machines and three buffers, with a single server at each machine. A crucial feature of this model is the re-entrant structure, which is typical of semiconductor manufacturing models [22].

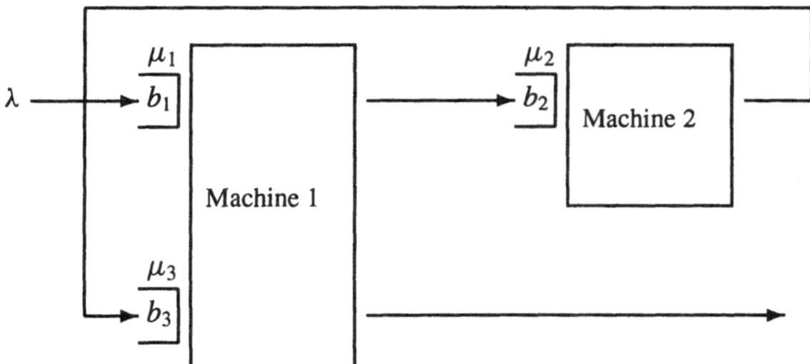

Fig. 1. A Re-entrant network model

The models which we consider consist of d single server stations, and a total of K buffers, which are located among the server stations. In the example illustrated in

Figure 1, $K = 3$ and $d = 2$. Customers at buffer k are serviced at station $s(k)$, with service times $\{\eta_k(n), n \geq 1\}$. The description of the network is simplified since routing is deterministic: customers arrive to buffer one, with interarrival times $\{\xi(n), n \geq 1\}$, where they wait in queue until service. After service is completed, a customer moves on to buffer two, and so forth, until it finally reaches buffer K. After service is completed at this final queue, the customer leaves the network.

It is assumed that service disciplines are "state dependent" and non-preemptive. Hence, after a service is completed at a server station, a new buffer is chosen for service at this station based solely upon the current buffer levels in the system. This is a rather strong condition which is required due to lack of space. The construction of a Markovian state process for more general service disciplines is treated in [9], and once such a state process is constructed, analysis follows in a manner analogous to that presented here (see for example [30] for a treatment of the FIFO policy).

Throughout this paper, we make the following three assumptions on the network. We first assume that

$$\xi, \eta_1, \ldots, \eta_K \quad \text{are mutually independent, and i.i.d. sequences.} \tag{1}$$

Next, we place moment assumptions on interarrival and service times: For some integer $p \geq 1$,

$$\mathsf{E}[|\eta_k(1)|^p] < \infty \text{ for } k = 1, \ldots, K, \quad \text{and} \quad \mathsf{E}[|\xi(1)|^p] < \infty \ . \tag{2}$$

Finally, we assume that interarrival times are unbounded, with a spread-out distribution. That is, for some positive function p on $\mathbb{R}_+$, and some integer j_0,

$$\begin{gathered} \mathsf{P}(\xi(1) \geq x) > 0 \quad \text{for all} \quad x > 0, \\ \mathsf{P}(\xi(1) + \cdots + \xi(j_0) \in dx) \geq p(x)\,dx \quad \text{and} \quad \textstyle\int_0^\infty p(x)\,dx > 0. \end{gathered} \tag{3}$$

Condition (1) allows us to construct a Markov state process, and the other two assumptions facilitate the formulation of conditions for ergodicity. We will henceforth set $\mu_k = 1/\mathsf{E}[\eta_k(1)]$ and $\lambda = 1/\mathsf{E}[\xi(1)]$.

For each $i = 1, \ldots, d$ we define the *nominal load* at station i as $\rho_i = \sum_{k:s(k)=i} \lambda/\mu_k$. To guarantee any practical form of stability for the network it is obviously necessary that the capacity constraint $\rho_i < 1$ be satisfied for each $i = 1, \ldots, d$. However, it is now known that this condition is not sufficient for stability.

2.2 A Markovian State Process

We now define a Markov process $\boldsymbol{\Phi} = \{\Phi_t : t \geq 0\}$ which describes the dynamics of the queueing network. For each t, $\Phi_t = (X_1(t), \ldots, X_K(t))^\top$, where $X_i(t)$ summarizes all the network information at buffer i. If all the interarrival time and service time distributions are exponential, then one may take $X_i(t) = b_i(t) =$ the queue length at buffer i, so that $\boldsymbol{\Phi}$ is a countable state space Markov chain with state space $\mathbb{Z}_+^K$. If the distributions are general, then it is necessary to augment this state description with the residual exogenous interarrival times and residual service times to preserve the Markov property. At time t, let $u(t)$ be the residual exogenous interarrival time, and $v_k(t)$ be the residual service time at buffer k. These residual processes are taken to be right continuous.

For the state process $\mathbf{\Phi}$, we then take $X_1(t) = (u(t), b_1(t), v_1(t))^\top$, and $X_i(t) = (b_i(t), v_i(t))^\top$ for $i \geq 2$. The state space X in this case will be equal to some subset of Euclidean space $\mathbb{R}^{2K+1}$, and will not be countable in this more general framework. For example, take the tandem queue illustrated in Figure 2. For this model, the state process $\mathbf{\Phi}$ becomes

$$\Phi_t = (u(t), b_1(t), v_1(t), b_2(t), v_2(t))^\top, \tag{4}$$

and the state space X is evidently equal to a subset of $\mathbb{R}^5_+$.

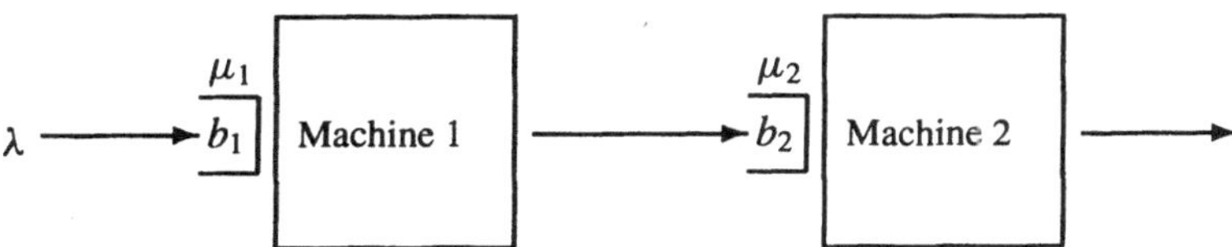

Fig. 2. Two queues in tandem

Under the assumptions (1–3), the state process $\mathbf{\Phi}$ has several desirable topological as well as measure-theoretic structural properties. As argued in [9], the process $\mathbf{\Phi}$ is a strong Markov process. It may also be shown using (3) as in [30] that all compact sets are *petite* for some skeleton chain $\mathbf{\Phi}^\delta = \{\Phi_0, \Phi_\delta, \Phi_{2\delta}, \dots\}$, and hence $\mathbf{\Phi}$ is a ψ-irreducible and aperiodic T-process [32]. These properties will simplify several of the results presented below.

We can now give a formal definition of stability. For the network model satisfying (1–3), the process $\mathbf{\Phi}$ is *Harris recurrent* if there exists one compact set $C \subset \mathsf{X}$ such that $\mathsf{P}_x\{\tau_C < \infty\} = 1$ for all initial states $x \in \mathsf{X}$, where τ_C is the hitting time $\tau_C = \inf\{t \geq 0 : \Phi_t \in C\}$. For the process $\mathbf{\Phi}$, Harris recurrence is equivalent to *non-evanesence*, the condition that $\mathsf{P}_x\{\mathbf{\Phi} \to \infty\} = 0$ for all x [32, Theorem 3.2]. This means that the buffer levels will return to reasonable levels infinitely often, but is far too weak to serve as a useful definition of stability. Rather than assuming merely finiteness of the hitting time τ_C, *positive Harris recurrence* requires that the *mean* return time to a bounded set be finite. For any timepoint $\delta \geq 0$ and any compact set C, define $\tau_C(\delta) := \delta + \theta^\delta \tau_C$ as the first hitting time on C after δ, where θ^δ is the usual backwards shift operator. The state process $\mathbf{\Phi}$ is called *positive Harris recurrent* if it is Harris recurrent, and there exists a compact set $C \subset \mathsf{X}$ and a $\delta > 0$ such that the expectation $\mathsf{E}_x[\tau_C(\delta)]$ is bounded on C.

It is shown in [32, Theorem 3.2] that positive Harris recurrence is also equivalent to *tightness* of the distributions of $\mathbf{\Phi}$ for each initial condition. In this case an invariant probability π exists, and it then follows that $\mathsf{P}_x\{\Phi_t \in A\} \to \pi(A)$, as $t \to \infty$, for any measurable set $A \subset \mathsf{X}$. These properties make positive Harris recurrence a useful form of stability for use in analyzing complex queueing networks. For such models, positivity can frequently be strengthened to far stronger forms of stability, such as geometric convergence of the expectation $\mathsf{E}_x[f(\Phi_t)]$ to its steady state value $\int f\,d\pi$ for sub-exponential functions f on X, and various sample path limit theorems also hold [33, 18].

We now describe the stochastic Foster-Lyapunov method for determining these various forms of stability of the network model with state process $\mathbf{\Phi}$.

3 Stochastic Lyapunov Functions

All of the methods for addressing stability which we describe in this paper build upon the application of stochastic Lyapunov functions. This method is also known as *Foster's Criterion*, although it is related to sixty year old dynamical systems concepts, and was developed for diffusions in the Russian literature prior to Foster (see [20]).

To describe this approach and its generalizations for continuous time operations research models, we first define a version of the generator for the process $\mathbf{\Phi}$.

We denote by $D(\widetilde{\mathcal{A}})$ the set of all functions $f : \mathsf{X} \to \mathbb{R}$ for which there exists a measurable function $g : \mathsf{X} \to \mathbb{R}$ such that for each $x \in \mathsf{X}$, $t > 0$,

$$\mathsf{E}_x[f(\Phi_t)] = f(x) + \mathsf{E}_x\left[\int_0^t g(\Phi_s)\,ds\right] , \tag{5}$$

$$\int_0^t \mathsf{E}_x[|g(\Phi_s)|]\,ds < \infty . \tag{6}$$

We write $\widetilde{\mathcal{A}} f := g$ and call $\widetilde{\mathcal{A}}$ the *extended generator* of $\mathbf{\Phi}$; $D(\widetilde{\mathcal{A}})$ is called the domain of $\widetilde{\mathcal{A}}$. In many cases we can write

$$\widetilde{\mathcal{A}} V(x) = \lim_{h \downarrow 0} \frac{\int P^h(x, dy) V(y) - V(x)}{h} .$$

In this section we show if $V \in D(\widetilde{\mathcal{A}})$ is a positive function on X, and if the *drift* $\widetilde{\mathcal{A}} V\,(x)$ is suitably negative off some compact subset of X, then the network model will be positive recurrent. We begin with the most general results of this kind.

3.1 Criteria for Stability

We state here one of the simplest stability tests, based on the existence of a stochastic Lyapunov function V. This result gives conditions for positive Harris recurrence in terms of the drift $\widetilde{\mathcal{A}} V$. For a proof and various extensions and converse results, see [34].

Theorem 1. *For the Markovian network with state process $\mathbf{\Phi}$ satisfying (1–3), suppose that the following conditions hold: There exists some compact set C, some $\gamma > 0$, and some non-negative test function V such that*

$$\widetilde{\mathcal{A}} V(x) \le -\gamma + b \mathbb{1}_C(x) . \tag{7}$$

Then $\mathbf{\Phi}$ is positive Harris recurrent.

While this result holds for a vast class of models – it is only ψ-irreducibility that is really required – because of the simpler state description we will restrict our attention largely to exponential network models. In §5 we will show how these methods extend themselves to more general queueing network models.

For virtually all examples in the operations research literature, Theorem 1 can be considerably strengthened. If V satisfies the drift criterion (7), then one can set $V_*(x) = \exp(\epsilon V(x))$, and frequently one then obtains, for sufficiently small ϵ, the stronger drift

$$\widetilde{\mathcal{A}} V_*(x) \leq -\gamma_* V_*(x) + b_* \mathbb{1}_C(x) \ .$$

It then follows that there exist constants $\varepsilon > 0$, $B < \infty$, such that

$$\left| \mathsf{E}_x[f(\Phi_t)] - \int f\, d\pi \right| \leq B V_*(x) e^{-\varepsilon t}, \qquad t \geq 0 \ ,$$

for any function f satisfying the bound $|f| \leq V_*$. See for example [40, 31, 30, 24].

Even though for network models the operator $\widetilde{\mathcal{A}}$ is rarely a differential operator, calculus plays an important role in analyzing the drift $\widetilde{\mathcal{A}} V$. This is because typically the function V will be a norm, or a polynomial function of a norm, on the state space X, and we will see that we can typically assume that V can be extended to form a smooth (C^∞) function on a suitable subset of $\mathbb{R}^n$. For instance, in [30] a linear function is constructed, and Dupuis and Williams [14] explicitly construct such a test function for a class of reflected Brownian motions. Other examples will be given below.

With X equal to a subset of $\mathbb{R}^n$, define the coordinate variables $f_i : \mathsf{X} \to \mathbb{R}$ so that $\Phi_t = (f_1(\Phi_t), \dots, f_n(\Phi_t))^\top$, and define for $x \in \mathsf{X}$, the vector $\Delta(x) \in \mathbb{R}^n$ as

$$\Delta(x) := (\widetilde{\mathcal{A}} f_1(x), \dots, \widetilde{\mathcal{A}} f_n(x))^\top \ .$$

In the examples we consider we will perform a Taylor series expansion to write

$$\widetilde{\mathcal{A}} V(x) = \Delta(x)^\top V'(x) + o(\widetilde{\mathcal{A}} V(x)) \ ,$$

where V' denotes the gradient of V, and $o(\widetilde{\mathcal{A}} V(x)) \ll |\widetilde{\mathcal{A}} V(x)|$ for large x. This gives a geometric approach to stability, which was first described in this operations research context by Kingman in [21]. If the above condition holds and if one can verify that

$$\Delta(x)^\top V'(x) < -\gamma \qquad x \neq 0 \ , \tag{8}$$

then positive recurrence is guaranteed by Theorem 1. We will illustrate this approach in several examples below.

For many of our examples a stability analysis is superfluous, since an invariant measure can be explicitly constructed. However, it is worthwhile illustrating the techniques first with simple models.

We begin by considering a specific class of linear Lyapunov functions to test for stability.

3.2 "Work" as a Lyapunov Function

For a large class of models one can take $V(x)$ equal to the *mean work in the system*, measured in units of time, which is a natural first order guess to approximate the mean time at which the network first empties, which is always a valid Lyapunov function for a stable system [34, Theorem 6]. We will illustrate this method with two examples.

First take the M/M/1 queue with service and arrival rates respectively μ and λ, where we can take the state space $\mathsf{X} = \mathbb{Z}_+$. We define the *work* $V(x)$ as

$$V(x) = x/\mu = \mathsf{E}_x \left[\begin{array}{l} \text{mean time to empty,} \\ \text{given no new} \\ \text{arrivals occur} \end{array} \right] .$$

Using the expression $\Delta(x) = (\lambda - \mu) + \mu \mathbf{1}(x = 0)$ we then obtain for $x \neq 0$,

$$\widetilde{\mathcal{A}}V(x) = \Delta(x)^{\mathsf{T}} V'(x) = -(1 - \lambda/\mu) .$$

Under the capacity constraint $\rho = \lambda/\mu < 1$, we see that the model is positive recurrent.

A similar argument is used in [30] as a first step in an inductive proof of positive recurrence for a class of generalized Jackson networks. To see how this approach can be generalized, consider the case of two queues in tandem as illustrated in Figure 2, where again we assume exponential inter-arrival times, and service times, with means $1/\lambda$, $1/\mu_1$, and $1/\mu_2$.

For this model the mean work in the system is defined to be $V(x) = x_1(1/\mu_1 + 1/\mu_2) + x_2/\mu_2$, and the drift vector is

$$\Delta(x) = \begin{pmatrix} \lambda - \mu_1 \mathbf{1}(x_1 > 0) \\ \mu_1 \mathbf{1}(x_1 > 0) - \mu_2 \mathbf{1}(x_2 > 0) \end{pmatrix} . \tag{9}$$

Applying the generator to the Lyapunov function V gives for all x,

$$\widetilde{\mathcal{A}}V(x) = \Delta(x)^{\mathsf{T}} V'(x) = -(1 - \rho_1) - (1 - \rho_2) + \mathbf{1}(x_1 = 0) + \mathbf{1}(x_2 = 0) .$$

We now face difficulties because the negative drift does not hold on the boundary $\partial\mathsf{X} = \{x : x_1 = 0 \text{ or } x_2 = 0\}$. Since the boundary is not compact, we cannot apply Theorem 1 directly.

One solution is to search for a drift over a finite period of time, rather than the instantaneous drift (7). This amounts to *averaging* the drift $\widetilde{\mathcal{A}}V(x)$ as follows: From (5) we have

$$P^T V(x) = V(x) + \int_0^{\mathsf{T}} P^s \widetilde{\mathcal{A}}V(x)\, ds$$

$$= V(x) + \int_0^{\mathsf{T}} [\mathsf{P}_x(x_1(s) = 0) - (1 - \rho_1)]\, ds + \int_0^{\mathsf{T}} [\mathsf{P}_x(x_2(s) = 0) - (1 - \rho_2)]\, ds.$$

One may show that for $x_1 + x_2$ is large, there exists a T such that the right hand side is less than or equal to $V(x) - 1$. This is proved by induction in [30], and may be seen in this simple example using two properties of the M/M/1 queue: (i) For large x, $\mathsf{P}_x(x(s) = 0) \approx 0$; (ii) $\mathsf{P}_x(x(s) = 0) - (1 - \rho) \to 0$ as $s \to \infty$, uniformly on any bounded subset of X. This drift property for the skeleton chain $\mathbf{\Phi}^T = \{\Phi_0, \Phi_T, \Phi_{2T}, \ldots\}$

then implies positive recurrence for the process Φ. To justify this comparison with an isolated M/M/1 queue it is necessary to combine stability of the single queue, with certain monotone properties of these single class networks [30].

Although as shown in [30] this is an effective approach for stability for single class networks, it does appear to be a rather model specific technique since the average of the drift must be separately analyzed. Moreover, it may be seen that for multiclass models with a buffer priority service discipline, linear functions are grossly ineffective as test functions using either the criterion (7), or the averaged criterion described here.

To see why the linear function V used in this example does not exhibit an instantaneous drift, and why in fact no linear function satisfies (7), we must look at the geometry of the problem in greater detail. This will lead us to consider more general classes of test functions.

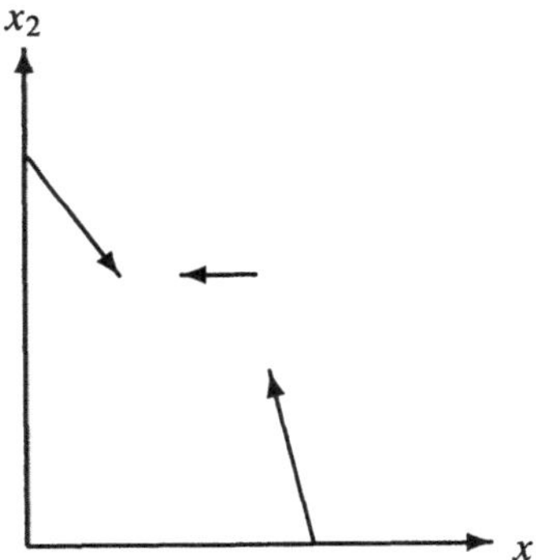

Fig. 3. The drift vector field $\Delta(x)$ for the tandem queue.

In Figure 3 we have illustrated the drift vector $\Delta(x)$ defined in (8) in the special case $\lambda = 1$, $\mu_1 = \mu_2 = 1.1$. These vectors have been skewed to exaggerate their relative angles. To satisfy (7), the entire vector $\Delta(x_0)$ with origin at x_0 must lie within the half space $\{x \in \mathbb{R} : c^{\top} x \leq c^{\top} x_0\}$ for all $x_0 \in \mathsf{X}$. It is clear from Figure 3 that this is impossible in this example for any fixed c.

3.3 Quadratics

Fayolle in [15] avoids the difficulties described above by considering a general quadratic Lyapunov function $V(x) = x_1^2 + 2\alpha x_1 x_2 + \beta^2 x_2^2$. Applying the generator to this function we have for the tandem pair of queues,

$$\begin{aligned}
\widetilde{\mathcal{A}}V(x) &= \lambda[2x_1 + 2\alpha x_2 + 1] \\
&\quad + \mu_1[-2x_1 + 1 + 2\alpha(-x_2 + x_1 - 1) + \beta^2(2x_2 + 1)]\mathbb{1}(x_1 > 0) \\
&\quad + \mu_2[-2\alpha x_1 - 2\beta^2 x_2 + \beta^2]\mathbb{1}(x_2 > 0) \\
&= 2[\lambda + \mu_1(\alpha - 1) - \alpha\mu_2 \mathbb{1}(x_2 > 0)]x_1 \\
&\quad + 2[\lambda\alpha + \mu_1(\beta^2 - \alpha)\mathbb{1}(x_1 > 0) - \mu_2\beta^2]x_2 \qquad + b(x)
\end{aligned}$$

where $b(x) = \mu_1(1-2\alpha+\beta^2)\mathbf{1}(x_1>0)+\mu_2\beta^2\mathbf{1}(x_2>0)+\lambda$, a bounded function of x. Thus $\widetilde{\mathcal{A}}V(x) = \Delta(x)^\top V'(x)+b(x)$, and again we see that the model will be stable if $\Delta(x)^\top V'(x) < 0$ for some choice of α and β, and all $x \neq 0$. With $\alpha \geq 0$, the conditions which must be satisfied to ensure this are

$$\begin{aligned} \lambda + \mu_1(\alpha-1) &< 0, \\ \lambda\alpha - \mu_2\beta^2 &< 0, \\ \lambda\alpha + \mu_1(\beta^2-\alpha) - \mu_2\beta^2 &< 0, \end{aligned} \qquad \text{or, equivalently,} \qquad \begin{aligned} \alpha &< 1-\rho_1, \\ \rho_2\alpha &< \beta^2, \\ \beta^2 &< \frac{1-\rho_1}{1-(\rho_1/\rho_2)}\alpha. \end{aligned}$$

When $\rho_2 \leq \rho_1 < 1$ it is immediate that $\alpha = 0$ and $\beta = 1$ satisfies these conditions. Otherwise, under the load condition $\rho_1 < 1$ we see that a positive α exists satisfying the first inequality. The second two inequalities have a solution β^2 for a given $\alpha > 0$ if and only if $\rho_2 < 1$. Hence we see from Theorem 1 that this model is positive Harris recurrent whenever the capacity conditions are satisfied.

Solving these equations by hand is a tedious process in complex models. In §4 we will show how this can be automated by formulating drift inequalities such as (10) as a linear programming problem.

3.4 Piecewise Linear Functions

An alternative approach to stability that can also be computer automated is to search for a *piecewise linear* function V which satisfies the geometric condition (8), where $V'(x)$ now denotes any subgradient vector at x. We do not know if the function V then satisfies (7), however the following result shows that this is irrelevant:

Theorem 2 [13]. *Consider an exponential network with state space $\mathsf{X} = \mathbb{Z}_+^n$, and suppose that the drift vector $\Delta(x)$ can be extended to all of $\mathbb{R}_+^n$ so that it is a radially homogeneous function of x. That is, $\Delta(\alpha x) = \Delta(x)$ for all $x \in \mathbb{R}_+^n$, and all $\alpha > 0$.*

Suppose that $V : \mathbb{R}^n \to \mathbb{R}_+$ is a piecewise linear function of the form

$$V(x) = \max(d_i(x) : 1 \leq i \leq m) \ ,$$

where each function d_i is of the form

$$d_i(x) = \sum_{j=1}^{n} a_{ij}x_j \ ,$$

with $a_{ij} \geq 0$, and suppose that $V(x) > 0$ for all $x \in \mathsf{X}$, $x \neq 0$. If (8) is satisfied for all $x \neq 0$ and all subgradient vectors $V'(x)$ at x, then there exists a smooth, uniformly Lipschitz function $V_s : \mathbb{R}^n \to \mathbb{R}_+$ which satisfies (7).

Hence, under these conditions the model is positive Harris recurrent.

For the tandem pair of queues with state space $\mathsf{X} = \mathbb{Z}_+^2$, let $V(x) = \max(d_1(x), d_2(x))$, where $d_i(x)$ is a linear function of the form described above. We will also write $d_i(x) = x^\top d^i$ where $d^i = \binom{a_{i1}}{a_{i2}} \in \mathbb{R}_+^2$. By examining the drift vectors $\Delta(x)$ for the specific model illustrated in Figure 3 one can see that the choice $d_1(x) = 2x_1$, $d_2(x) = x_1 + x_2$ gives rise to a function which satisfies (8). The resulting function V is in fact equal to $\max(aW_1(x), bW_2(x))$ where W_i is the work at station i, or the mean time at station i to process customers initially in the system. The scaling factors $a, b > 0$ are chosen to ensure that the desired drift does occur on the boundary of X. To see that the inequality (8) holds in general under the capacity constraints, observe that for $x \neq 0$,

$$\Delta(x)^{\mathsf{T}} d^1 = -2(\mu_1 \mathbb{1}(x_1 > 0) - \lambda) \ ,$$

which is negative when $x_1 > 0$; and

$$\Delta(x)^{\mathsf{T}} d^2 = -(\mu_2 \mathbb{1}(x_2 > 0) - \lambda) \ ,$$

which is negative when $x_2 > 0$. It follows that (8) does hold, and by Theorem 2 we again see that the network is positive Harris recurrent.

4 The LP Approach

The search for either a quadratic or piecewise linear Lyapunov function is made feasible in complex models by posing the problem of satisfying the desired drift condition as an LP. This is developed for quadratic Lyapunov functions in [24] and for piecewise linear Lyapunov functions in [13]. We demonstrate both methods by considering the model given in Figure 1.

4.1 Quadratics

The quadratic function considered is of the form

$$V(x) = q_{11}x_1^2 + q_{22}x_2^2 + q_{33}x_3^2 + 2q_{12}x_1x_2 + 2q_{13}x_1x_3 + 2q_{23}x_2x_3 \ ,$$

where $q_{ij} \geq 0$. We would like to solve for the coefficients $\{q_{ij}\}$ such that (8) is satisfied for all $x \neq 0$, and some $\gamma > 0$.

The problem of solving for $\{q_{ij}\}$ may be posed as a linear programming problem, with the objective function $\gamma \leq 1$. If the value of the LP is $\gamma^* = 1$, then by (8) the network is positive Harris recurrent. Let $\rho_1 = \lambda(1/\mu_1 + 1/\mu_3)$, $\rho_2 = \lambda/\mu_2$, and consider the case when $\mu_1 = \mu_3$. The set of pairs (ρ_1, ρ_2) for which $\gamma^* = 1$ was found numerically, and the results are illustrated graphically in Figure 4. While a large amount of the stability region is captured, there is a region where no conclusion about stability may be made. This is remedied by using piecewise linear functions, which we now consider.

4.2 Piecewise Linear Functions

As in §3.4, we search for a function of the form $V(x) = \max(d_i(x) : 1 \leq i \leq 2)$, which satisfy the following drift conditions:

$$\Delta(x)^{\mathsf{T}} d^1 < -\gamma \qquad \text{when } x_1 + x_3 > 0, \tag{10}$$

$$\Delta(x)^{\mathsf{T}} d^2 < -\gamma \qquad \text{when } x_2 > 0, \tag{11}$$

where $\gamma > 0$. We also require that the vectors d^i be normalized so that the appropriate linear function dominates on the boundaries of X: If $x_1 + x_3 = 0$, and $x_2 \neq 0$, then

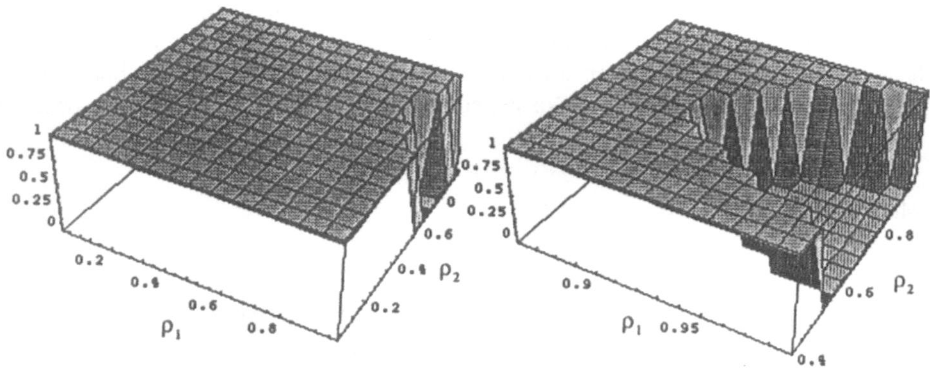

Fig. 4. Value of the quadratic LP for the network of Figure 1. The figure on the right is a more detailed view of one corner of the figure on the left.

we require that $V(x)' = d^2$ or, equivalently, $V(x) > x^\top d^1$. A similar constraint must hold if $x_2 = 0$. For the model under consideration, the condition (10) yields the linear constraints

$$\begin{aligned} a_{11}\lambda - a_{12}\mu_2 + a_{13}(\mu_2 - \mu_3) &< -\gamma \ , \\ a_{11}(\lambda - \mu_1) + a_{12}(\mu_1 - \mu_2) + a_{13}\mu_2 &< -\gamma \ , \\ a_{11}\lambda - a_{13}\mu_3 &< -\gamma \ , \\ a_{11}(\lambda - \mu_1) + a_{12}\mu_1 &< -\gamma \ . \end{aligned}$$

Similar linear constraints may be developed for (11). To ensure that the dominance conditions hold, choose $a_{22} > a_{12}$, $a_{11} > a_{21}$, $a_{13} > a_{23}$, and $a_{ij} \geq 0$ for all i, j. As when we considered quadratics, the problem of solving for $\{a_{ij}\}$ while maximizing $\gamma \leq 1$ may be formulated as a linear programming problem. In this example we find that a solution to the LP may be found, and hence the network is stable, whenever the nominal load is less than unity at each machine.

Since in this simple example a solution to the LP may be constructed whenever the capacity conditions are satisfied, we find that the network is stable for any non-idling policy. Also, since for certain arrival and service rates it is impossible to construct a quadratic Lyapunov function, even though the network is known to be stable, we see that piecewise linear functions improve upon quadratics in this example.

In [24] and in [13], substantially more complex network models are analyzed using quadratic and piecewise linear functions, respectively, using this LP approach. One such example is the network introduced by Dai and Wang in [11], which is a re-entrant line with two machines and five buffers. In [24] it is found that for certain service rates, no quadratic Lyapunov function can be constructed, although it was not known whether these parameter values would give rise to stable or unstable systems. When these same examples were analyzed in [13] using piecewise linear Lyapunov functions, it was found that the LP does give a suitable Lyapunov function for a larger range of parameter values, and that for parameters outside of this region, strong evidence from simulations suggests that the system is in fact unstable.

We conjecture that this method is tight, so that a piecewise linear function always exists for a network satisfying certain spacial homogeneity conditions. The proof of this result probably rests on proving that the mean return time to the origin is a Lipschitz function of x, so that for large x, it may be approximated by a piecewise linear function. Since the mean emptying time always serves as a Lyapunov function, this would give the desired converse result. This is similar to the approach followed in [14]. The key to this construction may lie in analyzing a fluid model, which is the next and final approach to stability that we will consider in this paper.

5 Fluid Models

To conclude the paper we now describe a recent method for assessing stability based upon an analysis of a fluid model approximation [9]. Up to now, all of our examples have focused on exponential networks. We will see from Dai's result that the distributions of the interarrival and service processes are largely irrelevant when addressing stability. As one consequence, this allows the application of the LP method to more general classes of networks.

For queueing network models it has been known for some time that a functional law of large numbers holds, which allows a given model to be approximated by an associated fluid model (see e.g. [6, 7]). Remarkably, if the fluid model is stable (i.e. empties in finite time) then the original process is positive Harris recurrent, and bounds on polynomial moments as well as useful ergodic theorems may be obtained [9, 10].

We will illustrate here the construction and analysis of a fluid model through an example. Again we will consider the pair of queues in tandem, but we will remove the restrictive distributional assumptions on the arrival and service processes.

The idea of the fluid model construction is to scale both space and time to smooth out the local behavior of the network. Specifically, take an initial condition $x \in \mathsf{X}$ which we assume is *large*, let $|x|$ denote the usual l_1 norm of x, and define the scaled process ϕ^x as

$$\phi^x(t) = \frac{1}{|x|}\Phi(|x|t), \qquad t \in \mathbb{R}_+ \ .$$

For the tandem queue (4) under consideration, we have

$$\phi^x(t) = \frac{1}{|x|}(u(|x|t), b_1(|x|t), v_1(|x|t), b_2(|x|t), v_2(|x|t))^{\top} \ .$$

We now let the initial condition $x \to \infty$ in such a way that $x/|x|$ converges, so that $\phi^x(0) \to \bar{\phi}(0) \in \mathbb{R}^5_+$. Dai shows in [9] that the distributions of the stochastic processes $\{\phi^x : x \in \mathsf{X}\}$ are tight, and that any weak limit must be a solution to a certain integral equation. In the present example, since $\phi^x(0)$ converges, it is not difficult to see that $\phi^x(\,\cdot\,)$ does in fact converge almost surely to a deterministic process $\bar{\phi}(\,\cdot\,)$ which we call the *fluid model*.

It is easy to describe the limit process. The limiting residual arrival process is $\bar{u}(t) = (\bar{u}(0) - t)^+$, where $x^+ = \max(x, 0)$. The limiting residual service processes have the similar form $\bar{v}_i(t) = (\bar{v}_i(0) - t)^+$. Since $\bar{u}(0) + \bar{v}_1(0) + \bar{v}_2(0) \leq |\bar{\phi}(0)| = 1$, we see that the residual processes are identically zero for all $t \geq 1$. The buffer levels $\bar{b}_i(t)$ are only slightly more complex. Both are continuous and finitely piecewise linear. For the first buffer, after a delay of $\bar{u}(0)$ units of time, "fluid" flows in at rate λ, and after the residual service time $\bar{v}_1(0)$, fluid flows out at rate μ_1. Hence we have the expression

$$\bar{b}_1(t) = [\bar{b}_1(0) + \lambda(t - \bar{u}(0))^+ - \mu_1(t - \bar{v}_1(0))^+]^+ \ .$$

A simple calculation shows that $\bar{b}_1(t) = 0$ for all $t \geq \tau_1 = (\mu_1 + 1)/(\mu_1 - \lambda)$, assuming $\rho_1 < 1$. The behavior of the second queue is made more complex because of the changes in the internal arrival stream to this buffer. However, after time τ_1, this internal arrival stream has rate λ, and hence the second queue can be analyzed as if it were in isolation. After some finite time $\tau_2 \geq \tau_1$ we again have $\bar{b}_2(t) = 0, t \geq \tau_2$, provided $\rho_2 < 1$.

Under the capacity conditions we see that the fluid model is stable in the sense that $\bar{\phi}(t) \to 0, t \to \infty$. It follows that for any $t \geq \tau_2$ we have weak convergence of $\frac{1}{|x|}\Phi(t|x|)$ to 0, and under (2) one can then show that

$$\mathsf{E}\left[\frac{1}{|x|^p}|\Phi(t|x|)|^p\right] \to 0, \qquad t \to \infty \ ,$$

where p is the integer in the moment condition (2). Thus, there exists a finite constant L such that

$$\mathsf{E}[|\Phi(t|x|)|^p] \leq \frac{1}{2}|x|^p, \qquad |x| \geq L \ ,$$

and this may be interpreted as a *drift property* for the network. Letting $V(x) = |x|^p$, and $n(x) = \tau_2|x|$, we have for some $\epsilon > 0$,

$$\int P^{n(x)}(x, dy)V(y) \leq V(x) - \epsilon|n(x)|^p + b\mathbb{1}_C(x), \qquad x \in \mathsf{X} \ , \tag{12}$$

where $C = \{x \in \mathsf{X} : |x| \leq L\}$. With $p = 1$ it follows from (12) and [31, Theorem 19.1.2] that the process is positive Harris. For $p \geq 2$, further analysis based on the Comparison Theorem of [31, p. 337] shows that in steady state $\mathsf{E}_\pi[|\Phi(t)|^{p-1}] < \infty$. In particular, in this case the mean queue size is finite in steady state [10].

We saw in this example that one simplification when analyzing the fluid model comes from the fact that the residual arrival and service processes may be discarded, since they eventually vanish. The resulting model is similar to the exponential network, for which the LP approach leads to a quick test for stability. Following these ideas, it has recently been shown in [5] that the LP methods described in §4 do indeed extend naturally to give an algorithmic approach to stability, even when the interarrival and service distributions are not exponentially distributed. Other examples are given in [37, 12, 13]. The fluid model approach thus allows the use of Lyapunov function methods for complex models using techniques which have been devised for relatively simple exponential networks.

Acknowledgements Much of this material is based upon joint work with Jim Dai, P.R. Kumar, and Richard Tweedie. We thank them for allowing us to reproduce some of these results here. Jim Dai, Sunil Kumar, and Richard Tweedie read earlier versions of the manuscript, and their comments are very much appreciated.

This research was supported in part by the NSF under Grant No. 1-5-30679.

References

1. F. Baccelli and S. Foss. Stability of Jackson-type queueing networks, I. To appear in *QUESTA*, 1993.
2. A. A. Borovkov. Limit theorems for queueing networks. *Theory Probab. Appl.*, 31:413–427, 1986.
3. D.D. Botvich and A.A. Zamyatin. Ergodicity of conservative communication networks. Technical report, Institut National de Recherche en Informatique et en Automatique, Rocquencourt, 1993. Technical Report.
4. M. Bramson. Instability of FIFO queueing networks. Technical report, Mathematics Department, University of Wisconsin, Madison, WI, 1993.
5. H. Chen. Fluid approximations and stability of multiclass queueing networks I: Work conserving disciplines. Technical Report, 1994.
6. H. Chen and A. Mandelbaum. Discrete flow networks: Bottlenecks analysis and fluid approximations. *Mathematics of Operations Research*, 16:408–446, 1991.
7. H. Chen and A. Mandelbaum. Heterogeneous fluid networks, with applications to multi-type queues. Preprint, 1993.
8. D. Bertsimas, I. Ch. Paschalidis and J. N. Tsitsiklis. Scheduling of multiclass queueing networks: Bounds on achievable performance. In *Workshop on Hierarchical Control for Real–Time Scheduling of Manufacturing Systems*, Lincoln, New Hampshire, October 16–18, 1992.
9. J. G. Dai. On the positive Harris recurrence for multiclass queueing networks: A unified approach via fluid limit models. Technical report, Georgia Institute of Technology, 1993.
10. J. G. Dai and S.P. Meyn. Stability and convergence of moments for multiclass queueing networks via fluid limit models. Technical report, University of Illinois and Georgia Institute of Technology, 1993. (In preparation).
11. J. G. Dai and Y. Wang. Nonexistence of Brownian models of certain multiclass queueing networks. *Queueing Systems: Theory and Applications*, 13:41–46, May 1993.
12. J. G. Dai and G. Weiss. Stability and instability of fluid models for certain re-entrant lines. Technical report, Georgia Institute of Technology, 1994.
13. D.G. Down and S. P. Meyn. Piecewise linear test functions for stability of queueing networks. In *Proceedings of the 33rd Conference on Decision and Control*, Buena Vista, FL, December 1994. (submitted).
14. P. Dupuis and R. J. Williams. Lyapunov functions for semimartingale reflecting Brownian motions. Technical report, Brown University, Providence, RI, 1993.
15. G. Fayolle. On random walks arising in queueing systems: ergodicity and transience via quadratic forms as Lyapounov functions - part I. *Queueing Systems*, 5:167–183, 1989.
16. G. Fayolle, V. A. Malyšhev, M.V. Men'šikov, and A. F. Sidorenko. Lyapunov functions for Jackson networks. Technical Report, INRIA Rocquencourt, 1991.
17. F. G. Foster. On the stochastic matrices associated with certain queueing processes. *Ann. Math. Statist.*, 24:355–360, 1953.
18. P. W. Glynn and S. P. Meyn. A Lyapunov bound for solutions of Poisson's equation. *Ann. Probab.*, 1993 (to appear).
19. J. M. Harrison and V. Nguyen. Brownian models of multiclass queueing networks: Current status and open problems. *Queueing Systems: Theory and Applications*, 13:5–40, 1993.
20. R. Z. Khas'minskii. *Stochastic Stability of Differential Equations*. Sijthoff & Noordhoff, Netherlands, 1980.
21. J. F. C. Kingman. The ergodic behaviour of random walks. *Biometrika*, 48:391–396, 1961.
22. P. R. Kumar. Re–entrant lines. *Queueing Systems: Theory and Applications*, 13:87–110, 1993.

23. P. R. Kumar and S. P. Meyn. Duality and linear programs for stability and performance analysis analysis of queueing netwroks and scheduling policies. Technical report, C. S. L., University of Illinois, 1993. submitted to *IEEE Transactions on Automatic Control.*
24. P. R. Kumar and S. P. Meyn. Stability of queueing networks and scheduling policies. Technical report, C. S. L., University of Illinois, 1993. To appear *IEEE Transactions on Automatic Control.*
25. S. H. Lu and P. R. Kumar. Distributed scheduling based on due dates and buffer priorities. *IEEE Transactions on Automatic Control*, 36(12):1406–1416, December 1991.
26. V. A. Malyšhev. Classification of two-dimensional positive random walks and almost linear semi-martingales. *Soviet Math. Dokl.*, 13:136–139, 1972.
27. V. A. Malyšhev. Networks and dynamical systems. *Adv. Appl. Probab.*, 25:140–175, 1993.
28. V. A. Malyšhev and M.V. Men'šikov. Ergodicity, continuity and analyticity of countable Markov chains. *Trans. Moscow Math. Soc.*, 1:1–48, 1982.
29. M. V. Men'šikov. Ergodicity and transience conditions for random walks in the positive octant of space. *Soviet Math. Dokl.*, 15:1118–1121, 1974.
30. S. P. Meyn and D. Down. Stability of generalized Jackson networks. *Ann. Appl. Probab.*, 1993 (to appear).
31. S. P. Meyn and R. L. Tweedie. *Markov Chains and Stochastic Stability*. Springer-Verlag, London, 1993.
32. S. P. Meyn and R. L. Tweedie. Stability of Markovian processes II: Continuous time processes and sampled chains. *Adv. Appl. Probab.*, 25:487–517, 1993.
33. S. P. Meyn and R. L. Tweedie. Stability of Markovian processes III: Foster-Lyapunov criteria for continuous time processes. *Adv. Appl. Probab.*, 25:518–548, 1993.
34. S. P. Meyn and R.L. Tweedie. A survey of Foster-Lyapunov techniques for general state space markov processes. In *Proceedings of the Workshop on Stochastic Stability and Stochastic Stabilization, Metz, France, June 1993*, 1994.
35. A. N. Rybko and A. L. Stolyar. On the ergodicity of stohastic processes describing open queueing networks. Technical report, Moscow, 1992.
36. S. Kumar and P. R. Kumar. Performance bounds for queueing networks and scheduling policies. Technical report, Coordinated Science Laboratory, University of Illinois, Urbana, IL, 1992.
37. S. Kumar and P. R. Kumar. The last buffer first serve priority policy is stable for stochastic re-entrant lines. Technical report, Coordinated Science Laboratory, University of Illinois, Urbana, IL, 1994.
38. T.I. Seidman. 'First come, first served' is unstable! Technical report, Department of Mathematics and Statistics, University of Maryland, 1993. Technical Report.
39. K. Sigman. The stability of open queueing networks. *Stoch. Proc. Applns.*, 35:11–25, 1990.
40. F. M. Spieksma and R. L. Tweedie. Strengthening ergodicity to geometric ergodicity for Markov chains. *Stochastic Models*, 10, 1994.

Stationary Regime and Stability of Free-Choice Petri Nets

F. Baccelli and B. Gaujal

INRIA Sophia-Antipolis, BP 93, 06902 Sophia-Antipolis, France

1 Introduction

The aim of this paper is to give conditions under which a class of stochastic Petri nets called free choice nets satisfy a set of monotonicity and separability conditions ensuring the existence of a finite stationary regime for the marking process (or a finite periodic regime in the deterministic case). The scheme of conflict resolution is via a stochastic routing sequence. This assumption is essential for ensuring basic monotonicity properties. The main tools for proving these properties is ergodic theory.

2 The Separable-Monotone Framework for Counters and Daters

We consider a discrete event system composed of a set of nodes, and submitted to an input point process $N = \{T_n\}_{n\in\mathbb{Z}}$. Let $c.N$ be the c-dilation of N, namely the point process with arrivals $c.N = \{cT_n\}_{n\in\mathbb{Z}}$. Let $N_{[m,n]}$ be the $[m,n]$-restriction of the point process N, namely the point process $\{T_l\}_{m\le l\le n}$. We will say that $N_{[m,n]} \le N'_{[m,n]}$ if $T_l \le T'_l$ for all $m \le l \le n$. In what follows, this point process will also be characterized through its counting measure $R_{[m,n]} : \mathbb{R} \to \mathbb{N}$, where $R_{[m,n]}(t)$ counts the number of points of $N_{[m,n]}$ which are less than t.

The discrete event system is characterized by two equivalent sets of variables:

- The daters: $\mathcal{X}^i_{[m,n]}(k) \in \mathbb{R}$ will denote the epoch of the k-th event on node i, when the system is submitted to $N_{[m,n]}$ (here, we take $k \in \mathbb{N}$ and $\mathcal{X}^i[m,n](k) = \infty$ if there are less than k events on node i).
- The counters: $X^i_{[m,n]}(t) \in \mathbb{N}$ will denote the number of events which took place on node i before time t (we will take this function left-continuous).

Note that counters and daters are related by

$$X^i_{[m,n]}(t) = \sum_{k\in\mathbb{N}} \mathbf{1}_{\{\mathcal{X}^i_{[m,n]}(k)\le t\}} \ . \tag{1}$$

The separable-monotone framework consists of the following set of assumptions:

External Monotonicity If $N_{[m,n]} \leq N'_{[m,n]}$, then for all k and i then (with obvious notations) $\mathcal{X}^i_{[m,n]}(k) \leq \{\mathcal{X}'\}^i_{[m,n]}(k)$, which is equivalent to the property that for all t and i $X^i_{[m,n]}(t) \geq \{X'\}^i_{[m,n]}(t)$.

Conservation Let

$$X^i_{[m,n]} \equiv \lim_{t\to\infty} X^i_{[m,n]}(t) \ . \tag{2}$$

This limit exists since the function is non-decreasing. In words, $X^i_{[m,n]}$ counts the *total* number of events on node i for $N_{[m,n]}$. We assume that $X^i_{[m,n]}$ is *finite* and *independent* of the values taken by the variables T_l, $n \leq l \leq m$ (provided m, n and $\{T_l\}$ are finite of course). Of particular interest to us will be the *maximal dater* defined by:

$$\mathcal{X}_{[m,n]} = \max_i \mathcal{X}^i_{[m,n]}\left(X^i_{[m,n]}\right) \ . \tag{3}$$

Separability The separability assumption states that if $T_{l+1} \geq \mathcal{X}_{[m,l]} + M$, for some non-negative M, then

$$\begin{aligned} \mathcal{X}^i_{[m,n]}(k) &= \mathcal{X}^i_{[m,l]}(k) \ , \quad k \leq X^i[m,l] \ , \\ \mathcal{X}^i_{[m,n]}(k + X^i_{[m,l]}) &= \mathcal{X}^i_{[l+1,m]}(k) \ , \quad k \geq 1 \ , \end{aligned} \tag{4}$$

or equivalently

$$\begin{aligned} X^i_{[m,n]}(t) &= X^i_{[m,l]}(t) \ , \quad t < T_{l+1} \ , \\ X^i_{[m,n]}(t) &= X^i_{[m,l]} + X^i_{[l+1,m]}(t) \ , \quad t \geq T_{l+1} \ . \end{aligned} \tag{5}$$

It is easy to check that the separation and the conservation properties imply that for all $m \leq l < n$, $X^i_{[m,n]} = X^i_{[m,l]} + X^i_{[l+1,n]}$ regardless of $\{T_l\}$.

Homogeneity The homogeneity assumption states that if $T'_l = T_l + c$, then $\{\mathcal{X}'\}^i_{[m,n]}(k) = \mathcal{X}^i_{[m,n]}(k) + c$ for all k and i or equivalently that $\{X'\}^i_{[m,n]}(t+c) = X^i_{[m,n]}(t)$ for all t and i.

Let

$$\begin{aligned} \mathcal{W}_{[m,n]} &\equiv \mathcal{X}_{[m,n]} - T_n \ , \\ W^i_{[m,n]}(t) &\equiv X^i_{[m,n]} - X^i_{[m,n]}(T_n + t) \ , \quad t \geq 0 \ . \end{aligned}$$

The following theorems are proved in [4]:

Theorem 1. *Under the above properties, for all n, $\mathcal{W}_{[m-1,n]} \geq \mathcal{W}_{[m,n]}$; for all $t \geq 0$, i and n, $W^i_{[m-1,n]}(t) \geq W^i_{[m,n]}(t)$, so that*

$$\exists \lim_{m\to-\infty} \uparrow \mathcal{W}_{[m,n]} \equiv \mathcal{W}_{[-\infty,n]} \ , \tag{6}$$

$$\exists \lim_{m\to-\infty} \uparrow W^i_{[m,n]}(t) \equiv W^i_{[-\infty,n]}(t) \ . \tag{7}$$

Theorem 2. *If the system has stationary ergodic input point process defined on some probability space $(\Omega, \mathcal{F}, P, \theta)$, where θ is a shift on Ω which is ergodic and leaves P invariant, and if it is such that $\mathcal{W}_{[n,n+k]} = \mathcal{W}_{[0,k]} \circ \theta^n$, then the following a.s. limit takes place for all $c \geq 0$:*

$$\exists \lim_{n \to \infty} \frac{\mathcal{W}_{[0,n]}(c.N)}{n} = \gamma_c \ . \tag{8}$$

where γ_c is a constant. If the intensity λ of the input point process is such that $\lambda\gamma_0 < 1$, then $\mathcal{W}_{[-\infty,n]}(1.N) \equiv \mathcal{W}_{[-\infty,n]} < \infty$, for all n. If in addition $W^i_{[n,n+k]}(t) = W^i_{[0,k]}(t) \circ \theta^n$ for all n, i and t and

$$\left\{\mathcal{W}_{[-n,0]} \to_{n\to\infty} \infty\right\} \stackrel{\text{a.s.}}{=} \left\{\exists i \mid W^i_{[-n,0]} \to_{n\to\infty} \infty\right\} \ , \tag{9}$$

then, if $\lambda\gamma_0 < 1$, $W^i_{[-\infty,n]}(t) < \infty$ for all n, i and t.

Remark. Often, one also defines classes of events like $\mathcal{X}^{i,j}(k)$, which count departures from node i to node j, and the above framework extends naturally to this type of variables (and the associated counters). In this case, it is natural to define the variables:

$$Q^i_{[m,n]}(t) = W^i_{[m,n]}(t) - \sum_j W^{j,i}_{[m,n]}(t) \ , \tag{10}$$

which represent the number of objects (customers, tokens) on node i at time $t + T_n$ for $N_{[m,n]}$. So, if $\lambda\gamma(0) < 1$, we have constructed a stationary (θ-compatible) version of the $Q^i(t)$ process. For instance, $Q^i_{[-\infty,n]}(t)$ for $t \in [T_n, T_{n+1})$ provides a stationary process Q for the $[-\infty, +\infty]$-restriction of N, namely N.

3 Timed Petri Net

A Petri net is a t-uple $(\mathcal{P}, \mathcal{T}, \mathcal{C}, \mathcal{M}_0)$ where $\mathcal{P}$ is the set of places, $\mathcal{T}$ is the set of transitions, $\mathcal{C}$ the set of arcs between places and transitions and between transitions and places ($\mathcal{C}$ is a subset of $\mathcal{P} \times \mathcal{T} \cup \mathcal{T} \times \mathcal{P}$). $\mathcal{M}_0$ is the initial marking in the places. We denote by ${}^\bullet t$ the set $\{p \in \mathcal{P} : (p, t) \in \mathcal{C}\}$ (i.e. the set of all the input places of t). We define similarly the sets $t^\bullet, {}^\bullet p, p^\bullet$ as the set of the output places of t, the set of the input transitions of p and the set of the output transitions of p, respectively.

A timed Petri net is a Petri net with timings attached to transition firings: $\sigma^t(n)$ is the duration of the n-th firing of transition t. This means that if transition t begins to fire for the n-th time at epoch e, this firing will end at epoch $e + \sigma^t(n)$ and tokens are taken out of input places and put in output places of t according to the firing rule of a Petri net.

3.1 Free Choice Nets

Free choice nets (FCN) are Petri nets verifying the following conditions: $\forall p \in \mathcal{P}, t_1, t_2 \in p^\bullet, t_1 \neq t_2, {}^\bullet t_1 = {}^\bullet t_2 = \{p\}$. In other words, whenever two transitions share an input place, they have no other input place.

Free choice nets have been extensively studied in the 70's [6] and have regained interest recently [7], [9] because they constitute a nice compromise between power of description and tractability of problems.

For any place p with several output transitions, the dynamic is characterized by a routing function $\nu^p : \mathbb{N} \to p^\bullet$ associated with place p, where $\nu^p(k)$ is the transition to which the k-th token to enter place p is routed. The routing function can be a deterministic or a random sequence. For comments on this type of routing policy, see [1].

3.2 Decomposition into Marked Graphs Components

A place p in a FCN F is *serial* if $|^\bullet p| = |p^\bullet| = 1$.

First we define a relation $\mathcal{L}$ by: $t, t' \in T, t\mathcal{L}t'$ if there is a serial place p verifying $\{^\bullet p, p^\bullet\} = \{t, t'\}$. Let $\mathcal{K}$ be the transitive closure of $\mathcal{L}$. $\mathcal{K}$ is a parallelism relation. We partition the set of transitions $\mathcal{T}$ into its maximal $\mathcal{K}$-classes, $\mathcal{T}_1, \ldots, \mathcal{T}_n$. We construct a decomposition of F in the following way: $\mathcal{P}_i = \{p \in \mathcal{P} \mid p$ serial and $^\bullet p, p^\bullet \in \mathcal{T}_i\}$. for all $1 \leq i \leq n$.

The marked graph component (MGC) G_i of F is the sub-Petri net of F $(\mathcal{P}_i, \mathcal{T}_i, \mathcal{C} \cap (\mathcal{P}_i \times \mathcal{T}_i \cup \mathcal{T}_i \times \mathcal{P}_i))$. One can easily check that G_i is a marked graph and is maximal in the sense that no marked graph included in F contains G_i, except G_i itself.

The places which do not belong to any component G_i are the places with several input transitions and/or several output transitions. These places will be called *routing* places in the following. The set of the routing places is denoted $\mathcal{R}$.

3.3 Classification of Free Choice Nets

We propose a classification of the marked graph components (MGC) of a FCN based upon its links with the routing places.

A MGC G_i is said *Single Input* (SI) if $\#\{t \in \mathcal{T}_i, {}^\bullet t \not\subset \mathcal{P}_i\} = 1$. G_i is said *Multiple Input* (MI) if $\#\{t \in \mathcal{T}_i, {}^\bullet t \not\subset \mathcal{P}_i\} > 1$. A MGC G_i is said *Single Output* (SO) if $\#\{t \in \mathcal{T}_i, t^\bullet \not\subset \mathcal{P}_i\} = 1$. A MGC G_i is said *Multiple Output* (MO) if $\#\{t \in \mathcal{T}_i, t^\bullet \not\subset \mathcal{P}_i\} > 1$.

Thus all the MGC of a FCN can be put in one of the four classes, SISO, SIMO, MISO, MIMO. A FCN is said SI (resp. SO, MI, MO , SISO , SIMO, MISO, MIMO) if all its MGC are SI (resp. SO, MI, MO , SISO, SIMO, MISO, MIMO).

4 Evolution Equations for Counters

Consider a FC net satisfying the following assumption: a transition t with more than one incoming arc (i.e. with an and-convergence) is never preceded by a place p with more than one incoming arc (i.e. with an or-convergence). This restriction introduces no loss of generality: because of the FC constraint, a transition t as above cannot be preceded by a place with multiple outcoming arcs; in addition, each place p as above can be replaced by a triple p', t', p'', where $^\bullet p' = {}^\bullet p$, $p''^\bullet = p^\bullet = t$, and where $p'^\bullet = t'$, $^\bullet t' = p'$, $t'^\bullet = p''$, $^\bullet p'' = t'$, without altering the evolution of the net.

Let $X^t(u)$ denote the counter associated with t, namely, the number of firings initiated by transition t by time u. We will consider the version of this process that is continuous to the right. Let $\mathcal{A}$ be the set of transitions such that all their upstream places are serial, and let $\mathcal{B}$ be the set of transitions which do not belong to $\mathcal{A}$. Let $Y(u)$ be the vector $\{X^t(u),\ t \in \mathcal{A}\}$, where the transitions are arranged in some order, and let $Z(u)$ be the vector $\{X^t(u),\ t \in \mathcal{B}\}$. Note that each transition of $\mathcal{B}$ has at most one non-serial upstream place due to the FC constraint. However, this place may precede several transitions of $\mathcal{B}$.

Constant Firing Times We shall first consider the case when firing times are constant, positive, and all multiple of a common number, which will be taken equal to 1 without loss of generality. These assumptions are essentially for the sake of easy exposition. We will in particular show in §7 how to address the case with stochastic times, which can be treated with a similar method. We will denote M the (integer-valued) upper bound on the firing times. We will denote $\nu^p(m)$ the m-th routing decision from place p ($\nu^p(m) \in p^\bullet$) and $\Pi^t(m)$ the sum

$$\Pi^t(m) = \sum_{l=1}^{m} \mathbf{1}_{\nu^{\bullet t}(l)=t} \ , \quad t \in \mathcal{B} \ . \tag{11}$$

Similarly, for sake of easy exposition, we will limit ourselves to the case when the jump times T_n of R are integer-valued.

Lemma 3. *Under the above assumptions, for all integers n, the counting vectors* $\{Y(k), Z(k)\}$ *satisfy the following evolution equation, which is valid for* $n > M$*:*

$$Y(k) = \bigoplus_{l=1}^{M} (A_l \otimes Y(k-l) \oplus B_l \otimes Z(k-l)) \ , \tag{12}$$

$$Z(k) = \Pi\left(\sum_{l=1}^{M} (P_l \times Z(k-l) + Q_l \times Y(k-l)) + R(k)\right) \ , \tag{13}$$

where $(\oplus, \otimes)$ *is* (min, +) *(see [2]),* $(+, \times)$ *is the usual algebra, and*

- *The matrix* A_l *on* $\mathcal{A} \times \mathcal{A}$ *is defined by* $A_l(t, t') = c$ *if the firing time of* $t \in \mathcal{A}$ *is* l *and there is a serial place with* c *initial tokens between* $t' \in \mathcal{A}$ *and* t*;* ∞ *otherwise. If there are more than one serial place between* t' *and* t*, we take* c *equal to the minimum of their initial markings.*
- *The matrix* B_l *on* $\mathcal{A} \times \mathcal{B}$ *is defined by* $B_l(t, t') = c$ *if the firing time of* t *is* l *and there is a serial place with* c *initial tokens between* $t' \in \mathcal{B}$ *and* $t \in \mathcal{A}$*;* ∞ *otherwise.*
- *The matrix* P_l *on* $\mathcal{B} \times \mathcal{B}$ *is defined by* $P_l(t, t') = 1$ *if the firing time of* $t \in \mathcal{B}$ *is* l *and there is place connecting* t' *to* t*;* 0 *otherwise.*
- *The matrix* Q_l *on* $\mathcal{B} \times \mathcal{A}$ *is defined by* $Q_l(t, t') = 1$ *if the firing time of* $t \in \mathcal{B}$ *is* l *and there is place connecting* t' *to* t*;* 0 *otherwise.*
- $R(0)$ *is the matrix of initial markings:* $R(t', t) = c$ *if* $t \in \mathcal{B}$ *is such that* ${}^\bullet t$ *has an initial marking of* c*. More generally,* $R(k)(t', t)$ *is the cumulated external input in that place up to time* n*.* $R(k)(t', t) = R(k-1)(t', t) + I(k)(t, t')$*. (thus* $I(k) \neq 0$ *iff* $k \in \{T_n\}_n$*).*
- *For all vectors of integers* $Z = (Z_1, \ldots, Z_q)$*, where* $q = |\mathcal{B}|$*,* $\Pi(Z)$ *is the vector of integers* $(\pi_1(Z_1), \ldots, \pi_q(Z_q))$*.*

Proof. Immediate from definitions, the key observation being that due to our preliminary assumption, a transition which belongs to $\mathcal{B}$ will never have more than one input arc, which allows us to write (13).

Total Number of Firings Let $Z = Z(\infty)$ and $Y = Y(\infty)$ denote the vectors counting the total number of firings of the transitions. One can easily check the absence of deadlocks (a deadlock is a marking where no transition can fire) and related properties, from Y and Z: for instance, the systems is deadlocked for the initial marking ($\forall k,\ R(k) = R(0)$) if and only if Z and Y are finite.

Lemma 4. *The integer-valued vectors Z and Y satisfy the system of equations*

$$Y = A \otimes Y \oplus B \otimes Z \ , \tag{14}$$

$$Z = \Pi\,(P \times Z + Q \times Y + R) \ , \tag{15}$$

where $A = \bigoplus_{l=1}^{M} A_l(k)$, $B = \bigoplus_{l=1}^{M} B_l(k)$ $P = \sum_{l=1}^{M} P_l(k)$, $Q = \sum_{l=1}^{M} Q_l(k)$, $R = \lim_{k\to\infty} R(k)$ are independent of n.

A notable property is that this system does not depend on the variables σ^t anymore: in other words, all properties like liveness, deadlock and intermediates are associated with the switching functions, the topology and the initial marking only, and not with timing variables.

4.1 Assumptions on the Function Π

We assume that at the origin of time, the network is in a configuration where no transition can fire. Such a marking is called the *original deadlock.*

In the following, we add two assumptions on the function Π.

Assumption 1. The total firing vectors Z and Y are finite if R is finite.

This assumption is a property of the function Π since the total firing vector does not depend on the timing variables σ^t of the system. This assumption says that if the number of arrivals is finite, the system reaches a deadlock after a finite number of firings.

Assumption 2. If the network reaches a deadlock, then this deadlock is the original deadlock.

We can give a characterization of this property in terms of the vector $X = (Z, Y)$. If the network reaches a deadlock, then X is finite. If this is the original deadlock, then in any MGC G, all the transitions have fired the same number of times, $\forall t_1, t_2 \in G,\ X^{t_1} = X^{t_2}$. Conversely if X is finite and for all MGC G, $\forall t_1, t_2 \in G,\ X^{t_1} = X^{t_2}$, then the network has reached the original deadlock.

4.2 Restriction of the Arrival Process

Let $R^t_{[m,n]}$ be the counting measure of $N_{[m,n]}$ on place ${}^\bullet t$.

Assumption 1 allows us to say that the network with the input process $\left(R_{[0,n]}(k)\right)_{k\in\mathbb{N}}$ reaches a deadlock. We denote by $Z_{[0,n]}$ and $Y_{[0,n]}$ the total firing vectors for this system. With $\Pi_{[0,\infty]} \equiv \Pi$, they verify:

$$Y_{[0,n]} = A \otimes Y_{[0,n]} \oplus B \otimes Z_{[0,n]} \ ,$$

$$Z_{[0,n]} = \Pi_{[0,\infty]}\left(P \times Z_{[0,n]} + Q \times Y_{[0,n]} + R_{[0,n]}\right) \ .$$

If $t \in \mathcal{B}$, we denote by $U^{\bullet t}_{[0,n]}$ the total number of tokens that entered the place $^\bullet t$ and $U_{[0,n]} = \{U^{\bullet t}_{[0,n]},\ t \in \mathcal{B}\}$. We have

$$U_{[0,n]} = P \times Z_{[0,n]} + Q \times Y_{[0,n]} + R_{[0,n]} \ . \tag{16}$$

Since $Z_{[0,n]}$ and $Y_{[0,n]}$ are finite, $U_{[0,n]}$ is also finite.

Now, we introduce the system generated by the restricted input process $(R_{[m,n]}(k))_{k \in \mathbb{N}}$. We connect this system with the original one by taking

$$\nu^p_{[m,\infty]}(k) \equiv \nu^p \left(k + U^t_{[0,m-1]}\right) \ , \tag{17}$$

for all $t \in \mathcal{T}$ and for all $p \in \mathcal{R}$. Thus the function $\Pi_{[m,\infty]}$ is defined by

$$\Pi^t_{[m,\infty]}(k) \equiv \sum_{l=1}^{k} \mathbf{1}_{\nu^{\bullet t}_{[m,\infty]}(l)=t} = \Pi^t \left(k + U^{\bullet t}_{[0,m-1]}\right) - \Pi^t \left(U^{\bullet t}_{[0,m-1]}\right) \ .$$

Finally we define the vectors $Y_{[m,n]}(k)$ and $Z_{[m,n]}(k)$, which verify the equations

$$Y_{[m,n]}(k) = \bigoplus_{l=1}^{M} \left(A_l \otimes Y_{[m,n]}(k-l) \oplus B_l \otimes Z_{[m,n]}(k-l)\right) \ ,$$

$$Z_{[m,n]}(k) = \Pi_{[m,\infty]}\left(\sum_{l=1}^{M} \left(P_l Z_{[m,n]}(k-l) + Q_l Y_{[m,n]}(k-l)\right) + R_{[m,n]}(k)) \right) \ ,$$

with the initial conditions : $\forall k < T_m$, $Y_{[m,n]}(k+1) = 0$ and $Z_{[m,n]}(k) = 0$.

5 The Saturation Rule

We prove that the variables $X_{[m,n]}(k)$ satisfy the extented saturation conditions. We need a preliminary lemma.

Lemma 5. *If $T_m > \mathcal{X}_{[0,m-1]} + M$ then for all $k \geq T_m$, $X_{[0,n]}(k) = X_{[0,m-1]} + X_{[m,n]}(k)$.*

Proof. The proof holds by induction on k. For $k = T_m$, $X_{[1,m-1]} = X_{[1,m-1]}(k-l)\ \forall l > 1$ yields

$$Y_{[0,n]}(k) = Y_{[0,m-1]} \ ,$$
$$Z_{[0,n]}(k) = \Pi\left((P Z_{[0,m-1]} + Q Y_{[0,m-1]} + R_{[0,m-1]} + I(T_m)\right) \ .$$

So, by definition of $\Pi_{[m,\infty]}$:

$$Y_{[0,n]}(k) = Y_{[0,m-1]} + Y_{[m,n]}(k) \ ,$$
$$Z_{[0,n]}(k) = Z_{[0,m-1]} + Z_{[m,n]}(k) \ .$$

For the case $k > T_m$, the induction property yields

$$Y_{[0,n]}(k) = \bigoplus_{l=1}^{M} \left(A_l \otimes \left(Y_{[0,m-1]} + Y_{[m,n]}(k-l)\right) \oplus B_l \otimes \left(Z_{[0,m-1]} + Z_{[m,n]}(k-l)\right)\right) ,$$

$$Z_{[0,n]}(k) = \Pi\Bigg(\sum_{l=1}^{M} \left(P_l \left(Z_{[0,m-1]} + Z_{[m,n]}(k-l)\right)\right.$$
$$\left. + Q_l \left(Y_{[0,m-1]} + Y_{[m,n]}(k-l)\right)\right) + R_{[0,m-1]} + R_{[m,n]}(k)\Bigg) .$$

Using the characterization of assumption (A_2) for $Y_{[0,m-1]}$,

$$Y_{[0,n]}(k) = Y_{[0,m-1]} + \bigoplus_{l=1}^{M} \left(A_l \otimes Y_{[m,n]}(k-l) \oplus B_l \otimes Z_{[m,n]}(k-l)\right) ,$$

$$Z_{[0,n]}(k) = \Pi\left(\sum_{l=1}^{M} (P_l Z_{[m,n]}(k-l) + Q_l Y_{[m,n]}(k-l)) + U_{[0,m-1]} + R_{[m,n]}(k)\right) .$$

This yields

$$Y_{[0,n]}(k) = Y_{[0,m-1]} + Y_{[m,n]}(k) ,$$
$$Z_{[0,n]}(k) = Z_{[0,m-1]} + Z_{[m,n]}(k) .$$

Corollary 6. $X_{[0,n]} = X_{[0,m-1]} + X_{[m,n]}$.

Proof. This is an immediate corollary of the previous lemma considering the fact that $X_{[0,n]}$ does not depend on T_m.

Lemma 7 (External Monotonicity). *With two arrival counting measures $R'_{[m,n]}(k) \geq R_{[m,n]}(k)$, then $X'_{[m,n]}(k) \geq X_{[m,n]}(k)$.*

Proof. The vector $X_{[m,n]}(k)$ is an increasing function of $\left(X([m,n](k-l)\right)_{l=1,\dots,M}$ and of $R_{[m,n]}(k)$. The proof follows by a straightforward induction.

Lemma 8 (Conservation). *$X_{[m,n]}$ is finite and independent of the arrival times.*

Proof. Corollary 6 says that $X_{[m,n]} = X_{[0,n]} - X_{[0,m-1]}$. Therefore, $X_{[m,n]}$ is finite and independent of the arrival times.

Lemma 9 (Separability). *Suppose $T_r > \mathcal{W}_{[m,n]} + M$. Then for all $m \leq n$,*

- *if $k < T_r$, then $X_{[m,n]}(k) = X_{[m,r-1]}(k)$,*
- *if $k \geq T_r$, then $X_{[m,n]}(k) = X_{[m,r-1]} + X_{[r,n]}(k)$.*

Proof. The case $t < T_r$ is trivial. For the case $t \geq T_r$, the proof holds by induction on k. It is very similar to the proof of lemma 5 and is not reported here.

Lemma 10 (Homogeneity). *Let $R'_{[m,n]}$ be the arrival process shifted by a constant C, $R'_{[m,n]}(k) = R_{[m,n]}(k+C)$. Then, $X'_{[m,n]}(k) = X_{[m,n]}(k+C)$.*

Proof. This holds by immediate induction on k.

5.1 Stochastic Assumptions

All the random variables defined in what follows are assumed to be carried by some probability space $(\Omega, \mathcal{F}, P, \theta)$, where θ is an ergodic shift which leaves P invariant. We assume that the point process associated with the counting measure $R_{[-\infty,+\infty]}(k)$ is θ-stationary and ergodic, and that it has a finite intensity. When taking $\{T_0 = 0\}$, this θ-stationarity assumption here means that

$$R_{[n,\infty]}(T_n + k) = R_{[0,\infty]}(k) \circ \theta^n \quad , \tag{18}$$

for all $k \in \mathbb{R}$ and n. Consider the $\mathcal{T}$-valued sequences $\nu_{[0,\infty]} = \{\nu^p_{[0,\infty]}(k)\},\ p \in \mathcal{R}, k \in \mathbb{N}\}$ describing the routing decisions; we also assume that the following compatibility relation holds for all n.

$$\nu_{[n,\infty]} = \nu_{[0,\infty]} \circ \theta^n \quad . \tag{19}$$

If $\mu = \{\mu(k)\}_{k\geq 0}$ denotes a sequence, for all integers $V \geq 0$, we will denote $\tau_V \mu$ the shifted sequence $\{\mu(V + k)\}_{k\geq 0}$. If $U = (U^p,\ p \in \mathcal{R})$ is a vector of non-negative integer, we will denote $\tau_U \nu$ the sequences $(\tau_{U^p} \nu^p,\ p \in \mathcal{R})$. Eq. (17) and the above relation imply that

$$\nu_{[0,\infty]} \circ \theta^n = \tau_{U_{[0,n-1]}} \nu_{[0,\infty]} \quad , \tag{20}$$

where $U^{\bullet t}_{[0,n]}$ is the function defined in Eq. (16).

It should be clear that under the above assumptions, the functions $\Pi^t_{[n,\infty]}$ satisfy the compatibility property

$$\Pi_{[n,\infty]} = \Pi_{[0,\infty]} \circ \theta^n \quad , \tag{21}$$

so that the compatibility relations of Theorem 2 hold for the counters and the daters of the FCN. Besides this, since all the firing times are positive, one can easily check that condition (9) is satisfied. Thus, Theorem 2 holds for this class of systems under the above assumptions.

Remark. Note that since $\nu^p_{[0,\infty]}(k) = \nu^p_{[n,n]}(l)$ for $l = k - U^p_{[0,n-1]} \leq U^p_{[n,n]}$, the infinite sequences $\nu_{[0,\infty]}$ are fully determined by the finite sub-sequences

$$\left(\nu^p_{[n,n]}(k),\quad p \in \mathcal{R},\quad 0 \leq k \leq U^p_{[n,n]}\right)_{n\geq 0} \quad .$$

Thus, with our framework, for all nodes p, the whole routing sequence $\nu^p_{[0,\infty]}(k)$ is simply the concatenation of the routing sequences $\nu^p_{[n,n]}(l)$, $1 \leq l \leq U^p_{[n,n]}$.

6 Analysis of an Example: The SI-FCN

The aim of this section is to give sufficient conditions for the assumptions of the preceding section to hold. We will limit ourselves to the SI-FCN case, where certain simplifications take place.

A MGC G_i is *input-connected* if for each transition in G_i, there is a path from its input transition to t. This is equivalent to: B has no lines which values are all $+\infty$.

Lemma 11. *Let F be a SI-FCN with all its MGC input-connected. If F can reach a deadlock, then this deadlock is unique.*

Proof. If a routing place p contains a token, then one of the transitions in $p^\bullet$ is enabled, thus this marking is not a deadlock. Let t be a transition in G_i, let us follow the longest path in G_i without tokens. This path leads to a transition which is enabled except if it is the input transition. Now, a marking verifying these conditions is necessarily unique.

Therefore, for SI-FCN with input-connected components, assumption (A_2) is redundant.

For all $t \in \mathcal{B} \cap G_i$, let O^t be the set

$$O^t = \{q \in \mathcal{R} \mid \exists t \in G_i \text{ s.t. } q = t^\bullet\} \ ,$$

where q is counted with multiplicity n if there are n arcs going from G_i to q. We will then say that q is an *offspring* of t with multiplicity n.

We now describe the dynamics of a pseudo marking process (this marking process is different from the one in the real system) on the set of places of $\mathcal{R}$, which is driven by the routing functions only. Fix an arbitrary priority order on the nodes of $\mathcal{R}$: p_j has priority over p_{j+1} etc. Assume that the jump of R at T_0 brings $m^{{}^\bullet t}$ tokens to place ${}^\bullet t$, for all $t \in \mathcal{B}$. If $m^{p_1} > 0$, one token of p_1 is *moved* following the routing decision $t = \nu^{p_1}_{[0,0]}(1)$. Let n^{p_i} be the multiplicity of the offspring p_i of t (this multiplicity is zero if p_i does not belong to O^t). By definition, such a move leads to the new marking defined by $m^{p_i} + n^{p_i}$ for all $i > 1$ and $m^{p_1} + n^{p_1} - 1$ for p_1. If the new marking of p_1 is still positive, we move one token of p_1 as above, but according to the routing decision $\nu^{p_1}_{[0,0]}(2)$, which leads to a new marking; the procedure is repeated up to the time when no tokens are left in p_1 (this may never happen in which case this first step of the procedure never stops). We then move one token of type p_2 according to the routing decision $\nu^{p_2}_{[0,0]}(1)$, provided there is at least one token in place p_2 in the last obtained marking. This may possibly create new tokens of type p_1. The general rule is actually to move the token with highest priority at each step, according to the residual routing decisions. The procedure stops whenever there are no tokens left in the routing places.

Lemma 12. *The assumptions 1 (and therefore 2) are satisfied if and only if the above procedure stops after an almost surely finite number of steps.*

The proof is omitted. It is based on a generalization of the Euler property for directed graphs called the Euler-Ordered property, which is introduced in [3]. Note that if the above stopping property holds for this specific ordering of the moves, it will hold for any other ordering.

In the particular case of i.i.d. routing decisions, independent on different nodes, one can naturally associate a multitype branching process with the set $\mathcal{R}$ by saying that an individual of type p has a set of offspring O^t with probability $P = P(\nu^p = t)$. Properties expressed by Assumption 1 (and 2) will then a.s. hold whenever this multitype branching process is subcritical (namely whenever its population dies out a.s. for all finite initial conditions). This property boils down to checking that the maximal eigenvalue of the branching matrix is strictly less than 1 ([8]).

7 Generalization to Variable Firing Times

Variable Firing Times We now consider the case when firing times are still integer-valued and bounded, but variable with time. Let $\sigma^t(m)$ be the firing time of the m-th firing of transition t. Let $\zeta^t(k)$ be the minimum of M and the time which elapsed since the last time t has started firing before time k. If we consider the variables to be left-continuous, we have:

Lemma 13.

$$Y(k) = \bigoplus_{l=1}^{M} (A_l(k) \otimes Y(k-l) \oplus B_l(k) \otimes Z(k-l)) \ , \tag{22}$$

$$Z(k) = \Pi \left(\sum_{l=1}^{M} (P_l(k) \times Z(k-l) + Q_l(k) \times Y(k-l)) + R(k) \right) \ , \tag{23}$$

with $A_l(k)(t,t') = c$, the number of tokens in the initial marking of the place between t' and t if $\zeta^t(k) = l$, ∞ otherwise (with a similar definition for B) and $P_l(k)(t,t') = c$, the number of tokens in the initial marking of the place between t' and t if $\zeta^t(k) = l$, 0 otherwise (with a similar definition for Q).

A system with variable firing times also falls within the monotone separable framework and the same method as in the constant case can be applied (see [5]).

7.1 Future Research

The constant γ_0 was compute for Jackson networks ([3]) which happen to be a simple case of SI-FCN. Its computation within the more general framework of this paper is considered in [5]. Other results can also be obtained along the lines of [3] including conditions for coupling convergence with (and uniqueness of) the stationary regime.

References

1. Baccelli F., Cohen G., Gaujal B., *Recursive Equations and Basic Properties of Timed Petri Nets*, DEDS: Theory and Applications, 1(4)415–439, 1992.
2. Baccelli F., Cohen G., Olsder G.J, Quadrat J.P. *Synchronization and Linearity*, Wiley 1992.
3. Baccelli F., Foss S., *Stability of Jackson-type Queueing Networks, I*, INRIA Report n. 1845, To appear in QUESTA.
4. Baccelli F., Foss S., *On the Saturation Rule for the Stability of Queues*, INRIA Report n. 2015, 1993.
5. Baccelli F., Foss S., Gaujal B., *On the stability region for a class of stochastic Petri nets*, in preparation.
6. Commoner F., *Deadlocks in Petri Nets*, Applied Data Research, CA-7606-2311, 1972.
7. Esparza L., Silva M., *On the Analysis and Synthesis of Free Choice Systems*, G. Rozenberg editor, Advances in Petri Nets, Vol. 483 of LNCS, 243–286, 1990.
8. Harris T.E., *The Theory of Branching Processes*, Springer, Berlin, 1969.
9. Thiagarajan P.S., Voss K., *A Fresh Look at Free Choice Nets*, Information and Control, (62) 85–113, 1984.

Allocation Sequences of Two Processes Sharing a Resource

Bruno Gaujal

INRIA Sophia-Antipolis, BP 93, 06902 Sophia-Antipolis, France

1 Introduction

We consider a system where two processes A and B are sharing a common resource. Each process undergoes a number of activities (operations) in a cyclic manner. There are two types of activities depending on whether or not the common resource is needed. The activities utilizing the resource are followed by the set of activities not utilizing that resource. This system is typical in manufacturing applications.

Without any external control, the system follows its natural evolution which is periodic and does not depend on the initial conditions. If during one period, the resource is allocated p times to process A and q times to process B, then the allocation sequence is in some sense the "most regular" sequence among all the periodic sequences with p times A and q times B.

Now, if the frequencies of allocation are imposed, the optimal allocation sequence is once again the most regular sequence among all the periodic sequences respecting the imposed frequencies of allocation.

This result can be partially generalized to non periodic allocation sequences and non rational frequencies. We also expose a heuristic to deal with the case of N processes sharing the resource.

Figure 1 illustrates the system to be studied in this article modeled as a Petri net. Places represent the activities and transitions the events taking place in the system. Associated with places are temporizations: α_1 and α_2 for process A and β_1 and β_2 for process B. We denote by T the quadruplet $(\alpha_1, \alpha_2, \beta_1, \beta_2)$. The temporized system is denoted (S, T). As it can be seen from Figure 1, during α_1 (respectively β_1) the common resource is being used by process A (respectively B). During α_2 (respectively β_2), process A (respectively B) performs activities not involving the resource.

2 A System with No Control on the Resource

The resource is given to the first process which is waiting for it. If the two processes become ready simultaneously to use the resource, we call this situation a conflict. In the following, B is given a higher priority than A and gets the resource in a conflict situation.

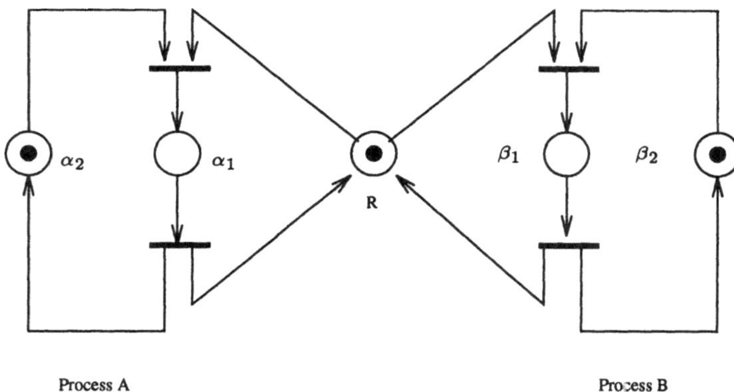

Fig. 1. General net modeling the system.

Let f and g be two periodic functions with respective periods $\alpha = \alpha_1 + \alpha_2$ and $\beta = \beta_1 + \beta_2$. On $[0, \alpha)$, f (respectively g) is given by :

$$f(x) = 1 \text{ if } 0 \leq x < \alpha_1 \quad \text{and} \quad g(x) = 1 \text{ if } 0 \leq x < \beta_1$$
$$f(x) = 0 \text{ if } \alpha_1 \leq x < \alpha_2 + \alpha_1 \qquad g(x) = 0 \text{ if } \beta_1 \leq x < \beta_2 + \beta_1.$$

These two functions will represent the use of the resource by the two processes. $f(x) = 1$ means that process A holds the resource at time x. $f(x) = 0$ means that A does not hold the resource at time x, and respectively with g and B. Now we define $f_{i_A}(x) = f(x - i_A)$ and $g_{i_B}(x) = g(x - i_B)$ to represent the two processes with their initial phases i_A and i_B. If the two processes are superposed, they behave as if they were alone as long as the intervals α_1 and β_1 do not overlap. If they overlap, then one of the processes has to wait for the resource. After a waiting point, two cases may happen: either A has been waiting and in which case the system restarts with a phase ϕ_A : *A begins its interval α_1 and B its interval β_2* and this phase corresponds to $i_A = 0$ and $i_B = \beta_2$; or B has been waiting for the resource in which case the system restarts with a phase ϕ_B between the two processes: *B starts its interval β_1 and A starts its interval α_2*. This phase corresponds to $i_A = \alpha_2$ and $i_B = 0$.

Theorem 1. *The system is periodic following a transient period and the periodic regime does not depend on the initial phase of the two processes.*

The proof of this theorem (not given here) is rather classical. It suggests that it is very unlikely to derive a close formula for the duration of the transient regime, the length of one period or the allocation sequence of the resource during one period. However, we can verify that this allocation sequence is the "most regular" sequence given the number of allocations of the resource to the two processes during one period. Let us assume that during one period, the resource is given n times to A and k times to B. n and k are the "natural" frequencies of the system (S, T) within $n + k$ allocations.

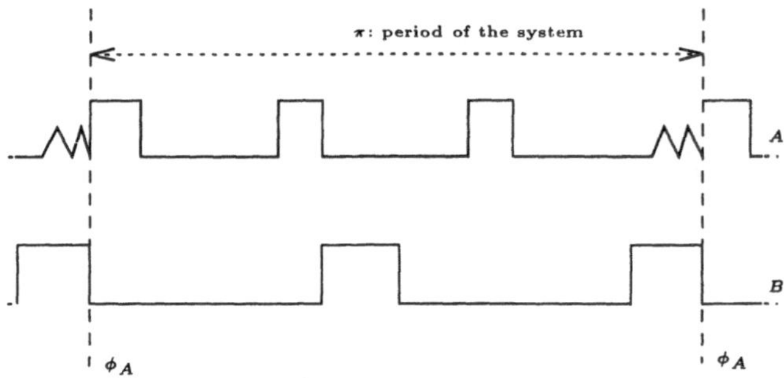

Fig. 2. First type of period: only one process waits for the resource.

In the case where the period contains waiting states, we can assume that the period begins with phase ϕ_A and ends with phase ϕ_A. Only two situations are possible: The first waiting process during one period is A in which case the period of the system is between these two waiting times (figure 2), or the next waiting process is B and the third waiting process is A (figure 3).

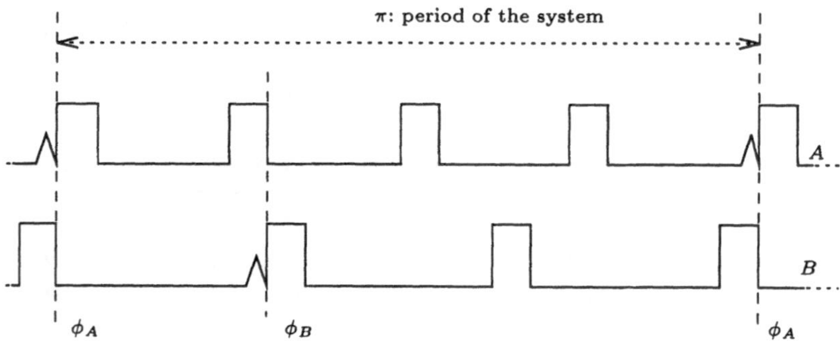

Fig. 3. Second type of period waiting states: both processes wait for the resource during one period.

In the first case, the allocation sequence is : $w = A^{c_1} B A^{c_2} B \cdots A^{c_k} B$ where $c_i = \lceil i\frac{n}{k} \rceil - \lceil (i-1)\frac{n}{k} \rceil$. In the case where the period also contains a waiting time for process B (see figure 3), this waiting time is smaller than α_1 and therefore cannot alter the allocation sequence within one period which remains the same as before.

The sequence w is the most regular integer sequence associated with the natural frequencies (n, k) of the system. This result is also true when the frequencies are imposed by a control on the resource as it is shown in the following.

3 A System with Imposed Frequency

Here, the objective is to find the allocation sequence minimizing the idle time of the resource given the following constraint on the frequency of resource allocation: Within N allocations of the resource, process A must receive the resource p times and process B must receive it q times with $p+q = N$. The couple (p, q) will be called the frequencies of the system.

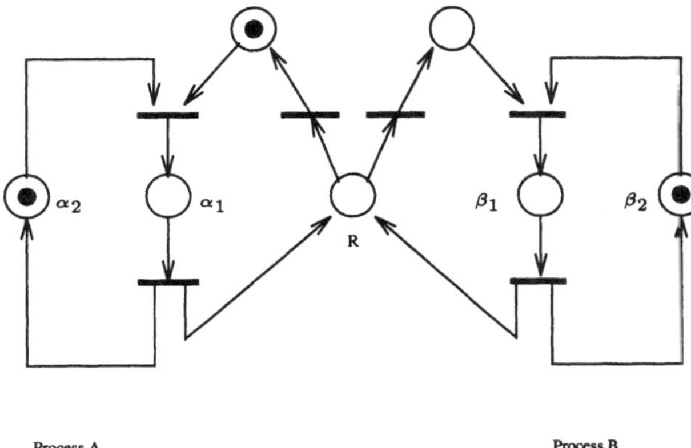

Fig. 4. Free choice net modeling the system S.

We are still free to choose any resource allocation sequence satisfying this constraint. Therefore, the general system of figure 1 reduces to that of figure 4 which depicts a free choice net. We further restrict ourselves to allocation sequences which are periodic of period N.

Under these conditions, minimizing the idle time of the resource is equivalent to minimizing the time required to go through N allocations called the cycle time of the system.

3.1 Allocation Sequences and Regular Word

A periodic allocation sequence can be considered as a word of N letters in the alphabet $\{A, B\}$.

Let $E(p, q)$ be the set of all the words in $\mathcal{N}$ formed with p A's and q B's. An element w in $E(p, q)$ corresponds to a period of a possible allocation sequence of the resource satisfying the given frequencies (p, q). The time that it takes for the system to execute periodically the sequence w is called the cycle time of the system (S, T, w) and is denoted M_w.

In $E(p, q)$, the most regular word in the word $r = r_1 r_2 \cdots r_N$ defined by:

$$r_i = \begin{cases} A \text{ if} \lceil i\frac{p}{N} \rceil - \lceil (i-1)\frac{p}{N} \rceil = 1, \\ B \text{ if} \lceil i\frac{p}{N} \rceil - \lceil (i-1)\frac{p}{N} \rceil = 0. \end{cases}$$

The word r distributes the A's and the B's as evenly as possible. These sorts of sequences are well known. If $\frac{p}{N}$ is replaced by an irrational number x, the most regular word is also called *word of Sturm* (see section 5.3).

Theorem 2. *r is an optimal allocation sequence in the sense that for any temporizations* $(\alpha_1, \alpha_2, \beta_1, \beta_2)$, $M_r \leq M_w$, $\forall w \in E(p, q)$.

This optimal sequence does not depend on the temporizations but only on the values of p and q. In [Haj], B. Hajek used a similar regular sequence to optimize the routing of customers into a queue. This suggest that the most regular world may help to find the optimal scheduling in many other problems. In our case, The worst sequence is known to be $w = A^p B^q$ which corresponds to a near complete sequentialization of the system (see [Cha]).

Proof. We give only a sketch of the proof. A general allocation sequence in $E(p, q)$ can be written $w = AB^{k_1}AB^{k_2}\cdots AB^{k_p}$, where for all i, $k_i \geq 0$ and $k_1 + \cdots + k_p = q$. We denote $w = (k_1, \cdots, k_p)$. The new system (S, T, w) can be modeled as a marked graph ($MG(S, T, w)$). If C is a circuit in the graph $MG(S, T, w)$, let $t(C)$ be the sum of the temporizations of the places belonging to the circuit. $t(C)$ is also called the temporization of the circuit. The cycle time of the system is then given by the critical circuit in the net $MG(S, T, w)$ (see [Bac]), where the critical circuit is the circuit with the maximum temporization.

If the critical circuit C_w "jumps" i_w times from process A to process B and if it contains s places from the circuit modeling B, it has a temporization of the form:

$$t(C_w) = F_{i_w,s_w}(w)(\beta_1 + \beta_2) + (p - s_w)(\alpha_1 + \alpha_2) + i_w(\beta_1 + \beta_2),$$

where $F_{i,s}(w) = \max\{e.w,\ e$ is a {0,1}-sequence with s 1's in i clusters}.

Lemma 3. *For any* (i, s), $\min_{w \in E(p,q)} F_{i,s}(w) = F_{i,s}(r)$.

This lemma is essential and constitute the heart of the proof. It is not necessary to show this result for all the couples (i, s) but only for the ones which do not "degenerate" to simpler cases. Since its proof is technical it is not given here.

The proof of the theorem follows quite directly from the lemma.

Using the lemma, $t(C_r) \leq t(C)$ for any circuit C in $MG(S, T, w)$ with p_r jumps and l_r places in B. Since $t(C_w) \geq t(C)$, we can conclude $t(C_w) \geq t(C_r)$.

4 n Processes Sharing a Resource

In the case of n processes sharing a resource with $n \geq 3$, theorem 2 cannot be extended. Indeed, in general, no strongly optimal allocation sequence exists. The optimal sequence depends on the temporizations of the processes.

To illustrate this negative result, we give an example where the frequencies of the $n = 3$ processors are fixed but if we change the temporizations, then the optimal sequence changes as well. During $N = 6$ allocations, the resource must be given $k_A = 3$ times to process A, $k_B = 2$ times to process B and $k_C = 1$ times to process C. With the temporizations: $(1, 1)$ for A, $(1, 1)$ for B and $(1, 1)$ for C, then the optimal sequence is *cababa* This sequence gives a cycle time of 6. But with the temporizations: $(1, 1)$ for A, $(1, 3)$ for B and $(2, 1)$ for C, then the optimal sequence is *cbaaba*

This negative result does not put to an end the search for an optimal sequence in the case where we have more than 2 processes sharing a resource. Although finding the optimal allocation of a token in a general Petri Net is an NP-Hard problem, see [Ram], we believe that a "even distribution" of the heaviest processes in terms of (processing time).(frequency) could help to solve the problem or at least is a good heuristic to start with.

5 Several Extensions

5.1 Non Periodic Sequences

In the previous section we have been interested only in periodic sequences. Here, we show that the results obtained in the periodic case can be extended in more general cases. We denote by $p_m(w)$ the prefix of w of length m. $V(p, q) = \{w, \forall k \in \mathbb{N}\ |p_{kN}w|_A = kp,\ |p_{kN}w|_B = kq\}$ We also denote $T_m(w)$ the time to go through the first m allocations of w. A general sequence $w \in V(p, q)$ is not periodic in general (of any period).

Theorem 4. *$\exists a_0$ s.t. $\forall a \geq a_0$, $\forall w \in V(p, q)$, $T_{aN}(w) \geq T_{aN}(r^\omega)$.*

Proof. $T_{aN}(w) = T_{aN}(p_{aN}(w)^\omega)$ because $T_{aN}(w)$ depends only on $p_{aN}(w)$. But now, $p_{aN}(w)^\omega$ is periodic of period aN. We get: $T_{aN}(p_{aN}(w)^\omega) \geq T_{aN}(r^\omega)$ if $a \geq a_0$ for some a_0 since the cycle time of the system $MG(S, T, p_{aN}(w)^\omega)$ is bigger than the cycle 356 time of $MG(S, T, r\omega)$.

5.2 Asymptotic Constraints

An even more general set of sequences would be sequences with asymptotic constraints only. Let $A(p, q)$ be the set of the infinite sequences w verifying: $\lim_{m\to\infty} \frac{|p_m(w)|_A}{m} = \frac{p}{N}$ and $\lim_{m\to\infty} \frac{|p_m(w)|_B}{m} = \frac{q}{N}$. In this case, we can state an asymptotic inequality:

Theorem 5. *For all $w \in A(p, q)$, $T_{aN}(w) \geq T_{aN}(r^\omega) + o(a)$.*

Proof. Let a be an arbitrary integer. $|p_{aN}(w)|_A = ap + \epsilon_A$ and $|p_{aN}(w)|_B = aq + \epsilon_B$ where $\epsilon_A = o(a)$ and $\epsilon_B = o(a)$.

If a is big enough,

$$
\begin{aligned}
T_{aN}(p_{aN}(w)) &\geq T_{aN}(r^a) - |\epsilon_A|(\alpha_1 + \alpha_2) - |\epsilon_B|(\beta_1 + \beta_2) \\
&= aT_N(r) + o(a).
\end{aligned}
$$

5.3 Irrational Frequencies

Suppose now that we impose irrational frequencies for the allocation of the resource. Namely, we consider the set $I(x)$ of all the sequences w verifying $\lim_{m\to\infty} \frac{|p_m(w)|_A}{m} = x$ where x is irrational, $0 < x < 1$.

Theorem 6. *For all $w \in I(x)$, $T_n(w) \geq T_n(r) + o(a)$.*

Proof. We build a sequence of rational numbers p_n, $\lim_{n\to\infty} p_n = x$ such that for all $i \leq n$, $\lceil i.p_n \rceil = \lceil i.x \rceil$. The sequence (p_n) is defined such that r_n, the most regular word with the frequency p_n and r, the most regular word associated with x have the same n first letters. Now, if w is a word in $I(x)$, $|p_n(w)|_A = |r_n|_A + \eta_A(n)$, $|p_n(w)|_B = |r_n|_B + \eta_B(n)$, where $\lim_{n\to\infty} \eta_A(n) = 0$, $\lim_{n\to\infty} \eta_B(n) = 0$. Now ,

$$\begin{aligned} T_n(w) &\geq T_n(r_n) - |\eta_A + 1|(\alpha_1 + \alpha_2) - |\eta_B + 1|(\beta_1 + \beta_2) \\ &= T_n(r_n) + o(n) \\ &= T_n(r) + o(n). \end{aligned}$$

References

[Bac] Baccelli, F., Cohen, G., Olsder, G.J., Quadrat, J.P.: *Synchronisation and Linearity.* Springer-Fverlag, New York, 1992.

[Cha] Chaouiya, C.: *Outils pour l'analyse de systèmes synchronisés.* PhD Thesis, University of Nice-Sophia-Antipolis, France, 1992.

[Haj] Hajek, B.: Extremal Splittings of Point Processes. *Mathematics of operation research* (1985), 10:543-556.

[Ram] Ramamoorthy, C.V., Ho, G.S.: Performance Evaluation of Asynchronous Systems Using Petri Nets. *IEEE Trans. on Software Eng.* (1980), SE-6:440-449.

Stability Criteria for yet Another Class of Multidimensional Distributed Systems*

Leonidas Georgiadis[1] *and Wojciech Szpankowski*[2]

[1] IBM T. J. Watson Research Center, Yorktown Heights, NY 10598, U.S.A
[2] Department of Computer Science, Purdue University, W. Lafayette, IN 47907, U.S.A

1 Introduction

Recently, in our paper [3] we have established rigorously stability conditions for the ℓ-limited token passing rings, and indicated that the technique can be extended to other disciplines. In this paper, we apply our methodology to derive stability criteria for *time-limited* token passing rings introduced recently by Leung and Eisenberg [6], [7] (see Theorem 1). In such a system each station transmits messages for at most an amount of time τ. If the transmission time exceeds τ, the station completes the transmission of the message in progress and sends the token to the next station (the so called nonpreemptive time limited discipline). While we study this system as an important application, the technique can be applied almost without modification to a class of monotonic and contractive policies (see Theorem 2).

Our approach to the stability of token passing rings follows the idea discussed in our paper [3], and differs from the standard methodology of the Lyapunov test function (see [12], [18], [14]). (For other than test function approaches see also [2], [11] [14, 15].) It resembles the general idea of Malyshev's *faces* and *induced Markov chains* [10]. Our method is based on a simple idea of stochastic dominance technique, and application of Loynes [9] stability criteria for an isolated queue. We note that this approach is *not* restricted to token passing rings, and stability of several other distributed systems can be assessed by this methodology (see [14, 15]).

We now summarize our main results. We shall analyze the token passing ring with Poisson arrivals with parameter λ_i for the ith station, general distribution of service times $\{S_i^k\}_{k=1}^{\infty}$ and switchover times $\{U_i^k\}_{k=1}^{\infty}$. We consider the gated version of τ-limited token ring. Define $\widetilde{L}_i = \min\{k : \sum_{j=1}^{k} S_i^j \geq \tau_i\}$ and $\ell_i = E\widetilde{L}_i$. Clearly, ℓ_i is the average of the maximum number of customers served during τ_i.

Theorem 1. *Consider a token passing ring consisting of M stations with τ-limited service schedule for the ith station, and Poisson arrivals. Then the system is stable if and only if $\sum_{j=1}^{M} \rho_j < 1$ and*

$$\lambda_j < \frac{\ell_j}{u_0}(1 - \rho_0) \ , \quad \text{for all } j \in \mathcal{M} = \{1, \dots M\} \ ,$$

* This research was supported in part by the NSF grants NCR-9206315, CCR-9201078, and INT-8912631, by AFOSR grant 90-0107, and in part by NATO Collaborative Grant 0057/89. This paper was revised while the author was visiting INRIA, Rocquencourt, France, and he wishes to thank INRIA (projects ALGO, MEVAL and REFLECS) for a generous support.

where $u_0 = \sum_{i=1}^{M} EU_i$ *is the average total switchover time, and* $\rho_0 = \sum_{i=1}^{M} \rho_i$ *with* $\rho_i = \lambda_i s_i$ *and* $s_i = ES_i$ *being the average service time at the* i*th station.*

It will be seen that the method of the proof can be applied virtually unchanged to the following class of monotonic and contractive policies. We assume that the number of customers served from queue i when there are n queued messages at the instant of token arrival to queue i is $f_i(n, \mathbf{X})$, where $\mathbf{X}$ is a possibly random quantity that depends on the policy. We assume that $f_i(\cdot, \mathbf{X})$ is a *nondecreasing* function of the number of customers in the ith queue. In addition, the following relation holds

$$f_i(n_1, \mathbf{X}) - f_i(n_2, \mathbf{X}) \leq n_1 - n_2 \quad \text{if} \quad n_1 > n_2 .$$

At the kth token arrival to queue i the quantity $\mathbf{X}$ takes the value $\mathbf{A}_k$. The random variables $\{\mathbf{A}_k\}_{k=1}^{\infty}$ are i.i.d., independent of the past arrival times, past service times and past switchover times. Note that $f_i(n, \mathbf{A}_k)$ tends to a random variable, $f_i^*(\mathbf{A}_k)$, when $n \to \infty$. We assume that $f_i^*(\mathbf{A}_k)$ is finite and $\ell_i \stackrel{\text{def}}{=} Ef_i^*(\mathbf{A}_k) < \infty$.

We will see in the next section that the time-limited token ring policy falls within the class just described. Other special cases are the ℓ-limited gated discipline and Bernoulli gated disciplines. In the last discipline, $f_i(n, X) = \min\{n, X\}$, where X is a geometrically distributed random variable. The following result follows directly from the arguments that will be presented in this paper.

Theorem 2. *Consider a token passing ring that employs a monotonic and contractive service discipline described above. Then, the system is stable if and only if*

$$\lambda_i < \frac{\ell_i}{u_0}(1 - \rho_0) \ , \qquad \text{for all } i \in \mathcal{M} \ ,$$

where $\rho_0 < 1$.

One policy that does not fall in the previously described category is the *preemptive* τ-limited token ring. In such as system the token interrupts his service immediately after the time limit τ expires and continues servicing the same customer in the next round. While the formal arguments of our methodology can be applied for this system, there are technical difficulties that need to be overcome for a complete proof. We believe that these technical arguments can be provided, but we do not have a complete proof yet, and therefore we express the following conjecture.

Conjecture. *The preemptive* τ*-limited token ring is stable if and only if*

$$\lambda_j < \frac{\tau_j}{s_j u_0}(1 - \rho_0) \ , \qquad \text{for all } j \in \mathcal{M} = \{1, \ldots M\} \ ,$$

and $\rho_0 < 1$.

2 Preliminary Results

We start with a precise definition of our stochastic model. We shall adopt the following assumptions.

(A1) There are M stations (queues) on a loop, each having infinite capacity buffer.

(A2) The maximum time customers are served during the token visit at a queue is limited to $\tau_i < \infty$ units of time. Only customers that are present at the instant of token arrival can be served. Moreover, we have nonpreemptive discipline, that is, the customer that is in the server when the time limit τ_i is reached, is served to completion before the token moves to the next queue.

(A3) The arrival process A_i^t, $t \in [0, \infty)$ to the ith queue is a Poisson process with parameter $\lambda_i > 0$. Here, A_i^t is the number of arrivals at queue i up to time t. The arrival process at a queue is independent of the arrival processes to other queues.

(A4) The service time process $\{S_i^k\}_{k=1}^{\infty}$ at queue i consists of i.i.d. random variables with mean $s_i = ES_i^1 > 0$. The service time process at a queue is independent of the arrival processes at all queues and independent of the service time processes at other queues.

(A5) The switchover times between i and $i+1 \bmod M$ queue, $\{U_i^k\}_{k=1}^{\infty}$, are i.i.d., independent of the switchover times $\{U_j^k\}_{k=1}^{\infty}$ for $j \neq i$, and independent of the arrival and the service time processes. The average total switchover time is defined as $u_0 = \sum_{i=1}^{M} EU_i^1$. To avoid inessential complications we assume that $P(U_i^n > 0) = 1$, $i = 1, \ldots, M$.

Let $\{T_n\}_{n=1}^{\infty}$ be the time instant of the nth visit of the token to a queue. Assumptions (A1)–(A5) imply that the service times of the customers at queue i at instant T_n are i.i.d. independent of the history up to time T_n. This permits us to consider a new model of the system which is stochastically equivalent to the original one and has the advantage that under this new model, many of the arguments that follow become simpler. Specifically, in the new system assumptions (A1)–(A3) and (A5) are the same, while assumption (A4) is replaced with

(A4') Service times are assigned to the customers at queue i upon beginning of service as follows. We consider a doubly infinite sequence of i.i.d random variables $\{S_i^{n,k}\}_{n,k=1}^{\infty}$ with $s_i = ES_i^{n,k} > 0$. The customers that are served during the nth arrival of the token to queue i are assigned the service times $S_i^{n,1}, S_i^{n,2}, \ldots$. The sequence $\{S_i^{n,k}\}_{n,k=1}^{\infty}$ is independent of the sequence $\{S_j^{n,k}\}_{n,k=1}^{\infty}$ for $i \neq j$, independent of the interarrival processes to the queues and of the switchover times.

Next, we need some relationships between the average number of customers served per token visit L_i^n and the average cycle time C_i^n. The former quantity is defined as follows. Let $\widetilde{L}_i^n = \min\{k : \sum_{j=1}^{k} S_i^{n,j} \geq \tau_i\}$. Then $L_i^n = \min\{\widetilde{L}_i^n, N_i^n\}$. The latter quantity is the length of time between the nth and $n+1$st visits of the token to the reference queue i. By EL_i and EC we denote the limiting averages of L_i^n and C_i^n. It turns out that the relations holding for the ℓ-limited case, continue to hold under the more general policies we consider here. Specifically, we have the following result which can be proved along the same lines as in our paper [3].

Proposition 3. *Let the Markov chain* $\mathbf{N}^n(i)$ *be positive recurrent (ergodic) for some* $i \in \mathcal{M}$. *Then,* $\mathbf{N}^n(j)$ *is ergodic for all* $j \in \mathcal{M}$, *and* $\rho_0 = \sum_{j=1}^{M} \rho_j < 1$. *In addition,*

$$EL_j = \lambda_j EC \ , \quad j \in \mathcal{M} \ , \tag{1}$$

$$EC = \frac{u_0}{1 - \sum_{j=1}^{M} \rho_j} \ , \tag{2}$$

where u_0 *is the total average switchover time (see assumption (A5)) and* $\rho_j = \lambda_j s_j$ *is the utilization coefficient for the* i*th queue.*

The next result is one of the key elements of our stability analysis. We state it in a general form, since it is needed to prove Theorem 2. Specifically, in the terminology of [8], we consider the class of monotonic, contractive policies. This amounts to replacing assumption (A2) with the following more general one.

(A2') Let $\mathbf{A}$ denote a sequence of real numbers $\{a_1, a_2, \dots\}$. Let $f_i(m, \mathbf{A})$ be the number of customers served from queue i when there are m queued messages at the instant of the nth token arrival at queue i and $\{S_i^{n,1}, S_i^{n,2}, \dots\} = \mathbf{A}$. We assume that for fixed $\mathbf{A}$, $f_i(m, \mathbf{A})$ is a nondecreasing function of m. In addition, for a fixed $\mathbf{A}$, the following relation holds

$$f_i(m_1, \mathbf{A}) - f_i(m_2, \mathbf{A}) \leq m_1 - m_2 \quad \text{if} \quad m_1 > m_2 . \tag{3}$$

For the case of τ-limited policy we have that $f_i(m, \{S_i^{n,1}, S_i^{n,2}, \dots\}) = \min\{m, \tilde{L}_i^n\}$.

Now we are ready to formulate our result. Consider two token passing rings, say θ and Θ. Both satisfy assumptions (A1), (A2'), (A3), (A4'), (A5). The system θ represents our original token passing ring. The system Θ differs only in the switchover times, namely, we assume that the switchover time for Θ is replaced by $\{\Delta_i^k + U_i^k\}_{k=1}^{\infty}$ for $i = 1, \dots, M$. We assume that for every $i \in \mathcal{M}$ and every $k \geq 0$ we have $\Delta_i^k \geq 0$. We make the following assumption for the process Δ_i^k.

(A6) The random variable Δ_i^k is independent of the service times, switchover times and the Poisson increments of the arrival processes to all stations after time $T_{M(k-1)+(i+1)} - U_i^k$ (see Fig. 1).

Proposition 4. *Let* $\widetilde{\mathbf{N}}^n(\theta)$ *and* $\widetilde{\mathbf{N}}^n(\Theta)$ *denote the process of queue lengths at times* T_n *in systems* θ *and* Θ *respectively. Then, under the above assumptions, and under the condition that the token starts from the same queue, say queue number one, and with the same number of initial customers in both systems, the following holds*

$$\widetilde{\mathbf{N}}^n(\theta) \leq_{\text{st}} \widetilde{\mathbf{N}}^n(\Theta), \tag{4}$$

where $\leq_{\text{st}}$ *means stochastically smaller.*

Proof. The proof is along the lines of Theorem 4 our paper [3]. For completeness in the presentation, we provide some details. To avoid cumbersome notation we present the proof only for $M = 2$ users.

We define some new variables. For a system θ let T_n^θ and D_n^θ denote the instances of the nth visit and the nth token departure from any queue respectively. Let also J_n^θ denote the queue number visited at the nth visit of the token to any queue. Finally, let $L_i^n(\theta)$ be the number of customers served from queue i at the nth visit of the token to any queue. Clearly, for our two station system $L_1^n(\theta) = 0$ for n even, and $L_2^n(\theta) = 0$ for n odd. In a similar manner we define respective quantities in the Θ system.

We will construct from the system Θ a token passing ring $\bar{\theta}$, which is stochastically equivalent to the system θ and for which we have that $\widetilde{\mathbf{N}}^n(\bar{\theta}) \leq \widetilde{\mathbf{N}}^n(\Theta)$. Fig. 1 should help to understand our construction. Assume $\widetilde{N}_i^1(\bar{\theta}) = \widetilde{N}_i^1(\Theta)$ for $i = 1, 2$. The service times in system $\bar{\theta}$ are assigned from the same sequences $S_i^{n,k}$ as in Θ (according to assumption (A4').) Also, the same functions $f_i(m, \mathbf{A})$, $i = 1, 2$ are used in both systems. Therefore, the decision to switch to queue 2 will occur at the same time, namely $D_1^{\bar{\theta}} = D_1^\Theta$. The switchover time for $\bar{\theta}$ becomes now U_1^1, and of course $T_2^{\bar{\theta}} \leq T_2^\Theta$ since $\Delta_1^1 \geq 0$ (see Fig. 1).

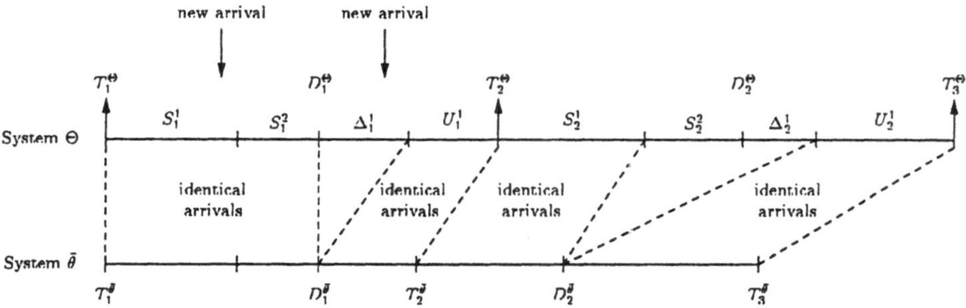

Fig. 1. Illustration to the proof of Theorem 4

The arrivals in the system $\bar{\theta}$ in $[D_1^{\bar{\theta}}, T_2^{\bar{\theta}})$ *are now assumed to be identical* to the arrivals in $[D_1^{\Theta} + \Delta_1^1, T_2^{\Theta})$ in Θ system. Therefore clearly $\widetilde{N}_i^2(\bar{\theta}) \leq \widetilde{N}_i^2(\Theta)$ for $i = 1, 2$. The arrivals to system $\bar{\theta}$ in $[T_2^{\bar{\theta}}\ ,\ T_2^{\bar{\theta}} + S_2^1 + \cdots + S_2^{L_2^1(\bar{\theta})})$ are taken to be identical to the arrivals in $[T_2^{\Theta}\ ,\ T_2^{\Theta} + S_2^1 + \cdots + S_2^{L_2^1(\bar{\theta})})$. Note that this can be done since by (A2') $L_2^1(\bar{\theta}) \leq L_2^1(\Theta)$. Observe also that $T_2^{\bar{\theta}} + S_2^1 + \cdots + S_2^{L_2^1(\bar{\theta})} = D_2^{\bar{\theta}}$ (Fig. 1).

To complete the description of the system $\bar{\theta}$ we have to specify the arrivals in $[D_2^{\bar{\theta}}\ ,\ D_2^{\bar{\theta}} + U_2^1)$. These are taken to be exactly the arrivals in $[D_2^{\Theta} + \Delta_2^1\ ,\ D_2^{\Theta} + \Delta_2^1 + U_2^1)$ in the dominant system Θ (see Fig. 1). Note from the construction that

$$\widetilde{N}_1^3(\bar{\theta}) = \widetilde{N}_1^2(\bar{\theta}) + A_{[T_2^{\bar{\theta}}, T_3^{\bar{\theta}}]} \leq \widetilde{N}_1^2(\Theta) + A_{[T_2^{\Theta}, T_3^{\Theta}]} = \widetilde{N}_1^3(\Theta)$$

and also by (3), $\widetilde{N}_2^3(\bar{\theta}) = \widetilde{N}_2^2(\bar{\theta}) - L_2^2(\bar{\theta}) \leq \widetilde{N}_2^2(\Theta) - L_2^2(\Theta) = \widetilde{N}_2^3(\Theta)$. We can now repeat exactly the same procedure to construct $\bar{\theta}$ in the interval $[T_n^{\bar{\theta}}, T_{n+1}^{\bar{\theta}})$, $n \geq 3$, in the same manner as it was constructed in the interval $[T_2, T_3)$. By construction the service times and switchover times of system $\bar{\theta}$ are identically distributed to the corresponding variables of system θ and are independent of the interarrival process. In addition, assumption (A6) and the fact that the servicing policy is nonanticipative assures that the times $T_{n+1}^{\Theta} - U_{J_n}^{k_n}$ are stopping times for the Poisson arrival processes to all stations. The independence of the increments of the Poisson process implies now, that the constructed interarrival process in system $\bar{\theta}$ is Poisson with rate λ_i for queue i. Moreover, by construction $\tilde{\mathbf{N}}^n(\bar{\theta}) \leq \tilde{\mathbf{N}}^n(\Theta)$. Since $\bar{\theta}$ is stochastically equivalent to θ, we have that the distribution of $\tilde{\mathbf{N}}^n(\theta)$ is identical to the distribution of $\tilde{\mathbf{N}}^n(\bar{\theta})$. This completes the proof of the proposition. □

3 Main Results

We start by showing a preliminary result. As in S2 consider a doubly infinite sequence of i.i.d random variables $\{S^{n,k}\}_{n,k=1}^{\infty}$ and recall the definition $\tilde{L}^n = \min\{k : \sum_{j=1}^{k} S^{n,j} \geq \tau\}$, $0 \leq \tau < \infty$. Consider further a queue with vacations such that upon the nth arrival of the server to the queue, $\tilde{L}^n$ customers (dummy if necessary) are served and then the server goes for a vacation. The service times of the customers served during the nth visit of the server are the random variables $S^{n,k}$, $k = 1, \ldots, \tilde{L}^n$. Let $\{C^n\}_{n=1}^{\infty}$ be the process of cycle times (time intervals between successive visits to the queue). It is assumed that the processes $\{C^n, \tilde{L}^n\}_{n=1}^{\infty}$ are jointly *stationary and ergodic* (no independence is required). The arrival process A^t to this queue is a Poisson process with parameter λ, independent of the processes $\{C^n, \tilde{L}^n\}_{n=1}^{\infty}$. Let N^n represent the queue length at the beginning of the nth cycle. By X^n we denote the number of customers arrived during the nth cycle. Since A^t is Poisson and independent of the processes $\{C^n, \tilde{L}^n\}_{n=1}^{\infty}$, the processes $\{X^n, \tilde{L}^n\}_{n=1}^{\infty}$ are jointly stationary and ergodic, and $EX = \lambda EC$ where $EC = EC^1$. Clearly, the process of queue lengths at the instants of the visits of the server to the queue satisfies the following recurrence

$$N^{n+1} = \max\{N^n + X^n - \tilde{L}^n \ , \ X^n\} \ , \ n = 1, 2, \ldots \tag{5}$$

Let $\ell = E\tilde{L}^n$. We prove the following stability result.

Lemma 5. *Consider the queueing system just described. If $\lambda EC < \ell$, then the queue is stable.*

Proof. Since the processes $\{X^n, \tilde{L}^n\}_{n=1}^{\infty}$ are stationary we can assume without loss of generality that they are defined for $n < 0$ as well. By telescoping the recurrence (5) we immediately obtain for $n \geq 2$, $N^n = \max_{1 \leq r \leq n-1}\{X^r + \sum_{k=r+1}^{n-1} \overline{X}^k\}$ where $\overline{X}^k = X^k - \tilde{L}^k$, provided $N^1 = 0$. Arguing as in Loynes [9] we have that N^n converges in distribution to an honest random variable if and only if

$$\limsup_{r \to \infty} \{X^{-r} + \sum_{k=1}^{r} \overline{X}^{-k}\} < \infty \, . \tag{6}$$

Observe now that $X^{-r} + \sum_{k=1}^{r-1} \overline{X}^{-k} = r(\sum_{k=1}^{r} X^{-k}/r - \frac{r-1}{r} \sum_{k=1}^{r-1} \tilde{L}^{-k}/(r-1))$ for $r = 2, 3, \ldots$. Since by the ergodicity of the sequences X^n, $\tilde{L}^n$ we have that $\lim_{r \to \infty} \sum_{k=1}^{r} X^{-k}/r = \lambda EC$ and $\lim_{r \to \infty} \sum_{k=1}^{r} \tilde{L}^{-k}/r = \ell$, the condition $\lambda EC < \ell$ assures the validity of (6). The assumption that $N^1 = 0$ can be removed as in [9]. □

Now we are ready to prove our main result described in Theorem 1. In the next proposition we show that the conditions of the theorem are sufficient.

Proposition 6. *The Markov chain $\mathbf{N}^n(i)$ representing the queue lengths in the token passing ring when it visits queue $i \in \mathcal{M}$ is ergodic if*

$$\lambda_j < \frac{\ell_j}{u_0}(1 - \rho_0) \quad \text{for all } j \in \mathcal{M} \tag{7}$$

where $\rho_0 = \sum_{j=1}^{M} \rho_j$ and $\ell_i = E\tilde{L}_i$.

Proof. We use mathematical induction. For $M = 1$ the proof is simple since it directly follows from Foster's criterion (see [12], [18]).

Now we assume that the theorem is true for $M - 1$ and prove that it can be extended to the M queue case. Let $(\mathcal{U}, \mathcal{S}), \mathcal{U} \neq \emptyset$, be a partition of the set $\mathcal{M}$ of M queues into *persistent* and *nonpersistent* queues. A persistent queue in $\mathcal{U}$ transmits $\tilde{L}_i^n$ messages during the nth visit of the token by transmitting dummy packets if necessary. A nonpersistent queue in $\mathcal{S}$ operates in the normal fashion. Note that the cardinality $|\mathcal{S}|$ of $\mathcal{S}$ is not larger than $M-1$. Let $\overline{\mathbf{N}}^n(i) = \{\overline{N}_1^n(i), \ldots, \overline{N}_M^n(i)\}$ be the queue lengths when the token visits the ith queue for the nth time in the $(\mathcal{U}, \mathcal{S})$ system. Observe that the modified system differs from the original token ring system only in the switchover time from a persistent queue to the successor of that queue in the ring. Specifically, if $i \in \mathcal{U}$, then the switchover times become $\overline{U}_i^k = \Delta_i^k + U_i^k$ where Δ_i^k is the time needed to service the dummy messages at node i (if any), and Δ_i^k satisfies condition (A6) of S2. Therefore, according to Proposition 4, if $\mathbf{N}^1(1) = \overline{\mathbf{N}}^1(1)$, then $\mathbf{N}^n(j) \leq_{\text{st}} \overline{\mathbf{N}}^n(j)$, for all n, $j \in \mathcal{M}$.

Note now that the queues in $\mathcal{S}$ constitute a token passing ring with $|\mathcal{S}|$ stations satisfying conditions (A1)–(A3), (A4'), (A5) of S2, whose operation is independent of the interarrival processes in the persistent queues. The total average switchover time $\overline{u}_0$ to this ring is equal to $\overline{u}_0 = u_0 + \sum_{i \in \mathcal{U}} \ell_i s_i$. Let the queue lengths in such a system be denoted as $\{\overline{\mathbf{N}}_{\mathcal{S}}^n(i)\}_{i \in \mathcal{S}}$. Clearly, $\overline{\mathbf{N}}_{\mathcal{S}}^n$ is a Markov chain, and since $|\mathcal{S}| \leq M-1$ we can apply the induction hypothesis. Hence, for $i \in \mathcal{S}$, $\overline{\mathbf{N}}_{\mathcal{S}}^n$ is ergodic if

$$\lambda_i < \frac{\ell_i}{u_0 + \sum_{i \in \mathcal{U}} \ell_i s_i} \left(1 - \sum_{i \in \mathcal{S}} \rho_i\right), \quad i \in \mathcal{S}. \tag{8}$$

Assume now that (8) holds, and consider a queue in $\mathcal{S}$, say queue 1 and let $C_{\mathcal{S}}^n(1)$ be the process of cycle lengths (successive visits to queue 1). The process $(\mathbf{N}_{\mathcal{S}}^n(1), C_{\mathcal{S}}^n(1))$ is a Markov chain, as easy to notice. Since $\mathbf{N}_{\mathcal{S}}^n(1)$ is ergodic, it is easy to see that also the process $\{\overline{\mathbf{N}}_{\mathcal{S}}^n(1), C_{\mathcal{S}}^{n-1}(1)\}_{n=2}^{\infty}$ is stationary and ergodic (for details see [3]). Now we are in position to use use Lemma 5 which implies that the process $\overline{N}_i^n(1)$, $i \in \mathcal{U}$ is stable provided that

$$\lambda_i < \frac{\ell_i}{EC_{\mathcal{S}}^1(1)} = \frac{\ell_i}{u_0 + \sum_{i \in \mathcal{U}} \ell_i s_i} \left(1 - \sum_{i \in \mathcal{S}} \rho_i\right) \quad i \in \mathcal{U}, \tag{9}$$

where the equality in (9) follows from the fact that by Proposition 4,

$$EC_{\mathcal{S}}^1(1) = \frac{\overline{u}_0}{1 - \sum_{i \in \mathcal{S}} \rho_i} = \frac{u_0 + \sum_{j \in \mathcal{U}} \ell_j s_j}{1 - \sum_{j \in \mathcal{S}} \rho_j}. \tag{10}$$

Since the process $\overline{N}_i^n(1)$, $i \in \mathcal{S}$ is stable by construction, it follows from the stochastic dominance that the irreducible, aperiodic Markov chain $\mathbf{N}^n(1)$ is substable and therefore, ergodic. The fact that $\mathbf{N}^n(j)$ is ergodic for all $j \in \mathcal{M}$ follows from Proposition 4.

Putting everything together, from (8) and (9) we finally have that the Markov chain $\mathbf{N}^n(j)$ is ergodic for every $j \in \mathcal{M}$ if

$$\lambda_i < \frac{\ell_i}{u_0 + \sum_{i \in \mathcal{U}} \ell_i s_i} \left(1 - \sum_{i \in \mathcal{S}} \rho_i\right) \quad i \in \mathcal{M}. \tag{11}$$

Since (11) holds for every partition $\mathcal{P} = (\mathcal{S}, \mathcal{U})$ of the set $\mathcal{M}$ such that $\mathcal{S} \neq \mathcal{M}$, we conclude that the sufficient condition for stability of the system is $\mathcal{R} = \bigcup_{\mathcal{S} \subset \mathcal{M}} \mathcal{R}_{\mathcal{S}}$ where $\mathcal{R}_{\mathcal{S}} = \{\lambda = (\lambda_1, \ldots, \lambda_M) : \text{condition (11) holds}\}$. Finally, to complete the proof we need to show that $\bigcup_{\mathcal{S} \subset \mathcal{M}} \mathcal{R}_{\mathcal{S}} = \{\lambda = (\lambda_1, \ldots, \lambda_M) : \lambda_i < \frac{\ell_i}{u_0}(1 - \sum_{i=1}^{M} \rho_i), \quad i \in \mathcal{M}\}$. This requires only algebraic manipulations which are almost identical to the ones in [3], therefore we omit them here. □

The fact that the conditions of Theorem 1 are also necessary can be proved along the lines of the corresponding proof in [3].

References

1. S. Asmussen, *Applied Probability and Queues*, John Wiley&Sons, Chichester 1987.
2. Chang C., Thomas J., and Kiang S., On the Stability of Open Networks: an Unified Approach by Stochastic Dominance, IBM RC 18343, 1992.
3. L. Georgiadis and W. Szpankowski, Stability of Token Passing Rings, *Queueing Systems*, 11 (1992) pp. 7-33.
4. O. Ibe and X. Cheng, Stability Conditions for Multiqueue Systems with Cycle Service, *IEEE Trans. Automatic Control*, 33 (1988) 102-103.
5. P. Kuehn, Multiqueue Systems with Nonexhaustive Cycle Service, *The bell System Technical Journal*, 58 (1979) 671-698.
6. K.K. Leung and M. Eisenberg, A Single-Server Queue with Vacations and Gated Time-Limited Service, *IEEE Trans. Commun.*, 38 (1990) 1454-1462.
7. K.K. Leung and M. Eisenberg, A Single-Server Queue with Vacations and Non-Gated Time-Limited Service, *Perf. Evaluation J.*, 12 (1991) 115-125.
8. H. Levy, M. Sidi and O. Boxma, Dominance Relations in Polling Systems, *Queueing Systems*, 6 (1990) 155-172.
9. R. Loynes, The stability of a queue with non-independent inter-arrival and service times, *Proc. Camb. Philos. Soc.*, 58 (1962), 497–520.
10. V.A. Malyshev and M.V Mensikov, Ergodicity, continuity and analyticity of countable Markov chains, *Trans. Moscov Math. Soc.* (1981), 1-18.
11. S. Meyn and D. Down, Stability of Generalized Jackson Networks, *Ann. Appl. Probab.*, to appear.
12. S. Meyn and R.L. Tweedie, Criteria for Stability of Markovian Processes I: Discrete Time Chains, *Adv. Appl. Prob.*, 24, 542-574, 1992.
13. S. Stidham, Jr., A last word on $L = \lambda W$. *Oper. Res.*, 22 (1974) , pp.417–421.
14. W. Szpankowski, Towards computable stability criteria for some multidimensional stochastic processes, in *Stochastic Analysis of Computer and Communication Systems* (ed. H. Takagi), Elsevier Science Publications B. V. (North-Holland), (1990), pp. 131–172.
15. W. Szpankowski, Stability Conditions for Some Distributed Systems: Buffered Random Access Systems, *Adv. Appl. Probab.*, 26 (1994).
16. H. Takagi, *Analysis of Polling Systems*, The MIT Press, Cambridge, Mass. 1988.
17. H. Takagi, Queueing Analysis of Polling Models, *ACM Computing Surveys*, 20 (1988) 5–28.
18. R. L. Tweedie, Criteria for classifying general Markov chains, *Adv. Appl. Prob.*, 8 (1976), 737–771.
19. J. Walrand, *An Introduction to Queueing Networks*, Prentice Hall, New Jersey 1988.
20. K. Watson, Performance Evaluation of Cyclic Service Strategies – A Survey. *Proc. PERFORMANCE'84*, E. Gelenbe Ed. (Paris 1984).

Flow Control of a Virtual Circuit*

Ashok K. Agrawala, Dheeraj Sanghi and Leyuan Shi

University of Maryland, College Park, MD 20742, USA

1 Introduction

Flow control schemes are used to control the sending of packets by the source node. Many schemes for flow control have been proposed in the literature [4],[11], [7], [12], [17], [3], [6], [13], [14]. In this paper we consider the transfer of packets from a given source host to a given destination. All packets between the given source-destination host pair are said to correspond to a virtual circuit. The flow of packets over a virtual circuit is described using the deterministic model proposed in [1]. In this model recursive equations are used to express the departure time as a function of inter-packet send time. This function is used to propose a send time based flow control scheme that results in minimum transit delay and/or maximum throughput. Under some simplifying assumptions we extend that model to take into account the cross traffic. A scheme to estimate critical parameters of the function is proposed. This estimation technique reflects the characteristics of network dynamics which have been observed on the Internet. The results show that the send time based flow control scheme performs much better then the TCP window based scheme in all the scenarios considered.

The rest of this paper is organized as follows: In §2, a deterministic model for a virtual circuit is introduced and the cross traffic through the virtual circuit is considered. A send-time based flow control scheme is used such that the performance measure of the virtual circuit is optimized. In §3, an estimation technique is proposed for those critical parameters used in the send-time based flow control scheme. Section 4 gives our conclusions.

2 Send-time Control Scheme

We first consider the model of a virtual circuit with no cross traffic. A virtual circuit is modeled as a system with n servers, each of which processes packets using a FIFO work-conserving discipline. Packets, $j = 1, 2, \dots$ arrive at server i, $i = 1, 2, \dots, n$, at

* This work is supported in part by Rome Labs and DARPA under contract F30602-90-C-0010 to UNIACS at the University of Maryland. Computer facilities were provided in part by NSF grant CCR-8811954. A long version of the paper has been sent to *Journal of Internetworking*.

time instant a_j^i and obtain service for time duration τ_j^i. Let d_j^i be the time instant when packet j is departs from server i. Then for a server i,

$$d_j^i = \max(d_j^{i-1} + \pi^i, d_{j-1}^i) + \tau_j^i, \tag{1}$$

where π^i is propagation delay from server $i - 1$ to server i.

We assume that, for any given server, all packets on a virtual circuit have the same service requirements. The assumption will be valid if, for example, all packets are of same length. This would imply, $\tau_j^i = \tau^i$, independent of packet number j.

We define the total transit time, τ, as

$$\tau = \sum_{i=1}^{n} (\tau^i + \pi^i),$$

and the bottleneck service time, τ^b, as

$$\tau^b = \max_i \{\tau^i\}.$$

If a packet does not have to wait at any server, its arrival and departure instances are related as follows.

$$d_j^n = d_j^0 + \tau. \tag{2}$$

In this case[2], the packet encounters minimum transit time. If the packet did wait at some server, it must wait at the bottleneck server [16]. Then,

$$d_j^n = d_{j-1}^n + \tau^b. \tag{3}$$

Note that Eq. (3) holds because of our assumption that for all j, $\tau_j^i = \tau^i$. Clearly, keeping the bottleneck server busy yields the maximum throughput for the virtual circuit.

In general, one of the two cases will hold, and the packet departure instance can be computed as follows.

$$d_j^n = \max(d_j^0 + \tau, d_{j-1}^n + \tau^b). \tag{4}$$

For a packet j, we define transit time T_j as the time spent by the packet in the network, Inter-packet time R_j is defined as the time interval between the departure from the network of packet j and packet $j - 1$. We define *Packet Performance Index* (PPI) for the packet j as

$$P_j = T_j * R_j = (d_j^n - d_j^0) * (d_j^n - d_{j-1}^n). \tag{5}$$

This measure is defined for each packet (except the first packet) and can be used to characterize the transient behavior of the connection. As we can see that a smaller transit time implies that packets are spending less time in the network. This indicates smaller waiting times at the store-and-forward nodes implying reduced congestion. A smaller inter-packet departure time implies that packets are reaching the destination at

[2] We consider the source node to be the node 0.

a faster rate. This indicates higher throughput. PPI, therefore is used as the performance measure to evaluate flow control schemes reflecting both transit time and throughput.

In the absence of cross traffic, send-time may be computed so that the performance measure PPI is minimized. The equations that are used to determine send-times are derived below.

In Eq. (4), d_j^0 is the time at which the packet is sent from the source. Note that d_j^0 is controlled by the sender. If we calculate d_j^0 by equating the right hand sides of Eq. (2) and (3), the jth packet is sent at a time which assures the minimum transit time as well as the maximum throughput for the virtual circuit. Thus,

$$d_j^0 = d_{j-1}^n + \tau^b - \tau. \tag{6}$$

Repeated substitutions from Eq. (3) in the above equation yields,

$$d_j^0 = d_{j-k}^n + k\tau^b - \tau. \tag{7}$$

From Eq. (2) and (6) we get

$$d_j^0 = d_{j-1}^0 + \tau^b. \tag{8}$$

Repeated expansion of the above equation yields

$$d_j^0 = d_{j-k}^0 + k\tau^b. \tag{9}$$

If we were to calculate d_j^0 from Eq. (6) or (7), we need information on departure times of earlier packets. If it were to be calculated from Eq. (8) or (9), it is strictly dependent on the send-times of earlier packets.

Next let us consider the model of a virtual circuit with cross traffic. Assume that each packet j will experience certain delay because of cross traffic. The observed effect of the presence of cross traffic is that the packet departure time may be later than that given by Eq. (4). If we represent this additional delay encountered by packet j to be δ_j, we may rewrite Eq. (4) as

$$d_j^n = \max(d_j^0 + \tau, d_{j-1}^n + \tau^b) + \delta_j. \tag{10}$$

Let $\tau_j = \tau + \delta_j$ and $\tau_j^b = \tau^b + \delta_j$. Viewing cross traffic this way, we are able to write the packet departure instance as follows:

$$d_j^n = \max(d_j^0 + \tau_j, d_{j-1}^n + \tau_j^b). \tag{11}$$

Under the assumption that τ_j and τ_j^b are independent of d_j^0, which means each packet j will experience a constant cross traffic regardless of the sending time, P_j is minimized if d_j^0 satisfies

$$d_j^0 + \tau_j = d_{j-1}^n + \tau_j^b = d_j^n. \tag{12}$$

We now, from Eq. (12) derive a send-time control scheme. Rewrite (12), we have

$$d_j^0 = d_{j-1}^n + \tau_j^b - \tau_j \tag{13}$$

$$= d_{j-2}^n + \tau_{j-1}^b + \tau_j^b - \tau_j \tag{14}$$

$$\vdots \tag{15}$$

$$= d_{j-k}^n + \tau_{j-k+1}^b + \cdots : + \tau_j^b - \tau_j. \tag{16}$$

Equation (16) leads the send-time control scheme: when an acknowledgement for packet $j-k$ is received, time to send the next packet, j, is determined as:

$$d_j^0 = d_{j-k}^n + k * \tau_{j-k}^b - \tau_{j-k}. \tag{17}$$

3 Parameter Estimate in Send-Time Control Scheme

In the last section we discussed a model of a virtual circuit in which the effect of cross traffic is reflected in two important parameters, τ_j and τ_j^b. Usually, τ_j represents increased transit time due to cross traffic and τ_j^b may reflect increased bottleneck service time. Both τ_j and τ_j^b are functions of cross traffic and sending time d_j^0. Explicit expressions of these two parameters in terms of cross traffic and d_j^0 are difficult. Nevertheless, if we can estimate these two parameters, the control scheme we derived in the previous section can be used.

In the following we provide an estimation technique to calculate these two key parameters.

Define $\hat{\tau}_j$ as the estimate of τ_j. Similarly, $\hat{\tau}_j^b$ is the estimate of τ_j^b. As before, T_j is defined to be the actual transit time of packet j. R_j is defined to to the inter-packet departure time for packet j.

From Eq. (11), if $d_j^n = d_j^0 + \tau_j$, then $\tau_j = d_j^n - d_j^0$, and if $d_j^n > d_j^0 + \tau_j$, then $\tau_j < d_j^n - d_j^0$. Thus,

$$\tau_j \leq d_j^n - d_j^0 = T_j. \tag{18}$$

Similarly,

$$\tau_j^b \leq d_j^n - d_{j-1}^n = R_j. \tag{19}$$

When

$$d_j^0 - d_{j-1}^n = d_j^0 - d_j^n + d_j^n - d_{j-1}^n = R_j - T_j, \geq 0 \tag{20}$$

it implies that the system is empty and $\tau_j = T_j$. If we draw a $R-T$ plot (see Fig. 1), we can see that in the region $T \leq R$, τ_j should be always set to equal T_j. But when $T \geq R$, from (18) and (19), τ_j and τ_j^b should be estimated in the shaded area in Fig. 1. Especially, given current transit time T_j and inter-packet departure time R_j, if we compare T_j and R_j with the previous estimates $\hat{\tau}_{j-1}$ and $\hat{\tau}_{j-1}^b$, four subregions can be identified (see Fig. 1).

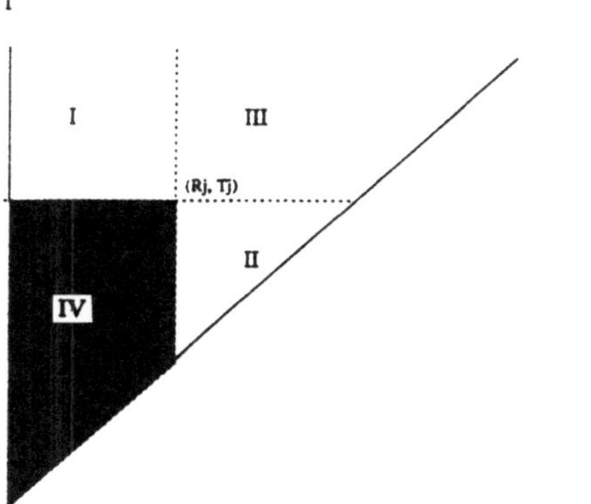

Fig. 1. R-T plot

Region I. $T_j < \hat{\tau}_{j-1}, \quad R_j \geq \hat{\tau}^b_{j-1}$.
From (18), we observe that τ_j should be less than or equal to T_j. But the previous estimate of τ, $\hat{\tau}_{j-1}$, is larger than T_j. This implies that the estimate is too high (or transit time has decreased). Therefore, the new estimate, $\hat{\tau}_j$, will be smaller than the previous estimate, $\hat{\tau}_{j-1}$ (see Fig. 1).

Region II. $T_j \geq \hat{\tau}_{j-1}, \quad R_j < \hat{\tau}^b_{j-1}$.
From (19), we observe that τ^b_j should be less than or equal to R_j. But the previous estimate of τ^b, $\hat{\tau}^b_{j-1}$, is larger than R_j. This implies that the estimate is too high (or bottleneck service time has decreased). Therefore, the new estimate, $\hat{\tau}^b_j$, will be smaller than the previous estimate, $\hat{\tau}^b_{j-1}$.

Region III. $T_j < \hat{\tau}_{j-1}, \quad R_j < \hat{\tau}^b_{j-1}$.
This case is combination of the above two cases. Both the estimates are very high and should be reduced.

Region IV. $T_j \geq \hat{\tau}_{j-1}, \quad R_j \geq \hat{\tau}^b_{j-1}$.
This is the most difficult case. We can only state with certainty that at least one of the estimates need to be increased (possibly both). We cannot say, which one. We will use heuristics to estimate the values properly.

Whenever we reduce τ or τ^b estimate, we use an auto-regressive filter to do that. In cases 1 and 3,

$$\hat{\tau}_j = \alpha_0 \hat{\tau}_{j-1} + (1 - \alpha_0) T_j,$$
$$\hat{\tau}_j = \hat{\tau}_{j-1} + (1 - \alpha_0)(T_j - \hat{\tau}_{j-1}).$$

In cases 2 and 3,

$$\hat{\tau}^b_j = \beta_0 \hat{\tau}^b_{j-1} + (1 - \beta_0) R_j,$$
$$\hat{\tau}^b_j = \hat{\tau}^b_{j-1} + (1 - \beta_0)(R_j - \hat{\tau}^b_{j-1}).$$

In case 4, we define

$$dr_j = d^n_{j-1} + \hat{\tau}^b_{j-1}, \quad \text{departure Time under wait condition,}$$
$$dt_j = d^0_j + \hat{\tau}_{j-1}, \quad \text{departure Time with no wait,}$$
$$\hat{d}^n_j = \max(dr_j, dt_j), \quad \text{estimated time of departure.}$$

In this case, $d^n_j \geq \hat{d}^n_j$. We increment both the estimates using auto-regressive filter.

$$\hat{\tau}^b_j = \hat{\tau}^b_{j-1} + (1 - \beta_1)(d^n_j - \hat{d}^n_j),$$
$$\hat{\tau}_j = \hat{\tau}_{j-1} + (1 - \alpha_1)(d^n_j - \hat{d}^n_j).$$

4 Conclusions

In this paper, we have developed an estimation technique for send-time flow control scheme. Based on our network dynamics observations, we have refined the estimation techniques used in send-time control. We have developed heuristics to reflect these observations in the estimation process. These heuristics have been validated using simulation. We have used simulation to compare TCP and DTP performance under a variety of network conditions. DTP performs much better in all the scenarios considered. It needs much fewer buffers in the intermediate nodes compared TCP. It has much smaller loss rate than TCP when they have the same buffer capacity. It has smaller average round-trip time and such average PPI than TCP.

Our results indicate that a significant improvement in performance can be obtained when using DTP. The impact of using DTP on all connections of a network needs further study. We believe, however, that the adaptive schemes of estimation which are used with DTP will result in stable, better performing network.

References

1. Ashok K. Agrawala and B.N. Jain, "Deterministic Analysis of Queues in Tandem with Applications to Computer Networks," Technical Report CS-TR-2511, UMIACS-TR-90-102, Department of Computer Science, University of Maryland, College Park, July 1990.
2. Ashok K. Agrawala and B.N. Jain, "Deterministic Model and Transient Analysis of Virtual Circuits," *IEEE Transactions on Software Engineering*, **19**, 2, 187–197, (1993).
3. J. Atkins, "Path Control: The Transport Network of SNA," *IEEE Transactions on Communication"*, **28**, 4, 527–538, (1980).
4. J.J. Bae and T.Suda, "Survey of Traffic Control Schemes and Protocols in ATM Network," *Proceedings of the IEEE*, **79**, 2, 170–189, (1991).
5. David R. Cheriton and C.L. Williamson, "VMTP as the Transport Layer for High-performance Distributed System," *IEEE Communications Magazine*, **27**, 6, 37–44, (1989).
6. David D. Clark, M.L. Lambert, and Lixia Zhang, "NETBLT: A High Throughput Transport Protocol," *Proceedings of ACM SIGCOMM '87*, 353–359, Stowe, VT, August 1987.
7. M.Gerla and L. Kleinrock, "Flow Control: A Comparative Survey," *IEEE Transactions on Communication*, **28**, 4, 553–574, (1980).
8. V. Jacobson, "Congestion Avoidance and Control", *Proceedings of ACM SIGCOMM '88*, 314–329, Stanford, CA, August 1988.
9. Raj Jain and Shawn A. Routhier, "Packet Trains-Measurements and a New Model for Computer Network Traffic," *IEEE Journal on Selected Areas in Communications*, **4**, 986–995, September 1986.

10. S. Keshav, "A Control-Theoretic Approach to Flow Control," *Proceedings ACM SIGCOMM '91*, Zurich, Switzerland, September, 1991.
11. N.F. Maxemchuck and M.E. Zarki, "Routing and Flow Control in High-Speed, Wide-Area Networks," *Proceedings of the IEEE*, **78**, 204–221, (1990).
12. J.B. Postel, "Transmission Control Protocol," Request for Comment: RFC-768, Network Information Center, SRI International, September 1981.
13. Robert M. Sanders and Alfred C. Weaver, "The Xpress Transfer Protocol (XTP) — A Tutorial," *Computer Communication Review*, **20**, 5, 67–80, (1990).
14. Djeeraj Sanghi and Ashok K. Agrawala, "DTP: An efficient Transport Protocol," *Proceedings of the IFIP TC6 International Conference on Computer Networks, Architecture and Applications (Networks '92)*, Elsevier North-Holland, Trivandrum, India, 1992.
15. Djeeraj Sanghi, Ashok K. Agrawala, Ó. Gudmundsson, and B.N. Jain, "Experimental Assessment of end-to-end behavior on Internet," *Proceedings of IEEE INFOCOM '93*, 867–874, San Francisco, CA, March 1993.
16. John Waclawsky, *Window Dynamics*, PhD thesis, Department of Computer Science, University of Maryland, College Park, May 1990.
17. Stuart Wecker, "DNA: The Digital Network Architecture," *IEEE Transactions on Communication*, **28**, 4, 510–526, (1980).

Stochastic Scheduling of Precedence Graphs

Lucian Finta and Zhen Liu

INRIA
Centre Sophia Antipolis
06902 Sophia-Antipolis Cedex
France

1 Introduction

Consider the following scheduling problem. We are given a set of tasks to be run in a system consisting of parallel and identical processors. The executions of these tasks have to satisfy some precedence constraints which are described by a directed acyclic graph, referred to as the task graph. A class of fork-join graphs will be considered, obtained from combination of in-forests and out-forests. The task running times are independent and identically distributed (i.i.d.) random variables with a common exponential distribution. The goal of scheduling is to find optimal preemptive schedule that stochastically minimizes the makespan, i.e., the completion time of all the tasks.

When the task graph is an in-forest and there are two processors, Chandy and Reynolds [2] proved that the Highest Level first (HL) policy minimizes the expected makespan. Here, the level of a task is simply the distance from it to the root of the tree it appears. They also have given one counterexample for in-trees under HL policy with three processors. Bruno [1] subsequently showed that HL stochastically minimizes the makespan when the system has two identical parallel processors. Pinedo and Weiss [6] extended this last result to the case where tasks at different levels may have different expected task running times. Frostig [4] further generalized the result of Pinedo and Weiss to include increasing likelihood ratio distributions for the task running times. When the number of identical parallel processors in the system is arbitrarily fixed, and the task running times have a common exponential distribution, Papadimitriou and Tsitsiklis [5] proved that HL is asymptotically optimal as the number of tasks tends to infinity.

Coffman and Liu [3] investigated the stochastic scheduling of out-forest on identical parallel processors. For the uniform out-forests where all the subtrees are ordered by an embedding relation, they showed that an intuitive priority scheduling policy induced by the embedding relation, referred to as the Largest Tree (LT) policy in this paper, stochastically minimizes the makespan when there are two processors. For a more restrictive graph, the r-uniform out-forests, they showed the optimality of LT policy for arbitrary number of processors.

In this paper, we investigate scheduling of fork-join graphs obtained from combination of in-forests and out-forests. We show that the heuristic Highest Level Most Successors stochastically minimizes the makespan on two processors. This result generalizes known results on stochastic scheduling of forests on two processors.

2 Problem Description and Preliminaries

A *task graph* $G = (V, E)$ is a directed acyclic graph, where $V = \{1, 2, \cdots, |V|\}$ is the set of vertices representing the tasks, $E \subset V \times V$ is the set of edges representing the *precedence constraints*: $(i, j) \in E$ if and only if task i must be completed before task j can start. Denote by $p(i)$ and $s(i)$ the sets of immediate *predecessors* and *successors* of $i \in V$, respectively, i.e.,

$$p(i) = \{j : (j, i) \in E\}, \qquad s(i) = \{j : (i, j) \in E\}.$$

Graph G is an in-forest (resp. out-forest) if $|s(i)| \leq 1$ (resp. $|p(i)| \leq 1$) for all $i \in G$. Let $S(i)$ be the set of (not necessarily immediate) successors of $i \in V$, i.e.,

$$\forall i: \quad \text{if} \quad s(i) = \emptyset \quad \text{then} \quad S(i) = \emptyset \quad \text{else} \quad S(i) = s(i) \bigcup \left(\bigcup_{j \in s(i)} S(j) \right).$$

The set of successors $S(i)$ and the vertex i form a subgraph of G denoted by $S_i(G)$.

All vertices $i \in V$ that have no successor are final vertices. The level of final vertices is zero by convention. The level of a vertex $i \in V$, denoted by $L(i)$, is the longest distance from it to final vertices in its set of successors. Let $T_l(G) = (V_l^{\mathrm{t}}, E_l^{\mathrm{t}})$ and $D_l(G) = (V_l^{\mathrm{d}}, E_l^{\mathrm{d}})$ be two subgraphs of G consisting of top and down vertices (separated by level l), respectively. More precisely,

$$V_l^{\mathrm{t}} = \{v \mid v \in V, \ L(v) \geq l\}, \quad V_l^{\mathrm{d}} = \{v \mid v \in V, \ L(v) \leq l\},$$

and E_l^{t} and E_l^{d} are restrictions of E on V_l^{t} and V_l^{d}.

Let $N_k(G)$ be the number of tasks of G at level k, $k \geq 0$. Clearly $N_k(G) = 0$ for all $k > L(G)$, where $L(G)$ is the level of graph G. For two graphs F and G, we say that F is *flatter than* G, denoted by $F \prec_{\mathrm{f}} G$, if

$$\forall i, \ i \geq 0: \qquad \sum_{k \geq i} N_k(F) \leq \sum_{k \geq i} N_k(G).$$

Let G_1, G_2 be two out-trees. Out-tree G_1 is said to embed out-tree G_2, or G_2 is embedded in G_1, denoted by $G_1 \succ_{\mathrm{e}} G_2$ or $G_2 \prec_{\mathrm{e}} G_1$, if G_2 is isomorphic to a subgraph of G_1. Formally, G_1 embeds G_2 if there exists an injective function f from G_2 into G_1 such that $\forall u, v \in G_2$, $v \in s(u)$ implies $f(v) \in s(f(u))$. Function f is called an embedding function. An out-forest G is said to be uniform if all its subtrees $\{S_v(G), \ v \in G\}$ can be ordered by the embedding relation. The embedding relation is extended to uniform out-forests: Let $G^1 = (V^1, E^1)$ and $G^2 = (V^2, E^2)$ be two uniform out-forests. Assume that the vertices of G^1 and G^2 are indexed in such a way that

$$S_1(G^1) \succ_{\mathrm{e}} S_2(G^1) \succ_{\mathrm{e}} \cdots \succ_{\mathrm{e}} S_{|V^1|}(G^1),$$
$$S_1(G^2) \succ_{\mathrm{e}} S_2(G^2) \succ_{\mathrm{e}} \cdots \succ_{\mathrm{e}} S_{|V^2|}(G^2).$$

Out-forest G^1 is embedded in G^2, referred to as $G^1 \prec^{\mathrm{e}} G^2$, if and only if

$$|V^1| \leq |V^2|, \quad \text{and} \quad \forall i, \ 1 \leq i \leq |V^1|: \quad S_i(G^1) \prec_{\mathrm{e}} S_i(G^2).$$

We will study a class $\mathcal{G}$ of task graphs, referred to as forest cut graphs, which can be cut into out-forest and in-forest. More precisely, $\mathcal{G}$ is the class of graphs satisfying the following conditions.

Definition 1. For any graph $G = (V, E) \in \mathcal{G}$, there exists a level l such that

1. $D_l(G)$ is an in-forest;
2. $T_l(G)$ is a uniform out-forest, whose vertices can be labelled in such a way that

$$S_1(T_l(G)) \succ^{\mathrm{e}} S_2(T_l(G)) \succ^{\mathrm{e}} \cdots \succ^{\mathrm{e}} S_{|V(T_l(G))|}(T_l(G)),$$

3. and that the subgraphs are in the *flatness* relation

$$S_1(G) \succ_{\mathrm{f}} S_2(G) \succ_{\mathrm{f}} \cdots \succ_{\mathrm{f}} S_{|V(T_l(G))|}(G).$$

The level l will be called the cut level.

Note that these class of graphs have the following closure property: if $G = (V, E) \in \mathcal{G}$ then $G-\{v\} \in \mathcal{G}$ for all $v \in V$ such that $p(v) = \varnothing$, where $G-\{v\}$ is the graph obtained by deleting vertex v and its adjacent edges. Such a closure property (by deletion) will be used in establishing our results.

We extend the notion of flatness as follows. Let $F, G \in \mathcal{G}$ be two forest cut graphs with the same cut level l, the relation of *extended flatness*, denoted by $F \prec^{\mathrm{f}} G$, means that $|V(T_l(F))| \leq |V(T_l(G))|$ and

$$S_1(F) \prec_{\mathrm{f}} S_1(G), \quad \cdots \quad S_{|V(T_l(F))|}(F) \prec_{\mathrm{f}} S_{|V(T_l(G))|}(G).$$

Definition 2. Two forest cut graphs $F, G \in \mathcal{G}$ with the same cut level l are in the relation $F \prec G$ if and only if they satisfy

$$F \prec_{\mathrm{f}} G, \qquad T_l(F) \prec^{\mathrm{e}} T_l(G), \qquad F \prec^{\mathrm{f}} G.$$

3 Stochastic Scheduling of Forest Cut Graphs

We define the Highest Level Most Successors (HLMS) policy as follows. The tasks of $G \in \mathcal{G}$ are labelled by $1, 2, \cdots, |V|$ in such a way that for any $1 \leq i \leq |V| - 1$, $L(i) \geq L(i+1)$, and if $L(i) = L(i+1)$ then $|S(i)| \geq |S(i+1)|$. At any time, only the enabled tasks (those without any unfinished predecessors) with the smallest labels are executed on the parallel processors. In other words, the HLMS assigns the tasks at the highest level among the enabled tasks to available processors. If there are several enabled tasks are at the same level, the priority is given to the task with most successors.

Let Ψ denote the class of policies which use no information on task running times. For any policy π, let $\pi(G)$ denote the makespan of G, i.e. the completion time of all tasks of G under policy π. Let γ denote policy HLMS. We will show that for two processors, a policy is optimal for the class $\mathcal{G}$ of task graphs if and only if the policy is HLMS.

Theorem 3. *Assume there are two parallel processors and task running times are i.i.d. with a common exponential distribution. Then for any graph $G \in \mathcal{G}$, and any $\pi \in \Psi$,*

$$\gamma(G) \leq_{\mathrm{st}} \pi(G). \tag{1}$$

Moreover, HL is the only policy that satisfies relation (1).

In order to simplify the proofs of this result, we make some reductions on the class of policies Ψ. Observe first that due to the memoryless property of the exponential distributions, the distribution of the remaining running time of a task running on processor k is still exponential with the same parameter. If we represent the state of the system by the remaining task graph and the distributions of the remaining running times of the tasks, then the state does not change between the instants of task completions. Therefore, we can, without loss of generality, confine ourselves to the class of policies where preemptions and new task assignments occur only at the instants of task completions. These instants are referred to as the *decision epochs*. Hence, we assume that all the policies in Ψ make their scheduling decisions at these time instants only.

The proof of Theorem 3 makes use of the following lemmas.

Lemma 4. *Let F and G be two subgraphs of a uniform out-forest H, obtained by successively deleting the vertices of H having no predecessor in H or in the previously obtained subgraphs. Let F' (resp. G') be the remaining subgraph when the second highest vertex without predecessor in F (resp. G), if any, is deleted. If $F \prec^{\mathrm{e}} G$, then $F' \prec^{\mathrm{e}} G'$.*

Lemma 5. *Let $H \in \mathcal{G}$ be an forest cut graph with cut level l. Let F and G be two subgraphs of graph $H \in \mathcal{G}$ obtained by successively deleting the vertices of H having no predecessor in H or in the previously obtained subgraphs. If $T_l(F) \prec^{\mathrm{e}} T_l(G)$, then $F \prec^{\mathrm{f}} G$.*

Lemma 6. *Let F and G be two subgraphs of forest cut graph $H \in \mathcal{G}$ obtained by successively deleting the vertices of H having no predecessor in H or in the previously obtained subgraphs. If $F \prec G$, then*

$$\gamma(F) \leq_{\mathrm{st}} \gamma(G).$$

Lemma 7. *Let $G \in \mathcal{G}$ be a forest cut graph. Let $\pi \in \Psi$ be a policy which follows the HLMS rule all the time except at the first decision epoch. Then*

$$\gamma(G) \leq_{\mathrm{st}} \pi(G).$$

Proof of Theorem 3. In order to show the theorem, we construct a finite sequence of policies $\{\pi_n\}_{n=1}^{N}$, $N = |V|$. Let $0 = c_1 < c_2 < \cdots < c_N$ be the decision times. For $1 \leq n \leq N$, policy π_n takes exactly the same scheduling decisions as π at times $c_1, c_2, \cdots, c_{n-1}$, i.e., the same tasks are assigned to the same processors at these time instants under policies π and π_n. From time c_n, policy π_n follows the HLMS rule. By definition, π_1 is an HLMS policy.

We use backward induction on n to show that for all n, $1 \leq n \leq N$,

$$\pi_n(G) \leq_{\mathrm{st}} \pi(G). \tag{2}$$

For $1 \leq n \leq N$, let G_n be the remaining task graph of G at time c_n under π. It is clear that $G_1 = G$ and G_N has a single vertex. Applying Lemma 7 to the task graph G_N implies that

$$\pi_N(G) = c_N + HLMS(G_N) \leq_{\mathrm{st}} c_N + \pi(G_N) = \pi(G).$$

Therefore, (2) holds for $n = N$. Assume that relation (2) holds for some $2 \leq n \leq N$. Then

$$\pi_{n-1}(G) = c_{n-1} + \gamma(G_{n-1}) \leq_{\text{st}} c_{n-1} + \pi_n(G_{n-1}) = \pi_n(G),$$

where we used Lemma 7 to obtain the inequality. Thus, by induction, relation (2) holds for all $1 \leq n \leq N$. Taking $n = 1$ immediately implies relation (1).

Moreover, it is not difficult to see that HLMS is the only policy that satisfies relation (1). Indeed, one can show that if a policy π violates the HLMS rule, then, with positive probability $\gamma(G) < \pi(G)$. □

References

1. J. Bruno, "On Scheduling Tasks with Exponential Service Times and In-Tree Precedence Constraints", *Acta Informatica*, **22** (1985), pp. 139–148.
2. K. M. Chandy, P. F. Reynolds, "Scheduling Partially Ordered Tasks with Probabilistic Execution Times", *Operating System Review*, **9** (1975), pp. 169–177.
3. E. G. Coffman, Z. Liu, "On the Optimal Stochastic Scheduling of Out-Forests", Rapport de Recherche INRIA No. 1156, 1990, *Operations Research*, **40** (1992), pp. S67–S75.
4. E. Frostig, "A Stochastic Scheduling Problem with Intree Precedence Constraints", *Operations Research*, **36** (1988), pp. 937–943.
5. C. H. Papadimitriou, J. N. Tsitsiklis, "On Stochastic Scheduling with In-Tree Precedence Constraints", *SIAM J. Comput.*, **16** (1987), pp. 1–6.
6. M. Pinedo, G. Weiss, "Scheduling Jobs with Exponentially Distributed Processing Times and Intree Precedence Constraints on Two Parallel Machines", *Operations Research*, **33** (1985), pp. 1381–1388.

Manufacturing Systems

Management of Manufacturing Systems based on Petri nets

Jean-Marie PROTH

INRIA-Lorraine, Technopôle Metz 2000, 4 rue Marconi, 57070 Metz, FRANCE
and
Institute for Systems Research, University of Maryland, College Park, MD 20742, USA.

1 Introduction

As far as production management is concerned, we usually consider two types of manufacturing systems, namely cyclic manufacturing systems (also known as off-line or ratio-driven systems), and non-cyclic manufacturing systems (also called on-line systems).
Cyclic manufacturing systems are highly automated manufacturing systems, such as flexible manufacturing systems, in which the part requirements are expressed in terms of ratios. In this kind of system, the goal is to maximise productivity while satisfying these ratios. It has been shown that an optimal cyclic control exists, and that it can be expressed by cyclic sequences at the entrance of each machine. The PN model of such a system is an event graph (also called a marked graph), and it models both the physical and the cyclic control. The cyclic systems have been extensively studied when the manufacturing times are deterministic or stochastic (see [Be1], [Ch1] and [Mu1]).
Non-cyclic manufacturing systems need a completely different approach. In this case, the PN model concerns only the physical system and the manufacturing constraints, but it influences the control in the sense that the decision making system (DMS) makes use of some properties of the PN model to make sure that the desirable qualitative properties will hold no matter what control is issued.
The second section of this paper presents the problem. In section three, we introduce the notion of decomposable net (DN) which is suitable for the problem at hand. In the same section, we also introduce the qualitative properties of the decomposable nets which are desirable as far as manufacturing systems are concerned. Based on the previous two sections, we present the short-term planning problem in section four. Section five is devoted to the scheduling problem. In particular, it will be shown that the scheduling PN model is derived from the short-term planning model by introducing some resource sharing places. Finally, section six contains some concluding remarks.

2 Presentation of the Problem

We are interested in a job-shop composed of n machines $M_1, M_2, \ldots, M_n$ able to manufacture q types of parts denoted by $P_1, P_2, \ldots, P_q$. The demand for each part type is known at the end of each of R consecutive elementary periods. For instance, an elementary period could be a day and R = 5: in this case, we are interested in managing the system over a working week on a day-to-day basis. In the remainder of the paper, the union of the R elementary periods is referred to as the sub-period. Let us denote by d_i^j, i = 1,2,.,q, j = 1,2,...,R, the demand for part type P_i at the end of the j-th period.

s_i^0 is the inventory level of part type i at the beginning of the first elementary period. We also define $\mathcal{M}(p_i)$ as the manufacturing process of a part of type P_i. $\mathcal{M}(p_i)$ provides:

(i) the sequence of operations to be performed on a part of type P_i;
(ii) the type of each operation, which can be either "assembly" or "regular"; disassembly operations are not considered;
(iii) the list of machines on which each operation can be performed;
(iv) the times required to perform each operation on its eligible machines.

The first problem to be solved is the short-term planning (STP) problem. Knowing the capacity of the system (i.e. the amount of time assigned to each elementary period), we aim at defining the number of parts of each type to be manufactured during each elementary period in order to optimize some criterion. Some commonly used criteria are the number of delayed products, the maximal delay, the weighted sum of delays, and the sum of the inventory and the backlogging costs. In the remainder of this paper, we aim at minimizing the sum of the inventory and the backlogging costs to illustrate the proposed approach.

Note that the amount of time assigned to each elementary period is bounded above by the duration of the period. It represents the total amount of time available for manufacturing parts. The difference between the duration of an elementary period and the time assigned to this period is the maximal idle time of the machines during each elementary period. It represents the flexibility of the system: the smaller the time assigned to an elementary period, the more likely is the existence of a feasible schedule, but the productivity of the system is smaller.

Starting from the number of parts of each type to be manufactured during the first elementary period (provided by the solution of the STP problem), the scheduling (S) process consists of assigning operations to their eligible machines and computing the beginning time of each operation in order to satisfy the usual manufacturing constraints, namely:

(i) operations should be performed according to the partial order issued from the manufacturing processes: two operations belonging to the same manufacturing process should be performed according to their order in this process;
(ii) a given machine performs at most one operation at a time.

We try not to optimize some criterion, but just to find a feasible schedule, i.e. a schedule which meets the requirements of the STP for the first elementary period. If such a feasible schedule does not exist, the only solution is to reduce the time assigned to the elementary periods, and to re-compute the STP: since such a change reduces the amount of parts to be manufactured during each elementary period, a feasible schedule more likely exists.

In figure 1, we summarise the process which leads to a feasible schedule. The STP problem can usually be solved using classical optimization software tools the choice of which depends upon the type of criterion to be optimized. But the S process is heuristic, since the problem is NP-hard.

In the next section, we introduce the basics which will be used to re-visit the previous approach in the light of PNs.

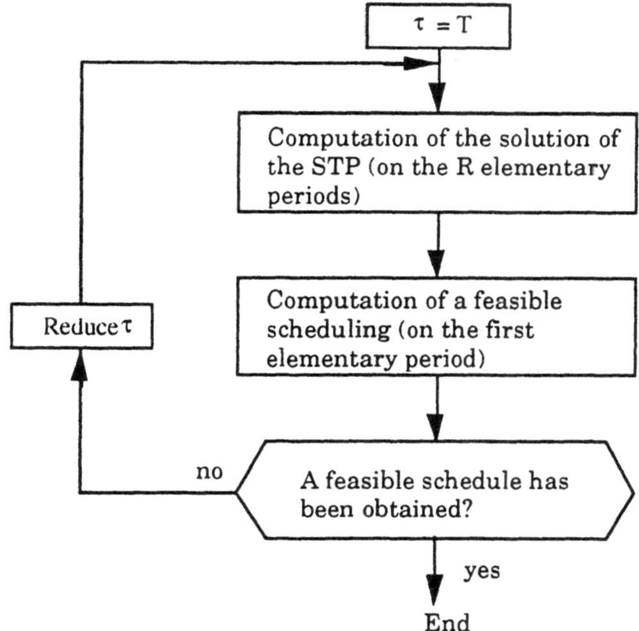

Fig. 1: General flow-chart

3 Decomposable Nets and their Qualitative Properties

4.1 Basic Definitions

We assume that the basics of PNs are known by the reader. More information can be found in [Be1], [HP1], [Ra1] and [RH1]. In this sub-section, we only recall some aspects which are critical for this presentation.

Let N = <P,T,F,W> be a PN where:

$P = \{p_1, p_2, \ldots, p_u\}$ is a finite set of places,

$T = \{t_1, t_2, \ldots, t_m\}$ is a finite set of transitions,

$F \subseteq (P \times T) \cup (T \times P)$ is the set of arcs,

$W: F \rightarrow \mathbb{N}^+$ is a weight function which assigns a strictly positive value to each arc; in this paper, the weight of any arc is 1.

We also assign a deterministic firing time to each transition: $Z = T \rightarrow Q^+$, where Q^+ is the set of non-negative rational numbers. Z(t) is called the firing time of transition t.

Let U be the incidence matrix of N. A t-invariant is a row-vector $X = [x_1, \ldots, x_m]$ such that x_i is a non-negative integer, at least one of the x_i being strictly positive, and such that $U.X^t = 0$. The set $T_X \subset T$ of transitions such that $t_i \in T_X$ if and only if $x_i > 0$ is called the support of the t-invariant X and denoted by $\|X\|$ (i.e. $T_X = \|X\|$).

Since parts enter and leave the system, the PN model of a manufacturing system contains source transitions (also called input transitions), which are used to model the entrance of the parts in the system, and sink transitions (or output transitions), which are used to model the exit from the system.

Let N_X be the PN derived from the t-invariant X as follows:

(i) $\|X\|$ are the transitions of N_X,

(ii) For any $t \in \|X\|$, t° and $^\circ t$ are the same in N and N_X. N_X is referred to as X-related subnet of N.

In addition, let us assume that:

N_X is structurally conflict-free (i.e. p° is unique whatever $p \in N_X$),

N_X has input and output transitions (which are input and output transitions of N).,

N_X does not contain cycles.

In that case, N_X will be referred to as X-CFIO of N, or as the CFIO derived from X, where CFIO stands for Conflict Free net with Input and Output transitions. We thus have the following definition.

Definition 1

Let X be a t-invariant of the PN model N = (P,T,F,W) of a manufacturing system. Let $N_X \subset N$ be a PN such that:

$$N_X = (P_X, \|X\|, F_X, W_X)$$

where $P_X = \{p \,/\, p \in P \text{ and } \exists\, t_1, t_2 \in \|X\| \text{ s.t. } p \in {}^\circ t_1 \text{ and } p \in t_2^\circ \text{ in } N\}$, $F_X = F \cap \{(P_X \times \|X\|) \cup (\|X\| \times P_X)\}$ and W_X is the restriction of W to F_X.

N_X is referred to as **X-related subnet of N**.

If, in addition:

(i) p° is unique for every $p \in P_X$,

(ii) there exists at least one $t_1 \in \|X\|$ and one $t_2 \in \|X\|$ such that $^\circ t_1 = \emptyset$ and $t_2^\circ = \emptyset$,

(iii) N_X is acyclic,

then N_X is referred to as **X-CFIO of N**.

4.2 Decomposability of a Net

The notion of decomposability is the key notion for the planning process we are proposing in this paper.

Definition 2

Let N be a PN model of a manufacturing system. Let $\{X_1, \ldots, X_r\}$ be a set of t-invariants of N such that:

(i) N_{X_i} is the X_i-CFIO of N for i = 1,2,...,r,

(ii) $N = \bigcup_{i=1}^{r} N_{X_i}$ (we will say that the set of X_i-CFIOs N_{X_i} covers N).

Then N is said to be decomposable.

4.3 Qualitative Properties

We consider the PN model of a manufacturing system, say N. To be acceptable as the model of a manufacturing system, N should be structurally live, and reversible and with home states. It should also be possible to keep it bounded. Let us recall these definitions to the reader and outline their importance in manufacturing.

a. Structural Liveness

Definition 3

A PN N = (P,T,F,W) is structurally live if their exists an initial marking M_0 such that, for any $t \in T$ and $M \in R(M_0)$ (i.e. M reachable from M_0) there exists a firing sequence which leads from M to a marking which enables t.

In terms of manufacturing systems, structural liveness means that it will be always possible to perform any operation for which the system was designed, assuming that the initial state of the system is adequately chosen. In other words, any operation which can be performed by the system will remain possible in the future, whatever the past sequence of decisions.

Result 1

A decomposable net N is structurally live for any initial marking M_0.

Proof

a. Let $M \in R(M_0)$ and σ a firing sequence which leads to M starting from M_0. N being a decomposable PN, there exists a set $\{X_1, \ldots, X_r\}$ of t-invariants of N such that the corresponding X-CFIO N_{X_i} covers N. Thus, it is possible to assign each element of σ to one of the N_{X_i}, designing firable sequences which keep the order the elements have in σ. Let us call σ_{X_i} (i = 1,2,...,r) the sequence related to N_{X_i}. Note that some of these sequences may be empty.

b. Let us now consider $t \in T$ and let N_{X_k} be a X-CFIO containing t. Due to the definition of X-CFIOs, there exists $\tilde{\sigma}_{X_k}$ such that firing $\sigma_{X_k} \circ \tilde{\sigma}_{X_k}$ from M_0^k (restriction of M_0 to N_{X_k}) leads to M_0^k again, and this sequence contains each transition at least once ($\circ$ represents the concatenation). As a consequence, $\sigma_k^* = \tilde{\sigma}_{X_k} \circ \sigma_{X_k} \circ \tilde{\sigma}_{X_k}$ is a firing sequence which applies to M and contains t.

This proof holds for any M_0. Q.E.D.

b. Reversibility and Home States

Definition 4

A PN N = (P,T,F,W) is reversible for a marking M_0 if, for any $M \in R(M_0)$,there exists a firing sequence σ_M which leads to M_0 from M.

This property is also very important in manufacturing. It guarantees that it is always possible to come back to the initial state, no matter what the current state is. This is often necessary for maintenance, tool adjustments or changes in production. The next definition generalizes the previous one.

Definition 5

A marking M_k of a Petri net N = (P,T,F,W) is a home state for the making M_0 if it can be reached from any marking reachable from M_0, i.e. $M_k \in R(M)$ for any $M \in R(M_0)$.

The use of a home state is the same as the one of M_0 in definition 4.
According to these definitions, any marking reachable from the initial marking in an reversible PN is a home state, but a PN with a home state may be not reversible.

Result 2

A decomposable PN N is reversible for any initial marking M_0.

Proof

a. This first part of the proof is identical to the first part of the proof of result 1.

b. Due to the definition of X-CFIOs, there exists $\tilde{\sigma}_{X_i}$ such that $\sigma_{X_i} \circ \tilde{\sigma}_{X_i}$ leads to M_0^i from M_0^i for i = 1,2,...,r. Thus $\tilde{\sigma} = \underset{i=1,2,\dots r}{0} \tilde{\sigma}_{X_i}$ is such that $\sigma \circ \tilde{\sigma}$ leads to M_0 from M_0.

Q.E.D.

c. ***Boundedness***

A decomposable net is not structurally bounded: for instance, the number of tokens in such a system increases indefinitely if we keep firing only the input transitions. Nevertheless, the following result holds.

Result 3

A decomposable net N = (P,T,F,W) can be kept bounded, no matter how many times the output transitions are fired, if this firings are finite.

Proof

Let k_1, ..., k_s be the minimal numbers of times the output transitions $t_{0,1}$, ...$t_{0,s}$, must be fired. Let X_1, ..., X_r be the t-invariants such that the corresponding X-CFIOs N_{X_i} cover N. Let n_j^i be the component of the t-invariant X_i which corresponds to $t_{0,j}$. At least one of the n_j^i, i = 1,2,...,r, is strictly positive for any $j \in \{1,\dots,s\}$, and at least one of the n_j^i, j = 1,2,...,s, is strictly positive for any $i \in \{1,\dots,r\}$. Thus the integer linear programming problem

$$Min \sum_{i=1}^{r} y_i$$

s.t.

$$\sum_{i=1}^{r} y_i . n_j^i \geq k_j, \quad j = 1,2,\dots,s$$
$$y_i \in \{0,1,\dots,\}, \quad i = 1,2,\dots,r$$

has a finite solution. The integer y_i is the number of times the set $\|X_i\|$ of transitions must be fired according to X_i in order to fire the output transitions the required number of times. Thus, the marking M(p) of $p \in t_v^\circ$ is bounded above by:

$$M_0(p) + \sum_{i=1}^{r} y_i x_v^i, \quad \text{where } x_v^i \text{ is the v} - \text{th componant of } X_i$$

This relation holds for any $p \in P$.

Q.E.D.

4 Short-Term Planning

4.1 Modelling Examples

Most of the manufacturing systems, when regarded from the short-term planning point of view, can be modelled using decomposable nets. For instance, let us consider three part-types P_1, P_2 and P_3 which can be performed on a set of five machines denoted by M_1, M_2, M_3, M_4 and M_5. The manufacturing processes of these part-types are given in figure 2. Each box represents one operation and contains the list of machines which can perform this operation.

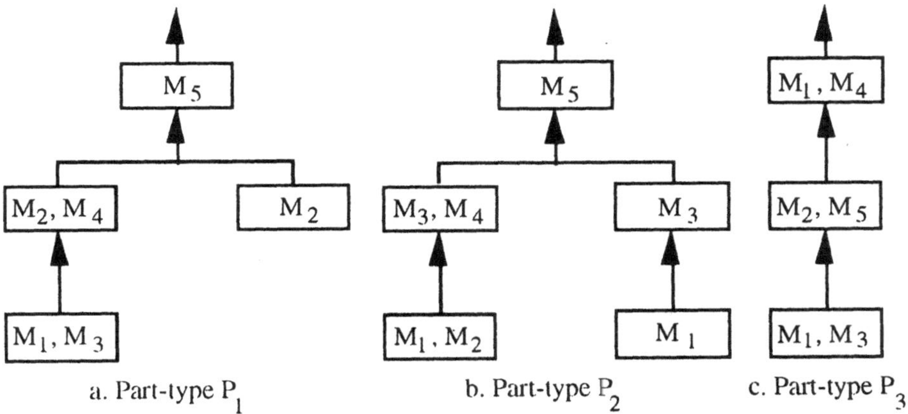

Fig. 2: Manufacturing processes

Note that the last operation required to complete parts P_1 and P_2 is an assembly operation. The models of these manufacturing processes are presented in figure 3. The numbers in squares are the manufacturing times of the operations (i.e. the firing times of the related transitions). The input transitions of each of these models are timed at zero and do not represent an operation, but model the launching of components in production. Each of the other transitions models an operation on a machine. The machines related to the transitions are in parentheses. At most one firing is in progress on each transition at a time, which means that a self-loop, with one token in the corresponding place, is associated to each transition. These self-loops are not represented in figure 3 for simplicity.

The PN model N presented in figure 3 is obviously a decomposable net. Several sets of t-invariants, the CFIOs derived from which cover N, can be proposed. We can, for instance, choose the minimal t-invariants of the PN, which are obtained by combining the minimal t-invariants of each of the manufacturing process models. Because the models corresponding to P_1, P_2 and P_3 have respectively 4, 4 and 8 minimal t-invariants, we would obtain $4 \times 4 \times 8 = 128$ minimal t-invariants fro the complete model. We can also choose some linear combinations of the minimal t-invariants provided that the derived CFIOs cover N.

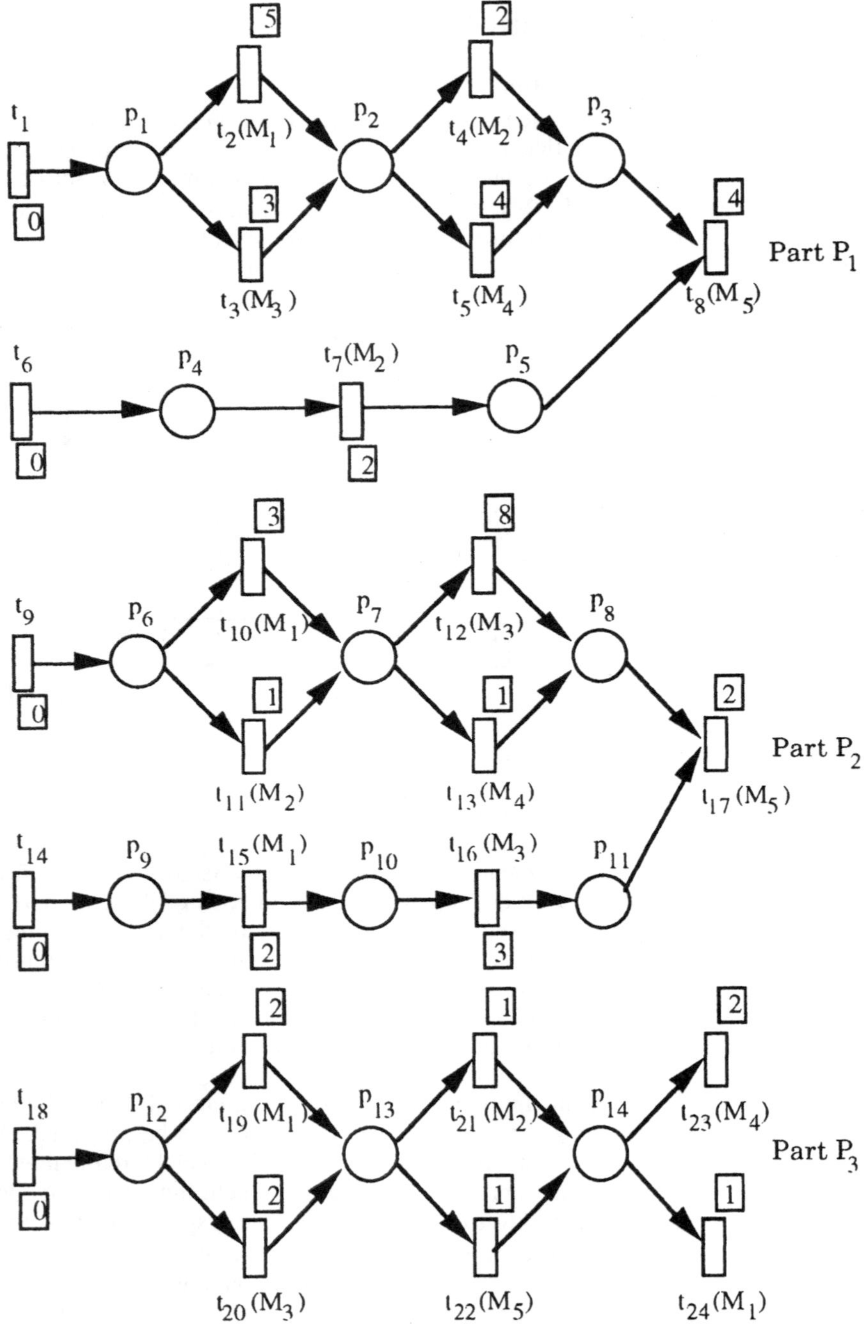

Fig. 3: The decomposable PN modelling the system from the short-term planning point of view

4.2 Planning Process

Let $\{X_1, \ldots, X_r\}$ be a set of t-invariants such that:

$$N = \bigcup_{i=1}^{r} N_{X_i}$$

Let O_k be the set of output transitions of the manufacturing process model of part type P_k, $k = 1,\ldots,q$. The demands for parts of type P_k are known at the end of each of R consecutive elementary periods. We denote, as in section 2, by d_k^j the number of parts of type P_k required at the end of the j-th elementary period; d_k^j is also the total number of times the transitions of O_k must be fired to meet the demand by the end of the j-th elementary period. T is the duration of an elementary period and τ is the time assigned to an elementary period (see section 2).

If y_i^j is the number of times the transitions of $\|X_i\|$ fire according to the components of X_i during the j-th elementary period, the following relations hold.

$$s_k^{j+1} = s_k^j - d_k^j + \sum_{i=1}^{r} \left[y_i^j \sum_{t \in O_k} x_t^i \right] \tag{1}$$

where x_t^i denotes the component of X_i corresponding to t in X_i.

This state equation holds for $k = 1,\ldots,q$ and $j = 0,\ldots,R-1$. s_k^0 is the initial inventory of parts of type P_k.
Furthermore, with Z(t) as the firing time of t:

$$\sum_{i=1}^{r} y_i^j x_t^i Z(t) \leq \tau, \quad j = 1,\ldots,R,\ t \in T \tag{2}$$

are the capacity constraints.
If the criterion to be minimized is the sum of the backlogging costs and inventory costs, it can be expressed as:

$$\mathit{Min} \sum_{k=1}^{q} \sum_{j=1}^{R} \left[b_k(-s_k^j)^+ + i_k(s_k^j)^+ \right] \tag{3}$$

where b_k (resp. i_k) is the backlogging cost (resp. the inventory cost) of one unit of part type P_k during one elementary period. Note that the problem which consists of minimizing (3) under constraints (1) and (2) can be re-written as a linear programming problem.

4.3 Remarks

The key to this approach is the choice of the set of t-invariants $\{X_1, \ldots, X_r\}$. Depending on this choice, N_{X_i} may be the model of one of the manufacturing processes or the model of a set of manufacturing processes of different part types. In the first case, the short-term planning process provides a better result, but requires a large amount of computation. In the second case, the number of y_i^j variables may be very small, and thus the amount of computation may be very limited, but the result will certainly be worst than in the previous case. In other words, selecting less t-invariants (assuming that the CFIOs derived from these t-invariants cover the PN model) offers the potential for reducing the computation burden at the expense of less use of the resources. But, in any case, the short-term planing process introduced in this section guarantees that the qualitative properties presented in the previous section hold.

5 THE SCHEDULING PROCESS

5.1 Problem Setting and Definitions

The goal of the scheduling process is to assign operations to resources (in the case that several resources are available) and to define the starting time of each operation for the first elementary period in order to meet the firing requirements of the short-term planning.

The model related to the scheduling process is obtained by extending the model used for the short-term planning using resource sharing places. Such a place:

(i) initially contains one token,

(ii) forms a self-loop with each transition related to the same given machine.

These places guarantee the non-reentrance. The notion of non-reentrance reflects the fact that machines can perform at most one operation at any given time. For instance, we could have five resource sharing places in the model given in figure 3, and one of these places would be linked in both ways to t_2; t_{10}, t_{15} and t_{19}.

Thus, there are two types of decision that have to be made, namely:

(i) The decisions related to the selection of a resource, when several resources are available to perform the same task. We denote such a decision by RU decision, where RU stands for Resource Use. A RU decision must be made in figure 4, where the next task to be performed on the part represented by the token can be performed either on M_1, M_2 or M_3.

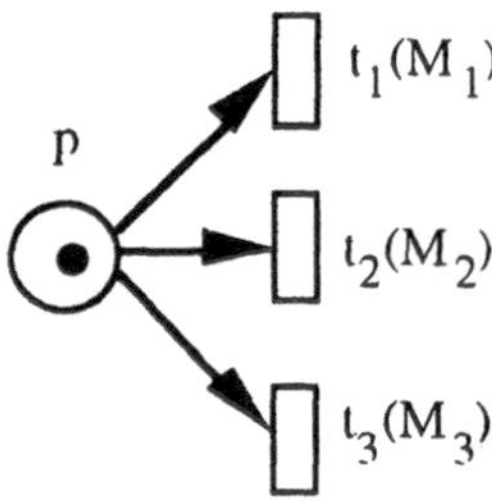

Fig. 4: Modelling of a RU type of decision

In figure 3, this situation can be observed in p_1, p_2, p_6, p_7, p_{12}, p_{13} and p_{14}.

(ii) The decisions related to the sequencing of the part types on the resources, called PS decisions, where PS stands for Product Sequencing. Figure 5 represents such a situation: we can fire either t_1, which represents the manufacturing of a part of type R_1 on M_1, or t_2, which represents the manufacturing of a part of type R_2 on M_1.

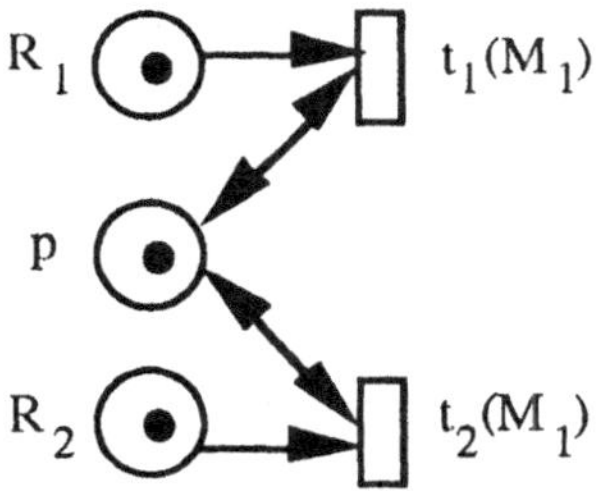

Fig. 5: Modelling of a PS type of decision

Hereafter, places related to RU (resp. PS) types of decision will be called RU (resp. PS) places.
We assume that transitions are fired as soon as they are enabled. As a consequence, a schedule is defined as soon as a sequence of transitions is assigned to each RU and PS place. The transitions belonging to such a sequence are the output transitions of the place, and they appear in the sequence as many times as specified by the short-term planning for the first elementary period. Let us for instance consider the RU place p in figure 4. If, according to the short-term planning, t_1, t_2 and t_3 have to be fired twice, once and three times respectively, a valid sequence could be $\sigma_p = < t_1, t_3, t_3, t_2, t_1, t_3 >$.
As we can see, the PN modelling of the scheduling problem exposes a well-separated modelling of RU and PS decisions, which makes easily understandable the decisions to be made. Nevertheless, it should be noticed that these decisions are not independent from each other and that non adequately designed sequences may lead to a blocking situation, as shown in figure 6 which contains two PS places, q_1 and q_2: if t_1, t_2, t_3 and t_4 must be fired once, then $\sigma_{q_1} = < t_4, t_1 >$ and $\sigma_{q_2} = < t_2, t_3 >$ leads to a blocking situation.

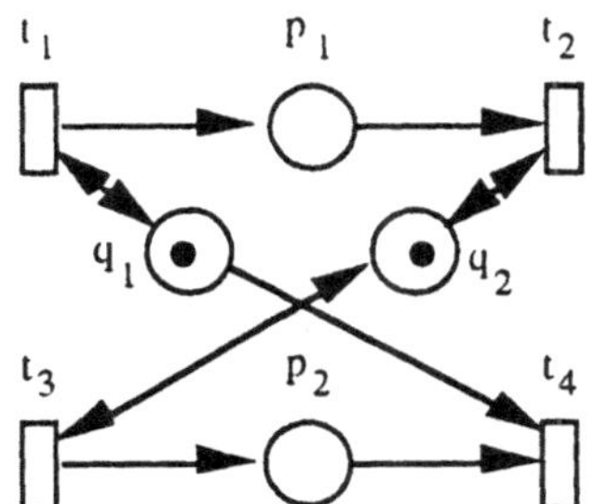

Fig. 6: Case where a blocking situation may occur

The objective of the problem can be expressed as follows:
determine the sequences to be assigned to each RU and PS places that make the makespan less than the duration of an elementary period, knowing that:

(i) the transitions belonging to a sequence are the output transitions of the related RU or PS place,
(ii) a transition appears in a sequence as many times as specified by the short-term planning for the first elementary period.

5.2 Scheduling Algorithm

It is well-known that the scheduling problem of the general job-shop case is NP-hard. As a consequence, only heuristic algorithms can be considered for solving large size problems. We developed two heuristic approaches: a simulated annealing approach and an approach based on the improvement of a critical circuit. For lack of place, only the second algorithm is presented in this section.

It starts with a feasible set of sequences, i.e. with a set of valid sequences which do not lead to a blocking situation. An easy way to build such a feasible set of sequences consists of assigning any valid sequence to each PS place, to simulate the system according to these sequences, the transitions being fired as soon as they are enabled, and to assign to the RU places the sequences issued from the simulation.

Let T be the duration of the elementary period, n_t the number of times t should be fired during the first elementary period, $S_t(k)$ the instant of the k-th firing initiation of t and $F_t(k) = S_t(k) + Z(t)$. These data refer to the initial feasible set of sequences, assuming that the starting time of the transition firing is 0 and that a transition fires as soon as it is enabled. A critical path is a sequence of pairs:

$$\mathcal{C} = < (t_{i_1}, k_{\alpha_1}), \ldots, (t_{i_r}, k_{\alpha_r}) >$$

such that:

(i) $S_{t_{i_1}}(k_{\alpha_1}) = 0$;

(ii) $F_{t_{i_r}}(k_{\alpha_r}) = \underset{(t,k)}{\mathbf{Max}}\{F_t(k)\}$. This value is the makespan;

(iii) $F_{t_{i_j}}(k_{\alpha_j}) = S_{t_{i_{j+1}}}(k_{\alpha_{j+1}})$, for j = 1,...,r-1.

A necessary condition to reduce the makespan is to reduce $F_{t_{i_r}}(k_{\alpha_r})$, and thus to bring forward the finishing time of one of the transitions firing that belongs to the critical path. To do this, we will have to postpone some transition firings which do not belong to the critical path. Note that it is not allowed to violate the machine precedence constraints (i.e. the constraints related to the manufacturing process, taking into account the initial marking). Furthermore, in order to reduce the computational burden, we introduce the **float time**, which is the maximal time a transition can be delayed without increasing the makespan. The calculation of the float time takes into account the scheduling and machine precedence dependencies of the operations. The objective is to forward a transition firing which belongs to the critical path if it results to postpone another transition firing for less than its float time. The two transition firings which are concerned in the scheduling perturbation are not taken into account in the computation of the float times. The following algorithm is derived from the above remarks.

Algorithm

1. Compute a feasible set of sequences and the initiation times of the transition firings.
2. Compute the critical path and the makespan.
3. Select (t_u, k_u) and (t_c, k_c) such that:
   * $t_u \neq t_c$,
   * (t_u, k_u) belongs to the critical path,
   * the k_c-th transition t_c just precedes the k_u-th transition t_u in one of the sequences of the feasible set of sequences,
   * swapping the k_u-th transition t_u and the k_c-th transition t_c in the sequence does not violate the machine precedence constraints,

* the float time associated to (t_c, k_c) is greater than or equal to the delay resulting from the swapping.

If several pairs (t_u, k_u) and (t_c, k_c) are candidates for the swapping, select the one for which the delay is the closest to the float time.

4. If a pair has been selected, go to 2, else stop.

6 Concluding Remarks

This paper proposes a class of Petri nets, called decomposable nets, which expose the desirable properties when manufacturing systems are concerned. They are used to model manufacturing systems from the short-term planning point of view. Such a model can be completed to obtain the model of the manufacturing system from the scheduling point of view. Thus, a comprehensive approach of production management is initiated in this paper.

This paper is part of our work on building a sound modular method for a modular approach of the management of complex manufacturing systems. The next step will be to examine which additional conditions have to be introduced in the decomposable nets in order to preserve the qualitative properties when integrating these nets.

Bibliography

[Be1] H.P. BENSON, "A Finite Algorithm for Concave Minimization over a Polyhedron," Nav. Res. Logis., vol. 32, pp. 165-177, 1985.

[Ch1] P. CHRETIENNE, "Les Réseaux de Petri Temporisés," Univ. Paris VI, Paris, France, Thèse d'Etat, 1983.

[DF1] M. DI MASCOLO, Y. FREIN, Y. DALLERY and R. DAVID, "A Unified Modeling of Kanban Systems Using Petri Nets," Technical Report No 89-06, LAG, Grenoble, France, September, 1989.

[HP1] H. HILLION and J.M. PROTH, "Performance Evaluation of Job-Shop Systems Using Timed Event-Graphs," IEEE Trans. Automat. Contr., vol. 34, No 1, pp. 3-9, January 1989.

[Mu1] T. MURATA, "Petri Nets: Properties, Analysis and Applications," Proceedings of the IEEE, vol. 77, No 4, pp. 541-580, April 1989.

[Ra1] C. RAMCHANDANI, "Analysis of Asynchronous Concurrent Systems by Timed Petri Nets," Lab. Comput. Sci., Mass. Inst. Technol. Cambridge, MA, Tech. Rep. 120, 1974.

[RH1] C.V. RAMAMOORTHY and G.S. HO, "Performance Evaluation of Asynchronous Concurrent Systems using Petri Nets," IEEE Trans. Software Eng., vol. SE-6, No 5, pp. 440-449, 1980.

[Si1] J. SIFAKIS, "A Unified Approach for Studying the properties of Transition Systems," Theoret. Comput. Sci. vol. 18, pp. 227-258, 1982.

[VT1] N. VAN THOAI and H. TUY, "Convergent Algorithms for Minimizing a Concave Function," Mathematics of Operations Research, vol. 5, No 4, pp. 556-566, 1980.

Approximate Closed Queueing Network Model of A Manufacturing Cell With A Batch Material Transporter *

S.Y. Ruan and M.A. Jafari

Department of Industrial Engineering, Rutgers University, P.O.Box 0909, Piscataway, NJ 08855

1 Introduction

We consider a manufacturing system composed of a work-cell, a material transporter (MT) and a central storage/retrieval system. The work-cell is composed of K workstations, a common buffer with capacity $N < \infty$, and an output buffer with unlimited capacity. Part routing within the work-cell follows a Markov chain. With probability p_{0i}, the newly arrived part goes to workstation i. With probability p_{ij}, the processed part by workstation i goes to workstation j; $j \neq i$. The probability that this part leaves the work-cell is p_{i0}. We assume that $p_{ii} = 0$, for $i = 0, \ldots, K$. The processing time S_i of a part on workstation i; $i = 1, \ldots, K$; follows a two-stage Coxian distribution with a known mean and a known squared coefficient of variation (SCV). Since the buffer within the work-cell is common to all workstations, there will not be any circumstances where a workstation is blocked by another workstation (see Jafari [1989]). When the number of finished parts in the output buffer reaches a threshold, say r_0, a signal is sent to the MT requesting for material transfer. Upon the receipt of this signal and given that it is available, the MT moves to the work-cell and starts unloading the output buffer. The maximum capacity of the MT is assumed to be R units. Upon the completion of the unloading operation, the MT moves to the storage/retrieval area, unloads its cargo, loads an equal number of parts, and then moves back to the work-cell. The total time, T_0, to complete these operations is assumed to follow a distribution with a known mean and a known SCV. In each round trip a random number B of finished parts are transferred from the output buffer to the storage/retrieval area. Also, the same number of new parts are transferred from the storage/retrieval area to the work-cell and stored in the common buffer or loaded to the starving workstations. The time it takes to store the finished parts and retrieve new parts is included in the total travel time.

In queueing terms, the above system is a closed network of K unit servers and one batch server. The unit servers process parts one by one, but the batch server processes parts in batches. The part routing resembles the customer routing in the classical central server queueing models, with a major difference that the operation of the batch server is somehow synchronized with the operation of the remaining K servers.

In real manufacturing systems with cellular layout, there may be more than one work-cells served by the same MT. In queueing terms, this is a large and complex network composed of a number of a closed subnetwork (chains) linked together by the MT. For

* This work is partially supported by the National Science Foundation under Grant DDM89-10007.

its performance analysis, one needs to decompose the system into its subnetworks. The source of interaction between these individual queueing networks is the common MT. To include this interaction in our model, we shall assume that the MT can also receive requests from some sources other than the work-cell. More specifically, we shall assume that the time between the arrival of two consecutive requests of this kind is exponentially distributed with rate λ. We also assume that there can be at most C outstanding requests of this kind at any time. Considering the whole manufacturing system, $C + 1$ could be the total number of cells in the system. λ could be considered as the aggregate request rate from all other cells. It is an input to our model. To avoid confusion, we shall denote by σ the specific work-cell being modeled here.

The number of outstanding material transfer requests from work-cell σ can be at most one at any time. But, if after unloading the output buffer, the number of remaining parts in there exceeds r_0, another request will be immediately generated. Thus, the number of requests awaiting processing by the MT can not exceed $C + 1$. The service policy of the MT is assumed to be FCFS.

In this article, we shall develop a queueing model for the above system. We believe that there are several important features of this system which make it quite challenging to analyze. These features are: (1) the batch processing nature of the MT, (2) the synchronous nature of the MT (3) the general processing time distributions on workstations, (4) the threshold policy by which the material transfer requests are generated, and (5) the finite capacity of the MT. As we shall discuss next, there are a number of queueing models in the literature that are either directly or indirectly for manufacturing applications. However, they lack one or more of these features.

The outline of the remainder of the paper is as follows. In Section 2, we discuss the queueing model, and in Section 3, we give numerical results and some discussion on the performance characteristics of the system.

2 Queueing Model

The CQNs with no blocking and general processing times have been studied by a number of researchers. Two basic approximations have been proposed (Baynat and Dallery [1991]): the aggregation method and Marie's method (Marie [1979]). Ruan and Jafari [1993] use the aggregation method for the same system as ours. Here we use a variation of Marie's method mainly because it can handle larger systems.

Following the Marie's procedure, we shall determine the arrival process to each queue (workstation or MT) by solving the CQN of the work-cell and MT with a product form solution. More specifically, we shall compute $\lambda_k(n)$ the load dependent arrival rate, and C_{ak}^2 the SCV of the arrival process to each queue, $k = 0, \ldots, K$. For the batch server we use k=0. Then we will study each node in isolation as a $C_2^{(n)}/C_2/1/\Omega/\Omega$ queue. Using the steady state probabilities $P_k(l), l = 0, \ldots, \Omega$, we shall then compute $\mu_k(n)$ the load dependent service rate for node k, $k = 0, \ldots, K$. These will then be used to again solve the CQN of the work-cell and MT with a product form solution. The process continues until some termination criteria (see Marie [1979]) are satisfied. The reader notes that the above procedure is quite straight-forward as long as the MT-node processes parts one by one. The complexity comes when that node becomes a batch processor and follows a threshold policy as we described earlier.

The batch server in isolation is a $C_2^{(n)}/C_2^{[B]}/1/\Omega/\Omega$ queue where $C_2^{(n)}$ indicates a load dependent arrival process and $C_2^{[B]}$ is a batch service process. Each process follows a general distribution approximated here by a two-stage Coxian. The maximum number of parts in this queue can never exceed Ω. Furthermore, the population from which the arrivals are generated is also finite of size Ω. Following Marie's procedure requires us to replace $C_2^{(n)}/C_2^{[B]}/1/\Omega/\Omega$ by an equivalent queue $C_2^{(n)}/C_2/1/\Omega/\Omega$ where parts are processed one by one. The equivalence must be defined such that the average flow of parts into the work-cell σ (i.e., workstation-nodes 1 through K) under the two configurations are as close as possible.

For the $C_2^{(n)}/C_2^{[B]}/1/\Omega/\Omega$ queue, let us define the random variable D_B to be the time between any two consecutive departures and the random variable B to be the batch size. For the $C_2^{(n)}/C_2/1/\Omega/\Omega$ queue, we define the random variable D_S to be the time between any two consecutive departures and the random variable A_S to be the service times. We note that under the batch processing configuration, on the average, $E(B)$ parts are delivered to the work-cell once every $E(D_B)$ time units. In the case of single processing configuration, the parts are delivered to the work-cell σ one by one every $E(D_S)$ time units. In order to keep an equivalent part flow under the two configurations, we must speed up the equivalent single processing server so that it also delivers, on the average, $E(B)$ parts to the work-cell during the time period $E(D_B)$. In other words, we have

$$E(D_S) = \frac{E(D_B)}{E(B)} , \qquad E(A_s) = \frac{E(D_s)}{1 - P_0} , \tag{1}$$

where P_0 is the proportion of times that the single processing server is idle. Similarly, one can compute the second moment of A_S. These two moments are used to compute the parameters of the service process in the equivalent $C_2^{(n)}/C_2/1/\Omega/\Omega$ queue. The only problem is that the distribution of D_B and B are unknown. We shall discuss these next.

Let us define three observation epochs: ξ_n, α_n, and e_n. ξ_n is the time epoch at which the MT-node starts processing the nth batch. In the real system, this is the time at which the MT leaves the output buffer for the nth time. α_n is the time epoch of the nth arrival of a batch to the work-cell σ (or workstation-nodes 1 through K), and e_n is the time epoch at which the number of parts in the output buffer (queue for the batch server) reaches r_0 for the nth time. In this network, let us also define I_n to be the inventory level in the output buffer at ξ_n, W_n the time duration from to e_n to ξ_n, and $\theta_n(r_0, I_n)$ the time period required to accumulate r_0 parts in the output buffer following ξ_n. Then, we have $\alpha_n - \xi_n = T_0$, $e_{n+1} - \xi_n = \theta_n(r_0, I_n)$, and $\xi_n - e_n = W_n$. Furthermore, we have

$$\alpha_{n+1} - \alpha_n = \xi_{n+1} - \xi_n = \begin{cases} W_{n+1} & \text{if } I_n \geq r_0 \\ W_{n+1} + \theta_n(r_0, I_n) & \text{if } I_n < r_0 \end{cases} \tag{2}$$

Under stationary conditions, $\alpha_{n+1} - \alpha_n$ defines D_B.

We note that $\theta_n(r_0, I_n)$ depends on the random variable I_n and on the departure process from the work-cell. We also note that W_n depends on the service process of the batch server. In general, W_n also depends on the queueing policy in the MT. But, here we fix the policy to FCFS.

2.1 Probability Distribution of I_n

The inventory levels I_n and I_{n+1} at ξ_n and ξ_{n+1} are related through

$$I_{n+1} = \max\{0, \max(0, r_0 - I_n) + I_n + N_{W_{n+1}} - R\} \tag{3}$$

where $\max(0, r_0 - I_n)$ and $N_{W_{n+1}}$ are, respectively, the number of parts arriving at the output buffer during $\theta_n(r_0, I_n)$ and W_{n+1}. Both $\theta_n(r_0, I_n)$ and $N_{W_{n+1}}$ depend on the departure process from the work-cell. This process depends on the number of parts on the work-cell among other things. Since W_{n+1} lies within $[\xi_n, \xi_{n+1}]$, the maximum number of parts within the work-cell (including the ones handled by the MT) will not change during W_{n+1}. Let $I(t)$ be inventory level in the output buffer at an arbitrary time t. Also, let

$$\begin{cases} X(0) = \Omega - I(\xi_n + \theta_n(r_0, I_n)) \\ X(w) = \Omega - I(\xi_n + \theta_n(r_0, I_n) + w), 0 < w \leq \xi_{n+1} - e_{n+1} \end{cases} \tag{4}$$

Then $X = \{X(w) \mid 0 \leq w \leq \xi_{n+1} - \xi_n\}$ defines a death process with death rate, $\gamma_n()$, dependent on the population size. Since a death could occur due to k different reasons, that is, a part may arrive to the output buffer from k different workstations, it is reasonable to assume exponential distribution for the time between any two consecutive occurrences of death.

An interesting problem here is that X(0) the initial population size is random variable. Distribution of X(0) can be determined using Eq. (4) and noting that

$$I(\xi_n + \theta_n(r_0, I_n)) = \begin{cases} I_n \text{ if } I_n \geq r_0 \\ r_0 \text{ if } I_n \leq r_0 \end{cases} \tag{5}$$

Given Eq. (5), we obtain,

$$P(N_{W_{n+1}} = m \mid W_{n+1} = w, I_n = i) = P(X(w) = \Omega - i - m \mid X(0) = \Omega - i), \quad i \geq r_0, 0 \leq m \leq \Omega - i \tag{6}$$

and

$$P(N_{W_{n+1}} = m \mid W_{n+1} = w, I_n = r_0) = P(X(w) = \Omega - r_0 - m \mid X(0) = \Omega - r_0), \quad 0 \leq m \leq \Omega - r_0 \tag{7}$$

Assuming that we have the death rates $\gamma_n()$ and $f_{W_{n+1}}(w)$ the density function of W_{n+1}, we can derive the conditional probability distribution of $N_{W_{n+1}}$ given that $I_n = i$. For $i \geq r_0$, we have

$$P(N_{W_{n+1}} = m \mid I_n = i) = \int_0^\infty P(N_{W_{n+1}} = m \mid W_{n+1}, I_n = i) f_{W_{n+1}}(w) dw \tag{8}$$

As we shall see later, we can only compute the moments of W_{n+1} and not its distribution. Using a two-moment approximation we approximate $f_{W_{n+1}}(w)$ by an appropriate distribution (e.g., two-stage Coxian or Erlang). Assuming a stationary process, using Eqs. (5), (6), (7), and (8), and

$$f_W(w) = c_1\psi_1 e^{-\psi_1 w} + c_2\psi_2 e^{-\psi_2 w}$$

we have, for $0 < m < \Omega - i$,

$$P(N_W = m \mid I_n = i) = \left[\prod_{r=l+1}^{L} \gamma(i)\right] \left\{\sum_{j=l}^{L} A_{j,l} \left[\frac{c_1\psi_1}{\gamma(j) + \psi_1} + \frac{c_2\psi_2}{\gamma(j) + \psi_2}\right]\right\} \tag{9}$$

and for $m = 0$

$$P(N_W = 0 \mid I_n = i) = \frac{c_1\psi_1}{\gamma(L) + \psi_1} + \frac{c_2\psi_2}{\gamma(L) + \psi_2} \tag{10}$$

where $l = \Omega - i - m$ and $L = \Omega - i$. For $i < r_0$, Eqs. (11) and (12) still hold except that $l = \Omega - r_0 - m$ and $L = \Omega - r_0$. Similarly, one can obtain the conditional distribution of N_W, given that W has Erlang distribution.

Given the probability distribution of N_W, we can now study the process $I = \{I_n \mid n \geq 0\}$. Based on the above formulation, it is easy to see that I is a Markov chain. Having solved the above Markov chain for its steady state probabilities, we compute $P(I = i)$.

2.2 Distribution of $\theta_n(r_0, I_n)$

$\theta_n(r_0, I_n)$ is defined according to: $\theta_n(r_0, I_n) = 0$ if $I_n \geq r_0$, or $\theta_n(r_0, I_n) = \sum_{j=1}^{r_0 - I_n} X_j$ if $I_n < r_0$, where X_1 is the time, measured from ξ_n, of the first arrival of a part to the output buffer, X_2 is the time between the first and the second arrivals, and so on. As discussed earlier, each X_j has an exponential distribution with rate $\gamma_n(\Omega - i - j + 1)$, given that $\{I_n = i, I_n < r_0\}$. Using the above equation we have

$$E(\theta_n(r_0, I_n) \mid I_n = i, I_n < r_0) = \sum_{j=1}^{r_0 - i} \frac{1}{\gamma_n(\Omega - i - j + 1)} \tag{11}$$

$$V(\theta_n(r_0, I_n) \mid I_n = i, I_n < r_0) = \sum_{j=1}^{r_0 - i} \frac{1}{\gamma_n^2(\Omega - i - j + 1)} \tag{12}$$

Assuming a stationary process and using Eqs. (11) and (12) and $P(I = i)$ we compute $E(\theta(r_0, I) \mid I < r_0)$ and $V(\theta(r_0, I) \mid I < r_0)$. These quantities are used to compute the parameters of an appropriate conditional distribution for $\theta(r_0, I)$, given that $I < r_0$.

2.3 Distribution of W

The computation of W is rather complex. We note that the batch server is associated with another queue where signals or requests from the the various work-cells wait. To compute the waiting time distribution, we need to know how many signals are already waiting in this queue at the time that the signal from this work-cell is generated. We shall not discuss this here. Interested readers are referred to Jafari and Ruan [1992].

2.4 Distribution of B

Given the probability distributions of N_W and the inventory level, we can easily compute the probability distribution of B. The formulation is not shown here for the lack of space.

Next, we present an algorithm summarizing the steps described above. $\mu_k(l)$ and $\lambda_k(l)$ are load dependent service and arrival rates to node k, $k = 0, \ldots, K$, with l being the number of parts in that node.

Algorithm

1. 1.0 Initialize $E(A_s)$ and $V(A_s)$. Also let $\mu_k(l) = \mu_k$, $k = 0, \ldots, K$, $l = 0, \ldots, \Omega$.
   1.1 Compute the steady state probabilities of the CQN of the work-cell and MT with closed product form solution.
2. 2.0 Initialize $p(I \geq r_0)$, $E(\theta(r_0, I) \mid I < r_0)$ and $V(\theta(r_0, I) \mid I < r_0)$.

2.1 Compute E(W) and V(W).

3. 3.1 Compute $P(N_W \mid I)$.

   3.2 Compute the transition probabilities of the process $\{I_n \mid n \geq 0\}$. Then compute $P(I = i)$.

   3.3 Compute new values for $E(\theta(r_0, I) \mid I < r_0)$ and $V(\theta(r_0, I) \mid I < r_0)$. Go back to step 2.1.

4. 4.1 Compute E(B) and V(B).

   4.2 Using Eq. (1) compute $E(A_s)$. Also compute $V(A_s)$.

5. 5.1 For each node k, $k = 0, \ldots, K$, form a $C_2^{(n)}/C_2/1/\Omega/\Omega$ queue where the parameters $\lambda_k(n)$ and C_{ak}^2 for the arrival process are obtained in Step 1.2. For $k = 0$ (MT-node) the parameters for the service process are computed using $E(A_s)$ and $V(A_s)$.

   5.2 Compute the steady state probabilities $P_k(l)$ for each of the above queues.

   5.3 Compute load dependent service rates $\mu_k(l)$ using $\mu_k(l) = \lambda_k(l-1)\frac{P_k(l-1)}{P_k(l)}$.

   5.4 Go back to Step 1.1.

The above algorithm has two loops, an outer loop and an inner loop. The inner loop stops when the value of $P(I \geq r_0)$ stabilizes. The outer loop stops when the total number of parts in the work-cell and MT is close to Ω, and the number of parts in the individual nodes from two consecutive iterations are close.

3 Numerical Results

The algorithm presented in the last section has been implemented and run for a number of cases some of which are shown here. The program requires input values for C, K, N, r_0, R, μ_0, μ_k, $k = 0, \ldots, K$, SCV for all workstations and the MT, λ, and $[p_{ij}]$. The program gives output for the following performance measures: Throughput rate (TH), utilization of each workstation (ut_i), utilization of the MT (ut_0), the average number of parts in each queue (L_{q_i}), the average number of parts in each workstation (L_{s_i}), the expected batch size ($E(B)$), and the expected time required to accumulate r_0 parts in the output buffer ($E(\theta(r_0, I) \mid I < r_0)$).

To determine the accuracy of our results we have also developed a simulation program with the same basic assumptions as in our analytical model. The simulation results are compared to the approximate results while changing the input parameters. We have tested many cases. Interested readers are referred to Jafari and Ruan [1992]. Here, we present the results of a few cases. In each case the approximation result is placed above the simulation result.

References

1. Baynat, B., and Dallery, Y.: "A Unified View Of Production-Form Solution Techniques For General Closed Queueing Networks," 1991.
2. Bitran, G. R., and Tirapati, D.: "Multi-Product Queueing Networks with Deterministic Routing: Decomposition Approach and the Notion of Interference," *Management Science*, 34, 1, 1988.
3. Jafari, M.A. and Ruan, S.Y.: "Approximate Closed Queueing Network Model of a Manufacturing Cell With a Material Handling System", Working Paper, Dept. of I.E., Rutgers University, 1992.
4. Ruan, S. and M.A. Jafari: "An Approximation Model for a Manufacturing Cell Attended By a Material Handling System", Queueing Networks with Finite Capacity, R.O. Onvural and I.F. Akyildiz (Editors), North-Holland, 225-238, 1993.

Table 1. Changing N

C	K	N	r_0	R	μ_0	μ_i	λ	TH	ut_i	ut_0	L_{q_i}	L_{s_i}	E(B)	$E(\theta)$
6	5	12	2	8	2.00	1.00	0.50	0.72	0.72	0.42	2.43	3.15	3.20	2.70
								0.74	0.74	0.42	2.43	3.16	2.17	2.72
6	5	10	2	8	2.00	1.00	0.50	0.70	0.70	0.41	2.005	2.75	3.16	2.79
								0.71	0.71	0.41	2.06	2.80	2.17	2.80
6	5	7	2	8	2.00	1.00	0.50	0.65	0.65	0.40	1.53	2.18	3.08	3.01
								0.66	0.66	0.40	1.49	2.15	2.15	3.01
6	5	5	2	8	2.00	1.00	0.50	0.61	0.61	0.39	1.18	1.79	3.01	3.18
								0.62	0.62	0.40	1.16	1.80	2.14	3.19

Table 2. Changing R

C	K	N	r_0	R	μ_0	μ_i	λ	TH	ut_i	ut_0	L_{q_i}	L_{s_i}	E(B)	$E(\theta)$
2	4	10	1	2	1.00	1.00	1.00	0.66	0.66	0.93	1.86	2.52	1.71	1.36
								0.67	0.67	0.94	1.71	2.38	1.66	1.40
2	4	10	1	3	1.00	1.00	1.00	0.68	0.68	0.90	1.96	2.64	2.02	1.36
								0.70	0.70	0.92	2.01	2.71	1.86	1.37
2	4	10	1	4	1.00	1.00	1.00	0.69	0.69	0.89	2.01	2.70	2.17	1.35
								0.71	0.71	0.92	2.07	2.78	1.92	1.37
2	4	10	1	5	1.00	1.00	1.00	0.69	0.69	0.88	2.03	2.73	2.24	1.35
								0.72	0.72	0.92	2.12	2.84	1.95	1.36

Table 3. Changing μ_0

C	K	N	r_0	R	μ_0	μ_i	λ	TH	ut_i	ut_0	L_{q_i}	L_{s_i}	E(B)	$E(\theta)$
4	4	10	2	5	1.00	1.00	1.00	0.65	0.65	0.90	1.79	2.43	3.52	2.63
								0.69	0.69	0.92	1.88	2.57	2.99	2.70
4	4	10	2	5	2.00	1.00	1.00	0.74	0.74	0.62	2.48	3.22	2.82	2.62
								0.74	0.74	0.63	2.42	3.16	2.31	2.70
4	4	10	2	5	3.00	1.00	1.00	0.75	0.75	0.44	2.62	3.37	2.49	2.62
								0.74	0.74	0.44	2.53	3.28	2.14	2.68
4	4	10	2	5	5.00	1.00	1.00	0.76	0.76	0.27	2.72	3.48	2.23	2.62
								0.75	0.75	0.27	2.59	3.35	2.05	2.67

Table 4. Changing μ_i

C	K	N	r_0	R	μ_0	μ_i	λ	TH	ut_i	ut_0	L_{q_i}	L_{s_i}	E(B)	$E(\theta)$
4	4	10	2	5	1.00	1.00	1.00	0.65	0.65	0.90	1.79	2.43	3.52	2.63
								0.69	0.69	0.92	1.88	2.57	2.99	2.70
4	4	10	2	5	1.00	0.50	1.00	0.37	0.73	0.85	2.45	3.18	3.01	5.21
								0.36	0.73	0.88	2.34	3.08	2.51	5.44
4	4	10	2	5	1.00	1/3	1.00	0.25	0.75	0.82	2.59	3.34	2.75	7.83
								0.24	0.74	0.85	2.40	3.15	2.34	8.14
4	4	10	2	5	1.00	0.2	1.00	0.15	0.76	0.80	2.70	3.45	2.49	13.06
								0.15	0.75	0.83	2.55	3.31	2.20	13.50

Optimizing the Transient Behavior of Hedging Control Policies in Manufacturing Systems

Sami El-Férik[1] *and Roland P. Malhamé*[1,2]

[1] Ecole Polytechnique de Montréal, Montréal, Québec H3C-3A7
[2] Gerad, Montréal, Québec
e-mail: malhame@auto.polymtl.ca

1 Introduction

Starting with the seminal work of Kimemia and Gershwin [3], the class of hedging type control policies in manufacturing systems was recognized as very likely to comprise in many situations the optimal control policy. In the single part multiple state manufacturing system case, hedging policies are characterized by a set of critical inventory levels -- one associated with each feasible machine state -- towards which one must converge as quickly as possible. Using a dynamic programming framework Akella and Kumar [1] established rigorously the optimality of such a policy in the two state, single part, constant demand manufacturing system case. They also solved for the critical inventory level. Subsequently, Bielecki and Kumar [2] and Sharifnia [7] recognized that it was easier to solve a parameter optimization problem within the class of hedging control policies, rather than a full-fledged optimal control problem, and did so in the two-state and multiple-state machine cases respectively. All existing results however were based on optimizing the statistical steady-state of an ergodic system, and thus essentially dealt with infinite horizon problems.

In this paper we use extensions of the renewal theoretic framework first developed in [4], to derive an integral equation associated with an integral type performance functional over finite horizon, for the single part, multiple machine state case. The kernel of that integral equation is the probability density function associated with the time to first return to a given hedging level (in the corresponding machine state). The main quantities in the integral equation are obtained as solutions of coupled partial differential equations with appropriate boundary conditions. This framework is used to derive a theorem which constitutes a rather intuitive bound on the speed with which the time average of the cost functional converges to its ergodic steady-state (based on which existing optimal solutions have been obtained). It states essentially that this speed is bounded above by the settling speed of the statistics of the underlying machine failure state Markov chain. Subsequently, closed form expressions are obtained for the asymptotic cost functionals associated with hedging control policies in the particular case of a two-state manufacturing system, over a possibly large but finite time horizon. These expressions are used to optimize the choice of hedging levels over a finite horizon. The minimal length of the horizon for the asymptotic analysis to hold is characterized in terms of the parameters of the failure mode Markov chain.

2 An integral equation for finite horizon cost

We use the modeling framework of Sharifnia [7]. The parts production process is represented by a fluid flow with random disruptions. More specifically:

$$\frac{dx}{dt} = u_\alpha - d$$

$x(t)$: parts surplus at time t

u_α : production rate in failure mode α

d : rate of demand for parts

- in mode α, u_α is bounded by $\hat{u}_\alpha$
- α evolves according to an irreducible finite state Markov chain with state transition intensity matrix Λ.
- $\alpha = 1, \ldots, m$ feasible (i.e. $\hat{u}_\alpha > d$)
  $\alpha = m + 1, \ldots, n$ infeasible.

A finite horizon integral cost J is considered:

$$J = E\left[\int_0^T \ell(x_t)\,dt \mid x_0, \alpha_0\right]$$

where $\ell(x_t)$ is the cost per unit time at surplus level x_t. This cost is to be minimized.

The hybrid probability density vector $\mathbf{f}(x,t) = [f_1(x,t), \ldots, f_n(x,t)]^T$ evolves according to [4]:

$$\frac{\partial \mathbf{f}(x,t)}{\partial t} = -V_i \frac{\partial \mathbf{f}(x,t)}{\partial x} + \Lambda' \mathbf{f}(x,t), \quad \forall Z_{i+1} < x < Z_i, \quad i = 1, \ldots, m, \tag{1}$$

at hedging levels $Z_1, Z_2, \ldots, Z_m$ and where $Z_{m+1} = -\infty$. The boundary conditions at hedging points are given by:

$$\lambda_{ij} P_{Z_i}(t) + v_j^i(t) f_j(Z_i^-, t) - v_j^{i-1}(t) f_j(Z_i^+, t) = 0 \text{ for } j = 1, 2, \ldots, n;\ j \neq i \tag{2a}$$

$$\frac{dP_{Z_i}(t)}{dt} = \lambda_{ii} P_{Z_i}(t) + v_i^i(t) f_i(Z_i^-, t) - v_i^{i-1}(t) f_i(Z_i^+, t) \text{ for } j = i \tag{2b}$$

where the P_{Z_i}'s are the probability masses at Z_i, $i = 1, 2, \ldots, m$. In addition

$$\mathbf{f}(x,t) = 0, \quad \forall x > Z_1; \qquad \lim_{x \to -\infty} \mathbf{f}(x,t) = 0. \tag{2c}$$

Finally note that:

$$v_\alpha^i(t) = u_\alpha^i(t) - d \text{ with } u_\alpha^i = \begin{cases} 0 & \text{if } \alpha > i \\ \hat{u}_\alpha & \text{if } \alpha \leq i \end{cases}, \qquad V^i = \operatorname{diag}\left[v_\alpha^i\right].$$

In [4], a Markov chain embedded into the hybrid state Markov process is defined by focusing on the successive instants where hedging points are reached and recording thereupon the corresponding machine state. When the machine state is in correspondance with the hedging point, the regeneration point will be called a hedging point state of the embedded Markov chain.

Theorem 1. *Let i be a hedging point regeneration state from the embedded Markov chain [4]. The cost functional conditional to the initial state i $C_i(T)$ satisfies the following integral equation:*

$$C_i(T) = \int_0^T E\left[\ell(x(\tau))|_{\alpha=i,x(0)=Z_i}\right] d\tau + \int_0^T C_i(T-\tau)g_{ii}(\tau)\,d\tau \qquad (3)$$

where in (3) $g_{ii}(t)$ is the p.d.f. of the first return of $x(t)$ to Z_i with machine state $\alpha = i$.

Thus the first term in the right-hand side of (3) represents the contribution of all the possible system trajectories up to the first passage-time, as well as those that never reached Z_i in state i at time t. In what follows, we show how one can obtain $g_{ii}(t)$ as the solution of (1)-(2) but subject to slight modifications in the boundary conditions. Computation of the p.d.f of the first return to Z_i in machine state i proceeds as follows.

- set the appropriate initial p.d.f's and probability $P_{Z_\alpha}(0)$. ex: $\mathbf{f}(x,0) = 0$, $P_{Z_i}(0) = 1$, $P_{Z_j}(0) = 0, \forall j \neq i$,
- set the appropriate absorbing boundary conditions relative to the initial state i by *eliminating the probability exchange term at the right-hand side of (2b).* Thus

$$\lambda_{ij}P_{Z_i}(t) + v_j^i(t)f_j(Z_i^-,t) - v_j^{i-1}(t)f_j(Z_i^+,t) = 0 \text{ for } j = 1,2,\ldots,n; \quad j \neq i$$

but:

$$\frac{dP_{Z_i}(t)}{dt} = \lambda_{ii}P_{Z_i}(t) \text{ for } j = i$$

and:

$$g_{ii}(t) = v_i^i(t)f_i(Z_i^-,t) - v_i^{i-1}(t)f_i(Z_i^+,t). \qquad (4)$$

Theorem 2. *The eigenvalues of the intensity matrix Λ of the machine state Markov chain are also solutions of the equation:*

$$1 - g_{ii}^*(s) = 0$$

where $g_{ii}^*(s)$ is the Laplace transform of $g_{ii}(t)$.

All proofs are omitted but can be found in [6].

Using integral equation (3) and theorem 2, it is possible to conclude that the eigenvalues of Λ are poles of the Laplace transform expression of the cost functional. This implies that if one considers the time average $\frac{1}{T}C_i(T)$ performance functional as in [2] or [1], it cannot reach steady-state any quicker than the statistics of the underlying machine state Markov chain. This puts a lower bound on the length of time horizon before one can hope that the ergodic optimum hedging points of [2] or [7] become acceptable approximations of the true optima. Thus we have:

Corollary 3. *In the ergodic case $\left(\frac{C_i(T)}{T} - C_\infty\right)$ is at least $O(\exp(\lambda_{\min}t))$ where $\lambda_{\min}$ is the dominant eigenvalue of the intensity matrix of the machine state Markov chain, and C_∞ is the steady-state average cost per unit time.*

We now give a complete treatement of the asymptotics of the linear cost functional for the two-state machine case. While for lack of space as well as simplicity, we use specific numerical values for the dynamic model, the analysis can be carried out in full generality (see [6]).

3 Asymptotic optimization of the two-state machine

We consider a sample example with two modes. The discrete system state evolves according to a Markov chain with intensity matrix:

$$\Lambda = \begin{pmatrix} -1 & 1 \\ 2 & -2 \end{pmatrix}.$$

Furthermore let: $\hat{u}_1 = 2$, $\hat{u}_2 = 0$, $d = 1$.

Because only one state is feasible, hedging policies are characterized by a single hedging point denoted Z. Let $\ell(x)$, the cost per unit time be c^+x if $x > 0$ and c^-x if $x < 0$. Application of the integral equation (3), together with the first passage-time computation in (4) yields:

$$\begin{aligned} C^*(s) = \frac{2s+5-\sqrt{1+4s(s+3)}}{2s^2(s+3)} &\left[\frac{2(c^+ + c^-)}{1+\sqrt{1+4s(s+3)}}\right. \\ &\exp\left(-(\frac{1+\sqrt{1+4s(s+3)}}{2}Z\right) + c^+Z \\ &\left. - \frac{1c^+}{1+\sqrt{1+4s(s+3)}}\right] + \frac{c^+Z}{s^2(s+3)}\frac{1+\sqrt{1+4s(s+3)}}{2} \end{aligned} \tag{5}$$

where it is assumed that at $t = 0$, $x(0) = Z$, the machine is in state 1, and $C^*(s)$ is the laplace transform of the cost functional.

The singularities of that Laplace transform are a mixture of poles at $s_1 = s_2 = 0$, $s_3 = -3$ and branching points due to the square root term $\sqrt{1+4s(s+3)}$ at $s_4 = -0.0857864$ and $s_5 = -2.9142$ respectively. The asymptotic behavior of $C(T)$ as T grows indefinitely can be obtained by considering the residues of the poles at zeros, as well as the most dominant branching point (i.e the closest to the $j\omega$ axis). Following [5], we neglect any integer powers in the asymptotic expansion of $C^*(s)$ around the branching point to obtain:

$$\begin{aligned} C^*(s) \approx &\left[0.6667(c^+ + c^-)\exp(-Z) + c^+(Z - 0.6667)\right]\frac{1}{s^2} \\ &- \left[2.88889(c^+ + c^-)\exp(-Z) + 2(c^+ + c^-)Z\exp(-Z) - 2.8889c^+\right]\frac{1}{s} \\ &+ \left[c^+(135.882Z - 225.137) + 225.17(c^+ + c^-)\exp\left(-\frac{Z}{2}\right)\right] \\ &- \left[914.103(c^- + c^+)(1 + 0.414214Z)\exp\left(-\frac{Z}{2}\right) - 914.103c^+\right]\sqrt{s + 0.0857864} \\ &+ 178.49\left[\left(-178.243(c^+ + c^-)\exp\left(-\frac{Z}{2}\right) + 178.243c^+\right) - \right. \\ &\left.(78.9522Z(c^+ + c^-) + 7.24264Z^2(c^+ + c^-) + Z^3(c^+ + c^-))\exp\left(-\frac{Z}{2}\right)\right] \\ &(s + 0.0857864)^{\frac{3}{2}} + O\left[(s + 0.0857864)^{\frac{5}{2}}\right] \end{aligned} \tag{6}$$

This in turn yields:

$$
\begin{aligned}
\frac{c(T)}{T} &= \left\{0.6667(c^+ + c^-)\exp(-Z) + c^+(z - 0.6667)\right\} \\
&- \left[2.88889(c^+ + c^-)\exp(-Z) + 2(c^+ + c^-)Z\exp(-Z) - 2.8889c^+\right]\frac{1}{T} \\
&+ \frac{\exp(-0.0857864T)}{T}\left\{c^+(135.882z - 225.137) + 225.17(c^+ + c^-)\exp\left(-\frac{Z}{2}\right)\right. \\
&- \left[914.103(c^- + c^+)(1 + 0.414214Z)\exp\left(-\frac{Z}{2}\right) - 914.103\right]\frac{1}{\sqrt{\pi T}} \\
&+ 178.49\left[\left(178.243(c^+ + c^-)\exp\left(-\frac{Z}{2}\right) - 178.243c^+\right)\right. \\
&+ \left.\left(78.9522z(c^+ + c^-) + 7.24264z^2(c^+ + c^-) + Z^3(c^+ + c^-)\right)\exp\left(-\frac{Z}{2}\right)\right] \\
&\left.\frac{1}{2\sqrt{\pi T^3}} + O\left[\frac{1}{\sqrt{\pi T^5}}\right]\right\} \qquad (7)
\end{aligned}
$$

where in (7) the constant term coïncides with the Bielecki-Kumar limit [2].

Using (7), for $T > \frac{2}{0.08576} = 23.3137$, we seek to optimize via a steepest descent algorithm the cost functional in (7). The results for the cost, and the optimal hedging point as a function of the length of the optimization horizon are shown in Fig. 1 and 2 below. Also, note that we include a figure (Fig. 3) illustrating the behavior of the closest branching point (from the jw axis) as the ratio $r = \frac{d}{\pi_1 \hat{u}_1}$ is varied from 0.01 to 0.9, where $\pi_1 \hat{u}_1$ is the average maximal production capacity. Recall that the position of this branching point governs the speed at which convergence to steady-state occurs.

4 Conclusion

We have extended the renewal theoretic framework of single part multiple state manufacturing systems under hedging policies first developed in [4]. This was used to asymptotically optimize hedging point policies over finite horizons for two-state systems. In future work, we will report on the theory for three state manufacturing systems.

References

1. R. Akella and P. R. Kumar. "Optimal control of production rate in a failure prone manufacturing system". *IEEE Transactions on Automatic Control*, AC-31:pp.116–126, February 1986.
2. T. Bielecki and P. R. Kumar. "Optimality of zero-inventory policies for unreliable manufacturing systems". *Operat. Res.*,, vol. 26:pp. 532–540, July-Aug. 1988.
3. J. B. Kimemia and S. B. Gershwin. "An algorithm for the computer control of production in flexible manufacturing systems". *IIE Transactions*,, vol. 15 (no 4):pp. 353–362, Dec. 1983.
4. R. P. Malhamé and E. K. Boukas. "A renewal theoretic analysis of a class of manufacturing systems". *IEEE Transactions on Automatic Control*,, AC-36 (no 5):pp. 353–362, May 1991.

5. B. Van Der Pol and H. Bremmer. *"Operational Calculus Based on the Two Sided Laplace Integral"*. The Sybdics of the Cambrige University Press, New York, 1964.
6. S. El-Férik and R. P. Malhamé. "optimizing the transient behavior of hedging control policies in failure prone manufacturing systems ". *to appear as a GERAD report.*
7. A. Sharifnia. "production control of manufacturing system with multiple machine states". *IEEE Transactions on Automatic Control,*, AC-33 (no 7):pp. 620–625, July 1988.

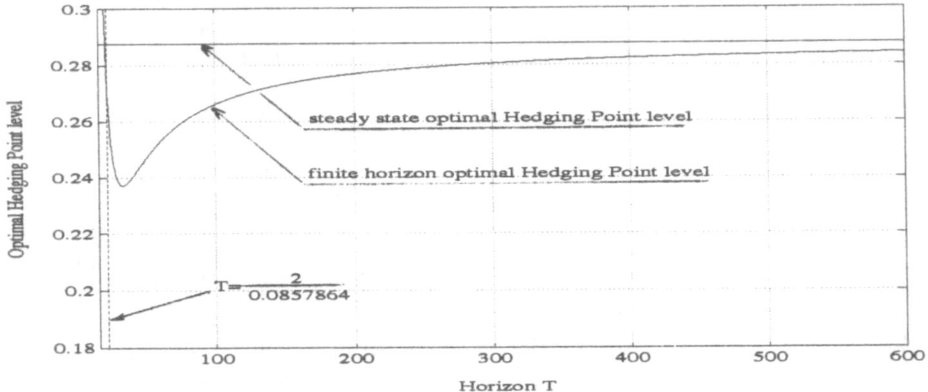

Fig. 1. Identification of the optimal hedging point level of the two-state machine case

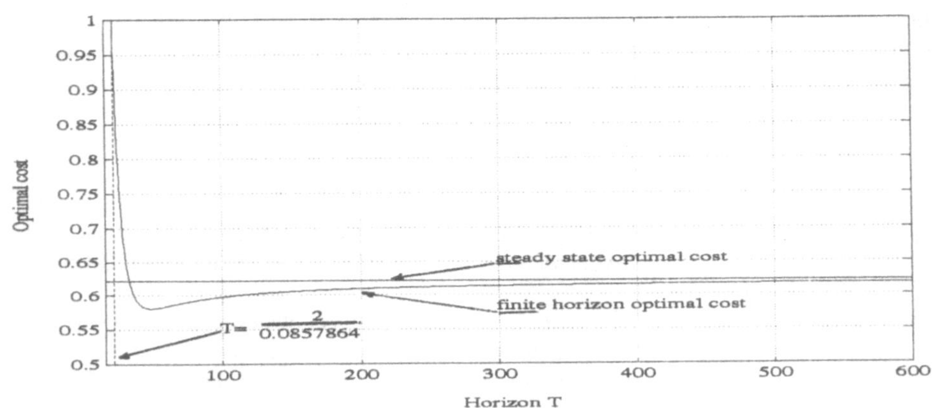

Fig. 2. The optimal finite horizon average cost of the two-state machine case

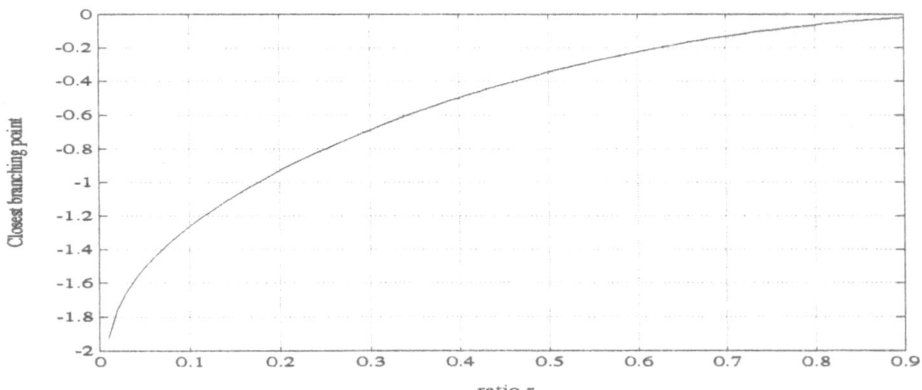

Fig. 3. The behavior of the closest branching point (from the jw axis) as $r = \frac{d}{\pi_1 \hat{u}_1}$ is varied

Finding Optimal Number of Kanbans in a Manufacturing System via Stochastic Approximation and Perturbation Analysis

Houmin Yan[1]*, *Xun Yu Zhou*[1]* and *G. Yin*[2]†

[1] Department of Systems Engineering, The Chinese University of Hong Kong, Hong Kong
[2] Department of Mathematics, Wayne State University, Detroit, MI 48202.

1 Introduction

Many manufacturing systems are large systems characterized by the presence of random interferences such as machine failures, demand fluctuations, stochastic setup times, uncertain yields, and so on. These uncertain factors have made it particularly difficult for managers to plan production, to regulate inventories, and to meet customer demand. However, by using the Just-In-time (JIT) system implemented with Kanbans, Toyota has managed to reduce work-in-process (WIP) and cycle time even in the presence of the uncertainties mentioned above. The success of this easily implementable approach and the desire for manufacturing competitiveness have stimulated a great deal of research in this area. An important issue in the implementation of the Kanban system is to design both the location and the number of circulating Kanbans, especially when the manufacturing system experiences random interferences such as random machine failures and stochastic demands. This problem has not been solved.

In order to reduce computational efforts, a technique called perturbation analysis which enables gradient estimates to be calculated based on a single simulation run was developed by Ho and his co-workers (see Ho and Cao [HC1] and Glasserman [GL1], see also Caramanis and Liberopoulos [CL1], Suri and Fu [SF1, SF2] and Yan, Yin and Lou [YYL, YY1]). In this paper, we investigate a manufacturing system with two tandem failure-prone machines. First, we illustrate how to devise gradient estimators based on the observations in a single simulation run and then design an iterative algorithm, a constant step-size stochastic approximation procedure, for finding the optimal number of circulating Kanbans. It is shown that the gradient estimators converge in probability and the iterative algorithm converges to the optimal value in an appropriate sense. A numerical example is presented to demonstrate the performance of the proposed algorithms.

* Research was supported in part by The Chinese University of Hong Kong under Direct Grant 220500850 and 220500660.

† Research was supported in part by the National Science Foundation under grant DMS-9224372, and in part by Wayne State University.

2 A Two-Machine Tandem System

Consider a manufacturing system consisting of two machines in tandem producing a single product. We use $u_1(t)$ and $u_2(t)$ to denote the production rates of the first and the second machine, respectively. The demand rate for the product is d. The machines are subject to breakdown and repair over an infinite horizon. Each machine is assumed to have two states: up and down. When a machine is down, its production rate is zero; when it is up, its production rate can be adjusted to any value between zero and a maximum rate $u_{i\max}$, $i = 1, 2$. Define the machine capacity process by $\alpha_i(t) = 1$, if machine i is up; $\alpha_i(t) = 0$, otherwise. The production rates (controls in this problem) must then satisfy $0 \le u_i(t) \le u_{i\max}$, $i = 1, 2$. As is customary, assume that the maximum capacity of each of the machines is greater than the demand, i.e., $\min\{u_{1\max}, u_{2\max}\} > d$. Without loss of generality, assume that $u_{1\max} > u_{2\max}$.

Denote the inventory level in the buffer following the first machine by $x_1(t)$ and the surplus (inventory/backlog) level of the single product by $x_2(t)$. Then the system dynamics and the state constraints can be described as

$$\dot{x}_1(t) = u_1(t) - u_2(t) \ , \quad \dot{x}_2(t) = u_2(t) - d \ , \quad x_1(t) \ge 0 \ . \tag{1}$$

Note that no constraint is imposed on $x_2(t)$. It can be either positive or negative, corresponding to an inventory or a backlog, respectively. A Kanban-control is defined as a function of the system state (x_1, x_2) and the machine capacity (α_1, α_2) in the following way:

$$u_2(x_1, x_2, \alpha_1, \alpha_2) = \begin{cases} d\alpha_2 & \text{if } x_2 = B_2 \text{ and } (x_1 > 0 \text{ or } \alpha_1 = 1); \\ u_{2\max}\alpha_2 & \text{if } x_2 < B_2 \text{ and } (x_1 > 0 \text{ or } \alpha_1 = 1); \\ 0 & \text{otherwise}; \end{cases} \tag{2}$$

$$u_1(x_1, x_2, \alpha_1, \alpha_2) = \begin{cases} u_2(x_1, x_2, \alpha_1, \alpha_2)\alpha_1 & \text{if } x_1 = B_1; \\ u_{1\max}\alpha_1 & \text{if } x_1 < B_1; \\ 0 & \text{otherwise}, \end{cases} \tag{3}$$

where $B_i \ge 0$, $i = 1, 2$, are called " threshold levels". They are also referred to as "threshold values", "numbers of circulating Kanbans" or " numbers of Kanbans". These terms will be used interchangeably in the sequel.

Remark. We assume that the production rate can be adjust between zero and the maximum capacity. Therefore, in such a setting, *numbers of Kanbans* may not be an integer. The relationship between the continuous and discrete production lines has been investigated by Suri and Fu [SF1]. They demonstrated that continuous production models can be used to approximate discrete lines.

The objective of the system is to find the optimal number of circulating Kanbans $B = (B_1, B_2) \in \mathbb{R}^2$ that minimizes the cost function

$$J(B) = \lim_{T\to\infty} \frac{1}{T} E \int_0^\infty [c_1 x_1(t) + c_2^+ x_2^+(t) + c_2^- x_2^-(t)]\, dt \ , \tag{4}$$

where $x_2^+ = \max\{0, x_2\}$ and $x_2^- = \max\{0, -x_2\}$.

Define a joint random process $\xi(t) = (x(t), \alpha(t))$. As in (Bielecki and Kumar [BK1] and Yan, Yin and Lou [YYL, YY1]), throughout the paper, it will be assumed that there exists an invariant measure $P^B(\cdot)$ for the joint process $\xi(t)$. By using $P^B(\cdot)$, the cost function can be written as

$$J(B) = \int [c_1 x_1 + c_2^+ x_2^+ + c_2^- x_2^-] P^B(dx) \ . \tag{5}$$

In the following sections, we first investigate how to calculate the gradient estimates of the cost functional $J(B)$ with respect to B, and then find the optimal value B^* by using a stochastic approximation algorithm.

3 Sample Gradients

First, the representation of sample gradients is given. The sample gradients are obtained by studying the difference between the nominal paths and the perturbed paths. As in [YYL] for each T and B, define

$$L_T(B) = \frac{1}{T} \int_0^T [c_1 x_1(t) + c_2^+ x_2^+(t) + c_2^- x_2^-(t)]\, dt \ .$$

The sample gradient is defined by

$$h_{T,i}(B) = \lim_{\Delta B_i \to 0} \frac{L_T(B + \Delta B_i e_i) - L_T(B)}{\Delta B_i} \ , \quad i = 1, 2 \ . \tag{6}$$

Notice that both $L_T(\cdot)$ and $h_T(\cdot)$ depend on $\xi(s)$, for $0 < s \leq T$. Although it is not really needed to figure out the form of h_T for simulation or computation purpose, it is necessary for studying the asymptotic properties of the algorithms in the sequel. In Fig. 1, nominal paths $x_1(t)$ and $x_2(t)$ (solid lines) and perturbed paths $x_1^{\Delta B_1}(t)$ and $x_2^{\Delta B_1}(t)$ (dashed lines) arising from a perturbation $\Delta B_1 > 0$ on the threshold value B_1 are depicted. We point out two important features of this perturbation generation-propagation process.

1. If two or more propagation periods overlap, the effect of perturbation propagation may be cumulative. During a time interval in which k propagation periods overlap, the perturbed paths will be shifted by the amount $k\Delta B_1$. One such example is the period between τ_3 and τ_4, where two perturbation propagation periods overlap.
2. The interactions between the sample paths $x_1(t)$ and $x_2(t)$ can be cumulative as well. That is, an up (or down) shift of $x_2(t)$ can be propagated to $x_1(t)$ to cause the same amount of up (or down) shift of $x_1(t)$, which in turn can be propagated back to $x_2(t)$, and so on. But this "chain" will eventually be broken. For example, the chain breaks at τ_7 in Fig. 1.

Definition 1. $p_1(t)$ and $p_2(t)$ are piecewise constant functions that are right continuous, and have left limits such that

1. $p_1(0) = p_2(0) = 0$;
2. If boundary A is hit by $\mathbf{x}(t)$ at time t_A, then $p_1(t_A) = \Delta B_1$, $p_2(t_a) = p_2(t_A^-)$;

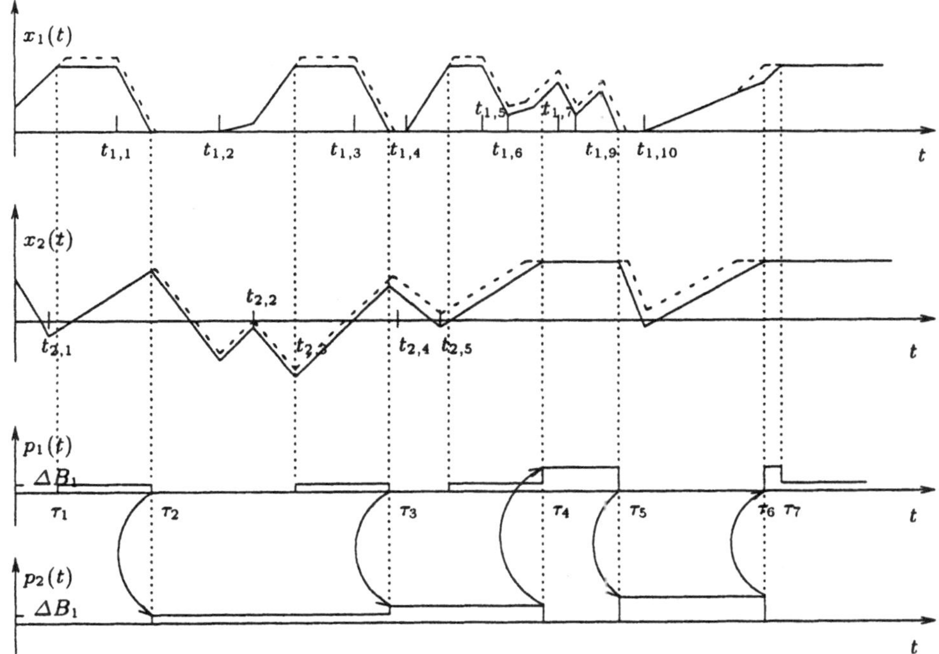

Fig. 1. Sample paths for tandem two-machine system (Kanban-control)

3. If boundary B is hit by $\mathbf{x}(t)$ at time t_B, then $p_1(t_B) = 0$, $p_2(t_B) = p_2(t_B^-) + p_1(t_B^-)$;
4. if boundary C is hit by $\mathbf{x}(t)$ at time t_C, then $p_1(t_C) = p_1(t_C^-) + p_2(t_C^-)I_{\{x_1(t_C)<B_1\}}$, $p_2(t_C) = 0$.

For $k = 1, 2, \ldots,$ define

$$\begin{aligned} \mathcal{A}^k &= \{t \mid p_1(t) = k\Delta B_1\} \ , & \mathcal{B}^k &= \{t \mid p_2(t) = k\Delta B_1\} \ , \\ \mathcal{B}^{k+} &= \{t \mid t \in \mathcal{B}^k, x_2(t) \geq 0\} \ , & \mathcal{B}^{k-} &= \{t \mid t \in \mathcal{B}^k, x_2(t) < 0\} \ . \end{aligned} \tag{7}$$

In view of the discussion above, by assuming that the longest perturbation chain M is finite, we arrive at

Lemma 2. *Suppose that $M < \infty$ is the length of the longest perturbation propagation chain. With probability one, for each fixed $T > 0$,*

$$\begin{aligned} \bar{h}_{T,1}(B, \xi(t)) &= \sum_{k=0}^{M} \left[c_1 k I_{\{t \in \mathcal{A}^k\}} + c_2^+ k I_{\{t \in \mathcal{B}^{k+}\}} - c_2^- k I_{\{t \in \mathcal{B}^{k-}\}} \right] \ , \\ \bar{h}_{T,2}(B, \xi(t)) &= c_2^+ I_{\{x_2(t) \geq 0\}} - c_2^- I_{\{x_2(t) < 0\}} \ . \end{aligned} \tag{8}$$

A stochastic optimization technique will be used to develop iterative algorithms to search for the optimal threshold value B^* as follows. A sequence $\{B^n\}$ is generated and the recursive algorithm is given by:

$$B^{n+1} = B^n - \frac{\epsilon}{T} \int_{nT}^{nT+T} \bar{h}_T(B^n, \xi(s))\, ds \ , \tag{9}$$

where $\epsilon > 0$ is a small constant step size, $T = T^\epsilon \to \infty$ as $\epsilon \to 0$, $B^n = (B_1^n, B_2^n)$ and $\bar{h}_T = (\bar{h}_{T,1}, \bar{h}_{T,2})$.

4 Recurrency, Consistency and Convergence

In this section, we demonstrate that the auxiliary processes $p_1(t)$ and $p_2(t)$, which are defined by the nominal paths only, can be used to represent the sample gradient estimators. it has been shown that

$$\lim_{\Delta B_1 \to 0} \lim_{T \to \infty} \frac{1}{T} E \int_0^T \left(\frac{c_1 p_1(t)}{\Delta B_1} + \frac{c_2^+ p_2(t) \cdot I_{\{x_2(t)>0\}}}{\Delta B_1} \right.$$
$$\left. + \frac{(c_2^+ - c_2^-) p_2(t) \cdot I_{\{0>x_2(t)>-\Delta B_1\}}}{\Delta B_1} - \frac{c_2^- p_2(t) \cdot I_{\{x_2(t) \le -\Delta B_1\}}}{\Delta B_1} \right) dt \quad (10)$$

can be used to replace $\frac{\partial J}{\partial B_1}$ if the following capacity condition satisfies [YZY].

Lemma 3. *Inventory-state* $(x_1(t) = B_1, x_2(t) = B_2)$ *is recurrent the following holds:*

$$\frac{u_{2\max} - d}{\lambda_1 + \lambda_2} - \frac{d}{\mu_1 + \mu_2} > 0, \quad \textit{and similarly} \quad \frac{u_{1\max} - u_{2\max}}{\lambda_1} - \frac{u_{2\max}}{\mu_1} > 0 .$$

Theorem 4. *Let* $\pi(B, \xi)$ *denote a function which can be any one of the indicator functions in (8). Under the condition of Lemma 3, suppose the limit in probability of* $(1/T) \int_0^T \pi(B, \xi(t)) dt$ *exists. Then* $\partial L_T(B)/\partial B \xrightarrow{T} \partial J(B)/\partial B$ *in probability.*

Theorem 5. *Let the conditions of Theorem 4 be satisfied and let the differential equation*

$$\dot{B} = -\nabla J(B) \quad (11)$$

have a unique solution for each initial condition $B(0)$. *Define* $\{B^\epsilon(\cdot)\}$ *by* $B^\epsilon(t) = B^n$ *for* $t \in [\epsilon n, \epsilon n + \epsilon)$. *Suppose* $B^{\epsilon,0} \to B(0)$. *Then* $\{B^\epsilon(\cdot)\}$ *is tight in* $D^2[0, \infty)$ *and any weakly convergent subsequence has limit* $B(\cdot)$ *which is a solution of (11).*

Theorem 6. *In addition to the conditions of Theorem5, suppose that* $V(\cdot)$ *is a Liapunov function for (11) such that the ODE (11) has a unique asymptotically stable point* B^* *(in the sense of Liapunov stability), and* $V_B'(B)\nabla J(B) > \lambda V(B)$ *for some* $\lambda > 0$ *and all* $B \neq B^*$. $V_{BB}(\cdot)$ *is bounded,* $|V_B(B)| \le K(1 + V^{1/2}(B))$ *such that* $V(B) \xrightarrow{|B| \to \infty} \infty$. *Then we have*

- *The set* $\{B^n;\ n < \infty, \epsilon > 0\}$ *is tight.*
- *Let* $t_\epsilon \to \infty$ *as* $\epsilon \to 0$. *Then,* $B^\epsilon(t_\epsilon + \cdot)$ *is tight in* $D^2[0, \infty)$ *and any weak limit is equal to the optimal threshold value* B^*.

Proofs of these theorems are in [YZY].

5 Numerical Example

A numerical example is provided to demonstrate our algorithm. The simulation horizon is divided into $N = 100$ segments, each having length $T = 10,000$. The step size ϵ is set at 0.5. At the end of each segment, sample gradients are calculated, and threshold values are updated according to (8) and (9). The system parameters are $\lambda_1 = 0.1$, $\lambda_2 = 0.125$, $\mu_1 = 0.2$, $\mu_2 = 1$, $u_{1\max} = u_{2\max} = 2$, and $d = 1$ with the initial threshold values being $B_1^0 = B_2^0 = 2$, $B_1^0 = 20$, and $B_2^0 = 10$. Since there is no analytical result available to date, contour curves are generated by conducting simulations for each set of control parameters (threshold values), and the costs under the Kanban-control policy are computed. These curves are displayed in Fig. 2. The algorithms seem to be converging very fast. After a few iterations, the average costs are already close to the optimal ones.

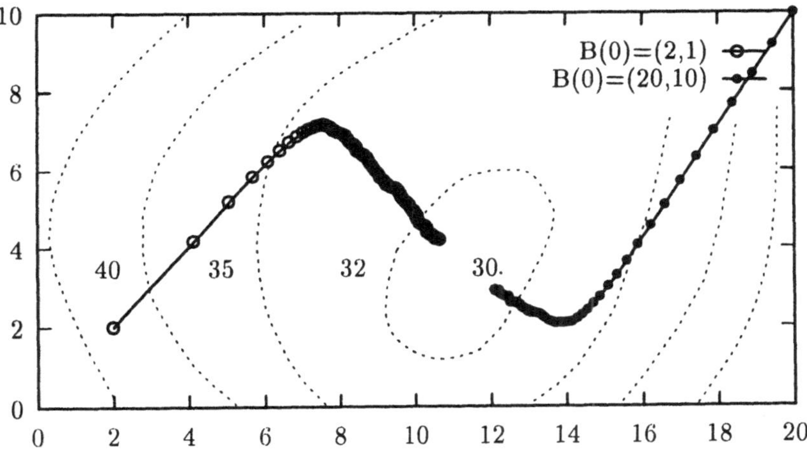

Fig. 2. Contour curves and convergence trajectories

Acknowledgment

The authors would like to thank Professor Harold J. Kushner for stimulating discussions, and for pointing out the possibilities of weakening the conditions on convergence analysis.

References

[BK1] Bielecki, T. and P.R. Kumar: Optimality of zero-inventory policies for unreliable manufacturing systems, *Oper. Res.* **36**, (1988) 532–541.

[CL1] Caramanis, M. and G. Liberopoulos: Perturbation analysis for the design of flexible manufacturing system flow controllers, *Oper. Res.* **40**, (1992) 1107–1125.

[HC1] Ho, Y.C. and X.R. Cao: *Perturbation Analysis of Discrete Event Dynamic Systems*, Kluwer, Boston, MA. 1991.

[GL1] Glasserman, P.: *Gradient estimation via perturbation analysis*, Kluwer, Boston, MA. 1991.

[SF1] Suri, R., and B.R. Fu: On using continuous flow lines for performance estimation of discrete production lines, *Proc. 1991 Winter Simulation Conf.* (1991) 968–977.

[SF2] Suri, R., and B.R. Fu: Using continuous flow models to enable rapid analysis and optimization of discrete production lines — a progress report, *Proc. 19th NFS Grantees Conf. Design and Manufacturing Syst. Res.* Charlotte, NC, Jan. 1993.

[YYL] Yan, H., G. Yin, and S. Lou: Using stochastic optimization to determine threshold levels for control of unreliable manufacturing systems, *J. Optim. Theory Appl.* forthcoming.

[YZY] Yan, H., Zhou, X.Y., and Yin G.: Optimal numbers of circulating Kanbans in a two-machine flowshop with uncertainty, Preprint, 1994.

[YY1] Yin, G., H. Yan and X. Lou: On a class of stochastic optimization algorithms with applications to manufacturing models, in *Model-Oriented Data Analysis*, W.G. Müller, H.P. Wynn and A.A. Zhigljavsky Eds., Physica-Verlag, (1993) 213–226.

Performance Evaluation of a Generalized Kanban System with a General Arrival Process of Demands

Maria Di Mascolo

Laboratoire d'Automatique de Grenoble, URA 228 CNRS, ENSIEG/INPG
BP 46, 38402 Saint Martin d'Hères, France.
email: maria@lag.grenet.fr, Telefax +33 76.82.63.88, Tel: +33 76.82.64.16

1 Introduction

The generalized kanban system is a variant of the kanban system, that has been introduced by Buzacott [3] and includes the classical kanban system as a special case. This paper is concerned with the analysis of such a generalized kanban system when the arrival process of demands is general. It is an extension of our previous work presented in [5] dealing with a generalized kanban system with a Poisson arrival process of demands. This extension will enable to study the influence of the variability of demands arrival on the performance of a generalized kanban system.

Let us first recall that the kanban system is a simple way to implement the pull control policy by transferring the information from the downstream to the upstream of a production system. The principle is to decompose the production system into several stages, and to associate a fixed number of kanbans (cards) with each stage. These kanbans are used as production orders: a part can proceed to a given stage only if there is a kanban available to be attached on it. In that case, the kanban will be detached only when the part is transferred to the next stage.

In the generalized kanban system, which is the subject of this paper, we have the same decomposition into stages, with their associated kanbans, but the way the information is transferred from downstream to upstream is different. Indeed, in the classical kanban system, it is the consumption of a finished part that is transferred, whereas in the generalized kanban system, the consumption of a finished part and the transfer of information are partly dissociated and it is the effective demand that is transferred from downstream to upstream. Moreover, in a classical kanban system, each part always has a kanban attached on it and thus, the number of kanbans associated with a stage is an upper bound on the number of parts in this stage, whereas in the generalized kanban system, the kanbans and the finished parts are not always linked (upon completion of its processing, a finished part is put into an output buffer, but its associated kanban is immediately detached and stored elsewhere). Thus, the maximum number of finished parts in a stage can be different from the number of kanbans associated with the stage.

The kanban system has attracted wide interest in recent years and much work has been devoted to the modeling of such systems, as well as to methods to evaluate their performance (see [4] and [9] for a list of references). The generalized kanban system, however, has been less studied. In [5], we propose an analytical method to evaluate the performance of a generalized kanban system with stages in series and receiving demands according to a Poisson process. The generalized kanban system is modeled by a queueing network with synchronization mechanisms. This queueing network is then analyzed thanks to an approximation method based on decomposition. In the present paper, we show how to extend this method in order to obtain the performance of a generalized kanban system when the arrival process of external demands is no more Poisson but general. In fact, we assume that the interarrival time distribution is a phase-type (or PH) distribution. Indeed, the PH distribution, introduced by Neuts [8], is a good approximation for any general distribution. It consists in a mixture of exponential distributions and enables to characterize the behavior of the station by a Markov chain. Some of the performance of interest (those related to the demands) are obtained thanks to the analysis in isolation of a station representing the linkage between the system and the external demands, and it is their computation which is different from the case presented in [5]. The station is analyzed by solving the underlying Markov chain, which has an infinite number of states. But, by regrouping them into sets, it can be seen that we have a quasi-birth-and-death process with a very regular structure, that leads to a matrix-geometric formulation of the problem. We can thus use the results developed by Neuts [8], as we did in [6] for the classical kanban system with a general arrival process of demands.

The paper is organized as follows: in §2, we give more details on the generalized kanban system, and we summarize the method used to evaluate its performance. For easier presentation, we mainly restrict our attention to the case of a generalized kanban system consisting of a single-stage. The extension to a multi-stage kanban system can be done exactly like in [5]. In §3, we deal with the computation of the performance related to the demands, together with the analysis in isolation of the concerned station. Finally, some numerical results are presented in §4.

2 Modeling and Analysis of a Single-Stage Generalized Kanban System

2.1 Assumptions

We consider a single-stage generalized kanban system producing only one type of parts. It consists of a manufacturing process, which contains parts that are either waiting for or receiving a service, and an output buffer, which contains the finished parts of the stage. The manufacturing process consists of m machines. We assume that: all the queues have infinite capacity; the routing throughout the subnetwork is probabilistic; the service time distribution of each server is represented by a PH distribution. With the stage is associated a fixed number of kanbans, say K, and a maximum number of finished parts, say S. We consider here that the kanbans are associated with individual parts and that there is no return delay for a kanban from the output to the input of a stage. We also assume that there is an infinite supply of raw parts at the input of the production system, whereas demands arrive at the system according to a fixed arrival process such that the interarrival time is distributed according to a PH distribution. Finally, demands that are not immediately satisfied are backordered.

2.2 Queueing Network Model

Figure 1 represents the queueing network for a single-stage generalized kanban system consisting of four machines in tandem. The manufacturing process is modeled by a subnetwork of $m = 4$ stations. The linkage between the system and the external demands is represented by two synchronization stations, each with two upstream queues, P and

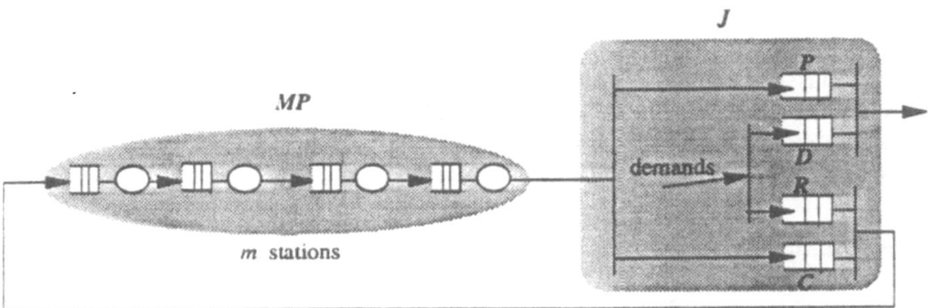

Fig. 1. Queueing network model for a single-stage generalized kanban system with four machines in tandem

D on the one hand, and R and C on the other hand. This linkage is denoted by J, and will be studied in §3.

Let us describe this queueing network of Fig. 1 more precisely. A part which has finished its service in the manufacturing system joins an output buffer, but its associated kanban is immediately detached and stored elsewhere. In the queueing network, this is represented by a split of an entity leaving subnetwork MP into two entities, one of them joining queue P (finished Part) and the other joining queue C (free kanban or Card). Similarly, an arriving external demand splits into two entities, one joining queue D (Demand for a finished part) and enabling the consumption of a finished part waiting in P, and the other joining queue R (production Request for a new part) and enabling the transfer of information to the upstream of the stage. An arriving external demand is immediately satisfied if there is at least one finished part in the output buffer (i.e. an entity in queue P). Otherwise, the demand cannot be immediately satisfied and is then backordered until a finished part is available. The backordered demands are represented by entities waiting in queue D, and the synchronization between the finished parts of the system and the external demands is represented by the synchronization station whose upstream queues are P and D. When a demand is satisfied (either immediately or after being backordered), a finished part leaves the system (i.e. as soon as there is an entity in P and one in D, they both leave the queueing network). The transfer of information to the upstream of the stage is done independently from the consumption of a part, thanks to the synchronization station whose two upstream queues are R and C. The production request is immediately transferred to the upstream of the stage if there is an entity representing a free kanban waiting in C. Otherwise, the entity representing the request waits in R until an entity is available in C. As soon as there is an entity in R and one in C, they merge together and join the upstream of the stage and a new raw part is allowed to enter the manufacturing process (remember that there are always raw parts at the input of the system).

We can point out two important features of the generalized kanban system. First, we have seen that the kanban is detached from its associated part as soon as this part has been processed, and not when the part has been consumed by the following stage, as it is the case for the classical kanban system. Thus, the kanbans and the finished parts are no longer linked and the maximum number of finished parts of the stage, S, can be different from the number of kanbans, K. Moreover, according to the relative values of K and S, we have quite different behaviors for the system, and three classes of generalized kanban systems can be considered [3]: special kanban system ($S = K$), reserve stock kanban system ($S > K$), and backordered kanban system ($S < K$). Note that the special kanban system is equivalent to the classical kanban system [5]. The second important feature is that instead of transferring the information of the consumption of a part as it is done in the classical kanban system, the generalized kanban system enables to transfer the effective demands from downstream to upstream. Indeed, we have just seen that the production request can be transferred upstream even if there is no finished part available (i.e. no entity in P).

2.3 Performance Analysis

The queueing network presented in the preceding section is a closed queueing network with only one class of customers, if we consider the kanbans as the customers. In order to analyze this queueing network, we use a method based on the product-form approximation technique, as we did in [5]. The idea of such a technique (see [1] for details and references) is to partition the closed queueing network into a set of subsystems (in Fig. 1, we have 5 subsystems: the four stations of the manufacturing process and station J). Each subsystem is then approximated by an exponential station with a load-dependent service rate. The resulting network is then a product-form network. In order to obtain the parameters of the equivalent network, we use Marie's approach [1]: each subsystem is analyzed in isolation as an open system fed by a state-dependent markovian process, using an iterative procedure. We can then derive all the performance parameters of the generalized kanban system using an efficient algorithm for product-form networks (see e.g. [2]). These performance are: the throughput and the average work in process (i.e. the number of parts in the manufacturing process). Other performance parameters of high interest (namely, the average number of finished parts, the average number of backordered demands, the mean waiting time of a backordered demand, and the proportion of backordered demands) can be obtained from the analysis in isolation of station J under a load dependent markovian arrival process. Thus, we will now focus on the way they are obtained.

3 Computation of the Performance Related to Station J

3.1 Behavior of Station J

The subsystem under study is shown in Fig. 2. It is a set of two synchronization stations, each with two upstream queues, P and D on the one hand, and R and C on the other hand. The number of entities in the queues are respectively denoted by n_P, n_D, n_R, and n_C. Station J is fed by two processes, one representing the arrival of the finished parts and their associated kanban from the manufacturing process, and the other representing the arrival of external demands. The finished parts and their associated kanbans arrive at the station according to a markovian arrival process with state dependent arrival rate $\lambda_C(n_C)$. Remember that the kanbans are the customers of the closed queueing network representing the single-stage generalized kanban system and thus n_C is such that $0 \leq n_C \leq K$, which implies that $\lambda_C(K) = 0$. The arriving entity splits into two entities, one joining queue P and the other joining queue C. The external demands arrive according to a markovian process such that the distribution of the time between two arrivals is represented by a PH distribution [8]. For sake of simplicity, we only present here the special case of coxian distribution, but everything remains true when we have a more general PH distribution. We denote by η the number of stages of the coxian distribution, by μ_i the processing rate at stage i, for $i = 1, ..., \eta$, by a_i the probability to go from stage i to stage $i + 1$, for $i = 1, ..., \eta - 1$, and by $b_i = 1 - a_i$

the probability to end with the coxian distribution from stage i, for $i = 1, ..., \eta - 1$, which corresponds to the arrival of a demand at station J. Again, the arriving entity

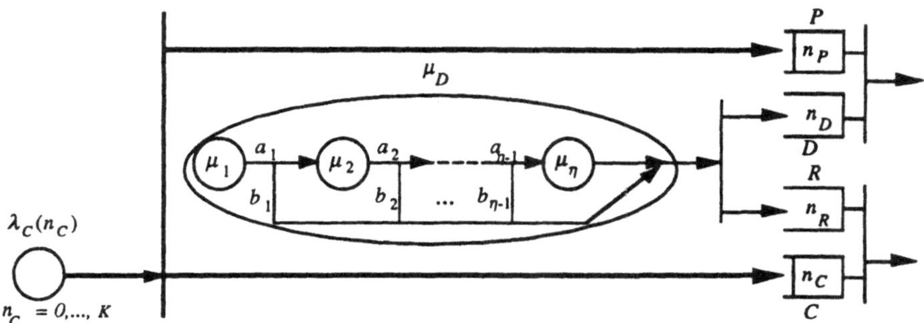

Fig. 2. Station J fed by two arrival processes

splits into two entities, one joining queue D and the other joining queue R. Note that n_D and n_R are not limited. The behavior of station J is the following: as soon as there is an entity in queue P (resp. R) and an entity in queue D (resp. C), they both leave the station. Thus, n_P and n_D, on the one hand, and n_R and n_C, on the other hand, are not simultaneously non null. Moreover, if we consider the situation where all the free kanbans and the finished parts of the stage are in station J (i.e. $n_C = K$ and $n_P = S$, which implies that $n_R = n_D = 0$), then network MP is empty and thus n_C and n_P can only decrease (due to an arrival of external demands, i.e. an entity in D and one in R). Thus, we see that n_P is limited by S.

3.2 Analysis of the Station

An exact solution of the open queueing system of Fig. 2 is obtained by solving the underlying Markov chain. The system state is defined as the following vector (n_C, n_R, n_D, n_P, i), where we have $0 \leq n_C \leq K$, $n_R \geq 0$, $n_D \geq 0$, $0 \leq n_P \leq S$, and $i = 1, ..., \eta$. The structure of the Markov chain is depicted in Fig. 3, for a backordered kanban system ($S < K$). The states have been regrouped into sets such that the only non-zero transition rates out of a given set of states are to its two neighboring sets of states. On Fig. 3, the arcs between two sets of states represent all the transitions between the states of one of them to the states of the other one (and correspond either to a demand arrival or to the arrival of a part and its associated card at station J). From a given state, there are also non-zero transition rates to some states inside the same set (corresponding to the progression of an entity from stage to stage in the coxian law). Note that, according to the relative values of K and S, we can observe a different evolution of the state, but the structure of the Markov chain is similar [7]. Especially, when we have $S = K$, the Markov chain is, as expected, exactly the one we had for the classical kanban system [6].

The Markov chain of Fig. 3 represents a quasi-birth-and-death-process, with an infinite number of states, but with a repetitive structure. We can then use the results developed by Neuts [8] for the analysis of this Markov chain, in order to find the steady-state probabilities $p(n_C, n_R, n_D, n_P, i)$. We can then derive the performance of interest, namely the proportion of backordered demands (probability that queue P is empty when

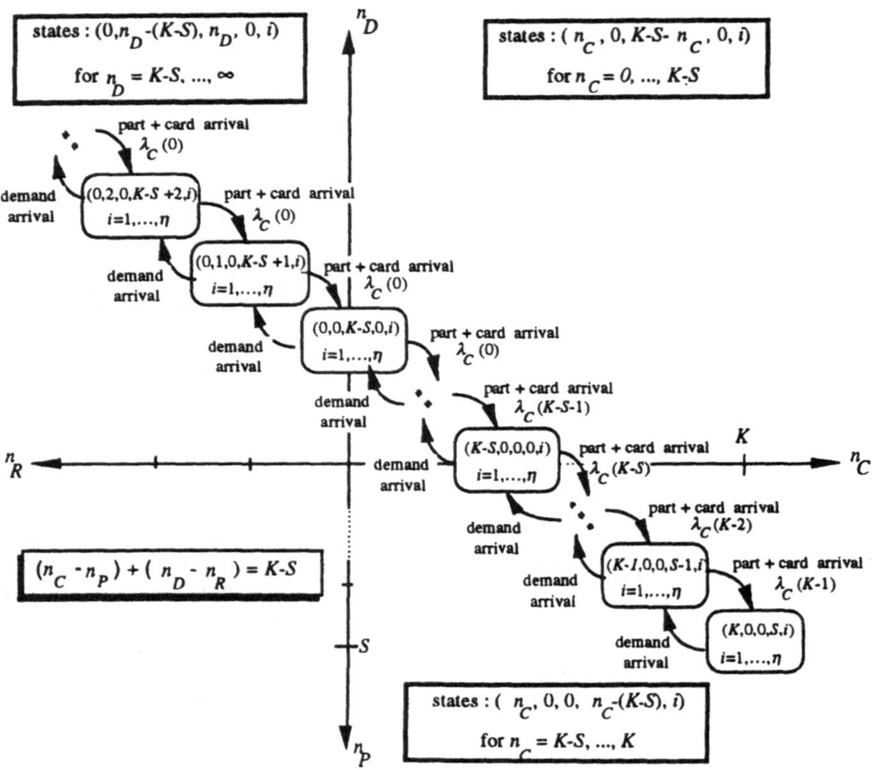

Fig. 3. Structure of the state-diagram representing the behavior of the subsystem presented in Fig. 2, when we have $S < K$. The state is (n_C, n_R, n_D, n_P, i)

a demand arrives), the average number of backordered demands (queue D), the mean waiting time of a demand, and the average number of finished products (queue P). See [7] for details.

4 Some Numerical Results

In this section, we briefly illustrate the kind of results we obtain with our method Because of space limitation, we just present here a simple example consisting in the generalized kanban system of Fig. 1 with all the service times exponentially distributed, with the same mean equal to 1. We fix the mean arrival rate of demands to $\mu_D = 0.5$, while varying the squared coefficient of variation of the distribution (cv^2) from $cv^2 = 0.25$ (4-stage Erlang distribution) to $cv^2 = 10$ (2-stage Coxian distribution). We consider a backordered kanban system with $K = 25$ and $S = 15$. The results obtained thanks to our approximation method are compared to simulation results obtained using the SIMAN simulation language. Fig(s). 4-a,-b, and -c give the mean waiting time of a backordered demand (W_B), the average number of finished parts (FP), and the average work in process (WIP) as functions of the cv^2. The approximation method is fairly accurate, but the error tends to increase with the cv^2. Moreover, this method is very rapid since the results are always obtained in less than one second on a SUN SPARC IPC station, whereas the simulation results were obtained in around 2 hours. Note also that although

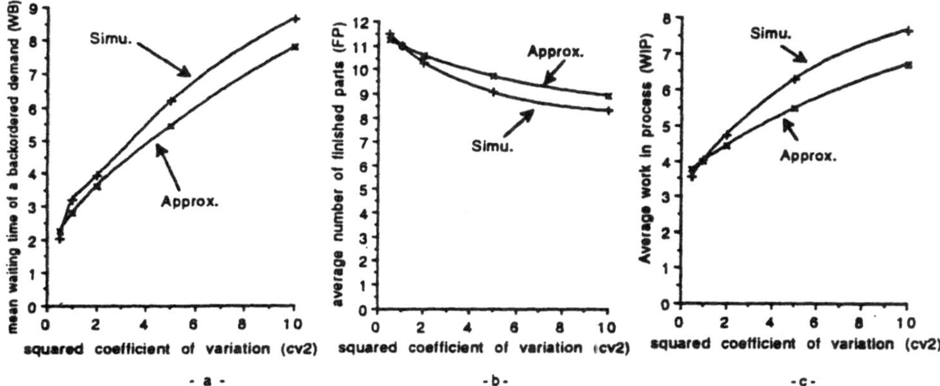

Fig. 4. Influence of the variability of the demand arrival process on W_B, FP, WIP, for a backordered kanban system with $K = 25$ and $S = 15$.

the simulation was long (corresponding to the circulation of 200000 parts in the system), the confidence intervals are not always very tight, especially for W_B for which they are in general around 10%.

On Fig. 4, we can also notice that the variability of the demand has a great influence on the performance, which shows the interest of the extension proposed here in order to take a general arrival process of demands into account.

References

1. B. Baynat and Y. Dallery: "A Unified View of Product-form Approximation Techniques for General Closed Queueing Networks". To appear in Perf. Evaluation.
2. S.C. Bruell and G. Balbo: "Computational Algorithms for Closed Queueing Networks". Elsevier North Holland, (1980).
3. J.A. Buzacott: "Queueing Models of Kanban and MRP Controlled Production Systems". Engineering Cost and Production Economics, (1989).
4. M. Di Mascolo, Y. Frein, and Y. Dallery: "Queueing Network Modeling and Analysis of Kanban Systems". Third International Conference on Computer Integrated Manufacturing, Rensselaer Polytechnic Institute, Troy (NY), IEEE Computer Society Press, pp. 202–211, (May 1992).
5. M. Di Mascolo, Y. Frein, B. Baynat, Y. Dallery: "Queueing Network Modeling and Analysis of Generalized Kanban Systems". Second European Control Conference (ECC93), Groningen, The Netherlands, pp.170–175, (June 1993).
6. M. Di Mascolo: "Analysis of a Synchronization Station for the Performance Evaluation of a Kanban System with a General Arrival Process of Demands", Tech. Rep. LAG, **93–187**, (1993).
7. M. Di Mascolo: "Performance Evaluation of a Generalized Kanban System Receiving Demands According to a General Process". Tech. Rep. LAG, (1994)
8. M.F. Neuts:, "Matrix Geometric Solutions in Stochastic Models". The John Hopkins University Press, Baltimore, London, (1981).
9. R. Uzsoy and L.A. Martin-Vega: "Modelling Kanban-based Demand-Pull Systems: a Survey and Critique". Manufacturing Review, **3–3**, (September 1990).

HOIST SCHEDULING PROBLEM IN A REAL TIME CONTEXT

J.Lamothe[1], M.Correge[1], J.Delmas[2]

1 Centre d'Etudes et de Recherches de Toulouse / Département d'Etudes et de Recherches en Automatique
2 Avenue Edouard Belin BP 4025, 31055 TOULOUSE Cédex France
Phone:(33)61.55.71.11,Fax:(33)61.55.71.94, E-Mail lamothe@saturne.cert.fr

2 also professor at Ecole Nationale Supérieure de l'Aéronautique et de l'Espace

1. INTRODUCTION

An electroplating, or chemical treatment, line is composed of (see fig 1) : tanks containing each a chemical reagent, an input and an output buffer, hoists transporting the products from tank to tank.

Therefore, a treatment results from the successive immersion of a product in several tanks. Tanks and hoists can only process one job at a time. Specific chemical constraints exist :

• once a product finishes its processing in a tank, it must go as fast as possible to the next tank. This means there is no in-process inventory.

• the soaktime, as it corresponds to the time needed to complete a chemical reaction, is not precisely given. But, in order to ensure the quality of the reactions (and, then, of the whole treatment) chemists give minimum and maximum limits. Those quality constraints are considered absolute in printed circuit boards or airplane parts production. As no inventories are allowed, this soaktime tolerance is the only source of flexibility.

The control of the hoist's movements with respect to those constraints is known as the Hoist Scheduling Problem (HSP).

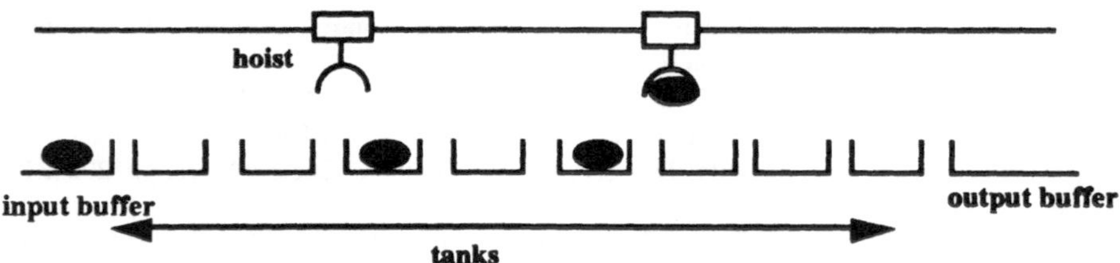

fig 1 : a chemical treatment line

We consider a multi-product production in a real time context : jobs to be carried out in the shop arrive one by one and randomly. The main difficulty is to schedule the hoist's movements with respect to the soaktime and no-inventory constraints. So, the only way to solve this real time problem consists in computing a new schedule of the hoist's movements, any time a new job arrives. This new schedule must ensure the feasibility of the new job and of the not yet finished ones. We will refer to this scheduling problem that must be repeated successively as the Dynamic Hoist Scheduling Problem (DHSP).

In this article, we will consider the case in which one hoist must carry out all the products.

2. PREVIOUS RESEARCH

First, Philipps and Unger in 1976 introduced a HSP in which all jobs were identical. Their goal was to find a cyclic schedule in which one job enters the line periodically (this problem is called the 1 periodic HSP). They developed a mixed integer model and proposed a resolution on a 13-tank benchmark. Shappiro and Nuttle ([SHA88]) using a branch and bound procedure and linear programming to bound the search, solved this benchmark and proved the optimality. Their algorithm introduces successively the jobs with the same period (this point of view is close to the DHSP but it includes the fact that a solution must be periodic). Algorithms that use other methods to bound the search and other enumeration procedures have also been proposed : Baptiste and Legeard ([BAP92]) use CLP (Constraint Logic Programming), Amstrong and Lei ([ARM92]) introduce a minimum time span variable easy to compute.
More generally, Lei and Wang ([LEI89]a) propose a model and a resolution to find a schedule in which k jobs enter the line in a cycle (k-periodic cycle). But the computation time happens to become very long when k grows. They also proved that the problem is NP-hard ([LEI89]b).

Another point of view opposed to the cyclic HSP can also be found in literature : soaktime maximum limits are not considered. Then a truly real time approach is adopted : at any instant the problem is to decide which hoist and which product is going to move next. Thesen and Lei ([THE86]) proposed an architecture for an expert scheduling system. Yih ([YIH90]) improves the rule selection rule with the Trace Driven Knowledge Acquisition method.

The DHSP, exposed in this paper, is a trade off between those two points of view. We Compute a forecasting schedule with an analytical model, and an optimal resolution method; so it looks like the cyclic HSP ones. But, as the DHSP must be solved in real time, rules and heuristics are introduced in the resolution and their efficiency must be analyzed. A first strategy, introduced by [SON93] and [YIN92], consists in keeping the processing jobs sequence and, insert the new job inside. It leads to a really quick computation (some seconds). In order to quantify the production rate reachable with the DHSP method and also the first strategy efficiency, another type of strategy, a more general one, is presented. It consists in choosing the feasible solution that optimizes a criterion (this criterion is a parameter to be specified). In order to compare those strategies, we simulate a line on a given horizon with data from the Philipps and Unger benchmark problem to measure the global production rate and the computation time.

3. MODEL

In order to find a schedule for the DHSP, a mixed integer model is presented.
A new job arrives at the input buffer, at an initial instant called t0, while (J-1) jobs are in-process and not yet finished. So, at instant t0, the position and the remaining soaktime of all the products in the line is known.
Then, the problem consists in finding a schedule that respects these initialization constraints and also soaktime, no in-process inventory.and resources constraints.

We also assume that one hoist must do all moves, that the travelling times from tank to tank are given, and that an operation can only take place in one tank.

3.1. Hoist travelling time

A chemical constraint imposes that a product must go as far as possible from a tank to another. As a consequence, the in-charge travelling time (the time to travel a product from a tank to another) is supposed given. But, between two successive in-charge travels, an empty move is necessary. It depends on the sequence of the hoist's movements and may take long. The hoist being sometimes a bottleneck, the duration of those empty moves must be precisely taken ito account.
So, two kinds of hoist's movements are considered : in-charge and empty movements.

3.2. Notations

The following notations are used :

- J : number of jobs
- nb_op(j) and init(j) : number of operations and the initial operation of job j.
- $O_{i,j}$: operation i of job j. $1 \leq j \leq J$, $1 \leq init(j) \leq i \leq nb\_op(j)$.
- $m_{i,j}$ and $M_{i,j}$: minimum and maximum soaktime for operation $O_{i,j}$.
- $b_{i,j}$ characterizes the bath corresponding to operation $O_{i,j}$.
- $t_{i,j}$: starting time of operation $O_{i,j}$ in bath $b_{i,j}$.
- $r_{i,j}$: in-charge travelling time from bath $b_{i-1,j}$ to bath $b_{i,j}$.
- r'(u,v) : empty travel time from bath u to bath v.
- $r_{i,j/k,l}$: time so that the hoist travels product j in bath $b_{i,j}$ knowing the

previous movement finished at bath $b_{k,l}$.:

$$r_{i,j/k,l} = r_{i,j} + r'(b_{k,l},b_{i-1,j}).$$

Then, $t_{i,j}$ are the decision variables of the DHSP.

3.3. Initialization constraints

At instant t0, two cases can be considered :

- Either the job j is processing in the bath $b_{init(j),j}$.
- Or the hoist is moving the job j in to the bath $b_{init(j),j}$.

In all cases, one can assume that : $t_{init(j),j}$ -t0 is given for all j ($1 \leq j \leq J$). (C1)

3.4. Specific chemical constraints

As a product must be travelled as far as possible from a tank to another, the soaktime of job j in bath bi,j can be expressed by : $t_{i+1,j} - t_{i,j} - r_{i+1,j}$.
Then, the window constraint concerning soaktime can be expressed by :

$$m_{i,j} \leq t_{i+1,j} - t_{i,j} - r_{i+1,j} \leq M_{i,j} \qquad \text{(C2)}$$

$1 \leq j \leq J$, $1 \leq init(j) \leq i \leq nb\_op(j)$.

3.5. Resource allocation constraints

The hoist and the tanks are indivisible resources. Moreover, their capacity is 1 and pre-emption is not allowed. Those constraints are usually expressed by disjunctions. In our application, hoist disjunctions and bath disjunctions are considered separately. Between the starting time of two operations there must be a sufficient time so that the hoist can move. Then, hoist disjunction between two operations, $O_{i,j}$ and $O_{k,l}$, can be expressed by :

$$t_{k,l} - t_{i,j} \geq r_{k,l/i,j} \qquad (O_{i,j} \text{ before } O_{k,l})$$

or (C3)

$$t_{i,j} - t_{k,l} \geq r_{i,j/k,l} \qquad (O_{k,l} \text{ before } O_{i,j})$$

$1 \leq j \leq l \leq J$, $init(l) \leq k \leq nb\_op(l)$, $init(i) \leq j \leq nb\_op(i)$

As one hoist carries out all the products, if operation $O_{i,j}$ is processed before operation $O_{k,l}$ in a tank, the hoist moves job j to tank $b_{i+1,j}$ before operation $O_{k,l}$ starts. So the bath disjunction between $O_{i,j}$ and $O_{k,l}$ can be expressed by :

$$t_{k,l} - t_{i+1,j} \geq r_{k,l/i+1,j} \qquad (O_{i,j} \text{ before } O_{k,l})$$

or (C4)

$$t_{i,j} - t_{k+1,l} \geq r_{i,j/k+1,l} \qquad (O_{k,l} \text{ before } O_{i,j})$$

$1 \leq j \leq l \leq J$, $init(l) \leq k \leq nb\_op(l)$, $init(i) \leq j \leq nb\_op(i)$, $b_{i,j} = b_{k,l}$.

3.6. Criterion

The general objective is to maximize the whole throughput rate. Obviously, this optimum cannot be reached by optimizing a criterion in each DHSP. But, the criterion helps to distinguish a good DHSPsolution.

The following criteria are, then, considered :

- makespan : $MIN \{ \max_{1 \leq j \leq J} t_{nb\_op(j),j} \}$
- criterion 2 : $MIN \{ \sum_{1 \leq j \leq J} t_{nb\_op(j),j} \}$
- criterion 3 : $MIN \{ t_{nb\_op(J),J} + \sum_{1 \leq j \leq J} t_{nb\_op(j),J} \}$

The criterion characterizes the solution retained in a DHSP. Then, when DHSPs are successively solved in a real time context, the throughput rate is a consequence of the criterion definition.

4. RESOLUTION

Constraints (C1) and (C2) are linear. Constraint (C3) and (C4) are disjunctions, but disjunctions between two linear propositions.

To treat those disjunctions, a branch and bound procedure with a depth search priority is adopted. The resolution is initialized with the only linear constraints (C1) and (C2). At each node of the search, a set of linear constraints is given and an optimal solution corresponding to this set is obtained. For this local solution, we look for conflicts : a conflict concerns two operations using the same resource (hoist or tanks) during a same period of time. If no conflict is detected, a feasible solution of the DHSP is found. Otherwise the set of all conflicts is determined and a branching heuristic selects one conflict from this set. Then two possibilities to add a new linear constraint exist to branch the search (see.fig 2).

At each node of the search, the branching heuristic, that leads to the choice of a disjunction, characterizes the speed of thc algorithm. A good heuristic must lead quickly to a near optimal solution so that the bound should be more effective. Then different branching heuristics will be compared in terms of speed of the resolution.

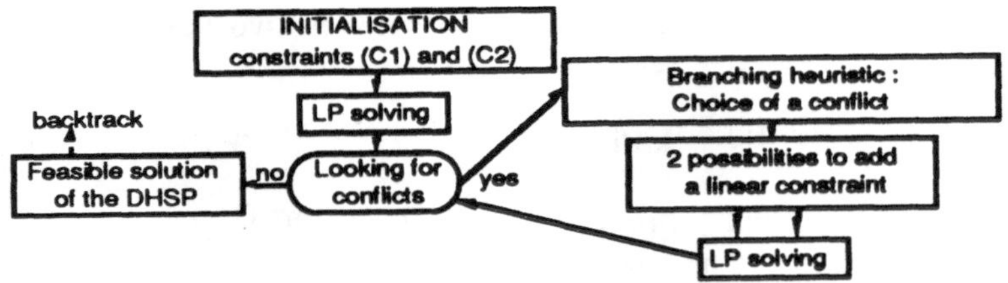

fig 2 : branching principle

5. COMPUTATIONAL RESULTS

The objective of this computational study is to analyze the consequences of the criterion definition on the whole throughput, and to compare our optimizing strategy with the simpler one that keeps the sequencing of the not yet finished jobs operations for each DHSP.

5.1. Branching heuristic implementation

At each node of the branch and bound resolution, the branching heuristic selects one conflict into a set.
In the first runs FIFO rule was used and led to excessive computation times.
But, industrial definition of multi-product is rather specific. Because of the chemical processes, the problem of scheduling the jobs into tanks looks like a flow-shop problem with no wait in-process. Moreover the job starting order is given because of the real time approach So, the branching heuristic considers first the tanks conflicts and selects them in the decreasing order of tanks utilization (see fig 3).

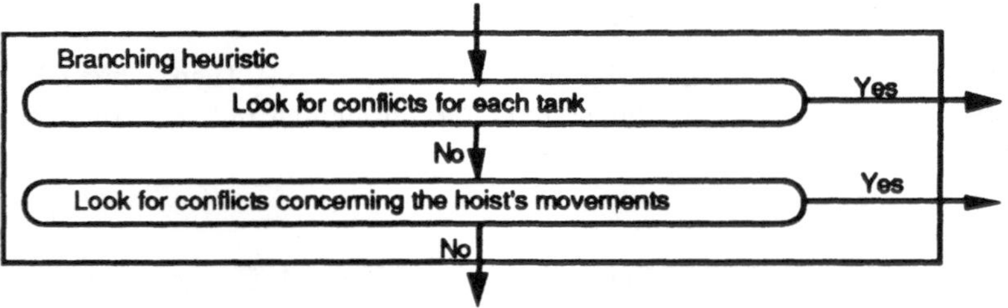

fig 3 : the branching heuristic

At the present time, FIFO rule is used for the hoist conflict selection. Some tests have been made with other rules (SPT, Shortest Remaining Process,...) with no significant improvement on the computation time.

5.2. throughput rate

5.2.1. solving the real time HSP

In order to compare the different strategies, we simulate the Phillips and Unger benchmark line (13 tanks) in real time. Jobs are all identical. We suppose that a new job arrives at the input buffer any time another one leaves it. Then, depending on the way the DHSPs are solved, a cyclic production is reached. We define the average period as the cycle time divided by the cycle length (the number of jobs entered during this cycle). It characterizes the average throughput of a simulation.

The three criteria introduced above are compared. With each one we also compare the application, or not, of the simple heuristic strategy (see table 1).

table 1. DHSP on Philipps and Unger benchmark problem

simple strategy	criterion	average period	average computation time per DHSP	cycle length
yes	makespan	652	7 cpu sec	1 periodic
no	makespan	561	143 cpu sec	3 periodic
yes	criterion 1	746	2 cpu sec	1 periodic
no	criterion 1	564	70 cpu sec	3 periodic
yes	criterion 2	731	1.1 cpu sec	1 periodic
no	criterion 2	564	67 cpu sec	3 periodic

computation time are given on a Sparc station 1

This table shows that the criterion optimizing strategy leads to an improvement of the average period. Fortunately the computation time per DHSP remains satisfactory.

5.2.2. solving the cyclic HSP

The real time HSP, as simulated just above, can be considered as a heuristic to find a solution for the cyclic HSP. The advantage of this method is that, contrary to the classical resolution of the cyclic HSP, the cycle length is not fixed.

To solve the real time HSP, only one job has been considered at the input buffer in each DHSP. But the model and the resolution proposed can take into account more than one job at the input buffer in each DHSP.

Results of the simulation when two jobs are taken into account at the input buffer are presented in the following table 2.

table 2. looking for the best cyclic HSP solution

criterion	average period	mean computation time per DHSP	cycle length
makespan	521	700 cpu sec	1 periodic
criterion 1	502	800 cpu sec	2 periodic
criterion 2	502	1200 cpu sec	2 periodic

521(respectively 502) have been shown to be the optimal average period to the 1-periodic (2-periodic) cyclic HSP. (see [SHA88] and [LEI89]a).

Table 2 shows that when the size of the problem (the number of independent variables) grows the algorithm can no longer respect the real time limit on computation time . But, when a cyclic schedule solution is wanted no more real time constraint exists on this computation time.

6. CONCLUSION

The purpose of this paper was to introduce the scheduling problem of a chemical treatment line in a multi-product production and a real time context. This problem implies the iterative computation of a schedule with a limit on the computation time. A mixed integer linear programming model of the scheduling problem is presented.

A simulation on the Philipps and Unger benchmark problem reveals the algorithm ability to repeatedly compute schedules with respect to computation time limits. It also shows our criterion optimizing strategy efficiency on throughput rate compared to the simple heuristic strategy usually introduced. Considered as a heuristic to solve the cyclic Hoist Scheduling Problem, the real time approach makes it possible to find again classical results without any assumption on the cycle length.
However, our algorithm needs long computation times. A computation rule is presented to improve it. But, more efficient rules and good heuristics are still to be found.
Finally, the approach presented to compute a schedule, does not take into account any information that could be recorded from previous computations. This point of view also needs to be developed.

References

[ARM 92] Armstrong,R. ,Lei,L and Shanhong,G.: a Bounding Scheme for Deriving the Minimal Cycle Time of a Single-Transporter N-stage Process with Time-Window Constraints. Working paper 92-07, Rutgers University (feb 1992)

[BAP 92] Baptiste,P.,Legeard,B.,Varnier,C.: Hoist Scheduling Problem: an Approach Based on Constraints Logic Programming. Proceedings of the 1992 IEEE, International Conference on Robotics and Automation, Nice-France, (may 1992)

[LEI 89]a Lei,L and Wang,T,J : On the Optimal Cyclic Schedules of a Single Hoist Electroplating Processes. Working paper 89-0016, Rutgers University (april 1989).

[LEI 89]b Lei,L and Wang,T,J : A Proof : The Cyclic Hoist Scheduling Problem is NP-Complete. Working paper 89-0016, Rutgers University (august 1989).

[LEI 91] Lei,L., and Wang, T-J : The Minimum Common-Cycle Algorithm for Cyclic Scheduling of two Material Handling Hoists with Time Window Constraints. Management Science **37 n°12**, (december 1991)

[PHI 76] Philipps,L.W.,and Unger,P.S. :Mathematical programming solution of a Hoist Scheduling Program. AIIE transactions,**28 n°2**(june 1976) 219-225

[SHA 88] Shapiro, G.W., and Nuttle, H.L.W : Hoist Scheduling for a PCB Electroplating Facility. IIE Transactions, **20 n° 2**, (june 1988).

[SON 93] Song W., Zabinsky Z., Storch R.L. : An Algorithm for Scheduling a Chemical processing Tank Line. IProduction Planning and Control,**4 n°4**(April 1993) 323-332

[THE 90] Thesen,A. and Lei.L : An expert scheduling system for material handling hoist. Journal of Manufacturing Systems **9 n° 3** .

[YIH 90] Yih, Y. : Trace-driven knowledge acquisition for a rule-based real time scheduling systems. Journal of Intelligent Manufacturing **1** (july 1990) 217-230 .

[YIN 92] Yin,N.C. Yih,Y. : Crane Scheduling in a Flexible Electroplating line : a Tolerance Based Approach. Journal of Electronic Manufacturing 2 (july 1992)137-144

Annex

Datas of the benchmark Philipps and Unger problem are summarized as follows :

tank i	0	1	2	3	4	5	6	7	8	9	10	11	12
m_i (s)	120	150	90	120	90	30	60	60	45	130	120	90	30
M_i (s)	∞	200	120	180	125	40	120	120	75	∞	∞	120	60
r_i (s)	30	31	22	22	22	25	23	22	22	22	47	27	22

Empty travelling time from tank to tank can take from 1 to 29 seconds

On Controlling a Class of Assembly System with Alternative Production Routeing

Jean-Marie PROTH, Liming WANG and Xiaolan XIE

INRIA-LORRAINE, Technopole Metz 2000, 4 rue Marconi, F-57070 Metz, France

1 Introduction

Petri net has been recognized as an important tool for studying FMSs[2,3]. Many of the previous works were on performance evaluation [2-7]. In this paper we focus on *controlling* a class of manufacturing systems based on the Timed Petri Net (TPN) modelling framework.

During the course of our investigation into the operational control of manufacturing systems, we have found it intractable (both in the sense of complexity theory and in common sense) to incorporate at the same time many real-world factors into one model. Thus we gradually reduce the problem complexity by adding a certain amount of assumptions, bearing in mind the belief that we will eventually return to our ultimate goal, that is, specification and implementation of effective operational controller for real-world manufacturing systems.

The first assumptions we made are; (A1) The systems under study are viewed as discrete event versus continuous variable systems. We are not interested in the detailed movements of material handling or the movement trajectories of robot arms. Rather we focus on system evolution driven by events occurred at discrete time instants. (A2) The system under study are deterministic in that they are reliable and that the time consumption of each operation is fixed.

In section 2, we make further assumptions to arrive at a description of the system at hand. We then model the system by a class of TPN which we shall define in section 3, and formulate the problem by a matrix sequence design problem under certain constraints. In section 4, we propose a control mechanism, reduce it to the finding of a sequence of maximal s-concurrent markings, and illustrate the idea by a numerical example. Finally, in section 5, we give a concluding remark and present some further research directions.

2 System Description

(A3) We confine ourselves to the case of one resource-type, i.e., machines, and assume that all machines are of different kind. Each machine can process several product-types. Each kind of processing on a machine required by a product-type is called an *operation*. The function of an operation of processing a product-type may be equally fulfilled by an alternative processing of the same product-type on another machine, namely, (A4) The system consists of alternative production routeing. We call an *action* the processing of a product-type on possibly several alternative machines to satisfy a given operational function, specifically we use input (output) action(s) to model load/unload (L/UL) area. (A5) There exists, for each product-type, a predetermined order of actions it must traverse, which we call *production process*. Successful production of some products may need assembly of some semi-finished products (SPs). For those product-types requiring assembly action, more than one sequences of actions are needed, each of which is called a *subprocess* for a given product-type.

Each execution of an operation consumes certain amount of time. If the time consumptions are the same for all the operations involved, then it is called homogeneous, otherwise, non-homogeneous. (A6) Homogeneous case is considered in this paper. As such it is reasonable to assume that one execution of each operation consumes one unit of time.

[Example 1] A manufacturing workshop consisting of five machines denoted by M1, M2, M3, M4, M5, is to process two types of product P1 and P2. P1 needs one production process which comprises three consecutive actions (except the input and output actions): a1, a2, and a3. a1 can be realized on either M1 or M3, a2 on M2 or M4, while a3 can only be realized on M5. Successful production of P2 requires an assembly action a8 using the SPs from earlier steps of the two subprocesses for P2. The two subprocesses comprise respectively actions a4, a6, a8 and a5, a7, a8 (excluding input and output actions). Actions a4, a5, a6, a7, and a8 can only be realized on machines M2, M1, M4, M3, and M5 respectively. Each realization of an action is an operation that consumes one unit of time. At the beginning ,there are 2 SPs of P1; one is waiting for realizing a1, the other is waiting for realizing a2. As for P2, 2 units of SPs are waiting for realizing a6 in one subprocess, 1 unit of SPs is waiting for realizing a7 in the other subprocess, while 2 units of SPs, one from each of the subprocesses for P2, are waiting for the assembly action a8.

3 Problem Formulation

3.1 A Brief Introduction of Timed Petri Net

Time parameters can be associated with a PN in various ways. In this paper, we assign time to each transition, and adapt to the following definition.

An *Open Timed Petri Net* (OTPN) is a 2-tuple OTPN=(PN, D), where PN is an ordinary Petri net: PN=(P, T, F, M_0), in which P is the union of four disjoint place sets P', P'', P''', and R the elements of which are called S-input, S-internal, S-output and resource places respectively (S stands for system). T is the union of three disjoint transition sets T', T'', and T''' the elements of which are called S-input, S-internal, and S-output transitions respectively. D is a function defined by (N^+ is denoted to be set of nonnegative integers). D(t) = $d(t) \in N^+$ if $t \in T''$, and 0 if $t \in T' \cup T'''$. S-input (S-output) transitions are those without input (output) places, while S-input (S-output) places are output (input) places of S-input (S-output) transitions. In this paper, d(t)=1 for all $t \in T''$. $M_0(p)=1$, $\forall\ p \in R$.

Remark 1: Roughly speaking, real-world manufacturing system management consists of two steps: strategic planning and operational control. We are concerned with the latter some constraints of which come from the requirements or commands of the former. One such constraint on the operational controller is the requirement from an OTPN-based planner that each transition should fire exactly how many times, i.e., each operation should be activated how many times within a certain planning horizon. The aim of operational level is to control the system operating in a mode of high productivity. (Other important functional aspects of this level have to be left to further research)

In homogeneous case, it is reasonable to require instead how many times each action should be realized. As such, we denote by Q the set of all actions excluding input and output ones, and define the function $\nabla : Q \to \{1,2,\ldots,|Q|\}$ by $\nabla(q) = k$, if $k \in \{1,2,\ldots,|Q|\}$ is the label of action $q \in Q$, otherwise undefined.

3.2 Formulating the Problem

Let N and M denote respectively the total number of product-types and the total number of machines. We have reason to view the homogeneous system to evolve stage by stage. A stage is characterized by the enablement of transitions at time s (or equivalently the token distribution at time s). By slightly abuse of notation we also denote by s the stage at time s. Let S be the total number of stages for the system evolution concerned. Denote by L_i the number of subprocesses for product-type i, $A_i^{(l)}$ the number of actions for the l-th ($l=1,2,\ldots,L_i$) subprocess of product-type i (including assembly action but excluding input and output actions). We also denote by K_k to be the required number of realizations of action $a_k \in Q$ $(k=1,2,\ldots,|Q|)$, and V_k the number of alternatives for the realization of action $a_k \in Q$.

Proposition 1: $$|Q| = \sum_{i=1}^{N}\left(\sum_{l=1}^{L_i} A_i^{(l)} - L_i + 1\right)$$

Indeed, if we count the actions in the system one by one, we have $\sum_{l=1}^{L_i} A_i^{(l)}$ actions for product-type i (i=1,2,...,N). Note in this way we have counted the assembly action L_i-1 times more for product-type i. Thus the total number of recounting is $\sum_{i=1}^{N}(L_i - 1)$. Denote $\Delta = \sum_{i=1}^{N}\sum_{l=1}^{L_i} A_i^{(l)}$, then the total number of actions in the system is $\Delta - \sum_{i=1}^{N}(L_i - 1) = \sum_{i=1}^{N}\left(\sum_{l=1}^{L_i} A_i^{(l)} - L_i + 1\right)$.

Once $|Q|$ is specified, we can define a matrix sequence $\{X(s)\}_s$, where

$X(s) = [X_1(s) \vdots X_2(s)]_{|Q|\times(M+1)}$ and

$$X_1(s) = \left[x_{kj}^1(s)\right] \in \{0,1\}^{|Q|\times M}, \quad X_2(s) = \left[x_k^2(s)\right] \in \{0,1,\ldots,MAX\}^{|Q|}$$

in which (*) $MAX = \max\{K_k : k=1,2,\ldots,|Q|\}$, (*) x_k^2 is the number of realizations of a_k up to stage s. (*) $x_{kj}^1(s)=1$, if a_k is realized on M_j at stage s, 0, otherwise.

[PROB1 —— Matrix Sequence Design Problem] Find a matrix sequence $\{X(s)\}_{s=0}^{S^*}$ such that S^* is the smallest S satisfying:

(1) At each stage $0 \le s \le S$, $\sum_{k=1}^{|Q|} x_{kj}^1(s) \le 1, \quad \forall\, j=1,2,\ldots,M$

(2) At each stage $0 \le s \le S$, $\sum_{k=1}^{M} x_{kj}^{1}(s) \le V_k$, $\forall\ a_k \in Q\ (k = 1,2,\ldots,|Q|)$

(3) For each $0 \le s \le S^*$, $[0] < \left[x_k^2(s) \right] \le [K_k]$, while for s=S*,

$$\left[x_k^2(s)\right] = \left[x_k^2(S^*)\right] = [0]$$

(4) $x_{kj}^{1}(s) \in \{0,1\}$, $x_k^2(s) \in \{0,1,\ldots,MAX\}$, MAX is specified as before.

[Example 2] For the workshop in example 1, the following parameters are known: N=2, M=5; L_1=1, L_2=2; A_1=3, $A_2^{(1)} = 3$, $A_2^{(2)} = 3$. Thus

$$|Q| = \sum_{i=1}^{2} \left(\sum_{l=1}^{L_i} A_i^{(l)} - L_i + 1 \right) = (A_1 - L_1 + 1) + (A_2^{(1)} + A_2^{(2)} - L_2 + 1) = 8.$$

The OTPN model of the manufacturing workshop is exhibited in Fig. 1. For convenience we use two-arrowed lines between resource places and relevant transitions to graphically represent the fact that resources are utilized non-preemptively and on an equal footing by all the relevant operations. Thus the workshop of example 1 is characterized by OTPN=(P, T, F, M_0; D), where $P = P' \cup P'' \cup P''' \cup R$, P'={p1, p2, p3}, P''={p4 — p9}, P'''={p10, p11}, R={R1—R5}, $T = T' \cup T'' \cup T'''$, T'={t1,t7,t8}, T''={t2—t6,t9-t13},T'''={t14,t15},$F \subseteq (P \times T) \cup (T \times P)$, M_0=(m_0(p1), ...,m_0(p11),m_0(R_1), ...,m_0(R_5))= (1,0,0,1,2,1,0,1,1,0,0,1,1,1,1,1). D(t) =1 if $t \in T''$, and 0 if $t \in T' \cup T'''$. We can define $\nabla: Q \to \{1,2,\ldots,8\}$ as follows:

$\nabla(a1) := \nabla(\{t2,t3\}) = 1$, $\nabla(a2) := \nabla(\{t4,t5\}) = 2$, $\nabla(a3) := \nabla(\{t6\}) = 3$,
$\nabla(a4) := \nabla(\{t9\}) = 4$, $\nabla(a5) := \nabla(\{t10\}) = 5$, $\nabla(a6) := \nabla(\{t11\}) = 6$,
$\nabla(a7) := \nabla(\{t12\}) = 7$, $\nabla(a8) := \nabla(\{t13\}) = 8$.

We also know that V_1=V_2=2, V_3=...=V_8=1. Moreover we are informed of the requirements from the planner that K_1=K_2=K_3=5, K_4=...=K_8=4.

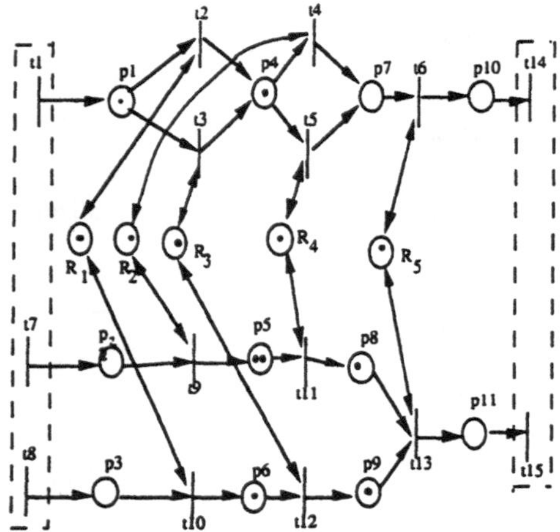

Fig. 1: The OTPN model of the manufacturing workshop in example 1

4 A Control Mechanism

Before getting on the exploration of a control mechanism, we make a slight adaptation on the OTPN model discussed in section III.

4.1 A-OTPN Model

In order to focus our attention on *operational action* rather than on how L/UL area works, we are lead to the following definition;

An A-OTPN is an OTPN in which all the S-input transitions, and S-output transitions together with S-output places are removed, while as many tokens as the required number of realizations of actions for each subprocess of a product-type are added to the corresponding S-input places. Formally, with the above notations, A-OTPN=($P' \cup P'' \cup R$, T'', F, W, M_0, D), in this paper, it is assumed that W and D are unit functions.

[Example 3] The A-OTPN model of the workshop in example 1 is derived from Fig. 1 and shown in Fig. 2.

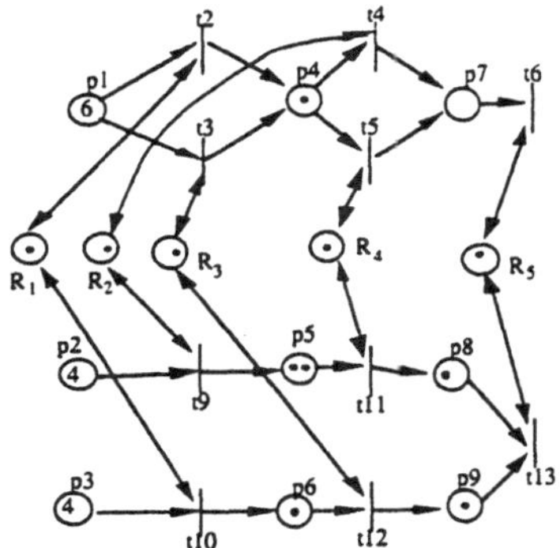

Fig. 2: The A-OTPN model of the workshop in example 1

4.2 Maximum Concurrency Policy

We base our discussion on a A-OTPN model A-OTPN=(P, T, F, W, M_0,D):=($P' \cup P'' \cup R$, T'', F, W, M_0, D) and use the following notations:

$$F_0 = \{(p,t) : (p,t) \in F,\ p \in \bigcup_{i=1}^{M} R_i\} \cup \{(t,p) : (t,p) \in F,\ p \in \bigcup_{i=1}^{M} R_i\}$$

$R_i^* = \{t : (p,t) \in F_0 \wedge (t,p) \in F_0,\ p \in R_i,\ t \in T\}$, $i = 1,2,\ldots,M$, and denote by REACH(M_0) the set of all markings reachable from M_0 (including M_0).

Definition 2: (g-concurrency) For a pair of markings M_1, $M_2 \in$REACH(M_0), if $M_1(p) \geq M_2(p)$, $\forall p \in {}^{\bullet}t$, $t \in R_i^*$ $(i = 1,2,\ldots,M)$, then M1 is called to be more g-concurrent than M2 .

Due to the existence of nondeterminism, for each stage s of the system evolution, several markings may occur. Let M^s be the set of all possible markings at stage s under a given initial marking M_0. Obviously $M^s \subset$ REACH(M_0).

Definition 3: (s-concurrency) For a pair of markings $M_1^S, M_2^S \in M^S$, if $M_1^S(p) \geq M_2^S(p)$, $\forall p \in {}^{\bullet}t$, $t \in R_i^*$ $(i = 1,2,\ldots,M)$, then M_1^S is called to be more s-concurrent than M_2^S.

Definition 4: For a marking $M_i^S \in M^S$ $(i \in \{1,2,\ldots,|M^S|\})$, if there does not exist any $M_j^S \in M^S$ $(j \neq i,\ j \in \{1,2,\ldots,|M^S|\})$, such that M_j^S is more s-concurrent than M_i^S, then M_i^S is called to be a maximal s-concurrent marking.

[PROB2 —— Maximal s-concurrency Tracking Problem]: Given an A-OTPN, find a sequence of maximal s-concurrent markings $\{M_{i_s}^S\}_{s \geq 0}$ such that (1) $M_{i_s}^S \in M^S$ $(i_s \in \{1,2,\ldots,|M^S|\})$, (2) $M_{i_s}^S$ is a maximal s-concurrent marking.

Theorem 5: The solution set of maximal s-concurrency tracking problem (PROB2) implies that of the matrix sequence design problem (PROB1).
(Due to lack of space, the proof is omitted.)

Once a sequence of maximal s-concurrent markings is given (as for the generation of a sequence of maximal s-concurrent markings, we leave it to forthcoming papers), we can readily construct a corresponding matrix sequence which is a solution to PROB1 that specifies the control mechanism.
[Example 4] For the workshop in example 1, based on its A-OTPN model (Fig. 2), we are given a sequence of maximal s-concurrent markings in the following format $M_i=(m_i(p1), \ldots, m_i(p9), m_i(R_1), \ldots, m_i(R_5))$.

M_0=(6,4,4,1,2,1,0,1,1,1,1,1,1,1), M_1=(6,3,3,1,2,1,0,1,1,1,1,1,1,1),
M_2=(6,2,2,1,2,1,0,1,1,1,1,1,1,1), M_3=(6,1,1,1,2,1,0,1,1,1,1,1,1,1),
M_4=(4,0,1,2,3,1,1,0,0,1,1,1,1,1), M_5=(4,0,0,1,2,1,1,1,1,1,1,1,1,1),
M_6=(2,0,0,2,2,1,1,1,1,1,1,1,1,1), M_7=(1,0,0,1,2,1,2,1,1,1,1,1,1,1),
M_8=(1,0,0,1,2,1,1,1,1,1,1,1,1,1), M_9=(1,0,0,1,2,1,0,1,1,1,1,1,1,1),
and the control mechanism is manifested in the matrix sequence as follows.
X(0), X(1), X(2), X(3), X(4); X(5), X(6), X(7), X(8); and X(9)=[0].

$$\begin{bmatrix}0&0&0&0&0&5\\0&0&0&0&0&5\\0&0&0&0&0&5\\0&1&0&0&0&4\\1&0&0&0&0&4\\0&0&0&1&0&4\\0&0&1&0&0&4\\0&0&0&0&1&4\end{bmatrix}\quad\begin{bmatrix}0&0&0&0&0&5\\0&0&0&0&0&5\\0&0&0&0&0&5\\0&1&0&0&0&3\\1&0&0&0&0&3\\0&0&0&1&0&3\\0&0&1&0&0&3\\0&0&0&0&1&3\end{bmatrix}\quad\begin{bmatrix}1&0&1&0&0&5\\0&0&0&1&0&5\\0&0&0&0&0&5\\0&1&0&0&0&1\\0&0&0&0&0&1\\0&0&0&0&0&1\\0&0&0&0&0&1\\0&0&0&0&1&1\end{bmatrix}\quad\begin{bmatrix}0&0&0&0&0&3\\0&1&0&0&0&4\\0&0&0&0&1&5\\0&0&0&0&0&0\\1&0&0&0&0&1\\0&0&0&1&0&1\\0&0&1&0&0&1\\0&0&0&0&0&0\end{bmatrix}$$

$$\begin{bmatrix}1&0&1&0&0&3\\0&1&0&0&0&3\\0&0&0&0&1&4\\0&0&0&0&0&0\\0&0&0&0&0&0\\0&0&0&0&0&0\\0&0&0&0&0&0\\0&0&0&0&0&0\end{bmatrix}\quad\begin{bmatrix}1&0&0&0&0&1\\0&1&0&1&0&2\\0&0&0&0&1&3\\0&0&0&0&0&0\\0&0&0&0&0&0\\0&0&0&0&0&0\\0&0&0&0&0&0\\0&0&0&0&0&0\end{bmatrix}\quad\begin{bmatrix}0&0&0&0&0&0\\0&0&0&0&0&0\\0&0&0&0&1&2\\0&0&0&0&0&0\\0&0&0&0&0&0\\0&0&0&0&0&0\\0&0&0&0&0&0\\0&0&0&0&0&0\end{bmatrix}\quad\begin{bmatrix}0&0&0&0&0&0\\0&0&0&0&0&0\\0&0&0&0&1&1\\0&0&0&0&0&0\\0&0&0&0&0&0\\0&0&0&0&0&0\\0&0&0&0&0&0\\0&0&0&0&0&0\end{bmatrix}$$

5 Concluding Remark

In this preliminary report, a class of deterministic discrete manufacturing systems with assembly action and alternative production routeing is discussed within the framework of TPN. The problem is formulated as a matrix sequence design problem under certain constraints. Based on the concepts of OTPN and A-OTPN, we introduce two notions of g-concurrency and s-concurrency, and reduce the problem to the finding of a sequence of maximal s-concurrent markings. As such, maximum concurrency mechanism is brought into the control of this class of systems.

One of the main directions for further research is towards an effective method for constructing a maximal s-concurrent markings. Much work needs to be done here.

Much more work awaiting to be done are the adaptation and extensions of the results in this paper to more realistic environment. It has been well noted that real-world systems are much more complicated than the simplified versions presented in this paper and many other papers. Variable requirements, uncertainty of machine breakdown and resource availability, random processing time, etc. all contribute to a dim picture of the decision space over a parameter or even structure time-variant system.

Acknowledgements

The authours would like to thank an anonymous referee for his or her helpful comments.

References

[1] S.B. Gershwin, R.R. Hilderbrant, R. Suri, S.K. Mitter, A control perspective on recent trends in manufacturing systems, *IEEE Control System Magazine*, April, 1986, 3-14

[2] R.Y. Al-Jaar, A.A. Desrochers, Petri nets in automation and manufacturing, preprint, 1987

[3] M. Silva, R. Valette, Petri nets and flexible manufacturing, *Advances in Petri Nets*, 1989, Springer-Verlag, 374-416

[4] A.A. Desrochers, Performance evaluation using Petri nets, *J of Intelligent and Robotic System*, 6, 1992, 65-79

[5] Y. Narahari, N. Viswanadham, A Petri net approach to the modelling and analysis of flexible manufacturing systems, *Ann. of Operations Research* 3, 1985, 449-472

[6] T. Murata, Petri nets: properties, analysis, and applications *Proceedings of IEEE*, vol.77, no.4, 1989, 541-580

[7] C.V. Ramamoorthy, G.S. Ho, Performance evaluation of asynchronous concurrent system using Petri nets, *IEEE Trans. Software Eng.* vol.SE-6, 1980, 440-449

[8] Y.C.Ho (ed.) Special Issue on Dynamics of Discrete Event System, *Proceedings of IEEE*, vol.77, no.1, 1989

[9] W.E. Wilhelm, Scheduling and logistics on electronics manufacturing and assembly: an introduction, *IIE Trans.*, vol.24, no.4, 1992

Planification Hiérarchisée de la Production: Agrégation du Temps et Cohérence

G. Fontan, G. Hetreux et C. Merce

Laboratoire d'Automatique et d'Analyse des Systèmes du CNRS
7, Avenue du Colonel Roche, 31077 Toulouse Cedex, France

1 Introduction

Cet article traite d'un problème de planification multi-produits et multi-étapes. Il s'intéresse à la mise en oeuvre d'une structure décisionnelle à deux niveaux basée sur une agrégation du temps. Il s'agit dans ce cadre d'établir des conditions de *cohérence* des décisions agrégées assurant l'existence au niveau inférieur de solutions détaillées faisables obtenues par désagrégation. L'expression de la cohérence s'appuie sur une modélisation graphique du processus de production. Les conditions obtenues peuvent être des conditions nécessaires et suffisantes de cohérence ou se limiter à des conditions nécessaires suivant la nature du processus de fabrication. Dans ce dernier cas, l'article propose également des conditions suffisantes de cohérence permettant une mise en oeuvre efficace des notions développées.

2 Présentation du Problème

Il s'agit de planifier la production de N produits i sur un horizon discrétisé en $\mathcal{T}$ périodes élémentaires notées τ, ($\tau \in [1, \mathcal{T}]$). Les demandes en produits finis relatives à chaque période détaillée $d_{i\tau}$ sont connues. Les demandes non satisfaites sont perdues. La fabrication d'un produit i s'effectue en J_i étapes *(opérations)* constituant une gamme linéaire. Cette dernière précise pour chaque opération j d'un produit i, la quantité w_{ij}^r de la ressource r requise. La capacité de chaque ressource r dans chaque période τ, notée cap_τ^r, est connue. Chaque produit est placé dans un stock intermédiaire à la fin de chaque opération. On note $s_{ij\tau}$ la quantité de produit i dans l'*état* j (c'est-à-dire, ayant subi l'opération j et en attente de l'opération $j+1$) en stock en fin de période τ.

Afin de faire apparaître une notion de durée opératoire, on suppose que dans une même période, un produit dans un état donné ne peut subir qu'une et une seule opération de sa gamme.

La planification définit pour chaque période, le volume de produits $x_{ij\tau}$ à réaliser à chaque étape du processus ainsi que les stocks $s_{ij\tau}$ associés. Ceci est effectué de manière à minimiser le coût lié au volume de produits en stock en fin de chaque période élémentaire à chaque étape du processus. On appelle $h_{ij\tau}$ le coût unitaire de stockage.

3 L'Approche Globale

Le **modèle global** $\mathcal{G}$ associé au problème considéré s'écrit: min $\sum_{\tau=1}^{\mathcal{T}} \sum_{i=1}^{N} \sum_{j=1}^{J_i} h_{ij\tau} s_{ij\tau}$ sous

$$\begin{cases} \mathcal{G}_1 : s_{ij\tau} = s_{ij\tau-1} + x_{ij\tau} - x_{ij+1\tau} & \text{pour } i \in [1, N], j \in [1, J_i - 1], \tau \in [1, \mathcal{T}] \\ \mathcal{G}_2 : s_{iJ_i\tau} = s_{iJ_i\tau-1} + x_{iJ_i\tau} - d_{i\tau} & \text{pour } i \in [1, N], \tau \in [1, \mathcal{T}] \\ \mathcal{G}_3 : \displaystyle\sum_{i=1}^{N} \sum_{j=1}^{J_i} w_{ij}^r x_{ij\tau} \le \text{cap}_\tau^r & \text{pour } r \in [1, R], \tau \in [1, \mathcal{T}] \\ \mathcal{G}_4 : x_{ij\tau} \ge 0, s_{ij\tau} \ge 0 & \text{pour } i \in [1, N], j \in [1, J_i], \tau \in [1, \mathcal{T}] \\ \mathcal{G}_5 : s_{ij\tau-1} \ge x_{ij+1\tau} & \text{pour } i \in [1, N], j \in [1, J_i - 1], \tau \in [1, \mathcal{T}] \end{cases}$$

La contrainte $\mathcal{G}_5$ concrétise la notion de durée opératoire présentée dans le §2. À partir de ce modèle linéaire, il est possible de définir un graphe associé aux différents flux des étapes de production d'un produit donné [JOHNSON 74]. Un tel graphe schématisé par la figure 1 est également appelé *graphe du processus de production*. Il modélise directement les contraintes $\mathcal{G}_1$, $\mathcal{G}_2$, $\mathcal{G}_4$, $\mathcal{G}_5$ du problème $\mathcal{G}$.

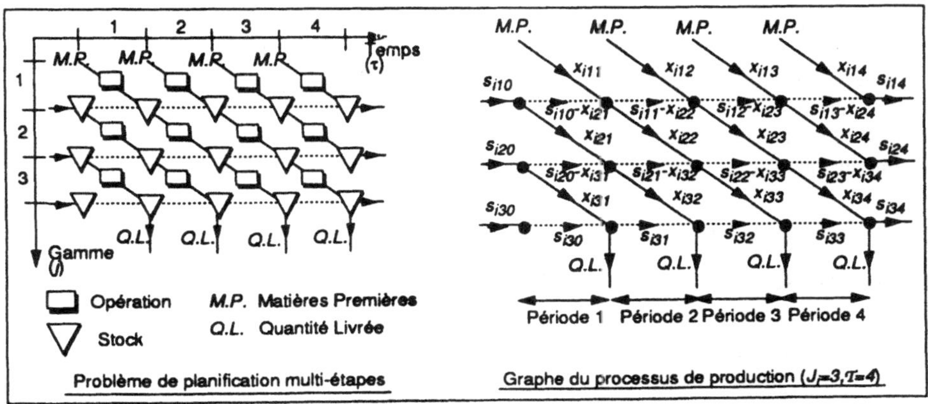

Fig. 1. Graphe du processus de production associé au produit i

Les nombres N_v de variables et N_c de contraintes sont pour ce modèle :

$$N_v = 2\mathcal{T}\sum_{i=1}^{N} J_i \quad \text{et} \quad N_c = \mathcal{T}\left(R - N + 2\sum_{i=1}^{N} J_i\right).$$

Les limites de l'approche globale, qui ont été largement présentées dans la littérature ([HAX 78], [AXSATER 84], ...) ne sont pas reprises ici.

4 L'Approche Hiérarchisée

On considère une structure décisionnelle à 2 niveaux basée sur une agrégation du temps. Le niveau supérieur travaille sur des *macro-périodes* encore appelées *périodes agrégées* qui correspondent à un regroupement de k périodes élémentaires. L'horizon de planification agrégé est donc discrétisé en $T = \mathcal{T}/k$ périodes agrégées notées t.

La planification s'effectue alors en 2 étapes. Le niveau agrégé définit pour chaque macro-période t, le volume de production par produit et par opération de manière à satisfaire les demandes agrégées. Le niveau inférieur *désagrège* les décisions du niveau supérieur: il définit pour chaque période détaillée τ, le volume de production par produit et par opération, de manière à satisfaire les demandes détaillées et les décisions agrégées. Soient :

X_{ijt} : quantité de produits i subissant l'opération j durant la période agrégée t,
S_{ijt} : quantité de produits i dans l'état j en stock en fin de période agrégée t.

Les variables agrégées et détaillées sont liées par les relations d'agrégation/dé-sagrégation suivantes:

$$\forall i \in [1, N], \forall j \in [1, J_i], \forall t \in [1, T] \quad \begin{cases} X_{ijt} = \sum_{\tau=k(t-1)+1}^{kt} x_{ij\tau}, & (1) \\ S_{ijt} = s_{ijkt}. & (2) \end{cases}$$

5 Modélisation

Les modèles associés aux 2 niveaux de la structure sont *dérivés* du modèle global $\mathcal{G}$ présenté précédemment.

5.1 Le Modèle Détaillé

Le niveau inférieur désagrège, période par période, la production agrégée. Un modèle détaillé $\mathcal{D}^t$ est associé à chaque période t. Il correspond au modèle global $\mathcal{G}$ auquel on rajoute les contraintes (1) et (2), en se limitant aux périodes τ concernées par la désagrégation, c'est à dire $\tau \in [k(t-1)+1, kt]$. Ces dernières contraintes matérialisent les relations d'agrégation/désagrégation entre modèles. Les états initiaux agrégés S_{ijt-1} et détaillés $s_{ijk(t-1)}$ sont donnés $\forall i \in [1, N], \forall j \in [1, J_i]$ et sont tels que $s_{ijk(t-1)} = S_{ijt-1}$.

5.2 Le Modèle Agrégé Direct

Le modèle agrégé direct $\mathcal{A}$ est le suivant : min $\sum_{\tau=1}^{T} \sum_{i=1}^{N} \sum_{j=1}^{J_i} H_{ijt} S_{ijt}$ sous

$$\begin{cases} \mathcal{A}_1 : S_{ijt} = S_{ijt-1} + X_{ijt} - X_{ij+1t} & \text{pour } i \in [1, N], j \in [1, J_i - 1], t \in [1, T] \\ \mathcal{A}_2 : S_{iJ_it} = S_{iJ_it-1} + X_{iJ_it} - D_{it} & \text{pour } i \in [1, N], t \in [1, T] \\ \mathcal{A}_3 : \sum_{i=1}^{N} \sum_{j=1}^{J_i} w_{ij}^r X_{ijt} \leq \mathrm{CAP}_t^r & \text{pour } r \in [1, R], t \in [1, T] \\ \mathcal{A}_4 : X_{ijt} \geq 0, S_{ijt} \geq 0 & \text{pour } i \in [1, N], j \in [1, J_i], t \in [1, T] \end{cases}$$

où :

$$H_{ijt} = \sum_{\tau=k(t-1)+1}^{kt} h_{ij\tau}, \tag{3}$$

$$D_{it} = \sum_{\tau=k(t-1)+1}^{kt} d_{i\tau}, \tag{4}$$

$$\mathrm{CAP}_t^r = \sum_{\tau=k(t-1)+1}^{kt} \mathrm{cap}_\tau^r. \tag{5}$$

Ce modèle est obtenu en *agrégeant* le modèle global $\mathcal{G}$ c'est à dire en sélectionnant dans l'ensemble des contraintes de $\mathcal{G}$, celles qui peuvent être exprimées en termes agrégés. Ainsi $\mathcal{A}_1$, $\mathcal{A}_2$, $\mathcal{A}_3$ et $\mathcal{A}_4$ résultent de l'agrégation directe de $\mathcal{G}_1, \mathcal{G}_2, \mathcal{G}_3$ et $\mathcal{G}_4$ compte tenu des relations (4) et (5). Avec cette logique, il n'est pas possible d'agréger directement la containte $\mathcal{G}_5$. Elle traduit une contrainte interne à la période agrégée (prise en compte des durées opératoires).

Par analogie avec le modèle global, le critère agrégé pénalise les stocks de fin de période, pour obtenir un plan agrégé au plus tard. Le coût de stockage agrégé retenu est fourni par la relation (3). Les dimensions de ce modèle sont alors :

$$N_v = 2T \sum_{i=1}^{N} J_i \quad \text{et} \quad N_c = T \left(R + \sum_{i=1}^{N} J_i \right).$$

5.3 Limites de cette Modélisation

Une décision agrégée est dite cohérente lorsqu'elle peut être désagrégée en une solution détaillée faisable [ERSCHLER 85]. Or, les plans agrégés fournis par le modèle direct ne sont pas toujours cohérents.

Exemple 1. Supposons que l'on désire planifier pour un horizon de $\mathcal{T} = 4$ périodes détaillées, la production d'un produit i unique. Sa gamme est constituée de 2 opérations $(i1)$ et $(i2)$ qui nécessitent respectivement une quantité $w_{i1}^{r_1} = 1$ de ressource r_1 et une quantité $w_{i2}^{r_2} = 1$ de ressource r_2. De plus, les capacités associées à chaque ressource sont $cap_\tau^{r_1} = 15, \forall\tau \in [1,4]$ et $cap_\tau^{r_2} = 10, \forall\tau \in [1,4]$. Les stocks initiaux sont nuls. Enfin, les demandes sont réparties de la façon suivante : $d_{i1} = 0$, $d_{i2} = 0$, $d_{i3} = 15$, $d_{i4} = 5$. Le graphe du processus de production est celui de la Fig. 2.a. Les nombres figurant sur les arcs représentent les capacités de production associées aux flux, ou les demandes. Une solution du problème global est donnée par la Fig. 2.b. Les nombres figurant sur les arcs sont les flux de produits. Pour une agrégation où $k = 2$, (regroupement en

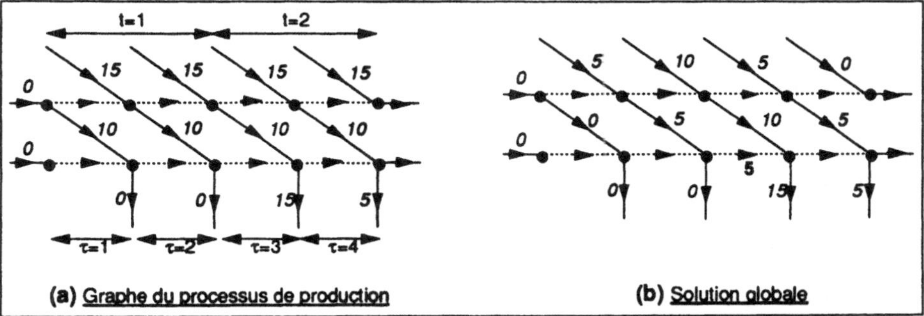

Fig. 2. Cas d'un problème où : $N = 1$, $J_1 = 2$, $R = 2$, $\mathcal{T} = 4$

$T = 2$ périodes agrégées), les données agrégées sont :

$$\begin{cases} CAP_t^{r_1} = 30, & \forall t \in [1,2] \\ CAP_t^{r_2} = 20, & \forall t \in [1,2] \end{cases} \quad \text{et} \quad \begin{cases} D_{i1} = 0 \\ D_{i2} = 20 \end{cases}$$

Le plan agrégé $X_{i11} = 0$, $X_{i21} = 0$, $X_{i12} = 20$, $X_{i22} = 20$ conduisant aux stocks $S_{i11} = S_{i12} = S_{i21} = S_{i22} = 0$ satisfait l'ensemble des contraintes du modèle agrégé direct. Il n'est cependant pas cohérent car il n'est pas possible de désagréger la production de la période agrégée 2. En effet, l'absence de stock ($S_{i11} = S_{i21} = 0$) ainsi que la répartition détaillée des capacités et des demandes ne permet pas de construire une solution détaillée faisable.

Différents cas de non cohérence peuvent être mis en évidence [HETREUX 93]. Ils sont liés au fait que le modèle agrégé direct ignore certaines caractéristiques détaillées du problème de planification: durées opératoires, contraintes d'enchaînement des opérations dans la gamme, répartition détaillée des capacités et des demandes. Ceci entraine une surévaluation par le niveau agrégé des flux de produits pouvant subir une opération (ij) dans la période t, et peut générer des infaisabilités lors de la désagrégation.

Des contraintes supplémentaires doivent donc être intégrées au modèle agrégé pour assurer la cohérence des décisions.

6 Conditions de Cohérence

Le paragraphe précédent a mis en évidence la faiblesse du modèle agrégé direct qui ne peut assurer la cohérence des décisions agrégées. Les conditions de cohérence visent à prendre en compte dans le modèle agrégé les caractéristiques du problème détaillé de manière à assurer l'existence d'une solution au niveau inférieur.

Ce type de problématique a déjà été développé dans le cadre d'une structure hiérarchisée basée sur une agrégation de produits [HAX 78], [AXSATER 84], [ERSCHLER 85], ainsi que dans le cadre d'une démarche liant planification et ordonnancement [DAUZERE 94].

Dans cette étude, l'exploitation du graphe du processus de production permet d'établir des conditions de cohérence, nécessaires et suffisantes, seulement nécessaires, ou seulement suffisantes selon la nature du processus de production.

On distingue 2 cas :

- Une ressource est dite *non partagée* lorsqu'elle est utilisée par une et une seule opération d'un et un seul produit.
- Une ressource est dite *partagée* dans les autres cas.

6.1 Cas de Ressources Non Partagées

Lorsque toutes les ressources sont non partagées, les produits peuvent être planifiés indépendamment les uns des autres (absence de couplage par les ressour-ces).

Conditions de Cohérence et Conditions d'Existence d'un Flot dans un Réseau Pour une période agrégée donnée t et un produit i, les flux de production détaillés peuvent être représentés à l'aide du réseau $\mathcal{R}_{it}$, construit à partir du graphe du processus de production complété d'une source O reliant les flux entrants et d'un puits M reliant les flux sortants (cf. Fig. 3). Les arcs *horizontaux* matérialisant les flux de produit restant dans le même état, ont une capacité infinie. Les arcs *obliques* symbolisant les flux $x_{ij\tau}$ ont une capacité $\text{cap}'_{ij\tau} = \text{cap}^r_\tau / w^r_{ij}$, si r est la ressource utilisée par l'opération (ij).

Les conditions de cohérence visent à limiter les flux agrégés de manière à ce qu'ils soient *désagrégeables* en flux détaillés. Elles correspondent donc aux conditions d'existence, dans ce réseau, d'un flot saturant les demandes (et compatible avec les grandeurs agrégées). De manière classique, la valeur de ce flot est limitée par la capacité des coupes séparant le puits M de la source O. Les contraintes sont donc de la forme:

$$\sum_{\tau=k(t-1)+1}^{kt} d_{i\tau} + \sum_{j=1}^{J_i} S_{ijt} \leq C(\delta), \quad \forall \delta \in \Delta \tag{6}$$

où δ est une coupe du réseau $\mathcal{R}_{it}$, Δ est un ensemble des coupes du réseau $\mathcal{R}_{it}$ et $C(\delta)$ est la capacité de la coupe δ.

Description des Coupes δ Candidates à Être Minimales La structure particulière du réseau (en niveaux) facilite l'énumération des coupes et le calcul de leur capacité. On peut remarquer sur cette classe de réseau 2 types de coupes:

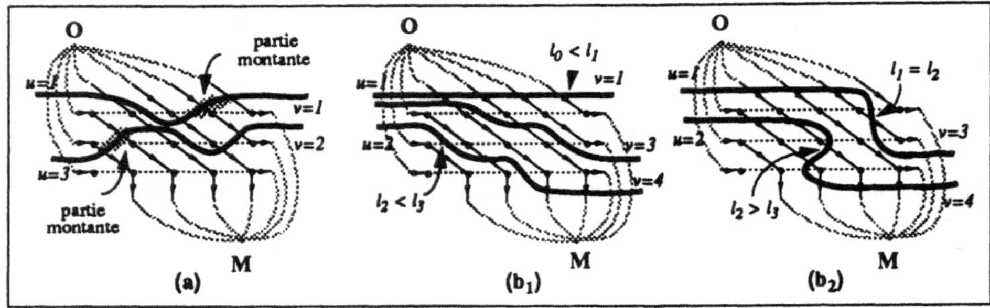

Fig. 3. Les différents types de coupes

(a) les coupes ayant une partie *"montante"* (cf. Fig. 3.a) : ces coupes contiennent des arcs horizontaux à capacité infinie, elles ont donc elles-même une capacité infinie; elles ne sont donc pas à considérer;

(b) les coupes *"descendantes"* : chaque coupe descendante est caractérisée par un triplet (u, v, l), tel que :

- u : le niveau de "départ" d'une coupe (extrêmité gauche),
- v : le niveau "d'arrivée" d'une coupe (extrêmité droite),
- l : le vecteur $(l_{u-1}, l_u, l_{u+1}, \ldots, l_p, \ldots, l_{v-1}, l_v)$ où :
  - * chaque l_p repère la période détaillée dans laquelle la coupe passe du niveau p au niveau $p+1$, pour $p \in [u, v-1]$,
  - * $l_{u-1} = k(t-1)$ et $l_v = kt + 1$.

On peut distinguer dans ce cadre:

(b_1) *les coupes descendantes* $(u \leq v)$ *t.q.* $l_{p-1} < l_p \; \forall p \in [u, v]$ (cf. Fig. 3.b_1),

(b_2) *les coupes descendantes* $(u \leq v)$ *t.q.* $\exists p \in [u, v]$ *t.q.* $l_{p-1} \geq l_p$ (cf. Fig. 3.b_2). Il est clair que la capacité de ce type de coupe est toujours supérieure à la capacité de coupes descendantes de type (b_1).

En conséquence, les seules coupes candidates à être minimales sont celles de type (b_1), constituant un ensemble $\widehat{\Delta} \subset \Delta$. Ainsi, tout triplet $(u, v, l) \in \widehat{\Delta}$ vérifie : δ est une coupe *descendante* où $u \leq v$ et $l_{u-1} < l_u < \ldots l_p \ldots < l_{v-1} < l_v, \forall p \in [u, v]$. On note $\widehat{L}(u, v) = \{l \mid l_{u-1} < l_u < \ldots l_p \cdots < l_{v-1} < l_v, \forall p \in [u, v]\}$.

Évaluation de la Capacité des Coupes $\delta \in \widehat{\Delta}$

– Si $v \leq J_i$:

$$C(u, v, l) = \sum_{j=u}^{J_i} S_{ijt-1} + \sum_{\tau=l_{u-1}+1}^{l_u-1} \mathrm{cap}'_{iu\tau} + \sum_{\tau=l_u+1}^{l_{u+1}-1} \mathrm{cap}'_{iu+1\tau} + \sum_{\tau=l_{v-1}+1}^{l_v-1} \mathrm{cap}'_{iv\tau} + \sum_{j=1}^{v-1} S_{ijt}$$

$$\text{soit:} \quad C(u, v, l) = \sum_{j=u}^{J_i} S_{ijt-1} + \sum_{j=1}^{v-1} S_{ijt} + \sum_{m=0}^{v-u} \sum_{\tau=l_{u+m-1}+1}^{l_{u+m}-1} \mathrm{cap}'_{iu+m\tau}. \tag{7}$$

– Si $v = J_i + 1$: la capacité des arcs obliques du niveau $J_i + 1$ est la demande détaillée. On a:

$$C(u, J_i+1, l) = \sum_{j=u}^{J_i} S_{ijt-1} + \sum_{j=1}^{J_i} S_{ijt} + \sum_{m=0}^{J_i-u} \sum_{\tau=l_{u+m-1}+1}^{l_{u+m}-1} \mathrm{cap}'_{iu+m\tau} + \sum_{\tau=l_{J_i}}^{kt} d_{i\tau}. \tag{8}$$

Remarque. On pose $\sum_{i=a}^{b} \alpha_i = 0$ lorsque $b < a$.

Expression des Conditions de Cohérence Les relations (6) doivent être écrites pour toutes les coupes $\delta \in \widehat{\Delta}$ de chaque réseau $\mathcal{R}_{it}$. Ces relations, une fois simplifiées à partir de l'expression de la capacité des coupes fournissent les conditions de cohérence. Ainsi :

– lorsque $v \leq J_i$:

$$\sum_{\tau=k(t-1)+1}^{kt} d_{i\tau} + \sum_{j=1}^{J_i} S_{ijt} \leq \sum_{j=u}^{J_i} S_{ijt-1} + \sum_{j=1}^{v-1} S_{ijt} + \sum_{m=0}^{v-u} \sum_{\tau=l_{u+m-1}+1}^{l_{u+m}-1} \mathrm{cap}'_{iu+m\tau}$$

$$\text{d'où :} \quad \sum_{\tau=k(t-1)+1}^{kt} d_{i\tau} + \sum_{j=v}^{J_i} S_{ijt} \leq \sum_{j=u}^{J_i} S_{ijt-1} + \sum_{m=0}^{v-u} \sum_{\tau=l_{u+m-1}+1}^{l_{u+m}-1} \mathrm{cap}'_{iu+m\tau}.$$

Les relations (4), $\mathcal{A}_1$ et $\mathcal{A}_2$ permettent d'écrire:

$$\sum_{\tau=k(t-1)+1}^{kt} d_{i\tau} + \sum_{j=v}^{J_i} S_{ijt} = D_{it} + \sum_{j=v}^{J_i} S_{ijt} = \sum_{j=v}^{J_i} S_{ijt-1} + X_{ivt}.$$

Il vient alors : $$\sum_{j=v}^{J_i} S_{ijt-1} + X_{ivt} \leq \sum_{j=u}^{J_i} S_{ijt-1} + \sum_{m=0}^{v-u} \sum_{\tau=l_{u+m-1}+1}^{l_{u+m}-1} \mathrm{cap}'_{iu+m\tau},$$

soit, $\forall u \in [1, J_i], \forall v \in [u, J_i]$: $$X_{ivt} - \sum_{j=u}^{v-1} S_{ijt-1} \leq \min_{l \in \widehat{L}(u,v)} \left(\sum_{m=0}^{v-u} \sum_{\tau=l_{u+m-1}+1}^{l_{u+m}-1} \mathrm{cap}'_{iu+m\tau} \right). \quad (9)$$

Dans le cas de ressources non partagées, lorsque $u = v$ (coupes horizontales), la relation (9) devient $X_{ivt} \leq \sum_{\tau=k(t-1)+1}^{kt-1} \mathrm{cap}'_{iv\tau}$ et est redondante avec $\mathcal{A}_3$.

- lorsque $v = J_i + 1$, on a:

$$\sum_{\tau=k(t-1)+1}^{kt} d_{i\tau} + \sum_{j=1}^{J_i} S_{ijt} \leq \sum_{j=u}^{J_i} S_{ijt-1} + \sum_{j=1}^{J_i} S_{ijt} + \sum_{\tau=l_{J_i}}^{kt} d_{i\tau} + \sum_{m=0}^{J_i-u} \sum_{\tau=l_{u+m-1}+1}^{l_{u+m}-1} \mathrm{cap}'_{iu+m\tau},$$

soit, $\forall u \in [1, J_i]$: $$\sum_{j=u}^{J_i} S_{ijt-1} \geq \max_{l \in \widehat{L}(u, J_i+1)} \left(\sum_{\tau=k(t-1)+1}^{l_{J_i}-1} d_{i\tau} - \sum_{m=0}^{J_i-u} \sum_{\tau=l_{u+m-1}+1}^{l_{u+m}-1} \mathrm{cap}'_{iu+m\tau} \right). (10)$$

Ainsi, dans le cas de ressources non partagées, les relations (9) et (10) sont des conditions nécessaires et suffisantes de cohérence. Elles permettent d'*intégrer* au niveau agrégé les contraintes d'enchaînement d'opérations et de durées opératoires et prennent en compte la répartition détaillée des capacités et des demandes. Dans ces conditions, le nombre N_c^{sup} de contraintes supplémentaires est :

$$N_c^{\mathrm{sup}} = T \sum_{i=1}^{N} \frac{J_i(J_i+1)}{2} \approx \frac{TNJ^2}{2}$$

(J représente le nombre moyen d'opérations par produit).

Exemple 2. L'exemple du §5.3 est repris. L'écriture des conditions de cohérence en période agrégée 2 donne les contraintes supplémentaires suivantes (cf. Fig. 4):

$$\begin{cases} X_{i22} - S_{i11} \leq \mathrm{cap}_{i13} \\ X_{i22} - S_{i11} \leq \mathrm{cap}_{i24} \end{cases} \text{soit} \begin{cases} X_{i22} - S_{i11} \leq 15 & (\text{Coupe } \delta_1) \\ X_{i22} - S_{i11} \leq 10 & (\text{Coupe } \delta_2) \end{cases}$$

$$\text{et} \begin{cases} S_{i11} + S_{i21} \geq d_{i3} \\ S_{i21} \geq d_{i3} - \mathrm{cap}_{i23} \end{cases} \text{soit} \begin{cases} S_{i11} + S_{i21} \geq 15 & (\text{Coupe } \delta_3) \\ S_{i21} \geq 5 & (\text{Coupe } \delta_4) \end{cases}$$

Le plan agrégé proposé au §5.3 ne satisfait pas ces conditions de cohérence. Par contre, le plan agrégé $X_{i11} = 15$, $X_{i21} = 5$, $X_{i12} = 5$, $X_{i22} = 15$ vérifie ces conditions. Ce plan est donc cohérent et sa désagrégation donne la solution détaillée de la Fig. 2.b.

6.2 Cas des Ressources Partagées

Le problème est plus délicat à résoudre puisque les produits et les opérations sont couplés par la capacité des ressources qu'ils se partagent. Néanmoins, en exploitant les résultats précédents, il est possible d'obtenir soit des conditions nécessaires de cohérence, soit des conditions suffisantes.

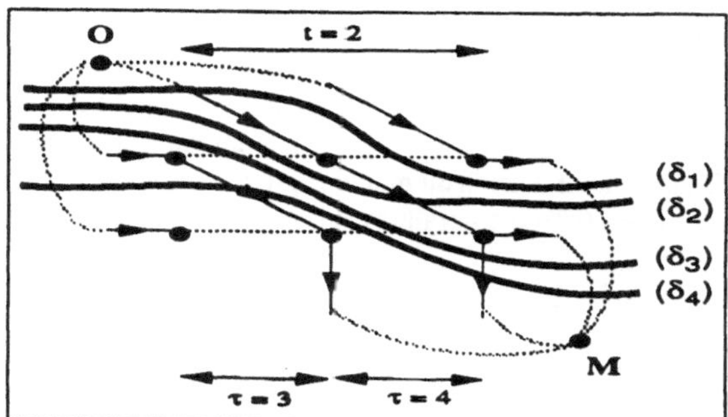

Fig. 4. Coupes sur le réseau $\mathcal{R}_{i2}$

Conditions Nécessaires de Cohérence Les conditions précédentes (9) et (10) constituent dans le cas de ressources partagées, des conditions seulement nécessaires. En effet, elles supposent que chaque opération de chaque produit dispose de la totalité de la capacité de la ressource correspondante, ce qui n'est pas le cas. L'efficacité de ces conditions pour se rapprocher de la cohérence est alors liée au *degré de partage* des ressources.

Conditions Suffisantes de Cohérence Le principe de base d'obtention de conditions de cohérence est le suivant: on effectue une pré-affectation de la capacité des ressources aux différentes opérations. Le problème devient *à ressources non partagées* et les résultats précédents peuvent être utilisés. Ainsi, le modèle agrégé fixe pour chaque période agrégée, une répartition des capacités entre les différentes opérations et définit simultanément des volumes de production agrégés cohérents.

Ceci est réalisé en remplaçant dans le modèle agrégé les termes $\text{cap}'_{ij\tau}$ par les termes variables $\beta_{ijt}\text{cap}'_{ij\tau}$. Les variables supplémentaires β_{ijt}, $(i \in [1, N], j \in [1, J_i], t \in [1, T])$ matérialisent cette répartition de capacité avec les contraintes suivantes :

$$0 \leq \beta_{ijt} \leq 1, \qquad \sum_{ij \in \mathcal{O}_{ij}^r} \beta_{ijt} = 1, \quad \forall r \in [1, R],$$

avec $\mathcal{O}_{ij}^r$ ensemble des opérations ij utilisant la ressource r. Les conditions de cohérence sont alors obtenues à partir des conditions (9) et (10) dans lesquelles les données $\text{cap}'_{ij\tau}$ deviennent des termes variables $\beta_{ijt}\text{cap}'_{ij\tau}$. D'où, $\forall u \in [1, J_i], \forall v \in [u, J_i], \forall l \in \widehat{L}(u, v)$:

$$X_{ivt} - \sum_{j=u}^{v-1} S_{ijt-1} - \sum_{m=0}^{v-u} \sum_{\tau=l_{u+m-1}+1}^{l_{u+m}-1} \beta_{iu+mt}\text{cap}'_{iu+m\tau} \leq 0, \tag{11}$$

et, $\forall u \in [1, J_i], \forall l \in \widehat{L}(u, J_i + 1)$:

$$\sum_{j=u}^{J_i} S_{ijt-1} + \sum_{m=0}^{J_i-u} \sum_{\tau=l_{u+m-1}+1}^{l_{u+m}-1} \beta_{iu+mt}\text{cap}'_{iu+m\tau} \geq \sum_{\tau=k(t-1)+1}^{l_{J_i}-1} d_{i\tau}. \tag{12}$$

Les relations (11) et (12) sont des conditions suffisantes de cohérence dans le cas de ressources partagées.

L'hypothèse supplémentaire d'une répartition constante des capacités peut dans certains cas limiter le nombre de solutions agrégées et même générer des infaisabilités (pas de solutions agrégées) alors que globalement, il existe une solution. Ceci explique le caractère suffisant de l'approche.

Le modèle détaillé qui reste celui présenté au §5.1 ne contient pas l'hypothèse de répartition constante de capacité retenue pour le niveau agrégé. Les seules variables à désagréger restent donc X_{ijt} et S_{ijt}. Il en découle que le niveau détaillé dispose d'autonomie pour générer une solution qui respecte ou ne respecte pas la répartition des capacités élaborée au niveau supérieur.

Quant à la dimension du modèle agrégé, le nombre N_v de variables devient: $N_v = 3T \sum_{i=1}^{N} J_i$ et le nombre N_c^{sup} de contraintes supplémentaires est :

$$N_c^{\text{sup}} = T\left(N + \sum_{i=1}^{N} J_i + \sum_{i=1}^{N}\sum_{j=2}^{J_i+1}(J_i + 2 - j)\left(\prod_{m=0}^{j-2}(k-m)\right)/(j-1)!\right).$$

Le nombre de contraintes supplémentaires est donc considérable dans le cas général. Il est cependant important de noter que ce nombre peut être grandement réduit lorsque les capacités de chaque niveau et les demandes sont constantes dans une période agrégée. On obtient alors :

$$N_c^{sup} = T\sum_{i=1}^{N}\frac{J_i(J_i+1)(J_i+3)}{6} \approx \frac{TNJ^3}{6} \ .$$

De plus, l'exploitation des propriétés particulières de certaines coupes permet de réduire de façon trés significative le nombre de conditions de cohérence.

7 Conclusion

Des expérimentations numériques actuellement en cours permettent de montrer la validité et la faisabilité des résultats présentés. Il faut cependant noter que la mise en oeuvre des conditions de cohérence peut conduire à rajouter de nombreuses contraintes supplémentaires dans le modèle agrégé. Pour minimiser l'accroissement de la complexité qui en découle, les études actuelles s'attachent à réduire le nombre de contraintes supplémentaires à prendre en compte. Deux voies de recherches sont envisagées:

- Étude a priori des propriétés des coupes associées au graphe du processus.
- Calcul heuristique hors ligne des grandeurs β ($N_c^{\text{sup}} \approx TNJ^2/2$).

Ces études permettront de mieux cerner les difficultés liées à la prise en compte de la cohérence, dans un souci constant d'applicabilité.

Références

[AXSATER 84] Axsater S., Jonsson H.: *Aggregation and desaggregation in hierarchical production planning*, European Journal of Operation Research **17** 338–350, 1984.

[DAUZERE 94] Dauzere-Peres S., Lasserre J.B.: *Integration ot lotsizing and scheduling decisions in a Job-Shop*, To appear in a special issue of European Journal of Operation Research on "Lotsizing Models for Production Planning", 1994.

[ERSCHLER 85] Erschler J., Fontan G., Merce C.: *Consistency of the disaggregation process in Hierarchical Planning*, Operation Research **14**, No. 3, May June 1985.

[HAX 78] Hax A.C., Golovin J.J.: *Hierarchical Production Planning System*, in Studies in Operation Management, AC Hax (Ed.) Amsterdam, North Holland, 1978.

[HETREUX 93] Hetreux G.: *Approche hièrarchisée pour la planification de la production*, Rapport de D.E.A Automatique/Informatique Industrielle, Institut National des Sciences Appliquées de Toulouse, Juin 1993.

[JOHNSON 74] Johnson A.J., Monygomery D.C.: *Operations Research in Production Planning, Scheduling, and Inventory Control*, Wiley, New York, 1974.

Multi-site planning: A centralized or a distributed approach ? [1]

*C. THIERRY**, P. LE PAGE*, N. CHAPEAUBLANC* and G. BEL **

1. *A part of this work has been done within PROMIP and is closely related to the DISCO ESPRIT Project with Magneti Marelli, Bull, AEG, ISA, Fraunhofer Institut Stuttgart(IAO).*
* *Centre d'Etudes et de Recherches de Toulouse/ Département d'Etudes et de Recherches en Automatique (CERT/DERA)*
** *CERT/DERA, Université Toulouse II Le Mirail*
CERT/DERA, 2 avenue Edouard Belin BP 4025, 31055 TOULOUSE Cedex FRANCE
Phone : (33) 61.55.71.11, Fax : (33) 62.25.25.25, E-Mail : thierry@saturne.cert.fr

1. Introduction

The increasing industrial competition in the world leads large companies to try to be more and more efficient as far as production planning control is concerned. Those big companies very often have several geographically independent production units with several production alternatives to produce customers orders. A plant oriented production, with local planning and control systems, exists in most of these companies, which induces a lot of inefficiencies at the company's level (cf [THI 94]).
This paper will present studies whose aim is to improve the performance of the company without any important updating of data, expensive hardware installation and long elaboration time. Consequently, the multi-site planning problem has been studied and two different approaches (a centralized approach and a distributed approach) have been tested and compared. Those two approaches appear to be interesting : in some companies a central department of logistics exists (so a centralized approach is advisable), on the other hand, other companies are broken up in several totaly autonomous entities with their own objectives (so a distributed approach is advisable).

2 Multi-site planning problem

The multi-site planning problem is represented on figure 1. It can be defined by the following set of data :

- a set of production units with their geographical site and their available aggregated resources (expressed in resources×time),
- a set of items (or components) produced with for each item, the available production units for the manufacture, the production duration, the production cost and the necessary quantity of resources in the different production units,
- a set of customer orders with the reference of each order, the final item required, the item required quantity, the delivery date required by the customer and the delivery production unit,
- the finished items bills of materials,
- the transportation duration between production units.

The problem is to determine items (WHAT?) quantities (HOW MUCH?) to be produced in the different production units (WHERE?) and transported between production units on different time periods (WHEN?). This must be done satisfying a

set of constraints (precedence constraints, resources limitation constraints, due dates constraints, transportation duration constraints) with a low global production cost and respecting a set of preferences.
The problem has been dealt with two different approaches, a centralized and a distributed approach. Those approaches will be introduced with both a modelling and a solving method.

3. The centralized approach

3.1. The model

The multi-site planning centralized approach is essentially based on the assumption that a central logistic department exists in the company and is in charge of coordinating all the activities of planning, programming and purchasing. Figure 1 represents the centralized approach organization.

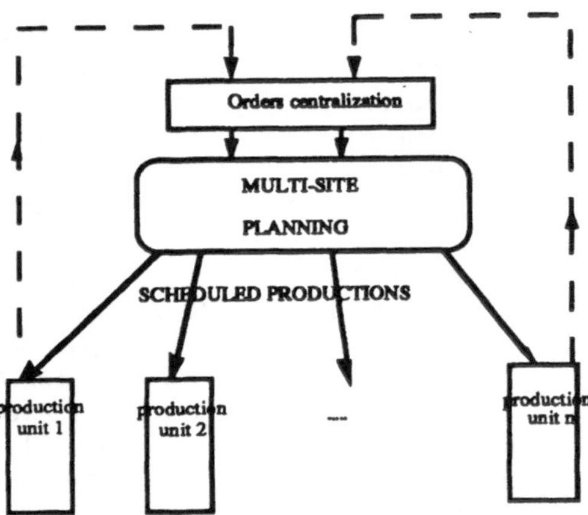

Figure 1. A centralized approach

The problem consists in a set of decision variables and dependant variables, connected with different constraints. We chose a so-called "flow model" with the following decision variables :

- $QP(i,t,p)$ = total quantity of item i to be produced on production unit p during the time period t,
- $QT(i,t,tp)$ = quantity of item i to be transported during the time period t through the transportation path tp,

The state variables which depend on the decision variables are the following :

- $QS(i,t,p)$ = quantity of item i stocked on production unit p at the beginning of the time period t,
- $QRU(t,p)$ = quantity of aggregated resources (expressed in resources×time) required on production unit p during the time period t,
- $QDC(i,t,p)$ = quantity of item i required on production unit p before the end of time period t.

This flow model includes several types of constraints :

• **The capacity constraints** are expressed as following : the consumed number of resources×time on each production unit, for each time period, must be smaller than the global resource capacity expressed in resources×time.

$\forall p, \forall t,\quad QRU(t,p) \leq QRA(t,p)$

$$\text{with } QRU(t,p) = \sum_i QP(i,t,p)*QNR(i,p) \qquad (1)$$

with : $QNR(i,p)$ = necessary quantity of resources×time to produce item i on production unit p,

$QRA(t,p)$ = available resources×time on production p during the time period t,

• **The due dates constraints** express the respect of the customers' due dates. A constraint states that the quantity of each ordered product must be available in stock in the delivery production unit at the right due-date (expressed in time period).

$\forall i, \forall p, \forall t,$

$$QS(i,t,p) = required\_quantity(i, t, p) \qquad (2)$$

with : $required\_quantity(i,t,p)$ = required quantity of item i on the production unit p during the period t.

• **The bill of material constraints** express the links between the different items in the bill of material. $QDC(i,t,p)$ is the quantity of item i required on production unit p before the end of time period t to produce items j.

$\forall i, \forall p, \forall t,$

$$QDC(i,t,p) = \sum_{j \in LIU(i,p)} QP(j,t,p)*QIU(i,j) \qquad (3)$$

with : $LIU(i,p)$ = list of items that can be produced on production unit p and which directly need item i to be produced (a view of the bill of material),

$QIU(i,j)$ = necessary quantity of item i to produce one unit of item j,

• **The conservation constraints (or state equations)** are the main feature of the flow model. They describe stocks level relations due to the production and transportation activities.

For each item, the following conservation equation must be respected :

$\forall i, p, t,$

$$QS(i,p,t+1) = QS(i,t,p) + QP(i,t,p) - QDC(i,t,p) + Q_{arrived}(i,t,p) - \sum_{tp \in from(p)} QT(i,t,tp) \qquad (4)$$

with $Q_{arrived}(i,t,p)$ = quantity of item i arrived on production unit p during the time period t which can be expressed as follows :

$$Q_{arrived}(i,t,p) = \sum_{tp \in to(p)} QT(i,t-TT(i,tp),tp) \qquad (5)$$

$TT(i,tp)$ = transportation time needed to transport item i through transportation path tp,

$from(p)$ = set of transportation paths which origin is p ,

$to(p)$ = set of transportation paths which destination is p ,

3.2. Solving method

The above considerations show that the multi-site scheduling problem can be stated as constraint satisfaction problem (CSP). Given a set of variables, a set of possible values for these variables (domain) and a set of constraints connecting the variables, a CSP consists in finding the values of the variables such that the constraints are satisfied (see [DEC 88]).
In the above multi-site scheduling problem formulation, the purpose is to find the values of integer variables (decision variables as QP) satisfying the constraints of the flow model. The key idea to solve CSP problems is to use constraints to prune the search space in an a priori way. This is done by a constraint manipulation mechanism called the constraint propagation. When some variables are not yet instanciated (the propagation mechanism only expresses necessary conditions), an heuristic knowledge has to be used in a decision process to find a solution of the problem. The decision process first chooses a variable then its value. This process can be viewed as a strategy to explore the set of feasible solutions. After each choice, the propagation mechanism deduces domain restrictions. Backtracking may occur if the decision leads to a constraint violation.
With this approach, various strategies can be easily implemented and the tuning of strategies is performed according to specific behaviour of the implementation site (cf [BEL93]). So this approach is very flexible as new constraints can be easily added and specific criteria or preferences can be used in the decision process.

4. The distributed approach

4.1. The model

A distributed approach respecting the natural distribution of the different production units in the company has been chosen [MAS 90], [DEM 90]. This approach can be represented as in figure 2 : the multi-site planning problem is solved thanks to the communications and dialogues between the entities.
An iterative method has been chosen to deal with the problem . That is to say that orders will be treated one by one. Each item production is labelled by an order.
Multi-site planning distributed model consists in [LEP 93]

- a set of entities, the different production units,
- a set of relations between those entities with :
  - requirements emissions from an entity to another,
  - answers to those requirements from the concerned entity to the requirement transmitter.
- a set of behaviours of the different entities,

The relations between entities can be described as in figure 3. Each entity can alternatively emit or answer a requirement. Nevertheless, for a specific order, there is a command transmitter (the production unit where the order has been put in) and several entities which answer this requirement (the entities that can communicate with the transmitter).
A requirement to an entity can be expressed by the transmitter to production unit p as follows : "can you produce quantity $QDC\ (i,o,t,p)$ of item i concerning order o for the due date t?". Entities answers will be an acceptance or a refusal or an acceptance under conditions (for instance "yes if I do not produce this other order").

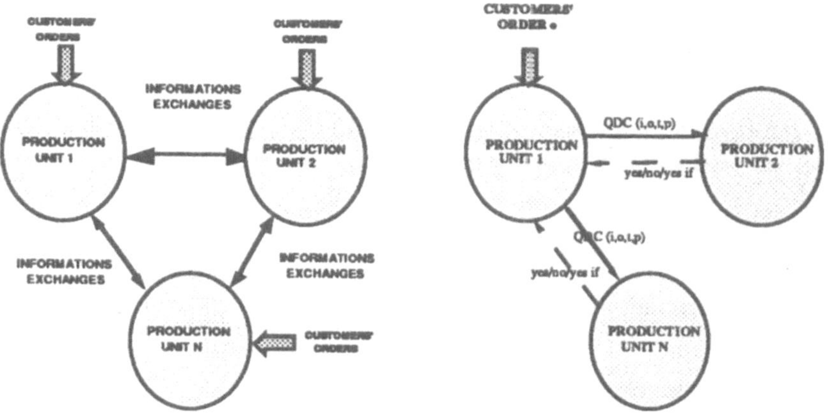

Figure 2. A distributed approach

Figure 3. Entities relations

Entities behaviours have to take into account a set of constraints. They elaborate their answers and requirements and they decide to give a value to all the decision variables concerning themselves ($QP(i,o,t,p)$=quantity of item i concerning order o produced on production unit p during t.).

• **The bill of material constraints** will be taken into account during the requirement elaboration. Let us consider an item i and an order o of item i which is produced in production unit p. Then, if item j is a component of item i, a quantity of item j corresponding to order o will be required to production unit $p1$ with the due date dd. This requirement can be calculate as following:

$$QDC(j,o,dd,p1)=QDC(i,o,t,p)\times QIU(j,i) \quad (6)$$

with $QIU(j,i)$=quantity of item j necessary to produce one item i.

As a production corresponding to a requirement is considered to be entirely achieved in a single time period, due date dd for the requirement concerning item j will be calculated as follows :

$$dd = PTP(i,o,t,p) - TT(j,p,p1) \quad (7)$$

TT= transportation time for item j between production unit p and production unit $p1$.
PTP(i,o,t,p) = production date (expressed in time periods) of item i concerning order o on production unit p.

• **The capacity and due date constraints** are taken into account during the answer elaboration concerning a requirement (for instance QDC(i,o,t,p)). Then t is the due date for the item i production concerning order o and both the constraints are expressed with the same constraint as follows : in each production unit,

$$\exists\ t0>t\ /QRU(i,o,t0)\leq QRA(t0) \quad (8)$$

with $QRU(i,o,t)$= required resources (expressed in resources×time) of the concerned production unit during time period t, for the production of item i concerning order o.

$$QRU(i,o,t)=QDC(i,o,t,p)*QNR(i,p) \quad (9)$$

QRA(t)=available resources (expressed in resources×time) on the concerned production unit during time period *t*.

$$QRA(t)=QRAi(t)-\sum_{o\in ATO}\sum_{i\in ATI} QP(i,o,t,p)\times QNR(i,p) \qquad (10)$$

QRAi(t)= initially available resources (expressed in resources×time) of the concerned production unit, at the beginning of the planning horizon during time period t.

ATO=orders already planned on production unit p when the requirement arrives in this production unit

ATI =items already planned on production unit *p* when the requirement arrives in this production unit p

QNR(i,p)=necessary quantity of (resources×time) for item *i*.

4.2. Solving method

In this approach, we propose a simple algorithm tested with a discrete events simulation. A requirement transmission is sent (containing a requirement due date) to a production units for items respecting bill of material constraints. An acceptance (or respectively a refusal) is sent back if constraints are respected (or respectively if they are not). When an acceptance is sent then we chose latest starting time (expressed in time period) for orders planning. When a refusal is sent the requirement is transmitted to the other production units.
New solving methods can be defined, as new relations types are added to the previous model. For instance we could choose earliest starting time (expressed in time periods) for orders planning.

5. Comparison

A comparison of both approaches can be made as far as the following issues are concerned:

• **optimization** :
With a centralized approach a global optimization criterium can be defined (for instance if global cost minimization is concerned). With a distributed approach we can only define local criteria even if a global cost minimization should emerge from those local criteria.
So if an optimization approach is used in the centralized approach we are sure to globally optimize this criteria whereas it is not true for the distributed approach. But if for calculus efficiency reasons, heuristics are used then this approach does not guarantee the global minimization of the criterium even if our aim is to find a "good solution"(as far as the global cost is concerned).

• **autonomy** :
In the centralized approach we are essentially interested in the global behaviour of the system that is to say with the whole company. In the distributed approach each entity (production unit) is studied individually as far as local objectives and behaviours are concerned. So each entity (production unit) has the possibility to take decisions such as "I can produce this item" or "I cannot produce this item", whereas in the centralized approach decisions are imposed on the different production units by the logistics department.

• **problem structuration and model evolution** :
Since the distributed approach keeps the natural distribution, different variables

definitions are more convenient than in the centralized approach. With a distributed approach, a problem structuration is necessary whereas in the centralized approach a structuration effort must be made independently of the approach.Moreover in the distributed approach, as the natural distribution is kept each entity can be easily described precisely, which is more difficult with a centralized approach. For instance, production units modelling (as entities) can be detailed modelising sub-entities inside those production units entities. Each sub-entities can be a different department. For instance a production unit can be modelled with a customers Department, a planning Department, a scheduling Department and a manufacturing Department There is no need of a big data base where all data are stocked. Each entity will has its own behaviour and new entities can be added easily.

• **dynamicity** :

In the distributed approach an iterative algorithm is used that is to say that requirements are treated one by one. On the contrary, in the centralized approach requirements are treated globally. So as far as our approaches are concerned, the distributed approach permit a react to unpredictable events (like arrival of new orders). Moreover the distributed approach enables to limit a perturbation influency (random requirements, capacities variations...) to the concerned entity neighbourhood.

6. Conclusion

The multi-site planning problem has been studied. Two approaches have been used to model and solve this problem. This lead us to compare principles of the two main approaches in term of optimization, autonomy, evolution and dynamicity. Up to now only the centralized approach has been validated on real data from industrial case (the lighting division of Magneti Marelli). A real case with about 30 production units, 5 time periods and 30 orders with 70 parts of final items (with about 10 bill of material levels) has been tested. The CPU time to solve these cases with 4000 decision variables and 16000 constraints is about 5 minutes on a SPARC station. The validation of the distributed approach will be done on these data to compare quantitatively both approaches efficiency in term of optimality and solving efficiency.

References

[BEL 93] G. BEL and C. THIERRY "A constraint based system for multi-site coordination and scheduling", IJCAI, Workshop on knowledge based production planning scheduling and control, Chambery, 29 Août 1993,

[DEC 88] DECHTER R. and PEARL J. (1988), "Network-based heuristics for constraints satisfaction problems", Artificial Intelligence, 1988, vol 34, pages 1-38,

[DEM 90] Y. DEMAZEAU and MULLER, *"Decentralized artificial Intelligence"*, Proceedings of the 1st modelling autonomous agents and multi-agents worlds, North Holland, 1990,

[LEP93] P. LE PAGE, Thèse de doctorat de l'ENSAE (Ecole Nationale Supérieure de l'Aéronautique et de l'Espace), décembre 1993,

[MAS 90] G. MASINI, A. NAPOLI, D. COLNET, D. LEONARD and K. TOMBE, *"Les langages à objets"*, Inter Edition 1990.

[THI 94] C. THIERRY, Thèse de doctorat de l'ENSAE (Ecole Nationale Supérieure de l'Aéronautique et de l'Espace), janvier 1994,

Algorithms for Simultaneous Scheduling of Machines and Vehicles in a FMS

Tadeusz Sawik

University of Mining and Metallurgy
Department of Computer Integrated Manufacturing
30-059 Cracow, Poland

1 Introduction

Production planning and scheduling in flexible manufacturing systems are concerned with short-term allocation of production resources so as to meet the requirements for produced part types and to best utilize the system capabilities. Two distinct approaches for solving the production planning and scheduling problem are used in practice.

The first approach, termed as multi-level or hierarchical, decomposes the problem into a hierarchy of subproblems to be solved sequentially. The solutions of subproblems from the upper hierarchy impose constraints on the lower level subproblems. A typical hierarchical decision structure for an FMS production planning and scheduling consists of two levels: the upper level which involves tactical planning and the lower level involving operational control, e.g., Gershwin et.al [2], Sawik [3,4].

The second approach, denoted as monolithic or single-level, usually formulates the problem as a large scale mathematical programming problem with a single level decision making structure required for its solution, e.g., Buzacott and Yao [1].

The purpose of this paper is to present and compare both the approaches for the problem of simultaneous scheduling of machines and vehicles in a flexible machining system.

2 Description of an FMS

Consider an FMS composed of m machine tools i, $(i = 1, \dots, m)$, one loading $(i = 0)$ and one unloading $(i = m + 1)$ stations. In the system n different (part) operations can be performed. Let $J = \{1, \dots, n\}$ be the set of all *part operations*, where each part operation $j \in J$ corresponds to a particular part type. The manufacture of part type k requires a sequence of part operations $j_{k0}, j_{k1}, \dots, j_{kn}, j_{k,n+1}$ to be performed in the listed order, beginning with loading operation j_{k0} on machine $i = 0$ and ending with unloading operation $j_{k,n+1}$ on machine $i = m+1$. Part operation $j \in J$ can be performed on any machine in the subset $I_j \subset I = \{0, 1, \dots, m, m+1\}$ of the appropriate machines. Let p_{ij} be the processing time on machine $i \in I_j$ of part operation j.

Each part operation requires one or more different types of cutting tools for processing. The tools for all part operations that can be performed by machine i are stored in its limited capacity tool magazine containing S_i tool slots, where each tool type $g \in G$ occupies s_g slots.

The material handling system consists of a AGVs that permit parts to move between any pair of machines. Let q_{il} be the transportation time required for a vehicle to move from machine i to machine l.

Finally, let r_j be the production requirement for part operation j (in number of parts of a particular type), and d_j - the associated due date for all r_j part operations. The production requirements must be satisfied with the primary objective of minimizing the maximum difference between the actual completion time and due date for all part types.

3 Short-term planning and scheduling in FMS: single-level versus multi-level approach

The short-term production planning and scheduling of an FMS includes machine loading, part routing, and machine and vehicle scheduling (e.g., Sawik [3,4]). The solution to all the above problems can be find simultaneously using a single-level approach or sequentially, if a multi-level approach is applied.

The single-level approach aims at direct scheduling of machines and vehicles so as to meet all part type requirements. The other subproblems need not to be considered separately using this approach since they are solved implicitely by making the detailed scheduling decisions.

In the multi-level approach however, solving of a lower level subproblem requires earlier solution to all higher level subproblems (see, Fig.1). In order to simultaneously schedule machines and vehicles, first the machine loading determines operation assignments z_{ij} so as to balance workloads on each machine. Then, the part routing selects flows of parts f_{ilj} required to meet the operations assignments z_{ij} in such a way as to balance transportation time on all paths of the network.

The machine loading problem and the part routing problem are formulated below (e.g., Sawik [3,4]).

Minimize

$$P_{\max} \tag{1}$$

subject to

$$\sum_{i \in I} z_{ij} = r_j; \quad j \in J; \tag{2}$$

$$\sum_{j \in J} p_{ij} z_{ij} \leq P_{\max}; \quad i \in I; \tag{3}$$

$$\sum_{g \in G} s_g t_{gi} \leq S_i; \quad i \in I; \tag{4}$$

$$\sum_{j \in J_{gi}} z_{ij} \leq \Big(\sum_{j \in J_{gi}} r_j\Big) t_{gi}; \quad g \in G; \quad i \in I; \tag{5}$$

$$z_{ij} \geq 0; \text{ integer}; \quad i \in I; \quad j \in J; \tag{6}$$

$$t_{gi} \in \{0, 1\}; \quad g \in G; \quad i \in I. \tag{7}$$

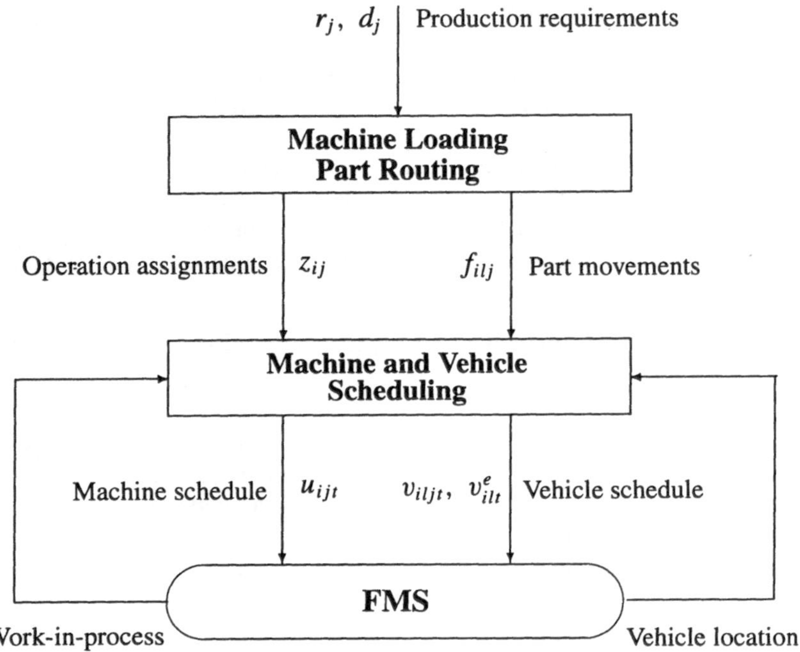

Fig. 1. Machine and vehicle scheduling in FMS: a multi-level approach

The objective (1) is a measure of system imbalance and represents workload of the bottleneck machine defined by constraint (3). Equation (2) ensures that all part operation requirements are allocated among the machines. Constraints (4) and (5) ensure loading of the required tools on appropriate machines with limited capacity tool magazines (J_{gi} is the set of part operations that require tool type g on machine i).

Minimize

$$Q_{\max} \tag{8}$$

subject to

$$\sum_{l=1}^{m+1} f_{ilj} = z_{ij}; \qquad i = 0, \dots, m, \qquad j \in J \setminus J_{UL}; \tag{9}$$

$$\sum_{i=0}^{m} f_{ilj} = z_{lj+1}; \quad l = 1, \dots, m+1, \quad j \in J \setminus J_{UL}; \tag{10}$$

$$\sum_{i=0}^{m} \sum_{j \in J} q_{il} f_{ilj} \le Q_{\max}; \qquad l \in I; \tag{11}$$

$$f_{ilj} \ge 0; \text{ integer}; \qquad i, l \in I; \qquad j \in J. \tag{12}$$

The objective (8) is a measure of transportation network imbalance. $Q_{\max}$ denotes total transportation time required to transfer parts directly to the bottleneck machine

Table 1. Decision Variables

Machine loading and part routing	
f_{ilj}	= number of part operations j to be transferred from machine i, to machine l to perform next part operation $j+1$;
t_{gi}	= 1, if tool type g is loaded on machine i; otherwise $t_{gi} = 0$
z_{ij}	= total number of part operations j assigned to machine i
Machine and vehicle scheduling	
u_{ijt}	= 1 if part operation j is assigned to machine i in period t; otherwise $u_{ijt} = 0$
v_{iljt}	= 1 if transferring of part operation j from machine i to machine l to perform next part operation $j+1$ starts at the beginning of period t; otherwise $v_{iljt} = 0$
v^e_{ilt}	= 1 if at the beginning of period t an empty vehicle starts moving from machine i to machine l to pick up a part waiting for transfer; otherwise $v^e_{ilt} = 0$
h_t	= length of assignment period t

defined by constraint (11). Constraints (9) and (10) ensure that the required operation assignments on each machine be met (J_{UL} is the set of unloading operations).

4 Machine and Vehicle Scheduling

The objective of the detailed machine and vehicle scheduling is to find an assignment of operations to machines for each part over a scheduling horizon as well as the associated time table for vehicle movements, where some are transferring parts between machines and others are moving empty to pick up parts waiting for transfer.

The scheduling horizon is assumed to be made up of T assignment periods (T is an unknown integer) which may have unequal time durations, during each of which the assignment of operations to machines and vehicles to routes (pair of machines) is considered fixed.

The algorithms presented below are period-by-period heuristics in which each assignment period is considered once, see also Sawik [3,4]. The assignment decisions in every period are supported by a family of dispatching rules. A rule helps an idle (or soon to be idle) machine to select its next part operation from the parts waiting at the machine's buffer. Similarly, another rule allows an idle (or soon to be idle) vehicle to select its next part for movement to one of the part's next machines as well as to select the next machine.

The decision variables (see, Table 1) for each assignment period t are determined in the following sequence:

1. Machine and parts unloading assignment, $u_{ijt},\ i = 1, \dots, m+1$;
2. Parts loading assignment, u_{0jt};
3. Vehicle assignment for parts transferring, v_{iljt};
4. Idle vehicle assignment, v^e_{ilt};
5. Length of assignment period, h_t.

In the scheduling algorithms described below the machine and vehicle assignments are made in the order of nondecreasing values of one of the following priority indices chosen for part operations $j \in J$.

STP& TT (Shortest Total Processing and Transportation Time)
EDD (Earliest Due Date)
MDD (Earliest Modified Due Date) *Original due date d_j or the latest estimated completion time of the remaining part operations j, whichever is greater,*
SCR (Smallest Critical Ratio) *Due date for all part operations j /total processing and transportation time (per one vehicle) required to complete remaining part operations j.*

Algorithm for the multi-level approach

Step 0. Choosing the priority indices and the initial conditions.
Step 1. Machine and parts unloading assignment in period t.
Part operations j for the remaining uncompleted tasks z_{ij} assign to idle machines i in the order of their priority.
Step 2. Vehicle assignment for parts transferring in period t.
For the remaining uncompleted transportation tasks f_{ilj}, part operations j assign to idle vehicles in the order of their priority.
Step 3. Idle vehicle assignment in period t.
An idle vehicle waiting at machine i send to machine l to pick up part operation j with the highest priority.
Step 4. Length h_t of assignment period t.
Set h_t equal to the longest possible period until the earliest completion of an operation or the earliest arrival of an AGV in the destination machine.
Step 5. The terminal conditions.
If all unloading operations are completed then set $T = t$ and terminate. Otherwise, calculate the in-process inventory and vehicles location at the end of period t, set $t = t + 1$ and go back to Step 1.

In the multi-level approach each machine must complete the assigned production tasks z_{ij}, whereas the vehicles must complete the associated transportation tasks f_{ilj}. In the single-level approach however, the production schedule is constructed based on the current system workload and remaining production requirements. Generally, the algorithm proceeds similarly as that for the multi-level approach with slight modifications in Steps 1 and 2, which now should be read as follows:

Step 1. Part operations j waiting at the idle machines capable of performing them assign to these machines in the order their priority.
Step 2. Part operation j with the highest priority after completing on machine i send to such a machine l on which $j + 1$ will be completed by the earliest time.

5 Computational Experience

In order to compare the two approaches proposed, 100 test problems have been solved with number of vehicles $a = 1, 2$, number of machine tools $m = 3, 4, 5$, and number of part operations $n = 10, 20, 30$.

The operation assignments and the part movements for the multi-level approach have been determined by solving the sequence of machine loading and part routing problems using discrete optimizer LINGO. The CPU time was varied from 30 seconds to 15 minutes on an IBM 486DX/25MHz computer.

The results of the computational experiments have indicated that the best dispatching rules for machine scheduling and for vehicle scheduling are rarely the same and are usually different for the single-level and the multi-level approach. In most cases the multi-level approach yields slightly better results.

In the single level approach the maximum lateness was greatly dependent on the dispatching rule chosen for loading operations and in most cases the best results were achieved with EDD or MDD rule. The multi-level approach, however does not show such a sensitivity of solution to any particular priority rule. On average the best results were obtained for $STP\&TT$ rule. The CPU time required to determine schedules for the test problems was not greater than 90 seconds.

6 Conclusion

The multi-level approach determines slightly better solutions in a longer computation time required to solve the integer programmes for the upper level problems. Both the approaches construct machine and vehicle schedules period-by-period, in a relatively short CPU run time, and therefore are capable of supporting the decision making in real time.

Acknowledgment

The work has been supported by KBN, grant nr 681/S5/93/05.

References

1. Buzacott J.A., Yao D.D., "Flexible manufacturing systems: a review of analytical models," *Management Science*, vol. 32, pp. 890–905 (1986).
2. Gershwin S.B., Akella R., Choong Y.F., "Short-term production scheduling of an automated manufacturing facility," *IBM Journal of Research and Development*, vol.29, pp. 392–400 (1985).
3. Sawik T., "Operation scheduling and vehicle routing in an FMS," *Proceedings of the Conference on the Practice and Theory of Operations Management*, AFCET, Paris, pp. 31–38 (1989).
4. Sawik T., "Modelling and scheduling of a flexible manufacturing system," *European Journal of Operational Research*, vol. 45, pp. 177–190 (1990).

Lecture Notes in Control and Information Sciences

Edited by M. Thoma

1989–1994 Published Titles:

Vol. 135: Nijmeijer, Hendrik; Schumacher, Johannes M. (Eds.)
Three Decades of Mathematical System Theory. A Collection of Surveys at the Occasion of the 50th Birthday of Jan C. Willems.
562 pp. 1989 [3-540-51605-0]

Vol. 136: Zabczyk, Jerzy W. (Ed.)
Stochastic Systems and Optimization. Proceedings of the 6th IFIP WG 7.1 Working Conference, Warsaw, Poland, September 12-16, 1988.
374 pp. 1989 [3-540-51619-0]

Vol. 137: Shah, Sirish L.; Dumont, Guy (Eds.)
Adaptive Control Strategies for Industrial Use. Proceedings of a Workshop held in Kananaskis, Canada, 1988.
360 pp. 1989 [3-540-51869-X]

Vol. 138: McFarlane, Duncan C.; Glover, Keith
Robust Controller Design Using Normalized Coprime Factor Plant Descriptions.
206 pp. 1990 [3-540-51851-7]

Vol. 139: Hayward, Vincent; Khatib, Oussama (Eds.)
Experimental Robotics I. The First International Symposium, Montreal, June 19-21, 1989.
613 pp. 1990 [3-540-52182-8]

Vol. 140: Gajic, Zoran; Petkovski, Djordjija; Shen, Xuemin (Eds.)
Singularly Perturbed and Weakly Coupled Linear Control Systems. A Recursive Approach.
202 pp. 1990 [3-540-52333-2]

Vol. 141: Gutman, Shaul
Root Clustering in Parameter Space.
153 pp. 1990 [3-540-52361-8]

Vol. 142: Gündes, A. Nazli; Desoer, Charles A.
Algebraic Theory of Linear Feedback Systems with Full and Decentralized Compensators.
176 pp. 1990 [3-540-52476-2]

Vol. 143: Sebastian, H.-J.; Tammer, K. (Eds.)
System Modelling and Optimizaton. Proceedings of the 14th IFIP Conference, Leipzig, GDR, July 3-7, 1989.
960 pp. 1990 [3-540-52659-5]

Vol. 144: Bensoussan, A.; Lions, J.L. (Eds.)
Analysis and Optimization of Systems. Proceedings of the 9th International Conference. Antibes, June 12-15, 1990.
992 pp. 1990 [3-540-52630-7]

Vol. 145: Subrahmanyam, M. Bala
Optimal Control with a Worst-Case Performance Criterion and Applications.
133 pp. 1990 [3-540-52822-9]

Vol. 146: Mustafa, Denis; Glover, Keith
Minimum Enthropy H Control.
144 pp. 1990 [3-540-52947-0]

Vol. 147: Zolesio, J.P. (Ed.)
Stabilization of Flexible Structures. Third Working Conference, Montpellier, France, January 1989.
327 pp. 1991 [3-540-53161-0]

Vol. 148: Not published

Vol. 149: Hoffmann, Karl H; Krabs, Werner (Eds.)
Optimal Control of Partial Differential Equations. Proceedings of IFIP WG 7.2 - International Conference. Irsee, April, 9-12, 1990.
245 pp. 1991 [3-540-53591-8]

Vol. 150: Habets, Luc C.
Robust Stabilization in the Gap-topology.
126 pp. 1991 [3-540-53466-0]

Vol. 151: Skowronski, J.M.; Flashner, H.; Guttalu, R.S. (Eds.)
Mechanics and Control. Proceedings of the 3rd Workshop on Control Mechanics, in Honor of the 65th Birthday of George Leitmann, January 22-24, 1990, University of Southern California.
497 pp. 1991 [3-540-53517-9]

Vol. 152: Aplevich, J. Dwight
Implicit Linear Systems.
176 pp. 1991 [3-540-53537-3]

Vol. 153: Hajek, Otomar
Control Theory in the Plane.
269 pp. 1991 [3-540-53553-5]

Vol. 154: Kurzhanski, Alexander; Laseicka, Irena (Eds.)
Modelling and Inverse Problems of Control for Distributed Parameter Systems. Proceedings of IFIP WG 7.2 - IIASA Conference, Laxenburg, Austria, July 1989.
170 pp. 1991 [3-540-53583-7]

Vol. 155: Bouvet, Michel; Bienvenu, Georges (Eds.)
High-Resolution Methods in Underwater Acoustics.
244 pp. 1991 [3-540-53716-3]

Vol. 156: Hämäläinen, Raimo P.; Ehtamo, Harri K. (Eds.)
Differential Games - Developments in Modelling and Computation. Proceedings of the Fourth International Symposium on Differential Games and Applications, August 9-10, 1990, Helsinki University of Technology, Finland.
292 pp. 1991 [3-540-53787-2]

Vol. 157: Hämäläinen, Raimo P.; Ehtamo, Harri K. (Eds.)
Dynamic Games in Economic Analysis. Proceedings of the Fourth International Symposium on Differential Games and Applications. August 9-10, 1990, Helsinki University of Technology, Finland.
311 pp. 1991 [3-540-53785-6]

Vol. 158: Warwick, Kevin; Karny, Miroslav; Halouskova, Alena (Eds.)
Advanced Methods in Adaptive Control for Industrial Applications.
331 pp. 1991 [3-540-53835-6]

Vol. 159: Li, Xunjing; Yong, Jiongmin (Eds.)
Control Theory of Distributed Parameter Systems and Applications. Proceedings of the IFIP WG 7.2 Working Conference, Shanghai, China, May 6-9, 1990.
219 pp. 1991 [3-540-53894-1]

Vol. 160: Kokotovic, Petar V. (Ed.)
Foundations of Adaptive Control.
525 pp. 1991 [3-540-54020-2]

Vol. 161: Gerencser, L.; Caines, P.E. (Eds.)
Topics in Stochastic Systems: Modelling, Estimation and Adaptive Control.
1991 [3-540-54133-0]

Vol. 162: Canudas de Wit, C. (Ed.)
Advanced Robot Control. Proceedings of the International Workshop on Nonlinear and Adaptive Control: Issues in Robotics, Grenoble, France, November 21-23, 1990.
Approx. 330 pp. 1991 [3-540-54169-1]

Vol. 163: Mehrmann, Volker L.
The Autonomous Linear Quadratic Control Problem. Theory and Numerical Solution.
177 pp. 1991 [3-540-54170-5]

Vol. 164: Lasiecka, Irena; Triggiani, Roberto
Differential and Algebraic Riccati Equations with Application to Boundary/Point Control Problems: Continuous Theory and Approximation Theory.
160 pp. 1991 [3-540-54339-2]

Vol. 165: Jacob, Gerard; Lamnabhi-Lagarrigue, F. (Eds.)
Algebraic Computing in Control. Proceedings of the First European Conference, Paris, March 13-15, 1991.
384 pp. 1991 [3-540-54408-9]

Vol. 166: Wegen, Leonardus L. van der
Local Disturbance Decoupling with Stability for Nonlinear Systems.
135 pp. 1991 [3-540-54543-3]

Vol. 167: Rao, Ming
Integrated System for Intelligent Control.
133 pp. 1992 [3-540-54913-7]

Vol. 168: Dorato, Peter; Fortuna, Luigi; Muscato, Giovanni
Robust Control for Unstructured Perturbations: An Introduction.
118 pp. 1992 [3-540-54920-X]

Vol. 169: Kuntzevich, Vsevolod M.; Lychak, Michael
Guaranteed Estimates, Adaptation and Robustness in Control Systems.
209 pp. 1992 [3-540-54925-0]

Vol. 170: Skowronski, Janislaw M.; Flashner, Henryk; Guttalu, Ramesh S. (Eds.)
Mechanics and Control. Proceedings of the 4th Workshop on Control Mechanics, January 21-23, 1991, University of Southern California, USA.
302 pp. 1992 [3-540-54954-4]

Vol. 171: Stefanidis, P.; Paplinski, A.P.; Gibbard, M.J.
Numerical Operations with Polynomial Matrices: Application to Multi-Variable Dynamic Compensator Design.
206 pp. 1992 [3-540-54992-7]

Vol. 172: Tolle, H.; Ersü, E.
Neurocontrol: Learning Control Systems Inspired by Neuronal Architectures and Human Problem Solving Strategies.
220 pp. 1992 [3-540-55057-7]

Vol. 173: Krabs, W.
On Moment Theory and Controllability of Non-Dimensional Vibrating Systems and Heating Processes.
174 pp. 1992 [3-540-55102-6]

Vol. 174: Beulens, A.J. (Ed.)
Optimization-Based Computer-Aided Modelling and Design. Proceedings of the First Working Conference of the New IFIP TC 7.6 Working Group, The Hague, The Netherlands, 1991.
268 pp. 1992 [3-540-55135-2]

Vol. 175: Rogers, E.T.A.; Owens, D.H.
Stability Analysis for Linear Repetitive Processes.
197 pp. 1992 [3-540-55264-2]

Vol. 176: Rozovskii, B.L.; Sowers, R.B. (Eds.)
Stochastic Partial Differential Equations and their Applications. Proceedings of IFIP WG 7.1 International Conference, June 6-8, 1991, University of North Carolina at Charlotte, USA.
251 pp. 1992 [3-540-55292-8]

Vol. 177: Karatzas, I.; Ocone, D. (Eds.)
Applied Stochastic Analysis. Proceedings of a US-French Workshop, Rutgers University, New Brunswick, N.J., April 29-May 2, 1991.
317 pp. 1992 [3-540-55296-0]

Vol. 178: Zolésio, J.P. (Ed.)
Boundary Control and Boundary Variation. Proceedings of IFIP WG 7.2 Conference, Sophia- Antipolis,France, October 15-17, 1990.
392 pp. 1992 [3-540-55351-7]

Vol. 179: Jiang, Z.H.; Schaufelberger, W.
Block Pulse Functions and Their Applications in Control Systems.
237 pp. 1992 [3-540-55369-X]

Vol. 180: Kall, P. (Ed.)
System Modelling and Optimization. Proceedings of the 15th IFIP Conference, Zurich, Switzerland, September 2-6, 1991.
969 pp. 1992 [3-540-55577-3]

Vol. 181: Drane, C.R.
Positioning Systems - A Unified Approach.
168 pp. 1992 [3-540-55850-0]

Vol. 182: Hagenauer, J. (Ed.)
Advanced Methods for Satellite and Deep Space Communications. Proceedings of an International Seminar Organized by Deutsche Forschungsanstalt für Luft-und Raumfahrt (DLR), Bonn, Germany, September 1992.
196 pp. 1992 [3-540-55851-9]

Vol. 183: Hosoe, S. (Ed.)
Robust Control. Proceesings of a Workshop held in Tokyo, Japan, June 23-24, 1991.
225 pp. 1992 [3-540-55961-2]

Vol. 184: Duncan, T.E.; Pasik-Duncan, B. (Eds.)
Stochastic Theory and Adaptive Control. Proceedings of a Workshop held in Lawrence, Kansas, September 26-28, 1991.
500 pages. 1992 [3-540-55962-0]

Vol. 185: Curtain, R.F. (Ed.); Bensoussan, A.; Lions, J.L.(Honorary Eds.)
Analysis and Optimization of Systems: State and Frequency Domain Approaches for Infinite-Dimensional Systems. Proceedings of the 10th International Conference, Sophia-Antipolis, France, June 9-12, 1992.
648 pp. 1993 [3-540-56155-2]

Vol. 186: Sreenath, N.
Systems Representation of Global Climate Change Models. Foundation for a Systems Science Approach.
288 pp. 1993 [3-540-19824-5]

Vol. 187: Morecki, A.; Bianchi, G.; Jaworeck, K. (Eds.)
RoManSy 9: Proceedings of the Ninth CISM-IFToMM Symposium on Theory and Practice of Robots and Manipulators.
476 pp. 1993 [3-540-19834-2]

Vol. 188: Naidu, D. Subbaram
Aeroassisted Orbital Transfer: Guidance and Control Strategies.
192 pp. 1993 [3-540-19819-9]

Vol. 189: Ilchmann, A.
Non-Identifier-Based High-Gain Adaptive Control.
220 pp. 1993 [3-540-19845-8]

Vol. 190: Chatila, R.; Hirzinger, G. (Eds.)
Experimental Robotics II: The 2nd International Symposium, Toulouse, France, June 25-27 1991.
580 pp. 1993 [3-540-19851-2]

Vol. 191: Blondel, V.
Simultaneous Stabilization of Linear Systems.
212 pp. 1993 [3-540-19862-8]

Vol. 192: Smith, R.S.; Dahleh, M. (Eds.)
The Modeling of Uncertainty in Control Systems.
412 pp. 1993 [3-540-19870-9]

Vol. 193: Zinober, A.S.I. (Ed.)
Variable Structure and Lyapunov Control
428 pp. 1993 [3-540-19869-5]

Vol. 194: Cao, Xi-Ren
Realization Probabilities: The Dynamics of Queuing Systems
336 pp. 1993 [3-540-19872-5]

Vol. 195: Liu, D.; Michel, A.N.
Dynamical Systems with Saturation Nonlinearities: Analysis and Design
212 pp. 1994 [3-540-19888-1]

Vol. 196: Battilotti, S.
Noninteracting Control with Stability for Nonlinear Systems
196 pp. 1994 [3-540-19891-1]

Vol. 197: Henry, J.; Yvon, J.P. (Eds.)
System Modelling and Optimization
975 pp approx. 1994 [3-540-19893-8]

Vol. 198: Winter, H.; Nüßer, H.-G. (Eds.)
Advanced Technologies for Air Traffic Flow Management
225 pp approx. 1994 [3-540-19895-4]